Integrated Pest Management in the Global Arena

This Handbook is dedicated to all the participants
of the Michigan State University's International
IPM Short Course and to the faculty members
who have provided support to this course.

Integrated Pest Management in the Global Arena

Edited by

K.M. Maredia

Professor
Institute of International Agriculture and Department of Entomology
Michigan State University
East Lansing
Michigan
USA

D. Dakouo

Senior Research Scientist
INERA
Bobo-Dioulasso
Burkina Faso

and

D. Mota-Sanchez

Research Associate
Department of Entomology
Michigan State University
East Lansing
Michigan
USA

CABI Publishing

CABI is a trading name of CAB International

CABI Head Office
Nosworthy Way
Wallingford
Oxfordshire OX10 8DE
UK

Tel: +44 (0)1491 832111
Fax: +44 (0)1491 833508
Email: cabi@cabi.org
Web site: www.cabi.org

CABI
745 Atlantic Avenue
8th Floor
Boston, MA 02111
USA

Tel: +1 617 682-9015
Email: cabi-nao@cabi.org

A catalogue record for this book is available from the British Library, London, UK

Library of Congress Cataloging-in-Publication Data
Integrated pest management in the global arena / edited by K.M. Maredia,
D. Dakouo, D. Mota-Sanchez
p.cm.
Includes bibliographical references.
ISBN 0-85199-652-3
1. Pests—Integrated control. I. Maredia, Karim M. II. Dakouo, D.
(Dona), 1951- III. Mota-Sanchez, D. (David), 1960-
SB950.I4575 2003
632'.9—dc21

2002154965

ISBN-13: 978-0-85199-652-3

First published 2003
Transferred to print on demand 2015

Printed and bound in the UK by Marston Book Services Ltd, Didcot, Oxon.

Contents

Contributors ix

Preface xv

Acknowledgments xvii

Foreword xix

Acronyms and Abbreviations xxi

1 **Introduction and Overview** 1
K.M. Maredia

PART I: EMERGING ISSUES IN INTEGRATED PEST MANAGEMENT

2 **Online Resources for Integrated Pest Management Information Delivery and Exchange** 9
W.I. Bajwa and M. Kogan

3 **Biological Control and Integrated Pest Management** 19
R.J. O'Neil, J.S. Yaninek, D.A. Landis and D.B. Orr

4 **The Influence of Biotechnology on Integrated Pest Management in Developing Countries** 31
K.V. Raman, E. Grafius and K.M. Maredia

5 **Pesticide Policy and Integrated Pest Management** 49
G. Fleischer and H. Waibel

6 **Industrial Perspective on Integrated Pest Management** 65
G. Head and J. Duan

7 **Role of Integrated Pest Management and Sustainable Development** 73
G.W. Bird

8 Social and Economic Considerations in the Design and Implementation of Integrated Pest Management in Developing Countries 87
B. Maumbe, R. Bernsten and G. Norton

9 Integrated Pest Management Adoption by the Global Community 97
W.I. Bajwa and M. Kogan

PART II: COUNTRY EXPERIENCES

EXPERIENCES FROM AFRICA

10 Integrated Pest Management in Burkina Faso 109
D. Dakouo, M.S. Bonzi, A. Sawadogo, C.L. Dabiré, S. Nacro and B. Thio

11 Ghana National Integrated Pest Management Program 119
K. Afreh-Nuamah

12 Development and Implementation of Integrated Pest Management in the Sudan 131
Y.G.A. Bashir, Elamin M. Elamin and Eltigani M. Elamin

13 Integrated Pest Management in Tanzania 145
B.T. Nyambo, A.M. Varela, Z. Seguni and G. Kirenga

14 Integration of Integrated Pest Management in Integrated Crop Management: Experiences from Malawi 157
S. Snapp and E. Minja

15 Integrated Pest Management in South Africa 169
D.S. Charleston, R. Kfir, N.J. van Rensburg, B.N. Barnes, V. Hattingh, D.E. Conlong, D.Visser and G.J. Prinsloo

EXPERIENCES FROM ASIA

16 Integrated Pest Management in China 197
Z.-Y. Wang, K.-L. He, J.-Z. Zhao and D.-R. Zhou

17 Integrated Pest Management in India 209
A. Singh, S. Singh and S.N. Rao

18 Integrated Pest Management in Indonesia: IPM by Farmers 223
I.P.G.N.J. Oka

19 Integrated Pest Management in the Philippines 239
P.A. Javier, M.L.Q. Sison and R.V. Ebora

EXPERIENCES FROM NORTH AND SOUTH AMERICA

20 Integrated Pest Management in the USA 249
L. Olsen, F. Zalom and P. Adkisson

21 Integrated Pest Management in Mexico 273
D. Mota-Sanchez, F.S. Gonzalez, B. Alvardo-Rodriguez, O. Diaz-Gomez, H.B. Mojica, G.S. Martinez and R. Bujanos

22 **Integrated Pest Management in Brazil** 285
C.B. Hoffmann-Campo, L.J. Oliveira, F. Moscardi, D.L. Gazzoni, B.S. Corrêa-Ferreira, I.A. Lorini, M. Borges, A.R. Panizzi, D.R. Sosa-Gomez and I.C. Corso

23 **Integrated Pest Management in Peru** 301
M.P. Lazo, A.L. Travaglini, R.V. Ochoa, E.C. Hernández and I. Segovia

24 **Integrated Pest Management in Argentina** 313
D. Carmona, M. Huarte, G. Arias, A. López, A.M. Vincini, H.A. Alvarez Castillo, P. Manetti, E. Chávez, M. Torres, J. Eyherabide, J. Mantecón, L. Cichón and D. Fernández

EXPERIENCES FROM EUROPE, AUSTRALIA AND NEW ZEALAND

25 **Integrated Pest Management in Greenhouses: Experiences in European Countries** 327
J.C. van Lenteren

26 **Integrated Pest Management in the Mediterranean Region: the Case of Catalonia, Spain** 341
R. Albajes, M.J. Sarasúa, J. Avilla, J. Arnó and R. Gabarra

27 **Integrated Plant Protection Management in Russia** 357
V.A. Zakharenko and A.L. Il'ichev

28 **Integrated Pest Management in Australia** 371
D.G. Williams and A.L. Il'ichev

29 **Integrated Pest Management in New Zealand Horticulture** 385
D.M. Suckling, C. McKenna and J.T.S. Walker

PART III: REGIONAL AND INTERNATIONAL EXPERIENCES

30 **FAO Integrated Pest Management Programs: Experiences of Participatory IPM in West Africa** 397
P. Stemerding and S. Nacro

31 **Integrated Pest Management Collaborative Research Support Program (USAID – IPM CRSP): Highlights of its Global Experience** 407
B. Gebrekidan

32 **Bridging the Gap with the CGIAR Systemwide Program on Integrated Pest Management** 419
B. James, P. Neuenschwander, R. Markham, P. Anderson, A. Braun, W.A. Overholt, Z.R. Khan, K. Makkouk and A. Emechebe

33 **The World Bank and Pest Management** 435
T.W. Schillhorn van Veen

34 **Integrated Pest Management Case Studies from ICIPE** 441
Z.R. Khan, W.A. Overholt and A. Ng'eny-Mengech

35 Integrated Pest Management Experiences of CIRAD-France in Developing Countries 453
A. Ratnadass, X. Mourichon, M. Vaissayre, S. Quilici and J.-P. Deguine

36 IPMEurope, the European Group for Integrated Pest Management in Development Cooperation: Adding Value to Research Effort 467
M. Iles

37 Building IPM Programs in Central America: Experiences of CATIE 477
C. Staver and F. Guharay

38 Integrated Pest Management at CAB International 493
D.R. Dent, M. Holderness and J.G.M. Vos

PART IV: CONCLUSIONS AND RECOMMENDATIONS

39 Making IPM Successful Globally: Research, Policy, Management and Networking Recommendations 501
K.M. Maredia, D. Dakouo, D. Mota-Sanchez and K.V. Raman

Index 507

Contributors

P. Adkisson, *Department of Entomology, Texas A&M University, College Station, TX 77843, USA.*

K. Afreh-Nuamah, *Faculty of Agriculture, University of Ghana, Legon, PO Box 25, Legon, Accra, Ghana.*

R. Albajes, *Universitat de Lleida, Centre UdL-Institute de Recerca i Tecnologia Agroalimentàries, Av. Alcalde Rovira Roure 177, Lleida 25006, Spain.*

B. Alvarado-Rodriguez, *Corregidora No. 529-2, Guasave, Sinaloa, México.*

H.A. Alvarez Castillo, *FCA, UNMdP-INTA Balcarce, Ruta Nac. 226 Km. 73,5 CC 276, 7620 Balcarce, Buenos Aires, Argentina.*

P. Anderson, *Centro Internacional de Agricultura Tropical (CIAT), AA 6713, Cali, Colombia.*

G. Arias, *INTA Saenz Peña, CC 64, Presidencia Saenz Peña (3700), Chaco, Argentina.*

J. Arnó, *Institut de Recerca i Tecnologia Agroalimentàries, Centre de Cabrils, Cabrils, Catolonia, Spain.*

J. Avilla, *Universitat de Lleida, Centre UdL-Institut de Recerca i Tecnologia Agroalimentàries, Av. Alcalde Rovira Roure 177 Lleida 25006, Spain.*

W.I. Bajwa, *Integrated Plant Protection Center (IPPC), Oregon State University, Corvallis, OR 97331, USA.*

B.N. Barnes, *ARC Infruitec-Nietvoorbij, Private Bag X5013, Stellenbosch, 7599, South Africa.*

Y.G.A. Bashir, *ElObeid Agricultural Research Station, PO Box 429, ElObeid, Sudan.*

R. Bernsten, *Department of Agricultural Economics, 205 Agriculture Hall, Michigan State University, East Lansing, MI 48824, USA.*

G.W. Bird, *Department of Entomology, 243 Natural Science Building, Michigan State University, East Lansing, MI 48824, USA.*

M.S. Bonzi, *INERA, Direction 01 BP 7192 Ouagadougou, Burkina Faso.*

M. Borges, *Embrapa Recursos Geneticos e Biotechnologia, Brazilia, DF.*

A. Braun, *Centro Internacional de Agricultura Tropical (CIAT), AA 6713, Cali, Colombia.*

R. Bujanos, *Instituto Nacional de Investigaciones Agropecuarias y Forestales (INIFAP), Guanajuato, Mexico.*

S. Capezio, *FCA, UNMdP-INTA Balcarce, Ruta Nac. 226 Km. 73,5 CC 276, 7620 Balcarce, Buenos Aires, Argentina.*

D. Carmona, *FCA, UNMdP-INTA Balcarce, Ruta Nac. 226 Km. 73,5 CC 276, 7620 Balcarce, Buenos Aires, Argentina.*

D.S. Charleston, *ARC–Plant Protection Research Institute, Rietondale Research Station, 600 Soutpansberg Road, Private Bag X134, Pretoria 0001, South Africa.*

E. Chávez, *FCA, UNMdP-INTA Balcarce, Ruta Nac. 226 Km. 73,5 CC 276, 7620 Balcarce, Buenos Aires, Argentina.*

L. Cichón, *INTA Alto Valle, CC 872, (8332) General Roca, Río Negro, Argentina.*

D.E. Conlong, *South African Sugar Association Experiment Station, Private Bag X02, Mount Edgecombe, 4300, South Africa.*

B.S. Corrêa-Ferreira, *Embrapa Soja, Caixa Postal 231, 86001-970 Londrina, PR, Brazil.*

I.C. Corso, *Embrapa Soja, Caixa Postal 231, 86001-970 Londrina, PR, Brazil.*

C.L. Dabiré, *INERA, Station de Kamboinsé, BP 476 Ouagadougou, Burkina Faso.*

D. Dakouo, *INERA, Station de Farako-ba, 01 BP 910 Bobo-Dioulasso, Burkino Faso.*

J.-P. Deguine, *Centre de Coopération Internationale en Recherche pour le Développement (CIRAD), Centre de Recherches de Montpellier, Avenue Agropolis, 34398 Montpellier, Cedex, France.*

D.R. Dent, *CABI Bioscience UK Centre, CAB International, Bakeham Lane, Egham, Surrey TW20 9TY, UK.*

O. Diaz-Gomez, *Alvaro Obregon 64, Facultad de Agronomia, Universidad Autonoma de San Luis, San Luis Potosi, CP 78000, México.*

J. Duan, *Monsanto Company V2C, 800 North Lindbergh Boulevard, St Louis, MO 63167, USA.*

R.V. Ebora, *National Institute of Molecular Biology and Biotechnology, University of the Philippines, Los Baños College, Laguna 4031, Philippines.*

Elamin M. Elamin, *ElObeid Agricultural Research Station, PO Box 429, ElObeid, Sudan.*

Eltigani M. Elamin, *ElObeid Agricultural Research Station, PO Box 429, ElObeid, Sudan.*

A. Emechebe, *International Institute of Tropical Agriculture (IITA), Kano, Nigeria.*

J. Eyherabide, *FCA, UNMdP-INTA Balcarce, Ruta Nac. 226 Km. 73,5 CC 276, 7620 Balcarce, Buenos Aires, Argentina.*

D. Fernández, *INTA Alto Valle, CC 872, (8332) General Roca, Río Negro, Argentina.*

G. Fleischer, *Rural Development Division, German Technical Cooperation (GTZ), OE 45, PO Box 5180, 65726 Eschborn, Germany.*

R. Gabarra, *Institut de Recerca i Technologia Agroalimentaries, Centre de Cabrils, Cabrils, Catalonia, Spain.*

D.L. Gazzoni, *Embrapa Soja, Caixa Postal 231, 86001-970 Londrina, PR, Brazil.*

B. Gebrekidan, *IPM CRSP, Office of International Research and Development, Virginia Tech, 1060 Litton Reaves Hall, Blacksburg, VA 24061-0334, USA.*

F.S. Gonzalez, *Department of Entomology, 243 Natural Science Building, Michigan State University, East Lansing, MI 48824, USA.*

F. Guharay, *Regional Program CATIE/IPM-AF (NORAD), Apartado Postal P-116, Managua, Nicaragua.*

E. Grafius, *Department of Entomology, 245 Natural Science Building, Michigan State University, East Lansing, MI 48824, USA.*

V. Hattingh, *Citrus Research International, PO Box 28, Nelspruit, South Africa.*

K.-L. He, *Institute of Plant Protection, Chinese Academy of Agricultural Sciences, Beijing 100094, China.*

G. Head, *Monsanto Company, 700 Chesterfield Parkway, GG3A, St Louis, MO 63198, USA.*

E.C. Hernández, *Psje. Francisco de Zela s/n., Piso 10 – Edificio del Ministerio de Trabajo, Servicio Nacional de Sanidad Agraria (SENASA), Lima 11, Perú.*

C.B. Hoffmann-Campo, *Embrapa Soja, Caixa Postal 231, 86001-970 Londrina, PR, Brazil.*

M. Holderness, *CABI Bioscience UK Centre, CAB International, Bakeham Lane, Egham, Surrey TW20 9TY, UK.*

M. Huarte, *FCA, UNMdP-INTA Balcarce, Ruta Nac. 226 Km. 73,5 CC 276, 7620 Balcarce, Buenos Aires, Argentina.*

A.L. Il'ichev, *Present address: Institute of Sustainable Irrigated Agriculture (ISIA), Tatura Centre, Private Bag 1, Ferguson Road, Tatura, Victoria, 3616, Australia.*

M. Iles, *IPMEurope, Natural Resources Institute (NRI), Chatham Maritime, Kent ME4 4TB, UK.*

B. James, *The SP-IPM Secretariat, International Institute of Tropical Agriculture, Benin, 08 BP 0932 Tri Postal, Cotonou, Benin.*

P.A. Javier, *National Crop Protection Center (NCPC), University of the Philippines, Los Baños College, Laguna 4031, Philippines.*

R. Kfir, *ARC–Plant Protection Research Institute, Rietondale Research Station, 600 Soutpansberg Road, Private Bag X134, Pretoria 0001, South Africa.*

Z.R. Khan, *International Center of Insect Physiology and Ecology (ICIPE), PO Box 30772 Nairobi, Kenya.*

G. Kirenga, *Ministry of Agriculture and Food Security, Plant Protection Services, PO Box 1519, Dar es Salaam, Tanzania.*

M. Kogan, *Integrated Plant Protection Center (IPPC), Oregon State University, Corvallis, OR 97331, USA.*

D.A. Landis, *Department of Entomology, 204 Center for Integrated Plant Systems, Michigan State University, East Lansing, MI 48824, USA.*

M.P. Lazo, *Centro Internacional de la Papa (CIP), PO Box 1558, Lima, Perú.*

J.C. van Lenteren, *Laboratory of Entomology, Wageningen University, PO Box 8031, 6700 EH Wageningen, The Netherlands.*

A. López, *FCA, UNMdP-INTA Balcarce, Ruta Nac. 226 Km. 73,5 CC 276, 7620 Balcarce, Buenos Aires, Argentina.*

I.A. Lorini, *Embrapa Trigo, Caixa Postal 451, 99001-970 Passo Fundo, RS, Brazil.*

K. Makkouk, *International Center for Agricultural Research in the Dry Areas (ICARDA), PO Box 5466, Aleppo, Syria.*

P. Manetti, *FCA, UNMdP-INTA Balcarce, Ruta Nac. 226 Km. 73,5 CC 276, 7620 Balcarce, Buenos Aires, Argentina.*

J. Mantecón, *FCA, UNMdP-INTA Balcarce, Ruta Nac. 226 Km. 73,5 CC 276, 7620 Balcarce, Buenos Aires, Argentina.*

K.M. Maredia, *Institute of International Agriculture and Department of Entomology, 416 Plant and Soil Science Building, Michigan State University, East Lansing, MI 48824, USA.*

R. Markham, *International Institute of Tropical Agriculture (IITA), PMB 5320, Ibadan, Nigeria.*

G.S. Martinez, *Direccion General de Sanidad Vegetal, Guillermo Perez Valenzuela 127, Col. El Carmen-Coyoacan, Mexico, DF, CP 04100, México.*

B. Maumbe, *Department of Agricultural Economics, 205 Agriculture Hall, Michigan State University, East Lansing, MI 48824, USA.*

C. McKenna, *Horticulture and Food Research Institute of New Zealand, RD2, 412 No 1 Road, Te Puke, New Zealand.*

E. Minja, *International Crop Research Institute for the Semi-Arid Tropics (ICRISAT), Nairobi, Kenya.*

H.B. Mojica, *Instituto de Fitosanidad, Colegio de Postgraduados en Ciencias Agricolas Km 36.5 Carretera Mexico-Texcoco, Montecillo, Estado de Mexico, CP 56230, México.*

F. Moscardi, *Embrapa Soja, Caixa Postal 231, 86001-970 Londrina, PR, Brazil.*

D. Mota-Sanchez, *Center for Integrated Plant Systems, Michigan State University, East Lansing, MI 48824, USA.*

X. Mourichon, *Centre de Coopération Internationale en Recherche pour le Développement (CIRAD), Centre de Recherches de Montpellier, Avenue Agropolis, 34398 Montpellier, Cedex, France.*

S. Nacro, *INERA, Station de Farako-ba, BP 910 Bobo-Dioulasso, Burkina Faso.*

P. Neuenschwander, *Biological Control Center for Africa, International Institute of Tropical Agriculture-Benin, 08 BP 0932 Tri Postal, Cotonou, Benin.*

A. Ng'eny-Mengech, *International Center of Insect Physiology and Ecology (ICIPE), PO Box 30772 Nairobi, Kenya.*

G. Norton, *Department of Agricultural Economics, Virginia Polytechnic Institute, 205-B Hutcheson Hall, Blacksburg, VA 24061, USA.*

B.T. Nyambo, *Integrated Crop Management Practitioner, PO Box 90, Village Market-, Nairobi, Kenya.*

R.V. Ochoa, *Carretera Huaral,Chancay Km 5.8, Instituto Nacional de Investigación Agraria (INIA), Lima, Perú.*

I.P.G.N.J. Oka, *c/o Dr Mahyuddin Syam, IRRI Liaison Office, Jalan Merdeka 147, Bogor 16111, Indonesia.*

L.J. Oliveira, *Embrapa Soja, Caixa Postal 231, 86001-970 Londrina, PR, Brazil.*

L. Olsen, *Department of Entomology, Michigan State University, East Lansing, MI 48824, USA.*

R.J. O'Neil, *Department of Entomology, Purdue University, West Lafayette, IN 47907, USA.*

D.B. Orr, *Department of Entomology, North Carolina State University, Raleigh, NC 27695, USA.*

W.A. Overholt, *International Center of Insect Physiology and Ecology (ICIPE), PO Box 30772 Nairobi, Kenya.*

A.R. Panizzi, *Embrapa Soja, Caixa Postal 231, 86001-970 Londrina, PR, Brazil.*

G.J. Prinsloo, *ARC–Small Grain Institute, Private Bag X29, Bethlehem, 9700, South Africa.*

S. Quilici, *Centre de Coopération Internationale en Recherche pour le Développement (CIRAD), Centre de Recherches de Montpellier, Avenue Agropolis, 34398 Montpellier, Cedex, France.*

K.V. Raman, *Department of Plant Breeding, 313 Bradfield Hall, Cornell University, Ithaca, NY 14853-1901, USA.*

S.N. Rao, *Acharya N. G. Ranga Agricultural University, Andhra Pradesh, India.*

A. Ratnadass, *Centre de Coopération Internationale en Recherche pour le Développement (CIRAD), Centre de Recherches de Montpellier, Avenue Agropolis, 34398 Montpellier, Cedex, France.*

N. J. van Rensburg, *ARC–Plant Protection Research Institute, Rietondale Research Station, 600 Soutpansberg Road, Private Bag X134, Pretoria 0001, South Africa.*

M.J. Sarasúa, *Universitat de Lleida, Centre UdL-Institut de Recerca i Tecnologia Agroalimentàries, Av. Alcalde Rovira Roure 177, Lleida 25006, Spain.*

A. Sawadogo, *INERA, Station de Farako-ba, BP 910 Bobo-Dioulasso, Burkina Faso.*

T.W. Schillhorn van Veen, *Natural Resources Unit, Europe and Central Asia Region, ECSSD, H5-503, The World Bank, 1818 H St NW, Washington, DC 20433, USA.*

I. Segovia, *Esteban Camere No 182, Urb. San Roque, Universidad Nacional Agraria La Molina, Lima 33, Perú.*

Z. Seguni, *Agricultural Research Institute, Mikocheni, PO Box 6226, Dar es Salaam, Tanzania.*

A. Singh, *National Centre for Integrated Pest Management (NCIPM), Pusa Campus, New Delhi 110 012, India.*

S. Singh, *National Centre for Integrated Pest Management (NCIPM), Pusa Campus, New Delhi 110 012, India.*

M.L.Q. Sison, *National Institute of Molecular Biology and Biotechnology (BIOTECH), University of the Philippines, Los Baños College, Laguna 4031, Philippines.*

S. Snapp, *Department of Horticulture, 440 Plant and Soil Sciences Building, Michigan State University, East Lansing, MI 48824, USA.*

D.R. Sosa-Gomez, *Embrapa Soja, Caixa Postal 231, 86001–970 Londrina, PR, Brazil.*

C. Staver, *Programa Regional CATIE/IPM-AF (NORAD), Apartado Postal P-116, Managua, Nicaragua.*

P. Stemerding, *Global IPM Facility, Food and Agriculture Organization of the United Nations (FAO), Viale delle Terme di Caracalla, 00100 Rome, Italy.*

D.M. Suckling, *Pipfruit Sustainability/Market Access, HortResearch, PO Box 51, Lincoln Canterbury, New Zealand.*

B. Thio, *INERA, Station de Farako-ba, 01 BP 910 Bobo-Dioulasso, Burkina Faso.*

A.L. Travaglini, *Julio Rodavero 682, Urb. Las Brisas, Red de Acción en Alternativas al Uso de Agroquímicos (RAAA) and Universidad Nacional Federico Villarreal, Lima 1, Perú.*

M. Torres, *FCA, UNMdP-INTA Balcarce, Ruta Nac. 226 Km. 73,5 CC 276, 7620 Balcarce, Buenos Aires, Argentina.*

M. Vaissayre, *Centre de Coopération Internationale en Recherche pour le Développement (CIRAD), Centre de Recherches de Montpellier, Avenue Agropolis, 34398 Montpellier, Cedex, France.*

A.M. Varela, *PO Box 41607, Nairobi, Kenya.*

A.M. Vincini, *FCA, UNMdP-INTA Balcarce, Ruta Nac. 226 Km. 73,5 CC 276, 7620 Balcarce, Buenos Aires, Argentina.*

D. Visser, *ARC–Vegetable and Ornamental Plant Institute (VOPI), Private Bag X293, Pretoria 0001, South Africa.*

J.G.M. Vos, *CABI Bioscience UK Centre, CAB International, Bakeham Lane, Egham, Surrey TW20 9TY, UK.*

H. Waibel, *Faculty of Economics, University of Hannover, Postfach 6009, D-30060 Hannover, Germany.*

J.T.S. Walker, *Horticulture and Food Research Institute of New Zealand, Private Bag 1401, Havelock North, New Zealand.*

Z.-Y. Wang, *Institute of Plant Protection, Chinese Academy of Agricultural Sciences, Beijing 100094, China.*

D.G. Williams, *Institute of Sustainable Irrigated Agriculture (ISIA), Tatura Centre, Private Bag 1, Ferguson Road, Tatura, Victoria, 3616, Australia.*

J.S. Yaninek, *Department of Entomology, Purdue University, West Lafayette, IN 47907, USA.*

V.A. Zakharenko, *Department of Plant Protection, Russian Academy of Agricultural Sciences, Krzhizhanovsky str.,15/2, Moscow 172189, Russia.*

F. Zalom, *Department of Entomology, University of California, Davis, CA 95616, USA.*

J.-Z. Zhao, *Department of Entomology, Cornell University, New York Agricultural Experiment Station, Geneva, NY 14456, USA.*

D.-R. Zhou, *Institute of Plant Protection, Chinese Academy of Agricultural Sciences, Beijing 100094, China.*

Preface

Michigan State University (MSU), a premier land grant university in the USA, is recognized as a center of excellence in international development. MSU has been organizing international short courses in Integrated Pest Management since 1995. Over 100 participants from more than 25 countries have attended these courses. In addition, MSU hosts visiting scientists, students and interns from around the world. MSU, through the IPM short course and other international collaborative projects, has established an excellent global IPM network.

The participants of MSU International IPM short course have always requested for references for IPM experiences of different countries. To our knowledge, there is no book that brings together IPM experiences of the global community. Both developing and developed countries have accumulated a wealth of information and experience based in IPM. This book brings together the unique case studies of IPM from developed and developing countries, and international centers and programs.

Many of the chapters in this book have been contributed by MSU's IPM short course participants and collaborating scientists from the national and international community. It was impossible to include IPM experiences of all countries in one book. However, efforts were made to include IPM experiences of several countries and key international programs around the world. The chapters included, illustrate the development stage of IPM programs in different geographic regions.

It is hoped that this book will serve as a useful reference in international courses, seminars and workshops around the world. International cooperation and collaboration is a hallmark of MSU. This handbook is one way of sharing information with the global community.

Karim M. Maredia
Michigan State University, USA

Dona Dakouo
INERA, Burkina Faso

and

David Mota-Sanchez
Michigan State University, USA

Acknowledgments

The editors would like to first thank CABI Publishing for publishing this book. The Michigan State University IPM Program Office, the Institute of International Agriculture, and the Department of Entomology and their faculty and staff members were very helpful in the publication of this book. Many thanks to all the international and US contributors for providing insight into the current status of IPM in their countries, regions or international programs.

The financial support from the USDA National IPM Program is greatly appreciated. Many thanks to Dr Larry Olsen and Dr Edward Grafius from Michigan State University for their encouragement and support. We sincerely appreciate the assistance of Ms Kirsten Fondren and Mr Bernard Gwekwerere, Ms Cathy Pisano and Mr Ken Fittig for their assistance in the editing process. Finally, help from Ms Patricia Sutherland in formatting all the chapters is greatly appreciated. Both Cathy and Patricia helped in keeping everything together.

It is difficult to describe the monumental efforts required to develop this book, so we must also thank our families for their understanding and support. We are grateful to Dr Jack Schwille and his wife Sharon for hosting Dr Dona Dakouo during the book editing process. Thanks to Mr Mahyuddin Syam, International Rice Research Institute's representative for Indonesia, Malaysia and Brunei Darussalam for facilitating contacts with the author of the Indonesia IPM chapter.

Karim M. Maredia
Michigan State University, USA

Dona Dakouo
INERA, Burkina Faso

and

David Mota-Sanchez
Michigan State University, USA

Foreword

As in most research areas, major advances have been made in integrated pest management (IPM) since it was developed some 60 years ago, in particular with the application of computational-, information- and biotechnologies. However, this does not mean that what has been there before has lost its relevance or should be disposed of; on the contrary, I believe that what is most important is to learn from the past and integrate experience with new knowledge, i.e. learn from the past and build on it.

This book on IPM in the global arena is attempting to address not only the new science that will undoubtedly influence pest management in the future, but also practical experiences from six continents with a national and international angle as presented in Parts II and III. It also looks critically, but with optimism, at the future with recommendations for the scientists and the practitioners.

The first part considers the emerging issues in pest management at many levels, from the latest developments and expectations from biotechnology to policy issues. One of the main constraints in gaining broad support for IPM from the national and international support agencies, as well as from the implementers, lies in the fact that IPM has been widely 'undersold'. This lies in the fact that IPM has not, generally, been well evaluated and documented for its deliverables. This is unlike the case in other disciplines that have contributed to the Green Revolution and also to the general agricultural productivity increase, such as plant breeding, which has been credited for much or even most of the productivity increases. When we look at the sustainability contribution of IPM and also some of its main components like biological control (both classical and through better use of the endemic natural enemies), there is no doubt that IPM has had a major role in the recorded productivity increase. It is now up to the social scientists to get to the task and place some figures on the table of the government agencies responsible for the promotion and support of IPM.

New research, based not only on entomological parameters but now also on meteorological, agronomic and economic ones, using the new power of information and communication technologies, needs to address the development of increasingly important decision-support tools. The farmers and health officials need these new tools to assist in the prevention of outbreaks, which should remain the keystone of any IPM strategy. 'Effective prevention and smart cures' is and should remain the bottom line for IPM.

This book will serve a broad readership, with its many examples from the four corners of the world, as well as the experiences from international organizations in IPM implementation. Although we live in the information age, it is refreshing and useful to have so much information in one place.

Hans R. Herren
World Food Prize Laureate (1995)
Director General
International Center for Insect Physiology and Ecology (ICIPE)
Nairobi, Kenya

Acronyms and Abbreviations

AAACU: Asian Association of Agricultural Colleges and Universities
AAIS: African Association of Insect Scientists
ABSP: Agricultural Biotechnology Support Project (Michigan State University, USA)
ACIAR: Australian Center for International Agricultural Research
AESA: agroecosystem analysis
AFFI: African Fruit Fly Initiative
AIL: Africa Integrated Pest Management (IPM) Link
AMEWG: Africa–Middle East Working Group
APCPA: Asia–Pacific Crop Protection Association
APHIS: Animal and Plant Health Inspection Service (USA)
APO: Asian Productivity Organization
APRTC: Asian–Pacific Regional Technology Center
ARC: Agricultural Research Corporation (Sudan)
ARC: Agricultural Research Council (South Africa)
ARC–SGI: Agricultural Research Council – Small Grain Institute (South Africa)
ARMCANZ: Agriculture and Resource Management Council of Australia and New Zealand
ARPPIS: African Regional Postgraduate Programme in Insect Science
ARS: Agricultural Research Service (USA)
ASSINSEL: International Association of Plant Breeders for the Protection of Plant Varieties
AVRDC: Asian Vegetables Research and Development Center (Taiwan)
BAPPENAS: National Agency for Planning and Development (Indonesia)
BBSRC: Biotechnology and Biological Sciences Research Council
BBTs: biologically based technologies
BC: biological control
BINAS: Biotechnology Information Network and Advisory Service (UNIDO-Italy)
BIOTECH: National Institute of Molecular Biology and Biotechnology (Philippines)
Bt: *Bacillus thuringiensis*
CAAS: Chinese Academy of Agricultural Sciences
CABI: CAB International
CAMBIA: Center for the Application of Molecular Biology to International Agriculture (Australia)
CASAFE: Agricultural Protection Chamber and Fertilizers (Argentina)
CATIE: Tropical Agriculture Research and Higher Education Center (Costa Rica)
CBD: Convention of Biological Diversity
CDPR: California Department of Pesticide Regulation

CGIAR: Consultative Group on International Agricultural Research
CIAT: International Center for Tropical Agriculture (Colombia)
CIBC: Commonwealth Institute of Biological Control
CICP: Consortium for International Crop Protection (USA)
CID: Compendium of IPM definitions
CILSS: Comité Permanent Inter-Etats de Lutte contre la Sécheresse dans le Sahel
CILSS: Inter-States Committee for Drought Management in the Sahel
CIMMYT: International Maize and Wheat Improvement Center (Mexico)
CIP: International Potato Center (Peru)
CIRAD: Centre de Coopération Internationale en Recherche pour le Développement (France)
CNESOLER: Centre National de l'Energie Solaire et des Energies Renouvelables (Mali)
CNRST: Centre National de Recherche Scientifique et Technologique (Burkina Faso)
CPI/OUA: Conseil phytosanitaire Inter-Africain/Organisation Unité Africaine
CPITT: Centre for Pest Information Technology and Transfer (Australia)*
CRIs: Crown Research Institutes (New Zealand)
CSIR: Council for Scientific and Industrial Research (Ghana)
CSIRO: Commonwealth Scientific and Industrial Research Organization (Australia)
CSREES: Cooperative State Research, Education, and Extension Service
CTA: Center for Technical Cooperation in Agriculture (The Netherlands)
DANIDA: Danish International Development Assistance
DEAT: Department of Environmental Affairs and Tourism (South Africa)
DFID: Department For International Development (UK)
EBPM: Ecological Based Pest Management
EC: European Commission
ECASARD: Ecumenical Association of Sustainable Agriculture and Relief Development
EDWIP: ecological database of the world's insect pathogens
EMATER-PR: Extension Service of the Paraná State
EMBRAPA: Brazilian Agricultural Research Corporation
EPA: Environmental Protection Agency (USA)
EQIP: Environmental Quality Incentive Program
ESA: Entomological Society of America
ETL: economic threshold level
FAO: Food and Agriculture Organization of the United Nations
FCA: College of Agricultural Sciences (Argentina)
FDA: Food and Drug Administration (USA)
FECOTRIGO: an agricultural cooperative (Brazil)
FFS: farmers' field schools
FPA: Fertilizer and Pesticide Authority (The Philippines)
FPR: Farmer Participatory Research
FQPA: Food Quality Protection Act (USA)
FSIPM: Farming Systems Integrated Pest Management
FTP: file transfer protocols
GAO: US General Accounting Office
GATT: General Agreement on Tariffs and Trade
GCPF: Global Crop Protection Federation
GIS: geographic information systems
GMOs: genetically modified organisms
GOAN: Ghana Organic Agricultural Network

* Name change in 2003 to Centre for Biological Information Technology (CBIT).

GPS: global positioning systems
GREEEN: generating research and education to meet economic and environmental needs
GTZ: Deustche Gesellschaft für Technishe Zusammenarbeit (German Agency)
HACCP: hazard analysis critical control point
IAPAR: Instituto Agronômico do Paraná (Brazil)
IARC: International Agricultural Research Center
IASCAV: Argentinean Institute of Animal and Vegetal Health and Quality
IBS: ISNAR Biotechnology Service
ICAR: Indian Council of Agricultural Research
ICARDA: International Center for Agricultural Research in the Dry Areas (Syria)
ICEAPS: Southern Africa Pigeonpea Improvement Program
ICGEB: International Center for Genetic Engineering and Biotechnology
ICIPE: International Centre for Insect Physiology and Ecology (Kenya)
ICM: Integrated Crop Management
ICPM: Integrated Crop and Pest Management
ICRAF: International Center for Research in Agroforestry (Kenya)
ICRISAT: International Crops Research Institute for the Semi-Arid Tropics (India)
IDEA: Investment in Developing Export Agriculture (USAID)
IDRC: International Development Research Center (Canada)
IER: Institut d'Economie Rurale (Mali)
IFAD: International Fund for Agricultural Development (Italy)
IFPRI: International Food Policy Research Institute (USA)
IGO: inter-governmental organization
IGRs: insect growth regulators
IIBC: International Institute of Biological Control
IICA: Interamerican Institute for Cooperation in Agriculture
IITA: International Institute of Tropical Agriculture (Nigeria)
INASE: National Seed Institute (Argentina)
INERA: Institut de l'Environnement et de Recherches Agricoles (Burkina Faso)
INGEBI: Institute of Genetic and Molecular Biology of Buenos Aires (Argentina)
INIA: National Institute for Agricultural Research (Peru)
INIAA: National Institute for Plant and Animal Husbandry Research (Peru)
INIBAP: International Network for Improvement of Banana and Plantain
INIFAP: National Institute of Research in Forestry, Agriculture and Animal Sciences (Mexico)
INTA: National Institute of Agricultural Research (Argentina)
IOBC: International Organisation for Biological and Integrated Control of Noxious Animals and Plants
IOPRM: International Organization for Resistance Pest Management
IPC: Integrated Pest Control
IPM: Integrated Pest Management
IPM-CRSP: Integrated Pest Management Collaborative Research Support Program (USAID)
IPPC: International Plant Protection Congress
IPPC: Integrated Plant Protection Center (Oregon State University, USA)
IPPM: Integrated Plant Protection Management
IPR: intellectual property rights
IPVM: Integrated Pest and Vector Management
IRAC: Insecticide Resistance Action Committee
IRD: Institut de Reserche pour le Développement (France)
IRM: Insect Resistance Management
IRRI: International Rice Research Institute (Philippines)
ISAAA: International Service for Acquisition of Agri-Biotech Applications

ISDMS: Inter-States Committee for Drought Management in the Sahel
ISNAR: International Service for National Agricultural Research
ISTA: International Seed Testing Association
ITESM: Tecnologico of Monterrey (Mexico)
ITL: injury threshold levels
KARI: Kenya Agricultural Research Institute
KZN: Kwazulu Natal (South Africa)
LACPA: Latin American Crop Protection Association
LDCs: less developed countries
LISA: Low Input Sustainable Agriculture Program (USA)
LMOs: living modified organisms
LUBILOSA: Lutte Biologique contre les Locustes et Sautériaux (IITA)
MARDI: Malaysian Agricultural Research and Development Institute
MFAT: Ministry of Foreign Affairs and Trade (New Zealand)
NAFTA: North American Free Trade Agreement
NARO: National Agricultural Research Organization (Uganda)
NARS: National Agricultural Research Systems
NATESC: National Agriculture Technology Extension and Service Center (China).
NBCC: National Biological Control Committee (Ghana)
NCIPM: National Center for Integrated Pest Management (India)
NCPC: National Crop Protection Center (Philippines)
NDA: National Department of Agriculture (South Africa)
NGEBI: Institute of Genetic and Molecular Biology (Argentina)
NGOs: non-governmental organizations
NORAD: Norwegian Agency for Development Cooperation
NPVs: nuclear polyhedrosis viruses
NRC: National Research Council (USA)
NRI: Natural Resources Institute (UK)
NSESD: National Strategy for Ecologically Sustainable Development (Australia)
NSF: National Science Foundation (USA)
NZKMB: New Zealand Kiwi fruit Marketing Board
OBEPAP: Organisation Bénoise pour la Promotion de l'Agriculture Biologique, Benin
OCLALAV: Organisation Commune de Lutte Anti-acridienne et de Lutte Antiaviaire.
OECD: Organization of Economic Cooperation and Development (France)
OPMP: Office of Pest Management Policy
PAN: Pesticide Action Network (FAO)
PCARRD: Philippine Council for Agriculture, Forestry and Natural Resources and Development
PEDUNE: Protection Ecologiquement Durable du Niebé (IITA)
PLRV: Potato Leaf Roll Virus
PPRI: Plant Protection Research Institute (South Africa)
PROINPA: Foundation for the Promotion and Investigation of Andean products
PRONAF: Cowpea Project for Africa
PRONAMACHCS: National Program for the Management of Soils and Watersheds (Peru)
PTD: participatory technology development
PVX: Potato Virus X
PVY: Potato Virus Y
QTL: quantitative trait loci
RAAA: Action Network for Alternatives to Agrochemicals (Peru)
REDBIO: Technical Co-operation Network on Plant Biotechnology in Latin America and the Caribbean (FAO)
RENACO: West and Central Africa Cowpea Research Network

SADCC: Southern African Development Co-ordination Conference
SAGARPA: Secretary of Agriculture, Animal, Rural Development, Fish and Food (Mexico)
SARE: Sustainable Agriculture Research and Education Program (USA)
SAUs: State agricultural universities (India)
SCARM: Standing Committee on Agriculture and Resource Management (Australia)
SCRI: Scottish Crop Research Institute
SEARCA: Regional Center for Graduate Study and Research in Agriculture (Philippines)
SENASA: National Service for Plant and Animal Health (Peru)
SENASICA: National Service for Agricultural and Food Health, Safety and Quality (Mexico)
SIDA: Swedish International Development Agency
SPFS: Special Program for Food Security (FAO)
SP-IPM: Systemwide Program on Integrated Pest Management (CGIAR)
TGPPP: Thai–German Plant Protection Program
TSWV: Tomato Spotted Wild Virus
TYLCV: Tomato Yellow Leaf Curl Virus
UCIPM: University of California Statewide IPM Project
UNALM: Universidad National Agraria, La Molina (Peru)
UNCED: United Nations Conference on Environment and Development
UNDP: United Nations Development Programme
UNEP: United Nations Environment Programme
UNPCB: Union Nationale des Paysans Producteurs de Coton du Burkina
UPLB : University of the Philippines Los Baños
UPOV: International Union for the Protection of New Varieties of Plants
UPWARD: Users Perspectives with Agricultural Research and Development
USAID: United States Agency for International Development
USDA: United States Department of Agriculture
USSR: Union of Soviet Socialist Republics
VIZR A: Russian Institute of Plant Protection
VIZR A11: Scientific Research Institute of Plant Protection (Russia)
VNIIBZR: Scientific Research Institute of Biological Plant Protection (Russia)
VNIIF: Scientific Research Institute of Phytopathology (Russia)
VOPI: Vegetable and Ornamental Plant Institute (South Africa)
WARDA: West Africa Rice Development Association (Côte d'Ivoire)
WCASRN: West and Central African Sorghum Research Network
WHO: World Health Organization
WIPO: World Intellectual Property Organization
WPRS: West Palaeartic Regional Section (of IOBC)
WRI: World Resources Institute
WTO: World Trade Organization

Chapter 1
Introduction and Overview

Karim M. Maredia
Institute of International Agriculture and Department of Entomology, Michigan State University, East Lansing, Michigan, USA

Chemical control of agricultural pests has dominated the scene, but its overuse has adverse effects on farm budgets, human health and the environment, as well as on international trade. New pest problems continue to develop. Integrated pest management, which combines biological control, host plant resistance and appropriate farming practices, and minimizes the use of pesticides, is the best option for the future, as it guarantees yields, reduces costs, is environmentally friendly and contributes to the sustainability of agriculture.

Agenda 21 UNCED

Background on IPM

Globally, approximately half of all food and fiber produced is lost to field and storage pests (insects, pathogens, nematodes, weeds, and vertebrate pests) (Pimentel, 1997). These losses threaten global food security and are a serious economic and nutritional burden to the farmers and consumers around the world. The national governments, private industry, universities, NGOs, and international centers/programs have all been working to manage pest problems to improve the agricultural productivity on a sustainable basis to feed the growing populations. Not all organisms are pests and a majority of them are beneficial to the global society and environment. Many cultures use insects as a food source (Fasoranti and Ajiboye, 1993; ESA Newsletter, 1994). Development of sustainable agricultural systems depends on discovering means of managing pests in an environmentally friendly manner while conserving natural resources and protecting biodiversity, human and animal health.

Overuse, misuse and improper use of pesticides endanger health of farm workers and consumers of agricultural products worldwide (Goodell, 1984). Many examples exist that document the environmental and health risks from the indiscriminate use of pesticides. The global community has expressed a willingness to reduce its reliance on chemical pesticides and to move towards a more balanced approach to pest management that relies on cultural, biological and biorational control measures. The shift is driven by the high cost of pesticides, increased pest resistance to pesticides and the negative impacts of pesticides on biodiversity, food and water quality, human and animal health (Rola and Pingali, 1993), and the environment. Also as the global economy moves towards a free market economy and free trade, strict pesticide residue regulations in European and North American markets are forcing many exporting countries to redesign their pest

 Integrated Pest Management in the Global Arena (eds K.M. Maredia, D. Dakouo and D. Mota-Sanchez)

management strategies to remain globally competitive (Schillhorn van Veen *et al.*, 1997; Henson and Loader, 2000). The forces driving the shift from chemical-based pest management paradigm to ecologically/biologically based pest management paradigm are listed in Table 1.1.

IPM has emerged as a science-based approach to minimize the risks associated with the use of pesticides (Nagrajan, 1990; NRI, 1992). IPM is gaining increased attention as a potential means of reducing food and fiber losses to pests, reducing reliance on chemical pest control, and therefore fostering the long-term sustainability of agricultural systems (World Bank, 1994). IPM is a knowledge-intensive, farmer-based approach that encourages natural control of pest populations by anticipating pest problems and preventing pests from reaching economically damaging levels (Indonesian IPM Secretariat, 1997). Control of pest populations is achieved using techniques such as enhancing natural enemies, planting pest resistant crops, cultural management and using pesticides as a last resort (Leslie and Cuperus, 1993; Maredia and Mihm, 1994; Maredia, 1997; Schillhorn van Veen *et al.*, 1997).

IPM is a systems approach to the design, use, and continued evaluation of pest management procedures that result in favorable socioeconomic and environmental consequences (Isley, 1957; Ruesink, 1976; Bird *et al.*, 1990). The fundamental importance of IPM is evidenced in its recent adoption as a basic tenet of the sustainable agriculture movement around the world (ASSINSEL, 1997). Different strategies used under IPM and salient features of IPM are listed in Tables 1.2 and 1.3. Historical perspective and contemporary development in IPM have been presented in an excellent review by Kogan (1998). Radcliffe's IPM World Textbook (http://ipmworld.umn.edu/textbook.htm) and Kennedy and Sutton's book (2000) also serve as additional resources for IPM related information.

Rapid changes in the implementation of IPM are anticipated in the future. During the last decade many new developments have taken place to reduce chemical input in agriculture. New knowledge, policies, technologies and strategies have emerged to improve/reduce pesticide usage or develop alternative strategies to manage pests in an environmentally friendly way. The following section discusses some of the new developments.

Table 1.1. Forces driving the shift from chemical-based pest management paradigm to biologically/ecologically based pest management paradigm.

- Environmental concerns
- Sustainability
- Human and animal health
- Food safety
- Biodiversity
- Pest resistance
- Global trade

Table 1.2. Different strategies used under IPM.

- Behavioral control
- Biological control
- Biopesticides
- Botanical pesticides
- Chemical pesticides
- Cultural control
- Host plant resistance
- Mechanical control
- Transgenic plants (GMOs)
- Quarantine and regulations

Table 1.3. Salient features of IPM.

- Multi-disciplinary approach
- Integration of multiple strategies
- Knowledge and information intensive
- Systems approach
- Risk minimization (safety, profitability and durability)
- Links agriculture with environment, biodiversity, human health and sustainability
- Combines sophisticated high technologies and low conventional technologies
- Useful environmental educational tool for extension workers, farmers and general public
- Integral part of an overall ICM program

IPM – Most sustainable solution to pest problems worldwide

New Developments in IPM

Policy framework

National and regional pesticide use policies are rapidly changing worldwide to reduce the reliance and availability of chemical pesticides. Treaties and conferences, such as GATT/WTO, NAFTA, CBD, and UNCED are driving such policy changes in both developed and developing countries (Henson and Loader, 2001). For example, IPM is the preferred strategy for pest management under Agenda 21 of the United Nations Conference on Environment and Development (UNCED, 1992). In the USA, the Food Quality Protection Act passed in 1996 will pressure many growers to implement IPM practices to cope with fewer pesticides available for use. Similar legislation to reduce dependence on chemical pesticides is under consideration in many countries that export agricultural products to the European Union, the USA, and Canada – this is a major shift from the former policies which subsidized the use of agricultural chemicals (Schillhorn van Veen *et al.*, 1997).

Pest diagnostic and monitoring tools/ techniques and services

Timely and efficient monitoring of pests is the foundation of sound IPM programs. Success in pest management operations depends on accurate timing of the occurrence of stages susceptible to control (Isely, 1957). Poor decisions can lead to the overuse of pesticides when 'insurance' applications are made to control potential pests. New tools and techniques for monitoring and pest identification are now available to assist in making appropriate pest management decisions. Computer modeling software helps to identify critical control times in pest lifecycles and to predict pest outbreaks. Pheromone baited traps allow efficient monitoring of selected pests. Immunoassay and DNA tests can be used to give species identification of pest organisms.

In many countries, government-supported extension services have provided scouting, monitoring and forecasting of pest outbreaks. They also have provided pest management education and training of pest management personnel and consultants. Many private companies now have emerged that provide these goods and services directly to farmers and, in some cases, groups of farmers hire a consultant to advise them on pest management matters. Private companies and independent consultants will play a more important role in the transfer and synthesis of information, and an increase in automated information sources will facilitate diagnostic and monitoring information transfer at all levels (NRC, 1996).

Biotechnology and biopesticides

Biotechnology offers opportunities to enhance agricultural productivity worldwide. During the last decade, large investments have been made by both the private and public sector in biotechnology research and development. Biotechnology involves the use of living organisms, or parts of organisms to improve plant and animal production (DaSilva *et al.*, 1992; Altman and Wantabe, 1995; Persley, 2001).

In agricultural biotechnology, efforts have been made to insert desired genes from one organism into another to produce genetically engineered transgenic plants resistant to specific insects, pathogens and herbicides (Whalon and Norris, 1996; Roush and Shelton, 1997). Examples include cotton transformed with genes of the bacterium Bt, resistant to attack by bollworms; maize (corn) resistant to stem borers; soybeans resistant to the herbicide glyphosate; tomatoes resistant to viral disease; and cold- and salinity-tolerant plants; some of these genetically engineered crop varieties are now commercially grown in developed countries. Emerging countries are acquiring these new technologies to enhance their agricultural productivity.

Bioprospecting and bacterial fermentation processes have allowed many new biopesticides to be developed. Environmentally-friendly biopesticides including Bt derivatives, neem trees, new IGRs, and NPVs with insecticidal properties are now used in many countries around the world. In addition, biotechnology is employed to develop new diagnostic tools that enable development of new pest/disease resistant varieties, determination of evolutionary status and population dynamics, prediction of pest outbreaks, accurate identification of plant diseases, assessment of control strategies, and production of disease-free planting materials.

Precision agriculture technology and GIS

Precision agriculture is moving IPM into the 21st century by combining computer and satellite technology into agricultural equipment. GPS and GIS are combined to allow manipulation and analysis of large amounts of field-specific data. Maps of pest infestations, pest movements and plant nutritional needs can be generated using soil, crop, pest scouting, and yield data. The site specificity of the map allows the farmer to only treat nutrient-deficient or pest-infested areas of the field. Precision agriculture technology, although potentially valuable, is currently a very expensive tool to add to IPM systems.

Biological pest management

Pests and their natural enemies are living organisms which move between habitats searching for the resources in order to survive, grow and reproduce. Many farmers apply pesticides without considering the pest's lifecycle or their impact on other organisms. In the search for alternatives to chemical pesticides, more emphasis is placed on the natural control of pests using cultural and biological means. The habitat surrounding crops plays an important role in supporting and sustaining important natural enemies (Maredia *et al.*, 1992; Landis *et al.*, 2000). The factors that influence habitat and landscape structure or the ecology and behavior of the pest and the natural enemy populations must be taken into account. Key components of the system are biological control, cover crops and mixed cropping to provide an alternate resource base for the pest controlling organisms, crop rotation and sanitation of crop residues.

Information, communication and education

Information is an integral part of successful IPM adoption and implementation worldwide. Information sharing will require cooperation among the global community. International organizations such as CABI, FAO, UNDP, CGIAR, CICP, IPM-CRSP and IPMEurope are developing and/or disseminating IPM information. Much of this information is available on the Internet, which also connects IPM practitioners to each other worldwide (see Chapter 2 in this book).

In various parts of the world, IPM education is being given in government farmer schools, e.g. in Ghana (Afreh-Nuamah *et al.*, 1996) and in Indonesia (Indonesian National IPM Program, 1993). IPM specialists, classes and policies are becoming a standard part of many institutions. IPM short courses, with topics ranging from IPM research and extension to pest resistance are offered worldwide by such institutions as Michigan State University, USA, University of Illinois, Urbana–Champaign, USA, and the Centre for Pest Information Technology and Transfer at the University of Queensland, Australia. Emerging information and communication technologies, including digital television, increasingly powerful computers, new software concepts and distance learning via satellite promise to revolutionize the practice of IPM and information exchange even more in the coming years.

Farmer empowerment through IPM

Farmers participation in IPM through FFS has significantly contributed in the empowerment of farmers in Indonesia and other countries in Asia (Indonesian IPM Secretariat, 1997; Van Huis and Meerman, 1997; Nelson *et al.*, 2001). This success of IPM through farmer empowerment programs has now been spreading in other countries and continents (e.g. Ghana, Burkina Faso and Sudan in Africa).

International Initiatives in IPM

During the last two decades, there have been many scientific, policy and technological developments that have tremendous potential for implementing IPM throughout the world. Today, IPM is the prevailing paradigm for crop protection worldwide. Many national, regional and international initiatives have contributed significantly in building capacity and a favorable environment for IPM. A few examples are as follow.

UN–FAO Plant Protection Service

The role of FAO in the development and diffusion of IPM has been well documented. The rice IPM program implemented by national governments of Indonesia and the Philippines serves as an excellent example of FAO's contributions in IPM (known as FAO Inter-country IPM Program in Rice). The latest development in the support of IPM at FAO has been the establishment of the global IPM facility. More information about plant protection services of FAO can be found on FAO's website at: www.fao.org/WAICENT/faoinfo/agricult/agp/agpp/ipm/

Global IPM facility

The global IPM facility was established in mid-1990s with co-sponsorship of FAO, UNDP, UNEP and the World Bank. This facility serves as a coordinating, consulting, advising and promoting entity for the advancement of IPM worldwide. More information about the Global IPM Facility can be found at: www.fao.org/ag/AGP/AGPP/IPM/gipmf/default.htm

USAID IPM Collaborative Research Support Program (CRSP)

The United States Agency for International Development has established a CRSP on IPM. The program includes a consortium of several public and private institutions, NGOs, and national programs of selected countries in Asia, Africa and Latin America. This is a research, education/training, and information exchange collaborative partnership among US and developing country institutions. More information on this program can be obtained at: www.ag.vt.edu/ipmcrsp/

CGIAR SP-IPM

The SP-IPM is an inter-Center initiative of the CGIAR. More information about this program can be found at: www.cgiar.org/spipm/

IPMEurope/IPMForum

IPMForum is an initiative of the Natural Resources Institute in UK. The aim of the IPMForum is to help poor farmers in developing-countries by strengthening the capacity of NGOs to promote and implement appropriate IPM approaches and techniques, as a component of sustainable agricultural development at the farm level. More information can be found at: www.nri.org/IPMForum/index.htm

CABI Bioscience

CABI Bioscience is an international organization with expertise in biological pest management. There are CABI Bioscience Centers in a number of countries, including Malaysia and Pakistan. More information on CABI Bioscience can be found at: www.cabi.org/bioscience/index.htm

CICP

The CICP, a non-profit organization, was formed in 1978 by a group of US universities. Its principal purpose was to assist developing nations reduce food crop losses caused by pests while also safeguarding the environment. CICP's basic goal is to advance economically efficient and environmentally sound protection practices in developing countries and to ensure the health of rural and urban communities. For more information, see their website: www.orst.edu/Dept/IPPC/ippcbroc.html or www.ipmnet.org/countries/international.html

CPITT at the University of Queensland (Australia)

The mission of this center is to develop high quality, innovative software products to inform, educate and train students, practitioners and others involved in agricultural and natural resource management, particularly pest management. More information on this center is available at: www.cpitt.uq.edu.au/

National IPM Centers/Programs

Many national programs have initiated IPM programs. Many countries have set-up national IPM centers for coordinating IPM activities. For example, the USDA in the USA has a US National IPM Network (www.reeusda.gov/agsys/nipmn/index.htm). India has also set-up an NCIPM in New Delhi (see Chapter 17). Indonesia has a National IPM Secretariat in Jakarta.

NSF Center for IPM

This virtual Center for IPM is an Internet-based information source from and about the NSF sponsored Industry/University Cooperative Research Center for Integrated Pest Management located at North Carolina State University, Raleigh. More information about NSF Center for IPM can be found at: www.ncsu.edu:8150/cipm/

ICIPE and Africa IPM Forum

The ICIPE located in Kenya coordinates the Africa IPM Forum. The primary mandate of ICIPE is research, capacity and institution building in integrated arthropod management. The Africa IPM Forum is a web-based forum for online IPM information sharing and discussion. More information about ICIPE's programs and activities can be found at: www.icipe.org

Goal and Purpose of the Global IPM Book

IPM programs have been developed and implemented worldwide. The purpose of this book is to present the experiences and successful case studies of IPM from developed and developing countries and international centers/programs. The developed and developing countries have accumulated a wealth of information and experience-base in IPM. This book brings together these unique case studies. The unique features and successful case studies of IPM are presented in different chapters. The chapters are organized in three sections with an additional section on recommendations.

- Section I focuses on the emerging issues of IPM encompassing the role of information technology, biotechnology, biological control, pesticide policy, private sector, socioeconomic issues. In addition, IPM adoption by the global community and integration of IPM into sustainable agriculture are discussed.

Table 1.4. Implications for IPM implementation in the future.

- Priority setting, favorable IPM policies
- Research, teaching and extension capacity building
- Public perception and communication
- Private sector cooperation
- Structural changes in the institutions, and food systems
- Global networking and information exchange
- Sustainable financing of IPM programs
- National, regional and global cooperation

- Section II presents country case studies from 20 countries in Asia, Africa, Latin America, Europe, North America, Australia and New Zealand.
- Section III reports experiences of international organizations and programs in IPM.

Many factors will have implications for the successful implementation of IPM in the future (Table 1.4). These factors are discussed in Chapter 39. This section discusses research, policy, management, education and networking recommendations for making IPM a successful strategy globally.

The book is a true collaboration between developing and developed countries. The co-editors come from Asia, Africa and Latin America, with various chapters written by authors from all over the world. Owing to limited space, we could not include experiences of some countries. However, efforts were made to include information from as many IPM programs as possible. We hope this book will serve as a useful resource for the IPM community around the world and foster cooperation/collaboration within the global IPM community.

References

Afreh-Nuamah, K., Boob, S.M., Korang-Amoakoh, S., Dixon, G., Poku, J.A., Poku-Mensah, E.O. and Anamoah, C.H. (1996) *Institutional Innovations in Integrated Pest Management (IPM) Technology Transfer and Adoption: Farmers Field Schools in Ghana.* Paper Presented in All Africa Crop Science Congress, University of Pretoria, South Africa, 13–17 January 1996.

Altman, D.W. and Watanbe, K.N. (1995) *Plant Biotechnology Transfer to Developing Countries.* Academic Press, New York, 300 pp.

[ASSINSEL] International Association of Plant Breeders (1997) *Feeding the 8 Billion and Preserving the Planet.* ASSINSEL Secretariat, NYON, Switzerland, 11 pp.

Bird, G.W., Edens, T., Drummond, F. and Groden, E. (1990) Design of pest management systems for sustainable agricultural systems. In: Francis, C.A., Flora, C.B. and King, L.D. (eds) *Sustainable Agriculture in Temperate Zones.* John Wiley & Sons, New York, pp. 55–109.

DaSilva, E.J., Rutledge, C. and Sasson, A. (1992) *Biotechnology: Economic and Social Aspects: Issues for Developing Countries.* Cambridge University Press, 388 pp.

ESA Newsletter (1994) Insects: international food. *Bees Wax.* The Newsletter for ESA's youth members, Entomological Society of America (ESA), 17(2), 2.

Fasoranti, J.O. and Ajiboye, D.O. (1993) Some edible insects of Kwara state, Nigeria. *American Entomologist,* Summer 1993, 113–116.

Goodell, G. (1984) Challenges to international pest management research and extension in the Third World: do we really want IPM to work? *Bulletin of the Entomological Society of America,* 30(3), 18–26.

Henson, S. and Loader, R. (2001) Barriers to agricultural exports from developing countries: the role of sanitary and phytosanitary requirements. *World Development* 29, 85–102.

Indonesian IPM Secretariat (1997) *IPM By Farmers: the Indonesian Integrated Pest Management Program.* Published by Republic of Indonesia, IPM Secretariat, Jakarta, 20 pp.

Indonesian National IPM Program (1993) IPM farmer training: the Indonesian case. FAO-IPM Secretariat. Yogyakarta, Indonesia, 94 pp.

Isely, D. (1957) *Methods of Insect Control: Part I and II,* revised, Braun-Brumfield, Ann Arbor, Michigan, 208 pp.

Kennedy, G.G. and Sutton, T.B. (2000) *Emerging Technologies for Integrated Pest Management: Concepts, Research and Implementation.* Proceedings of an IPM Conference held at Raleigh, North Carolina, USA (March 8–10, 1999). American Phytopathological Society (APS) Press, St Paul Minnesota, 526 pp.

Kogan, M. (1998) Integrated pest management: historical perspectives and contemporary developments. *Annual Review of Entomology* 43, 243–270.

Landis, D., Wratten, S.D. and Gurr, G. (2000) Habitat manipulation to conserve natural enemies of arthropod pest in agriculture. *Annual Review of Entomology* 45, 173–199.

Leslie, A.R. and Cuperus, G.W. (1993) *Successful Implementation of Integrated Pest Management for Agricultural Crops.* CRC Press, Boca Raton, Florida, 193 pp.

Maredia, K.M. (1997) Sustaining host plant resistance derived through conventional and biotechnological means. In: Mihm, J.A. (ed.) *Proceedings of an International Symposium on Insect Resistant Maize: Recent Advances and Utilization*, held at CIMMYT, Mexico (27 November to 3 December 1994), pp. 175–179.

Maredia, K.M. and Mihm, J.A. (1994) Integrated management of stored grain insect pests in developing countries. In: Jewell, D.C., Waddington, S.R., Ransom, J.K. and Pixley, K.V. (eds) *Proceedings of Fourth Eastern and Southern Africa Regional Maize Conference*, held in Harare, Zimbabwe (March 28 to April 1, 1994). CIMMYT, Mexico, pp. 205–210.

Maredia, K.M., Gage, S.H., Landis, D.A. and Wirth, T.M. (1992) Ecological observations on predatory Coccinellidae in southwestern Michigan. *The Great Lakes Entomologist* 25, 265–270.

Nagarajan, S.N. (1990) *Pest Management: Integrated Approach Vital.* The Hindu Survey of Indian Agriculture.

[NRC] National Research Council (1996) *Ecologically Based Pest Management: New Solutions for a New Century.* National Academy Press, Washington, DC, 144 pp.

Nelson, R., Orrego, R., Ortiz, O., Tenorio, J., Mundt, C., Fredrix, M. and Vien, N.V. (2001) Working with Resource-Poor Farmers to Manage Plant Diseases. The American Phytopathological Society. *Plant Diseases* 85, 684–695.

[NRI] Natural Resources Institute (1992) *Integrated Pest Management in Developing Countries: Experience and Prospects.* Natural Resources Institute, Chatham, UK.

Rola, A.C. and Pingali, P.A. (1993) *Pesticides, Rice Productivity, and Farmers' Health: an Economic Assessment.* International Rice Research Institute, Manila, Philippines, 100 pp.

Persley, G. (2001) Science and environmental effects of GM crops. *Agrolinks* September, 4–6, published by the Asia-Pacific Crop Protection Association.

Pimentel, D. (ed.) (1997) *Techniques for Reducing Pesticide Use: Economic and Environmental Benefits.* John Wiley & Sons, New York, 444 pp.

Roush, R.T. and Shelton, A.M. (1997) Assessing the odds: the emergence of resistance to Bt transgenic plants. *Nature Biotechnology* 15, 816–817.

Ruesink, W.G. (1976) Status of systems approach to pest management. *Annual Review of Entomology* 21, 27–46.

Schillhorn van Veen, T.W., Forno, D., Joffe, S., Umali-Deininger, D.L. and Cooke, S. (1997) Integrated pest management: strategies and policies for effective implementation. *Environmentally Sustainable Development Studies and Monograph Series No. 13.* The World Bank, Washington, DC.

UNCED (1992) Promoting sustainable agriculture and rural development. *Agenda 21*, Chapter 14. United Nations Conference on Environment and Development (UNCED), Switzerland.

Van Huis, A. and Meerman, F. (1997) Can we make IPM work for the resource-poor farmers in sub-Saharan Africa? *International Journal of Pest Management* 43, 313–320.

Whalon, M.E. and Norris, D.L. (1996) Resistance management for transgenic *Bacillus thuringiensis* plants. *Biotechnology and Development Monitor* 29, 8–12.

World Bank (1994) Integrated pest management (IPM): an environmentally sustainable approach to crop protection. *Agriculture Technology Notes No. 2.* The Agricultural Technology and Services Division, The World Bank, Washington, DC. (Table 1. IPM resources on the Internet.)

Chapter 2

Online Resources for Integrated Pest Management Information Delivery and Exchange

Waheed I. Bajwa and Marcos Kogan
Integrated Plant Protection Center, Oregon State University, Corvallis, Oregon, USA

Introduction

The Internet is a network system that ties computers together using existing phone-lines, Ethernet, and Fiber Optic networks. The most popular Internet tools namely electronic mail (e-mail), Telnet, File Transfer Protocol (FTP), and the World Wide Web (WWW), operate as client/server systems. All these tools are basically interactive. In an interactive program, a user interacts with a client program (e.g. browser), which manages the details of how data are presented to the user. The client program interacts with one or more servers (e.g. web server). The server receives a request, processes it, and sends a result back to the client. The advantage of the client/server model lies in distributing the work so that each component focuses on a specialized task: the server distributes information to many clients while the client software for each user handles the individual user's interface and other details of the requests and results.

Knowledge and information are key to correct pest management decisions. IPM is information intensive and depends heavily on accurate and timely information for field implementation by practitioners. Additionally, it is critical to strengthen the communication links between researchers and extension professionals and their clientele to expedite multi-way exchange of information and technology transfer. In addition, researchers and extension specialists need the most up-to-date information to design new projects and set future research goals and directions. There is already a large volume of useful IPM information available on the Internet, however, the information is scattered all across the globe. These resources range from topics such as pest identification, biology, control tactics, IPM definitions and basic concepts, to modeling and systems analysis. As awareness of the Internet increases worldwide, more people are participating not only as users of the information but also as creators of new information; as a consequence, the number of both IPM Internet servers and clients is increasing rapidly, perhaps slightly lagging but generally accompanying the exponential growth of the Internet itself.

The Internet enables collaboration and information sharing on an unprecedented scale. It is becoming a prime medium for research and extension communication. The WWW – the Internet's hypertext, multimedia publishing protocol – makes it

 Integrated Pest Management in the Global Arena (eds K.M. Maredia, D. Dakouo and D. Mota-Sanchez)

possible to combine information from many different sites in a seamless fashion. The potential for using the web to integrate all types of static and interactive (dynamic) information is unique and unprecedented. The web provides excellent interfaces for all kinds of interactive network databases, and many kinds of online analyses and data processing. Web-based models and decision support systems (DSS) are becoming popular because little or no client software is required, thus reducing software management and distribution costs. No other medium offers such ability as simultaneous real-time weather information, multimedia, analytical processing and multi-way discussion and feedback.

This chapter focuses on the Internet as an information delivery and exchange tool, and describes how to retrieve IPM information online.

The Internet as an Information Delivery and Exchange Tool

An efficient information system requires rapid and accurate transmission of information at a minimal cost. Currently, different scientific communication media are being used such as printed materials, electronic archives on CD-ROM, and websites on the Internet. Printed paper in the form of books, scholarly journals, and newsletters are bulky, make delivery to remote locations expensive, and can quickly become outdated after finally reaching the user. Electronic publishing provides several advantages including the ability to search and index text, quick access to reference materials, and interactive multimedia capabilities. Information on CD-ROM has the advantage of being randomly accessed and is easy to store and ship, but is inherently static and unchanging (Bajwa and Kogan, 2000). Using the Internet as an information exchange tool has a number of advantages over both paper or CD-ROM media. Among various Internet tools, the WWW is the most rapidly growing medium for information exchange throughout the world. The web provides a cost-effective (Channin and Chang, 1997), multimedia means of delivering and exchanging quantitative and qualitative information via its user-friendly interactive interface. It makes information accessible globally to any person at any time (Jensen *et al.*, 1996a). The web provides a collaborative environment for the development and maintenance of electronic information that can occur among distant and dispersed developers and institutions (Gilman and Green, 1997). Since personal computers can be used and the browser software is low cost or free, the only requisite is an Internet connection. In academia and most research communities, the Internet connections are provided at no or low cost, and in the private sector Internet server providers (ISP) are reducing access charges to build up their clientele.

The web is advancing quickly toward mass-media status in the world. It has sustained double- and triple-digit annual growth in the USA and throughout the world. Having started with a small number of users with less than 1% of the USA population and less than 0.1% of the world population in 1990 (Gardner, 1999), the WWW currently has approximately 163 million users around the world (NUA Surveys, 1999). In the USA, 28.3% of the population used the Internet in January 1999, and the number is expected to grow to 48.6% in 2000 and 72.1% in 2005 (Gardner, 1999). The current share of different regions is as follow: North America 55.5%, Western Europe 23.3%, Asia Pacific 15.5%, Eastern Europe/Russia 2%, Latin America 1.8%, and Middle East/Africa 1.9% (Gardner, 1999). The Internet is projected to be four times bigger by 2005 with a total of 716 million users: 32.1% in North America, 28.2% in Western Europe, 23.8% in Asia Pacific, 6.1% in Russia, 6.1% in Latin America, and 3.7% in the Middle East/Africa (Gardner, 1999). In July 1999, it was estimated that, of their total hours using PCs, US households spent 53% of the time on the web; thus indicating increased reliance on Internet information (Strassmann, 2000). In fact, the Web and Internet provide a common platform that connects information servers in all geographic areas into a global information base. The Web may soon become a primary medium for exchange of scientific information replacing hardcopy books and serials. The libraries of the future will serve as electronic repositories of information (Lineberger, 1998).

The Internet and DSS

Web-based models and DSS are becoming popular because little or no client software is required, thus reducing software management and distribution costs (Power and Kaparthi, 1998). Several internet-based DSS have been developed for industry, medicine, business, meteorology, and agriculture. DSS have emerged as essential tools to bridge the gap between science-based technology and end-users who make day-to-day management decisions. A DSS integrates a user-friendly front end to often complex models, knowledgebases, expert systems, and database technologies (Coulson *et al.*, 1987; Jones 1989). A general DSS website is 'DSS Resources' http://www.dssresources.com (Power, 2000). This site provides information on basic concepts, development, deployment, and evaluation of DSSs. Also, it links to university and research DSS sites and case studies, various web-based DSSs and DSS related articles, websites of DSS companies, etc.

As a general rule, an IPM-DSS should provide users all necessary information including pest identification/disease diagnosis, pest/pathogen life histories (cycles), sampling and decision-making criteria, sampling threshold calculators, pest/disease developmental models linked to weather networks, biorational pest control methods, plus currently available pesticides, and their safety issues and environmental impacts. There are no true IPM-DSS online at this time, but many of the resources are available and waiting for proper integration. For example, various weather-based disease and insect pest models are available online for local forecasting of pest situations based on real time, near-real time, and/or historical weather data. For example, the phenology model database of the University of California Statewide IPM Project (UCIPM, 1998) contains information about, and models of, more than 100 plants, pests, and beneficial organisms (predators and parasitoids). This information can be utilized for developing web-based pest management DSS. The Integrated Plant Protection Center (IPPC) http://ippc.orst.edu of Oregon State University developed several online interactive resources including near real-time daily weather data, various degree-day products (calculators, phenology models, maps, and map calculator), and weather-based phenology models for pest management decision making in the four northwestern US states (Oregon, Washington, Idaho, and Montana) (Coop, 2000). Forecasting pest and disease incidence and development is highly valuable in planning and adjusting control measures. Another example is the Codling Moth Information Support System (CMISS) http://ippc.orst.edu/codlingmoth This site contains various knowledgebases, databases, phenology models, and links to worldwide resources on codling moth. Currently, there is an evolution of pest control recommendation resources towards online interactive, more comprehensive decision support tools. Examples include Cornell University vegetable IPM recommendations at http://www.nysaes.cornell.edu/recommends and Pacific Northwest Plant disease control guidelines at http://plant-disease.orst.edu

The Internet and Extension Services

Information exchange by electronic means has revitalized the role of extension services in providing information, education, and decision-making assistance to agricultural producers. Cooperative extension services in many countries have developed electronic information systems. For example, the states of Florida (http://edis.ifas.ufl.edu/) and Colorado (http://www.colostate.edu/Depts/CoopExt/index.html) (USA) offer the majority of their publications through the web or on CD-ROM. Now, it is possible to use an electronic mail or 'Ask an Expert' web page for requesting information from extension professionals on a specific topic. The web-based systems rely on e-mail servers/clients for responding to the queries; however, a record of each question and its answer is kept in a searchable database. Clients have access to these services 24 h a day to identify and contact an expert for answering questions. With these systems, extension professionals may respond in a more thoughtful manner by completing any necessary research before responding. These systems are better than using telephone to call a professional who may not have information readily available thus requiring additional phone calls and time delays. Many extension services offer searchable AAE databases. Examples include 'Ask an Expert' http://www1.agric.gov.ab.ca/staff/ate.nsf of Alberta Agriculture, Food and Rural Development (Canada), and 'Ask Our Experts!' http://www.ppdl.purdue.edu/ppdl/Ask_Expert.html of Virtual Plant & Pest Diagnostic Laboratory, Purdue University (USA).

A digital photograph of a plant problem can be sent to a crop consultant for proper advice and treatment recommendations. These services are readily accepted and greatly appreciated by the public (Gilman and Green, 1998). It seems that using electronic means of information exchange enhances the image of the extension services as a modern and effective source of information, education, and decision support for its clientele, thus strengthening its leadership role. One example is the Distance Diagnostic Identification System (DDIS) (http://edis.ifas.ufl.edu/MENU_DDIS) from the University of Florida, USA.

Web-based information systems and databases are freely accessible by users, whether a producer, a professional consultant, or an extension worker. These systems provide a solid base for exchanging information between experts and their clientele. They have proved to be efficient and cost-effective means of decision-support in agriculture. Online databases (e.g. for decision support in IPM) increase the ability of an extension professional to provide the latest information to the local public. In addition, they enable extension professionals to keep in touch with the technological advancements in areas inside and outside their personal expertise. Also, research and information needs may be identified by gaps in databases. The scope of web-based information delivery is not just limited to a particular (local) area/community. It is readily available to broader areas resulting in less duplication of effort by local experts. If appropriately coordinated, it can encourage collaborative efforts among professionals in the neighboring areas/states/countries, thus greatly enhancing the quality of information. With the Internet, specialists can participate cooperatively in a wide-area/national project with minimum travel and other expense involved. Electronic networking may ease and enhance extension specialists–researchers interactivity and cooperative development of comprehensive national and international databases. It may reduce overhead costs such as telephone, mail, printing, and storage costs.

A wealth of online, extension IPM resources exists including identification keys, diagnostic guides, predictive models, in-season pest alerts, pest and disease management guidelines, pest management alternatives, etc. Examples of some outstanding resources are given in Table 2.1.

The Internet and Research

Internet tools like e-mail and the Web are frequently used in the academic/scientific communities. Some research activity like literature search and acquisition, and research collaboration is the most impacted by the Internet (Leung, 1998). However, other uses are emerging. Most informational databases, previously available in the academic libraries or university's local area networks (LAN), are now online. The same is true for most journals and magazines. Some online databases provide information directly, but most are bibliographic,

Table 2.1. Some outstanding online IPM resources from different regions and perspectives.

Theme	Address
1. Information Retrieval and Referral Systems	
• Database of IPM Resources (DIR)	http://www.IPMnet.org/DIR/
• Acarology WWW Home Page	http://www.nhm.ac.uk/hosted_sites/acarology/
• AgNIC – a guide to online agricultural information	http://www.agnic.org/
• Agricultural Genome Information Server	http://ars-genome.cornell.edu/
• All the Virology on the WWW	http://www.tulane.edu/~dmsander/garryfavweb.html
• Arachnology Page (spiders and their relatives)	http://www.ufsia.ac.be/Arachnology/Arachnology.html
• Compendium of IPM Definitions (CID)	http://www.ippc.orst.edu/IPMdefinitions/home.html
• Entomology Index of Internet Resources	http://www.ent.iastate.edu/list/
• Internet Resources on Weeds & Their Control	http://www.ippc.orst.edu/cicp/gateway/weed.htm
• Internet Resources on Vertebrate Pests	http://www.ippc.orst.edu/cicp/pests/vertpest.htm
• IPMnet NEWS	http://ipmwww.ncsu.edu/cicp/IPMnet_NEWS/archives.html
• Nematology Sites on the Web	http://nematode.unl.edu/wormsite.htm
• Pesticide & Agrichemical Industry Information	http://www.bmckay.com/
• Pesticide Information Profiles (PIPs)	http://ace.ace.orst.edu/info/extoxnet/pips/pips.html
• Plant Pathology Internet Guide Book	http://www.ifgb.uni-hannover.de/extern/ppigb/ppigb.htm
• US National Pesticide Information Retrieval System	http://www.ceris.purdue.edu/npirs/npirs.html
2. Phenology, Models, and Pest Forecasting and Alert Systems	
• Blue Mold Forecast Website (USA)	http://www.ces.ncsu.edu/depts/pp/bluemold/
• Disease Model Database (USA)	http://www.ipm.ucdavis.edu/DISEASE/DATABASE/
• Models of Plants, Pests, and Beneficials Using Degree-Days (USA)	http://www.ipm.ucdavis.edu/PHENOLOGY/models.html
• Near Real-time Pest Alert Systems	http://ippc.orst.edu/pestalert/
• Online Weather Data and Degree-Days (USA)	http://www.orst.edu/Dept/IPPC/wea/
3. North America	
• Biocontrol of Plant Diseases	http://www.barc.usda.gov/psi/bpdl/bpdl.html
• BT (*Bacillus thuringiensis*) Toxin Resources	http://www.nalusda.gov/bic/BTTOX/bttoxin.htm
• Cornell University's Guide to Natural Enemies in North America	http://www.nysaes.cornell.edu/ent/biocontrol/
• Clemson Entomology – Insect Information	http://entweb.clemson.edu/cuentres/
• Crop Protection Guide (insects, disease and weeds)	http://www.agr.gov.sk.ca/Docs/crops/cropguide00.asp
• Diagnostic Key to Major Tree Fruit Diseases in the Mid-Atlantic Region	http://www.caf.wvu.edu/kearneysville/wvufarm6.html
• Electronic Resources on Lepidoptera	http://www.chebucto.ns.ca/Environment/NHR/lepidoptera.html
• Fungal Databases	http://nt.ars-grin.gov/fungaldatabases/databaseframe.cfm
• Gypsy Moth Server	http://www.gypsymoth.ento.vt.edu/vagm/index.html
• Insect Notes (North Carolina State University)	http://www.ces.ncsu.edu/depts/ent/notes/
• Northwest Berry & Grape InfoNet	http://www.orst.edu/dept/infonet/
• Overview of Organic Fruit Production	http://www.attra.org/attra-pub/fruitover.html
• Pest/Biocontrol Information	http://www.ceris.purdue.edu/napis/pests/index.html
• Pesticide Handling and Storage Tutorial	http://danpatch.ecn.purdue.edu/~epados/farmstead/pest/src/main.htm

continued

Table 2.1. *Continued.*

Theme	Address
• Photo Gallery of Insects and Mites	http://ipmwww.ncsu.edu/current_ipm/otimages.html
• Plant and Insect Parasitic Nematodes Homepage	http://nematode.unl.edu/wormhome.htm
• University of California Pest Management Guidelines	http://www.ipm.ucdavis.edu/
• Urban Integrated Pest Management	http://hammock.ifas.ufl.edu/en/en.html
• Weed Images and Descriptions	http://www.rce.rutgers.edu/weeddocuments/index.htm
4. Australasia	
• Insect and Allied Pests of Extensive Farming in Western Australia	http://www.agric.wa.gov.au/ento/allied1.htm
• Plant Viruses Online	http://biology.anu.edu.au/Groups/MES/vide/refs.htm
5. Asia	
• Japan's Pesticide Database	http://chrom.tutms.tut.ac.jp/JINNO/PESDATA/00database.html
• Malaysia's Crop Technology	http://agrolink.moa.my/doa/english/croptech/crop.html
6. Africa	
• Biological control of Cereal Stemborers in East and Southern Africa	http://nbo.icipe.org/agriculture/stemborers/default.html
7. South America	
• Brazilian National Fungal Catalogue	http://www.bdt.org.br
8. Europe	
• A Guide to the Use of Terms in Plant Pathology	http://www.bspp.org.uk/fbpp.htm
• Cereal Pathology at Scottish Crop Research Institute (SCRI), UK	http://www.scri.sari.ac.uk/mbn/cerpath/cerpath.htm
• Chemical Ecology (Sweden)	http://www.vsv.slu.se/cec/h.htm
• ExPASy – Molecular Biology Server (Switzerland)	http://www.expasy.ch/
• IPM Europe (UK)	http://www.nri.org/IPMEurope/homepage.htm
• The Pherolist (Sweden)	http://www-pherolist.slu.se/
9. International	
• FAO: Pesticide Management	http://www.fao.org/waicent/FaoInfo/Agricult/AGP/AGPP/Pesticid/
• Global Plant Protection Information System	http://pppis.fao.org/
• IMPnet	http://www.IPMnet.org/
• International Survey of Herbicide-Resistant Weeds	http://www.weedscience.com/
• The Universal Virus Database	http://life.anu.edu.au/viruses/canintro1.htm
10. Industry	
• American Crop Protection Association's IPM: The Quiet Evolution	http://www.acpa.org/public/pubs/quiteevol.html
• Cyanamid's Weed Identification Guide	http://www.cyanamid.com/tools/weedguide/index.shtml
• Integrated Pest Management (IPM) from Asia-Pacific Crop Protection Association	http://www.apcpa.org/ipm.htm

Table 2.1. *Continued.*

Theme	Address
11. Growers	
• Grape Grower's Notebook	http://users.erols.com/gmead/
• North American Fruit Explorers Website	http://www.nafex.org/
12. Books/Literature	
• AGRICOLA – The bibliographic database	http://www.nal.usda.gov/ag98/
• Florida Entomologist (Online Journal, USA)	http://www.fcla.edu/FlaEnt/
• Quantitative Population Ecology (A. Sharov, Department of Entomology, Virginia Tech, USA)	http://www.gypsymoth.ento.vt.edu/~sharov/PopEcol/popecol.html
• Radcliffe's IPM World Textbook	http://ipmworld.umn.edu/
• World Textbook of Ecotoxicology	http://ecotox.orst.edu/
• Texas Plant Disease Handbook (USA)	http://cygnus.tamu.edu/Texlab/tpdh.html

providing only references to the literature where information can be found; abstracts are sometimes given as an option. However, online searching has recently been undergoing a shift in focus, with full-text databases appearing. These databases offer access to primary sources through the complete text of articles and books, bypassing the bibliographic stage. Most online resources are free except for the most recent issues of journals/serials and commercial database services. Corporate subscriptions to these services and resources (online journals and other serials) permit employees/students/users to read, download or print the whole article using their own computer. Several publishers have already made, or are in a process of making, issues published more than 2 years ago available free of charge. The Internet is also being used by researchers in data collection, analysis, and as a tool for publishing scholarly journals (Oblinger and Maruyama, 1996). Electronic surveys (both http://www- and e-mail-based) have proved to be cost effective and a convenient method of collecting data for extension and agricultural specialists (Bajwa and Kogan, 2000). WWW provides a cost-effective conduit to disseminate research-based information. Research results can be published on the Web rapidly.

Authentic online databases provide up-to-date information on a given topic. They can be used to identify research and information needs by exploring the gaps in knowledge. Graduate students may use these resources to find new thesis topics. A few examples of these resources include genome databases for several plant, fungal, and other organisms available at: http://ars-genome.cornell.edu The site is a user friendly system with a variety of information on cotton, maize, wheat, barley, rye, beans (*Glycine*, *Phaseolus*, *Vigna*), pearl millet, rice, *Solanaceae*, *Rosaceae*, rice blast fungus (*Magnaporthe grisea*), and fungal pathogens of small-grain cereals. This site also hosts various botanical databases on plant ecological ranges, native American food plants, medicinal plants of native America, phytochemicals, plant variety protection, and worldwide plant uses. The site is very useful for finding information on a plant species and its associative organisms such as arthropod fauna and microbial flora. Another resource, NEMABASE (http://ucdnema.ucdavis.edu/imagemap/nemmap/nemabase.htm), is a database on the host status of plant species for plant-parasitic nematodes. This database contains information (for each host–parasite interaction) on nematode species, nematode subspecific designation, host species and cultivar, susceptibility to damage, damage functions and thresholds, geographic location, and fungal, bacterial or viral interactions. The 'Ecological Database of the World's Insect Pathogens (EDWIP)' (http://insectweb.inhs.uiuc.edu/pathogens/

EDWIP/) provides information on fungi, viruses, protozoa, mollicutes, nematodes, and bacteria (other than *Bacillus thuringiensis*) infectious to insects, mites, and other arthropods. This source provides information on host range, countries, and habitats where host–pathogen association can be observed. This database can be used for risk analysis and environmental impact assessment of the use of entomopathogens as control agents for insect and mite pests. Plant Viruses Online (http://biology.anu.edu.au/Groups/MES/vide/refs.htm) contains information on most species of virus known to infect plants including viruses with virions and those (e.g. umbraviruses) that have no virion protein genes of their own, and use the virion proteins of their symbiotic helper viruses instead. Resources such as International Survey of Herbicide-Resistant Weeds (http://www.weedscience.com), and Insecticide Resistance Action Committee (IRAC) (http://PlantProtection.org/IRAC), contain general and specific information about respective pesticide insecticide resistance, latest facts, and results of worldwide surveys.

The Retrieval of Internet-based IPM Information

IPM is an information-intensive system for the control of pest populations whenever they impinge on human activity and well being (or welfare). Its research and implementation depend on the reliable supply of timely information. IPM researchers, like other scientists, must rely on access to data from extant studies (Jensen *et al.*, 1996b). The Internet (particularly the Web) has opened a vast array of data resources for IPM research, extension, teaching, and learning that was not readily available before. The Web is fast becoming a critical component of IPM information exchange. Search tools on the Web, called search engines and directories, which index Internet sites, offer keyword searching and subject browsing of information. There are now thousands of IPM sites online from all over the world. The future of IPM delivery systems through the Internet is promising; internet-based information exchange is quickly becoming an absolute requirement for local, regional/areawide, and international implementation of IPM systems.

The Internet is a repository of all kinds of information. In fact, the amount of this information has overwhelmed its current information management and search technology. According to Lawrence and Giles (1998), the best search agents (e.g. Yahoo, HotBot, Alta Vista, Excite, Northern Light, Magellan, etc.) index only one-third of the total web pages which are expected to grow exponentially over the next few years. General search engines are typically of two types: evaluative or non-evaluative. Evaluative search engines include sites evaluated for quality by a person (generally a database manager, not a subject specialist). Generally, these search tools return fewer 'hits' as their databases are usually smaller because of the time necessary for evaluation. An example of an evaluative search engine is Magellan. It includes a summary, a link to the review, and a link to the site itself for each document retrieved. Non-evaluative search engines usually rely on automation. Now, search engines increasingly include a hierarchical directory structure for categorizing their web pages. Advanced search features are often available from most general search engines. These features includes Boolean searching, duplication detection, limiting retrieval by field (e.g. by specifying a keyword search in title words or heading words), proximity and/or phrase searching, retrieval display options, truncation, and wildcards (automatic or user-defined, e.g. entering 'behavio*' searches for 'behavior', 'behavioral', and 'behaviour', etc.).

It is a challenge to find the required information on a specific topic. The best known (and largest) search engine, Yahoo, searches its category words first rather than the text of its web pages. This search engine receives 56% of search referrals, yet it indexes only 10% of total Internet resources (Strassmann, 2000). Most popular engines normally turn up thousands of links, only a few (or none) of which may fit the user's needs. These search engine sites, at times of peak traffic, may be overloaded and attempts to connect may be refused. A good strategy is, however, to narrow the search by their advanced utilities using logical operators AND, OR, NEAR and NOT. Subject guides/directories and specialty search engines are preferred because they are more specific and quite easy to use. Subject guides can be used as a reference point for information retrieval on a given topic. Examples of some of the best IPM related guides are Plant Pathology Internet Guide Book (PPIGB) (http://www.ifgb.uni-hannover.de/extern/ppigb/ppigb.htm), Entomology Index of Internet Resources (http://www.ent.iastate.edu/list/) and Insects WWW (http://www.isis.vt.edu/~fanjun/text/ Links.html). Specialty Search Engines are subject specific searching utilities. Among them is DIR (http://www.IPMnet.org/DIR/; ippc.orst.edu/DIR), which is like the Yahoo of online IPM information. Infomine (a scholarly resource guide for biological, agricultural, and medical sciences) of the University of California designates DIR as 'a well organized, annotated Web virtual library of IPM information'. (Infomine is available at: http://infomine.ucr.edu/search/bioagsearch.phtml). Several other examples of specialty search engines are available from the IPM Informatics website (http://ippc.orst.edu/ipminformatics/) (Bajwa and Kogan, 1997).

Conclusion

Like the concept of a decision support system, the Internet's potential is an integrative system of static, dynamic, and real-time multi-way communications and information. To date, the examples of Internet-based IPM resources often represent extraordinary effort and creativity in exploiting this new medium. At present, most IPM information accessible via the Internet is static, such as electronic versions of informational brochures, fact sheets, extension guides and papers. Nevertheless, the number of online interactive (database-driven) resources is growing with time. It is promising that after only about 8 years since the Internet was first used for agriculture, that so many excellent resources are available, and that improvement and further adoption are seen on a daily basis during the new era of the Internet's exponential growth.

References

Bajwa, W.I. and Kogan, M. (1997) Online IPM informatics. Electronic Knowledgebase. Oregon State University. http://ippc.orst.edu/ipminformatics

Bajwa, W.I. and Kogan, M. (2000) Database management system for Internet IPM information. In: Shenk, M. and Kogan, M. (eds) *IPM in Oregon: Achievements and Future Directions*. Oregon State University, Corvallis, Oregon, pp. 216–220.

Channin, D.S. and Chang, P.J. (1997) Primer on computers and information technology. Part two: an introduction to computer networking. *Radiographics* 17(4), 988–992.

Coop, L. (2000) Online IPM weather data and degree-days for pest management decisions making in Oregon. Integrated Plant Protection Center, Oregon State University. http://osu.orst.edu/Dept/IPPC/wea

Coulson, R.N., Folse, L.J. and Loh, D.K. (1987) Artificial intelligence and natural resource management. *Science* 237, 262–267.

Gardner, E. (1999) Net is advancing quickly toward mass-media status in United States. *Internet World* 5(15), 13–14.

Gilman, E.F. and Green, J.L. (1998) Efficient, collaborative, inquiry-driven electronic information systems. *HortTechnology* 8(3), 297–300.

Jensen, A.L., Thysen, I., Hansen, J.G., Jensen, T., Secher, B.J.M. and Juhl, O. (1996a) An information system for crop production on World Wide Web, 6th International Congress for Computer Technology in Agriculture (ICCTA '96), Wageningen, The Netherlands, pp. 604–609.

Jensen, A.L., Thysen, I. and Secher, B.J.M. (1996b) Decision support in crop production via the Internet. In: Secher, B.J.M. and Frahm, J. (eds) *Proceedings of the Workshop on Decision Support Systems in Crop Protection*, Münster, Germany, 4–8 November 1996. *SP-Report, Danish Institute for Plant and Soil Science* 15, 39–47.

Jones, J.W. (1989) Integrating models with expert systems and data bases for decision making. In: Weiss, A. (ed.) *Climate and Agriculture – System Approaches to Decision Making*. Charleston, SC, 5–7 March 1989, pp. 194–211.

Lawrence, S. and Giles, C.L. (1998) Searching the world wide web. *Science* 280, 98–100.

Leung, Y. (1998) Using the Internet for natural resource research: results from an online user survey. *Journal of Natural Resources and Life Science Education* 27, 8–12.

Lineberger, R.D. (1998) Integrating the World Wide Web into existing extension and educational technology. *Hort Technology* 8(3), 313–315.

NUA Surveys (1999) Nua Internet survey. http://www.nua.ie/

Oblinger, D.G. and Maruyama, M.K. (1996) Distributed learning. *CAUSE Professional Paper Series*, No. 14. New York, 39 pp.

Power, D.J. (2000) Decision Support Systems Web Tour. World Wide Web, dssresources.com/tour, Version 3.1, 4 February 2000.

Power, D.J. and Kaparthi, S. (1998) The changing technological context of decision support systems. In: Berkeley, D., Widmeyer, G., Brezillion, P. and Rajkovic, V. (eds) *Context-Sensitive Decision Support Systems*. Chapman and Hall, London, pp. 41–54.

Strassmann, P.A. (2000) Knowledge metrics. *Knowledge Management* 3(4), 16.

UCIPM (1998) UC IPM online (electronic publication). http://axp.ipm.ucdavis.edu

Chapter 3
Biological Control and Integrated Pest Management

Robert J. O'Neil[1], J. Stephen Yaninek[1], Douglas A. Landis[2] and David B. Orr[3]

[1]Department of Entomology, Purdue University, West Lafayette, Indiana, USA; [2]Department of Entomology, Michigan State University, East Lansing, Michigan, USA; [3]Department of Entomology, North Carolina State University, Raleigh, North Carolina, USA

Introduction

Biological control of insects[1] is the use of natural enemies (parasitoids, predators and pathogens) to reduce or maintain insect pest populations below an economic, action or aesthetic threshold (DeBach and Rosen, 1991; Bellows and Fischer, 1999). The practice of biological control includes methods that reunite pests with their natural enemies, or recommend modifications of the environment (or the natural enemy itself) to favor natural enemy population growth and impact on pest dynamics. Biological control is practiced worldwide in almost every natural and human modified habitat. Because natural enemies are often key factors in the dynamics of pests, biological control should be the cornerstone of IPM (integrated pest management) practices. However, the resources dedicated to the study and implementation of biological control are frequently insufficient (DeBach and Rosen, 1991), and thus for most crop–pest systems, biological control remains an under-utilized option (DeBach and Rosen, 1991; Van Driesche and Bellows, 1996; Huffaker and Dahlsten, 1999).

Because biological control options differ depending on the ecological, agronomic and socioeconomic conditions of the pest situation, it is important to understand the principal practices of biological control to see how they can be applied to any given system. In this chapter we will review the major approaches to implement biological control and use case studies to illustrate their application. We use examples from both subsistence and production agriculture to highlight how biological control can be used in both systems. While we focus on agronomic and horticultural crops, we provide information on examples, institutions and programs working outside these production systems (Tables 3.1–3.3). Finally, by illustrating that biological control can be

[1] Biological control methods are also used against weeds and plant pathogens. While there are similarities of approach, the interested reader is referred to Harley and Forno (1992) for an introduction to weed biological control, and Cook and Baker (1983) for biological control in plant pathology.

 Integrated Pest Management in the Global Arena (eds K.M. Maredia, D. Dakouo and D. Mota-Sanchez)

Table 3.1. Government, university, commercial and non-profit websites on biological control.

United States
- National Biological Control Institute
www.aphis.usda.gov/ppq/nbci/
- USDA Insect Biological Control Laboratory
www.barc.usda.gov/psi/ibl/iblhome.htm
- Cornell University Biological Control Guide
www.nysaes.cornell.edu/ent/biocontrol/
- Purdue Entomology Biological Control
www.entm.purdue.edu/Entomology/research/bclab/BCMAC.HTML
- Biological Control at Michigan State University
www.cips.msu.edu/biocontrol/
- UC Berkeley Center for Biological Control
www.CNR.Berkeley.EDU/biocon/
- University of California Statewide IPM Project
www.ipm.ucdavis.edu/
- Midwest Institute for Biological Control
www.inhs.uiuc.edu/cee/biocontrol/home.html
- Biological Virtual Information Center
ipmwww.ncsu.edu/biocontrol/
- Biological Control News
www.entomology.wisc.edu/mbcn/mbcn.html
- Insect Parasitic Nematodes
www.oardc.ohio-state.edu/nematodes/
- The Association of Natural Bio-control Producers
www.anbp.org/
- List of Suppliers of Beneficial Insects in North America
cdpr.ca.gov/docs/ipminov/bensuppl.htm

International
- The Biotechnology and Biological Control Agency
www.e-bbca.net/main.htm
- International Organisation for Biological Control and Integrated Control of Noxious Animals and Plants (IOBC)
www.iobc-wprs.org/
- CAB International
www.cabi.org/
- Centre de Recherche de l'Est sur les cereals et oleagineux
res2.agr.ca/ecorc/isbi/biocont/libhomf.htm
- Institut National de la Recherche Agronomique (INRA)
inra.fr/Internet/Hebergement/OPIE-Insectes/luttebio.htm

Table 3.1. *Continued.*

- Centro de Control Biologico de Plagas y Enfermedades
www.usfq.edu.ec/1AGROEMPR/HOME.HTML
- CPL Worldwide Directory of Agrobiologicals
www.cplscientific.co.uk/press/wda-features.html
- Institute of Arable Crops Research (Great Britain)
www.iacr.bbsrc.ac.uk/iacr/tiacrhome.html
- Embrapa (Brazil)
www.embrapa.br/
- FAO – Community Integrated Pest Management
www.communityipm.org/
- The Consortium for International Crop Protection (CICP)
www.ipmnet.org/
- IPM Europe: IPM Working for Development Newsletter
www.nri.org/IPMForum/ipmwd.htm
- CIAT – Integrated Pest and Disease Management (IPDM)
www.ciat.cgiar.org/ipm/index.htm
- European and Mediterranean Plant Protection Organization (EPPO)
www.eppo.org/index.html
- The Center for Agroecology and Sustainable Food Systems
www.agroecology.org/index.html

used in areas once thought to be outside its purview, we hope to show that what limits its application are not agronomic, ecological or economic constraints, but a willingness to invest in biological control as a management option.

Methods of Biological Control

Biological control is practiced using *importation*, *augmentation* and *conservation*[2] methods. Each method is most appropriate for a particular crop–pest system, however,

[2] 'Habitat management' is another commonly used term to describe methods to alter the environment to favor natural enemies. Because it is considered a subset of conservation biological control (Landis *et al.*, 2000), we will use the term conservation biological control as it is more inclusive of practices that enhance the control effectiveness of natural enemies through manipulation of the environment or the characteristics of natural enemies themselves. Also, the important method is sometimes referred to as 'classical biological control'.

there are commonalities among methods, and the same crop–pest system can be targeted using more than one approach.

Conservation biological control

The goal of conservation biological control is to modify the environmental factor(s) that may limit the control effectiveness of natural enemies (Table 3.2). In general, conservation of natural enemies involves reducing factors that interfere with natural enemies or providing resources that natural enemies need in their environment (Rabb *et al.*, 1976; Barbosa, 1998). Many factors can interfere with the ecological requirements of natural enemies and reduce their effectiveness as control agents. Pesticide applications may directly kill natural enemies or have indirect effects through reduction in the numbers or availability of hosts (Croft, 1990). Various cultural practices such as tillage or burning of crop debris can kill natural enemies or make the crop habitat unsuitable (Gurr *et al.*, 2000). In orchards, repeated cultivation for weed control may create dust deposits on leaves, killing small predators and parasites and causing increases in certain insect and mite pests (DeBach and Rosen, 1991). Finally, host plant effects such as chemical or physical defenses may reduce the effectiveness of natural enemies by altering their search efficiencies or life history characteristics (Kogan *et al.*, 1999).

The goals and approaches of conservation biological control closely match those of IPM. In both, a fundamental understanding of the ecological mechanisms driving pest dynamics is key to success (Huffaker, 1980). Conserving natural enemies often requires modification of production practices that are similar to changes in practices recommended by IPM principles (e.g. increase diversification of crops, reduction in pesticide use, etc.). The use of thresholds to make decisions, common to IPM systems, is closely tied to the impact of natural enemies whose density, composition and impact on pest dynamics (and damage) are dependent on the crop cultivation practices and environmental milieu. The interdependence of farming practices, pest dynamics and the impact of natural enemies often requires farmers to modify practices. As such, farmer education is key to success. Examples of farmer education span a number of extension approaches that include bulletins, field days, grower meetings, electronic media and farmer field schools. Two case studies illustrate the importance of farmer education in conservation biological control, as well as the opportunities to use this method in pest management in subsistence crops.

Maize in Honduras[3]

Recently in Honduras, a validation study was conducted to determine if sugar solutions applied to maize (*Zea mays* L.) would

Table 3.2. Examples of conservation biological control methods (see Barbosa, 1998; Gurr and Wratten, 2000; Landis *et al.*, 2000; Stoll, 2000).

- Reduce pesticide application frequency and or rates
- Use 'softer' pesticides such as microbials, soaps and botanicals
- Plant flowers/using cultivars that provide pollen and nectar sources
- Apply sugar-water or protein sprays to attract/maintain natural enemies
- Provide shelters for or avoid destroying nests of social wasps
- Reduce dust in orchards that can hinder predaceous mites
- Plant 'banker plants' that harbor alternative (non-pest) prey
- Diversify crop plantings using intercropping, relay cropping, etc.
- Alter harvest and/or cultivation practices to maintain 'refuge strips' for natural enemies
- Use cover crops to increase overwintering survival of natural enemies
- Use tillage and fertilization practices that enhance natural enemy diversity and densities

[3] This section adapted from an article by RJO first published in *Midwest Biological Control News* II (3): March 1995 (now: Biological Control News: http://www.entomology.wisc.edu/mbcn/mbcn.html).

attract natural enemies of the key pest, *Spodoptera frugiperda* Smith (Cañas and O'Neil, 1998). Using a sugar solution as a conservation technique has been reported in a number of scientific journals and has been tried in the USA in lucerne and some vegetable systems (Ben Saad and Bishop, 1976; Hagen *et al.*, 1976; Evans and Swallow, 1993). However, the idea for using sugar solutions in Honduran maize did not arise from a scientific journal, but rather from a farmer who had invented a new (to her) technology of pest control. The pathway from farmer invention to testing by university scientists, to extension, to other farmers was predicated on a simple, yet profound idea. That idea, that farmers, like the rest of us, experiment with the familiar to gain insight on what they don't know, was used as the basis for an IPM program in Honduras (Bentley, 1989; Bentley *et al.*, 1994). In brief, field studies by crop protection specialists and the anthropologist, J.W. Bentley, at the Zamorano College in Honduras (then the Panamerican School of Agriculture) identified critical gaps in farmer understanding and use of IPM in subsistence crops (maize and beans). A key finding was that farmers did not appreciate the role of natural enemies (primarily ants, social wasps and parasitoids), and thus were not manipulating their practices to conserve natural enemies. A workshop was developed and offered to farmers, who participated in a number of role-playing exercises (on pest and natural enemy biology), field studies (seeing social wasps attacking pests), and discussions (classroom presentations were minimized). Our farmer[4] attended one of these workshops, which resulted in her invention of using sugar-water to attract natural enemies (other workshop farmers also invented this and other control technologies). It is important to note that farmers were taught that ants eat pests, and not: 'Use sugar-water to attract ants to control pests.' Our inventive farmer took what she knew, that ants like sugar (she owned a small store where ants were pests of sweet products she sold), and added it to what she learned, that ants are predators. She then began to experiment with using sugar-water in her *milpa* (small production plot), which lead to the validation work cited above. The repeated invention of this technology by workshop participants, and the validation study by Cañas and O'Neil (1998) led to the extension of this technology to thousands of farmers in Honduras, Nicaragua and El Salvador. Farmer innovation can be a powerful mechanism in conservation biological control, and programs that directly involve farmers in the development and testing of practices should increase the adoption and spread of this technology (Stoll, 2000).

Asian rice

Rice is grown by tens of millions of small farmers and is one of the most important food crops worldwide. A 'Green Revolution crop', rice production increased dramatically with implementation of new varieties, irrigation schedules, fertilization and pesticide use. These changes in production practices, while increasing yields, also increased plant protection problems, including the inducement of secondary pest outbreaks of the brown planthopper (BPH), *Nilaparvata lugens* (Stål) (Kenmore *et al.*, 1984; Way and Bowling, 1994). Initial efforts using resistant varieties were met with success, but resistance was short-lived, leading to cycles of varietal development, deployment and eventual failure. BPH management was further complicated by governmental policies that subsidized the purchase of insecticides causing widespread mis- and over-use further exasperating BPH (and other pests) management programs. The net result of these efforts was a decrease in rice production productivity in an unsustainable system reliant on chemical subsidies (Kiritani *et al.*, 1972; Hare, 1994). In reaction to these developments, agricultural scientists and policy makers in several national (e.g. Indonesia, Vietnam,

[4] Sra. Hublado Castro, Comayagua, Honduras.

India, the Philippines) and international research centers (notably, the IRRI) collaborated with US, Japanese and European scientists and donor agencies (e.g. US Agency for International Development, the United Nations, The World Bank), to develop a new rice IPM program (Matteson *et al.*, 1994). The hallmarks of the Asian Rice IPM program were a reliance on conservation biological control and the use of FFS to investigate, implement and extend IPM technologies. FFS are organized at the village level and involve farmers in the development and implementation of pest management practices[5]. Because rice IPM is built upon a foundation of conserving endemic natural enemies (principally predatory bugs, egg parasitoids and spiders), farmers are involved in observing natural enemy attacks on rice pests, and investigating relationships between production practices and natural enemy diversity, dynamics and effectiveness. FFS are assisted by trainers (often NGO groups) whose role is to facilitate discussion by posing questions to stimulate participation, without lecturing as the 'expert'. Farmers are encouraged to develop investigations that test insights they themselves have proposed (similar to the maize IPM example). The success of this approach is evident not only through its implementation in the rice production system, but also in its application to other crop–pest systems in Asia and other parts of the world (Raj and Suresh, 2000).

Augmentation biological control[6]

Augmentation is the direct manipulation of natural enemy populations to increase their effectiveness as biological control agents (Debach and Rosen, 1991; Barbosa and Braxton, 1993; Elzen and King, 1999) (Table 3.3). In augmentation, natural enemies are typically reared in insectaries then released into the target environment where pest suppression is desired (Ridgway *et al.*, 1998). There are two ways in which such periodic colonization is conducted; inundative and inoculative releases. In inoculative releases, the natural enemy is intended to establish and reproduce on the pest population with future generations of the natural enemy essential in achieving pest control. Alternatively, inundative releases involve the initial release of large numbers of a natural enemy such that the released population overwhelms the pest. In inundative releases reproduction and persistence of the natural

Table 3.3. Selected references on augmentative biological control, mass rearing of natural enemies, non-target impacts of biological control and biology of major natural enemy taxa used in augmentation.

Topic	Reference
Overview of augmentation	Elzen and King, 1999
Mass rearing of natural enemies	Ridgway *et al.*, 1998
Non-target impacts	Follett and Duan, 2000
	Wajnberg *et al.*, 2001
Augmentation in glasshouses	van Lenteren and Woets, 1988
	Parella *et al.*, 1999
Biology and use of egg parasitoids	Wajnberg and Hassan, 1994
Biology and use of Coccinellidae	Obrycki and Kring, 1998
Biology and use of Chrysopidae	Canard *et al.*, 1984

[5] The initial FFS focus on insect IPM in rice has expanded to other crops, production issues and pests. The existence of a cadre of farmers experienced with research protocols, teamwork and organization has made farmer–researcher–extension efforts more common and successful (Matteson *et al.*, 1994).
[6] Adapted from: Landis, D.A. and Orr, D.B. (1996) Biological control: approaches and applications. In: Radcliffe, E.B. and Hutchison, W.D. (eds) *Radcliffe's IPM World Textbook*. University of Minnesota, St Paul, Minnesota. http://ipmworld.umn.edu

enemy is not required. Genetic enhancement of natural enemies to improve their survival or effectiveness has also become an important component of some modern augmentation efforts (Whitten and Hoy, 1999). As with other methods of biological control, an understanding of the basic biology of the natural enemy is key to the effectiveness of any given program (e.g. see Canard *et al.*, 1984; Wajnberg and Hassan, 1994; Obrycki and Kring, 1998).

An early example of the inoculative release method is the use of the parasitoid wasp, *Encarsia formosa* Gahan, to suppress populations of the greenhouse whitefly, *Trialeurodes vaporariorum* (Westwood) (Hussey and Scope, 1985; Parrella, 1990). The greenhouse whitefly is a worldwide pest of vegetable and floriculture crops that is difficult to manage with pesticides. Releases of relatively low densities (typically 0.25–2 per plant, depending on the crop) of *E. formosa* immediately after the first whiteflies are detected can effectively prevent populations from developing to damaging levels. Augmentative methods have been used against a wide diversity of pests and estimates of the number of hectares under augmentative protection are in the tens of thousands worldwide (van Lenteren and Woets, 1988; Parrella *et al.*, 1999).

A common form of inundative release is the use of *Trichogramma* wasps to control lepidopteran pests in a number of crop and forest systems. These minute endoparasitoids of insect eggs are released in crops or forests in large numbers (up to several million per hectare) timed to the presence of pest eggs. *Trichogramma* are the most widely augmented species of natural enemy, having been mass-produced and field released for almost 70 years in biological control efforts. Worldwide, over 32 million ha of agricultural crops and forests are treated annually with *Trichogramma* in 19 countries, mostly in China and republics of the former Soviet Union (Li, 1994). As illustrated below, research into *Trichogramma* biology and release technology was key to successful implementation (Wajnberg and Hassan, 1994).

Genetic enhancement of natural enemies has proven to be an important key to success in several augmentation programs (Whitten and Hoy, 1999). For example, cultures of the parasitic wasp *Trioxys pallidus* Haliday were artificially selected for resistance to azinphosmethyl, an insecticide commonly used in an integrated management of the walnut aphid, *Chromaphis juglandicola* Kaltenbach, in California walnut orchards (Hoy *et al.*, 1990). Following mass release of the resistant parasitoids into walnut blocks, they established and rapidly became the dominant strain in four of five release sites, demonstrating the potential of the technique.

Trichogramma in China

In China, agricultural production and pest management systems capitalize on low labor costs and generally follow highly innovative yet technologically simple processes. For example, *Trichogramma* inundatively released to suppress sugarcane borer, Chilo spp., populations in sugarcane are protected from rain and predators inside emergence packets. Insectary-reared parasitized eggs are wrapped in sections of leaves that are manually placed on the sugarcane plants. Most *Trichogramma* production in China takes place in facilities producing material for a localized area. These facilities range from open-air insectaries to mechanized facilities that are leading the world in development of artificial host eggs.

Rearing of natural enemies for augmentative release can be labor intensive, particularly when it involves the rearing of the host insect. Chinese scientists were the first to develop *in vitro* methods for the culture of *Trichogramma* (Guan *et al.*, 1978; Liu *et al.*, 1979). Initially, these artificial diets incorporated insect-derived ingredients to support normal development, while later diets devoid of insect material were developed (Wu *et al.*, 1980; Liu and Wu, 1982; Wu *et al.*, 1982). These early efforts were responsible, in part, for the current worldwide interest in development of *in vitro* rearing techniques for natural enemies.

Trichogramma in Western Europe

One of the barriers to wider implementation of biological control in Western agriculture has been socioeconomic constraints (van Lenteren, 1988). Currently, large-scale production agricultural systems place a premium on efficiency and economy of scale. In Western Europe, almost two decades of intensive research resulted in the commercial marketing of three products utilizing the European native, *Trichogramma brassicae* Bezdenko, to suppress the European corn borer, *Ostrina nubilalis* Hübner, in corn fields (Bigler *et al.*, 1989). These products are annually applied to over 70,000 ha in France, Switzerland, Germany and Austria. All three products are based on manufactured plastic or paper packets designed to provide protection for the wasps against weather extremes and predation until emergence in the field.

As in the Chinese example above, European *Trichogramma* products are, for the most part, applied to crop fields by hand. One product called Trichocaps® consists of hollow walnut-shaped cardboard capsules (approximately 2 cm diameter) each containing 500–1000 parasitized eggs of the Mediterranean flour moth, *Ephestia kuehniella* Zwolfer (Kabiri *et al.*, 1990). Developing *Trichogramma* inside capsules are induced into an overwintering (diapause) state in the insectary, and then are refrigerated for up to 9 months. This system allows for production during winter months for later distribution to growers when needed in the summer.

Once removed from cold storage, *Trichogramma* inside the capsules begin development and emerge approximately 100 (°C) degree-days later. This reactivation process can be manipulated so that capsules containing *Trichogramma* at different developmental stages can be applied to fields at the same time, extending the emergence period of parasitoids and increasing the residual activity of a single application to approximately 1 week. Preparation of Trichocaps® for application is done by the supplying company, with growers responsible for applying the product to crop fields.

Importation biological control

Importation biological control is the purposeful reuniting of natural enemies with their hosts/prey that have become pests in areas outside their original geographic distribution[7]. The first major success of this method, the complete control of the cottony cushion scale, *Icerya purchasii* Maskell, in California in the late 1800s set the stage for worldwide application. The central elements of the approach include, identification of the 'area of origin' of the exotic pest (requiring detailed taxonomic and biogeographical study), and travel to collect natural enemies (a process called 'foreign exploration'). Following collection, putative natural enemies and their hosts are held in quarantine laboratories (most run by state or federal governments) where their basic biology is initially studied and the natural enemies are evaluated for host specificity and contamination by insect pathogens and other parasites. Field releases are then made and their impact on the target hosts (sometimes including economic impact) is determined. Worldwide this method has been practiced on 415 target pest insects, with some level of control success achieved in 40% of cases (Huffaker and Dahlsten, 1999).

Cassava green mite in Africa

In the early 1970s, the neotropical spider mite *Mononychellus tanajoa* (Bondar) was discovered attacking cassava in East Africa. This exotic mite quickly spread throughout the 'cassava belt' (an area larger than the continental USA) causing up to 80% reduction in yield. Cassava green mite threatened production in many marginal areas where cassava was often the last crop available for

[7] The importation method has also been practiced using natural enemies of different hosts, an approach referred to as the 'novel association method' (see Hokkanen and Pimental, 1984).

harvest when all other crops had failed (Yaninek and Herren, 1988).

The exotic nature of the pest and its host plant in Africa prompted scientists in 1984 to initiate a classical biological control program to complement the efforts in resistance breeding. The control campaign focused on phytoseiid (predaceous mites) predators of neotropical origin. Initially, natural enemies were selected and shipped to Africa for experimental releases because of their abundance and frequency on cassava. Between 1984 and 1988, more than 5.2 million phytoseiids belonging to seven species of Colombian origin were imported to Africa and released in 348 sites in ten countries. None of these species and populations ever became established in the wide range of agronomic and ecological conditions tested, apparently because of inadequate alternative food sources when *M. tanajoa* densities were low and there were extended periods of low relative humidity (Yaninek *et al.*, 1993).

Foreign exploration was adjusted in 1988 to focus on neotropical regions that were agrometeorologically homologous to areas in Africa where the potential for severe *M. tanajoa* damage existed. Natural enemies associated temporally and spatially with *M. tanajoa* and capable of surviving periods of low *M. tanajoa* densities on alternative food sources in the new exploration sites were given selection priority. Several natural enemy candidates were immediately identified from northeast Brazil and shipped to Africa. Approximately 6.1 million phytoseiids of the species *Neoseiulus idaeus* (Denmark and Muma), *Typhlodromalus manihoti* Moraes, and *T. aripo* DeLeon, of Brazilian origin were released in 358 sites in 16 countries between 1989 and 1997. *Neoseiulus idaeus* became established in Benin and Kenya (Yaninek *et al.*, 1992), *T. manihoti* became established in Benin, Burundi, Ghana and Nigeria (Yaninek *et al.*, 1999), and *T. aripo* became established in 20 countries spanning the cassava belt of Africa (unpublished data).

Prospects for control of *M. tanajoa* were initially inferred from the impact of natural enemies in their area of origin, but eventually field results in Africa told the real story. *Neoseiulus idaeus* never spread beyond the two original release and 'establishment' sites, and has probably been extirpated due to insufficient mite prey on cassava and associated host plant species. *Typhlodromalus manihoti* has become established and continues to spread. However, its impact on *M. tanajoa* has been difficult to quantify at the low predator densities found in the field. *Typhlodromalus aripo*, a predator confined to the growing tip of cassava plants, has been the big surprise. This predator has become established in 20 countries and rapidly spread beyond most of the original release sites. *Mononychellus tanajoa* populations have been reduced by more than 50% and yields increased by more than 35% where *T. aripo* is present. The economic return for this predator has been estimated to be equivalent to millions of dollars in food aid each and every growing season if yield losses had to be replaced.

Several constraints were overcome to successfully implement the cassava green mite program. The exotic phytoseiids had to be available in sufficient numbers and at the right time to assure worthwhile experimental releases. This was partly resolved by new mass production technology that was developed to facilitate decentralized rearing of phytoseiid predators. A cadre of highly trained and dedicated core project staff was needed to properly handle exotic phytoseiid mites. Most national programs had no experience working with mites of any kind. Thus, capable individuals were identified and trained in basic acarology and biological control applications. Release fields were judiciously selected and faithfully monitored to understand the conditions needed to achieve widespread establishment. Post-release follow-up monitoring was one of the most difficult tasks for national program staff to accomplish because of many constraints. Continued training and close mentoring by experienced collaborators helped resolve the problem. Ultimately, close and continuous contact between national programs with specific local capacities and needs, international programs with unique expertise and implementation resources and a donor community committed to long-term support, set the stage for the success achieved in this campaign.

The ecological risks associated with biological control, particularly importation biological control, has been the topic of considerable discussion in recent years (Follett and Duan, 2000). Evidence of non-target affects (e.g. impacts on non-target species, associated habitats and surrounding ecosystems) by a few natural enemies introduced before more stringent taxonomic specificity standards were adopted, and increasing concern about these introductions as potential invasive species, has put importation biological control practices under considerable scrutiny. In the cassava green mite project, candidate natural enemies were systematically evaluated for potential multi-trophic, intra-guild, community and ecosystem impact as part of the selection and prerelease procedures. There has been no evidence that any released natural enemy has had an unanticipated impact or has posed an ecological risk anywhere in the African cassava belt (Yaninek *et al.*, 1993). Biological control will continue to be a desirable and appropriate intervention tactic in the future. However, success and impact will be measured, in part, by the risks that are mitigated in the process.

Biological control and transgenic crops

The advent of transgenic crops introduced an entirely new category of pest management intervention technologies into the environmental risk equation. In 1988, the National Academy of Sciences proposed three principles when evaluating the environmental effects of transgenic organisms: (i) the act of transforming an organism poses no special threat to the environment; (ii) the environmental risks of releasing a transgenic organism should be evaluated the same way any conventional release would be treated; and (iii) consider the characteristics of the transgenic organism and the environment, not how the organism was produced when evaluating risk (NRC, 1989). Transgenic crops are increasingly important components of production agriculture with the 'next generation' biotech products already on the horizon (Pew, 2001). While transgenic crops may not pose unique environmental risks compared with traditional crops (NRC, 2001, 2002), there are risks to consider (Rieger *et al.*, 2002), and more empirical experiences will be needed to understand their potential impact on the environment and compatibility with biological control.

In 2002 the IOBC established a working group to develop guidelines for evaluating the environmental risk of transgenic plants in agricultural production systems (IOBC, 2002). These guidelines recommend five scientific activities as part of the risk assessment process including:

1. Establish a framework to evaluate the need for the transgenic crop;
2. Characterize the transgenic construct and phenotype in order to evaluate its stability and inheritance;
3. Assess non-target species effects and biodiversity impacts;
4. Determine resistance risk and develop appropriate management responses; and
5. Monitor gene flow and geographic and genetic spread of transgenes.

These guidelines will be tested in Kenya, Brazil and Vietnam which no doubt will improve our understanding of the relationships between transgenic crops and biological control. However, the flexibility of biological control to be used in diverse habitats, against a wide variety of pest target systems and in combination with other IPM tactics suggests that it will be an integral component in a transgenic cropping system.

Summary

Biological control has enjoyed a rich history of success and, as the above examples illustrate, its application is not bounded by crop, target pest or geography. The case histories also illustrate that the keys to success lie not so much in the ecological or economic constraints of the particular crop–pest system, but in the effort expended to institute biological control as a pest management option. The methods of biological control require different levels of technical and farmer input, ranging from high levels of technical input in importation biological control to high levels of farmer input in conservation programs. The rice and maize examples show that farmer involvement is key in those systems in which biological control is integrated in IPM programs. Not only does farmer involvement avoid conflicts (e.g. pesticide use following release of natural enemies), it maximizes adoption and spread of IPM technologies by other farmers (Stoll, 2000). The future of biological control therefore rests on foundations of basic sciences such as systematics, population ecology and predator–prey theory, and applied efforts in pest management, sampling and other quantitative methods. The application of biological control in the global arena will grow as networks of scientists, farmers, extension specialists and national and international institutions coalesce around critical pest situations. The opportunities that biological control offers as an environmentally sound, safe and cost-effective control technology are therefore limitless.

Acknowledgments

We appreciate the extreme patience of Dr Karim Maredia in waiting for the initial draft of this document. Kim Rebek and Farah Heraux helped with editing. Drs Cliff Sadof and T. Gibb provided helpful reviews of the manuscript.

References

Barbosa, P. (ed.) (1998) *Conservation Biological Control.* Academic Press, San Diego, California, 396 pp.

Barbosa, P. and Braxton, S. (1993) A proposed definition of biological control and its relationship to related control approaches. In: Lumsden, R.D. and Vaughn, J.L. (eds) *American Chemical Society Conference Proceedings Series. Pest Management: Biologically Based Technologies.* American Chemical Society, Washington, DC, pp. 21–27.

Bellows, T.S. and Fisher, T.W. (eds) (1999) *Handbook of Biological Control.* Academic Press, San Diego, California, 1046 pp.

Ben Saad, A. and Bishop, G.W. (1976) Attraction of insects to potato plants through use of artificial honeydews and aphid juice. *Entomophaga* 21, 49–57.

Bentley, J.W. (1989) What farmers don't know can't help them: the strengths and weaknesses of indigenous technical knowledge in Honduras. *Agriculture and Human Values* 6, 25–31.

Bentley, J.W., Rodríguez, G. and González, A. (1994) Science and people: Honduran campesinos and natural pest control inventions. *Agriculture and Human Values* 11, 178–182.

Bigler, F., Bosshart, S. and Walburger, M. (1989) Bisherige und neue Entwicklungen bei der biologischen Bekampfung des Maiszunslers, *Ostrinia nubilalis* Hbn., mit *Trichogramma maidis* Pint. et Voeg. in der Schweiz. *Landwirtscahft Schweiz* 2, 37–43.

Canard, M., Séméria, Y. and New, T.R. (1984) *Biology of Chrysopidae.* Junk, The Hague, 294 pp.

Cañas, L. and O'Neil, R.J. (1998) Applications of sugar solutions to maize, and the impact of natural enemies on fall armyworm. *International Journal of Pest Management* 44, 59–64.

Cook, R.J. and Baker, K.F. (1983) *The Nature and Practice of Biological Control of Plant Pathogens.* American Phytopathological Society, St Paul, Minnesota, 539 pp.

Croft, B.A. (1990) *Arthropod Biological Control Agents and Pesticides.* John Wiley & Sons, New York, 723 pp.

DeBach, P. and Rosen, D. (1991) *Biological Control by Natural Enemies.* Cambridge University Press, UK, 440 pp.

Elzen, G.W. and King, E.G. (1999) Periodical release and manipulation of natural enemies. In: Bellows, T.S. and Fisher, T.W. (eds) *Handbook of Biological Control.* Academic Press, San Diego, California, pp. 253–270.

Evans, E.W. and Swallow, J.G. (1993) Numerical response of natural enemies to artificial honeydew in Utah alfalfa. *Environmental Entomology* 22, 1392–1401.

Follett, P.A. and Duan, J.J. (2000) *Nontarget Effects of Biological Control.* Kluwer Academic Publishers, Boston, Massachusetts, 316 pp.

Guan, X.C., Wu, Z.X., Wu, T.N. and Feng, H. (1978) Studies on rearing *Trichogramma dendrolimi* Matsumura in vitro. *Acta Entomologica Sinica* 21, 122–126.

Gurr, G. and Wratten, S.D. (eds) (2000) *Biological Control: Measures of Success.* Kluwer Academic Publishers, Boston, Massachusetts, 429 pp.

Gurr, G., Wratten, S.D. and Barbosa, P. (2000) Success in conservation biological control. In: Gurr, G. and Wratten, S.D. (eds) *Biological Control: Measures of Success.* Kluwer Academic Publishers, Boston, Massachusetts, pp. 105–132.

Hagen, K.S., Greany, P., Sawall, E.F. Jr and Tassan, R.L. (1976) Tryptophan in artificial honeydews as a source of an attractant for adult *Chrysopa carnea. Environmental Entomology* 5, 458–468.

Hare, D.J. (1994) Status and prospects for an integrated approach to the control of rice planthoppers. In: Denno, R.F. and Perfect, T.J. (eds) *Planthoppers their Ecology and Management.* Chapman and Hall, New York, pp. 615–633.

Harley, K.L.S. and Forno, I.W. (1992) *Biological Control of Weeds.* Inkata Press, Sydney, Australia, 74 pp.

Hokkanen, H. and Pimental, D. (1984) New approach for selecting biological control agents. *Canadian Entomologist* 116, 1109–1121.

Hoy, M.A., Cave, F.E., Beede, R.H., Grant, J., Krueger, W.H., Olson, W.H., Spollen, K.M., Barnett, W.W. and Hendricks, L.C. (1990) Release, dispersal, and recovery of a laboratory-selected strain of the walnut aphid parasite *Trioxys pallidus* (Hymenoptera: Aphidiidae) resistant to azinphosmethyl. *Journal of Economic Entomology* 83, 89–96.

Huffaker, C.B. (ed.) (1980) *New Technology of Pest Control.* John Wiley & Sons, New York, 500 pp.

Huffaker, C.B. and Dahlsten, D.L. (1999) Scope and significance of biological control. In: Bellows, T.S and Fisher, T. (eds) *Handbook of Biological Control.* Academic Press, San Diego, California, pp.1–16.

Hussey, N.W. and Scopes, N. (1985) *Biological Pest Control: the Glasshouse Experience.* Cornell University Press, Ithaca, New York, 240 pp.

[IOBC] International Organization of Biological Control (2002) GMO Guidelines Project: Development of International Scientific Biosafety Testing Guidelines for Transgenic Plants. www.gmo-guidelines.info/

Kabiri, F., Frandon, J., Voegele, J., Hawlitzky, N. and Stengel, M. (1990) Evolution of a strategy for inundative releases of *Trichogramma brassicae* Bezd. (Hym Trichogrammatidae) against the European corn borer, *Ostrinia nubilalis* Hbn. (Lep.: Pyralidae). In: *Proceedings, ANPP-Second International Conference on Agricultural Pests,* Versailles, 4–6 December 1990.

Kenmore, P.E., Carino, F.O., Perez, C.A., Dyck, V.A. and Gutierrez, A.P. (1984) Population regulation of the rice brown planthopper (*Nilaparvata lugens*) within rice fields in the Philippines. *Journal of Plant Protection in the Tropics* 1, 19–37.

Kiritani, K., Kawahara, S., Sasaba, T. and Nakasuji, F. (1972) An approach to the integrated control of rice pests: control with selective low dosage insecticides by reduced number of applications. *Japanese Journal of Applied Entomology and Zoology* 16, 28–40.

Kogan, M., Gerlingand, D. and Maddox, J.V. (1999) Enhancement of biological control in annual agricultural environments. In: Bellows, T.S. and Fisher, T.W. (eds) *Handbook of Biological Control.* Academic Press, San Diego, California, pp. 789–818.

Landis, D.A., Wratten, S.S. and Gurr, G.M. (2000) Habitat management to conserve natural enemies of arthropod pests in agriculture. *Annual Review of Entomology* 45, 175–201.

Li, Li-Ying (1994) Worldwide use of *Trichogramma* for biological control on different crops: a survey. In: Wajnberg, E. and Hassan, S.A. (eds) *Biological Control with Egg Parasitoids.* CAB International, Wallingford, UK, pp. 37–54.

Liu, W.H. and Wu, Z.X. (1982) Recent results in rearing *Trichogramma* in vitro with artificial media devoid of insectan additives. *Acta Entomologica Sinica* 25, 160–163.

Liu, W.H., Zie, Z.N., Xiao, G.F., Zhou, Y.F., Ou Yang, D.H. and Li, L.Y. (1979) Rearing of the *Trichogramma dendrolimi* in artificial diets. *Acta Phytophylacia Sinica* 6, 17–25.

Matteson, C.P., Gallagher, K.D. and Kenmore, P.E. (1994) Extension of integrated pest management for planthoppers in Asian irrigated rice: empowering the user. In: Denno, R.F. and Perfect, T.J. (eds) *Planthoppers their Ecology and Management.* Chapman and Hall, New York, pp. 656–686.

[NRC] National Research Council (1989) *Field Testing Genetically Modified Organisms: Framework for Decisions.* National Academy Press, Washington, DC.

[NRC] National Research Council (2001) *Ecological Monitoring of Genetically Modified Crops: a Workshop Summary.* National Academy Press, Washington, DC.

[NRC] National Research Council (2002) *Environmental Effects of Transgenic Plants: the Scope and Adequacy of Regulation.* National Academy Press, Washington, DC.

Obrycki, J.J. and Kring, T.J. (1998) Predaceous Coccinellidae in biological control. *Annual Review of Entomology* 43, 295–321.

Parrella, M.P. (1990) Biological pest control in ornamentals: status and perspectives. *Bulletin of the Section Régionale Ouest Paléarctique/Western Palaearctic Regional Section* XIII/5, 161–168.

Parrella, M.P., Stengård Hansen, L. and van Lenteren, J. (1999) Glasshouse environments. In: Bellows, T.S. and Fisher, T.W. (eds) *Handbook of Biological Control.* Academic Press, San Diego, California, pp. 819–840.

Pew (2001) Harvest on the Horizon: Future Uses of Agricultural Biotechnology. Pew Initiative On Food and Biotechnology. http://pewagbiotech.org/research/harvest/

Rabb, R.L., Stinner, R.E. and van den Bosch, R. (1976) Conservation and augmentation of natural enemies. In: Huffaker, C.B. and Messenger, P.S. (eds) *Theory and Practice of Biological Control.* Academic Press, New York, pp. 233–254.

Raj, D.A. and Suresh, C. (2000) Farmers' research on natural crop protection. In: Stoll, G. (ed.) *Natural Crop Protection in the Tropics,* 2nd edn. AGRECO/CTA, Margrof Verlag, Weikersheim, Germany, pp. 291–296.

Ridgway, R.L., Hoffmann, M.P., Inscoe, M.N. and Glenister, C.S. (1998) *Mass-Reared Natural Enemies: Application, Regulation, and Needs.* Thomas Say Publications in Entomology, Entomological Society of America, Lanham, Maryland, 332 pp.

Rieger, M.A., Lamond, M., Preston, C., Powles, S.B. and Roush, R.T. (2002) Pollen-mediated movement of herbicide resistance between commercial canola fields. *Science* 296, 2386–2388.

Stoll, G. (2000) *Natural Crop Protection in the Tropics,* 2nd edn. AGRECOL/CTA, Margrof Verlag, Weikersheim, Germany, 376 pp.

Van Driesche, R.V. and Bellows, T.S. Jr (1996) *Biological Control.* Academic Press, New York, 539 pp.

van Lenteren, J.C. (1988) Implementation of biological control. *American Journal of Alternative Agriculture* 3, 102–109.

van Lenteren, J. and Woets, J. (1988) Biological and integrated control in greenhouses. *Annual Review of Entomology* 33, 239–269.

Wajnberg, E. and Hassan, S.A. (eds) (1994) *Biological Control with Egg Parasitoids.* CAB International, Wallingford, UK, pp. 37–54.

Wajnberg, E., Scott, J.K. and Quimby, P.C. (2001) *Evaluating Indirect Ecological Effects of Biological Control.* CAB International, Wallingford, UK, 261 pp.

Way, M.O. and Bowling, C.C. (1994) Insect pests of rice. In: Luh, S.B. (ed.) *Rice Production,* Vol. I. Van Nostrand Heinhold, New York, pp. 237–268.

Whitten, M.J. and Hoy, M.A. (1999) Genetic improvement and other genetic considerations for improving the efficacy and success rate of biological control. In: Bellows, T.S. and Fisher, T.W. (eds) *Handbook of Biological Control.* Academic Press, San Diego, California, pp. 271–296.

Wu, Z.X., Zhang, Z.P., Li, T.X. and Liu, D.M. (1980) Artificial media devoid of insect additives for rearing larvae of the endoparasitoid wasp *Trichogramma. Acta Entomologica Sinica* 23, 232.

Wu, Z.X., Qin, J.D., Li, T.X., Zhang, Z.P. and Liu, D.M. (1982) Culturing *Trichogramma dendrolimi* in vitro with artificial media devoid of insect materials. *Acta Entomologica Sinica* 25, 128–133.

Yaninek, J.S. and Herren, H.R. (1988) Introduction and spread of the cassava green mite, *Mononychellus tanajoa* (Bondar) (Acari: Tetranychidae) an exotic pest in Africa and the search for appropriate control methods: a review. *Bulletin of Entomological Research* 78, 1–13.

Yaninek, J.S., Mégevand, B., de Moraes, G.J., Bakker, F., Braun, A. and Herren, H.R. (1992) Establishment of the neotropical predator *Amblyseius idaeus* (Acari: Phytoseiidae) in Benin, West Africa. *Biocontrol Science and Technology* 1, 323–330.

Yaninek, J.S., Onzo, A. and Ojo, B. (1993) Continent-wide experiences releasing neotropical phytoseiids against the exotic cassava green mite in Africa. *Experimental and Applied Acaralogy* 16, 145–160.

Yaninek, J.S., Megevand, B., Ojo, B., Cudjoe, A.R., Abole, E., Onzo, A. and Gnanvossou, T.S. (1999) Establishment and spread of *Typhlodromalus manihoti,* an introduced phytoseiid predator of *Mononychellus tanajoa* in Africa. *Environmental Entomology* 27, 1496–1505.

Chapter 4

The Influence of Biotechnology on Integrated Pest Management in Developing Countries

K.V. Raman[1], Edward Grafius[2] and Karim M. Maredia[3]
[1]*Department of Plant Breeding, Cornell University, Ithaca, New York, USA;*
[2]*Department of Entomology, Michigan State University, East Lansing, Michigan, USA;*
[3]*Institute of International Agriculture and Department of Entomology, Michigan State University, East Lansing, Michigan, USA*

Introduction

Although the Green Revolution of the 1960s improved agricultural productivity, today an estimated billion people live in absolute poverty and suffer from chronic hunger. Seventy per cent of these individuals are farmers – men, women, and children – who make a living farming small plots of poor soils. These plots are mainly located in the tropical environments that are increasingly prone to drought, flood, bushfires, and hurricanes. Crop yields in these areas are stagnant and epidemics of pests and weeds often ruin crops (Persley and Doyle, 1999). Among the principal problems in this area is the widespread use of insecticides, fungicides, and herbicides that raises production costs and leads to contamination of soils and water sources, thus threatening the future of agriculture.

Increasing crop yields continues to be a challenge for many developing countries; whereas biotechnology and information technology improves the health, well-being, and the life styles of developing countries (Persley and Lantin, 2000). A wealth of information is available on biotechnology research and development issues on Internet websites (Table 4.1). In this chapter, we try to provide an overview of emerging new biotechnologies for pest management. We also examine specific studies involving the development and commercialization of transgenic plants with pest resistance. The policy issues affecting access to new biotechnologies, pest resistance management, public–private sector linkages and the future of biotechnology and its influence on IPM in the global arena are discussed.

Biotechnology Applications in IPM

Development of technologies using biotechnology is often time consuming and expensive. It can take about 10 years or more and many millions of dollars from the time of first identification of a novel gene to commercialization. Costs are expected to rise because society requires significant safety information on biotechnology-derived products before their release and acceptance. The private sector, a dominant player in commercialization, evaluates, early in its business plans, the size and

Table 4.1. List of selected websites related to agricultural biotechnology.

USAID – Agricultural Biotechnology Support Project (ABSP)
www.iia.msu.edu/absp
CABI – AgBiotechNet
www.agbiotechnet.com
International Service for Acquisition of Agri-Biotech Applications (ISAAA)
www.isaaa.org
Center for the Application of Molecular Biology to International Agriculture (CAMBIA)
www.cambia.org
Biotechnology Information Network and Advisory Service (BINAS)
www.binas.unido.org/binas/binas.html
Information Systems for Biotechnology
www.isb.vt.edu
ISNAR Biotechnology Service (IBS)
www.cgiar.org/isnar/ibs.htm
International Center for Genetic Engineering and Biotechnology (ICGEB – Biosafety)
www.icgeBtrieste.it/biosafety
Technical Co-operation Network on Plant Biotechnology in Latin America and the Caribbean (REDBIO/FAO)
www.rlc.fao.org/redes/redbio/html/home.htm
AgBioWorld
www.agbioworld.org
International Union for the Protection of New Varieties of Plants (UPOV)
www.upov.int
World Intellectual Property Organization (WIPO)
www.wipo.int
AfricaBio (NGO)
www.africabio.com

value of the market. If it is not sufficient to provide a return on investments made, the product is not developed.

In industrial countries, there is an increasing emphasis on reducing the reliance on the use of conventional pesticides using BBTs such as biological control, microbial control, pest behavior modifying chemicals, genetic manipulation of pest populations, plant immunizations, plant breeding and enhanced resistance to pests. The US Congress Office of Technology Assessment (1995) completed a study examining the current and potential role of BBTs for pest control.

Notable points are:

- There is a significant investment in the area of BBTs by the US Federal government – somewhere between US$150 million and US$200 million/year.
- The major use of BBTs at the moment is on insect pests of arable agricultural, forestry, and aquatic environments. The least use is on weed control in agriculture, even though herbicides account for 57% of the US chemical expenditures.
- BBTs are being adopted when conventional pesticides are unavailable, unacceptable, as in environmentally sensitive habitats, or where conventional pesticides are economically not feasible because costs are high relative to the value of resource.
- Increased knowledge of pests and ecological systems will hasten the transition from the current levels of pesticide use to BBTs.

The NRC that provides advice to the US government was recently given the task of determining the future of biological control agents for use in agriculture. The NRC examined questions such as: Why do we need new pest control methods in crop and forest productions systems? What can we realistically expect from investment in these new technologies? How do we develop effective and profitable pest control systems that rely on ecological processes of control? How should we oversee and commercialize biological control organisms and products? The consensus reached indicates that safety, profitability, and durability in pest management are the key issues for success in this area. The biotechnology related applications of importance to ecologically based pest management include: use of transgenic crops for insect and disease resistance, molecular markers for host plant resistance breeding, diagnostic tools, and the use of conventional and novel biological control agents. It is, therefore, timely for developing countries to consider the potential usefulness of these new biotechnologies for inclusion in their IPM programs. Persley (1996) edited an excellent book on the role biotechnology can play in IPM.

Transgenic plants with insect and disease resistance with emphasis on the use of Bt transgenic plants

The ability to cut and join DNA to create a new molecule (transgene) and to insert this transgene into the crop plant forms the basis of the most revolutionary crop improvement technology of the 20th century. The bulk of crop transgenes thus far commercialized were designed to aid in crop protection against insects, diseases, and weeds. The current value of the global market in transgenic crops is estimated at US$3 billion, increasing to US$8 billion in 2005, and US$25 billion by 2010 (Estruch *et al.*, 1997; James, 1999). Among the various disease and pest resistant transgenic crops released for cultivation, insect resistant crops are the ones which have received wide acceptance in many countries. The most extensively grown insect resistant transgenic crops today are: maize (also referred as corn) and cotton. According to James (2000), on a worldwide basis in 2000, Bt maize was grown on 6.8 million ha with an additional 1.4 million ha planted to Bt/herbicide tolerant maize. Bt cotton was grown on 1.5 million ha with an additional 1.7 million ha grown to Bt/ herbicide resistant cotton. Bt potatoes were grown on < 0.1 million ha. The adoption of Bt cotton is already bringing significant economic benefits to the small-scale farmers in South Africa (Gregory *et al.*, 2002).

China was the first country to commercialize transgenic plants in the early 1990s with the introduction of virus resistant tobacco, and later a virus resistant tomato. In 1995, the USA EPA approved the first registration of Bt maize, potato and cotton products; now more Bt crops are grown in the USA than any other country. Since this approval, there has been a dramatic increase in the use of Bt maize and Bt cotton in the USA (US EPA, 2000). The area of Bt maize in the USA was 0.2 million ha in 1996 and reached 8.2 million ha by 1999. The area of Bt cotton in the USA increased from 0.7 million ha in 1996, to 1.1 million in 1998. In 2000, 39% of the total cotton growing area was devoted to Bt cotton while 28% had stacked genes for Bt and herbicide resistance (Carpenter and Gianessi, 2001). Apart from the USA, Bt maize was also grown in: Canada, Argentina, South Africa, Spain, France, and Portugal. Similarly, Bt cotton was grown in: China, Australia, Mexico, South Africa, and Argentina.

Biosafety issues in field-testing and commercialization of transgenic plants received major attention in the industrialized nations. In these countries, governmental regulations for testing of transgenic crops in contained and field experiments for assessing the potential risks prior to the approval for commercialization are now in place (see Chapter 39). Researchers in industrialized countries such as the USA from both the private and public sectors now find it relatively easy to conduct field trials. The large-scale deployment of insect resistant crops using the Bt gene will be a major challenge in terms of the management and durability of Bt resistance. The future, however, looks promising because insect resistant products will cover a much broader range of pests that cause economic losses on crops in different regions of the world. Products with more than one Bt gene will increase the durability of Bt resistance and products with Bt and other mechanisms of resistance will provide further security and offer new possibilities for optimizing the durability of deployed genes. New Bt genes that confer resistance to many of the tropical pests damaging rice, soybean, sunflower, tomato, sugarcane, sweet potato, apple, and many other crops continue to be discovered. These new products, when commercialized, will significantly reduce the total use of insecticide. One study done in this area reports that of the US$8.1 billion spent annually on all insecticides worldwide, US$2.7 billion could be substituted with Bt biotechnology products (Krattiger, 1997).

It is also likely that, in the near future, transgenes for pest resistance are likely to become more sophisticated than the various genes for toxins employed today. Rather than killing pests, they may reduce the ability of pests to recognize and/or colonize the crop, or they may modify the damage-causing behavior of the pests. For example, transgenes are being developed that produce a protein, which inactivates the binding site in insect vectors where plant viral pathogens nominally bind. Anti-vectoring transgenes, for example, can be introduced into insects, and using genetics, the genes can be spread through the insect population. For a review on case studies involving the use of Bt in different crops including the economic, ecological, food safety, and social consequences of the deployment of Bt transgenic plants see the recently published work of Shelton *et al.* (2002).

Crop-specific issues

Major impact can be made for several crops grown in developing countries by deploying transgenic pest and disease resistant crops. The following three examples provide a clear idea of the benefits.

Rice

Unfortunately many developing countries still face pest problems such as weeds (red rice), bacterial, fungal, viral diseases, and insect pests like the stem borer and plant hoppers. Plant biotechnologies using resistant transgenes have now been researched and are available. For example, herbicide resistance genes genetically engineered into conventionally grown rice may allow for selective use of herbicides to control weeds such as red rice, a major weed problem in many regions of South America. Similarly, Bt resistance genes have now been tested for resistance to rice stem borer, a major insect pest of rice in Asia, and this technology can be combined with genes for rice sheath blight resistance, and the tungro virus. By developing such rice with multiple resistances to pathogens and pests, one would be able to reduce the current estimated yield loss of 20–30% annually in Asia. Additional information on the laboratories around the world which have transformed rice with Bt genes and evaluated them in greenhouse and field trials are presented by Tu *et al.*, 2000 and Ye *et al.*, 2001. In the near future, there is a good possibility for the release of transgenic Bt rice for farmers in developing countries (Cohen *et al.*, 2000).

Maize

Technologies for transformation and regeneration of tropical maize allows many maize-growing developing countries to engineer traits of national importance. For example, currently severe losses of over 40% occur in several tropical countries due to damage by pests (corn borers), and viruses such as the corn stunt, and others. Gene technologies (Bt, coat protein, replicase) provide an immediate short-term opportunity to develop superior pest and disease resistant maize germplasm with these genes. Such materials should have a major impact in increasing yields by over 30% (James, 2000).

Soybeans

Plant breeders are now able to back-cross elite soybean cultivars grown in developing countries such as Brazil with genetically engineered pest and disease resistant soybean varieties and create locally adapted germplasm for commercial release. Countries such as Brazil and China have access to some of the best soybean germplasm whereas the private sector has the gene technology. By combining efforts it is possible to develop transgenic soybeans with pest and disease resistance. Specific genes of interest are: herbicide resistance to the herbicide Roundup (glyphosate), and use of Bt genes for resistance to important lepidopteran insects damaging this crop. Yield gains of 30% and more can be expected due to the use of such technology.

Potatoes

Genotype independent transformation systems are now available to use in combination with *Agrobacterium tumefaciens*-mediated transformation procedures to obtain transgenic potatoes with a frequency of transgenic plant recovery of up to 50% (Douches *et al.*, 1998; Coombs, *et al.*, 2002). Potatoes with a codon-modified *Bt-cry5* (*Bt-cry1Ia1*) gene conferring resistance to potato tuber moth, *Phthorimaea operculella* Zeller, have been developed (Li *et al.*, 1999). 'Spunta', a variety grown in Egypt for local consumption, was used as the parent line because of its adaptation to tropical growing conditions. The highest expression was obtained in transgenic lines with the CaMV 35S promoter without the *gus* reporter gene. Laboratory trials in the USA and field and storage tests in Egypt in cooperation with the Agricultural Genetic Engineering Research Institute in Cairo and the International Potato Center station in Kafr-el-Zayat showed up to 100% control of leaf and tuber damage (Douches *et al.*, 2003). Lagnaoui *et al.* (2001) evaluated the *Bt-cry5* potatoes against both *P. operculella* and *Symmetrischema tangolias* (Lepidoptera: Gelechiidae) using detached leaf bioassays. All *Bt-cry5* Spunta lines caused high levels of mortality to both species (80–98%). The *Bt-cry5* Spunta lines are also effective against European corn borer, *Ostrina nubilalis* (Hübner), but not the cabbage looper, *Trichoplusia ni* (Hübner) in detached leaf bioassays (Santos, Grafius and Douches, unpublished).

The protection of potatoes in storage from potato tuber moth and related species is especially important in developing countries, where refrigerated storage may not be available. Even with insecticides for control of tuber moth in the field and careful sorting and removal of damaged tubers before putting the crop into storage, eggs and newly hatched larvae cannot be detected and some level of infestation is almost certain to enter the storage. Without treatment, tuber moth and the associated bacterial rotting organisms can rapidly spread throughout a storage. Regular removal of tubers from storage, sorting out damaged tubers, and treatment with insecticides directly on the remaining tubers before returning them to storage is commonly practiced to reduce injury and losses. Within 2–3 months 100% of susceptible tubers in a traditional Egyptian storage were damaged by tuber moth and infected with bacterial rot (Douches *et al.*, 2003). In the same storage, 98–100% of transgenic Spunta tubers were undamaged after 3 months (Douches *et al.*, 2003). Tuber moth damage in storage can also destroy tubers retained as seed for future crops, requiring purchase of commercial seed. Eliminating the need for treatment of stored tubers with insecticide will significantly reduce the risk of pesticide exposure to consumers, reduce the cost, and allow longer storage of potatoes for future consumption or seed.

Transgenic potatoes with resistance to Colorado potato beetle, *Leptinotarsa decemlineata* (Say), Potato Virus Y, and Potato Leafroll Virus have been produced and marketed in the USA by NatureMark (NatureMark, 2002), a subsidiary of Monsanto Corp. However, adoption was never widespread due to limited availability of seed, the additional cost of the transgenic seed, and consumer and buyer concerns about genetically engineered crops. As the result, these varieties are no longer available commercially. Research at Michigan State University and elsewhere continues on development of potato varieties resistant to Colorado potato beetle incorporating Bt-cry 3A toxin via genetic engineering and using traditional breeding techniques to incorporate additional mechanisms of resistance from other *Solanum* species (e.g. Coombs *et al.*, 2002).

Gene technologies to develop plants with resistance to pests and diseases can also be applied to other crops of importance in the tropics. These include, wheat, chickpea, oilseed (mustard), fruits and vegetables such as papaya, cucurbits, potatoes, tomatoes, aubergine, and other crops of significance to the economy of developing countries.

New biological control agents

Three main approaches for producing new biological control agents using genetic engineering are:

1. The engineering of natural enemies of pests to be more effective agents of biological control;
2. The engineering of plants or their microbial associates with genes that protect them from pests; and
3. The engineering of natural enemies of pests with genes for pesticide resistance, that increases options for pesticide use.

Among these three approaches, the creation and use of GMOs to improve plant protection receives the major attention. For additional details on current trends in biotechnology for pest management, see Waage (1996).

With rapid advances in molecular biology significant developments are occurring in the development of biologically produced pesticides (bio-pesticides). These advances usually include: naturally occurring organisms (such as fungi, baculoviruses, bacteria, nematodes), their by-products (such as chemicals derived from microbial organisms such as *Actinomyces* and *Streptomyces*), products derived from insects (such as pheromones), and products derived from trees (such as azadirachtin). The global market for these products is estimated at over US$500 million (Georgis, 1996). However, there are still a number of technological hurdles to overcome for mass-production, formulation, product safety, and persistence to reach this level of market sales.

The future of biological pesticides will depend largely on the financial strength and interest of the companies involved in producing these products. Initially, bio-pesticides were produced by pharmaceutical companies such as Sandoz and Abbott. However, in the last 2–3 years, the industry has begun to change. Firms such as Syngenta, Dow Chemicals, Zeneca, and DuPont have acquired many of the companies producing bio-pesticides. Additional research on genetic engineering and manufacturing processes (fermentation, synthesis, formulation) should strengthen the position of bio-pesticides in the market place.

The use of biotechnology will enable these bio-pesticides to be an even more effective component of future IPM programs in developing countries. Production of such agents could be developed as a small-scale or large-scale biotechnology industry in the tropics. Using biotechnology tools, researchers in these regions will be able to produce less expensive, and effective formulations for pest control. Transgenic bio-pesticides may prove faster acting and broader-spectrum than conventional products, but care must be taken not to sacrifice the valuable, self-renewing nature of biological control agents in developing more bio-pesticide-like products that require more repeated applications.

There is on-going work on developing more effective transgenic arthropods as biological control agents (Hoy, 1996). Currently, there is a lack of an example that demonstrates that such transgenic arthropods can be effective in a pest management program or that they can control a pest population. Until this has been achieved, adequate resources and funding will be difficult to obtain because it is considered to be high-risk research.

New diagnostic tools

Progress in the development of new diagnostic tools based on monoclonal antibodies and molecular markers has been rapid. Using techniques such as restriction fragment length polymorphism (RFLP), polymerase chain reaction (PCR) in random amplified polymorphic DNA (RAPD), and a new powerful synthesis of RFLP and PCR, called amplified fragment length polymorphisms (AFLP) enable breeders to develop new varieties. These diagnostic tools also allow IPM practitioners to determine: the evolutionary status and population dynamics of pests, pathogens and of the natural enemies of pests; to monitor changes in pest population and of the beneficial organisms; to predict pest outbreaks and detect and measure pesticides; and to identify products produced from transgenic plants.

For details on the use of these novel technology options in integrated pest management systems see Whitten *et al.* (1996).

Developments in human health care allow these tools to be modified for use in plant health diagnostics. In developing countries the development of these tools will be useful for:

1. Pest identification;
2. Enabling farmers to visualize the situation in the field, with respect to the presence of the pest or the disease;
3. Monitoring pest movement in time and space;
4. In resistance breeding by simplifying or eliminating the need for disease bioassay;
5. In monitoring pesticide contamination in the postharvest produce; and
6. In detecting pesticide resistant pests.

These tools are fairly easy to develop. For example, many developing countries such as Brazil, India, China, South Africa, and Egypt use enzyme-linked immunosorbent assays (ELISA) for the detection of important plant viruses. Similarly the quarantine programs of many developing countries use ELISA to prevent the entry of plant viruses of quarantine significance.

Developing countries are in need of additional diagnostic tools for detecting pesticide residues. The use of appropriate diagnostic technology in this area will promote exports and allow for pesticide-free certification. Other technologies such as the use of PCR to detect transgenic food products and biochemical kits for detection of resistant pests to transgenic plants and pesticides are also needed. It is in the interest of developing countries to identify key areas where diagnostic technologies can be best utilized in the short, medium, and long term.

DNA markers, mapping and application for developing pest and disease resistant plants

Many neutral-genetic markers are used as tags to identify the specific genes of interest to crop improvement. Once a statistical gene-marker association is established for a given cross, the marker phenotype of a progeny becomes a genetic shorthand for its individual gene makeup, allowing genotypic selection without, or in addition to, phenotypic selection. For the purpose of crop breeding, the major recent advances in genetic markers have consisted of finding ways to generate and assay large numbers of markers. The main physical tools are DNA cleavage via restriction enzymes, homologous DNA–DNA hybridization, electrophoretic DNA separation technology, and DNA amplification via DNA PCR. Various combinations of these have produced the genetic marker types known as RFLP and AFLP, RAPD, and simple sequence repeats. The practical consequence is that one can come arbitrarily close to knowing the actual genetic composition of any plant in comparison to its parents or siblings. This affords better estimates of the number, contributions and genomic locations of genes controlling a complex character (say, drought tolerance or pest resistance controlled by many genes) than could ever have been obtained without DNA genotyping. With this information one can identify, without expensive field testing, plants that are likely to possess the favorable alleles of these genes, and discard those that do not. The practical benefit of marker-assisted selection to plant breeding is expected to be large savings in the land, labor, and time required for variety development; estimates of potential time-saving vary from upward from 2 years for an annual crop. Marker-assisted selection is practical for single-gene traits whose assay is tedious (nematode resistance in wheat, tomato, potato), and horticultural crops of high value, and in long cycle species such as trees.

Current marker technology is not the end point in molecular breeding. Even the best markers are still relatively expensive (in time, labor, skill and materials) to assay. Although exact cost analysis is highly dependent on site, scale, and goals of a project, the cost of a data point (of the tens of thousands required for many mapping experiments) was estimated to be from US$ 0.10 to > 1.00 (Ragot and Hosington, 1993), and has not decreased radically since then. The short-term trend is toward increasing throughput of marker–genotyping labs via multiplex PCR reactions, fast DNA fragment-sizing methods (thin-gel or capillary electrophoresis), or methods that require separations such as allele-specific PCR (Gu *et al.*, 1995), robotics for DNA handling, and automated data acquisition involving optical scanning and image analysis. New tools have been developed for fast, cheap molecular genotyping using the oligo microarray (Chee *et al.*, 1996), e.g. a small chip of glass slide on which are fastened many thousands of DNA oligonucleotide fragments for simultaneous hybridization with a fluorescent labeled DNA probe. The resulting patterns, read via microscope, are processed with computer image-analysis software. This technology (pioneered by the Santa Clara, California based biotechnology company Affymetrix, among others) allows the simultaneous assay of large number of DNA polymorphisms with one or a few hybridizations.

Specific case studies involving the use of molecular marker technology for analysis and manipulation of resistance to crop pests such as the bacterial blight of rice, and rice blast illustrate the value of these tools for developing rice varieties with resistance to these diseases (Nelson, 1996). In the next few years, molecular techniques of various kinds are likely to become universal in plant breeding. However, the rate at which molecular data can now be acquired has overtaken that at which it can be synthesized, understood, and applied. Developing countries should focus on expanding their ability not only to collect molecular data but also to develop bio-informatics tools and capabilities to synthesize, analyze, and apply it for crop improvement. A database system unifying crop performance with molecular and pedigree data for a range of tropical crops should be created, and linked, where possible, to established databases for the same or related crops in other countries. Expertise should be acquired in computational biology as applied to database model building, data mining and visualization, and the statistical synthesis of complex laboratory and field data sets to arrive at decisions for marker-assisted crop breeding. A core of national expertise will then be able to supplement or replace foreign centers in training new developing country molecular breeders and acquainting established breeders with new tools. Application of this expertise to existing technology is expected to result in varieties with broad adaptation and stable performance under stress. This is in contrast to transformation, which will be used for making major and abrupt alterations in single-gene traits, but whose long-term impact is uncertain especially in smallholder agriculture.

Policy Issues Affecting Access to New Biotechnologies

The majority of new and emerging biotechnologies are proprietary and reside in the private sector of the developed world. The access, use and management of these proprietary technologies require a sound policy framework in IPR, biosafety, food safety and technology transfer (Maredia and Erbisch, 1998; Maredia *et al.*, 1999).

Intellectual property is a broad term used to cover patents, plant variety protection, trademarks, trade secrets and copyrights. All of these have a part to play in the use and management of biotechnologies. The intellectual property rights provide legal protection for genetic resources, genetically transformed organisms, and related products and technologies as well as access to technologies from local, regional and international sources. Biosafety encompasses policies and procedures adopted to ensure environmentally safe applications of new technologies. The environmental safety issues surrounding the use of biotechnology include gene flow/gene transfer, weediness, pest/pathogen effects, impacts on non-target and beneficial organisms, and development of pest resistance. The food safety issues encompass allergenecity, toxicity, and altered nutritional content of the genetically modified food products and their impact on human, and animal health.

The IPR, biosafety, food safety and technology-transfer policies and their implementation are important for safe and legal exchange and commercialization of biotechnology. They provide incentives for local researchers and firms, they are required by international laws, and they can assist in the international transfer of technology and international trade. In order to access new biotechnologies, many countries are putting together appropriate policy frameworks and building their capacity to implement these policies. For example, the government in Egypt has developed national Biosafety guidelines and IPR policies as they relate to the use and management of biotechnology in Egypt.

Pest Resistance Management

Pests and diseases have a remarkable capacity to adapt to several control agents. Development of resistance in insect pests to control methods using resistant plant varieties, cultural or biological controls, insect controlling pathogens, and insecticides is common. More than 4000 examples of resistance to insecticides have been documented in populations of about 500 species of insects. Insect pests have evolved resistance to all major categories of insecticides including insecticidal crystal proteins from the bacterium Bt, and some major insect pests have evolved resistance to new insecticides within 1–3 years (Metcalf, 1989; Georghiou and Lagunes-Tejeda, 1991; Gould *et al.*, 1997).

The tragic failures are not isolated incidences. Every major crop – cotton, rice, maize, fruits, vegetables, and ornamentals – has one or more resistant pests. This problem is not isolated to developing countries. Cotton producers in the USA, Australia, China, Turkey, Pakistan and India are battling resistance in more than a dozen species of insects and mites; so are fruit growers in Europe, Canada, the northwestern USA, Australia and Japan (among others).

The price of insecticide resistance in lost yields and higher insect control costs is staggering – more than US$1 billion in the USA from the budworm/bollworm complex alone. Once a crop protection product is rendered ineffective by resistance, it may be lost from the IPM toolbox forever. There are several projects in many developing countries that focus on pesticide resistance management.

With the wide adoption of transgenic crops with insect resistance, the issue of pests developing resistance is of major concern in both developed and developing countries. It is, therefore, important to have information on resistance management strategies, regulatory options and other deployment strategies adopted in the USA that may have significance in developing countries. In the USA, the EPA Office of Pesticide Programs has formed the scientific Pesticide Resistance Management Working Group to consider EPA's role concerning resistance management of insect resistant transgenic plants. The seven key elements identified by this group for developing an adequate Resistance Management Strategy (RMS) are:

1. Knowledge of pest biology and ecology;
2. Appropriate gene deployment strategy;
3. Appropriate refugia (primarily for insecticidal genes such as Bt);
4. Monitoring and reporting of incidents of pesticide resistance development;
5. Employment of IPM;
6. Communication and educational strategies on product use; and
7 Development of alternative modes of action. For more details see: Lewis and Matten (1995).

Many of the above elements can be achieved by relying on four key strategies: (i) diversification of mortality sources; (ii) reduction of selection pressure; (iii) maintaining a susceptible population via refugia or immigration; and (iv) prediction of resistance development. Significant research developments are in progress at several private and public sector institutions to implement these strategies at the field level. RMS is further enhanced by the development of several tactics. These include the use of gene strategies. It is possible to use single or multiple genes in a pyramid fashion. Multiple gene deployment reduces the likelihood of resistance development since multiple mutations would have to occur concurrently in individual insects. This tactic is still in the development phase and no cultivars with multiple genes have yet been released for commercial production. The second involves the use of gene promoters which are either constitutive, or tissue specific or inducible (e.g. wounding). Targeted expression is aimed at reducing the time period of insect exposure to a toxin by expressing it only in vulnerable parts of the plant or in a certain part of the plant simultaneously with a particularly critical time in the development of the plant. Several of the private companies now have the technology to express the Bt toxin gene in selected parts of the maize plant. The third tactic includes the use of high-dose gene expression including the expression of genes in mixtures. The high level of expression of a single toxin in all plants is aimed at killing the highest possible percentage of insect populations and is generally implemented in conjunction with refugia. Low levels of expression of a single gene toxin in all plants is expected to produce a sub-lethal dose that would reduce fertility and growth of insect populations and also make the affected insects prone to predators and parasites. The fourth tactic that is employed at the field level includes the use of a uniform single gene, mixtures of genes, gene rotation, mosaic planting or providing refuges to delay the development of resistant insects. The use of refugia is now a common practice that is recommended as a field tactic to delay resistance in insects by several of the private companies involved with commercialization of insect resistant transgenic crops. A refuge enables insects to breed and thus provide a steady supply of wild-type insects (or non-resistant ones) that would be most likely to mate with potentially resistant insects. This would reduce the chances of an increase in the frequency of resistance genes. Further explanations on the various gene strategies to delay resistance are provided in McGaughey and Whalon (1992).

Specific RMS used in the USA for transgenic plants containing Bt include:

1. Establishing baselines and monitoring shifts in Bt susceptibility;
2. Research on ecological and genetic factors of Bt resistance;
3. Experimentally validating resistance management strategies;
4. Integrating Bt with other pest control tactics;
5. Assuring an appropriate regulatory environment;
6. Characterizing Bt cross resistance;
7. Estimating Bt resistance gene frequencies;
8. Mapping and cloning Bt resistance genes;
9. Establishing a national scientific advisory group for management as a part of the national biosafety program; and
10. Including resistance management as a part of the national biosafety program.

The private sector in the USA involved with the commercialization of insect resistant transgenic crops has developed refuge management strategies for different transgenic crops to prevent the likelihood of insect resistance (Shelton, 2002). These deployment strategies already implemented show that a high-dose strategy with refuges has been adopted in cotton, maize, and potato. These are all in the USA where five Bt transgenic products are already in the market. The exact refuge depends on the crop and on the selected treatment of the refuge area. Monsanto, in the commercialization of its Bt transgenic cotton under the trade name of Bollgard, provides two options. The first is to provide 20% unprotected cotton (non-transgenic) as a refuge using non-transgenic as a total refuge using no insect control whatsoever. The private sector in the USA tends to favor the 20% unprotected refuge strategy as an acceptable optimum.

Along with the refuge strategy, several of the private sector firms have established a monitoring program to sample insects. The Novartis company (now Syngenta) monitors maize insects such as the European corn borer and corn earworm populations annually. The goal of these monitoring efforts is to identify as early as possible suspected changes in insect susceptibility to Bt maize hybrids, allowing time to modify RMS recommendations. Such monitoring of insects for early identification of possible resistant insects is critical in all the strategies outlined above. This poses formidable challenges because collecting insects at random may not necessarily allow early enough detection of resistance to allow remedial actions to be implemented. Requesting farmers to monitor insect populations has limitations, particularly in developing countries with resource-poor small-scale agriculture, where the extension efforts required for such a system to work are tremendous. Many farmers lack knowledge of the simple tools to identify and collect the right insects. The cooperation of many different groups is needed to establish an effective monitoring program.

Regulatory policy options for RMS

It is clear that the success of any policy will depend on the number of entities or people regulated as well as the particular crop and production system involved. Obviously, the fewer the number of people that are targeted for regulation, the easier it will be to enforce. The current regulatory apparatus, the particular transgenic crop under consideration, and the existing seed-handling system are important factors in determining effective regulatory options in each country. Whalon and Norris (1997) have identified six main regulatory options for Bt transgenic plant deployment. These include: (i) licensing; (ii) central control of seed; (iii) regulation of seed distribution; (iv) labeling; (v) monitoring use; and (vi) monitoring resistance development. The easiest and most straightforward procedure to implement and enforce appears to be licensing and labeling, followed by monitoring resistance development.

Licensing

Commercialization of new inventions such as the development of transgenic plants with the Bt toxin gene is a high priority of the private sector in the developed nations. This is usually done by licensing their proprietary technology to other parties with pre-agreed terms regarding its use or sale. The private US company, Monsanto, used licensing to achieve compliance on utilization, production and sales of transgenic Bt crops like potato, cotton, and maize. Growers who partner in such agreements have an opportunity to purchase Bt crops with appropriate information on gaining maximum benefits from the use of this technology. Under this agreement growers agree to not reuse or save seed containing the Monsanto gene technology, and to set up an adequate refuge when planting Bt crops. Additional details are given in the 1997 Monsanto technology agreement (Monsanto, 1997).

Licensing strategies can be implemented at various levels. It can be done at the point where seeds are used by including seed companies, seed distributors, sales representatives, or end users. The enforcement apparatus would then vary based on the particular target. For example, a national government agency could enforce licensing agreements at all target levels, although government licensing of seed companies would be easier to implement than government licensing of individual farmers. Seed companies or local grower organization could also enforce licensing at the local level by acting as a licensor. Strict penalties are usually enforced if the license agreement is infringed upon or is not followed. In all cases, the establishment of such a policy requires the input of many resources such as capital, personnel, facilities, and educational campaigns to train in the license application process and proper use of the seed.

Labeling

Most biotechnology companies now provide ‘proper use labels’ on most products. Industry cooperation and compliance with this type of regulation can be high. For example, all pesticides currently sold carry labels that give precise information on the chemical name of the pesticide, its concentration, the pests it controls, and at what rate it should be applied. In the case of transgenic seed with the *Bt* toxin gene the label should include:

1. Identification of transgenic status;
2. Recommended planting ratio (percentage of transgenic versus non-transgenic seed, or recommended refugia for RMS);
3. A warning against misuse or over planting of the transgenic seed; and
4. An agency or industry contact in the event that resistance to the transgenic crop becomes evident in the pest population.

The local extension agencies in developing countries should provide grower education programs on how to interpret information on the labels and how to implement the appropriate proper use recommendations. Private companies should also be willing to participate in such efforts.

Use of IPM to delay pest resistance

IPM is a widely accepted strategy to reduce over-dependence on chemical insecticides and their potential negative environmental and economic effects. The use of IPM reduces the selection pressure on pests and thereby delays resistance development. IPM has been adopted widely for the control of rice pests in Asia. It is also practiced widely in many other crops grown in the developed and developing world. The insecticide resistance action committee in the USA recommends using IPM by the participating growers to delay resistance development. Five resistance guidelines are recommended to strengthen the durability of pesticides and to help keep grower costs down. These are:

1. Consulting with an agricultural advisor or extension agent for regional insecticide resistance and IPM strategies. This allows one to consider the appropriate pest management options available and map out a season-long plan to avoid unnecessary applications of insecticides.
2. Consideration, before planting, of the options for minimizing pesticide use by selecting pest-resistant crops, and by managing the crop for early maturity.
3. Regular monitoring of fields during the season to identify pests and natural enemies properly and track their stages of development. Pesticides should be used only if pest counts go over the local economic threshold – the point at which economic losses exceed the cost of the insecticide, plus application costs. Timing applications against the most susceptible life stages brings maximum benefit from the product.
4. Selection of insecticides that have minimum effect on beneficial insects and other natural enemies. This will reduce the number of applications and thereby reduce selection pressure on insects to develop resistance. Resistance grows under continued pressure.
5. The use of only one control component such as Bt crops can enhance resistance. If one combines this technology with others, such as the ones described above, the resistance developed will be delayed allowing for better protection, and increased profits.

It is, therefore, essential that during the deployment of insect resistant crops developed through genetic engineering, all available IPM control methods are used at the field level. National governments, regulatory agencies, the private and public sectors, and growers should conduct on-farm demonstrations on the use of such approaches. Such simple trials will demonstrate the value of IPM, which in turn will ensure the long-term durability of insect resistant crops.

Public–Private Sector Linkage Considerations

In most instances, conventional breeding and IPM technologies developed in the public sector are useful and should be continued. There are, of course, opportunities for using well-proven and tested gene technologies and other proprietary biological controls for increasing the durability of pest control. Many of these technologies being investigated by the public and private sector firms in the developed world are proprietary and mostly patented by the private sector. The participation of private sector is therefore critical to the delivery of effective biotechnologies for effective plant protection. The public sector needs to create a favorable environment to encourage private-sector participation by providing a regulatory system that accurately informs the public of the benefits and risks involved in the use of new technologies; a legal framework for protecting IPR; adequate infrastructure for power, transport and telecommunications; a fair tax system and investment incentives; a skilled workforce, including a well-supported university sector; public funding for research and development; and incentives to establish innovative public–private collaboration, and joint ventures at the national and international levels.

For NARS to access proprietary technologies developed in the private or public sector requires the resolution of intellectual property and ownership issues. This issue affects patents, plant variety protection, seed certification and access to biodiversity. Capacity building in the IPR area is crucial so that NARS can effectively collaborate with the owners of technology. Human resource development in the area of intellectual property management, legal, technical and business expertise including database management are important. A few NARS are now creating IPR management focal points to serve as technology transfer offices. The personnel in these offices are trained in the area of IP policy development, protection and licensing mechanisms, networking, and creating new start-up companies. IPM researchers who do not have time to go through the negotiation process with the private sector should consider using these focal points to help them work with the private sector. The main purpose of such focal points should be to protect local inventions and enable access to technologies developed elsewhere. The capacity building in the IPR area is a continuing process and requires institutional and financial commitment. It also needs to work in collaboration with other regional and global bodies that have the expertise in the area of IPR. Gaining knowledge on how to develop effective confidentiality agreements, material transfer agreements, collaborative research agreements, research-only license agreement, commercial license agreement, visiting scholar agreement, and confidential disclosure of invention will help NARS to promote partnerships with the private sector.

Several US universities and others in Europe provide training in all areas of IPR (Maredia *et al.*, 1997; Maredia and Erbisch, 1998). A few NARS are using these opportunities to develop their human resource base so that they are in a strong position to negotiate with the private sector. It is also apparent that once genetically modified organisms are developed, NARS will have to field test them under their own conditions following appropriate biosafety and food safety guidelines. Biosafety and the regulation of biotechnology activities have been at the forefront of the biotechnology debate for almost a decade, especially in developing countries. In the USA there are three federal agencies that have regulatory responsibility and authority to maintain the safety and nutritional value of America's food supply. They are the EPA, the FDA and the USDA. Each agency has specific authority. However, such biosafety mechanisms are limited in developing nations. It is only during the last 3 years that many developing nations have started developing their own biosafety guidelines, and regulations for field-testing transgenic plants. The ultimate aim of these programs is to develop regulations that are science based; transparent; harmonized with international protocols, domestic legislation, and import–export requirements; and implemented by credible institutions. This is a tall order and progress will be slow, as many NARS have limited financial and human resources to implement successful biosafety and food safety programs. Success will only come by creating human resource capacity for biosafety regulation through programs such as training, workshops, seminars, and technical meetings that will continue to build capacity in biosafety and food safety. In order for GMOs to reach developing NARS, other areas such as public participation and awareness programs should be established from the outset to communicate effectively with the public about the rationale for decisions, and the risks and the benefits of crop biotechnology.

The program should also encourage public participation in the decisions regarding the use of transgenic products. The benefits such as direct saving to growers and any other indirect savings and its impact on processors and consumers should be demonstrated to the public.

Information exchange and experience sharing with developed countries such as the USA, which has now commercialized several transgenic crops, including potatoes, would be useful. Through partnerships and knowledge-sharing developing country NARS will be able to develop their own IPR, biosafety and food safety regulatory structure and procedures for implementing policy. Once such guidelines are in place, it becomes easy for the private sector to become a partner with NARS to promote technologies such as the use of genetically engineered plants with resistance to pests and diseases. For additional information see: Raman, 1995, 1998; Ives *et al.*, 1998; Maredia *et al.*, 1999.

Future of Biotechnology in IPM

The vast majority of transgenic pest and disease resistant crops developed and commercialized are the result of single-gene transfers, in which one or more genes coding for desired characteristics such as insect, disease, virus or herbicide resistance are introduced. In some cases both insect and herbicide resistance were combined. Such efforts, although important for controlling specific pests, are unlikely to control other important pests and diseases of importance in the tropics. To create plants with multiple resistance, the plants would have to be thoroughly re-engineered with more than one or two genes.

The technological advances in the area of genomics may provide clues to identifying novel genes that may provide broad-based resistance to many pests. Since the early 1980s, a succession of technological advances has made it possible to perform large-scale DNA sequencing. New genomic data has been generated at an explosive rate: the volume has doubled every 15 months since about 1984. Extracting biological insight from this raw data is one of the central challenges of biology today.

The ability to relate genomic data to biological function would constitute a major step in understanding the process of life. Such an advance would likely have a significant impact on medicine and agriculture. There are international efforts to establish the full DNA sequence of every gene required to produce a plant. This is a pioneering initiative for crop breeding. The subsequent steps involve interpretation of genes' structures and patterns of expression in each organism. Once a gene is identified in one species, its functional value can be found in other species to aid breeding of any crop for pest, and disease resistance.

Progress in the genomics area will significantly influence the future of biotechnology research, and the speed with which we can develop plants with resistances to many pests and diseases. New developments could include the ability of the plant's own cells to fight pests and diseases. Global stakeholders in the life sciences sector have already committed a significant level of internal, and collaborative investments in genomics research with significant contribution flowing from universities and other world-class research bodies who have expertise in genomics. In the future, new advances in the area of genomics and allied areas such as proteomics will further stimulate research and commercialization opportunities for the biotechnology sector.

The convergence of classical biology, genomic tools, and sequence information and the availability of sophisticated model genetic systems are enabling unique opportunities for the agricultural scientists to both identify and commercialize new pest control technologies. The emergence of new scientific fields, such as nano-biotechnology that fuses non/microfabrication and biosystems to develop microscopic tools that can extract and explore genetic information, is expected to fuel major breakthroughs as it will allow more precise information to be gathered from biological systems in a more rapid manner. New ultra performance DNA sequencing instruments based on novel fluorescent detection technologies will have the capability of sequencing large amounts of genomic data, over 340,000 base pairs/hour or 5000/min. At the same time these novel techniques will enhance the resolution of mixed DNA applications leading to significant improvements in the development of DNA-based diagnostic tools.

There is now a growing interaction between fundamental researchers, the product developers, and marketers in the USA, Europe, and to some extent, in Japan. This is helping to accelerate the realization of benefits to society from the scientific advancements in the life sciences area far more rapidly during the last decades. The growing interaction between multidisciplinary researchers will be essential to stimulate new discoveries to strengthen existing IPM. In the developed world, issues related to IPM labeling, consumer education, and the role of the retail sector in providing a way for consumers to buy food that is produced in environmentally friendly ways continues to be a high priority. In the developing world, implementation of these and other IPM strategies have been limited. NARS will have to continually develop and integrate existing IPM control components to offset the risks to farmers due to the loss of conventional chemical control, and the problems of pest resistance. For example, in many areas of South and Central America, the leaf miner fly, white flies, aphids, and potato tuber moth are resistant to the commonly applied insecticides. There is, therefore, a need for continuing investments from national governments, NGOs, private sector and international development agencies to support IPM. Successful IPM labeling programs need to be promoted as this will allow consumers to differentiate between food produced using IPM and conventional systems. Such foods could enhance exports to other environmentally concerned countries.

In many countries, important food crops are genetically engineered for virus, fungus, and insect resistance. The challenge is to integrate these products into IPM and crop management programs to delay or prevent pest resistance. The IPM therefore should become part of the national biosafety and biotechnology policy. There are also issues related to intellectual property rights and regulatory hurdles, that one needs to consider for the responsible use of biotechnology. Continuing efforts and investments to build capacity in biotechnology research, policy, and management are required. Education and awareness on the risk/benefits of the products of modern biotechnology should be continuously promoted at the national and international level, for the greater acceptance of biotechnology products.

References

Carpenter, J.E. and Gianessi, L.P. (2001) Agricultural biotechnology: updated benefit estimates. National Center for Food and Agricultural Policy. http://ncfap.org/pup/biotech/updatedbenefits.pdf Accessed 30 August 2001.

Chee, M., Yang, R., Hubbell, E., Berno, A., Huang, X.C., Stern, D., Winkler, J., Lockhart, D.J., Morris, M.S. and Fodor, S.P. (1996) Accessing genetic information with high-density DNA arrays. *Science* 274, 610–614.

Cohen, M.B., Gould, F. and Bentur, J.S. (2000) *Bt* rice: practical steps to sustainable use. *International Rice Research Notes* 25(2), 4–10.

Coombs, J., Douches, D., Li, W., Grafius, E. and Pett, W. (2002) Combining engineered (Bt-cry3A) and natural resistance mechanisms in potato for control of Colorado potato beetle. *Journal of the American Society for Horticultural Science* 127, 62–68.

Douches, D.S., Westedt, A.L., Zarka, K. and Grafius, E.J. (1998) Potato transformation to combine natural and engineered resistance for controlling tuber moth. *HortScience* 33, 1053–1056.

Douches, D., Santos, F., Coombs, J., Grafius, E., Li, W., Metry, E., Nasr El-Din, T. and Madkour, M. (2003) Field and storage testing Bt-potatoes for resistance to potato tuber moth (Lepidoptera: Gelichiidae). *Journal of Economic Entomology* (in press).

Estruch, J.J., Carozzi, N.B., Desai, N., Duck, N.B., Warren, G.W. and Koziel, M.G. (1997) Transgenic plants: an emerging approach to pest control. *Nature Biotechnology* 15, 137–141.

Georgis, R. (1996) Present and Future Prospects of Biological Insecticides. Cornell Community conference on Biological Control, 11–13

April 1996, Cornell University, Ithaca, New York, USA. http://www.nysaes.cornell.edu/ent/bcconf/talks/georgis.html

Georghiou, G.P. and Lagunes-Tejeda, A. (1991) *The occurrence of Resistance to Pesticides in Arthropods.* FAO, Rome, Italy.

Gould, F., Anderson, A., Jones, A., Sumerford, D., Heckel, D.G., Lopez, J., Micinski, S., Leonard, R. and Laster, M. (1997) Initial frequency of alleles for resistance to *Bacillus thuringiensis* toxins in field populations of *Heliothis virescens. Proceedings of the National Academy of Science USA* 94, 3519–3523.

Gregory, P., Stewart, R. and Satvrou, S. (2002) Adoption of Bt Cotton by Small-Scale Farmers in South Africa. *Pesticide Outlook*, February, 31–34.

Gu, W.K., Weeden, N.F., Yu, J. and Wallace, D.H. (1995) Large-scale, cost effective screening of PCR products in marker-assisted selection applications. *Theoretical and Applied Genetics* 91, 465–470.

Hoy, M.A. (1996) Novel-arthropod biological control agents. In: Persley G.J. (ed.) *Biotechnology and Integrated Pest Management.* Biotechnology in Agriculture No. 15. CAB International, Wallingford, UK, pp. 164–185.

Ives, C.L., Bedford, B.M. and Maredia, K.M. (1998) The agricultural biotechnology for sustainable productivity project: a new model in collaborative development. In: Ives, C.L. and Bedford, B.M. (eds) *Agricultural Biotechnology in International Development.* CAB International, Wallingford, UK, pp. 1–14.

James, C. (1999) Global status of commercialized transgenic crops: 1999. *ISAAA Briefs* No. 17. ISAAA, Ithaca, New York, 65 pp.

James, C. (2000) Global status of commercialized transgenic crops: 2000. *ISAAA Briefs* No. 21: Preview. ISAAA, Ithaca, New York.

Krattiger, A.F. (1997) Insect resistance in crops: a case study of *Bacillus thuringiensis* (Bt) and its transfer to developing countries. *ISAAA Briefs* No. 2, ISAAA, Ithaca, New York, 42 pp.

Lagnaoui, A., Canedo, V. and Douches, D. (2001) Evaluation of Bt-cry1IA1 (cry V) transgenic potatoes on two species of potato tuber moth, *Phthorimaea operculella* and *Symmetrischema tangolias* (Lepidoptera: Gelichiidae) in Peru. *CIP Program Report 1999–2000.* International Potato Center, Lima, Peru.

Lewis, P.I. and Matten, S.R. (1995) Consideration and management of pesticide resistance by the U.S. Environmental Protection Agency. *Resistant Pest Management Newsletter* 7(2), 10–13. Pesticide Research Center, Michigan State University, East Lansing, Michigan.

Li, W., Zarka, K., Douches, D.S., Coombs, J.J., Pett, W.L. and Grafius, E.J. (1999) Co-expression of potato PVY coat protein gene and cry V-Bt genes in potato (*Solanum tuberosum* L.). *Journal of American Society of Horticultural Science* 124, 218–223.

Maredia, K.M. and Erbisch, F.H. (1998) Capacity building in intellectual property management in agricultural biotechnology. In: *Intellectual Property Rights in Agricultural Biotechnology, Biotechnology in Agriculture Series, No. 20.* CAB International, Wallingford, UK.

Maredia, K.M., Erbisch, F.H. and Dodds, J.H. (1997) Strengthening the technology transfer framework in developing countries: a Michigan State University Internship Program. *Industry and Higher Education*, June, 145–149.

Maredia, K.M., Erbisch, F.H., Ives, C.L. and Fischer, A.J. (1999) Technology transfer and licensing of agricultural biotechnology in the international arena. *AgBiotechNet* Vol. 1, May. ABN 017.

McGaughey, W.H. and Whalon, M.E. (1992) Managing insect resistance to *Bacillus thuringiensis* toxins. *Science* 258, 1451–1455.

Metcalf, R.L. (1989) Insect resistance to insecticides. *Pesticide Science* 26, 333–358.

Monsanto (1997) *Monsanto Technology Agreement Bollgard Cotton, Roundup Ready Cotton, Roundup Ready Soybeans.* Special Publication of Monsanto Consumer Relations Center, PO Box 938, Langhorne, Pennsylvania, 19047–0938.

NatureMark (2002) http://www.naturemark.com/pages/Home.html

Nelson, R.J. (1996) Marker-assisted Plant Breeding. In: Persley, G.J. (ed.) *Biotechnology and Integrated Pest Management.* Biotechnology in Agriculture No. 15. CAB International, Wallingford, UK, pp. 186–204.

Persley, G.J. (1996) *Biotechnology and Integrated Pest Management.* Biotechnology and Agriculture Series No. 15. CAB International, Wallingford, UK, 475 pp.

Persley, G.J. and Doyle, J.J. (1999) *Biotechnology for Developing Country Agriculture: Problems and Opportunities.* IFPRI Brief 1, Focus 2 Washington, DC.

Persley, G.J. and Lantin, M.M. (2000) *Agricultural Biotechnology and the Poor.* Proceedings of an International Conference, Washington, DC, 21–22 October 1999. Consultative Group on International Agricultural Research, Washington, DC.

Raman, K.V. (1995) Facilitating plant biotechnology transfer to developing countries. In: Altman, D.W. and Watanabe, K. (eds) *Plant Biotechnology Transfer to Developing Countries.* Academic Press. R.G. Landes Company, Austin, Texas, pp. 267–275.

Raman, K.V. (1998) Public and private research promoting synergies. In: Chopra, V.L., Singh, R.B. and Varma, A. (eds) *Proceedings of the Second International Crop Science Congress, 1996, Crop Productivity and Sustainability, Shaping the Future.* Oxford & IBH Publishing, New Delhi, India, pp: 937–948.

Ragot, M. and Hoisington, D.A. (1993) Molecular markers for plant breeding: comparisons of RFLP and RAPD genotyping costs. *Theoretical and Applied Genetics* 86(8), 975–984.

Shelton, A.M., Zhao, J.Z. and Roush, R.T. (2002) Economic, ecological, food safety, and social consequences of the deployment of Bt transgenic plants. *Annual Review of Entomology* 47, 845–881.

Tu, J.M., Zhang, G.A., Datta, K., Xu, C.G., He, Y.Q. *et al.* (2000) Field performance of transgenic elite commercial hybrid rice expressing *Bacillus thuringiensis* delta-endotoxin. *Nature Biotechnology* 18, 1101–1104.

US Congress, Office of Technology Assessment (1995) *Biologically Based Technologies for Pest Control.* OTA-ENV-636. US Government Printing Office, Washington, DC.

US EPA, Office of Pesticide Programs, Biopesticides and Pollution Prevention Division (2000) *Biopesticides Registration Document; Preliminary Risks and Benefits Sections;* Bacillus thuringiensis *plant-pesticides.* US Environmental Protection Agency, Washington, DC.

Waage, J. (1996) Integrated pest management and biotechnology: an analysis of their potential for integration. In: Persley, G.J. (ed.) *Biotechnology and Integrated Pest Management.* Biotechnology in Agriculture No. 15. CAB International, Wallingford, UK, pp. 37–60.

Whitten, M.J., Jefferson, R.A. and Dall, D. (1996) Needs and opportunities. In: Persley G.J. (ed.) *Biotechnology and Integrated Pest Management.* Biotechnology in Agriculture No. 15. CAB International, Wallingford, UK, pp. 1–36.

Whalon, M.E. and Norris, D.L. (1997) Pest resistance management and transgenic plant deployment: perspectives and policy recommendations for Mesoamerica. In: *Transgenic* Bacillus thuringiensis *plants in Mesoamerican Agriculture.* Zamorrano Academic Press, Escuela Agricola Panamericana, Zamorano, Honduras, pp. 62–85.

Ye, G.Y., Shu, Q.Y., Yao, H.W., Cui, H.R., Cheng, X.Y., *et al.* (2001) Field evaluation of resistance of transgenic rice containing a synthetic cry1Ab gene from *Bacillus thuringiensis* (Berliner) to two stem borers. *Journal of Economic Entomology* 94, 271–276.

Chapter 5

Pesticide Policy and Integrated Pest Management

Gerd Fleischer[1] and Hermann Waibel[2]

[1]Rural Development Department Division, German Technical Cooperation (GTZ), Eschborn, Germany; [2]Faculty of Economics, University of Hannover, Hannover, Germany

Introduction

Chemical pesticide use in developing countries has grown substantially during the past four decades. The so-called Green Revolution technological package of high-yielding varieties and mineral fertilizer demanded the use of chemical pesticides to secure the full achievement of the yield potential. National governments, bilateral and multilateral donor agencies promoted pesticide use in order to achieve national food security and match the production targets for agricultural export commodities. The orientation towards unilateral use of chemical pesticides was challenged when widespread negative externalities became apparent. IPM has been developed to overcome the dependency of cropping systems on chemical pesticides and to reduce environmental and health risks. However, the diffusion of IPM is severely impeded because in many developing countries pesticide use is subsidized or favored through non-price instruments when compared with non-chemical pest control measures. The reform of the legal, fiscal, economic, and institutional framework governing pesticide use can thus be a contribution toward reversing the pro-pesticide bias in agricultural policies.

The aim of this chapter is to outline the rationale and the importance of policy analysis for designing rational decision making in the area of crop protection. The first section describes the most important trends in crop protection that shape the current policy agenda and exert pressure to change the pro-pesticide bias. Although the global policy environment has become more important in the last decade, actual implementation of IPM policies is still mainly a task for decision makers at the level of individual countries. A concept for country level studies has been used in a joint pilot project by the University of Hannover and GTZ (German Technical Cooperation). The analysis uses a welfare economic concept that approximates the socially optimal level of pesticide use by integrating external costs. Factors that drive a wedge between the private and social optimum are identified.

Conclusions for policy intervention can only be made when the driving forces of pesticide use and opportunities and constraints of influencing use trends are known.

Global Pesticide Use: Trends and Policy Challenges

It is estimated that about one-third of chemical pesticides is used in developing countries, 16% thereof in Latin America, and 22% in East Asia (including Japan and Korea) (Farm Chemicals International, 2002). Nearly all regions of the developing world saw a higher average annual increase between 1993 and 1998 than in the ten years before (see Fig. 5.1). World sales of chemical pesticides dropped from US$27.8 billion in 2000 to US$25.2 billion in 2002.

However, forecasts of market research companies suggest a further increase of sales in major regions of the developing world. While pesticide use in industrialized countries, especially Europe and Japan, has stabilized due to internal adjustment of agricultural policies, major developing country markets in Asia and Latin America show high predicted growth rates. While agricultural intensification is generally considered necessary to meet the expected demand increases for food and fiber due to population and income growth, there is a less clear picture about the role of pesticides in these strategies. Researchers of the IFPRI state that there is a 'paradox of increased pesticide use and increased losses from pests' (Yudelman *et al.*, 1998).

According to global estimates of Oerke *et al.* (1994) an increasing proportion of crop output is lost to pests after a period when the use of chemical pesticides has rapidly increased. Increased vulnerability of crops and production systems, reduction in crop diversity and crop rotation, and emergence of resistance of pests to pesticides are among the factors that hint at the lack of sustainability of crop protection strategies relying exclusively on chemical pesticide use. These concerns are a driving force towards adoption of an IPM strategy that come from within the agricultural sector itself.

The mounting evidence on the negative externalities of pesticides is a factor that has pushed demands from a large number of actors of the global society for reconsideration of policies and strategies with regard to pesticides. As globalization of markets, institutions and information increased, this trend is increasingly affecting developing countries. The last decade has witnessed the emergence of vocal and powerful global civil society organizations that argue for the full

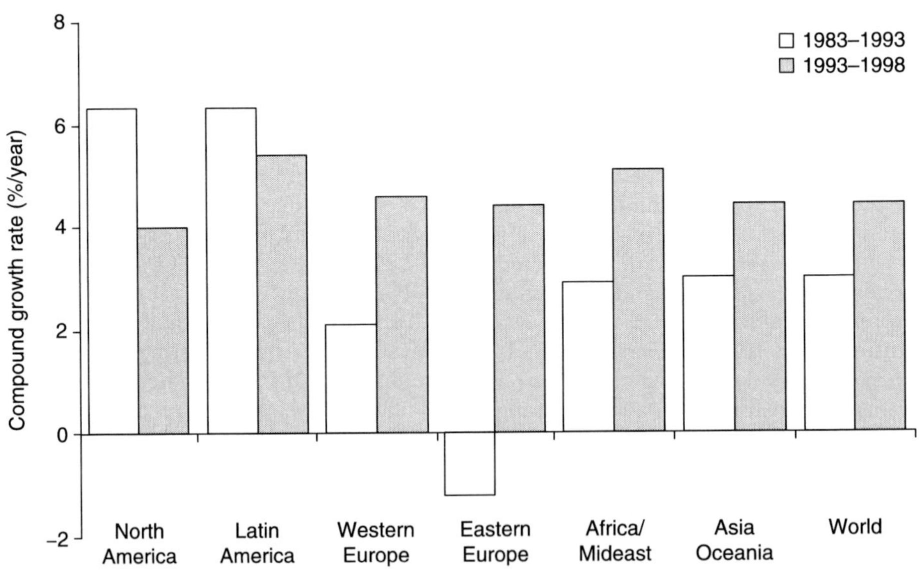

Fig. 5.1. Growth rates of pesticide use in world regions, 1983–1998. Source: After Wood *et al.*, 2001.

recognition of previously neglected impacts of agricultural development and inclusion of disadvantaged groups. Pesticides and their effects on farmers' health, natural resources and biodiversity have been part of that debate. At the same time, a remarkable shift in global food demand has started. Based on increasing consumer awareness, food supply chain operators, mainly in Europe, put more stringent standards on reducing pesticide residues in food and meeting requirements of environmentally and socially responsible production methods (see protocol of European retailer initiative, EUREP-GAP, 2001). Market access for high-value crops of developing countries is affected by private and public standards. This forces institutions in developing countries to respond effectively.

Decision makers in developing countries find themselves in a policy dilemma. An appropriate balance has to be found between a careful reversal of proactive pesticide support in the traditional agricultural development agenda, which nevertheless does not endanger food security and rural development objectives, and the mounting forces that call for reduced use of pesticides, particularly highly toxic and persistent substances. The policy dilemma is reinforced through conflicting signals from international institutions. On the one hand, liberalization and deregulation of government intervention, which are part of the structural adjustment programs that aim at facilitating integration into world markets, frequently have weakened capacities to monitor and enforce health and environmental precautions with regard to pesticides. Government services were scaled down, which had a negative impact on delivery of public goods such as extension and information on IPM. On the other hand, a number of global conventions, which require national governments to upgrade capacities for stricter pesticide control, have been put in place, e.g. the Rotterdam convention on Prior Informed Consent (UNEP, 1998) and the recent Stockholm convention on Persistent Organic Pollutants.

Theoretical Framework for Pesticide Policy Analysis

The objective for policies on pesticide use should be not to eliminate yield losses from pests but to reduce them towards a socially and economically optimal level. Welfare economic theory suggests that the marginal utility of a unit of loss reduction should be equal to the marginal costs. Because of the need to account for external costs, decision making about the extent of pesticide use in pest management strategies has to be made both at the farm level and for the society as a whole (Pearce and Tinch, 1998).

Farmers' decision making on the type and amount of pesticide use depends on several considerations, i.e. type of pest, expected crop loss, ratio of output and input prices, risk attitude and availability of input resources. However, subsidies and other institutional factors distorting use levels may contribute to pesticide use that exceeds the optimum from the society's point of view. At the same time, alternatives in pest management are under-utilized which leads to productivity loss for the national economy.

From the viewpoint of the society, pesticide use is often excessive, because the external effects caused by pesticides are usually not included in the pesticide price. The individual farmer's objective is to maximize profit. With respect to pesticide use, profit maximization implies that the marginal value product of pesticides equals their marginal costs. Taking other inputs as given, the farmer will seek to obtain the maximum level of crop loss prevention subject to the cost of pest control. Figure 5.2 shows how different types of costs are related to the optimal level of pesticide use. The x-axis shows units of prevented crop loss due to the application of pesticides. The line (OD) represents the benefit from pesticides. The benefit of pesticide use may be a linear function of crop loss based on the assumption that the producer is a price taker. The costs that users perceive as direct costs include the costs of pesticides plus other farm-level costs such as application, storage. Those costs will be represented by the cost

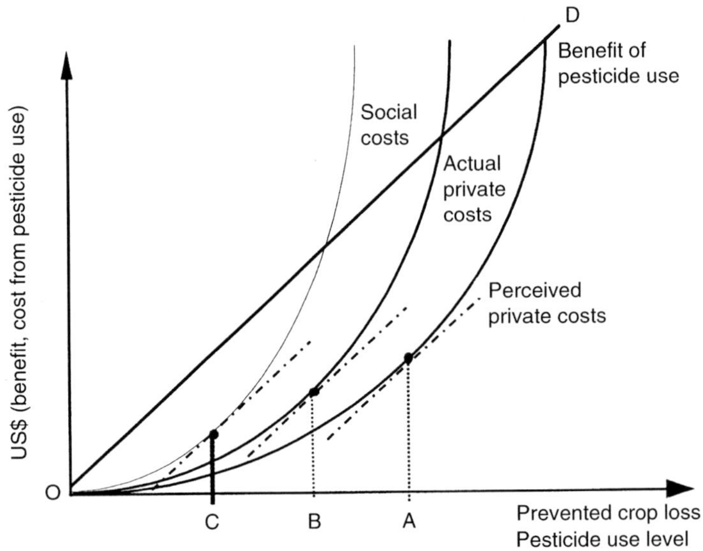

Fig. 5.2. Private and social optimum of pesticide use. (Source: Waibel, 1994.)

function labeled as 'perceived private costs'. The optimal level of pesticide use will be attained at (A). However, because pesticides often affect the health of farmers (WHO, 1990), these can result in health costs, which are normally not considered in the production costs. By internalizing these costs, the cost curve shifts from the perceived private costs to the cost function specified as 'actual private costs' in Fig. 5.2. The economic optimum of pesticide use decreases to (B).

From the society's point of view there are additional costs related to crop protection. External costs of pesticide use occur for example through the contamination of groundwater or food. Including these costs in the computation leads to a third cost function labeled 'social costs'.

The optimal level of pesticide use from the society's point of view would be reached at (C). This optimal level for society differs from the current level of pesticide use (A) if the external costs are not internalized.

The exact determination of the socially optimal level may not always be feasible due to uncertainty about the magnitude of effects and lack of data (Oskam, 1994). However, it appears particularly useful to determine the extent of the deviation of different private and social optima and the relative importance of distorting factors (Waibel, 1994).

In analyzing effects that contribute to the distortion of pesticide use from its socially optimal level, several groups of subsidizing and promoting factors can be distinguished (see Table 5.1). The farm-gate price of pesticides can be lowered by direct transfer payments to pesticide industries or retailers, or by administering controlled prices through government distribution. Concessions on taxes and import duties as well as interest rate subsidies may be in place. Also, external costs are in most cases not yet internalized in market prices. These factors can be classified as price factors. The decision making of the actual pesticide user whether to apply pesticides or to use alternative crop protection methods is influenced also by some other reasons which are acting indirectly and frequently hidden. Biases towards chemical solutions in institutional settings such as the agricultural education system, priorities in the research programs and organization of the extension service, have an important influence on the generation and the direction of technical progress and its implementation on the field level. With regard to human resources, the type and level of information about different crop protection strategies is decisive for the over- and misuse of chemicals of pesticides as well as the under-utilization of non-chemical

Table 5.1. Factors causing excessive pesticide use. (Source: Waibel, 1991.)

		Price factors		Non-price factors
Obvious factors	I	Government sells or gives pesticides Donors provide pesticides at low or no costs Government refunds pesticide companies costs Subsidized credit for pesticides Preferential rates for tax and exchange rate	III	Misguided use of governments' activities in reducing pesticide damage Governments' investments in pesticide research Inadequate government research in environmentally benign pest management
Hidden factors	II	Plant Protection Service outbreak budget Pesticide production externalities Pesticide use externalities	IV	Lack of adequate procedures for • pest definition • crop loss definition Lack of information on agroecological parameters Lack of transparency in regulatory decision making Curricula of agricultural education and extension Dominance of pesticide industry in the market for crop protection information

alternatives. Inappropriate government intervention in case of occurrence of external effects can also play an important role in keeping pesticide use levels high.

Pesticide subsidies have been first analyzed by Repetto (1985). He pointed at the high amount of direct price support, government distribution at low or no cost for the user and the effects of exchange rate regimes in nine developing countries. A literature review of the World Bank (Farah, 1994) used an extended framework for the identification of subsidizing factors (Table 5.1) based on research in Thailand (Waibel, 1991). Despite the decisive impact of structural adjustment programs on abolishing direct price subsidies it appeared that both hidden price factors and non-price factors play still an important role in many parts of the world. This stands in sharp contrast to the declaration of IPM as a national policy goal in many countries (Fleischer and Waibel, 1993).

In order to get country-specific information on the extent of direct and indirect subsidies, a methodological framework for country studies on pesticide policies was elaborated (Agne *et al.*, 1995). The framework aims at an in-depth overview on the current status of pesticide use and policies by revealing the economic and institutional factors that contribute to the deviation of the private costs from the social costs of pesticide use. A comprehensive analysis in the following areas is included:

1. Characteristics of the agricultural sector, e.g. relative importance of the sector, farm size structure, dependence of the sector on external economies.
2. Analysis of pest problems, pest management practices and pesticide use trends.
3. Externalities of pesticide use (e.g. occupational health impacts, water pollution, damage to natural resources, pesticide resistance).
4. Evaluation of agricultural policy, i.e. market and input price policies, trade and exchange rate policies, commodity price support measures.
5. Regulatory intervention, i.e. registration requirements and procedures.
6. Perception of pesticide use and regulation in the society, e.g. perception of crop loss by different groups of the society, perception of health and environmental risks.
7. Farm and crop characteristics of pesticide use, e.g. biophysical and socio-economic characteristics of farming system, crop profile, characteristics of and information level on plant protection strategies, awareness of health risks.
8. Economic evaluation of different pest control strategies from the viewpoint of the private user and the society.

So far, the framework has been applied to case studies in eight countries in cooperation with local, regional and international organizations: Benin (Affognon, 2003), Costa Rica (Agne, 1996), Côte d'Ivoire (Fleischer *et al.*, 1998), Ghana (Gerken *et al.*, 2001), Mali (Camara *et al.*, 2001), Pakistan (NARC, 2001), Thailand (Jungbluth, 1996), and Zimbabwe (Mudimu *et al.*, 1999). In most cases, the entry point for the policy analysis was the interest of national or sub-regional research institutions to become involved in the ongoing international discussion on pesticide policy research. The theoretical concept was adapted to local conditions in a structured awareness raising and training workshop. The GTZ/University of Hannover Pesticide Policy Project provided the link to the international debate especially with regard to methods for economic evaluation of pesticide externalities.

The situation analysis of the crop protection sub-sector is a first step for establishing a common information basis, which can be used as a starting point for achieving consensus about the extent of the problems and the underlying trends. Appropriate policies and strategies for pesticides and crop protection have to be formulated by taking into account the broader framework of agricultural, environmental, and health and consumer protection policy. Policy change will produce winners and losers among the interest groups, depending on the objectives and instruments of proposed policy instruments. Therefore relevant stakeholders should negotiate policy reform.

The process involved several steps (see Fig. 5.3) before policy instruments were actually implemented. In several of the case-study countries it has been observed that providing local economic research institutions with a link to the international debate about pesticide policy was a crucial element for starting the policy reform process. The process of developing a consensus started already at the beginning by exposing stakeholder representatives to the approach for economic evaluation to be used in the situation analysis. The workshops at the end of the study were in most cases convened by either the agricultural ministry or a policy research institute. In almost all countries,

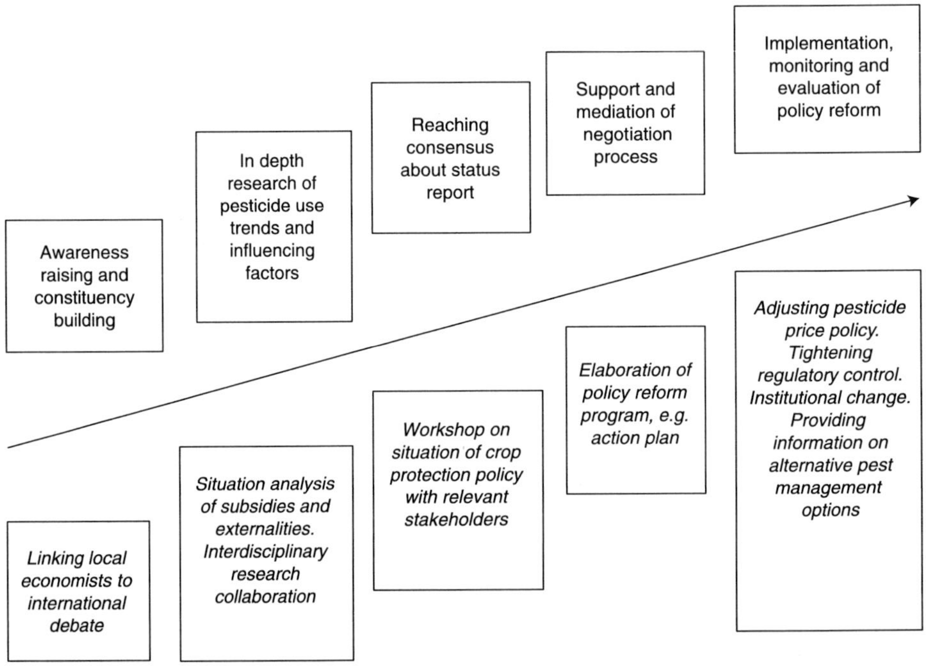

Fig. 5.3. Strategy for policy reform.

those workshops brought together for the first time a broad range of stakeholders at the national level in an open discussion forum. The provision of reliable information to all stakeholders is a vital step towards improving information access for those actors, which are structurally disadvantaged in the political decision-making process. It is thus a contribution towards more rational policy making.

Case Studies on Policy Distortions of Pesticide Use

The detailed account for pesticide use trends in the country proved to be the most important starting point for the policy analysis. Although import statistics have been updated in many countries for a considerable period, information about the trends of pesticide use at national level and for the major cropping systems was generally not readily available to most of the stakeholders in pesticide policy.

Pesticide use trend

Results for those countries where a time-series analysis was available show high growth rates in recent years (see Table 5.2) which generally exceed the increase in agricultural production. The increase is mainly caused by general intensification of agricultural production and the shift towards crops that consume high amounts of chemical pesticide per area unit. The latter holds true especially for fruit, vegetable and plantation crops. Countries with a large share of pesticides going to cotton production (Benin, Mali, Pakistan, and to a lesser extent Côte d'Ivoire) showed particularly high growth rates in the last decade which hints at widespread problems of declining productivity and emergence of pest resistance in cotton production systems.

Pesticide subsidies

The analysis of the driving forces for the increase in pesticide use shows that various economic and institutional mechanisms stimulate farm-level use patterns. Direct and indirect government subsidies distort farm level decision making and provide incentives for inefficiencies. Direct price subsidies are provided through two mechanisms. Many African countries have programs with financial assistance from the Japanese government to provide subsidized inputs, among them pesticides, for the production of rice and other food crops. A second line of support has been found in countries where the parastatal cotton agencies still hold monopolies for setting commodity and input prices. Indirect price distortions, i.e. by import duty and sales tax exemptions, are present in all countries where data were available (see Table 5.3). They play a major role in lowering the

Table 5.2. Pesticide use level in case study countries. (Source: own compilation based on results of eight country studies.)

Country	Current pesticide use level, at the time of study completion (US$ million)	Average growth rate of pesticide use per year
Benin	11.4 (1999)	n.d.
Costa Rica	90.1 (1996)	9.8% (1990–1994)
Thailand	247 (1994)	18.8% (1992–1994)
Côte d'Ivoire	41.1 (1997)	8.1% (1994–1997)
Mali	25 (1999)	19% (1994–1999)
Zimbabwe	56.3 (1997)	13.3% (1991–1997)
Pakistan	200 (1999)	n.d.
Ghana	25 (1998)[a]	n.d.

[a]Includes veterinary drugs. n.d. not determined due to lack of data.

Table 5.3. Direct and indirect pesticide price subsidies in case study countries. (Source: own compilation based on results of eight country studies.)

Country (year)	Type of price subsidy
Benin (1999)	Distribution of donor-financed pesticides through government at prices below costs Direct price and interest rate subsidy for cotton growers Exemption from import duty (29%) and sales tax (18%) Compound subsidy rate for cotton insecticides: 44%
Costa Rica (1996)	Exemption from import duty and sales tax (rebate ranging from 6% to 28%)
Thailand (1995)	Exemption from import duty, business and local tax (import duty on fertilizer = 10%, for agricultural machinery = 28%) Government budget for pesticide campaigns ('outbreak budget') = US$3 million
Côte d'Ivoire (1998)	Distribution of donor-financed pesticides through government at prices below costs (subsidy of over 50%) Exemption from import duty (18%) for cotton, banana and pineapple growers (2/3 of total market value)
Mali (2000)	Distribution of donor-financed pesticides through government at prices below costs Reduction of import duty (7.5% instead of 12%), exemption from sales tax (20%)
Pakistan (2000)	Import duty concessions for local formulators
Ghana (2000)	Distribution of donor-financed pesticides through government at prices below cost (budget of about US$1.2 million) Exemption from import duty (10%) and sales tax

pesticide price to the user. In Benin, total subsidies amount to 44% of the calculated full costs of cotton insecticides whose share is about 90% of the total use in the country. This is equivalent to a subsidy of over US$9 million or 0.4% of total GDP.

Eliminating direct and indirect price subsidies would not only improve farm level efficiency, but would also improve the government's financial position. Farmers could be compensated for income losses by funding the development and spread of alternative crop protection technologies.

Pro-pesticide biases in the institutional setting are more difficult to identify and can generally not be assessed in a quantitative manner. However, the entrenchment of biased support to pesticides in agricultural research, extension and education institutions as well as in the regulatory framework is a truly global phenomenon as all of the surveyed countries have been affected by agricultural modernization strategies which were pursued during the times of the Green Revolution. Priorities and resources in extension services and in research programs are still predominately geared towards the adoption of chemical solutions to pest problems. Some countries, such as Thailand, Pakistan and others, have allocated funds to research for IPM. However, weak research–extension linkages constrain the adoption of research results by farmers. Costa Rica and Ghana have established IPM farmer extension and training programs, while other countries such as Mali, Thailand, and Zimbabwe started pilot activities. Despite these efforts, IPM training programs currently reach only a small number of farmers and have still a negligible impact on national use levels. Educational and training curricula lack the adequate consideration of non-chemical crop protection strategies.

Government promotion of pesticide-intensive production systems also plays an important role. Those systems are supported by priority setting in agricultural development programs and hence indirectly subsidized and favored against other crops. In some crops such as cotton in Côte d'Ivoire, Mali, and Benin, credit programs are bound to obligatory pesticide use.

The informational environment for the decision making at farm level is almost exclusively dominated by chemical solutions for pest problems. Farmers lack, both, adequate information on alternatives to chemical products as well as the knowledge on feasible pesticide-use reduction measures such as a full understanding of agroecological principles.

One of the most challenging tasks for the status analysis is the economic assessment of pesticide externalities. Overall assessments are available for some industrialized countries (Pimentel *et al.*, 1993; Waibel *et al.*, 1999; Pretty *et al.*, 2000). They have demonstrated that aggregated information about the social costs of pesticide use is a useful tool for underpinning the notion that net benefits of pesticides to society are lower than their contribution to agricultural productivity. However, in most developing countries negative impacts frequently go unnoticed as the capacities for long-term monitoring of health and environmental effects are underdeveloped. For example, there is considerable underreporting of occupational poisoning cases. A case study in Thailand showed that the number of poisoning cases is likely to be 13 times higher than official records (Jungbluth, 1996).

Human wealth costs and pesticide use

The most visible and pronounced externality problem arises from occupational health impacts for farmers and farm laborers. According to WHO (1990) estimates, there are more than 20,000 fatal poisoning cases due to occupational pesticide exposure that occur annually worldwide. In the Pakistan study, an attempt for a comprehensive assessment of the costs of occupational poisoning was made. Based on evidence from case studies conducted in selected areas, an extrapolation for nine districts in the Punjab province was made where about 60% of the total cotton area of the country is located (NARC, 2001). Health costs have to be borne by three sections of the population (see Table 5.4). The largest groups are women from rural areas who are employed for picking cotton in the fields. Due to increasing intensity of pesticide use, mainly caused by pest resistance problems, pesticide exposure at harvest time has become a serious problem. Since in many cases cotton picking is the only source of income for landless families, exposure-related productivity loss has a direct bearing on total household income. Farmers and farm

Table 5.4. Cost estimate for health effects due to acute pesticide poisoning in nine cotton districts of Punjab Province, Pakistan. (Source: NARC, 2001.)

	No. of persons affected	Estimated costs per year (US$ million)
1. Exposure during application at farm level[a]		
Farmers and farm workers suffering sickness	1.08 million	
Hospital treatment	0.02 million	0.45
Work loss	0.24 million days	0.35
Accidental fatalities	271	4.25
2. Exposure during cotton harvest[b]		
Female cotton pickers suffering sickness	2.23 million	
Medical treatment		2
Work loss	11 million days	12.5
3. Exposure at local pesticide refilling facilities[c]		
Laborers suffering sickness	500	
Medical treatment		0.01
Work loss	450 days	0.002
Total		20

[a]Estimated for 2.17 million households of 9 major cotton growing districts.
[b]Estimated for 5,127,000 tons of cotton picked by 2.6 million women in nine cotton growing districts of cotton zone.
[c]Estimated for 1000 laborers working at 25 plants in Multan city.

workers are exposed to pesticides during application. Safety precautions are generally low and inadequate for the majority of insecticides used. A third group is workers in repackaging and refilling plants that are exposed to pesticide health hazards due to substandard working conditions. Total health costs for the nine cotton districts alone equal about 10% of the total value of pesticide imports into the country.

Environmental costs of pesticide use

External costs of pesticide use occur also due to contamination of the environment with residues. Pesticides contaminate the environment by leaching residues into ground and surface water and by soil accumulation. For example, in Thailand a comparatively large survey found residues in over 90% of soil, sediment and fish samples as well as in over 50% of water samples (Sinhaseni, 1992, cited in Jungbluth, 1996). However, contamination may only have an economic impact in the long run when avoidance, mitigation and damage abatement measures have to be undertaken, e.g. for securing access to safe drinking water resources.

When governments spend resources from general revenues in mitigation of pesticide damage, the polluter-pays-principle tends to be neglected. External costs are not internalized into private user's decision making. For example, pesticide use decisions of farmers are not affected by the costs that users of contaminated drinking water resources pay for clean-up or finding new sources. Thus, the price of pesticides does not reflect costs borne by the society as a whole.

A comprehensive estimation of external costs has been made in the Thailand study (see Table 5.5). External costs are between 9% and 93% of the market value of pesticides. This may well be an underestimation of the true costs since important effects such as costs of pest resistance development have not yet been quantified. Pimentel *et al.* (1993), estimate that the USA has a ratio of external costs to private costs that comes to about 200%. The study of Rola and Pingali (1993) on health costs of insecticides in rice farming in the Philippines estimates a similar ratio for health costs alone.

Achieving Stakeholder Consensus on Policy Change

Although the economic impact of different subsidy factors was clearly determined by

Table 5.5. External costs of pesticide use in Thailand. (Source: modified from Jungbluth, 1996.)

Cost type	Method of assessment	Estimated annual costs (million Baht[a])
Human health damage	Case study on acute occupational health in citrus growing area	13.0
Resistance and resurgence of pests	Outbreak budget of plant protection service for control of rice pests and support of rice farmers	57.4
Market produce loss due to residues[b]	Market value of contaminated fruits and vegetables	5037
Government regulation, control and extension	Budget for pesticide quality and food residue monitoring, for pesticide regulation and market control, and for research and extension of chemical control	404.4
External costs of pesticide use (range)[c]	(excludes other externalities, e.g. water contamination, loss of wild life etc.)	462.8 to 5491.8

[a]Exchange rate at the time of the survey: 1 US$ = 25.6 Baht.
[b]10% of samples in fruit and vegetables exceed maximum residue limits. Therefore, produce should be withdrawn, but regulatory control is inefficient. Therefore, other costs, e.g. chronic human health hazard may occur. More detailed research is needed in this area.
[c]Lower boundary excludes the market value of contaminated produce.

the research results in the different countries, the available information did not automatically trigger the removal of inefficient subsidies. Policy change is determined not only by the facts, but also by the perceptions, interests, and strategies of policymakers and affected groups. Therefore, study results were communicated to a broad range of stakeholders to become a dominant point of reference for the dialog among concerned groups. The objective was to reach consensus among experts from different fields and among the interest groups in order to enable further action towards policy change.

The results of the situation analysis were discussed with an expert forum where representatives from government ministries and agencies in agriculture, environment, and health protection, as well as non-governmental organizations, e.g. pesticide industry, farmers' associations and environmental groups participated. Qualitative indicators for the extent to which each factor influences pesticide use levels were used in a number of case-study countries. The experts did an overall ranking of stimulating and discouraging factors of chemical pesticide use. The example of Costa Rica shows that the overwhelming majority of influencing factors act towards pesticide use levels that exceed the social optimum (see Fig. 5.4). This clearly indicates that pesticide use is above the socially desired level when all direct and indirect costs are taken into account. Only few measures counterbalance the effects of indirect subsidies and other promoting influences. For example, farmer extension on IPM presently reaches only a small number of farmers in Costa Rica. IPM adoption shows little impact on overall pesticide use levels because of the large number of factors that favor chemical pesticide use.

Experience with policy reform workshops were, among other countries, made in Thailand (Poapongsakorn *et al.*, 1999), the Central America region (Reiche, 2000), Mali (Camara *et al.*, 2001), and Pakistan (NARC, 2001). The workshops were instrumental in developing a reform agenda. For example, a common perspective on the trends of pesticide use and its externalities was reached. The success of the effort in terms of pushing decision makers towards actual adoption of policy measures depended on several factors. In countries, where there was strong evidence of negative externalities, pressure from global markets to react to food residue problems and growing levels of civil society concern about pesticide-related issues, policy makers were more likely to endorse a pro-IPM policy reform agenda. In some countries, factors like the high stakes of agricultural agencies (ministry, research, extension service) in chemical crop protection, the sales strategies of pesticide industry, or general political instability hampered the adoption of a reform path which required a long-term vision for the improving sustainability of the agricultural sector.

The Central American workshop explored the potential role of economic instruments in crop protection policy based on the findings of pesticide policy reports from five countries. A common understanding was reached that legislative measures should be harmonized in the region but are not sufficient for effective regulatory control of pesticide externalities. Reduction of pesticide use and risks should be achieved by adding economic instruments, such as polluter-pay-taxes to the policy toolbox. The process of establishing a common market in the region demands also a harmonized approach for the withdrawal of tax exemption and for taxation of pesticides (Reiche, 2000).

Policy change may be triggered if decision makers are convinced that a crisis situation demands corrective action. In the case of Pakistan and Mali, this point was reached when there was unanimous consensus that continuing the present trend of pesticide use, especially in cotton, would have serious economic, health or environmental impacts (Camara *et al.*, 2001; NARC, 2001).

In Thailand, a national workshop among stakeholders from all relevant government organizations and societal interest groups recommended a master plan on sustainable agriculture (Poapongsakorn *et al.*, 1999). A new pesticide policy is currently under discussion and expected to be officially adopted in early 2002 (TGPPP, 2001). Since the country has been known for its too liberal pesticide market a need for tightening the regulatory policy was expressed, especially with a view on adhering to pesticide residue standards in export markets. The master plan contains instruments for a stepwise reduction of the most toxic

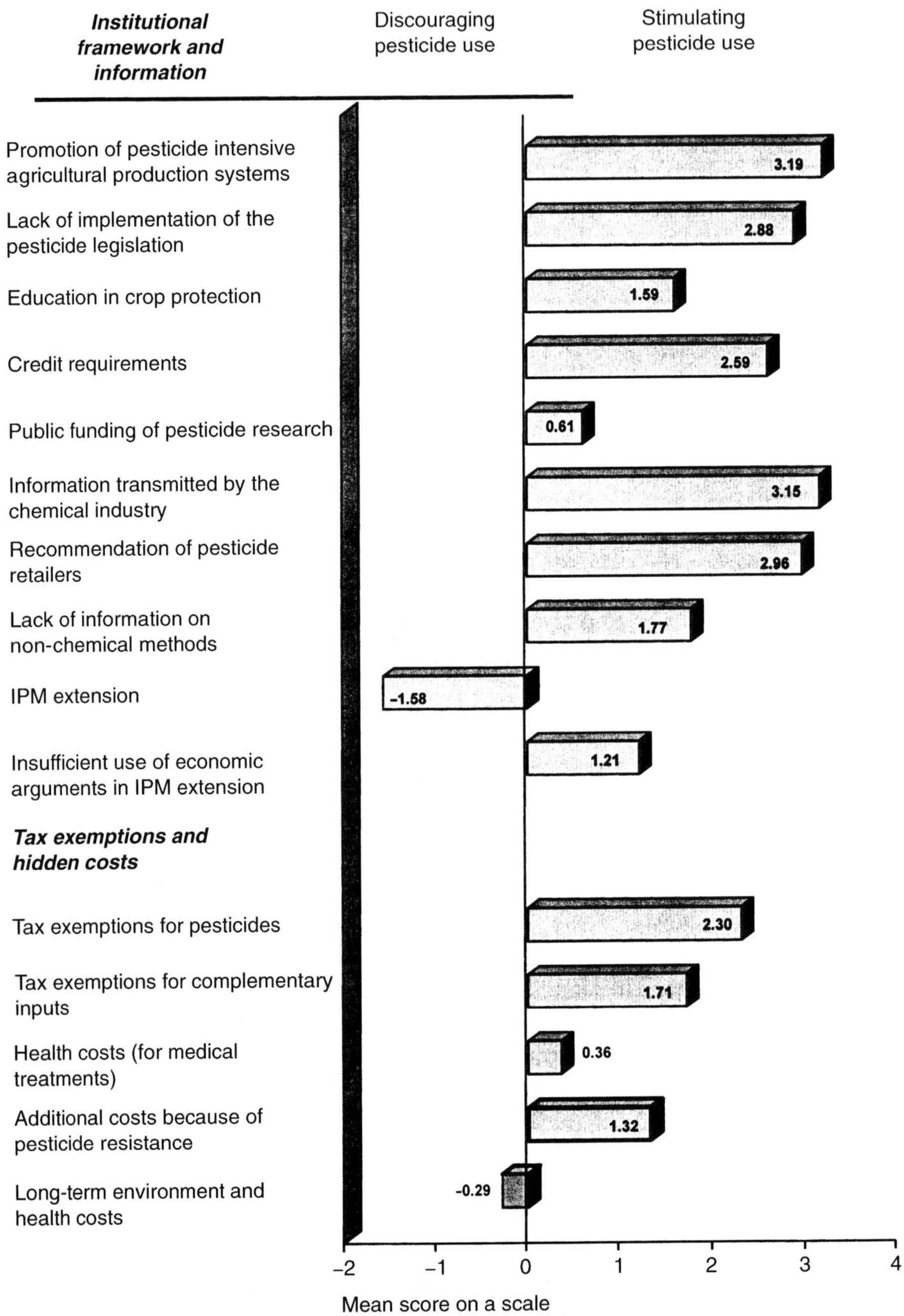

Fig. 5.4. Determinants of pesticide use and their impact (Agne, 1996).

pesticides. In a comparatively advanced economic situation like Thailand, changes in the institutional environment of crop protection decisions play a major role. Workshop participants demanded increased efforts supporting the adoption of the IPM paradigm in research, education and extension. It was stressed that obligations for pesticide use in credit schemes as well as special government budgets for subsidies in outbreak situations should be abolished.

Both, the Mali and the Pakistan pesticide policy workshops were held at a time when the crisis in the cotton sector provided favorable conditions for launching a policy reform strategy. In Mali, recommendations of the policy workshop were directly included into an agricultural lending program, financed by the World Bank and other donors. The component includes farmer training on IPM, strengthening of regulatory control, and adjusting the fiscal and economic framework related to pesticides. In Pakistan, escalating pesticide use trends in cotton have become a liability for the national economy. The federal agricultural ministry took the initiative to develop a national IPM program. Within this context, the policy study was an opportunity to win the support of the provinces and to draw the attention of the international lending agencies on this area.

Further continuation of the policy reform process depends on the awareness of major actors as well as on the relative position of the interest groups in the political process. Comprehensive pesticide use reduction plans have been used in several European countries to reconcile conflicting interests among groups and achieve a framework for the elaboration of a package of instruments (Reus *et al.*, 1994).

Conclusion

Pesticide policy studies are an important element in strategies that aim at rapid IPM implementation. Analyzing the trends in pesticide use and the economic impacts of influencing factors plays an important role for providing information to policy makers. Interaction between economists and natural scientists is particularly important to improve the assessment of the impacts of pesticide use for the society at large.

The analysis of pesticide use in the eight countries in Latin America, Asia and Africa demonstrates that a number of factors distort pesticide use levels from its social optimum. There is a consistent pattern of economic and institutional support measures for unilateral pesticide use which hamper the diffusion and adoption of IPM approaches in developing countries. Once the distorting factors are identified, appropriate regulatory and economic instruments for correcting the imbalances can be designed.

The second pillar of a policy reform strategy is the establishment of a dialog forum. The situation analysis of the crop protection sector is an appropriate entrance point for awareness creation among different stakeholders. Typically, there exist conflicts between different ministries and agencies on the extent of problems related to pesticide use and about suitable measures to be taken. Differing perceptions and interests block inter-agency committees. In this case, a policy study covering all aspects of the pesticide use problem in a manner that is as complete and as transparent as possible improves the common information base and helps to identify action areas of high priority.

More studies will be needed to substantiate the emerging evidence about negative long-term impacts of pesticide use on the agroecological resource base. Pest resistance and resurgence are compromising the long-term productivity of crop protection. To date, these effects have been only sporadically assessed in economic terms. More emphasis should be placed on the economic evaluation of externalities of pesticides as well as of other crop protection products and technologies. This information is essential to quantify the difference between the private and the social costs of crop protection technologies. Reliable information from both areas will help to establish better arguments for IPM that can be used in the political process.

References

Affognon, H. (2003) *Crop Protection Policy in Benin – Factors Influencing Pesticide Use. Pesticide Policy Project, Publication Series No. 11.* GTZ/University of Hannover, Pesticide Policy Project, Hannover, Germany (in press).

Agne, S. (1996) *Economic Analysis of Crop Protection Policy in Costa Rica. Pesticide Policy Project, Publications Series No. 4.* GTZ/University of Hannover Pesticide Policy Project, Hannover, Germany.

Agne, S., Fleischer, G., Jungbluth, F. and Waibel, H. (1995) *Guidelines for Pesticide Policy Studies – a Framework for Analyzing Economic and Political Factors of Pesticide Use in Developing Countries. Pesticide Policy Project, Publication Series No. 1.* GTZ/University of Hannover Pesticide Policy Project, Hannover, Germany.

Camara, M., Haidara, F. and Traoré, A. (2001) *Etude socio-economique de l'utlisation des pesticides au Mali.* Institut du Sahel, Bamako.

EUREP-GAP (2001) *EUREP-GAP Protocol for Fresh Fruit and Vegetables.* EUREP-GAP/FoodPlus GmbH, Cologne, Germany.

Farah, J. (1994) Pesticide policies in developing countries – do they encourage excessive pesticide use? *World Bank Technical Paper Series* 238. World Bank, Washington, DC.

Farm Chemicals International (2002) Meister Publishing Co., Willoughby, Ohio, December issue.

Fleischer, G. and Waibel, H. (1993) *Survey Among Plant Protection Officers During the Global IPM Conference, Bangkok, August 1993.* Report to the FAO/UNEP Panel of Experts on Integrated Pest Control, Göttingen/Rome.

Fleischer, G., Andoli, V., Coulibaly, M. and Randolph, T. (1998) *Analyse Socio-économique de la Filière des Pesticides en Côte d'Ivoire. Pesticide Policy Project Publication Series No. 6.* GTZ/University of Hannover Pesticide Policy Project, Hannover, Germany.

Gerken, A., Suglo, J. and Braun, M. (2001) *Pesticides Use and Policies in Ghana – an Economic and Institutional Analysis of Current Practice and Factors Influencing Pesticide Use. Pesticide Policy Publications Series No. 10.* GTZ/University of Hannover Pesticide Policy Project, Hannover, Germany.

Jungbluth, F. (1996) *Pesticide Policy in Thailand. Pesticide Policy Project Publications Series No. 5.* University of Hannover, Germany.

Mudimu, G.D., Waibel, H. and Fleischer, G. (1999) *Pesticide Policies in Zimbabwe – Status and Implications for Change. Pesticide Policy Project, Publications Series Special Issue No. 2.* GTZ/University of Hannover Pesticide Policy Project, Hannover, Germany.

NARC (2001) *Policy and Strategy for Rational Use of Pesticides in Pakistan.* National Agricultural Research Centre, Islamabad.

Oerke, E.C., Dehne, H.W., Schönbeck, F. and Weber, A. (1994) *Crop Production and Crop Protection. Estimated Losses in Major Food and Cash Crops.* Elsevier Science, Amsterdam.

Oskam, A. (1994) Pesticides – issues at stake for economists. In: Michalek, J. and Hanf, C.H. (eds) *The Economic Consequences of a Drastic Reduction in Pesticide Use in the European Union.* Revised papers of a workshop held in Tannenfelde (Schleswig-Holstein), 13th–14th November 1993, Kiel, pp. 81–108.

Pearce, D. and Tinch, R. (1998) The true price of pesticides. In: Vorley, W. and Keeney, D. (eds) *Bugs in the System. Redesigning the Pesticide Industry for Sustainable Agriculture.* Earthscan, London.

Pimentel, D., Acquay, H., Biltonen, M., Rice, P., Silva, M., Nelson, J., Lipner, V., Giordano, S., Horowitz, A. and D'Amore, M. (1993) Assessment of environmental and economic impacts of pesticide use. In: Pimentel, D. and Lehman, H. (eds) *The Pesticide Question – Environment, Economics, and Ethics.* Chapman and Hall, New York, pp. 47–84.

Poapongsakorn, N., Meenakanit, L., Jungbluth, F. and Waibel, H. (eds) (1999) Approaches to pesticide policy reform – building consensus for future action. A Policy Workshop in Hua Hin, Thailand, 3–5 July 1997. *Pesticide Policy Project, Publication Series No. 7.* GTZ/University of Hannover Pesticide Policy Project, Hannover, Germany.

Pretty, J., Brett, C., Gee, D., Hine, R., Mason, C., Morison, J., Raven, H., Rayment, M. and van den Bijl, G. (2000) An assessment of the external costs of UK agriculture. *Agricultural Systems* 65(2), 113–136.

Reiche, C. (ed.) (2000) *Memoria – Taller Sobre el Uso de Plaguicidas an America Central, 17–19 de Setiembre de 1997.* Proyecto IICA/GTZ sobre Agricultura Recurso Naturales y Desarrollo Sostenible. Instituto Interamericano de Cooperación para la Agricultura, San José, Costa Rica.

Repetto, R. (1985) *Paying the Price – Pesticide Subsidies in Developing Countries.* World Resources Institute, Washington, DC.

Reus, J., Weckseler, H. and Pak, G. (1994) *Towards a Future EC Pesticide Policy – an Inventory of Risks of Pesticide Use, Possible Solutions and Policy Instruments.* Centre for Agriculture and Environment, Utrecht, The Netherlands.

Rola, A. and Pingali, P. (1993) *Pesticides, Rice Productivity, and Farmers' Health – an Economic Assessment.* International Rice Research Institute (IRRI), Los Baños, Philippines.

TGPPP (2001) *New Pesticide Policy for Health Environment and Export Workshop, November 16, 2001.* Thai–German Plant Protection Programme, Bangkok. http://www.tgppp.org Accessed 20 December 2001.

UNEP (1998) *PIC – Rotterdam Convention on the Prior Informed Consent Procedure for Certain Hazardous Chemicals and Pesticides in International Trade.* Text and Annexes. United Nations Environment Programme, Nairobi.

Waibel, H. (1991) Pesticide subsidies in Southeast Asia. *FAO Plant Protection Bulletin* 38(2), 111–120.

Waibel, H. (1994) Towards an economic framework of pesticide policy studies. In: Agne, S., Fleischer, G. and Waibel, H. (eds) *Proceedings of the Göttingen Workshop on Pesticide Policies, Göttinger Schriften zur Agrarökonomie,* Vol. 66. Institute of Agricultural Economics, University of Göttingen, Germany, pp. 1–10.

Waibel, H., Fleischer, G. and Becker, H. (1999) The economic benefits of pesticides: a case study from Germany. *Agrarwirtschaft* 48(6), 219–230.

WHO (1990) *Public Health Impact of Pesticides Used in Agriculture.* World Health Organization, Geneva.

Wood, S., Sebastian, K. and Scherr, S. (2001) *Pilot Analysis of Global Ecosystems. Agroecosystems.* IFPRI and World Resources Institute, Washington, DC.

Yudelman, M., Ratta, A. and Nygaard, D. (1998) *Pest Management and Food Production. Looking to the Future. Food, Agriculture, and the Environment Discussion Paper 25.* International Food Policy Research Institute, Washington, DC.

Chapter 6

Industrial Perspective on Integrated Pest Management

Graham Head and Jian Duan
Monsanto Company, St Louis, Missouri, USA

Introduction

IPM is generally thought of as a management system that uses all available tools and techniques to maintain pest populations at levels below economical thresholds (Kogan, 1998). IPM aims to control insects, weeds, diseases, and other pests in an environmentally safe and cost-effective manner. IPM tools may be either preventative or interventional. The former category includes host-plant resistance traits, many methods of cultural control (for example, crop rotation), and genetically modified crops. The latter includes chemical treatments that are applied according to thresholds, some forms of biological control (for example, inundative and augmentative natural enemy releases), and sanitation methods that reduce pest populations and suitable pest habitat. Ideally these methods also work in a complementary fashion to natural biological control agents.

IPM is not a new concept, and its obvious value to farmers and the environment has long been recognized. Governments and academic scientists have had a longstanding commitment to IPM (e.g. Wallace, 1993; Kogan, 1998; Fitt, 2000). For example, in the USA, the CSREES-Land Grant University IPM Program involves a network of research and extension staff located in all 50 States and six Territories. This network broadly influences the management practices throughout the USA by encouraging research on IPM-related problems and disseminates the results of this research using media that include workshops, scouting programs, consultations, and a wide variety of printed and electronic materials (see http://www.reeusda.gov/ipm). Industry also recognizes this value and supports both the individual principles and IPM as a framework for defining good agricultural practices. This chapter describes the unique role that industry plays in developing, supporting and implementing IPM programs. Industry's influence can be seen in three basic areas:

1. Industry's primary role lies in providing pest control tools that can form the basis for IPM programs. We will discuss how IPM considerations are integrated into the new product development process.

2. In the product development process, industry develops a large amount of data on product performance and appropriate product use. These data are valuable in designing effective IPM programs. We will describe examples of industry working with other stakeholders, including academic scientists

 Integrated Pest Management in the Global Arena (eds K.M. Maredia, D. Dakouo and D. Mota-Sanchez)

and farmer organizations, to design such programs.

3. Industry, as the provider of technologies to farmers and thus trusted by them, has an opportunity to communicate and educate farmers in relation to desirable practices. IPM is one important topic dealt with in this way. We will talk about how this is accomplished and give examples of successful educational programs.

Product Development and IPM

Companies recognize that certain criteria must be met for a new product to be successful: the product must provide a highly effective solution to a particular pest problem of economic significance, and must also complement IPM practices and natural biological control (Briggs and Koziel, 1998). The design process, and subsequent testing, ensures that new products have desirable environmental properties and a good potential fit with IPM practices.

Product design

In designing a new product, considerable effort goes into ensuring that the active ingredient is both highly effective and specific. This is equally true of a conventional insecticide and the choice of an insecticidal protein to be expressed in a genetically engineered crop. The candidate molecule is screened against important target and non-target animal species to determine the range of activity. In addition, the mode of action of the molecule is examined so that predictions can be made about its activity against select species. The aim is to find proteins with high activity against the target pest insects and little or no activity against other taxa. As a consequence of this selection process, proteins that might cause adverse environmental impacts because of either broad toxicity or activity against key non-target groups are eliminated.

In many cases, similar molecules, or even the same molecule, have been used previously for crop protection. In such cases, the history of use can be studied. Where possible, insecticidal molecules that have been previously used in comparable ways without environmental problems are preferred. For example, the crystalline (Cry) proteins that have been incorporated into genetically modified, insect-protected maize, cotton and potatoes were used safely in foliar sprays for almost 40 years (McClintock *et al.*, 1995; EPA, 1998; Betz *et al.*, 2000).

Product testing

Initial testing of novel insecticidal products begins up to a decade before commercialization. The testing process follows a standard risk-assessment process and involves multiple tiers of laboratory and field testing (Sharples, 1991). The tests used are shaped by the requirements of regulatory agencies as well as by product stewardship considerations. In the USA, the EPA and the USDA–APHIS regulate the environmental safety of new pest control technologies (see, for example, EPA RED, 1998). In other countries, ministries of agriculture and the environment have divisions that play a comparable role. The tests performed on any new product include laboratory and field-based studies, and typically proceed in a tiered fashion. These tests address regulatory requirements regarding environmental safety and the agricultural role of the new product. Testing includes an assessment of the impact of the new product on important ecological guilds within agricultural systems such as predators, parasitoids and pollinators. Ultimately, potential new products are compared with reasonable agronomic alternatives with respect to agronomic and ecological impacts. Regulatory agencies, in granting product approvals, must make a determination that the benefits a new product can bring to farmers and society clearly outweigh any risks.

Even after commercialization, work on ecological and agronomic impacts continues, now in commercial-sized fields managed with standard farmer practices. At this stage, the emphasis is on larger scale effects and emergent properties (system functioning). Aspects like energy flow are examined and impacts on agronomic systems are assessed. These impacts can be direct or indirect, an example of the latter being the facilitation of a shift toward reduced tillage by genetically modified, herbicide-tolerant crops. This shift should positively affect both soil and water quality in areas where these products are used. In general, such studies are a part of product development and stewardship, as well as being regulatory requirements in some countries.

Through the initial product design process and subsequent safety tests, products with minimal impacts on non-target organisms, particularly biological control agents, are selected. This ensures that future products will complement the biological control capacity of agricultural systems, thereby fitting into existing IPM systems.

An example – genetically modified Bt crops and IPM

Bt is a common species of soil bacterium. Many strains exist which produce various combinations of insecticidal proteins. The Cry proteins are of particular interest because of their specificity and effectiveness (see review in Schnepf *et al.*, 1998). Each protein only affects a relatively small set of related insect species; for example, Cry1-type proteins control various Lepidoptera (moths and butterflies), while Cry3 proteins control certain Coleoptera (beetles). Unrelated non-target species are unaffected. The genes for Bt Cry proteins have been genetically engineered into certain crop plants. These so-called Bt crops, because of their effectiveness, can replace insecticides with undesirable environmental characteristics. Tropical systems and crops like cotton in which pest pressure is very high, and thus broad-spectrum insecticide use would otherwise be very high, are particularly benefited.

The incompatibility of broad-spectrum insecticide use and biological control, particularly the survival of natural enemies, has been a major stumbling block in the attempt to initiate IPM programs. Many of the widely used classes of conventional insecticides, including organophosphates and pyrethroids, have been shown to adversely affect a broad range of non-target species, including species of economic importance (e.g. Badawy and El-Arnaouty, 1999; Amano and Haseeb, 2001). In Indian cotton, over 600 such species have disappeared altogether (Sundaramurthy and Gahukar, 1998). These impacts on natural enemies have been shown to lead to flare-ups in non-target pest species, some of which were not previously economically important. Replacing these chemistries with Bt crops allows natural populations of predators and parasitoids to increase, and the opportunity for successfully introducing other suitable beneficial species also is improved. Through removing major negative impacts on natural enemy populations, the use of Bt crops and concomitant decrease in broad-spectrum insecticide use can contribute to improved pest control even of species not directly impacted by Bt proteins (Turnipseed *et al.*, 2001).

In the case of Bt cotton in the USA, Roof and DuRant (1997) demonstrated that beneficial arthropod numbers were greater in Bt cotton fields than conventional cotton fields in South Carolina. Similarly, in a multi-state study, Head *et al.* (2001) found equivalent or greater numbers of many generalist predators in commercial Bt cotton fields compared with commercial sprayed conventional cotton fields (see Table 6.1). Similarly, in a multiple-year field study, Reed *et al.* (2001) demonstrated that transgenic Bt potato eliminated the need for weekly Bt microbial sprays, biweekly permethrin sprays, or in-furrow application of systemic insecticides for Colorado potato beetle control. In doing so, the use of Bt potato enhanced the survival and reproduction of naturally occurring generalist predators such as big-eyed bugs (*Geocoris* spp.), damsel bugs (*Nabid* spp.), pirate bugs (*Orius* spp.), and spiders (Araneae) relative to

Table 6.1. Average abundance (number of individuals observed on beating-cloths) of various natural enemies in commercial conventional cotton (Conv) and Bt cotton (BG) fields in southeastern USA. Insecticide applications for lepidopteran pests began between 4 July and 11 July on the conventional cotton fields. The final row has the ANOVA results for each species or group of species. (From Head *et al.*, 2001.)

	Geocoris		*Orius*		Spiders		Ants	
Date	Conv	BG	Conv	BG	Conv	BG	Conv	BG
20 Jun	4.0	13	0	1.0	6.3	6.8	97	64
27 Jun	8.0	10	0.3	0.5	7.3	8.5	64	38
04 Jul	8.3	18	1.8	2.5	20	20	45	73
11 Jul	11	17	5.3	3.8	24	27	33	53
18 Jul	6.8	19	8.5	30	16	47	21	46
25 Jul	17	34	3.5	39	8.5	36	15	130
02 Aug	10	8.3	1.3	1.8	2.3	9.8	11	40
F, Prob.	$F = 7.75$, $P = 0.007$;		$F = 6.21$, $P = 0.016$;		$F = 7.51$, $P = 0.008$;		$F = 7.73$, $P = 0.08$	

conventional potato fields treated with broad-spectrum insecticides.

The impact of these increased non-target natural enemy populations (relative to fields sprayed with conventional insecticides) can be improved biological control of pest species not directly controlled by the Bt protein. This, in turn, means that the number of insecticide sprays used for these other pests are reduced. For example, in Alabama in 1996, beet armyworms were less likely to occur at economic levels in Bt cotton, apparently in part because of higher numbers of beneficial insects imparting natural protection (Smith, 1997). In the case of Bt potato fields, generalist predators effectively suppressed the population of green peach aphid, *Myzus persicae* – vector of potato viral diseases (Reed *et al.*, 2001).

Work in other countries has produced comparable results. With the introduction of Bt cotton in China, broad-spectrum insecticide use has been reduced by up to 80%, and this has been accompanied by a 24% increase in generalist predator populations (Xia *et al.*, 1999). In an agricultural system where a wide variety of crops are grown in close proximity, such as in many parts of India and Africa, Bt cotton could even act as a natural enemy reservoir for other crops, thereby increasing the role of biological control in these crops too. This process may require several years to produce its full effect because of the time that some of these natural enemy populations require to build up, particularly after the severe impacts of consistent broad-spectrum insecticide use.

Working with New Products

Industry, as the developer of new pest control tools, also has the greatest knowledge of these products. During the development process, companies create large data sets around the characteristics of each new product, and how these products should be used in the context of current practices and needs. As a consequence, companies can provide unique and valuable input into IPM programs; industry's information can form the basis for defining appropriate use practices for a new product.

Industry's role in promoting proper product use

Industry has a responsibility and vested interest in ensuring that use practices for a new product are optimized and that they fit with current IPM practices. Companies accomplish these goals through internal research and external collaborations. Industry routinely works with groups that include academics, regulators, scientists with government research institutions, and farmer organizations. The resulting use guidelines can include application practices, establishing windows for use, combining or rotating products, and establishing scouting thresholds for target pest species.

IRM practices form a subset of IPM programs in which industry has particular interest and where industry's information can be particularly useful. The aim of IRM is to maximize the durability of pest control tools by limiting the selection for resistance. Some common IPM practices that serve this dual purpose include rotation of different pest control tools, the establishment of spatial refuges for transgenic Bt crops, and the use of economic thresholds to determine when and where pest control tools are used.

Examples – the Center for Integrated Pest Management

The industry/university Center for Integrated Pest Management was established in 1991, and now consists of companies (including Monsanto, Syngenta, and Dow AgroSciences), government agencies (including the USDA Forest Service, USDA/APHIS, USDA/ARS/OPMP, and USDA/CSREES), commodity organizations (including groups associated with cotton, soybean, sweet potato, and turfgrass), grower groups, and an organization representing consultants (see http://ipmwww.ncsu.edu/cipm). The CIPM funds a variety of projects ranging from small, 1-year 'seed money' efforts to longer-term regional programs. The research funded involves multi-state problems and solutions, and works as a provider of unbiased IPM. The CIPM is working to include more grower associations, food processors, companies, government agencies, and other agricultural groups with an interest in affordable and safe food production.

IRAC and Insect Resistance Management programs

Industry has worked collectively with a variety of stakeholders to create IRM programs through the IRAC, a group that contains representatives of agricultural chemical companies (see http://www.plantprotection.org/irac). IRAC serves as a coordinating group for methodologies and resistance surveys and provides seed money to fund research in problem areas. IRAC also plays a role in IRM education by developing literature and videos and by placing articles in the popular press. IRAC has worked with academic scientists, extension services, farmer organizations, and commodity groups to achieve some outstanding successes (Thompson and Head, 2001).

For cotton-growing areas of Asia, IRAC has worked with academic scientists from universities in India, Pakistan, China, and the UK, and the NRI, to develop and disseminate effective and sustainable practices for cotton bollworm control. This has involved synthesizing existing knowledge, surveying insecticide use and resistance, developing practical tools for farmers to use in evaluating different insecticides, producing a handbook for use in smallholder systems in Asia and Africa, and organizing workshops in these areas.

In the southwestern USA, the local IRAC group cooperated with farmers, a commodity-related organization (Cotton Incorporated), and university research and extension personnel to research and communicate a plan to combat the whitefly, *Bemisia argentifolii* (Dennehy and Williams, 1997). This program has been successful for 4 years and is continuing with monitoring and refinement.

Influencing and Educating Farmers

IPM practices and programs will only be adopted and successfully implemented by farmers if they are seen as valuable and practical. In this sense, the success of IPM ultimately rests with farmers. Once appropriate IPM-related practices are developed, farmers must be educated on these practices, and industry plays an important role in this educational process (Jutsum *et al.*,

1998). With respect to industry participation, the education of farmers occurs through information being placed on product labels, through separate technical documents being developed and distributed, and through the development of special materials like informational websites and videotapes. Some of these communications are required by regulatory agencies, including the details placed on all product labels for agricultural chemicals and the technical guides developed for the use of transgenic crops. In addition, companies have made IPM and IRM key components of product stewardship (for example, Urech *et al.*, 1997). In some cases, industry has worked to fill broader needs where they exist. In developing countries, academic institutions often have limited resources and there may be no group that fills the traditional extension role. Under such conditions, coalitions of companies may step in to address the need for basic farmer education around agriculture, appropriate agricultural practices, and how IPM programs are assembled.

Obviously, with the introduction of any new pest control product, farmers must learn specific details on how that product should be used. Consequently, farmers are receptive to information at this stage, and this represents an opportunity to refine IPM programs and educate farmers broadly in relation to agricultural practices. Effective pest control technologies provide a special opportunity to influence farmers. Because they are effective, these technologies are particularly attractive to farmers, and they provide an opportunity to influence and possibly alter farmer practices in positive ways.

Examples – farmer practices and genetically modified crops

As described earlier, the first genetically modified crops were commercialized in the mid 1990s. These products have required a variety of changes in farmer practices. For example, Bt crops require unique scouting practices and economic thresholds because of their effectiveness and mode of action. These same products also require farmers to employ novel practices as part of IRM (Sherrick and Head, 2000). In the various countries in which they have been commercialized, farmers growing Bt crops must ensure that an adequate area of non-transgenic crop exists as a refuge for susceptible pest insects. This notion of farmers providing and maintaining fields of crops to support a pest population is not only new but also counter to the instincts of many farmers. Nevertheless, this practice has been successfully implemented on a global basis. This success reflects several factors. First, educational efforts in the relevant countries have been intense, and co-ordinated across industry and public sector scientists. Second, farmers recognize the value of these technologies and are highly satisfied with their performance, and thus are willing to go to exceptional lengths to preserve them. Third, other mechanisms have been introduced to ensure that farmers follow the practices required of them. In the USA and Australia, for example, farmers must sign license agreements to gain access to these products, and these agreements spell out their responsibilities with respect to IPM and IRM practices. Genetically modified, herbicide-tolerant crops have met with similar approval and also have brought changes in farmer practices (Sherrick and Head, 2000).

Web-based resources and distance learning programs

The CropLife International federation is a network of 75 national and regional associations (see http://www.croplife.org). This overarching organization has identified sustainable agriculture, including the development of IPM programs, as one of its five strategic pillars of action. To this end, an international Integrated Crop/Pest Management project team has been created. This team works to build private–public partnerships to shape IPM policies and facilitates training programs for farmers. Within the six regional nodes of CropLife

International, the AMEWG, the LACPA, and the APCPA are all working on separate, specific regional training and education programs for farmers and other interested groups.

APCPA efforts will be used to illustrate the sorts of programs that industry can and has created around farmer education and IPM (see http://www.apcpa.org). APCPA consists of eight companies and 14 national association affiliates. At the broadest level, APCPA has worked with non-governmental organizations like ISAAA to establish Information Centers in the region and has held workshops on safe use targeted at smallholder farmers. APCPA also has worked with non-governmental groups and governmental agencies to establish the APRTC (see http://www.aprtc.org). The APRTC is an educational network that uses distance learning tools and other Web-based resources to promote IPM and good agricultural practices. Courses are being offered on:

- digital literacy for agricultural professionals;
- English for agriculture;
- safe and effective use of agrochemicals;
- introduction to IPM;
- IPM for cotton;
- IPM for irrigated rice;
- IPM for vegetables.

The APRTC is actively working to develop new initiatives with two SEARCA-headed organizations – the AAACU and the Southeast Asian University Consortium for Graduate Education in Agriculture and Natural Resources. In such partnerships with the APCPA, academic institutions can benefit through improved course content, the opportunity to interact with a network of educated and experienced agricultural professionals, advanced training for faculty, and the opportunity to generate funding.

Conclusions

Agricultural chemical and biotechnological companies have an important function within IPM because of their role as technology providers. These companies are the source of new and better pest management tools (critical building blocks for IPM), they are the primary repositories for information on these tools (information that defines how these tools can fit into IPM programs), and they maintain strong relationships with their customers, the farmers, which can be useful in IPM education. Through the formation of partnerships within the industry, and with public sector groups, companies have played an important role in the development of IPM programs on a global basis, and they are committed to continuing to do so in the future.

References

Amano, H. and Haseeb, M. (2001) Recently-proposed methods and concepts of testing the effects of pesticides on the beneficial mite and insect species: study limitations and implications in IPM. *Applied Entomology and Zoology* 36, 1–11.

Badawy, H.M.A. and El-Arnaouty, S.A. (1999) Direct and indirect effects of some insecticides on *Chrysoperla carnea* (Stephens) s.L. (Neuroptera: Chrysopidae). *Journal of Neuropterology* 2, 67–76.

Betz, S., Hammond, B.G. and Fuchs, R.L. (2000) Safety and advantages of *Bacillus thuringiensis*-protected plants to control insect pests. *Regulatory Toxicology and Pharmacology* 32, 156–173.

Briggs, S.P. and Koziel, M. (1998) Engineering new plant strains for commercial markets. *Current Opinion in Biotechnology* 9, 233–235.

Dennehy, T.J. and Williams, L. III (1997) Management of resistance in *Bemisia* in Arizona cotton. *Pesticide Science* 51, 398–406.

EPA Registration Eligibility Decision (RED) (1998) *Bacillus thuringiensis*. EPA 738-R-98-004, March 1998.

Fitt, G.P. (2000) An Australian approach to IPM in cotton: integrating new technologies to minimize insecticide dependence. *Crop Protection* 19, 793–800.

Head, G., Freeman, B., Moar, W., Ruberson, J. and Turnipseed, S. (2001) Natural enemy abundance in commercial Bollgard and conventional cotton fields. *Proceedings of the Beltwide Cotton Conference*, National Cotton Council, Memphis, Tennessee, pp. 796–798.

Jutsum, A.R., Heaney, S.P., Perrin, B.M. and Wege, P.J. (1998) Pesticide resistance: assessment of risk and the development and implementation of effective management strategies. *Pesticide Science* 54, 435–446.

Kogan, M. (1998) Integrated pest management: historical perspectives and contemporary developments. *Annual Review of Entomology* 43, 243–270.

McClintock, J.T., Schaffer, C.R. and Sjoblad, R.D. (1995) A comparative review of the mammalian toxicity of *Bacillus thuringiensis*-based pesticides. *Pesticide Science* 45, 95–105.

Reed, G.L., Jensen, A.S., Riebe, J., Head, G. and Duan, J.J. (2001) Transgenic Bt potato and conventional insecticides for Colorado potato beetle (Coleoptera: Chrysomelidae) management: comparative efficacy and non-target impacts. *Entomologia Experimentalis et Applicata* 100, 89–100.

Roof, M.E. and DuRant, J.A. (1997) On-farm experiences with Bt cotton in South Carolina. *Proceedings of the Beltwide Cotton Conference.* National Cotton Council, Memphis, Tennessee.

Schnepf, E., Crickmore, N., van Rie, J., Lereclus, D., Baum, J., Feitelson, J., Zeigler, D.R. and Dean, D.H. (1998) *Bacillus thuringiensis* and its pesticidal crystal proteins. *Microbiology and Molecular Biology Review* 62, 775–806.

Sharples, F.E. (1991) Ecological aspects of hazard identification for environmental uses of genetically engineered organisms. In: Levin, M.A. and Strauss, H.S. (eds) *Risk Assessment in Genetic Engineering.* McGraw-Hill, New York, pp. 18–31.

Sherrick, S. and Head, G. (2000) General concepts, status, and potential of transgenic plants in IPM. In: Kennedy, G.G. and Sutton, T.B. (eds) *Emerging Technologies for Integrated Pest Management.* APS Press, St Paul, Minnesota, pp. 96–100.

Smith, R.H. (1997) An extension entomologist's 1996 observations of Bollgard technology. *Proceedings of the Beltwide Cotton Conference.* National Cotton Council, Memphis, Tennessee, pp. 856–858.

Sundaramurthy, V.T. and Gahukar, R.T. (1998) Integrated pest management of cotton insect pests in India. *Outlook on Agriculture* 27, 261–269.

Thompson, G.D. and Head, G. (2001) Implications of regulating insect resistance management. *American Entomologist* 47, 6–10.

Turnipseed, S.G., Sullivan, M.J., Hagerty, A. and Ridge, R. (2001) Cotton as a model IPM system in the southeast: a dream or potential reality. *Proceedings of the Beltwide Cotton Conference.* National Cotton Council, Memphis, Tennessee.

Urech, P.A., Staub, T. and Voss, G. (1997) Resistance as a concomitant of modern crop protection. *Pesticide Science* 51, 227–225.

Wallace, M. (1993) The National Coalition on IPM: working for safer food, cleaner water, and wildlife conservation through expanded implementation of IPM. In: Leslie, A.R. and Cuperus, G.W. (eds) *Successful Implementation of IPM for Agricultural Crops.* Lewis Publishers, Boca Raton, Florida, p. 193.

Xia, J.Y., Cui-Jin, J., Ma, L.H., Dong, S.L. and Cui, X.F. (1999) The role of transgenic Bt cotton in integrated insect pest management. *Acta Gossypii Sinica* 11, 57–64.

Chapter 7
Role of Integrated Pest Management and Sustainable Development

George W. Bird
Department of Entomology, Michigan State University, East Lansing, Michigan, USA

Introduction

IPM and sustainable development evolved as processes of significance during the last three decades of the 20th century. This chapter is designed to: (i) summarize the histories, practices, systems and philosophies of these topics; and (ii) illustrate how IPM can be of assistance in the quest for a sustainable future based on a reasonably high quality of life for all of our planet's people. A comprehensive understanding of the concept of alternative world-views, however, is imperative for use of IPM as a catalyst in the achievement of the goals of sustainable development. The contribution relies heavily on many IPM and sustainability science contributions, including two recent books published by the US Academy of Sciences, National Research Council (*Our Common Journey: a Transition Toward Sustainability*, 1999 and *Drama of the Commons*, Ostrom, 2002). Other works of major emphasis include, Capra's *The Web of Life: a New Scientific Understanding of Living Systems* (1996), Horne and McDermott's *The Next Green Revolution: Essential Steps to a Healthy, Sustainable Agriculture* (2001), Bird *et al.* (1990) *Design of Pest Management Systems for Sustainable Agriculture* and a presentation to the Michigan Agriculture Commission entitled, *Agriculture: Observations, Prognosis and Recommendations* (Bird, 1994).

Mechanistic and Ecological World-views

The current dominant world-view is a mechanistic world-view. It is based on linear relationships and the assumption that the whole represents the sum of the parts. Resources are considered as infinite or it is assumed that replacement technologies are continually available. There are relatively few direct feedback loops, and only a small number of system components with overlapping functions. This world-view does not mandate existence within a vibrant community of local ecological interdependence and partnerships (Capra, 1996). The quantitative phenomenon of growth is a fundamental component of this world-view. There is an increasing flow or throughput of matter and energy for production of goods and services for an economy based on both population and consumption growth (Miller, 2000).

An alternative world-view is an ecological world-view. It is a cyclic system based on the assumption that resources are finite and that the whole is greater than the sum of its parts. Local ecological interdependence and partnerships, cyclic patterns of organization, system components with overlapping functions and multiple feedback loops, and existence within a vibrant community are the characteristics of an ecological world-view (Capra, 1996). This world-view is based on self-organizing, interdependent and interconnected networks of living organisms that are autopoietic (self-replicating), dissipative (requiring energy inputs and providing residual outputs) and cognitive (responsive to their environment).

The mechanistic and ecological world-views are fundamentally different! What is commonly known as conventional agriculture is based on the mechanist world-view (Bird and Ikerd, 1993). It is a highly productive system of food, feed and fiber production that relies heavily on external system inputs. Sustainable development (the process of maintaining a system at a fuller or better state) is based on the ecological world-view. Although IPM originated within the boundaries of the mechanist world-view, many of its fundamental properties are based on attributes of the ecological world-view. During the last half of the 20th century, however, there were relatively few formal research and implementation initiatives related to management systems based on the ecological world-view (Bird, 1995; Lipson, 1997).

Sustainable Development

Potter *et al.* (1970) introduced the concept of sustainability through recognition of the significance of intergenerational equity and quality of life. In *Our Next Frontier* (1984), Rodale indicated that the first phase in the development of a society relates to the discovery of its natural resources, the second phase deals with learning how to use the resources to enhance quality of life, and the third phase is the challenge of sustainability. Sustainable development was defined in the 1987 World Commission on Environment and Development (*Our Common Future*) as, '. . . development that meets the needs of the present without compromising the ability of future generations to meet their own needs'. The report's recommendations were global in nature and involved progressive transformation of economy and society. During the subsequent decade, the topic of sustainable development emerged as a major imperative for all of global society (UNCED, 1992). Meadows *et al.* (1992), Wright and Nebel (2002) and Miller (2000) described visions of the concept of sustainable development (Fig. 7.1). More recently, the US Academy of Sciences, National Research Council described the transition to sustainability as *Our Common Journey* (1999). The report focuses on nature, life support systems and community as things that must be sustained; and people, economy and society as what needs to be developed.

To understand the concept of sustainable development, however, it is imperative to differentiate between *growth* and *development* (Meadows *et al.*, 1992)! *Growth is a quantitative phenomenon characterized by size increase through assimilation of matter.* Growth has distinct limits! *Development, however, is a qualitative phenomenon in which an entity realizes potential or is brought to a fuller or better state.* There are no known limits to development! The concept of sustainable development mandates that moral and ethical value judgments be made. These are often outside the normal boundaries of science and scientific method.

Current State

The tool revolution, estimated to have taken place about 2.4 mya, was followed in relatively recent times, *c.* 10,000 ya, by the beginning of the agricultural revolution; an event of major significance in relation to the evolution of modern society (Diamond, 1999). Only yesterday, about 250 ya, society entered into an age of growth initiated by the European Industrial Revolution. In response to these events, human population reached the 1.0 billion level *c.* AD 1830.

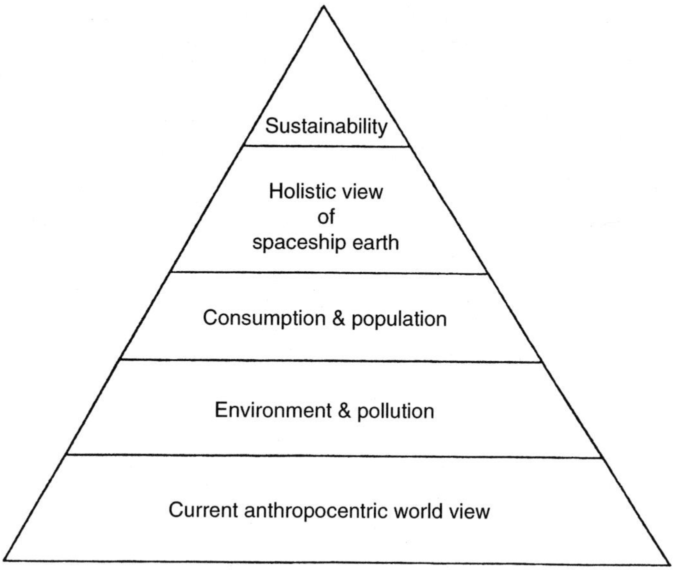

Fig. 7.1. Hierarchical levels of awareness (after Miller, 2000).

Table 7.1. Major societal events.

Major societal event	Time
Tool revolution	2.4 million to 100,000 years ago
Agricultural revolution	10,000 years ago
Industrial revolution	250 years ago
Chemotechnology era	55 years ago
Electronic era	15 years ago
Biotechnology era	10 years ago
Sustainable agriculture era	Unknown future date
Industrial growth age	1750–unknown future date
Age of sustainable development	Unknown future date–unknown development distant future date

This was followed in the 20th century by the chemical technology, electronic and biotechnology eras (Table 7.1).

During 10,000 years of development of western civilization, a single dominant mechanistic world-view emerged. As indicated above, it is based on the assumption that all components of the environment (air, water, soil, minerals, and all microbial, plant and animal species, including nematodes) are natural resources to be exploited for the advantage of humankind. It assumes that natural resources are essentially infinite, and that if a resource becomes extinct, another will be substituted as an alternative. It is a taker world-view and is supported by government policy, multinational corporations, current practice of economics and the heritage of science (Daly and Cobb, 1989; Goldsmith, 1992). Although the technologies associated with this world-view have resulted in highly significant increases in human population growth, and numerous amazing advances in quality of life, there have also been consequences that were either unexpected, or have the potential for major long-term detrimental impacts on society and the biosphere. These impacts are similar throughout most sectors of society.

The unexpected consequences associated with US agriculture resulted in both the evolution of IPM (Bird *et al.*, 1990) and the more recent activities concerning the sustainability of agriculture. These consequences include a decrease in the number of farms, increase in farm size, high dependency on off-farm purchased inputs, increase in risk of farm failure, decrease in system diversity, decrease in biological diversity, unacceptable risks associated with environmental quality, increase in risks associated with human health, decrease in reliance on rural communities, and decrease in direct contact between the farm sector and urban–suburban communities. This resulted in the evolution of a single dominant system of US agriculture. It is known as conventional agriculture or the industrial agribusiness model. It is frequently even referred to as *Traditional Agriculture*!

The US food system can be subdivided into three components: (i) market; (ii) input; and (iii) farm sectors. Between 1910 and 1990 the market sector grew 627% in absolute dollars, while the input sector increased 460%, and the farm sector declined 8% (Smith, 1992). The benefits of the increases in farm sector productivity were reaped by the off-farm sectors of the food system. Today, about 85% of the food and fiber produced in the USA comes from about 15% or 300,000 of the 2,000,000 farms. The vast majority of these enterprises are operated under the structural attributes of the conventional farm model (Table 7.2).

The narrow profit margin associated with the conventional farm model usually mandates growth in farm size as a strategy for economic viability. This fosters the continued decline in the number of viable full-time conventional farms. Recently, there has been increased interest in on-farm value added initiatives.

Another major component of the US farm sector is the part-time farm. There are about 1.2 million part-time farms, representing 60% of the total number of farm enterprises. The off-farm income of part-time farms exceeds the net farm income. Many part-time farms are not part-time farms by choice. They have adopted this type of farming as a default function designed to protect a desire for an agrarian style of life. The viability of the part-time farm usually depends on factors outside the farm sector, and in many cases factors outside of agriculture.

Table 7.2. Structural attributes of the conventional agriculture farm. (After Bird and Ikerd, 1993; Strange, 1988.)

- Centralized management
- Emphasis on specialization
- Hired worker days exceed owner on-farm work days
- Separation of management and labor
- Technology used to minimize labor inputs (limited education required)
- Heavy reliance on purchased inputs
- Technology designed to minimize real-time in-field decision-making
- Emphasis on standard farming practices

Miller (2000) outlined a philosophy of sustainable development based on hierarchical levels of awareness (Fig. 7.1). The first-order of awareness after recognition of the unexpected consequences of the mechanistic world-view is an understanding of the need to address environmental quality and pollution issues. IPM is an example of a philosophy and set of technologies resulting from this first-order level of awareness. The second-order level of awareness takes into consideration the issues of over-consumption and over-population. These are topics that are frequently outside the boundaries of discussion. The third-order level of awareness deals with a holistic approach to *spaceship earth*. This is addressed in detail by Goldsmith in his 1992 publication entitled, *The Way: an Ecological World-View*, and by Capra (1996). Only after development of appropriate strategies to deal with the first three levels of awareness is it possible to evaluate seriously the temporal nature of issues of sustainability in relation to decades, centuries or millennia. The original Michigan State University concept of IPM was based firmly on the techniques of systems science.

The philosophies of sustainable development presented by Meadows *et al.* (1992), and Wright and Nebel (2002) are very similar. Meadows *et al.* (1992) used five criteria to describe sustainable development as operating within both natural resource and social subsystems. All five of the criteria must be met for the system to be sustainable (Table 7.3).

The approach of Wright and Nebel (2002) uses four principles of sustainability. It also includes elements of both natural and social systems (Table 7.4).

Table 7.3. Attributes of sustainable development (Meadows *et al.*, 1992).

- Renewable resources must not be used at a rate greater than the regenerative capacity of the system.
- Non-renewable resources must not be used at a rate greater than the development of substitute resources.
- System residuals must not be produced at a rate greater than the assimilation capacity of the system.
- The system must meet the ecological world-view quality of life mandates associated with rural, regional and urban human living environments.
- The system must provide for intergenerational equity.

Table 7.4. Attributes of sustainable development (Wright and Nebel, 2002).

- The foundation of the system must be based on solar energy.
- The system must make optimal use of natural cycles.
- The system must be designed to prevent overconsumption–overpopulation.
- The system must promote biodiversity.

Sustainable Agriculture

During the next to last decade of the 20th century, a coalition of US environmental advocates, organic farmers and ecologists worked with the US Congress and the US Department of Agriculture to obtain funding for research and education programs in alternative agriculture systems. In 1988, appropriations were approved for the LISA. In 1990, the US Food, Agriculture, Conservation and Trade Act expanded the program. Today it is known as the SARE. The Act authorizing SARE defined sustainable agriculture as:

> an integrated system of plant and animal production practices having a site specific application that will, over the long-term: satisfy human food and fiber needs; enhance environmental quality and the natural resource base upon which the agriculture economy depends; make the most efficient use of nonrenewable resources and integrate where appropriate, natural biological cycles and controls; sustain the economic viability of farm operations; and enhance the quality of life for farmers and society as a whole.

Numerous individuals representing widely diverse sectors of US agriculture indicate that this definition represents a long-term goal for US agriculture. How to convert this goal into practical realities, however, is a major challenge. During the last decade of the 20th century, SARE served as a catalyst for building new coalitions between US farmers and ranchers, and representatives of non-profit private organizations, government academia and some agribusiness. It resulted in development of an alternative vision of US agriculture for the 21st century.

The vast majority of individuals associated with US conventional agriculture have limited or no experience with alternative farming systems. Because of this and other societal reasons, they frequently have serious doubts about and often strong bias against alternative systems. Evidence indicates, however, that this is beginning to change. Research data indicate various alternative systems, including organic agriculture, can be highly productive and economically profitable (Peterson, *et al.*, 2000). The types of technology associated with alternative systems of agriculture, however, are usually very different from those associated with our current dominant world-view.

One vision of sustainable agriculture for the 21st century includes an environment that would allow alternative agriculture systems to thrive. An example might be a 21st-century diversified farm (Table 7.5). Although evidence exists that there has been an increase in the number of enterprises in this category, it is highly probable that specific policy, research initiatives and education programs targeted for alternative systems are necessary to enhance the transformation.

For the 21st-century diversified farm to become a major reality, the farm sector will have to recapture a small portion,

Table 7.5. Structural attributes of the 21st-century diversified farm. (After Strange, 1988; Bird and Ikerd, 1993.)

- The farm is owner operated.
- Hired-worker days usually do not exceed farm-family worker days.
- The farm is a partnership of, usually, not more than three families.
- The farm is structured as a joint management–labor relationship.
- The operation places major emphasis on biological diversity.
- There is an emphasis on the use of on-farm resources.
- Site-specific and real-time decision-making are important components of the system.
- A diverse set of enterprise statements include environmental goals, natural resource conservation objectives, economic priorities, production system goals, family quality of life objectives, local community quality of life activities, and urban–suburban community interfacing mandates.

approximately 10%, of the resources currently controlled by the market and input sectors. It is believed that this should be possible to achieve through on-farm and local value-added initiatives. The results would include a significant increase in the number of viable farm-sector opportunities, a revitalization of rural communities, greatly enhanced environmental quality, and an overall improvement in quality of life for the farm-sector and society as a whole.

Pests and Pest Management

Humankind has always had to deal with the detrimental impacts of pests in its search for food, fiber, shelter or space. Other pests function as vectors of disease-causing or nuisance organisms in relation to human comfort or welfare. As hunters and gatherers, humans often migrated as a means of resolving pest and other natural resource issues. With the advent of agriculture and permanent settlements, the detrimental impact of pests became associated with all components of society, including urban, suburban, and rural environments, as well as industrial, agricultural, forest, and aquatic systems (Diamond, 1999). Pests are included among many of the 23 currently recognized kingdoms of living organisms. Numerous species of insects, nematodes, bacteria, fungi, viruses, vertebrates etc. become pests under specific environmental conditions.

Pest control procedures in the pre-synthetic pesticide era were diverse. Sulfur was used for controlling insects and mites as early as 2500 BC. Cultural procedures and various forms of habitat modification were part of early pest control programs. Biological control was used in citrus orchards in China as early as AD 307. Botanical pesticides such as pyrethrin and other toxicants including arsenic, mercury, Paris green, and Bordeaux mixture were discovered and used in Europe during the late 18th and 19th centuries. The vedalia beetle was imported from Australia to the USA in 1888 to control cottony cushion scale of citrus in California. Some pre-synthetic pest control tactics were very successful, while others left much room for improvement.

Between 1850 and 1925, scientists working with agricultural pests identified new pest problems, developed improved pest management strategies, and discovered basic principles that have served as catalysts for important developments in other areas of science and technology, including human medicine. A farmers bulletin on root-knot nematode management, published by E.A. Bessey in 1911, illustrated that a truly integrated approach to pest management was available shortly after the turn of the 20th century.

Today, conventional agriculture is perceived primarily in the context of gains in increased substitution of capital for labor and the associated productivity of labor and expansion of total product. It is often overlooked that agricultural output and pest management are closely tied to the availability and cost of synthetic inputs and their derivatives. The post World War II chemotechnology era resulted in a vast array of inexpensive pesticides; including acaricides, fungicides, herbicides, nematicides, bactericides, and rodenticides. Both weed science and the science of nematology evolved as direct results of chemical technology development.

The economics of scale evident in post World War II agriculture were made possible by inexpensive and abundant supplies of natural resources. Some of the direct spin-offs of conventional agriculture include increased specialization in the production process, reduced heterogeneity of cropping systems, and an associated decline in redundancy of natural feedback loops. This resulted in a decrease in system resiliency to perturbations and the movement toward larger production units. Pest management practices are directly related to prevailing agricultural technologies, which, in turn are determined by the cost and availability of existing energy inputs. The availability of inexpensive high-energy technology has led to the development of unique interactions among living organisms that would likely be vastly different in more labor-intensive and diversified systems. The development and broad-scale rapid adoption of herbicide–cyst nematode resistant soybean varieties is an excellent example.

In the USA, publication of Rachel Carson's *Silent Spring* catalyzed a general awareness of potential human health and environmental risks associated with some uses of pesticides. As a result, the US Federal Insecticide, Fungicide, and Rodenticide Act was amended in 1972. This changed the orientation of national pest control legislation from consumer protection for pesticide users to environmental protection. Concomitantly, the scientific community began to place increased research emphasis on the use of multiple pest control tactics and on potential environmental and human health hazards associated with pesticides. In addition to the highly significant benefits of the chemotechnology era to pest management, at least six unexpected consequences took place. These included: (i) human health risks; (ii) environmental risks; (iii) development of pest resistance to pesticides; (iv) impacts on non-target organisms; (v) pest population resurgence; and (vi) development of new pest problems. They resulted in the evolution of what has become known as IPM.

IPM

IPM is recognized as the development, use, and evaluation of pest control procedures that result in favorable socioeconomic and environmental consequences. In a 1979 US Presidential Message to Congress, IPM was defined as 'a systems approach to reduce pest damage to tolerable levels through a variety of techniques, including predators and parasites, genetically resistant hosts, natural environmental modifications and, when necessary and appropriate, chemical pesticides'. Although the development and utilization of IPM is far from complete, it can be conceptualized as a process involving seven core components (Fig. 7.2).

Biological monitoring is one of the core components of IPM. This is frequently referred to as 'scouting'. Biological monitoring consists of sampling procedures designed to estimate the stages and population densities of both pests and beneficial organisms. It also involves monitoring the stage of development and symptomatology of the associated crop, animal or other entity such as a human living environment. Biological monitoring is a very knowledge-intensive procedure, and requires highly trained individuals. The system manager or decision maker is responsible for the current state of the system, and is hypothetically best suited for this role. Private sector scouts from pest management associations or private consulting firms, however, are often hired to do the biological monitoring. In the USA, biological monitoring specialists are frequently trained by Land Grant Universities.

Pest populations and the growth and development of crop plants are governed by environmental parameters such as air temperature, soil temperature, soil moisture, light intensity, and relative humidity. This mandates that environmental monitoring be another core component of IPM. Weather

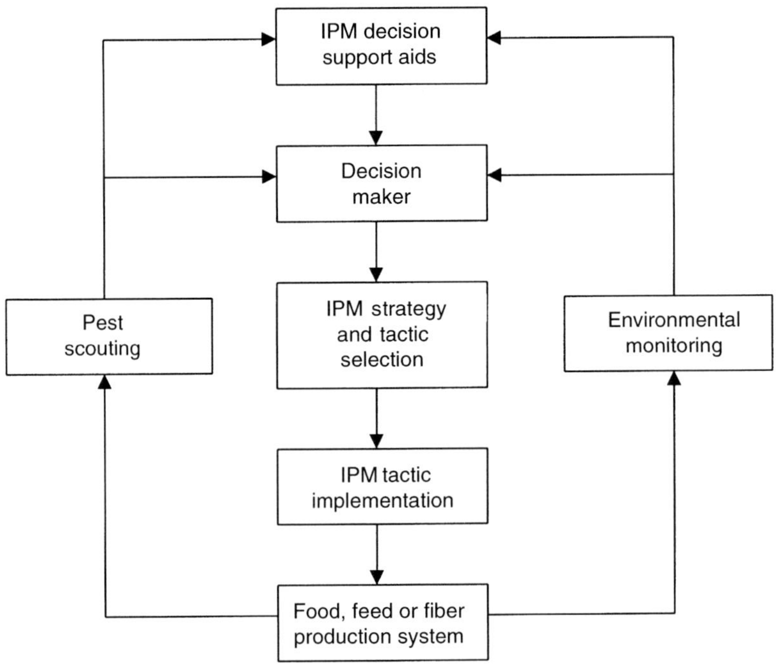

Fig. 7.2. Conceptual model of the components and process of IPM (Bird *et al.*, 1990).

monitoring information for IPM must be available on both a regional, local and sometimes microhabitat basis. This information must be readily available, user friendly, and as close to real-time as possible. Weather monitoring systems for use in IPM programs have improved greatly during the past decade. In addition to macro-climatological information, farm-level micro-climatological data is frequently imperative for predicting diseases caused by fungi and bacteria. Dedicated microcomputers are available for use in environmental monitoring in a significant number of agricultural systems.

Because of the complexity and the large number of potentially significant interactions between pests, beneficial organisms, crops, agricultural animals, and the environment, decision support aids are important aspects of IPM systems. These may be relatively simple look-up tables or computerized systems. A decision support system may be based on simulation models from research data or developed as expert systems. Elementary aspects of artificial intelligence have been investigated for a few systems. Human experience and wisdom are still, and will likely always remain, as essential elements of successful IPM programs.

The system manager or designated representative is responsible for pest management decisions. This aspect of IPM is a very knowledge-intensive process. An individual within the specific enterprise may be assigned the IPM decision-making responsibility, or a private consultant hired. Individuals trained only in pest scouting should never be delegated the responsibility of IPM strategy or tactic decision-making.

A fundamental difference between pest control and IPM is the use of population density thresholds (Fig. 7.3). The first threshold that needs to be considered is the action threshold. When the population density of a pest reaches an action threshold, or is predicted to reach this level in the near future, it is time to select appropriate IPM strategies and tactics for implementation. An economic threshold must be estimated when selecting IPM procedures. IPM procedures should usually be implemented when the marginal revenue derived from the management input is equal to or exceeds the marginal cost. The economic threshold is a dynamic concept

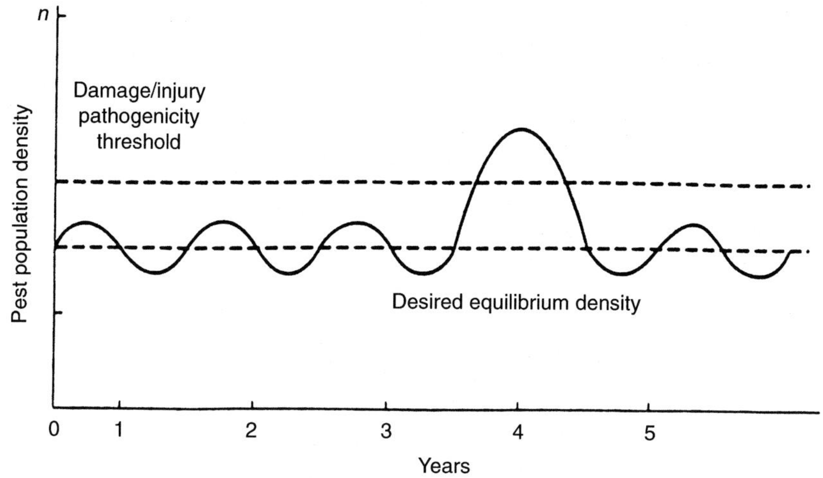

Fig. 7.3. Population dynamics of a potentially major pest in a sustainable agricultural system in which the population density is maintained below the damage/injury/pathogenicity threshold in 4 out of 5 years (Bird *et al.*, 1990).

and depends on the cost and efficacy of the management input, production system economics, nature of the pest and population density, and other environmental parameters. Although IPM strives to reduce environmental and human health risks, the costs associated with these factors are usually not available for incorporation into the economic threshold. Where the criteria of the economic threshold are met, an appropriate pest management procedure should be implemented. This will usually consist of manipulation of the pest, crop or animal, or regulation of the associated interactions.

After implementation of an IPM procedure, it is imperative that the biological and environmental monitoring programs be continued to determine if the desired pest management objectives were achieved or if there is need for additional action. IPM is a highly dynamic process. It has been successfully implemented in a significant number of important systems on a worldwide basis. Even with significant institutional, educational, and social constraints, IPM has become a well-established practice. The principles of IPM appear to be ideally suited for use in conjunction with the concepts of sustainability science and sustainable development.

Global agriculture and agribusiness appear to be changing rapidly. Although the potential benefits of IPM have become widely recognized, and adopted in a limited number of agricultural systems, resources for the design, research, education, and facilitation required for broader adoption have not developed quickly. In some cases, production system managers and other decision makers have received mixed signals about IPM, and have not invested in the additional educational and support resources required by this knowledge-intensive system. Several of the important success stories in IPM have evolved from crisis and not from the planned change through procedures of education, facilitation, and persuasion. These lessons should be studied in detail as plans are made for future systems of sustainable development. While IPM has been successful, it is not designed for solving all of the issues of sustainability. IPM is usually implemented as a strategy to deal with features of a system that are the causes of pest problems. IPM is designed to make incremental adjustments to the system's trajectory.

Soil quality and pest management are important components of sustainable agriculture. The nature of pest management in sustainable agriculture, however, is not well defined. The most important pest management strategy in sustainable agriculture may be pest problem avoidance through the use of alternative system designs. The primary objective is to exclude pests (including major pests of conventional agriculture) from the area of concern, or to maintain pest populations at a population density below the damage, injury, or pathogenicity threshold. For an agricultural system to be sustainable, it is suggested that this pest management objective be achieved in at least 4 out of every 5 years. This will mandate the use of system design as a management tool.

Norris and Caswell-Chen (2003) and Flint and Gouveia (2001) published comprehensive books on IPM. These need to be part of the library of all current IPM practitioners. Benbrook (1996) authored a book entitled, *Pest Management at the Crossroads*, containing substantial references to organic food and farming systems.

IPM represents both a vision and a methodology. As James Kendrick Jr (Vice President for Agriculture, University of California) succinctly put it in 1988, IPM '. . . is an ecological approach to maintaining plant health. It is an attitude evolving into a concept of controlling pest and disease damage to plants. It is based on an understanding of the entire ecological system to which the host that we are interested in keeping healthy belongs'. The concepts of sustainable development take IPM a step further and recognize it as a continuous journey.

Role of IPM in Sustainable Development

The practices, systems and philosophies developed under the rubric of IPM should be extremely useful in the journey towards an era of sustainable development (Fig. 7.4). Although the issues of sustainability in an era of 6–11 billion people on a planet with limited resources are immense, there are certain similarities with the issues encountered during the last half of the 20th century. In 1973, Rachel Carson observed, 'The entomologist, whose specialty is insects, is not so qualified by training, and is not psychologically disposed to look for undesirable side effects of his control program.' To generalize this observation, one need only point to the writings of Barry Commoner in

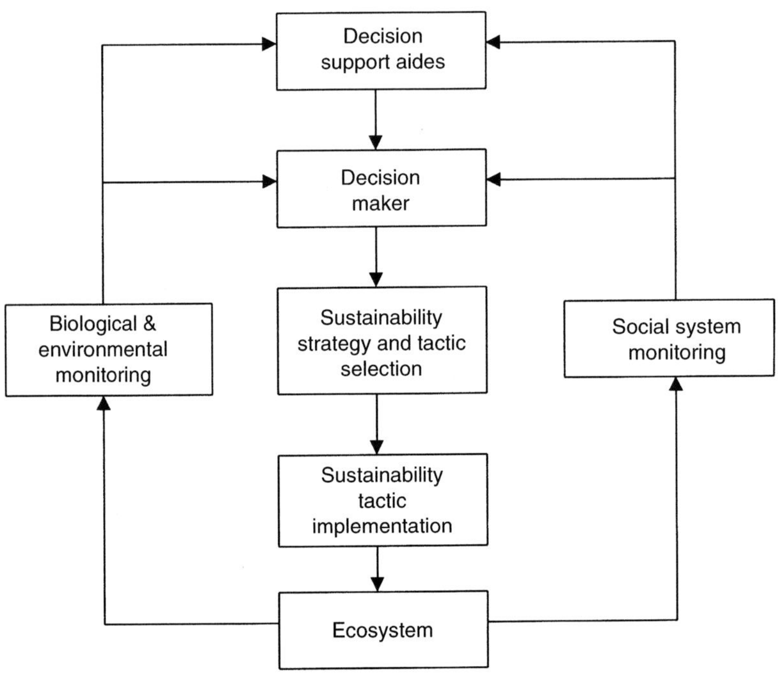

Fig. 7.4. Conceptual model of the components and process of sustainable development (after Bird *et al.*, 1990).

1969: 'We have become not less dependent on the balance of nature, but more dependent on it. Modern technology has so stressed the web of processes in the living environment at its most vulnerable points that there is little leeway left in the system.'

Commoner proceeds to point out the rules of environmental biology indicating that '. . . everything has to go somewhere' and 'everything is connected to everything else'. Indeed, this interconnectedness is the crux of our problem. Capra's 1996 treatise on *The Web of Life: a New Scientific Understanding of Living Systems*, is a truly interdisciplinary and holistic approach.

IPM is an interdisciplinary process. It cannot be stronger than its weakest disciplinary component. The same is true for sustainability science. Both IPM and sustainable development must be based on the ecological world-view. IPM has three decades of experience dealing with both the biological and social aspects of pest management. Sustainable development has a shorter but broader formal history. The procedures and lessons of IPM can provide a sound foundation for the future journey of sustainable development. The conceptual model of the process of IPM is easily convertible into one for sustainable development (Fig. 7.4).

At Michigan State University, the development of IPM was based on the principles of systems science and the legendary interactions among Dr Herman Koenig (system science), Dean Haynes (entomology), Thomas Edens (resource development), Lal Tummala (electrical engineering) and William Cooper (zoology). The process was similar to that described by Capra (1996) with his focus on the Macy Conferences and the concept for the search for patterns of sustainability. The principles and procedures of systems science are both useful and highly appropriate for the construction of conceptual models of sustainable development (Fig. 7.5).

Our willingness to place a high priority on understanding and mitigating our long-term impacts on our planet's ecosystems will dictate society's future state. A continuing issue that directly impacts our ability to understand and evolve towards sustainable agricultural and pest management practices is the time horizon adopted by various key participants in the political decision-making process. The very use of the term sustainable implies a temporal dimension. Economists and politicians frequently use rather short market or term-of-office oriented time frames in their analyses. Geologists or anthropologists, on the other hand, are accustomed to using decades, centuries or even millennia as a temporal frame of reference. This is perhaps why the 1999 National Research Council report speaks of the journey of moving toward a more sustainable system rather than attaining some conceptual level of sustainability.

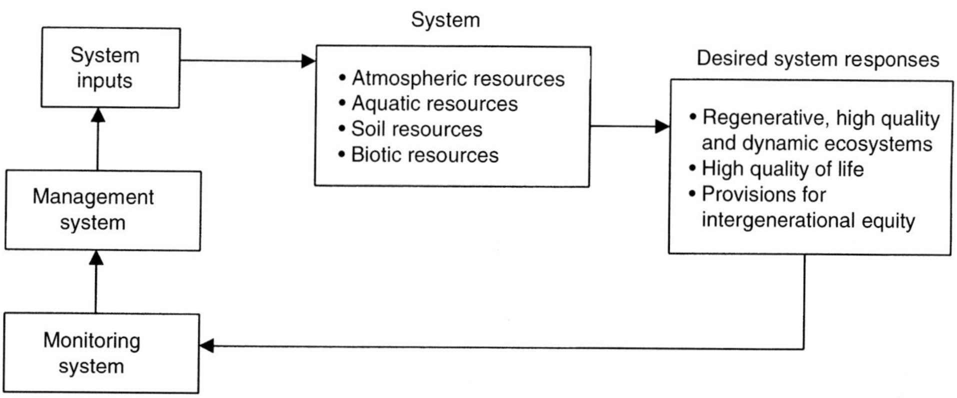

Fig. 7.5. Conceptual system model of sustainable development.

It is also essential to be aware of the political, economic, and social realities that play a major part in determining for whom sustainability is sought and for how long. In his Second Inaugural Address on 20 January 1937, President Franklin Delano Roosevelt stated that 'The test of our progress is not whether we add more to the abundance of those who have much; it is whether we provide enough for those who have too little.' Societies or individuals in a survival mode at the edge of starvation are not likely to concern themselves with or take actions to provide for the next generation. Even preparing for the next day or week is a major task. Many of our global commons problems are core factors in the issue of sustainability. These include, among others, rainforests destroyed, woodlands removed, and waters irrevocably contaminated. Individuals in low income nations and other stressed environments need and deserve the attention and support of the high income nations in order to move towards sustainability. Most wealthy nations, however, exist under the dominant mechanist world-view.

Sustainable development transcends the agricultural sector. It is useless to argue for the sustainability of anything outside the context of the total system within which it exists. This is one of the reasons it is imperative for sustainability science to rediscover the fundamentals of systems science as outlined at the legendary Macy Conferences and its second generation leaders like Herman Koenig and Fritjof Capra. IPM is a small but very important subsystem. The way in which this subsystem fits into the larger system in the context of a sustainable food system must be defined holistically. The determination and will to develop and nurture sustainable systems requires both a vision for the future and a goal for the present. The future of agricultural pest management is highly objective-dependent and presents several significant challenges. Unless there are major unlikely developments in the area of technological dominance of nature, agriculture will have to rely on an increasingly knowledge-dependent system of pest management. This must be designed to take advantage of multiple strategies and tactics for optimizing the long-term sustainability of ecosystems.

The objectives of both sustainable development and IPM can be viewed as the process of a journey: stated in terms of temporal stability, inter- and intra-group equity, and ecological impact along all perceivable time horizons. The problem quickly encountered is how, from a scientific view, does one measure whether or not a practice or process is sustainable or leads toward a more sustainable system. What are the indicators of sustainability? It is highly probable that isolated components will never be suitable for use as indicators of sustainability. It is more likely, however, that it will be possible to identify patterns of sustainability! Patterns result from the interactions of structure and process. An approach recommended within IPM is to focus on process and to evaluate elements of the process in a dynamic way so as to assess their contribution to sustainability in terms of their impact on the length or duration of subsystem cycles. In addition, it has been argued that a movement toward closure of cycles within subsystems is imperative for sustainable development (Edens and Haynes, 1982).

For an ecological world-view to become a dominant world-view, society must become ecologically literate. The Michigan State University International Education Initiative in IPM-Sustainable Agriculture is a small but very important contribution towards this effort. The participants are an important part of the key to global ecological literacy.

References

Benbrook, C.M. (1996) *Pest Management at the Crossroads.* Consumers Union, New York, 272 pp.

Bessey, E.A. (1911) *Root-knot and its Control.* Bureau of Plant Industry, USDA, Washington, DC, 89 pp.

Bird, G.W. (1989) The Integrated Pest Management Experience. In: *Reform and Innovation of Science and Education: Planning for the 1990 Farm Bill.* Committee on Agriculture, Nutrition and Forestry, United States Senate, pp. 31–41, 101–161.

Bird, G.W. (1992) Sustainable agriculture research and education program: with special reference to ecology. *Journal of Sustainable Agriculture* 2, 141–152.

Bird, G.W. (1994) Agriculture: Observation, Prognosis and Recommendations. Testimony to the Michigan Agriculture commission (USA), 7 June. East Lansing, Michigan, 7 pp.

Bird, G.W. (1995) *Sustainable Agriculture – a Case Study of Research Relevancy Classification.* Invited Presentation of the 1995 Annual Meeting of the American Association for the Advancement of Science, 20 February 1995. Atlanta, Georgia, USA, 3 pp.

Bird, G.W. and Ikerd, J. (1993) Sustainable agriculture: a twenty-first-century system. *Annals of the American Association of Political and Social Science* 529, 92–102.

Bird, G.W., Edens, T., Drummond, E. and Groden, E. (1990) Design of pest management systems for sustainable agriculture. In: Francis, C.A., Flora, C.B. and King, L.D. (ed.) *Sustainable Agriculture in Temperate Zones.* John Wiley & Sons, New York, pp. 55–110.

Capra, F. (1996) *The Web of Life.* Anchor Books, New York, 347 pp.

Daly, H.E. and Cobb, J.B. Jr (1989) *For the Common Good.* Beacon Press, Boston, 482 pp.

Diamond, J. (1999) *Guns, Ferms and Steel: the Fates of Human Societies.* W.W. Norton, New York, 480 pp.

Edens, T.C. and Haynes, D.L. (1982) Closed system agriculture: Resource constraints, management options, and design alternatives. *Annual Review of Phytopathology* 20, 363–395.

Flint, M.L. and Gouveia, P. (2001) *IPM in Practice: Principles and Methods of Integrated Pest Management.* University of California, Publication 3418, Davis, 296 pp.

Goldsmith, E. (1992) *The Way: an Ecological World-View.* Shambhala Publications, Boston, 442 pp.

Horne, J.E. and McDermott, M. (2001) *The Next Green Revolution: Essential Steps to a Healthy, Sustainable Agriculture.* Food Products Press, New York, 312 pp.

Lipson, M. (1997) *Searching for the O-Word: Analyzing the USDA Current Research Information System for Pertinence to Organic Farming.* Organic Farming Research Foundation, Santa Cruz, California, 82 pp.

Meadows, D.H., Meadows, D.L. and Randers, J. (1992) *Beyond the Limits.* Chelsea Green Publication Company, Post Mills, Vermont, 300 pp.

Miller, G.T. Jr (2000) *Living in the Environment: Principles, Connections and Solutions,* 7th edn. Brooks/Cole Publishing Company, New York, 815 pp.

National Research Council (1999) *Our Common Journey: a Transition Toward Sustainability.* National Academy Press, Washington, DC, 363 pp.

Norris, R.F. and Caswell-Chen, E.P. (2003) *Concepts in Integrated Pest Management.* Prentice Hall, Upper Saddle River, New Jersey, 586 pp.

Ostrom, E., Dietz, T., Dolsak, N., Stern, P.C., Stonich, S. and Weber, E.U. (2002) *The Drama of the Commons.* National Academy Press, Washington, DC, 521 pp.

Potter, V.R., Baerreis, D.A., Bryson, R.A., Currin, J.W., Johansen, G., McLeod, J., Rankin, J. and Symon, K.R. (1970) Purpose and Function of the University. *Science* 167, 1590–1593.

Peterson, C., Drinkwater, L.E. and Wagoner, P. (2000) *The Rodale Institute Farming Systems Trial: the First 15 Years.* Rodale Institute, Kutztown, Pennsylvania, 40 pp.

Rodale, R. (1984) *Our Next Frontier.* Rodale Press, Emmaus, Pennsylvania.

Smith, R.L. (1992) Farming activities and family farms: getting the concepts right. Joint Economic Committee Symposium. *Agricultural Industrialization and Family Farms: the Role of Federal Policy.* 21 October. United States Congress, Washington, DC.

Strange, M. (1988) *Family Farming: a New Economic Vision.* University of Nebraska Press, Lincoln, Nebraska.

UNCED (1992) *Agenda 21.* United Nations.

World Commission on Environment and Development (1987) *Our Common Future.* Oxford University Press, New York, 400 pp.

Wright, R.T. and Nebel, B.J. (2002) *Environmental Science: Towards a Sustainable Planet.* Prentice Hall, Upper Saddle River, New Jersey, 644 pp.

Chapter 8

Social and Economic Considerations in the Design and Implementation of Integrated Pest Management in Developing Countries

Blessing Maumbe[1], Richard Bernsten[1] and George Norton[2]
[1]Department of Agricultural Economics, Michigan State University, East Lansing, Michigan, USA; [2]Department of Agricultural Economics, Virginia Polytechnic Institute, Blacksburg, Virginia, USA

Introduction

Pest infestation represents one of the major agricultural production risks facing farmers worldwide. Although prophylactic pesticide sprays may act as a form of insurance against a pest attack, ineffective pest scouting, poor weather conditions, competing demands for the growers' time, and in some instances, lack of cash needed to purchase the chemicals, have often led to untimely pesticide treatment. An important question to ask is whether alternative IPM[1] strategies effectively insulate farmers from yield and income risks associated with pest damage. There is now a growing realization that IPM strategies provide viable alternatives to relying totally on pesticides to manage pests. By reducing farmers' dependence on chemicals, IPM programs potentially reduce environmental contamination from pesticide residue, the incidence of pesticide-related health risks, and the need for countries to spend scarce foreign exchange on importing pesticides.

Understanding why farmers in developing countries have failed to adopt IPM in a significant way, despite its proven success in other parts of the world, is central to motivating individual farmers to respond to IPM programs. Economists rely on several approaches, including benefit cost analysis (BCA), to weigh the benefits relative to the costs of adopting IPM practices. In its advanced form, BCA captures not only direct income benefits, but the value of intangible, non-market-based goods and services such as biodiversity, stable yields, and clean water. Understanding the behavioral characteristics of farmers adopting IPM is important for guiding extension education and the development of new IPM technologies worldwide. Both IPM farmers and agribusiness firms can benefit from such information, and it may help them to

[1] IPM is a strategy for reducing the level of economic damage to crops by relying on numerous user- and environmentally friendly technologies, including host plant resistance, pheromones, natural enemies, crop rotations, trap crops, synchronized planting, sanitation measures, and vegetative barriers.

make better decisions about investments in research and IPM-related products.

It is important to note that in most LDCs, the use of calendar-based spraying is common. Farmers have been advised by scientists, extension workers, and chemical companies to apply prophylactic, on-schedule applications of pesticides. In addition, many LDCs have had a long history of subsidizing agricultural chemicals, including pesticides. These subsidies to the chemical industry have been complemented by crop protection policies centered on chemical control of pests. Yet problems of pesticide misuse that include application of the wrong pesticides or the wrong amount of a pesticide, and/or even the wrong timing of pesticide application have been widespread (Tjornhom *et al.*, 1997).

In Europe and the USA, the search for alternatives to heavy reliance on chemical pesticides for controlling pests gained momentum in the 1960s after the publication of *Silent Spring* (Carson, 1962), which described health and environmental dangers of pervasive pesticides use (Norton *et al.*, 1999). In the 1980s, a shift from prescription-based systems to new methods of building farmer knowledge and decision-making capacity evolved in several LDCs. In the developing world, Asia has taken the lead, with notable IPM programs in place in several countries, including Indonesia, the Philippines, and Bangladesh. For example, the Indonesia FFS IPM program, considered one of the most successful cases, grew in a short space of time from a non-IPM mode in 1989 to one that reached more than 300,000 rice farmers and 10,000 vegetable growers by 1992 (Untung, 1995). However, it is disturbing that only a few countries have adopted a national IPM strategy, making lack of political commitment by LDC governments a major stumbling block to IPM uptake (World Bank, 1997). While the goal of IPM is to reduce the use of chemical control measures, most IPM programs allow the judicious use of chemicals as a last resort, if scouting or surveillance indicates that pest levels exceed a predetermined ETL[2] which can not be reduced using alternative methods.

The aim of this chapter is to: (i) highlight key socioeconomic factors that affect farmer adoption of IPM; (ii) identify economic considerations that explain farmers' behavior and thereby affect their ability and willingness to adopt IPM strategies; (iii) describe the type of socioeconomic data that are necessary to design a successful IPM program; and (iv) discuss how these data can be analyzed to increase the probability that an IPM program will be successful. Case studies in the following chapters illustrate successful IPM strategies as well as some of the problems and constraints encountered in their implementation. The case studies highlight technical aspects of IPM use, extension programs, and socioeconomic and institutional factors that drive the uptake of IPM in various regions of the world.

What Factors Affect the Success of IPM?

Understanding the nature of IPM technology

In contrast to pesticides that rely on a single technology to produce immediate results (e.g. dead insects), IPM programs typically incorporate several complementary components. It is also important to note that these pest management practices are both location and crop specific. Basically, IPM can be understood from two angles: the 'input oriented approach' which focuses on different IPM technology components, and the 'output oriented approach' which looks at IPM in terms of desired outcomes, including attainment of a certain level of profitability, human health, and environmental qualities (Swinton and Williams, 1998).

[2] This is the break-even point at which the dollar value for an increment of loss in yield quantity or quality is equal to the cost of a control method that successfully eliminates pest damage and yield loss (Kiss and Meerman, 1991).

Genetically modified crops that include herbicide or insecticide resistance, such as Roundup Ready soybeans or Bt cotton varieties, represent a new approach to pest management in the 21st century.

Concern for the design of effective IPM programs is now shifting from emphasis on the composition of technical components of IPM to institutional design. There is increasing realization that IPM is both a technical and social process that relies on well-functioning institutions (Waibel and Zadoks, 1995). Institutions are the rules of the game, which as in a sporting event, are an important guide to understanding human interaction (North, 1990). Institutions define and limit the opportunity set of farmers (i.e. institutions establish rules that create opportunities for some and place restrictions on others). Rapid adoption of IPM therefore requires sound knowledge and understanding of the institutional arrangements. In many LDCs, institutions are not conducive to IPM adoption and diffusion. Traditional train and visit extension programs have often failed to reach farmers with IPM methods they consider useful (Norton *et al.*, 1999).

In terms of IPM implementation, a holistic and farmer-driven IPM program may not realize its full potential if it conflicts with agricultural policy objectives such as intensification and food security. In the same vein, the World Bank has argued for the need to mobilize and develop strong local constituencies in support of environmental protection and public health issues (World Bank, 1997). A growing number of studies have clearly demonstrated the benefits of knowledge-based technologies such as IPM in significantly reducing over-application of pesticides, thus improving productivity, human health, and the environment (Antle and Pingali, 1994; Norton and Mullen, 1994; Fernandez-Cornejo, 1998; Swinton, *et al.*, 1999). To facilitate more extensive adoption of IPM, the USA has introduced the EQIP, under the 1996 Federal Agricultural Improvement and Reform Act, as a public cost-share program designed to reduce the cost of adopting IPM practices (Swinton and Day, 2000).

The FFS philosophy, originally developed as a strategy to fight the problem of pesticide resistance and farmer health risks in rice-based monocultures in Asia, is now increasingly being used to spread IPM in many parts of Africa (Gallagher, 1998). FFS is a participatory training approach that uses discovery-based learning techniques in pest and crop management with the overall aim of helping farmer-groups to understand agroecosystems analysis required to cope with biotic and abiotic stresses. It stresses the importance of farmers growing a healthy crop, observing their fields weekly, conserving natural enemies, and experimenting themselves using relevant science-based knowledge. Although this tool has not been extensively tested in all regions and in all cropping systems in Africa, initial results from a study of the smallholder cotton production system in Zimbabwe indicate that awareness of IPM technology through exposure to FFS is a major driving force in farmers' adoption of IPM (Maumbe, 2001). One of its main advantages is that it empowers limited-resource farmers to make independent pest management decisions. Some of its apparent drawbacks are the costliness of FFS-based IPM programs and the danger of program quality deterioration as training responsibilities are passed on successively to newly trained farmers (Seif and Lohr, 1998).

A key consideration in IPM use is the fact that it is difficult for farmers to observe the benefits of each specific IPM component, and the full impact of these benefits may be realized over a relatively long time horizon. The idea that the relationship between pest damage and a farmer's action is not obvious, suggests that IPM programs must include a significant educational component and farmers may adopt some program components relatively slowly. Evidence of selective step-wise adoption of interrelated packages has been highlighted in the literature (Byerlee and de Polanco, 1986). Equally important is the need to understand the core practices that may represent an ideal IPM technology package in each farming system, as these packages are still in the development stages in most regions. More importantly, trainers need to ensure that attendees at FFS events are directly involved in pest management themselves. In Zimbabwe, male farmers mainly apply pesticides. However, since most men work off-farm in the urban areas, women may be more appropriate participants in FFS training programs. This gender-related difference highlights the need for new pest management information to reach various stakeholders. Similar issues pertain to the case where hired workers, who may not be exposed to IPM knowledge, apply pesticide treatments.

The need for group versus individual actions

Many Green Revolution agricultural technologies such as fertilizers and high-yielding varieties are divisible – implying that they can be tested by individual farmers and, if successful, individual farmers can adopt them over their whole farm. In contrast, because many insects respect no boundaries, some IPM technologies are ineffective, unless adopted simultaneously by all farmers in a region. The challenge for a successful IPM program is to mobilize community support needed for simultaneous adoption. For that reason, the concept of FFS has grown as a strategy for spreading IPM in most parts of the world. For FFS or any other approach to successfully deliver IPM technology, farmers should have a common agenda. When farmers have a goal of reducing pests in more effective and safer ways, then the use of farmer group-based IPM may deliver bigger and earlier benefits to individual farmers than when IPM is viewed as component technologies that are only focused on higher yields and lower costs.

Macroeconomic determinants

At the national level, the removal of pesticide subsidies and the promulgation by governments of national IPM strategies is a major first step needed to usher in a new way of thinking among all stakeholders, including farmers, researchers, policy makers, extension workers, and the chemical industry – all of whom are critical for the adoption and diffusion of IPM. There is little doubt that the use of subsidies has contributed immensely to the overuse of pesticides. All forms of tax exemptions for pesticide imports and refunds by governments for pesticide sales taxes should be abolished because they stimulate artificial demand for pesticides. In contrast, pesticide taxation would reduce pesticide demand and generate additional funds that could be used to strengthen IPM research and training programs. Governmental commitment is essential to establish an enabling environment for abolishing policies that support environmentally unsustainable pest management and for strengthening regulatory institutions. Targeted support for IPM is crucial in the early stages of IPM diffusion. As has happened in the USA, governments of LDCs might consider subsidizing IPM technologies such as use of natural enemies in order to encourage farmers to adopt them more rapidly.

Another macroeconomic deterrent to IPM technology adoption is the prevalence of overvalued exchange rates (Tjornhom *et al.*, 1998). In most developing countries pesticides are imported, and overvalued exchange rates indirectly subsidize these imports.

The role of farmer's indigenous knowledge

Prior to the development and marketing of chemical pesticides, farmers relied on indigenous knowledge to help protect their crops from pests. Although there is growing recognition that farmers have considerable 'indigenous knowledge' of crop–pest ecology, they also have misconceptions that can act as barriers to the adoption of IPM (Bentley, 1994). For example, studies of IPM use in the Philippines and Uganda, have identified failure to recognize the difference between insect and disease damage and the stress caused by water deficiency and high temperatures as common problems facing farmers in FFS (Tjornhom *et al.*, 1997). They also fail to recognize pests that are difficult to observe such as nematodes (Norton *et al.*, 1999). In addition, it has been reported that farmers fail to identify the difference between pests and beneficial insects (natural predators), a key dimension of IPM technology (Kiss, 1995; Lazaro *et al.*, 1995). None the less, the World Bank (1997) argues that the effective integration of technical and social knowledge is one of the essential aspects of IPM diffusion. As a starting point, IPM programs need to know what farmers know in order to design a strategy for teaching them the 'new science' of IPM. Over time, scientists have discovered that traditional technologies such as the use of mud-sealed clay containers, and in some cases neem (*Azadirachta indica*)[3], to control post-harvest pests are not only quite effective, but are also relatively inexpensive – a key advantage to resource poor farmers – compared with some chemical alternatives. Challenges remain in the use of economic thresholds, which are still underdeveloped and require further refinement, if they are to be applied successfully in developing countries.

Farmer's resource endowments

The adoption of agricultural technology has been shown to differ across socioeconomic groups and over time (Feder *et al.*, 1985). In addition to economic status, farmer circumstances differ considerably in terms of their access to productive labor, educational levels, and land. These circumstances can be key factors affecting farmers' decisions to use IPM recommendations. For example, scouting[4] – a key component of IPM – requires considerable labor. Thus, farmers with limited labor may find it too time consuming, relative to the anticipated benefits. Second, because scouting is knowledge intensive, farmers without sufficient understanding of the rationale for counting pests may consider it a waste of time. Third, land tenure arrangements may affect the attractiveness of IPM. For example, in some countries tenant farmers and their landlords share the cost of purchased inputs, but the farmer must provide all of the required farm labor. Under this arrangement, there is less incentive for the tenant farmers to substitute labor-intensive IPM strategies for insecticides, compared to owner-operators.

What Social and Economic Concepts are Relevant?

In order to understand farmers' IPM technology adoption behavior, four concepts are critical: opportunity cost, uncertainty, marginal benefits, and ETLs.

Opportunity cost of pesticide use

Indiscriminate pesticide use has been associated with concerns about farmer health and environmental risks. Trade-offs between health and economic effects have been documented in recent studies in the Phillipines and Ecuador (Antle and Pingali, 1994, Crissman *et al.*, 1994, 1998; Cuyno, 1999). Pesticides threaten not only human health, but also the vitality of other species including beneficial insects. However, in deciding to adopt labor-intensive IPM technologies, farmers may have to delay other critical crop operations such as planting and weeding, and daily household chores such as gathering firewood, and fetching water. In diffusing IPM, proponents have to assess the new practices in terms of their opportunity cost – the activities that farmers

[3] In many parts of the developing world, farmers use an organic insecticide, which is made from the leaves of the neem tree, to protect their stored crop.

[4] Scouting involves periodically counting the number of harmful insects attacking the crop to determine if the insect population exceeds a predetermined threshold level, which requires the application of an insecticide.

may have to give up versus the benefits that farmers will gain from adopting IPM.

Uncertain benefits and positive marginal returns

If the production benefits justify the costs, then the spread of IPM will be relatively easy. Some technologies have performed very well under supervised experimental conditions and yet performed dismally under farmer field conditions. Further, while some 'research station' recommendations have worked well for some farmers, the same recommendations have not performed as well for farmers facing slightly different ecological and social circumstance. That farmers are relatively efficient but poor is well understood. Farmers will be hesitant to adopt a technological package unless it is clear to them that under their farming conditions (i.e. ecological and socioeconomic) each *element* of the package is beneficial, and it gives a sufficiently great marginal return to justify the time required to implement it. The current dilemma for farmers in many countries throughout the world is that production without pesticides is not profitable in the short run, even though further application of pesticides may lead to pesticide resistance, lower yields, higher pesticide expenses, and increased incidence of pesticide-related health risks.

Economic thresholds

A key component of many IPM programs is the recommendation that farmers to not apply chemicals unless the level of damage is sufficiently large to justify the cost of controlling the pest. Although this strategy is technically and economically sound, it often conflicts with farmers' current beliefs. For example, farmers may typically wait to take action until they observe adult insects (which may be too late) or spray as soon as they observe any leaf damage (which may be too early). The appropriateness of using an ETL in IPM programs in LDCs remains controversial. Some researchers believe that LDC farmers do not have sufficient information required to make the appropriate interventions. Others are concerned about the accuracy and precision needed to effectively apply an ETL in developing countries. In the USA and Europe, elaborate ETLs have been developed and are commonly used by farmers in pest management decisions. However, there is no single ETL for every crop and climatic condition in the world (Kiss and Meerman, 1991). None the less, others believe that it is in the best interest of LDC policy makers, farmers, researchers, and extension workers to invest the time and effort required to adapt ETLs to specific farming systems in LDCs, if pesticide interventions are to be meaningful. The bottom line is while the concept of ETLs is sound, at least for insect pests, getting farmers to follow the program's threshold recommendation requires a major education effort.

Socioeconomic Data Needed for IPM Program Design

Most IPM programs first attempt to understand the insect, disease, weed, and nematode complex, in order to set priorities in terms of which pests cause damage and what predators are present. For example, for insect pest problems, entomologists often conduct 'insect population surveys'. However, given the dynamic nature of pest populations due to climatic and other factors, these surveys must be carried out for several years to assess the relative importance of pest and disease constraints effectively. Too frequently, IPM and related agricultural research priorities are established based on an inadequate assessment of the pest constraints. Successful IPM programs also require data on: (i) farmers' knowledge of pests; (ii) farmers' current pest management practices and perceptions, against which the impact of the program can be measured; and (iii) farmers' resource endowments (Table 8.1). Such information

Table 8.1. Data needs for the design of successful IPM programs.

Farmers' knowledge of insect biology *Data needs* • Insect, nematode, weed, and disease types • Type of damage • Persistent versus sporadic pest problems • Farmers' knowledge of pest predators • Farmers' traditional pest control methods	**Farmer's knowledge of pesticides** *Data needs* • Types of pesticides applied • Pesticide prices • Rate of application per crop • Basis for spraying, crop height, pest pressure, etc. • Label literacy • Pesticide application technology • Safety precautions for handling pesticides • Training received on pesticides use and source • Benefits of current pesticide application strategy • Knowledge of 'safer pesticides'
Crop production and level of technology use *Data needs* (a) Varieties • Varieties grown (traditional versus improved) • Variety characteristics, yields, size, height, etc. • Rationale for choosing the specific varieties • Susceptibility to pests (b) Crop management • Cropping patterns, monoculture, intercropping • Planting dates for each crop • Irrigated versus dry land production • Use of trap crops • Use of field sanitation • Crop rotations • Major production constraints • Cash crops versus food crops produced	**Human and production resource endowments** *Data needs* • Farmers' educational levels • Household size and gender distribution • Labor sources (i.e. family versus hired) • Acreage cropped • Farm tenure arrangements • Access to credit • Market outlets and prices • IPM-related inputs available, sources and cost • Extension assistance • Farmers' perceptions about IPM practices • Farmers' preferences about IPM practices (rank) • Possible IPM adoption constraints • Farmers' views on health effects of pesticides • Farmers' views on environmental hazards • Training needs on pest management

can be obtained via participatory appraisals, group interviews, and farmer surveys.

Analysis of these data should focus on identifying the IPM components most likely to be accepted and those that are most likely to be rejected by farmers because they conflict with farmers' goals or preconceptions – due to significant information gaps in farmers' understanding of insect population dynamics, proposed IPM program components that are unfamiliar to the farmers, and misinformation that farmers will have to 'unlearn'.

Training to 'promote' key elements of an IPM program should be based on farmers' feedback. With insights gained regarding specific information gaps, the program can focus its attention on developing a curriculum to teach farmers what they need to know about IPM. Components that are most likely to be difficult to implement should be given priority during training.

At the planning stage, IPM programs should incorporate organizational innovations that provide continual feedback during project implementation. In the event that factors that condition farmers' adoption of IPM technology – such as lack of credit, non-availability of key inputs, especially improved varieties, and weak extension support – are highlighted as major constraints, then the IPM program can initiate a dialogue with policy makers to identify ways to remove these barriers to program success.

Finally, policy makers and donors now require agricultural research and development programs to document the impact of their programs. Impact analysis is important for two main reasons: (i) to measure program success and or failure; and (ii) to form the basis for guiding program redesign needed as new constraints emerge and new understanding is gained. Program redesign will benefit from farmer feedback on which components of the program they believe are ineffective and need modification. Evidence of program impact can be obtained by resurveying the farmers in the target area to measure the degree to which the goals of the program have been achieved, including the amount of change in farmers' attitudes, knowledge, and behavior following program implementation. In particular, the initial and follow-up surveys should ask questions about farmers' perceptions regarding levels of pest damage. Economic data should be collected to assess changes in cost of production, yields, returns to labor and profit. Economy-wide economic and environmental/health impacts can also be calculated.

Conclusion

Although crop protection specialists generally agree that IPM programs have been successful in many countries, the success of a specific IPM program depends on both the *validity* of the technical recommendations around which the program is built and the *compatibility* of these technical elements with the target farmer's ecological and socioeconomic circumstances. Thus, reliable information about both the agroecology and the socioeconomic environment of the target area is critical to successful implementation of IPM. At the farm level, recognition of key pests is a critical component of IPM training programs. Introduction of IPM is not easy when farmers are poorly motivated; hence the need to demonstrate clearly the opportunity cost and expected benefits and costs of adopting IPM. Rapid adoption of IPM can be stimulated by well-targeted economic incentives for farmers to adopt IPM technologies.

Government commitment to developing IPM programs and encouraging adoption can play a vital role in IPM diffusion. In designing IPM programs, social scientists as part of an interdisciplinary IPM team can provide insights into the social and economic environment that may contribute to the success of an IPM program. The diffusion of IPM requires early identification of a clear IPM technology package that is simple and effective under farmers' conditions. Such location-specific IPM programs are a win-win situation that will help solve pest problems, minimize pesticide costs, and raise agricultural productivity while reducing both environmental damage and farmer health risks. The FFS model is one approach that has been well received and is a growing force in the spread of IPM in Africa, Asia, and Latin America. However, it too can be a costly means for spreading IPM information and may not be the most cost-effective approach in every situation. The challenge remains in finding cost-effective ways of changing farmers' attitudes, given their cautionary approach to changing traditional farming practices.

References

Antle, J.M. and Pingali, P.L. (1994) Pesticides, productivity, and farmer health: a Philippine case study. *American Journal of Agricultural Economics* 76, 418–429.

Bentley, J.W., Rodriguez, G. and Gonzalez, A. (1994) Science and people: Honduran campesinos and natural pest control interventions. *Agriculture and Human Values*, Winter, 178–182.

Byerlee, D. and de Polanco, E.H. (1986) Farmers' step-wise adoption of technological packages: evidence from the Mexican altiplano. *American Journal of Agricultural Economics* 68, 519–527.

Carson, R. (1962) *Silent Spring*. Houghton Muffin Company, Boston.

Crissman, C.C., Cole, D.C. and Carpio, F. (1994) Pesticide use and farm worker health in Ecuadorian potato production. *American Journal of Agricultural Economics* 76, 593–597.

Crissman, C.C., Antle, J.M. and Capalbo, S.M. (1998) *Economic, Environmental, and Health Trade-Offs in Agriculture: Pesticides and the Sustainability of Andean Potato Production.* Kluwer Academic Publishers, Norwell, Massachusetts.

Cuyno, L.C.M. (1999) An economic evaluation of the health and environmental benefits of the IPM program in the Philippines. PhD dissertation, Virginia Tech., Blacksburg.

Feder, G., Just, R.E. and Zilberman, D. (1985). Adoption of agricultural innovations in developing countries: a survey. *Economic Development and Cultural Change* 33, 255–298.

Fernandez-Cornejo, J. (1998) Environmental and economic consequences of technology adoption: IPM in viticulture. *Journal of Agricultural Economics* 18, 145–155.

Gallagher, K.D. (1998) Farmer field schools for integrated pest management in Africa with special reference to East Africa. *Proceedings of the National Pre-Season Planning Workshop on the Implementation of Field School Groups for Integrated Production and Pest Management.* 31 August–1 September, ZIPAM, Darwendale, Government of Zimbabwe and FAO Global IPM Facility, Rome, Italy.

Kiss, A. (1995) Pest management needs and trends in Africa today. *International Food Policy Research Institute Workshop on Pest Management, Food Security, and the Environment: the Future to 2020.* May 10–11, Nairobi, Kenya.

Kiss, A. and Meerman, F. (1991) *Intergrated Pest Management and African Agriculture.* World Bank Technical Paper No. 142. African Technical Department Series, Washington, DC.

Lazaro, A.A., Heong, K.L., Canapi, B., Gapud, V. and Norton, G.W. (1995) *Farmers' Pest Management Knowledge, Attitude and Practices in San Jose, Philippines: a Baseline Survey.* Integrated Pest Management Collaborative Research Support Program (IPM CRSP) Working Paper 95-2. Virginia Tech., Blacksburg, Virginia.

Maumbe, B.M. (2001) Essays on the economics of cotton production in Zimbabwe: policy implications for technology adoption, farmer health and market liberalization. PhD dissertation, Michigan State University, East Lansing.

North, D.C. (1990) *Institutions, Institutional Change and Economic Performance. Political Economy of Institutions and Decisions.* Cambridge University Press, New York.

Norton, G.W. and Mullen, J. (1994) *Economic Evaluation of Integrated Pest Management Programs: a Literature Review.* VCE Publication 448–120, Virginia Tech., Blacksburg.

Norton, G.W., Rajotte, E.G. and Gapud, V. (1999) Participatory research in integrated pest management: lessons from the IPM CRSP. *Agriculture and Human Values* 16, 431–439.

Seif, A.A. and Lohr, B. (1998) Integrated Pest Management in horticultural crops in East and Southern Africa: activities and experiences of GTZ IPM horticulture project. *Proceedings of the Integrated Pest Management Communications Workshop: East and Southern Africa.* 1–6 March, Nairobi, Kenya.

Swinton, S.M. and Day, E. (2000) *Economics in the Design, Assessment, Adoption, and Policy Analysis of IPM.* Staff Paper, Department of Agricultural Economics, Michigan State University, East Lansing, Michigan.

Swinton, S.M. and Williams, M.B. (1998) *Assessing the Economic Impacts of Integratred Pest Management: Lessons from the Past, Directions For the Future.* Staff Paper 98-12, Department of Agricultural Economics, Michigan State University, East Lansing, Michigan.

Swinton, S.M., Batie, S.S. and Schulz, M. (1999) *FQPA Implementation to Reduce Pesticides Risks: Part II: Implementation Alternatives and Strategies.* Staff Paper 99-46, Department of Agricultural Economics, Michigan State University, East Lansing, Michigan.

Tjornhom, J.D., Norton, G.W., Heong, K.L., Talekar, N.S. and Gapud, V.P. (1997) Determinants of pesticide misuse in Phillipines onion production. *Philippine Entomology* 11, 139–149.

Tjornhom, J.D., Norton, G.W. and Gapud, V. (1998) Impacts of price and exchange rate policies on pesticide use in the Philippines. *Agricultural Economics* 18, 167–175.

Utung, K. (1995) Institutional constraints on IPM implementation in Indonesia. Institutional Constraints to IPM. Papers Presented at the *XIIIth International Plant Protection Congress (IPPC)*, The Hague, 2–7 July, 1995.

Waibel, H. and Zadoks, J.C. (1995) Institutional constraints to IPM. *Papers Presented at the XIIIth International Plant Protection Congress (IPPC)*, The Hague, 2–7 July, 1995.

World Bank (1997) *Integrated Pest Management: Strategies and Policies for Effective Implementation.* Environmentally Sustainable Development Studies and Monograph Series No.13. World Bank, Washington, DC.

Chapter 9
Integrated Pest Management Adoption by the Global Community

Waheed Ibrahim Bajwa and Marcos Kogan
Integrated Plant Protection Center, Oregon State University, Corvallis, Oregon, USA

Introduction

Nearly 33 years after introduction of IPM, pest control is still largely dependent on the use of pesticides. In many countries – developing countries in particular – pesticide consumption actually increased in the 1990s. For instance, pesticide use has increased by a factor of 39 between 1950 and 1992 and the developing countries now account for one quarter of the world's pesticide use (FAO Statistics). However, the industrial countries of North America and Western Europe still account for over half of the world's pesticide sales (Table 9.1). In the 1990s, pesticide use leveled off in the USA; declined in The Netherlands, Denmark, and Sweden; and slightly increased in the UK (Kogan and Bajwa, 1999). These countries have strong IPM programs, based on excellent research and outreach efforts. It is unfortunate that in the developed countries, 95% of pesticides is used to control the final 5% damage which is often cosmetic (van Emden and Peakall, 1996). In developing countries, most farmers still believe in high volumes, high pressure and high doses, as the most appropriate way to

Table 9.1. Sales of agrochemicals in the major regions of the world.

	1990[a]		1992[b]		1996[c]	
Region	US$ (billion)	% share	US$ (billion)	% share	US$ (billion)	% share
North America	5.4	21.9	7.3	29.2	9.2	29.4
Western Europe	6.6	26.7	6.7	26.7	8.2	26.2
Eastern Europe	1.9	7.7	1.2	4.6	na	na
Asia	6.8	27.5	6.1	24.4	7.7	24.5
Africa	1.2	4.9	na	na	na	na
Latin America	2.8	11.3	2.4	9.5	3.3	10.4
Rest of the world	na	na	1.4	5.6	3.0	9.5
Total	24.7	100	25.1	100	31.4	100

na, not available; [a]GIFAP, 1992. Asia Working Group. Publication of International Group of National Associations of Manufacturers of Agrochemical Products, Brussels. [b]*Chemistry & Industry*, 15 November 1993. [c]*Agrow: World Crop Protection News*, 13 December 1996, 14 February and 28 February 1997.

apply pesticides. Problems with pesticide-intensive pest control programs are the driving force behind IPM adoption or at least consideration for its adoption at the national or local level in most countries, and in many it has become official governmental policy.

IPM is an information-intensive, site-specific, multitactic approach to pest control. Rates and levels of adoption of IPM are determined by the resultant interplay of a regional producers culture and experience, influenced by promotional efforts of the agrochemical industry, moderated by public educational and outreach efforts and availability of extension support. In contrast to the rapid adoption of pesticide technology worldwide, adoption of a newly developed IPM approach or technology may take years. Table 9.2 provides a summary of possible reasons for the contrast in the rates of adoption of these crop protection technologies. In addition to the reasons suggested in Table 9.2, because of differences in climate, pests, soil, variety and other factors, a well-developed IPM program for a crop in a particular location may not necessarily work well in another situation. Farmers need site-specific information. Generally, they have to work with local IPM information providers (research and extension specialists, NGOs, private consultants) to acquire the information and knowledge necessary for developing an IPM program suited to their needs. Thus, IPM is a diffuse technology not amenable to generalized prescriptions. Decisions must be made at the local, or at best, at the regional level (Table 9.2). With its success in many parts of the world, IPM is viewed as an ecologically benign and cost-effective pest control strategy ideally suited for both small and large farmers around the world.

Levels of IPM Adoption

Adoption of IPM can be viewed under a continuum, starting with systems largely confined to using a single tactical approach such as using economic thresholds for better timing of pesticide applications. Along the continuum, additional non-chemical tactics such as cultural controls, biological controls, resistant crop varieties, mating disruption, sterile insect release, etc., may be integrated into the system. Above a certain level of tactical integration, a threshold is reached at which a previously

Table 9.2. Contrasting features of pesticide technology and IPM as possible reasons to explain the fast rate of adoption of the former and the slow rate of adoption of the latter.

Pesticides	IPM
Compact technology from acquisition to application. Easily incorporated into regular farming operations	**Diffuse technology** with multiple components. At times difficult to reconcile with normal farming operations
Promoted by the private sector	Promoted by the public sector
Strong economic interests. Large budgets for R&D	Budgets extremely limited for R&D
Aggressive sales promotion supported by professionally developed advertising campaigns	Promoted by extension personnel usually trained as educators not as salespersons
Skillful use of mass communications media	Limited support of trained communications media personnel. Educational programs of restricted scope
Capable of providing incentives to 'adoption' (free advice, slick publications, bonuses and small gifts)	Technical support usually provided, but limited by inadequate staffing. No material incentives
Results of applications usually immediately apparent.	Benefits often not apparent in the short run. Some difficult to demonstrate (e.g. results of biological control)
Consequently: pesticide technology was rapidly adopted	**Consequently: adoption of IPM technology has been slow**

pesticide-centered program, becomes an IPM system. At the other extreme of the continuum, higher levels of integration are reached including multiple pest impacts and consideration of ecosystem processes. Eventually a stage is reached at which pesticide use is minimized with a concurrent increase in the amount of time and management skills that are devoted to IPM operations (Fig. 9.1).

IPM is conceived at three levels of integration. Level I – is the integration of multiple control tactics into a control strategy for individual pest species or species complexes within the same pest category, i.e. arthropods, pathogens, or weeds. The operational unit is the crop field and the ecological scale is the pest population. Level II – is the integration of multiple and interactive impacts of all pests within the crop community and the tactics for their management. The operational unit is the individual farm or multiple farms within a region and the ecological scale is the crop biotic community. Level III – is the integration of multiple pests and controls within the context of the regional cropping system and surrounding natural vegetation. The operational unit is the regional agricultural production system and the ecological scale is the ecosystem (Kogan, 1988). Prokopy and Croft (1994) suggested the need for a fourth level, socio-political integration, but, in our view, regulatory and other issues determined by societal demands and political actions permeate all three levels of integration and thus should not be identified as yet an additional level.

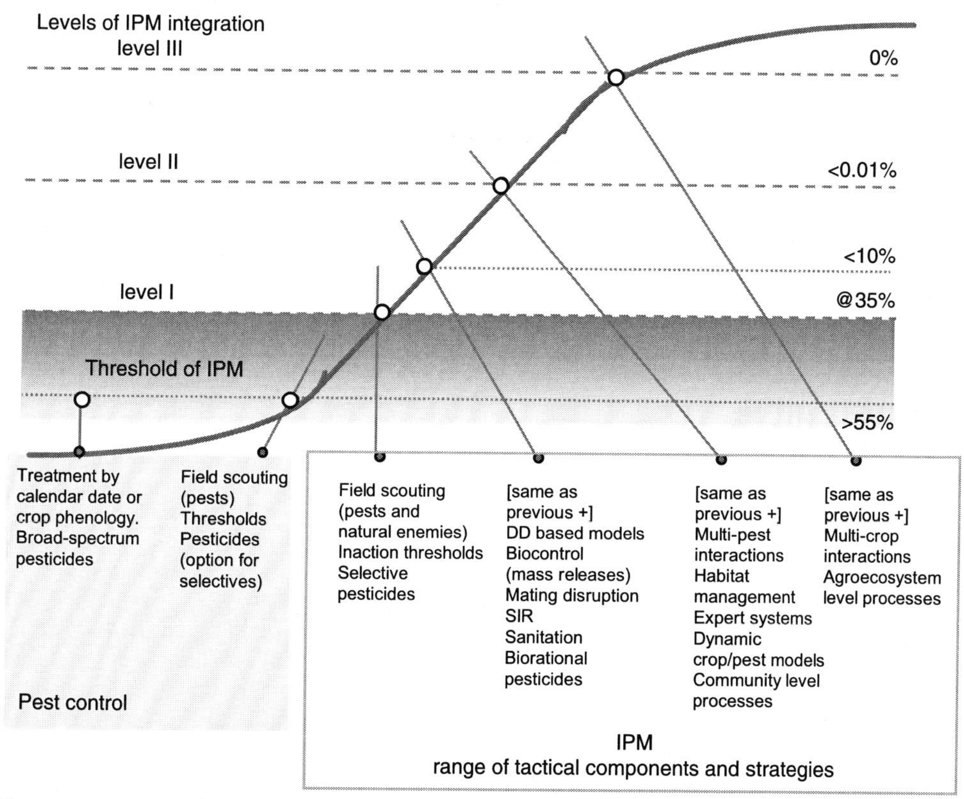

Fig. 9.1. Continuum from conventional pest control to level III IPM, as exemplified by Fruit IPM in USA. A minimum set of tactical components determines the 'threshold of IPM'. DD, degree-day; SIR, sterile insect release.

Criteria for the effectiveness of an IPM system

The following criteria have been suggested by Kogan and Bajwa (1999) for measuring performance of an IPM system:

- Ability of the system to maintain pest populations below established economic injury levels. As the stated goal of an IPM system is to eliminate or at least attenuate the economic impact of pests, efficacy in the reduction of pest populations is a primary indicator of effectiveness of the system.
- Measurable reduction of pest impact on crop yield and quality over a period of time leading to greater stability in the productivity of the system.
- Reduction in amounts of production and protection inputs of non-renewable resource origin (mainly pesticides) while maintaining stable productivity levels for the region.
- Level of adoption of the IPM system by producers.
- Preservation of environmental quality, as determined by measurable indicators.
- Increase in safety and comfort of rural workers and their families.
- Increase in the level of consumer confidence in the safety of agricultural products.

IPM adoption around the world

Measuring IPM adoption is a complex but much needed process because it provides information about the efficiency of an IPM program, identifies constraints to adoption, and identifies areas in need of improvement. Reliable estimates of IPM adoption are not available presently for most crops. In fact, the measurement of IPM adoption depends largely on the definition of IPM. IPM is often viewed as a strategy to integrate two or more control tactics. However, a decision to do nothing is perhaps the most desirable state of an IPM program in which forces of nature are identified as, *per se*, capable of achieving adequate pest population regulation. Such a stage, what Sterling (1984) called the inaction threshold, requires no integration of tactics, merely a profound understanding of the ecology of the agricultural system. Such was perhaps the nature of the most heralded IPM success in the world, the control of the brown planthopper, *Nilaparvata lugens*, in Southeast Asia (Kenmore, 1996).

In most countries, some form of IPM now exists with varying degrees of sophistication and adoption. Major effort has been directed to crops such as banana (Costa Rica), cotton (USA, many Asian, African, and South American countries), rice (many Asian and African countries), soybean (USA and South American countries), maize (USA, many Asian and African countries), vegetables (most countries), pome fruits (Europe, Australasia, and North America), citrus fruits (USA and Australia), and plantation crops (Malaysia) (see Tables 9.3–9.7). Several regional IPM programs have successfully been implemented on crops such as cassava (mealy bug in Africa), coconut (rhinoceros beetle in Asia/Pacific), crucifers (diamondback moth in Asia), and rice (brown planthopper in tropical Asia) (Mengech *et al.*, 1995; Soon, 1996). Experience from these examples has shown that IPM can work very well in both developed and developing countries; however, successful implementation requires raising general awareness of IPM and training at the research, extension, and farm levels. In many developing countries, IPM was found economically more efficient than conventional pest control approaches based on intensive use of pesticides. In these countries, a 50–100% reduction in pesticide use is possible with no detrimental effects on yield (Soon, 1996).

In the developed world (both in Europe and North America), the food industry (particularly baby food products) has recently accepted the concept of IPM and is actively encouraging its development and adoption. Several companies have hired IPM specialists to conduct IPM research and develop programs for their contract growers. Some companies help promote IPM by purchasing products from IPM/organic growers (Sorensen, 1998; Kogan and Bajwa, 1999).

Estimates on IPM adoption in some crops are given in the Tables 9.3–9.7. These tables contain information on direct measurement such as area under IPM or indirect measurement such as impact on reduction in pesticide use, application frequency, or treated area. Table 9.7 provides a list of other IPM programs mentioned in the literature as successful, but with no indication of levels of adoption.

USA and North America

Estimates on IPM adoption in USA are based on a survey carried out from 1990 to 1993 (Table 9.3). Overall, 50% or more of the crop acreage in fruits, nuts, vegetables, and field crops was under IPM for at least one of the three major pest types: insects, diseases, and weeds. About 5–15% of the area was under low level (just scouting and use of economic thresholds (ET) or at the level we call the IPM threshold), 23–35% under medium level (scouting, ET and one or two additional practices) and 20–30% under a high level (scouting, threshold and three or more additional practices) of IPM. In fruits, nuts and vegetables, about 50% of the area was under IPM with more than half classified as high-level IPM. Among field crops, 74% and 72% of planted area was under IPM for maize and fall potatoes, respectively. About 38% of the fall potato acres were classified at high level IPM for insects. Lack of crop consultants to deliver IPM services and the higher managerial input necessary for IPM implementation were the most frequent impediments to adoption. Adoption rate has recently improved, for example, in the state of Washington, where 100% of fruit growers are now using Level I IPM (Warner, 1998).

In Canada, rate of IPM adoption is 25–95% in different areas and commodities

Table 9.3. IPM adoption in field, vegetable, fruit and nut crops in major producing states of USA, 1991–1994.

	Production area (1000 ha or %)			Planted ha under IPM (%)		
Crops	Total US	Producing states reporting	% US ha	Insects	Diseases	Weeds
Field crops						
Maize	29,537	26,583	90	74 (22)	na	53 (51)
Soybean	25,760	21,638	84	na	na	59 (57)
Fall potatoes	482	453	94	72 (69)	63 (58)	66 (65)
Vegetables				52 (43)	41 (29)	35 (33)
Lettuce	108	105	97	81 (59)	80 (42)	41 (41)
Melons	165	132	80	56 (48)	52 (34)	47 (47)
Sweetcorn	305	259	85	43 (34)	34 (25)	46 (46)
Tomatoes	166	144	87	66 (55)	41 (36)	23 (23)
				All pests (insect, weed, disease)		
				High IPM	Total	
Fruits and nuts				31	50 (44)	
Almond	156	154	99	32	54 (53)	
Apple	188	154	82	27	42 (41)	
Grapes	299	296	99	37	54 (48)	
Oranges	248	248	100	26	64 (49)	
Pear	29	28	95	26	40 (37)	
Walnuts	74	74	100	31	43 (41)	

Values in parentheses are based on the IPM threshold concept by Kogan, 1998. na, not available. (Source: Vandeman *et al.*, 1994.)

(Table 9.6). In Nova Scotia, 95% of growers use Level I IPM and 25–40% Level II and III IPM. In British Columbia, rate of adoption is over 75% in pome and stone fruits.

Europe

In Western Europe, 35% of the total area (322,000 ha) of pome fruit production is under integrated fruit production (IFP), an approach in which IPM has a central role. The area has increased by 40% since 1991 (Schafermeyer, 1991). Area under pome IPM in Europe is given in Table 9.4. Adoption of IFP over a large area has led to promotion of higher standards of integrated pest management, up to 30% reduction in pesticide use, and use of environmentally benign pesticides (Cross *et al*., 1995). IFP has received a warm welcome from developed countries in other parts of the world, e.g. USA (Kogan and Bajwa, 1999), and New Zealand and Australia (Cross *et al*., 1995). Implementation of IFP is not confined to Western Europe or the developed world, it is spreading to many other fruit growing countries, e.g. Poland (Kogan and Bajwa, 1999), Hungary (Balázs *et al*., 1996), South Africa, and Argentina (Cross *et al*., 1995) have recently started initiatives to adopt IFP methods.

Asia and Australia

In Asia and Australia, various stages of IPM have been successfully adopted in Bangladesh (rice), China (cotton, fruit crops, maize, rice, soybean, vegetables), India (cotton, fruit crops, sugarcane, vegetables), Indonesia (rice), Korea (rice), Malaysia (vegetables, plantation crops), Pakistan (cotton, mango, sugarcane), the Philippines (rice), Tadjikistan (cotton), Thailand (cotton), Turkmenistan (cotton), Vietnam (rice), Australia (pome and citrus fruits) and New Zealand (pome fruit) (Tables 9.5 and 9.7). Information on IPM adoption is generally not available for many countries where emphasis was given to biological means of pest control as the major component of IPM. In these countries mass production and release of several biocontrol agents have occurred without subsequent study on the effect of the program. In China, where large-scale mass release of biocontrol agents has been adopted for many years, it was estimated that by the end of 1991, the area covered under the mass release program

Table 9.4. IPM adoption in Europe.

		Area (1000 ha)		Area under	Farmers adopted	
Country/region	Crop	Total crop	IPM ha	IPM (%)	IPM (%)	Reference
Western Europe	Apple & pear	–	–	50	–	Reed, 1995
Western Europe	Pome fruits	920	322	35 (1994)	–	Cross *et al*., 1995[a]
Western Europe	Fruit crops	–	–	29	–	Reed, 1993
Austria	Pome fruits	5.83	4.77	82	51	Cross *et al*., 1995[a]
Belgium	Pear	–	–	–	98	Schaetzen, 1996
Belgium	Pome fruits	20.00	4.51	23	31	Cross *et al*., 1995[a]
Denmark	Pome fruits	3.44	0.96	28	17	Cross *et al*., 1995[a]
France	Pome fruits	75.00	0.50	<1	~1	Cross *et al*., 1995[a]
Germany	Pome fruits	38.60	30.44	79	27	Cross *et al*., 1995[a]
UK	Pome fruits	17.00	13.00	76	77	Cross *et al*., 1995[a]
Italy	Pome fruits	71.24	38.00	53	47	Cross *et al*., 1995[a]
The Netherlands	Pome fruits	21.00	14.80	70	57	Cross *et al*., 1995[a]
Portugal	Pome fruits	25.50	1.10	~4	<1	Cross *et al*., 1995[a]
Spain	Pome fruits	56.00	0.40	<1	<1	Cross *et al*., 1995[a]
Switzerland	Pome fruits	6.08	4.35	72	39	Cross *et al*., 1995[a]
Hungary	Pome fruits	37.0	3	8	6	Valyi and Sallai, 1993[a]

[a]Values represent adoption of integrated fruit production.

was 25.8 million ha and 2.2 million ha were under microbial control (Raheja, 1995).

In Southeast Asia, a major breakthrough in IPM occurred in Indonesia in 1986–1987 when IPM was adopted as the national crop protection strategy. Of 66 broad-spectrum pesticides used on rice at the time 57 were banned by presidential decree (Morse and Buhler, 1997). This decree endorsed IPM as the official 'strategy' for rice production. Subsidies on pesticides were reduced from 75% in 1986 to 40% in 1987 and removed altogether by January 1989 (APO, 1993). Five years later, rice yields increased by 15%, while pesticide use dropped by 60% (Morse and Buhler, 1997). In the first 2 years alone the government saved US$120 million that it would have spent subsidizing chemicals (WRI, 1994). The overall economic impact of IPM has been estimated at US$1 billion (WRI, 1994). Part of this success came from field schools which allowed local farmers to harness their indigenous knowledge of natural pest control to IPM (Morse and Buhler, 1997). The schools were set up with the assistance of the FAO. In these schools, more than 250,000 farmers received IPM training from 1989 to 1994 (WRI, 1994). According to Wardhani (1991), Indonesian Rice IPM program represents a social movement that links the scientific development of ecological concepts with intensive field training of farmers on ecologically sound field management techniques. This success story has proved instrumental for IPM adoption by rice farmers in other Asian countries particularly in Vietnam (Soon, 1996) and China (Mangan and Mangan, 1998). Indeed, such a mass scale IPM adoption has influenced and motivated farmers all across the globe.

Table 9.5. IPM adoption and/or its impacts in Africa, Asia and Australia.

Country/ region	Crop	Area (1000 ha) Total crop	Area (1000 ha) IPM	Area under IPM (%)	Farmers adopted IPM (%)	Reduction in: pesticide use[a], AF[b] or TA[c]/CC[d] (%)	Reference
Asia	Rice	132,158	–	–	–	35–100[b]	FAO, 1993
		132,100	4,900	3.71	–	28[a] (5 Y)	Raheja, 1995
		133,000	6,600	4.96	–	–	Morse and Buhler, 1997
Australia	Cotton	270	–	–	90	–	Fitt, 1994
Bangladesh	Rice	9,919	–	–	–	95[c] (14 Y)	Raheja, 1995 van Emden and Peakall, 1996
China	Cotton	5,200	15,00	29 (1990)	–	–	Zhaohui *et al.*, 1991
	Cotton	–	–	–	–	85[b] (10 Y)	Raheja, 1995
	Maize	20,350	2,000	10	–	–	Zhaohui *et al.*, 1991
	Rice	32,500	10,000	31 (1990)	–	–	Zhaohui *et al.*, 1991
	Soybean	8,000	1,500	19 (1990)	–	–	Zhaohui *et al.*, 1991
	Vegetables[1]	–	–	–	–	d	Raheja, 1995
	Wheat	29,850	6,000	20 (1990)	–	–	Zhaohui *et al.*, 1991
India	Cotton[2]	–	–	–	–	70[b] (15 Y)	Raheja, 1995
Indonesia	Rice	10,734	–	–	–	60[a] (1994)	Morse and Buhler, 1997
Pakistan	Mango[3]	13	3.3	25	–	–	Soon, 1996
Sudan	Cotton	–	–	–	–	50[a]	Pretty 1995, Morse and Buhler, 1997
Tajikistan	Cotton	300	–	–	–	~85[a] (22 Y)	Sugonyaev, 1994
Turkmenistan	Cotton	620	–	–	–	~99[a,c] (24 Y)	Sugonyaev, 1994

[a]Pesticide use; [b]application frequency; [c]treated area; [d]pest control cost.
Y: year (s); M: million.
Reporting area: the whole country except for [1]200 cities in 22 provinces of China; [2]State of Andhra Pradesh, India; [3]Province of Punjab, Pakistan.

South America

In South America, IPM has been successfully implemented in Argentina (lucerne, citrus, soybean), Brazil (citrus, cotton, livestock, soybean, sugarcane, tomato, wheat), Chile (wheat), Colombia (cotton, ornamentals, soybean, sugarcane, tomato), Paraguay (cotton, soybean), Peru (cotton, sugarcane), and Venezuela (cotton, sugarcane) (Tables 9.6 and 9.7) (Campanhola *et al.*, 1995; Soon, 1996).

Table 9.6. IPM adoption and/or its impact in the Americas (countries other than USA).

Country	Crop	Area (1000 ha)		Farmers adopted IPM (%)	Reduction in: pesticide use[a], AF[b] or TA[c]/CC[d] (%)	Reference
		Total crop	IPM			
Argentina	Soybean	5,935	–	–	50[b]	Aragon, 1991
Canada	Field crops, fruits and vegetables[1]	–	–	–	30–50[a]	Surgeoner and Roberts, 1992
	Pome fruits (British Columbia)	–	–	75	–	CHC, 2002
	Fruit & field crops (Nova Scotia)	–	–	95 (Level I) 25–40 (Level II & III)	–	CHC, 2002
Brazil	Cassava	–	34[2]	50	80–90[d]	Braun *et al.*, 1993
	Citrus	1,000[3]	–	–	77[b] (1970 vs. 95)	Campanhola *et al.*, 1995
	Cotton	222[4]	–	70	–	Ramalho, 1994
	Soybean	5,935	–	–	85[b] (11 Y)	Campanhola *et al.*, 1995
		11,100	–	40 (1991)	60–80[b] (25 Y)	Campanhola *et al.*, 1995, Iles and Sweetmore, 1991
		10,728	–	40	60[a]	Moscardi and Sosa-Gomez, 1996
	Sugarcane	4,183	150	–	–	Campanhola *et al.*, 1995
	Wheat[5]	–	–	–	94[c] (5 Y, 1982)	Campanhola *et al.*, 1995
Chile	Wheat	–	–	–	(Annual saving US$20 M)[a]	Campanhola *et al.*, 1995
Colombia	Cotton	26[6]	26	–	85–90[b] (1995)	Campanhola *et al.*, 1995
	Cotton	–	6[7]	–	97[b] (7 Y)	Campanhola *et al.*, 1995
	Sugarcane	318[8]	318	100	–	Campanhola *et al.*, 1995, Escobar, 1986
	Soybean	–	–	–	80–90[d]	Garcia, 1990
	Tomato[9]	–	–	70	100[b]	Campanhola *et al.*, 1995
Costa Rica	Banana[10]	–	–	–	100[a] (1973)	Soon, 1996
Paraguay	Cotton	454	–	–	~80[a]	Servian de Cardozo, 1990
		19[11]	–	–	50[b]	Servian de Cardozo, 1990
Venezuela	Sugarcane	111	50	–	–	Campanhola *et al.*, 1995

[a]Pesticide use; [b]application frequency; [c]treated area; [d]pest control cost.
Y: year (s); M: million.
Reporting area: the whole country except for: [1]Province of Ontario, Canada; [2]State of Paraná, Brazil; [3 & 4]State of São Paulo, Brazil; [5]Wheat-growing areas of Rio Grande do Sul, Paraná and Santa Catarina, Brazil; [6, 8 & 9] Valle del Cauca, Colombia; [7]Municipality of Zarzal, Colombia; [10]Reduction in insecticide sprays; [11]By Cooperative Colonia Unidas, Paraguay.

Table 9.7. Other examples of reported IPM adoption.

Countries	Crop	Impact – reduction in/comments	Reference
Argentina	Citrus	Pesticide use & application frequency	Campanhola *et al.*, 1995
Bangladesh, Burkina Faso, India & Sri Lanka	Rice	Successfully implemented	Pretty, 1995
Egypt, Sudan, Togo, Zimbabwe	Cotton	Successfully used	Pretty, 1995
Malaysia	Plantation crops	Pesticide use (+ Environmental conservation)	Raheja, 1995
	Vegetables	Pesticide use	Raheja, 1995
Peru (Canete Valley)	Cotton	Pesticide use & control cost (1972)	Boza-Barducci, 1972
South Africa	Apple	Pesticide use & application frequency	Nel *et al.*, 1993
China, Malaysia, Indonesia, Taiwan, Thailand, the Philippines, Vietnam	Vegetables	Successfully used	van Emden and Peakall, 1996

Africa

The current status of IPM adoption is given in Tables 9.5 and 9.7. In Sudan, IPM produced good results with more than 50% reduction in insecticide use. Introduction of a parasitoid wasp, *Epidinocarsis lopezi*, spectacularly controlled the cassava mealybug, *Phenacoccus manihoti*, across the cassava belt (Zethner, 1995; Soon, 1996). This program started in 1979 and by 1990 *E. lopezi* had become established in 25 countries where cassava is cultivated (Zethner, 1995). IPM has been successfully used in South Africa on apple; Togo, Zimbabwe and Egypt on cotton; and Burkina Faso on rice. Ghana has recently launched IPM as the national crop protection strategy, which includes controls on the import of chemical pesticides (Zethner, 1995). Countries like Burkina Faso, Côte d'Ivoire, and Kenya are currently focusing more on capacity building as the initial step towards adopting IPM (Zethner, 1995).

Conclusion

IPM is a tangible reality in some privileged regions of the world, but still remains a distant dream for many others. Given the world demographic and social realities, however, adoption of IPM is not an option, it is a vital necessity for the conservation of the environment and for the very survival of the human race on earth. The robust conceptual foundation of IPM projects influences beyond the limits of crop protection. IPM has become a model for all other operational components of sustainable agriculture. It is just a matter of time and dedication from those who believe in its potential for IPM to become a global reality in practice, as well as in theory.

References

[APO] Asian Productivity Organization (1993) *Pest Control in Asia and the Pacific: Report of an APO Seminar 24th September–4th October, 1991.* Asian Productivity Organization, Tokyo, Japan, 409 pp.

Aragon, J. (1991) El manejo integrado de plagas en soja. *Desarrollo Argentino (Rosario)* 8, 9–12.

Balázs, K., Jenser, G. and Bujáki, G. (1996) Eight years' experience of IPM in Hungarian apple orchards. In: Polesny, F., Müller, W. and Olszak, R.W. (eds) Proceedings of the International Conference on Integrated Fruit Production, Cedzyna, Poland, 28 August–2 September 1995. *Acta Horticulturae* 422, 95–101.

Boza-Barducci, T. (1972) Ecological consequences of pesticides used for the control of cotton insects in Canete Vally, Peru. In: Farva, M.T. and Milton, J.P. (eds) *The Careless Technology.* Natural History Press, New York, pp. 423–438.

Braun, A.R., Bellotti, A.C. and Lozano, J.C. (1993) Implementation of IPM for small-scale cassava farmers. In: Altieri, M.A. (ed.) *Crop Protection Strategies for Subsistence Farmers*. Westview Press, San Francisco, pp. 103–115.

Campanhola, C., Moraes, G.J.D. and De Sa, L.A.N. (1995) Review of IPM in South America. In: Mengech, A.N., Saxena, K.N. and Gopalan, H.N.B. (eds) *Integrated Pest Management in the Tropics – Current Status and Future Prospects*. John Wiley & Sons, New York, pp. 121–151.

[CHC] Canadian Horticultural Council (2002) Crop protection – a better future for Canada. http://www.hortcouncil.ca/

Cross, J.V., Bonauer, A., Bondio, V., Clemente, J., Denis, J., Grauslund, J., Huguet, C., Jörg, E., Koning, S., Kvale, A., Malavota, C., Marcelle, R., Morandell, I., Oberhofer, H., Pontalti, M., Polesny, F., Rossini, M., Schenk, A., de Schaetzen, C. and Vilajeliu, M. (1995) The current status of integrated pome fruit production in western Europe and its achievements. In: Polesny F., Müller, W. and Olszak, R.W. (eds) Proceedings of the International Conference on Integrated Fruit Production, Cedzyna, Poland, 28 August–2 September 1995. *Acta Horticulturae* 422, 1–10.

Escobar, G.J.A. (1986) Criterios para la evaluación de programas de manejo integrado de plagas, con referencia especial al caso en caña de azucar. In: *Conferencias Especiales 13, Congresso Socolen*, 16–18 July 1986, Cali, Colombia, pp. 64–97.

FAO (1993) *Global IPM: a Report on the Global IPM Field Study Tour and Meeting August–September 1990*. FAO, Manila, 65 pp.

Fitt, G. (1994) An Australian perspective – cotton pest management. *Annual Review of Entomology* 39, 543–562.

Garcia, F. (1990) Avances y perspectivas del control biologico en Colombia. *Colombia Ciencia y Tecnologia* 8, 8–12.

Iles, M.J. and Sweetmore, A. (1991) *Constraints on the Adoption of IPM in Developing Countries – a Survey*. Natural Resources Institute, Kent, UK, 74 pp.

Kenmore, P.E. (1996) Integrated pest management in rice. In: Persley, G.J. (ed.) *Biotechnology and Pest Management*. CAB International, Wallingford, UK, pp. 76–97.

Kogan, M. (1988) Integrated pest management theory and practice. *Entomologia Experimentalis et Applicata* 49, 59–70.

Kogan, M. and Bajwa, W.I. (1999) Integrated Pest Management: a global reality. *Anais da Sociedade Entomologica do Brasil* 28, 1–25.

Mangan, J. and Mangan, M.S. (1998) A comparison of two IPM training strategies in China: the importance of concepts of the rice ecosystem for sustainable insect pest management. *Agriculture and Human Values* 15, 209–221.

Mengech, A.N., Saxena, K.N. and Gopalan, H.N.B. (eds) (1995) *Integrated Pest Management in the Tropics – Current Status and Future Prospects*. John Wiley & Sons, New York, 171 pp.

Morse, S. and Buhler, W. (1997) *Integrated Pest Management: Ideals and Realities in Developing Countries*. Lynne Rienner Publishers, London, 171 pp.

Moscardi, F. and Sosa-Gomez, D.R. (1996) Soybean in Brazil. In: Persley, G.J. (ed.) *Biotechnology and Integrated Pest Management*. CAB International, Wallingford, UK, pp. 98–112.

Nel, P.J., Addison, M.F., Schenk, A.M.E., Webster, A.D. and Wertheim, S.J. (1993) The development of an integrated pest management programme in apple orchards in Elgin, South Africa and the implications for integrated fruit production. *Second International Symposium on Integrated Fruit Production*, Veldhoven, The Netherlands, 24–28 August, 1992, pp. 323–326.

Pretty, J.N. (1995) *Regenerating Agriculture: Policies and Practice for Sustainability and Self-reliance*. Joseph Henry Press, Washington, DC, 320 pp.

Prokopy, R.J. and Croft, B.A. (1994) Apple insect management. In: Metcalf, R.L. and Luckmann, W.H. (eds) *Introduction to Insect Pest Management*, 3rd edn. John Wiley & Sons, New York, pp. 543–589,

Raheja, K. (1995) Practice of IPM in South and Southeast Asia. In: Mengech, A.N., Saxena, K.N. and Gopalan, H.N.B. (eds) *Integrated Pest Management in the Tropics – Current Status and Future Prospects*. John Wiley & Sons, New York, pp. 69–119.

Ramalho, F.S. (1994) Cotton pest management: Part 4. A Brazilian perspective. *Annual Review of Entomology* 39, 563–578.

Reed, A.N. (1993) Low or no chemical: will it work? In: *Proceedings of the Eighty-eighth Annual Meeting, Washington State Horticulture Association*, pp. 135–144.

Reed, A.N. (1995) Responsible choice: a systems approach to growing, packing, and marketing fruit. In: Hull, J. Jr and Perry, R. (1995) *The 125th Annual Report of the Secretary of the State Horticultural Society of Michigan for the year 1995*, pp. 68–78.

Schaetzen, C. de. (1996) Integrated pest management in Belgium orchards in 1995.

In: Polesny, F., Müller, W. and Olszak, R.W. (eds) Proceedings of the International Conference on Integrated Fruit Production, 28 August–2 September, 1995. *Acta Horticulturae* 422, 321–323.

Schafermeyer, S. (1991) *Vergleichende Untersuchungen zu Richtlinien fur die integrierte Kernobstproduktion in Europa. (Comparative Studies on Guidelines for Integrated Pome Fruit Production in Europe.)* P. Parey, Berlin, 110 pp.

Servian de Cardozo, J.F. (1990) Problematica de las plagas y los plaguicidas en el sector campesino. In: *Consulta Sudamericana sobre Manejo Integrado de Plagas en la Agricultura Campesina.* Instituto Agronomico Nacional, Caacupe, Paraguay, pp. 20–35.

Soon, L.G. (1996) Integrated pest management in developing countries. In: Persley, G.J. (ed.) *Biotechnology and Integrated Pest Management.* CAB International, Wallingford, UK, pp. 61–75.

Sorensen, A.A. (1998) Integrated pest management: future farming takes lessons from the past. http://ificinfo.health.org/insight/int-pest.htm

Sterling, W. (1984) *Action and Inaction Levels in Pest Management.* College Station, Texas, Texas Agriculture Experiment Station, B-1480, 20 pp.

Sugonyaev, E.S. (1994) Cotton pest management: Part 5. A Commonwealth of Independent States perspective. *Annual Review of Entomology* 39, 579–592.

Surgeoner, G.A. and Roberts, W. (1993) Reducing pesticide use by 50% in the province of Ontario: challenges and progress. In: Pimentel, D. and Lehman, H. (eds) *The Pesticide Question: Environment, Economics, and Ethics.* Chapman & Hall, New York, pp. 206–222.

van Emden, H.F. and Peakall, D.B. (1996) *Beyond Silent Spring: Integrated Pest Management and Chemical Safety.* Chapman & Hall, London, 322 pp.

Valyi, I. and Sallai, P. (1993) Integrated Fruit Production in Hungary. *Acta Horticulturae* (ISHS) 347, 371.

Vandeman, A., Fernandez-Cornejo, J., Jans, S. and Lin, B. (1994) *Adoption of Integrated Pest Management in U.S. Agriculture.* Resources and Technical Division, Economic Research Service, USDA. Agricultural Information Bulletin No. 707, 26 pp.

Wardhani, M.A. (1991) Development in IPM: the Indonesian case. In: Ooi, P.A.C., Lim, G.S., Ho, T.H., Manalo, P.L. and Waage, J. (eds) *Integrated Pest Management in the Asia–Pacific Region.* CAB International, Wallingford, UK, pp. 27–35.

Warner, G. (1998) People want IFP fruit, but are unwilling to pay more. *Good Fruit Grower* 49(15), 22–23.

[WRI] World Resource Institute (1994) *World Resources – a Guide to the Global Environment; People and the Environment; Resource Consumption; Population Growth; Women.* Oxford University Press, New York, 400 pp.

Zethner, O. (1995) Practice of integrated pest management in tropical and sub-tropical Africa: an overview of two decades (1970–1990). In: Mengech, A.N., Saxena, K.N. and Gopalan, H.N.B. (eds) *Integrated Pest Management in the Tropics – Current Status and Future Prospects.* John Wiley & Sons, New York, pp. 1–68.

Zhaohui, Z., Yongfan, P. and Juru, L. (1992) National integrated pest management of major crops in China. In: Ooi, P.A.C., Lim, G.S., Ho, T.H., Manalo, P.L. and Waage, J. (eds) *Integrated Pest Management in the Asia-Pacific Region.* CAB International, Wallingford, UK, pp. 255–266.

Chapter 10
Integrated Pest Management in Burkina Faso

Dona Dakouo[1], Marcel S. Bonzi[2], Abdoussalam Sawadogo[1], Clémentine L. Dabiré[3], Souleymane Nacro[1] and Bouma Thio[1]
[1]INERA, Bobo-Dioulasso, Burkina Faso; [2]INERA, Ouagadougou, Burkina Faso; [3]INERA, Station de Kamboinsé, Ouagadougou, Burkina Faso

Introduction

Burkina Faso is a landlocked country of 274,000 km^2 (105,869 square miles), located in the heart of Western Africa, 1000 km from the Atlantic Ocean. The country is bordered by Benin, Côte d'Ivoire, Ghana, Mali, Niger and Togo. With 10.9 million inhabitants and a density of 39.8 inhabitants/km^2 (13.1/square mile), Burkina Faso is one of the most highly populated states in West Africa.

Like other Sahelian countries, Burkina Faso suffers from drought and desertification, overgrazing, soil degradation, deforestation, and from the effects of uneven population distribution. The tropical weather in Burkina Faso is divided into two seasons: the dry season from November to May (with a cool and dry period from November to February, and hot weather from March to May), and the rainy season from June to October. The average temperature is 15°C at night, and 30°C during the day, except in the dry season when temperatures may rise to over 38°C. Average rainfall is approximately 1200 mm (47.2 inches) in the south, to less than 400 mm (15.7 inches) in the north and northeast.

Some 90% of the country's population is actively involved in agriculture. Agriculture accounts for 36.1% of the Gross Domestic Product and 65% of the country's exports. Cotton, shea nuts (karité), sesame, vegetables and tobacco are the major exports, while millet, sorghum, maize, rice, sweet potatoes and yams are the main food crops.

History of IPM in Burkina Faso

Chemical control

Plant protection started in Burkina Faso with the control of locusts and cotton pests in the early 1960s using chemicals. Chemical control was well known as a fast and easy way to control pests in field crops and stored products. However, chemical control rapidly showed its limits by causing problems of pest resistance and hazards to humans and the environment.

Based on the demonstrated need for new pest management strategies, IPM strategies have been developed by multidisciplinary teams in plant protection, breeding and agronomy. IPM requires adequate research, a protocol for decision making based on specified criteria, and the use of compatible control methods including natural enemies, habitat management, cultural practices, resistant varieties and pesticides (Bonzi, 1996).

IPM programs in Burkina Faso

The first successful IPM program in Burkina Faso was the Regional IPM Project (1980–1987) in the nine Sahelian countries. These countries are members of the CILSS (the Inter-States Committee for Drought Management in the Sahel). The project was funded by USAID and carried out with the technical assistance of FAO. In each country, the program relied on a multidisciplinary research team in food crops (cereals and vegetable crops). Regional meetings were held every year to discuss results among scientists and extension agents from the nine countries. The project worked well but ended before most of the results were implemented in farmers' fields.

In 1995, FAO began to experiment with the Asian FFS approach with pilot IPM projects in Ghana (1995), Côte d'Ivoire (1996), Burkina Faso (1997) and Mali (1998–1999). Other IPM programs in Burkina Faso included the cowpea IPM Project (1996–1999) and the sorghum IPM Project (1996–2000). A Regional IPM Project in rice, vegetables and cotton for three countries (Mali, Senegal and Burkina Faso) has been established and will be carried out over 3 years (2001–2003) with the assistance of FAO. Following the adoption of a Strategic Plan for Scientific Research in Burkina (Anonymous, 1995a), all research programs in plant protection are IPM oriented.

IPM Organizational Structure in Burkina Faso

Organizations involved in IPM

Organizations contributing to IPM include the Ministry of Higher Education and Scientific Research through its national research and higher education institutes, primarily the INERA and the Institut de Développement Rural (IDR). The Ministry of Agriculture is also involved through its technical services: Agricultural Extension Directorate, Regional Directions of Agriculture, and Plant Protection Service. Private sector groups also contribute to IPM. At the grassroots level, farmers' organizations in cotton, rice and vegetable crops participate in IPM in the field.

IPM research in Burkina Faso

Research focusing on IPM has primarily been carried out at INERA within the CNRST and at the University IDR both under the Ministry of Higher Education and Scientific Research. Regional or international research institutes such as CIRAD and IRD (French overseas research institutes), IITA, ICRISAT and WARDA (CGIAR International Centers) carry out research on plant protection in collaboration with national research institutes.

The development and adoption of the Strategic Plan for Scientific Research (Anonymous, 1995a) was done in collaboration with national and international organizations in research and extension, private sector, and farmers' organizations.

Agricultural research in Burkina Faso

Research activities within INERA are conducted according to five agricultural or ecological regions in Burkina Faso. Each of the five regions hosts a Regional Research Center where research is conducted according to the specific needs of the region. Such research is part of the activities of the 16 research programs of the Institute. Of these 16 research programs, five are focused on crops: traditional crops (sorghum, millet and maize), rice, cotton, vegetables and fruits, legume and oil crops. Research includes studies on ecology of pests, their relationships with biotic factors (host plants, natural enemies) and abiotic factors (climate, soils), and the development and implementation of IPM programs. The five regions include (Fig. 10.1):

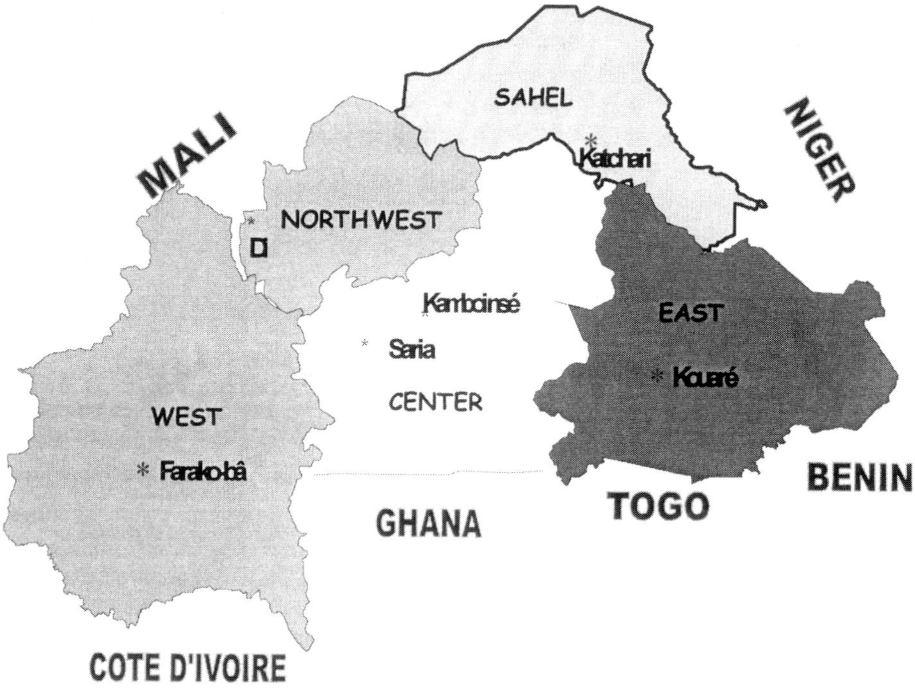

Fig. 10.1. The regional centers for environmental and agricultural research of INERA. *Indicates Regional Research Centers; Kamboinsé is a Regional Research and Training Center near Ouagadougou (Anonymous, 1995b).

- The Sahel Region, 36,896 km^2, population 662,169;
- The Center Region, 94,000 km^2, population 3,584,117;
- The Northwest Region, 30,817 km^2, population 1,280,933;
- The West Region, 52,000 km^2, population 1,184,000;
- The East Region, 60,000 km^2, population 1,500,000.

Agricultural extension

The Ministry of Agriculture administers this sector. Programs include development at the national level of extension methods, programming, and follow-up evaluation of animal and plant production. This sector is divided into 12 regional directions of agriculture.

Extension and research programs work together to develop IPM programs. Scientists train extension facilitators and farmers, while feedback on the successful or unsuccessful aspects of IPM programs is provided to research scientists by farmer organizations and extension services.

Other important groups in IPM

The Sofitex Cotton Company

Sofitex (Société Burkinabé des Fibres Textiles) is a parastatal company. It oversees the production, industrialization, sale and export of cotton. It also supplies inputs (fertilizers and insecticides) to cotton farmers through their organizations. Sofitex has a network of 130 cotton extension agents in plant protection throughout the country. The INERA research program in cotton trained these extension agents. They work with all aspects of cotton production and training of cotton farmers on the use of plant protection methods. The cotton plant protection program is based on chemical control. A total of 8,801,888 l of insecticides were used from 1997 to 2001 (Anonymous, 2001). However, in the past 3 years, progress has been made on the development of IPM methods in cotton.

NGOs

Two NGOs operating in Burkina Faso have a specific program in plant protection. The 'Albert Schweitzer Centre Ecologique' (a Swiss funded ecological center) promotes the use of local technologies in local environments. The Centre has an experimental farm where methods to protect vegetable crops and storage products are tested. Training in agroecology is also provided to farmers.

The 'Assistance Ecologique aux Projets de Développement' (Ecological Assistance to Rural Development Projects) also works in plant protection. It is based in Bobo-Dioulasso (in the southwest of the country) and was founded by a Catholic priest in 1981. Its mission is to inform and train farmers on the use of environmentally sound methods in agriculture.

The private sector

The private sector includes several chemical pesticide companies that conduct the distribution and sale of pesticides. Pesticide use in Burkina Faso is only common in cotton, rice and vegetable crops. Large quantities of pesticides have often been donated to the government of Burkina Faso by exterior aid for locust control. The private sector can play a greater role in environmental and IPM issues by improving farmers' knowledge on pesticides and their proper use.

Farmers' organizations

Farmers' organizations operate primarily at the village level, but a few work at regional and national levels. At the national level, the cotton farmers' organization, the 'Union national des paysans producteurs de coton du Burkina (UNPCB)', works with all aspects of cotton production and sale in connection with Sofitex. A similar organization for rice farmers, the 'Comité Interprofessionel du Riz du Burkina (CIR-B)' has been recently created. Such organizations are potential channels for the promotion of IPM methods, because they can most accurately identify constraints to production and sale of their products.

IPM and Pesticide Use Policy

In 1992, the members of CILSS adopted a common pesticide legislation (Diarra, 2000) and created 'The Sahelian Committee of Pesticides' in 1998. This body has the authority to regulate pesticide use, registration, import, export and local production in the nine member countries. The committee is made up of 24 members drawn from the nine countries and from regional and international organizations such as OCLALAV, CPI/OUA, FAO, and WHO.

The Sahelian Committee of Pesticides holds two annual meetings to examine requests for pesticide registration made by pesticide companies and to address all related matters such as issuing a list of registered pesticides in the nine countries (Diarra, 2000). In each country, the national committee administers the decisions and oversees matters related to pesticides in the country.

Each national committee consists of 16 members. In Burkina Faso, the members are drawn from the Ministries of Agriculture, Environment, Trade, Animal Resources, Finance, Higher Education and Scientific Research, Health, Labor, and Justice and from the private sector (pesticide companies, pesticide users and consumers' organizations). The National Committee meets twice a year.

The National Committee of Pesticides (NCP) is divided into four commissions: verification, control, pesticide management, and fraud. Each commission holds at least one meeting a year. The recommendations from the NCP are submitted to the Minister of Agriculture. Since August 2000, the NCP has held a training program for plant protection service officers in charge of pest and pesticide control at country borders.

Successful IPM in Burkina Faso

The Regional IPM Project in the Sahelian countries (1980–1987)

The first successful IPM project in Burkina Faso was the Regional IPM Project. Important results of this project included research-building capacity in plant protection, an inventory of food crop pests in the Sahel, training staff (masters and PhD) in several areas (entomology, plant pathology, weed science), and increasing the awareness of policy makers about IPM (Bonzi, 1996). In 1990, an international workshop was held in Bamako to highlight the major results of the project (Anonymous, 1990; Mattesson, 1990). Two major achievements of the project in Burkina Faso were the pilot program in millet and the IPM system in rice.

The pilot program in millet

This program took place in all the millet-producing CILSS countries from 1985 to 1987. The program strategy was based on evaluating all of the control methods against millet pests generated or collected by the Regional IPM Project since 1980, primarily against the millet downy mildew and striga, a parasitic plant. Methods used included hand weeding and burning striga and all millet infested by downy mildew. In addition to these, cultural practices such as plowing before sowing, use of fertilizer, sowing on lines, fixed spacing, and plant thinning were used (Bonzi, 1996).

The program began in the 1985/86 crop season with 15 farmers from three villages. By the 1986/87 season, 75 farmers from 12 villages were participating. The farmers reported increased yield in the pilot plots, effective control of millet pests, and an effective combination of control methods (Bonzi, 1996).

The IPM system in rice (1987–1990)

Irrigated rice covers only 7% of the total cropped acreage (40,000 ha) in Burkina Faso, but accounts for one third of the estimated rice production. Using modern rice production methods, two crop seasons a year may occur, with an average yield of 4 t/ha. Insect pests and rice blast caused by the fungus *Pyricularia oryzae* are of major importance. Important insect species include lepidopterous stem borers (*Chilo zacconius*, *C. diffusilineus*, *Maliarpha separatella* and *Sesamia calamistis*), the rice gall midge, *Orseolia oryzivora* and the stalk-eyed borer, *Diopis* sp.

Insect control in irrigated rice still largely depends on the use of insecticides. However, chemical control does not always meet the necessary criteria of efficiency and profitability. Systematic use of insecticides has proved to be prohibitively costly for small-scale farmers and disastrous for the environment (Matteson, 1990). Therefore, there was a need to develop an insect management system that minimized the use of chemicals.

An insect pest management system was developed (Dakouo *et al.*, 1992) using a procedure that relied on monitoring and threshold interventions based on levels of stem borer damage. The Vallée du Kou irrigated rice region (near Bobo-Dioulasso, in the southwest of the country) was selected for this case study because of its relatively large size (1100 ha), the possibility of two crop seasons per year and the high average yield of 4 t/ha. Lepidopterous stem borers were the major insect pests in the area. Damage symptoms included dying off of the central leaf during the vegetative stage (dead heart) or dying off of the panicle at the reproductive stage (white head). The efficiency and profitability of the insect pest management system was evaluated during two consecutive crop seasons with 30 randomly selected farmers. Twenty farmers were assigned to apply insecticides according to one of two threshold levels: (i) 5% of dead heart during vegetative stage; and (ii) 1% of white head during the reproductive stage. Ten of the 30 farmers were assigned to the control group, applying insecticides at pre-defined intervals without considering insect damage.

Table 10.1 illustrates the yield, average number of insecticide applications per crop season and the cost–benefit ratio. An average of 1.2 insecticide applications were recorded per crop season in the 20 pilot fields. In the control fields, 3 applications were done on average. In the pilot fields, yield increased by 10.5%, and the cost–benefit ratio was 1:17.4 (Dakouo *et al.*, 1995).

The cowpea IPM project (1994–1998)

Cowpea is an important staple grain legume in the lowland dry savanna and the Sahel regions of Africa. Unfortunately, it is heavily damaged by insect pests, diseases and parasitic weeds making it difficult to produce without pesticides. But effects of repeated applications of pesticides lead to the development of pest resistance, health hazards and environment pollution. In addition, these pesticides are not accessible to small-scale farmers. Alternative technologies for sustainable cowpea production including host-plant resistance, insecticides derived directly from plants, cultural control methods, and biological control, have been developed by IITA Cotonou (Benin) and the NARS. A regional IPM project, PEDUNE was initiated in mid-1994 for testing and adapting these technologies through a participatory approach and to make them available to farmers and rural communities. This project had two phases. The pilot phase, which lasted for 2½ years (1994–1996), was executed in five countries, namely, Benin, Burkina Faso, Mozambique, Niger and Nigeria. Very promising results from an extensive evaluation (Anonymous, 1999) were the basis for extending the project into an implementation phase (1997–1999) and expanding it to four new countries: Cameroon, Ghana, Mali and Senegal.

Main achievements of the project included the development of single and multiple resistant cultivars to bruchids, aphids, thrips, bugs, striga and drought, the development of methods to protect cowpea using botanical insecticides from plants such as *Hyptis spicigera*, *Cassia nigricans*, *Boscia senegalensis* and *Azadirachta indica* (Anonymous, 1999).

In 2000, PEDUNE and RENACO (Réseau de recherche sur le niébé pour l'Afrique occidentale et Centrale = West and Central Africa Cowpea Research Network) merged into a new project called Projet Niébé pour l'Afrique (Cowpea Project for Africa, PRONAF) (Anonymous, 2000) for a 2-year period (2000–2002). This project accomplished the completion of baseline surveys in eight PRONAF countries, improved storage techniques, and IPM training for farmer organizations and government institutions (Anonymous, 2000). Multilocational trials have been conducted across countries and different ecological regions using planting dates and intercroppings to control pod-sucking bugs. Findings in Burkina Faso showed that the population of the main sucking bug, *Clavigralla tomentosicollis* decreased with late plantings due to the activity of an egg parasitoid, *Gryon fulviventre*. Intercropping cowpea with cereals (sorghum or millet) reduced the bug population and its incidence on cowpea

Table 10.1. Profitability of the IPM system at the Vallée du Kou irrigated rice scheme in Burkina Faso.

Type of farmers	Yield (kg ha^{-1})	Yield increase (%)	Insecticide applications: Range	Average	Cost (US$)*	Cost–benefit ratio
Control	4145	–	3	3	17	–
Pilot	4581	10.5	0–2	1.2	7	1:17.4

*US$1 = 700 CFA Fr.

yield, the best result being with millet (Dabiré, 2001).

The Sorghum IPM Project (1996–2000)

The Sorghum IPM Project was funded by the European Union. It included five research partners: CIRAD-CA (France), the University of Heidelberg (Germany), CNESOLER (Mali), IER (Mali), and INERA (Burkina Faso). The objectives of the project were to reduce losses caused by insect pests, especially lepidopterous stem borers (*Busseola fusca*, *Eldana saccharina* and *Sesamia calamistis*), panicle-feeding pests (sorghum midge, *Stenodiplosis sorghicola*) and head bugs. The project focused on four IPM components: host plant resistance, plant-derived pesticides including extracts of physic nut (*Jatropha curcas*) and neem (*Azadirachta indica*), sex pheromones of the stem borer (*Busseola fusca*), and combining integrated control strategies. The final goal was to apply IPM packages in farmers' fields.

Several methodologies and research tools were developed in the sorghum project during its 4-year duration (Ratnadass, 2001). Some of the most important results in Burkina Faso included:

- identification of ten new sources of resistance to the sorghum midge (Dakouo *et al.*, 2000);
- development of three sorghum cultivars combining high yield potential and resistance or tolerance to midge and/or head bugs;
- improved efficiency of insecticides derived from neem seed powder against stem borers;
- an efficient trapping technique for the sorghum stem borer *Busseola fusca* using pheromones, that can be used for either direct or indirect control of the pest (Dakouo and Ratnadass, 1999);
- successful testing of IPM packages based on manipulation of planting dates; host-plant resistance and sorghum protection using neem in farmers' fields.

Parasitic nematodes of vegetable crops

Importance of parasitic nematodes on vegetable crops

Parasitic nematodes are major pests on several vegetable crops. A survey by Sawadogo (1990) identified 20 genera or species of nematodes associated with most of the cultivated vegetable crops in Burkina Faso (tomato, aubergine, potato, bean, okra, onion and sweetcorn). Vegetable crops are grown during the dry season, after rainy season crops such as cereals, or after the rice crop in irrigated rice areas. Nematode damage is highly correlated with soils, climatic conditions and farmer practices.

Parasitic nematodes are a limiting factor in maize and tomato productivity in upland growing areas (Sawadogo *et al.*, 1998). *Helicotylenchus*, *Scutellonema*, *Pratylenchus* and *Meloidogyne* are the most important genera of nematodes associated with vegetable crops. Five species of root-knot nematodes have been identified in Burkina Faso: *Meloidogyne incognita*, *M. javanica*, *M. mayaguensis*, *M. chitwoodi* and *M. hispanica*. *M. incognita* and *M. javanica* are the most common species.

The control of root-knot nematodes with crop rotation

High populations of root-knot nematode (*Meloidogyne* spp.) can occur in upland vegetable crops (Sawadogo *et al.*, 1998). *Meloidogyne* spp. are a threat to tomato production in West Africa. Recent research has focused on the effects of common rainy-season crops on nematode populations and tomato production. The goal was to identify alternative rainy-season crops that decrease the population of *Meloidogyne* spp. in soil, and lead to good tomato production in the dry season.

One study was conducted in two vegetable growing areas near Bobo-Dioulasso in 2 consecutive years. Seven treatments were tested as alternative crops in the rainy season: groundnut var. RMP12, maize var. Jaune de Fô, millet var. M12, sorghum var. Gnofing, fonio (*Digitaria exilis* Stapf.)

landrace, tomato var. Roma VF, and natural fallow. Nematode population levels at the end of the rainy season showed significant differences. Tomato, fallow and maize were highly infested, while groundnut, sorghum, millet and fonio were associated with low nematode population levels. These results indicated that crop rotation can be an effective strategy. Sorghum, *Digitaria exilis* and groundnut are the best alternative crops for tomato, helping to reduce nematode damage (Sawadogo *et al.*, 1995).

Parasitic nematodes of rice

Nematodes associated with rice

A survey of plant parasitic nematodes associated with irrigated rice was conducted in Burkina Faso and Mali in 1994 (Sawadogo *et al.*, 1994). *Hirschmanniella spinicaudata* and *H. oryzae* were the most abundant species. Yield loss assessments due to parasitic nematodes of *Hirschmanniella* genus were done in pot trials in greenhouse and on farm trials.

Pot trials showed a yield loss of 30% due to rice root nematode *H. spinicaudata* (Thio, 1998). In farm trials, the nematicides carbofuran and isazophos increased rice yield by 60%. The same result was observed with the nematicide dazomet applied in soils 2 weeks before transplanting.

A screening of 50 rice cultivars conducted in two locations indicated that some of them showed partial resistance to the *Hirschmanniella* and *Heterodera* genera (Anonymous, 2000). Resistance study carried on African rice species, *Oryza glaberrima*, and the Asian rice species *O. sativa* and their crosses showed good resistance of African rice species and partial resistance of hybrids (Thio, 1998).

A multidisciplinary approach to nematode control in irrigated rice

High costs and environmental hazards due to excessive use of pesticides led to the development of alternative pest control methods. A study comparing several control methods was conducted in 2000 in two locations (Banzon and Karfiguela) by Kabore *et al.* (2002). Preliminary results showed that *Hirschmanniella spinicaudata* and *H. oryzae* were the primary nematode species present in these locations. Three treatments were used in the study:

1. Untreated control.
2. Synthetic pesticides: Basudin (diazinon) at 1000 g a.i./ha, foliar sprays 20, 40 and 60 days after transplanting; a nematicide, Basamid (dazomet) at 100 kg/ha, incorporated into the soil 14 days before transplanting; and a fungicide, Kitazin (di-isopropyl-*s*-benzyl-thiophosphate), foliar spray at 720 g a.i./ha at panicle emergence.
3. Botanical pesticides (IPM): treatment with botanical pesticides derived from neem (*Azadirachta indica*) and organic matter against nematodes, kernel liquid extract of *A. indica* against insects, and rice stem ashes against fungi.

The synthetic pesticide treatments reduced nematode populations in both locations, but the botanical pesticides were efficient in root population reduction. The botanical pesticide treatments increased rice yields by 3% compared with 9.3% with synthetic pesticides at the Karfiguéla site. Rice yield was increased by 18.5% in both the synthetic and botanical pesticide treatments at the Banzon site.

Economic benefits of control

Table 10.2 illustrates the economic benefit of the botanical pesticide (IPM) treatments. The synthetic chemicals produced a benefit of US$142.10 at Banzon and US$23.90 at Karfiguéla, but the treatment cost was very high, at US$858.00. The botanical pesticide (IPM) treatments also produced a benefit of US$142.10 at Banzon, and a benefit of US$65.60 at Karfiguéla, but the treatment cost was much lower at US$42.80.

Expensive imported chemicals used in the synthetic chemicals treatment led to economic losses with a cost–benefit ratio of 1:0.16 at Banzon and 1:0.26 at Karfiguéla.

Table 10.2. Profitability of the integrated protection system combining organic manure, rice ashes and neem products against nematodes, blast and insects at Banzon and Karfiguéla during the 2000 wet season (Kaboré *et al.*, 2002).

	Yield (t/ha)		Benefit in US$*		Cost in US$		Cost–benefit ratio	
Treatments	Banzon	Karfiguéla	Banzon	Karfiguéla	Banzon	Karfiguéla	Banzon	Karfiguéla
Control	6.31[b]	5.78[b]	–	–	–	–	–	–
Synthetic chemicals	7.48[a]	7.69[a]	142.1	231.9	858.0	858.0	1:0.16	1:0.26
Botanical pesticides (IPM)	7.48[a]	6.32[a]	142.1	65.6	42.8	42.8	1:3.38	1:1.50

*US$1 = 700 CFA Fr.
Cost–benefit ratio = cost/benefit; the ratio expresses the return for 1 dollar invested.
Means followed by the same letters in a column are not significant at 5% level.

The IPM approach, using exclusively local natural products, provided a relevant cost–benefit ratio of 1:3.38 at Banzon and 1:1.50 at Karfiguéla, with the added advantage of better environmental protection.

Key Constraints of IPM

The following constraints are challenges to IPM in Burkina Faso and can explain the failure of some IPM projects:

- Inadequate awareness of the National IPM Policy: only public organizations (research and extension services) are aware of the policy.
- Lack of collaboration between research, extension services and NGOs: recommendations from research do not always reach farmers due to insufficient communication networks.
- Inadequate funding for research, IPM development and implementation.
- Low income of farmers: they cannot afford the use of costly imported inputs (fertilizers) on traditional crops (sorghum and millets), which results in low yields and poverty.

References

Anonymous (1990) *Food Crop Pests in the Sahel.* Regional Coordination Unit in Plant Protection (UCTR-PV), Institut du Sahel, Bamako Mali /Technical Centre for Agricultural and Rural Cooperation (CTA), Wageningen, The Netherlands, 128 pp.

Anonymous (1995a) *Strategic Plan for Scientific Research: Agricultural Research.* Ministère des Enseignements Secondaire, Supérieur et de la Recherche Scientifique, Ouagadougou, Burkina Faso, 90 pp.

Anonymous (1995b) *Agricultural Research Project – Formulation of Phase II (Preliminary report).* Centre National de la Recherche Scientifique et Technologique (CNRST), Ouagadougou, Burkina Faso.

Anonymous (1999) *PEDUNE Farmer Participatory IPM for Cowpea, 1997–1998 Progress Report.* ITTA /Swiss Agency for Development and Cooperation, Cotonou, Benin, 83 pp.

Anonymous (2000) Integrated Pest Management of legumes pests and diseases. In: *Annual Report 2000, Research Highlights.* International Institute of Tropical Agriculture (IITA), Ibadan, Nigeria, p. 8.

Anonymous (2001) *Statistics on Pesticides Importation over the 5 Past Years 1996–2000.* Société burkinabè des fibres textiles (Sofitex), Direction du Développement de la Production Cotonnière, Bobo-Dioulasso, Burkina Faso.

Bonzi, S.M. (1996) *Plant Protection in Burkina Faso: Part I – Diagnosis.* Consultant Report for the Institut du Sahel, Bamako, Mali.

Dabiré, L.C. (2001) Etude de quelques paramètres biologiques de *Clavigralla tomentosicollis* Stål. (Hemiptera: Coreidae), punaise suceuse des gousses du niébé (*Vigna unguiculata* (L.) Walp), dans une perspective de lutte durable contre l'insecte au Burkina Faso. Thèse de Doctorat d'Etat ès Sciences Naturelles, Université d'Abidjan, Côte d'Ivoire, 179 pp.

Dakouo, D. and Ratnadass, A. (1999) Les lépidoptères foreurs de tige en Afrique de l'Ouest,

fluctuations saisonnières, importance économique et perspectives de lutte intégrée. *Annales de la Société Entomologique de France* 35, 463–470.
Dakouo, D., Nacro, S., Post, R. and Traoré, Y. (1992) A system of rational control of insect pests in irrigated rice schemes in Burkina Faso: phytosanitary survey and threshold action. *Nuisibles Pests Pragas* 1, 1–9.
Dakouo, D., Nacro, S., Post, R., Traoré, Y., Nokoe, S. and Munyinyi, D.M. (1995) Evaluating Insect Pest Management in rice environnement. *Insect Science and its Application* 16, 93–101.
Dakouo, D., Trouche, G., Ratnadass, A., Ba, M. and Da, S. (2000) New sources of resistance to sorghum midge in Burkina Faso. *International Sorghum and Millet Newsletter* 41, 34–37.
Diarra, A. (2000) *Common Regulation of the Member States of CILSS on Pesticides Registration.* Inter-States Committee for Drought Management in the Sahel (CILSS), Ouagadougou, Burkina Faso, 7 pp.
Kaboré B., Dakouo, D., Thio, B. and Sawadogo, A. (2002) Efficiency and profitability of an IPM system against insects, diseases and nematodes at two irrigated rice schemes in Burkina Faso. *International Rice Research Notes* 27(1), 34–35.
Matteson, P.C. (1990) Lessons learned in the development of IPM in Asia. In: Institut du Sahel (ed.) *Proceedings of the International Symposium on IPM in the Sahelian Countries*, Bamako 4–9 January 1990, Institut du Sahel, Bamako, Mali, pp. 257–263,
Ratnadass, A. (2001) *Final Report of the INCO-DC Project – Sustainable Production of Sorghum in West Africa by Integrated Control of its Main Insect Pests.* CIRAD, Programme CALIM, Montpellier, France.
Sawadogo, A. (1990) Contribution à l'étude de la nématofaune et de la mycoflore antagoniste dans les cultures maraîchères du Burkina Faso. Thèse de Doctorat, Université des Sciences et Techniques du Langdoc, Montpellier, France.
Sawadogo, A., Thio, B., Konate, A. and Abdoulaye, A. (1994) Inventory of nematodes associated with rice in the Sahelian Burkina and at the Office du Niger (Mali). *Nuisibles Pests Pragas* 2(2), 130–148.
Sawadogo, A., Thio, B. and Kini, L. (1995) Crop rotation as a way to control tomato gall-nematodes in Western Burkina Faso *Nuisibles Pests Pragas* 3(2), 197–207.
Sawadogo, A., Thio, B. and Konate, A. (1998) Yield lossses due to nematodes in a maize–tomato crop rotation in Burkina Faso. *Cahiers Agricultures* 7, 323–325.
Thio, B. (1998) Studies on effects of *Hirschmanniella oryzae* (Pratylenchidae: Nematodes). MSc thesis, University of Ouagadougou, Burkina Faso.

Important Websites

Permanent Interstates Committee for Drought Control in the Sahel (CILSS)

www.cilssnet.org

Education and Research

Centre National de la Recherche Scientifique et Technologique (CNRST): www.cnrst.bf
Université de Ouagadougou: www. univ-ouaga.bf
Institute for Research and Development (IRD): www.ird.bf

NGOs

Albert Schweitzer Ecological Centre: www. ceas-ong.net

Chapter 11

Ghana National Integrated Pest Management Program

K. Afreh-Nuamah

Faculty of Agriculture, University of Ghana, Legon, Accra, Ghana

Brief History of IPM in Ghana

IPM has been recognized as one of the practical alternative measures that could be used to deal with many problems emanating from increasing pesticide use especially at the farm level. However, the implementation had been restricted to a few isolated crops in the developed world.

Recent developments, however, have shown that IPM could be more practical and field-oriented to the benefits of the ordinary farmer. Especially when it is adopted not as a technology, but as an approach and a strategy for developing technologies, to solving pest and disease problems as and when they occur (Kiss and Meerman, 1991).

Until 1991, research on IPM in Ghana was fragmented and lacked a focused approach. Until then most of the research work was undertaken within the institutes of the CSIR and the faculties of agriculture of the country's universities. These centered on developing control measures, which were usually pesticide oriented, and screening germplasm for resistance to insects pests/diseases for the various crop commodities.

Well-planned experiments to study population dynamics and seasonal distributions of pests and their natural enemies, and the nature and influence of interacting biotic factors on pest populations, had been lacking. Furthermore, contacts with farmers had been very minimal.

Following the West African IPM workshop at the Accra Conference Centre in Accra, Ghana, in 1991 (under the auspices of the IPM forum), the National Plant Protection and Pesticide Regulatory Committee of the National Agricultural Research Project (NARP) submitted a memorandum to the NARP Secretariat pushing adoption of IPM as a major component of Ghana's Plant Production/Protection Strategy. This recognized the excessive use of pesticides especially on crops like vegetables (tomato, cabbage and aubergines) had led to unacceptable residues in market produce resulting in risks to consumers and commodity rejection at the international market. Increasing incidence of farmer poisoning and long-term effects of pesticides on aquatic and terrestrial ecosystems were further causing concern to agriculturists and environmentalists.

The need to reduce dependence on chemical pesticides and the development and implementation of alternative pest control measures were therefore of urgent priority. In addition to the memorandum, a NBCC was formed with the assistance of the IITA, in September 1992.

The NBCC which was IPM oriented, established a number of multidisciplinary crop-based working groups. Members of these groups included leading scientists engaged in agricultural research (from Ghanaian Research Institutions), Technical Officers and Extensionists from the Ministry of Food and Agriculture and importantly, local farmers. These groups worked to identify pest problems and recommended environmentally friendly and sustainable strategies for controlling them. The targets were crops and pests known to be associated with overuse or abuse of pesticides and for which an adequate knowledge base was available (preferably with local research support). Major pests included variegated grasshoppers (*Zonocerus variegatus*), larger grain borer, mango mealybug, water hyacinth and the plantain/banana weevil (*Cosmopoliti sordidus*). Targeted crops were cereals and vegetables such as tomatoes, aubergines and cabbage.

Implementation of the FFS Methodology

In August/September 1993, the FAO Inter-Country IPM Program for South and Southeast Asia organized a global IPM meeting and study tour for representatives from 40 countries to Southeast Asia to expose participants to the IPM FFS training methodology and to sensitize participants to establish national IPM/FFS programs in their own countries.

After the study tour, proposals made by the Government of Ghana for an FAO/Technical Cooperation Programme project to establish an IPM Training of Trainers course in irrigated rice were approved for implementation at the Dawhenya Irrigated Rice Scheme from 31 May to 6 October 1995. The Training of Trainers course in healthy rice production was organized at Dawhenya for 24 staff of the Ministry of Food and Agriculture (MOFA) from Ghana, three from Côte d'Ivoire and one from Burkina Faso with the assistance of one IPM trainer from the FAO Inter-country IPM Program in Asia and two rice Master Trainers from the Philippines National IPM program. Technical inputs on special topics were provided by scientists and training specialists from WARDA and local research institutions and the University of Ghana. Overall policy guidance and supervision for this project was provided by an Oversight Committee chaired by the Director of Agricultural Extension Services Directorate of MOFA, Ghana and with representatives from three relevant technical departments of MOFA and the University of Ghana, Legon.

The Dawhenya experience showed that the IPM/FFS concept of training could also work in Africa. This observation was endorsed by participants of an FAO Technical Consultation Workshop on Participatory Training in IPM for Africa at Akosombo, Ghana, from 5 to 12 September 1995. Consequently, follow-up training programs for rice farmers were established at five irrigation sites (i.e. Ashaiman, Dawhenya, Afife, Botanga and Tono) in 1996, to extend the Dawhenya experience to other regions or ecologies and to train additional rice farmers. About 500 rice farmers were trained under this program.

At both the training of trainers (TOT) and post-TOT training, yields were increased from 3 t/ha to 6 t/ha compared with a reference group of farmers who had not been trained, and worked according to conventional farmers' practice, the average net returns of trained farmers were 138% higher.

After training, over 80% of farmers changed their practices and adopted the IPM strategies. In a number of communities, farmers formed groups after they completed their FFS training. These groups of FFS graduates meet and discuss their problems.

The results of these training activities convinced the relevant government authorities that IPM/FFS has the potential to complement the extension delivery in the country and, therefore, should be scaled up to cover other crops and made available to large groups of farmers.

Subsequently, the concept is now being progressively applied to Ghana's agriculture. To date, plantain, cassava, cowpea, sorghum/millet and vegetables have been handled. The UNDP through the FAO, in its National Poverty Reduction Program, is sponsoring the training of 1700 rice, vegetable, sorghum/millet and plantain farmers from five districts (viz. the Afram Plains, Accra Metropolitan Area (AMA), Bongo, Juabeso-Bia and Dangme West) using the IPM/FFS strategy.

The purpose of UNDP assistance to the National Action Program for Poverty Reduction, was to provide direct support to community-based organizations, NGOs, decentralized departments, the private sector and civil society operative at the district and community level in order to improve the living conditions of the poor.

The Poverty Reduction Program, therefore, focused on achieving rising real household incomes and possibilities for expanding opportunities for such increases permanently across the board.

IPM/FFS and TOT Course: a participatory training approach for both extension agents and farmers

The IPM/FFS is an experiential field-based training program, which encourages farmers to grow healthy crops in a more environmentally sustainable manner with little or no input of agrochemical pesticides. There are no standard recommendations or packages of technology offered.

Critical problems addressed during the training

1. *Low productivity and low farmer income due to poor knowledge and inadequate skills.* Intervention: provision of skills/scientific-based knowledge through experiential field oriented learning techniques using the FFS strategies. This makes crop production more sustainable and cost-effective.

2. *Excessive pesticide use stemming from ignorance about pests and pesticides, and leading to increasing unsustainable and cost-ineffective production, environmental contamination and unnecessary health hazards.* Intervention: ignorance about pests/diseases replaced by insight into the ecosystem and the interactions among pests and natural enemy populations, the causal relationship between crop agronomy and physiology, damage to plants and economic loss which form the basis for the farmers' capability to take crop management decisions based on findings from his/her field.

3. *Temporary and limited impact of extension services as a result of a top-down system which provides crop management instructions without giving farmers sufficient ownership of the information knowledge-base on which these instructions are based.* Intervention: the FFS approach offers a solution. The project seeks to establish a capacity to conduct FFS, thereby contributing to increased effectiveness of agricultural extension.

The training program, which is season-long, covers an entire crop-growing season from land preparation, nursery establishment and transplanting through to harvesting or postharvesting practices.

Principles of the training approach

The training approach is based on three main principles:

1. *Grow a healthy crop.* This involves all aspects of growing a healthy crop; soil and land preparation, agronomy, crop physiology, nutrient management, pest and disease management, irrigation and water management, and harvest/postharvest techniques.

2. *Monitor your field regularly.* Regular scheduled visits to the field, at least once a week for field crops and about twice a month for plantation crops. Field inspection and monitoring is done using the AESA technique – a tool for detailed field inspection and data collection, processing of data and informed decision making. Field observations and critical analysis of data/information gathered from the field are discussed by entire group, using previous knowledge of crop performance at particular crop growth stage. Trainees learn to become acquainted with monitoring and field inspection methods. At least 50% of the time is spent in the field.

3. *Conserve natural enemies.* This is an indirect way of encouraging trainees to reduce or avoid use of agrochemical pesticides. Training emphasizes the presence of predators and other beneficial organisms in the environment which all help to regulate pest population. Simple exercises to demonstrate harmful nature of agrochemicals in the environment, especially to predators are carried out. In addition, functions of pests and beneficial insects and their biology are demonstrated in insect zoos. Farmers find out for themselves that limited pest damage does not usually reduce yields, and that spraying against several pests increases both production costs and the risks of further pest outbreaks.

Farmers become experts after the training, and can properly understand the cause–effect relationship between pest/diseases and beneficial/predators and, therefore, learn to be careful in the use of agrochemical pesticides. Enhanced agronomic and soil nutrient management practices all lead to healthy crop growth and therefore higher yields and greater economic returns.

Full-time season-long TOT courses are organized to prepare Extension Staff to conduct FFS training. During the TOT they carry out comparative experiments, and grow and monitor the target crops to learn about the problems that farmers face throughout a cropping season.

A validation trial becomes necessary whenever FFS is to be introduced to new crops. This facilitates the testing and adaptation of theory and foreign experience to actual farm situations. This activity is usually conducted by future master trainers and selected farmers with involvement of researchers.

Training methodology

Baseline surveys during which ICPM trained Extension Agents determine traditional production constraints and agronomic practices, farmers' intervention to constraints and sociocultural information about the community, always precede any FFS training program. Survey results are processed and technological interventions from research put together in an experiential learning method as the training curriculum.

Usually, a training session involves a group of 30 or 25 farmers, who agree to meet regularly once a week for the entire crop cycle. These groups of farmers are subdivided into subgroups of five or six per subgroup. Each subgroup is led by a trained Extension Officer – facilitator trained in special TOT courses.

During the training session, two main plots – one depicting farmers' traditional farming practices based on information collected from a baseline survey in the community before the training – where pesticides use is on a calendar basis or as any insect/disease is observed on the field, use or little application of fertilizer, poor irrigation and land preparation, irregular random planting, untimely/irregular field visits and poor farm hygiene practices.

Then the ICPM plots, where the new farming strategies to overcome production constraints are introduced – nursing of good or clean seeds and transplanting of seedlings in rows, incorporation of organic manure or composting to enhance soil nutrient availability, regular visits for field monitoring using AESA. Informed decisions are made based on actual field situations, i.e. selective use of pesticides, usually environmentally friendly biopesticides, so as to encourage build-up of predators and other natural pest regulatory organisms.

During training, non-formal adult education exercises and group dynamic studies, like team building, group formation and cohesiveness, are all carried out.

Gender issues are also discussed and encouraged so that membership of groups is not discriminatory.

Through participation in the field schools, farmers quickly realize that the FFS can be effectively used to address other community development issues, such as improved health status due to drastically reduced pesticide poisoning, education for youth, etc. It could even be a medium for discussions on HIV/AIDS.

A few case studies will illustrate the potential of the IPM/FFS in managing crop pests.

Case Studies

Cabbage production at Weija Irrigation Scheme (AMA, 1998 season)

Because of high levels of pest damage from a variety of pests, e.g. *Plutella*, and lack of appropriate knowledge and skills in pest and crop management, farmers at the Weija irrigation scheme abandoned cabbage production for several years. In 1998, the project conducted season-long training for farmers at Weija in integrated pest and crop management in FFS for cabbage production. Marketable cabbage yields and net returns from IPM practice and farmers conventional fields were recorded and compared in three farmers' fields as follows (Table 11.1).

When farmers adopted their conventional cabbage production practices, they consistently made losses in their field. Using crop production skills acquired through training in FFS by the project, farmers at Weija are now able to resume cabbage production on a cost-effective and highly profitable and environmentally sound manner. They are able to generate additional revenue to reduce poverty and upgrade their standard of living.

Managing of plantain weevils and nematodes

On-station yield loss trials in Ghana showed the importance of nematodes and weevils as production constraints, particularly when in combination. Weevils alone (artificial infestation) gave a yield reduction of 35%; nematodes (natural population) reduced the yield by 64%; and combined, the pests led to a severe reduction of 85% in the plant crop.

ICPM Trial

An ICPM trial was carried out with the following objectives:

1. To determine the crop production and protection strategies that give the highest yield at lowest input cost.
2. To determine the best season of planting that gives the highest yield.

In the IPM plot the trainers made vital crop protection decisions as to whether to weed or prune old leaves or not based on monthly AESA, whilst in the farmers' practice (FP) plots weeding was done as and when necessary based on traditional local practices as reported during the baseline surveys.

Table 11.1. Marketable cabbage yields and net returns from IPM practice vs. conventional fields.

Field number	Marketable cabbage yields		% increase (IPM over FP)
	Farmers' practice (FP)	IPM practice	
Field 39			
Yield	4.9 t/ha	27.5 t/ha	
Revenue	–¢0.78 million	+¢3.9 million	550
Field 42			
Yield	14.4 t/ha	18.3 t/ha	
Revenue	–¢0.67 million	+¢1.7 million	354
Field 56			
Yield	1.8 t/ha	21.8 t/ha	
Revenue	–¢1.6 million	+¢3.75 million	334

US$1 = ¢7000 (Ghanaian Cedis).

No insecticides were applied in either the IPM or the FP plots and weeds were controlled manually in both plots. Yield data and net returns are presented in Table 11.2.

Percentage harvested plants for the major season were 60 and 7 for the IPM and FP respectively (Table 11.2). The corresponding figures for the minor season were 62% and 23%. Generally, yields were unexpectedly higher in the major season than in the minor season. This could be attributed to enhanced agronomic practices imposed and the healthy planting materials used as well as the unusually good rainfall distribution.

The net returns (Table 11.2) in terms of profit margins, followed the same trend as the yield data; while the IPM recorded a profit of about US$634.3 in the major season the FP recorded a loss of about US$51.42. However in the minor season planting, while IPM recorded a profit of about US$460, FP recorded a profit of about US$85.7. The loss incurred in the major season FP plot was due to the high incidence of toppling in the major season, caused by a high incidence of nematodes together with wind damage that followed the dry weather from January to March.

IPM Stakeholders in Ghana

IPM stakeholders in Ghana are considered to include the following:

- farmers
- extension agents of MOFA
- researchers
- intermediaries – NGOs, international centers, UN agencies and donors
- policy makers.

Research

Since 1990/91, Ghana Agricultural Research Systems has been restructured with the establishment of the NARP. NARP comprised all institutions engaged in agricultural research including the universities, research institutions under the CSIR, the MOFA and farmers. Under the NARP, commodity-based multidisciplinary and interinstitutional research teams are in place. Research is usually adaptive with farmers on farmers' fields, but there are also some basic research activities, especially in the universities. Each of these commodity-based research teams has some of the scientists concentrating on IPM issues, which usually emphasize pest/disease identification and control with pesticides on calendar or 'need-be' spray schedules, cultural practices and use of resistant varieties of crops.

Scientists on these commodity teams are the main local resource persons in IPM/FFS/TOT programs, with technical backstopping by the Global IPM Facility, through employment of consultants. A non-formal education workshop was organized for the local scientists who serve as resource persons in TOTs and curriculum development

Table 11.2. Percentage number of harvested plants for plant crop and first ratoon crop of Apantu Pa plantains grown under IPM and FP in the major and minor seasons. Net returns for major and minor seasons.

Season of planting	Plant crop	First ratoon	Mean	Net return (U$)
Major				
IPM	61	58	59.5	634.28
FP	9	4	6.5	–51.42
Minor				
IPM	62	61	61.5	460
FP	29	16	22.5	85.71
				Average: IPM US$547.1
				FP US$17.1

US$1 = ¢7000 (Ghanaian Cedis).

workshops to create awareness of adult education techniques and strengthen and enhance their participation in IPM. It was expected that the performance of the NARS would be further enhanced to be more responsive to demand-driven farmer production constraints through these IPM/FFS programs. This has been a very important output of the IPM program.

Policy makers

The Government of Ghana endorses the call from the UNCED to promote IPM at farm level in order to make crop production more sustainable and environmentally sound. Consequently, the MOFA is very keen to expand (IPM/FFS) training methodology to cover other crops after the successful IPM/FFS pilot activities in 1995/96 on rice and the positive experiments achieved for cassava and cowpea. A National Integrated Crop Protection Advisory Committee, chaired by a Deputy Minister, Ministry of Food and Agriculture was established by the Government in April 1995. This committee, which is an advisory body on all IPM issues, is made up of policy makers, researchers (from the universities and research institutions), the EPA, extension agents and farmer associations.

With time, the committee has fully endorsed the IPM/FFS training strategy and has recommended to the MOFA to adapt it as an extension tool in training of extension staff and farmers. Integrated crop management strategies have been the central theme of this committee's recommendations.

Institutional framework

In recognition of the validity of the IPM/FFS approach, the MOFA institutionalized the Project Oversight Committee (POC) in May 1997 to facilitate the expansion of IPM/FFS in Ghana. The POC is chaired by a Deputy Minister of MOFA and includes all the relevant Directors of MOFA, the EPA and NGOs. The reconstituted POC now has the Deputy Director General, Agricultural Research (NARP of CSIR) on it. A professor from the University of Ghana was appointed the National IPM Co-ordinator and acts as Secretary to the POC. The Senior Crop Protection Officer, at the FAO Regional Office for Africa, in Ghana, is an advisor to the POC.

Networking

In line with the Government of Ghana's recent decentralization exercise, the POC has recommended the establishment of Regional and District Oversight Committees throughout the country to supervise all IPM/FFS activities. Regional MOFA Directors are to chair these committees, with representatives from farmers, NGOs, researchers and extension agents. District Directors (MOFA) of all districts with IPM/FFS activities are to be members. The National IPM Co-ordinator will then link up with these committees to form a National IPM Network.

NGOs

The NGOs in Ghana have recently formed an action group, with a representative on the POC. We expect these groups to work closely with the National IPM Program. The POC accepted to train six NGOs during the vegetable IPM/TOT scheduled for August 1998. It was expected that after training of the respective NGOs they will mobilize funds on their own to enable their trained personnel to train farmers within their respective communities. I hope such collaboration will continue.

Coordination

In May 1997, a National IPM Co-ordinator was appointed. A secretariat was established at the Plant Protection and Regulatory Services Department (PPRSD) of MOFA. The secretariat is equipped with computers, an administrative clerk, telephone, one cross-country vehicle and a driver. Plans are in place to connect this secretariat to the Internet, so it could be the main Internet access to be linked with the regions/ districts, the universities and research institutions. Currently, coordination is achieved by regular visits to all regions, personal and telephone contacts with POC/National Integrated Crop production Advisory Committee members, and sometimes by mail correspondence. Workshops, occasional lectures and publicity in both the print and electronic media have also been used.

Overall strategy

The IPM/FFS strategy, though acknowledged to be potentially viable for enhancing crop production practices in the country, was developed in pilot cases. The idea was not to stop ongoing training approaches, nor to condemn existing practices, but to perfect and consolidate it in our system. For sustainability, emphasis was placed on use of local expertise and improvement of our own human resources.

Several other IPM projects have been ongoing in Ghana. These included projects on cowpea (USAID), striga control (GTZ), cassava and plantain (IITA), integrated crop protection (GTZ/MOFA), integrated food crop system project (NRI/NARP). The IPM/ FFS training methodology was expected to add value to these. A national mechanism for cooperation and coordination among IPM initiatives is envisaged to be established under the national IPM program.

The main target crop for IPM/FFS has been irrigated rice. Further pilot activities on IPM/FFS for cassava and cowpea conducted by the 1995 rice TOT graduates (cassava: Ecologically Sustainable Cassava Plant Protection – IITA, cowpea: USAID Bean-Cowpea Collaborative Research Support Program) in 1996, demonstrated that IPM/FFS has great potential to add value to these ongoing projects. Since October 1997, a pilot IPM/TOT/FFS scheme on plantain has been ongoing at the University of Ghana's ARS at Kade, with resource persons from ARS, Kade, Crop Research Institute and IITA/West African Plantain Project, and the PPRSD.

Achievements

National IPM policy environment

A very favorable policy environment has been created for the expansion and implementation of the IPM/FFS program to train more extension staff on more crops, so that a larger percentage of farmers could be reached.

At a recent workshop to review the IPM/ FFS program, the Communiqué released at the end of the workshop recommended that the IPM/FFS training methodology should be adopted as a normal training strategy in MOFA's extension system with the requisite budgetary allocation provided by the districts. As a result of this, it was decided that a more appropriate name to be adopted for the program is ICPM/FFS program, to take into account the holistic approach to healthy crop production that the program advocates.

Development of extension capacities

The project started with a series of studies commissioned to provide field-base information on farmers' cropping practices and constraints to efficient crop production. These studies were exhaustively discussed at technical workshops to define the components for conducting season-long training of extension agents in the use of integrated crop and pest management strategies in crop production. Some technologies, which did not directly address the production constraints, were validated to adjust them to a field school training and to build up the confidence of the trainers.

As a result of this, well established and tested guidelines for developing methodologies and curriculum for ICPM/TOT/FFS training have emerged which have already enabled us to develop curricula for training extension agents and farmers in new crops like plantain, sorghum and millet.

Benefits realized by trainers

The IPM trainers have achieved a high level of experience and competence in IPM training so much so that other African countries such as Malawi, Tanzania-Zanzibar and Tanzania-Mainland have requested assistance from Ghana for IPM trainers to facilitate the conduct of FFS in their countries. Other requests for assistance received include study tours from Senegal, Niger, Tanzania-Zanzibar (see Table 11.3).

The UNDP-supported project has therefore enabled Ghana to develop capacities within the agricultural extension service, to train farmers in the adoption of ICPM methods for sustainable, cost effective and environmentally sound crop production. In recognition of this expertise, the National IPM Oversight Committee has decided to create a cadre of extension agents known as IPM Master Trainers in order to further promote the development of agro-skills in smallholder farmers.

Benefits realized by farmers

Using the skills acquired from season-long training by the project, smallholders have begun to realize farm profits as well as economic and social benefits from their farming.

A recent evaluation and impact assessment of the program in all five poverty districts (June 2000) indicated the following:

- Farmers had enthusiastically adopted the ICPM technologies, because the methods give increased yields (reported by 55.8% respondents), are healthier for the farmer and even the consumer (33.7%), facilitate farming (33.7%) and reduce production costs (31.7%).
- Further, the FFS have given them the opportunity to form farmers' groups for common action to solve their problems. In some areas the program had introduced new crops helping them to diversify their productions (6.7%).

Impact of ICPM on crop production and quality

The impact assessment study observed that with ICPM, crop yield rose by a mean of 150%, crop losses dropped from 46.2% to 10.4% and production cost fell by about 40–58%. The beneficiary farmers indicated that ICPM vegetables are more wholesome with heavier, richer green, smoother skin free from insect holes. They are tastier and have a longer storage life than traditionally produced vegetables. ICPM-produced rice also gives a better-filled panicle with longer grains and very little breakage during

Table 11.3. Assistance requested from IPM project.

Country	Assistance requested
Tanzania-Zanzibar	Study tour for two trainers to study planning and implementation of TOT and FFS
Tanzania-Mainland	One IPM Trainer to backstop TOT and FFS activities
Senegal	Study tour for two plant protection technicians to study planning and implementation of TOT/FFS
CILSS/Departement de Formation en Protection des Végétaux Niger	Study tour for two senior IPM specialists to study field implementation of FFS
Malawi	Training of three extension agents on TOT/FFS training process
CGIAR/NGO	One IPM Master Trainer to facilitate FFS workshop
IITA/PEDUNE	IPM Master Trainers for cowpea technical backstopping and coordination of curriculum development workshop for a regional Cowpea TOT/FFS field training at Tamale

milling. Similarly, ICPM plantains have larger and heavier bunches and fingers.

Impact on farm revenue, lifestyle and household food security

With ICPM, beneficiary farmers' yearly earnings have increased by an average of 70–143%. This enabled them improve their housing, pay children's school fees, increase church contributions or acquire television sets, new furniture, utensils, new clothing for the family, etc.

Some farmers have been able to send their children to vocational or secondary schools. ICPM also facilitates family feeding during the lean season, by enabling farmers to store more food or buy the necessary ingredients. The longer shelf-life of ICPM foodstuffs also assists household food security.

Gender issues in FFS

During the conduct of TOT courses and FFS, every effort was made to encourage the participation of women or women's groups in these field activities. Women farmers continue to make significant and quality contributions to the planning of FFS and all the associated activities.

The participation of women in FFS was strongly location specific. In some communities, women were primarily home-bound taking care of the children, the home and marginally doing some subsistence farming just to supplement the family food requirements. In such locations, there were usually a low percentage of women in the FFS. Therefore, only a limited amount of time was devoted to discussions of specific gender issues in such schools.

Partnerships

The project has promoted and established collaborative programs with relevant agencies and related agricultural development projects in Ghana.

Six village worker NGOs were trained as trainers in vegetable production during the 1998 TOT course at the Weija Irrigation Project site.

Since then the GOAN and the ECASARD have adopted the ICPM/FFS strategy as the main component of extension training activities.

In addition, with assistance from the ICPM/FFS program, some of the existing related agricultural development projects such as the FAO-Ghana SPFS, the Root and Tuber Improvement, the Lowland Rice Development Project, have all incorporated ICPM/FFS training approach into their implementation activities.

Strong collaboration with and partnerships will also be established with the GTZ-supported PTD/FFS project and with the DFID Rural Sustainable Livelihood program.

Amex International of the USAID and the Vegetable Producers and Exporters Association have established linkages with the ICPM/FFS program training in healthy and sustainable production of crops like pineapples, mangoes and vegetables.

Emerging Issues and Constraints

1. *Workload and limited trained extension personnel.* Trained FFS facilitators have had to conduct FFS field activities in addition to their normal extension work, thus FFS training is seen as additional load. This affected quality of delivery as planning sessions prior to FFS days were usually left out.
2. *Marketing.* Marketing of produce was an initial problem, but improved as farmers explored other opportunities for marketing. This was one of the risks identified before start up of project.
3. *Erratic rainfall.* Over reliance on rainfall for crop production delayed and affected the training program. In addition crop production was restricted to only the rainy season. Thus farmers could not make as much money as should have been possible

were irrigation facilities available for the farming ventures.

4. *Access to credit.* Limited financial resources prevented trained farmers from expanding their farms and optimizing their farm operations. Another risk factor.

5. *Limited availability of land.* Land was a limiting factor in certain communities, e.g. Bongo, making it impossible for some of the trained farmers to take advantage of the knowledge gained in the FFS training.

Bibliography

Afreh-Nuamah, K. (1996) *Report on IPM Farmers' Field School in Ghana.* TCP/GHA/4553.

Afreh-Nuamah, K. (1999) *The Development and Implementation of Integrated Crop and Pest Management Farmers' Field School (ICPM/FFS) in Ghana.*

Anonymous (1996) *International Literature Search for IPM Interventions on Tomato and Cabbage Farming.* Project GHA/96/001, CAB International, Egham, UK.

Anonymous (1997) *Consultative Workshop on Integrated Pest Management in Ghana.* Sogakope, Ghana.

Anonymous (1998a) *Ghana IPM Programme, Preparatory Workshop for Design of IPM Validation Trials,* Project HA96/001, Consultancy Report. CAB International, Egham, UK.

Anonymous (1998b) *Ghana IPM Programme, Vegetable Curriculum Development Workshop in Dawhenya and Vegetable IPM/TOT course in Weija,* Consultancy Report. CAB International, Egham, UK.

Anonymous (1998c) *Ghana IPM Programme: Vegetable IPM Training of Trainers in Weija.* Project GHA/96/001, Consultancy Report. CAB International, Egham, UK.

Anonymous (1998d) *Vegetable IPM/TOT Curriculum Development Workshop, Proceedings.* Dawhenya, Ghana, Project GHA/96/001, August 1998.

Anonymous (1998e) *Report on Rice IPM Training of Trainers (TOT) Course at Dawhenya and Farmers' Field School (FFS) at the Kpong Irrigation Project Site, Asutsuare.*

Anonymous (1998f) *Report on Vegetable PM Training of Trainers (TOT) Course and Farmers' Field School (FFS) in Ghana, Weija (Tubakrom).*

Anonymous (2000) *Final Report of the National Integrated Crop and Pest Management Farmers' Field School Programme Review Workshop:* 16–17 December 1999.

Chowdhury, D. (1998) *Report on Integrated Crop Management Training: Rice-Fish-Dike Crop Production.*

Ghana IPM Secretariat (2000) *Report on Plantain IPM Training of Trainers Course.*

Keetelaar, J.W.H., Millomeda, G. and Pulmano, A. (1995) *Report on Training of Trainers and Farmers' Field Schools for Rice Integrated Pest Management in Ghana.* TCP/GHA/4553.

Kiss, A. and Meerman, F. (1991) Integrated Pest Management and African Agriculture. *World Bank Technical Paper number 142,* 122 pp.

Kwasivie, F. (2000) *An Impact Assessment of Integrated Crop and Pest Management (ICPM)/Farmers' Field School (FFS) in Rice, Vegetable and Plantain Farming at Five Sites – NPRP in Ghana.* Final Report June 2000.

Obeng-Ofori, D. (1998) *Data Base on Crop Health Problems of Vegetables in Ghana.*

Ofori-Budu, K.G., Timbilla, J.A. and Asante, V.O. (1998) *Survey Report on Tomato and Cabbage Production in the Vea/Tono, Akumadan, Weija and Damage Areas in Ghana.*

Vos, J.G.M. (2000) *Ghana IPM Programme, Vegetable Farmers' Field School Evaluation.* Consultancy Visit Report. CAB International, Egham, UK.

Youdeowei, A. (1999a) *Briefing Document to the UNDP Evaluation Mission.* September, 1999.

Youdeowei, A. (1999b) *Progress in Implementation of Integrated Pest Management in Irrigated Rice and Vegetables in Ghana.* Consultancy Report, 27 January 1999.

Youdeowei, A. (1999c) *Agroskills Development: Integrated Pest Management Farmers' Field Schools and Briefing Documents to the UNDP Mission,* September 1999. Ghana: UNDP Poverty Reduction Programme, GHA/96/001.

Youdeowei, A. and Amedzro, A. (1999) *Facilitators' Field Guide (2) on Non-Formal Education for IPM.*

UNDP/GoG/NPRP (1999) *Report of Evaluation Mission: September 1999.*

Chapter 12

Development and Implementation of Integrated Pest Management in the Sudan

Yasir G.A. Bashir, Elamin M. Elamin and Eltigani M. Elamin
ElObeid Agricultural Research Station, ElObeid, Sudan

Introduction

Agriculture in Sudan

Sudan, the largest country in Africa, occupies about 1 million square miles, extending from the desert in the north to the equatorial forests in the south. It lies between latitudes 3° and 22° north and longitudes 22° and 38° east, and shares its borders with nine countries. Sudan has a population of 25 million with an annual growth rate of 2.6%. The country contains rich natural resources, with vast areas suitable for agriculture and an adequate water supply.

Agricultural production in the country can be divided into three sectors. Irrigated crops compose one sector, with cotton, wheat, groundnuts, vegetables and sorghum as the most important crops. Another sector includes mechanized agricultural production of sorghum and sesame. The third sector, traditional rain-fed crop production, includes millet, groundnuts, sesame and field watermelon.

Agriculture is considered the backbone of the Sudanese economy. About 85% of the population depends on agriculture for a living. Agricultural production comprises 40% of the gross national product, and provides most of the raw materials for industry in Sudan. Most cultivated land in Sudan is on the clay plains that cover 10% of the country. In this region, the bulk of the cotton crop is grown on heavy black soil, in addition to sorghum, groundnuts and a variety of fruits and vegetables. Cotton is the most important cash crop in Sudan, representing nearly 40% of the total value of agricultural exports. About 350,000 ha of cotton are grown annually under irrigation or traditional rain-fed production. Most of the cotton crop is exported; only 20% is used for the Sudanese textile industry.

History of the IPM program in Sudan

The IPM program in Sudan was initiated in 1974 under the FAO/UNEP cooperative program on development and application of IPM. Top priority was given to cotton and rice (Elamin, 1997).

In 1975, the FAO global coordinator for the IPC programs informed the ministers of agriculture in Sudan, Egypt, Ethiopia, Somalia, Uganda, Kenya and Tanzania that FAO intended to select a cotton-growing country in Africa as a base for the African Regional IPC program. In 1976, Sudan was nominated by FAO/UNEP IPC Consultation Mission as a suitable country for the program. Major crop losses caused by insect pests in the Sudan, the vital importance of cotton to the Sudanese economy, the impacts of increased use of chemical insecticides, and the availability of well-developed facilities for research stations were all cited as reasons for choosing the Sudan for the program.

Cotton was one of the principal target crops of the program, partly because of its vital importance to many developing countries as a revenue source. Cotton was also a focus because of environmental problems and decreased effectiveness resulting from the over-reliance on chemical pesticides for control of cotton pests.

The program was titled 'The African Inter Country Program for the Development and Application of IPC in Cotton' and was based at Gezira Research Station in Wad Medani, Central Sudan. The negative consequences of over-reliance on pesticides in cotton were discussed in a symposium held on 'Crop Pest Management in Sudan' in February 1978 in Khartoum, Sudan, with special focus on the effect of DDT on cotton pests.

- *Phase 1 of the IPM program in Sudan.* In 1978, the program was changed to 'Development and Application of IPC on Cotton in the Sudan'. The FAO/UNEP Panel of IPC Experts closely monitored the developments on IPC on cotton. The project was titled 'The Development and Application of Integrated Pest Management in Cotton and Rotational Food Crops' and executed by the FAO as part of the global program on IPC. The Directorate General for International Cooperation of the Netherlands primarily financed it.

  The initial financial support extended to October 1982. Funds were allocated for background research. In 1982, the FAO and the Dutch government reviewed the progress of the project, and recommended its extension to 1990.
- *Phase II.* In the second phase, the program concentrated on preparation of an inventory of natural enemies, development of biological control strategies, analysis of cotton ecosystems and organization of large-scale demonstrations in commercial fields to show that fewer pesticides could be used without yield reduction.
- *Phase III.* The third phase began in July 1990 and ended in December 1992. During this phase, economic threshold levels for four major cotton pest species were revised after large-scale field experiments in cotton production schemes, improvement of the role of biological control agents in cotton crop protection, and initiation of research and demonstration activities for food crops in the cotton rotation.
- *Phase IV.* In 1992, the significant achievements of the program resulted in a recommendation to continue into a fourth phase, targeted towards the development of IPM in vegetables, where pesticide misuse and risk for health and environment are large. From 1993 to 1996, the program shifted in focus from cotton to vegetables and cereals. The project was titled 'Development and Application of IPM in Vegetables, Wheat and Cotton.'

During all phases of the IPM project, training was an important project component at both the local and international levels, in addition to organization of workshops on IPM for different target groups. Budgets allocated for the first, second, third and fourth phases were US$1 million, US$1.4 million, US$2.9 million and US$3 million, respectively.

IPM policy

The Sudanese Ministry of Agriculture is heavily involved in IPM activities, most notably by setting priorities for the project. The First Secretary of the Ministry of Agriculture, which determines the overall IPM strategy for the country and finds ways to reach IPM objectives, chairs the steering committee. In 1995, this committee became the National IPM Steering Committee, integrating all policy makers related to IPM in Sudan. Government support for IPM in Sudan is outlined in the recent publication, *Sudan Country Strategy Note 1997–2000.* At both the federal and state levels, the Ministry of Agriculture provides support for expanding IPM activities. Research done by the Agricultural Research Corporation, as well as universities, follows the IPM philosophy.

Pesticide regulations

To limit the use of pesticides and improve safety to humans, animals and environment, the Agricultural Research Corporation (ARC) issued Procedures and Regulations Governing Research on Agricultural Pesticides for Registration.

Research and Extension Focus in Sudan

Research

Cotton

The first three phases of the IPM project in Sudan focused exclusively on cotton. New economic threshold levels were adopted for the four major cotton pests: the cotton whitefly (*Bemisia tabaci*), jassid (*Jacobiasca lybica*), cotton aphid (*Aphis gossypii*), and the African bollworm (*Helicoverpa armigra*) (Table 12.1). The new thresholds dropped the average number of insecticide sprays from eight to three without yield reduction.

As part of the biological control program, an egg parasitoid (*Trichogramma pretiosum*) of the African bollworm was collected in Texas, reared in The Netherlands and released in cotton-growing areas.

Wheat

The second research focus was on wheat. The total area under wheat production in Sudan is 164,929 ha. Aphids, termites, stem borers, rodents and birds regularly attack wheat in Sudan, causing grain losses as high as 30%. Before the IPM program, chemical control was being used against two aphid species, *Rhopalosiphum maidis* and *Schizaphis graminum*, at a threshold level of 35% infested plants. More intensive scouting methods were developed to improve aphid monitoring procedures. A modified economic threshold level was developed, which considered the role of natural enemies. The new threshold for wheat aphid control was based on the degree of infestation (DOI) rather than the percentage of infested plants. The DOI was calculated by multiplying the number of plants infested in a sample of 100 tillers by the category (class) of infestation. Aphid population density was divided into three categories 1 = 1–9 aphids/tiller; 2 = 10–29 aphids/tiller and 3 = more than 30 aphids/tiller.

Benefits of introducing the improved scouting techniques in 1994/1995 season included a potential reduction of the area treated with insecticides by 35% due to the new economic thresholds. This would result in a savings of 51,402,934 Sudanese dinars at the rate of 897.6 dinars/ha.

Table 12.1. Recommended economic threshold levels (ETL) for four major cotton pests in the Sudan, since 1993.

Pest	Old ETL	New ETL
Cotton whitefly *Bemisia tabaci*	200 adults/100 leaves	600 adults/100 leaves
Jassid *Jacobiasca lybica*	50 nymphs/100 leaves	70 nymphs/100 leaves in *Gossypium hirsutum* and 100 nymphs/100 leaves in *G. barbadense*
Cotton aphid *Aphis gossypii*	20% infested plants	40% infested plants
African bollworm *Helicoverpa armigra*	5–10 eggs or larvae/ 100 plants	30 eggs or 10 larvae /100 plants. No insecticide spraying before advanced flowering

Vegetables

During the fourth phase (1993–1996) the focus of the program shifted to vegetables. Losses in vegetable crops due to pests in Sudan were estimated at 25% (Elwasila, 1981). A 1990 survey of vegetable farmers in the Khartoum and Gezira regions reported that most of the farmers used chemical insecticides and 90% sprayed their crops at weekly intervals (Siddig and Nasr El Din, 1990). The vegetable IPM program focused on tomato, aubergine, onion, potato and okra. A package of IPM options was developed for each crop with the participation of farmers. This included the development and testing of improved cultural practices for key pests with farmer participation.

IPM options for vegetables

A survey of the seasonal occurrence of pests, predators, and parasites on vegetable crops formed the basis for the development of IPM techniques to conserve natural enemies. Other options included evaluating varietal resistance of vegetables to pests and diseases, screening of selective insecticides, using botanical extracts (e.g. neem water extract), and using repellents and attractants (e.g. coriander and fenugreek).

Development of IPM on vegetables can be a more complex task than on cotton and wheat, because vegetables comprise a heterogeneous group of crops, that attract a variety of pests. In addition, vegetables are not produced on a large scale under close supervision of the plant protection specialist, but by individual small farmers or sharecroppers who are only infrequently reached by extension services (Dabrowski, 1994). In spite of this, the project succeeded in developing a solid basis for vegetable IPM.

Results of IPM pilot studies conducted in FFS demonstrated:

- A significant yield increase (151% on tomato) in the use of IPM options over traditional practices.
- Reduction in the number of chemical sprays, reducing the cost of crop production and saving of foreign exchange.
- Good quality products.
- Fewer pesticide applications result in fewer hazards to farmers, consumers and the environment.

Training and extension accomplishments

Training and extension were an important component of the program. Training included organization of workshops, conferences and international MSc and PhD fellowships, in addition to degrees offered in Sudan. Six higher degrees were offered in Sudan, as well as 11 degrees overseas and 18 study tours organized by the FAO/ARC IPM project.

The program also organized a series of workshops and seminars (Appendix 12.1) aimed at identifying farmers' needs for effective IPM transfer. In addition, the project reviewed the capabilities of the national agricultural extension service and their work in advising and training the farmers.

The program issued a number of extension and training materials (some are in press) to provide farmers and extension officers with easily understandable information on various aspects of crop management (see Appendix 12.2).

FFS

The IPM project in Sudan adopted the FFS approach in 1993 (Fig. 12.1). Sudan is the first African country to introduce this approach. The structure of the IPM FFS program is outlined in Table 12.2. The goal is to help farmers become experts in managing their fields. The FFS program was aided by close collaboration with researchers from the ARC, universities with extension services, and field plant protection staff.

In the 1995/96 season, seven pilot FFS were established. Each focused on a certain crop: one for wheat, one for cotton and five for vegetable production. Pilot field schools sought to introduce and evaluate IPM options with technical assistance from the IPM specialists and extension workers.

Fig. 12.1. An IPM FFS weekly session.

Overall 46 IPM FFS were established during the program to encourage hands-on research and extension by farmers. School activities included weekly meetings throughout the growing season with a group of 20–25 farmers.

Rural Womens' Schools

IPM Rural Womens' Schools were also established in 1995/96 with the objective of enabling rural women to understand the effect of pesticides on human life and environment. Other subjects included major diseases and how to protect their families, and how to establish small home gardens. Nutrition and home food preservation were also taught as part of the program.

The IPM program provided support through discussions and lectures during group training courses organized by other FAO/ARC/UNDP/IFAD projects coordinated from ElObeid (SUD/88/022) or Kosti (IFAD). Two leaders in participatory research and training were asked to administer an on-farm research project in traditional rainfed agriculture (FAO/UNDP/ARC) in ElObeid and the White Nile Project for Agricultural Development (IFAD/ARC) in Kosti to establish new FFS in those areas (Dabrowski, 1997). Total numbers of people reached directly by the program are presented in Table 12.3.

Successful IPM examples from Sudan: Cotton and Watermelon

Cotton IPM

Background on cotton production in Sudan

Cotton has been grown commercially in Sudan since 1867. In 1911, it was sown in central Sudan in the Gezira Scheme (Elamin, 1997). Before World War II and the introduction of the chemical insecticide DDT, pest control on cotton was based on cultural, physical and legislative approaches to reduce the damage of major cotton pests. These included flea beetle (*Podagarica puncticollis*), jassid (*Empoasca lybica*), pink bollworm (*Pectinophora gossypella*), and termites (*Microtermes thoracalis*). Minor pests of cotton included African bollworm, whitefly, and cotton aphid. A bacterial blight caused by *Xanthomonas malvacearum* was an important

Table 12.2. FFS structure, membership and activities (as recommended by Schulten and Meerman, 1995, cited from Dabrowski, 1997).

Role	Membership	Responsibilities
IPM planning and supervision team	IPM CTA IPM National Project Director IPM Extension and Training Expert IPM Biological Expert ARC DG Deputy	1. Approval of FFS plans, locations, structures and costs 2. Preparation of FFS program, incentives, inputs, fuel and transportation 3. Participate in field visits for supervision and problem solving 4. Approval of FFS final reports
IPM FFS coordination committee	IPM Planning and Supervision Team FFS coordinators Technical officers Horticulturist Protectionist Pathologist Agricultural mangers	1. Call for meetings 2. Program preparation and follow-up for FFS 3. Exchange experiences and opinions among members on program planning, supervision and execution
FFS area coordination team	Coordinator Entomologist Pathologist Horticulturist Extension worker Farmers' representative	1. Plan FFS programs and supervise their execution 2. Data collection, monitoring and evaluation of FFSs 3. Making available requirements and coordinate different concerned services
FFS organizer	Extension worker, horticulturist or entomologist	1. Conduct weekly meeting field sessions 2. Ensure that data are collected 3. Request the FFS area coordinator for assistance 4. Prepare monthly report for FFS area describing the activities undertaken, farmers' views on the IPM options and suggestions, further improvement and a listing of data collected 5. Responsible for organization and execution of all IPM FFS activities
FFS members	20 selected farmers who are willing and ready to learn and apply IPM principles	1. Attend weekly/monthly FFS sessions 2. Involved in all FFS activities 3. Learn IPM practices and teach others 4. Apply IPM principle

cotton disease, first reported in 1922. It is controlled by breeding resistant varieties. Other diseases include leaf curl, vectored by whiteflies, and cotton wilt caused by *Fusarium oxysporum* (Abu Salih and Kalifa, 1978).

Insecticide use in cotton

Cotton spraying with chemical insecticides began in 1945; an average of one spray or less was applied using DDT primarily against jassid. The continuous use of DDT and other chlorinated hydrocarbons until 1980, and the increasing use of organophosphates, led to the resurgence of secondary pests. The number of insecticide sprays per season gradually increased with an outbreak of the African bollworm in 1963, high infestation levels of whitefly, and abundance of the cotton aphid (Bashir, 1997). The average number of insecticide

Table 12.3. Participants in the FAO/ARC IPM Project, Sudan during 1993–1996.

	Number of participants				
Category	1993	1994	1995	1996	Total
Cadre	586	346	428	266	1626
Farmers (general session field days)	2161	625	264	230	3280
Farmers in the IPM FFS	108	140	221	448	917
Rural women	842	50	515	252	1659
Students	2071	146	284	80	2581

applications increased from one spray in 1945 to eight or nine sprays per season during the 1970s (Table 12.4). However, the cotton yield remained almost stagnant.

Another problem with aphids and whiteflies

The stickiness of the cotton fibers as a result of honeydew excreted by large numbers of aphids and whiteflies became a problem in the ginning mills; consequently, a number of complaints were raised by consumers of Sudanese cotton and prices fell still further.

The increasing use of insecticides on cotton eventually became a burden on the foreign exchange reserve of the Sudan. At about the same time, cotton prices fell sharply in the international markets. Prices of extra long staple cotton fell from about US$1.3/pound in 1951 in the Liverpool market to below US$0.40 in the early 1960s. Farmers' income followed a similar fall. The net income of farmers fell to less than US$2/acre in the 1963/1964 season (Elamin, 1997).

Cotton production in the Sudan was in a crisis, approaching the disaster point where Sudanese cotton would not sell on the world market. Field data indicated that rapidly developing strains of whiteflies were becoming resistant to organophosphate insecticides. For example, monocrotophos provided effective control of whitefly until 1977/78, when whiteflies were out of control in fields that had been sprayed as many as 11 times with this insecticide (Elamin, 1997).

Table 12.4. Number of insecticide sprays and cost of crop protection of cotton in the Gezira Scheme, since 1945.

Period/ year	Number of sprays	Crop protection cost as % of total production costs
1945–1954	0.5	–
1955–1964	1.6	–
1965–1974	5.0	14.5
1975–1978	6.9	25.5
1979/80	8.87	34
1980/81	8.61	32
1981/82	6.78	26
1982/83	5.22	24
1983/84	5.45	26
1984/85	4.14	23
1985/86	8.60	33
1986/87	5.20	30
1987/88	5.67	24
1988/89	5.27	22
1989/90	4.34	15
1990/91	3.72	10
1991/92	4.75	19
1992/93	4.93	35
1993/94	3.02	30
1994/95	2.85	22

Host plant resistance

During the first phase of the project for the 'Development and Application of IPM in Cotton and Rotational Crops' (1979–1983), research identified morphological and biochemical characters of a resistant cotton genotype such as gossypol content, okra leaf, hairiness and nectar. Consequently, a medium staple cotton variety Suda-K was released. The low humidity and high temperature in the microenvironment of the open canopy of the okra leaf variety created unsuitable conditions for the whitefly, better penetration of pesticide and reduced number of sprayings (two versus four to six for the normal Acala types) (Abdelrahman *et al.*, 1997).

Biological control

Heavy insecticide application at the beginning of the season often prevents the establishment of natural enemies (Bashir, 1992). The natural enemies of whitefly and aphid are important in central Sudan. In one experiment, when cotton was left without spraying throughout the season, both species were kept below the economic threshold level (Abdelrahman and Munir, 1989). During the second phase of the IPM program (1985–1989), the project concentrated on demonstration work in large-scale trials in addition to biological control activities. An egg parasitoid of the African bollworm, *Trichogramma pretiosum*, was introduced in commercial production areas. A large-scale program of mass release was carried out from 1986 to 1990 with the cooperation of the Dutch government, which agreed to compensate farmers for potential yield losses. The initial experiment involved 320 ha of cotton left unsprayed in the 1986/87 season. Initial studies showed that 45% of bollworm eggs were destroyed by the introduced parasitoid. In addition, Ahmed (1995) mentioned that the braconid wasp *Meteorus layphygmarum* destroyed about 20% of bollworm larvae in fields where early chemical application was avoided.

Economic benefits of IPM in cotton

In 1981, the Ministry of Agriculture reported that expenditures incurred for purchase and application of pesticides had risen to US$65 million. After the adoption of raised economic threshold levels and the introduction of IPM on cotton, pesticide spending dropped to an average of US$12 million in 1992–1995.

Raised economic threshold levels for four key insect pests in cotton and one key pest in wheat are now implemented in the Gezira and Rahad Schemes. As a result, the number of sprays in the Gezira Scheme has dropped from 5.6 to 3 in cotton and from 1.7 to 0.9 in wheat. This translates into savings of approximately US$2.6 million in cotton and US$1.3 million in wheat. In the Rahad Scheme the number of sprays in cotton dropped from 4.55 in 1993 to 3.1 in 1994. This saved US$700,000. The total donor contribution to the FAO/ARC IPM Project amounted to US$8.3 million from 1979 to 1995. The donor contributions turned out to be less than the savings in pesticide spending, after 2 years of IPM implementation in cotton and wheat in the Gezira and Rahad Schemes (Report of the Tripartite Review Mission 21–30 July 1994).

Field watermelon IPM

Field watermelon is one of the most important crops in the traditional rain-fed system in western Sudan. It can withstand the harsh climatic conditions prevailing in the area. It is grown as a sole crop or on a relay intercropping system with groundnut, millet, sorghum, sesame and okra. Watermelon is a strategic crop for small farmers in the traditional rain-fed sector. It is a multipurpose crop; the seeds are an important cash crop, the fruit is a source of drinking water, and fruit remains are used as feed for animals. In addition, field watermelon is the only crop that remains in the field after other crops are harvested (from July to March), helping to prevent soil erosion. During years of cereal shortages, melon seed flour is added to cereals in many rural areas in western Sudan. As a result, consumption of cereals such as millet and sorghum decreases by 60%. For this reason, watermelon is considered to be an important food security crop.

The production of watermelon in the state of North Kordofan has substantially decreased during the last ten years due to attack by the melon bug *Aspongopus viduatus*. This has led to negative economical, social and environmental impacts. Farmers use broad-spectrum pesticides such as malathion and carbaryl to control this pest. The large quantities of insecticides (10,000 kg and 4000 to 5000 l of liquid) required amount to a cost of 15 million Sudanese dinars, a difficult sum for the state to afford.

IPM control strategies for melon bug

After the biology and ecology of melon bug were studied, an IPM program was designed with community participation. The melon bug *Aspongopus viduatus* is oval, rather flat, and relatively large (1.8–2 cm long and 0.8–1 cm broad) (Schumeterer, 1969). From May to June, the bugs congregate in large numbers for a period of estivation.

As a result of this research, two IPM strategies were developed: (i) reduction of the number of adult bugs by hand picking; and (ii) the use of cultural practices and less hazardous insecticides to control the remaining bug population.

Hand picking

The hand picking campaign was started in 1998. As an incentive, 100 Sudanese dinars (US$0.39) were paid for each kilogram of bugs. The campaign began in May and ended in mid June prior to the rainy season. A total of 15 tons of bugs were collected and burned by farmers during the campaign. The rate of community participation was about 50%, highest among women and children (70%). Indirect benefits of hand picking included the raising of awareness among farmers of the advantage of cooperative pest control operations.

In 1999, another hand picking campaign was carried out in areas suffering a drinking water shortage. Hand picking was conducted in collaboration with some NGOs working in the area on the basis of issuing Food For Work (one sack of sorghum for one sack of melon bugs). A total of 226 tons of melon bugs were collected and burned by farmers (Table 12.5).

Table 12.5. Localities participating in hand picking campaign and the amount of melon bugs collected during May–June 1999.

Locality	No. of sacks	Weight (kg)
Al Mazroub	1722	86,100
Tayba	154	7,700
Um Kradim	2641	132,050
Elobeid	16	800
Total	4533	226,650

Hand picking is not often an effective method of pest control, but in the case of the melon bug, mass collection could be successful. Picking melon bugs does not require skilled labor because adult bugs are large and can easily be recognized and collected. Also, adult insects can be found in large numbers during their estivation period. The insects exhibit reduced movement and limited ability to fly during their estivation, and adult insects do not release odors or bite. Most importantly, the recommended time for hand picking (May–June) coincides with a let-up in other farm activities, leaving a large available workforce.

Cultural strategies

Watermelon planted early in the season is subject to attack by two consecutive generations of melon bug, which breed and reproduce on watermelon before they enter winter diapause. Therefore, delaying the planting date of watermelon to late August was recommended.

Insecticide trials

Botanical water extracts were evaluated in on-farm experiments after the hand picking of melon bugs. Botanical extracts were mixed with a small dose of malathion to control the remaining bug population. They were also targeted at the African melon beetle *Aulacophora africana*, which appears late in the season. Treatments were prepared as follows:

- 25 ml malathion in 4 l water (the farmers' usual practice).
- 200 g neem seed water extract and 10 ml malathion in 4 l of water.
- 200 g neem seed in 4 l of water.
- 250 g neem leaf in 4 l of water.
- 250 g neem leaf water extract and 10 ml malathion in 4 l of water.
- 100 g *Balanites egyptiaca* seed water extract and 10 ml malathion in 4 l of water.
- Control (water only).

Botanical extracts gave adequate control to both insects. Since heavy infestation by the two pests kills the watermelon plants, the botanical extracts increased the plant survival rate. About 4375 plants/ha survived in the neem seed/malathion treatment while 1650 plants/ha survived in the control.

Yield exceeded 8 tons/ha, 12.3 times higher than the average yield in North Kordofan during 1989–1994, and 50 times greater than the average production during 1995–1998.

Other benefits of the IPM program on field watermelon included:

1. A reduction in the malathion application rate reduced the amount of insecticide used by two-thirds, representing significant savings to the government and reducing the cost of production.
2. The Plant Protection Department reported that the total amount of insecticides used by farmers in watermelon production declined by 50% in 1999/2000. A drop of 75% is projected for 2000/01(annual report of State Ministry of Agriculture and Animal Wealth, North Kordofan, 1999/2000).
3. Better yield of watermelon during 1999/2000 encouraged settlement in rural villages and reduced migration of the population to cities searching for temporary jobs.
4. The State Ministry of Agriculture reported a decrease in the area cultivated with watermelon due to increase in productivity (annual report of Ministry of Agriculture and Animal Wealth, North Kordofan, 1999/2000).

References

Abdelrahman, A.A. and Munir, B. (1989) Sudanese experience in integrated pest management in cotton. *Insect Science and its Application* 10(6), 787–794.

Abdelrahman, A.A., ElJack, I., Banaga, A., Gendil, B., El Allam, E., Maragan, M. and Ahmed, A.M. (1997) Integrated pest management in cotton: production and protection. In: Dabrowski, Z.T. (ed.) *Integrated Pest Management in Vegetables, Wheat and Cotton in the Sudan: a Participatory Approach.* ICIPE Science Press, Nairobi, Kenya, pp. 173–176.

Abu Salih, H.S. and Khalifa, O. (1978) Problems of cotton diseases control. In: Bashir, S., El Tigani, K.B., Eltayeb, Y.M. and Khalifa, H. (eds) *Proceedings of the Symposium Crop Pest Management in the Sudan.* Khartoum University Press, Sudan, pp. 99–104.

Ahmed, M.A. (1995) Natural enemies in Gezira and Rahad. Annual Report 1994/1995. *National Insect Collection.* FAO/IPM Project GCP/SUD/025/NET, Wad Medani, Sudan. 4 pp.

Bashir, N.H.H. (1997) Cotton pest resistance in the Sudan: status quo. In: Dabrowski, Z.T. (ed.) *Integrated Pest Management in Vegetables, Wheat and Cotton in the Sudan: a Participatory Approach.* ICIPE Science Press, Nairobi, Kenya, pp. 21–24.

Bashir, Y.G.A. (1992) Some biotic and abiotic mortality factors affecting the population of *Chrysopa carnea* in cotton fields in the Gezira. MSc thesis, University of Gezira, Wad Medani, Sudan.

Dabrowski, Z.T. (1994) *Integrated Vegetable Management in the Sudan.* FAO/ Government of Sudan Cooperative Project GCP/SUD/025/ NET. ICIPE Science Press, Nairobi, Kenya, 71 pp.

Dabrowski, Z.T. (1997) The impact of the FAO/ARC Integrated Pest Management Project on vegetables, wheat and cotton in the Sudan. In: Dabrowski, Z.T. (ed.) *Integrated Pest Management in Vegetables, Wheat and Cotton in the Sudan: a Participatory Approach.* ICIPE Science Press, Nairobi, Kenya, pp. 25–39.

Elamin, E. Mohamed (1997) The development of Integrated Pest Management in the Sudan: a historical perspective. In: Dabrowski, Z.T. (ed.) *Integrated Pest Management in Vegetables, Wheat and Cotton in the Sudan: a Participatory Approach.* ICIPE Science Press, Nairobi, Kenya, pp. 13–20.

Elwasila, G. (1981) Pests of vegetables and fruits and their control strategy. *Proceedings of the Third Annual Meeting of the African Association of Insect Scientists (AAIS).* ICIPE Science Press, Nairobi, Kenya.

Ministry of Agriculture and Animal Wealth (1981) Annual report season 1980/81, Khartoum, Sudan.

MOA (2000) *Annual Report of the Ministry of Agriculture, Animal Wealth and Irrigation, Season 1999/2000.* ElObeid, Sudan.

Schmutterer, H. (1969) *Pests of Crops in Northeast and Central Africa.* Gustav Fisher, Stuttgart, 296 pp.

Siddig, A.S. and Nasr El Din, S. (1990) *Consultancy on Vegetable Production, Present Insect Status and their Control.* FAO Project GCP/SUD/025/NET, November 1990, Wad Medani, Sudan, 96 pp.

Appendix 12.1.

In-service training courses and workshops organized by the FAO/ARC IPM Project, Sudan since January 1993. KAP, Knowledge Attitude and Practice; SMS, Subject Matter Specialist. (Source: Dabrowski, 1997.)

No.	Title	Duration	Audience
1	IPM training course	4 days	59 extensionists and inspectors
2	IPM training course	2 days	18 project members
3	Training course on operation, maintenance and safety measures of knapsack sprays	2 days	14 extensionists and inspectors
4	IPM training course	13 days	54 extensionists and inspectors
5	Training in IPM	1 day	20 extensionists + 67 farmers
6	KAP survey training	1 day	19 cadres
7	Training course on operation, maintenance and safety measures of knapsack sprays	2 days	20 extensionists and technicians
8	National workshop on integrated vegetable production	3 days	101 cadres + 23 farmers
9	IPM training course	1 week	49 entomologists
10	IPM GTZ training course on vegetables	4 days	26 extensionists
11	Extension staff training course	2 months	5 cadres
12	Annual review and planning workshop	2 days	79 policy makers, researchers SMSs
13	National workshop on the role of agricultural extension in IPM	3 days	42 extensionists and researchers
14	IPM/FFS orientation day	1 day	47 FFS cadres
15	Training course on vegetable IPM	2 days	49 FFS cadres, SMS and researchers
16	IPM workshop	5 days	53 extensionists, SMSs and researchers
17	Biological control training course	8 days	20 IPM staff
18	Annual review and planning meeting	4 days	75 researchers, SMSs, policy makers and extensionists
19	Workshop on control of *Orobanche*	1 day	46 researchers, SMSs and extensionists
20	IPM workshop	5 days	40 extensionists and SMSs
21	Biological control workshop	2 days	45 IPM FFS trainers
22	Workshop for Gezira and Rahad schemes	5 days	48 inspectors
23	Workshop	6 days	48 inspectors
24	Vegetable diseases workshop	3 days	38 IPM FFS trainers
25	IPM workshop	5 days	75 inspectors
26	IPM workshop	5 days	73 inspectors
27	IPM workshop	5 days	50 inspectors
28	Group training on biology, ecology and control of nematodes in the Sudan	1 day	20 cadres
29	IPM agricultural extension workshop	3 days	8 extensionists and farmers
30	Annual review and planning workshop	3 days	95 researchers, extensionists, plant protectionist; cotton, wheat and vegetable agronomists

Total of 1359 individual trainees attended

subsidized and spraying results were easy and quick to demonstrate, chemical pesticides were popularized at farm level at the expense of the established IPM practices.

In the 1940s, only two sprays of copper-based fungicides were deemed necessary for the control of coffee leaf rust (CLR) in bad years. In most seasons, the disease could be effectively managed through good cultural practices. The situation changed dramatically in 1970s with the outbreak of the coffee berry disease (CBD). By 1983, the number of copper-based sprays had increased to nine per season (Kullaya, 1983). The application of copper-based fungicide was in the rate of 5.5–11 kg a.i./ha per spray by 1985 (Bujulu and Uronu, 1985). Since then, the dosage rate has continued to increase without necessarily an increase in efficacy and control of the diseases.

Coffee leaf miners can be very damaging to coffee but outbreaks are usually checked by a complex of its indigenous natural enemies and therefore spraying in most seasons was deemed unnecessary (Notely, 1948, 1956; Swynnerton *et al.*, 1948). However, in the early 1980s, coffee farmers began to complain that it was no longer possible to achieve effective control with recommended insecticides such as fenitrothion. It was later established that this was because the pest has developed resistance to recommended organophosphates due to an increase in spraying frequency (Bardner and Mcharo, 1988; Nyambo, 1993, unpublished).

Similarly, continued use of copper-based fungicides for the control of CBD and CLR has exacerbated the pest pressure on coffee and its companion crops. Increased frequency of spraying with copper-based fungicides is considered as one of the major factors contributing to the development and increased pressure of the coffee leaf miners, CLR, CBD and the African coffee root-knot nematodes (Furtado, 1969; Bridge, 1984; Masaba *et al.*, 1993). The yields of maize and beans, the two crops often grown in association with coffee has declined. The incidence and severity of root-knot nematodes on bananas, a crop often intercropped with coffee, has intensified. This is due to the toxicity of copper to a wide range of natural biocontrol agents including antagonists that inhabit the soil. Consequently, pest populations build up to damaging levels without their natural control. The exceptionally high copper levels in the coffee soils results in toxicity, this depressing the activity of potential indigenous biocontrol agents (Crowe, 1964; Kullaya, 1983; Bridge, 1984; Masaba, 1991; Ak'habuhaya and Rusibamayila, 2000).

The IPM approaches formulated and extended to coffee farmers in the 1930s and 1940s have been severely eroded and the system is now at a crisis. The increasing pest pressure despite increased spraying frequency has led to a critical situation in the small-scale coffee production systems in Northern Tanzania. Increased pesticide prices without parallel increases in cash returns from coffee sales have forced many of the small-scale farmers to neglect their coffee plantations and divert their resources to alternative cash generating crops, e.g. vegetables and beans (Nyambo *et al.*, 1994, 1996).

Cotton

Cotton is another important foreign exchange earner for Tanzania. Up and until the late 1980s, it ranked second to coffee as a major export crop. Some 90% of the crop is produced in the Western Cotton Growing Area (WCGA) (Mwanza, Shinyanga, Mara, Singida and Kigoma regions) by small-scale farmers and is solely rainfed. It is the main non-food cash crop in most areas where it is grown. The crop is grown over a wide range of ecological and climatic conditions and it is therefore subject to varying degree of attack by insect pests and diseases in different areas.

Although cotton was introduced in the region in 1904, research to improve the yield and quality of the crop did not begin until 1932 (Jones and Kapingu 1982; Nyambo, 1982). The main task at the beginning of the program was to identify key limiting factors to increased production.

Before the 1940s, jassids (*Empoasca* spp.), and bacterial blight (*Xanthomonas campestris* pv *malvacearum*), were identified as the major pests limiting production of quality cotton in the area (Peat and Brown, 1961; Arnold, 1965). An assessment of cotton crop damage and loss due to insect pests done between 1934 and 1957 revealed in addition that the American bollworm (*Helicoverpa armigera*), the spiny bollworm (*Earias* spp.) and the cotton stainers (*Dysdercus* spp. and *Calidea dregii*) were also important pests (Nyambo, 1982). As a basis for developing economically acceptable management options, studies on the biology and ecology of these insect pests and diseases were done to establish their seasonality and possible management strategies in cotton (Mackinlay and Geering, 1957; Perry, 1962; Reed, 1964, 1965, 1967, 1970, 1971, 1972; Wickens, 1964a; Robertson, 1970; Wood and Ebbles, 1972).

Before the 1940s, the lack of good effective insecticides and the fact that even if they were available they were not considered economic under the small-scale farming conditions, research efforts emphasized other viable alternatives. The discovery of the mechanism of jassid resistance in cottons in South Africa in 1935 (Parnell *et al.*, 1949) was an invaluable contribution to the industry. As a follow-up to this, breeding and selecting for jassid resistance started at the Ukiriguru Agricultural Research Institute, the main cotton research centre for the WCGA. The first real jassid-resistant variety (UK 46) was released in 1946. As a result of this early success, selecting for jassid-resistance became a continuous process in the cotton program and resistance levels were gradually improved with each release of the Ukiriguru (UK) commercial varieties. Similarly, breeding and selecting for bacterial blight resistance and improved yields and quality of cotton was emphasized. Dramatic increases in production followed the release of varieties with good resistance to jassids and bacterial blight in the 1960s (Arnold, 1957, 1963; Cross, 1963; Cross and Innes, 1963; Cross, 1964a,b; Cross and Hayward, 1964). The genetic improvement in yield and quality potential plus jassid and bacterial blight resistance played a major role in the increase in annual cotton production from 190,000 bales to about 390,000 bales of lint between 1960 and the mid-1970s (Peat and Brown, 1961). In addition, cotton production became possible in areas where production was previously impossible due to damage caused by jassids and blight (Spence, 1967).

The achievements made in the production of jassid and bacterial blight resistance opened an avenue to search for resistance to other pests as well. The outbreak of Fusarium wilt (*Fusarium oxysporum* f.sp. *vasinfectum* (Alk) Snyder and Hansen) in 1953 on the shores of Lake Victoria was taken as yet another challenge to the research program. Breeding and selection of potential varieties from the local varieties and crossbreeding with some resistant varieties from West Africa and America (Brown, 1964; Wickens, 1964b; Ebbles, 1975; Hillocks, 1984) produced UK 77, which was released in 1977. UK 77 carries a level of resistance equal to that found in most Fusarium wilt resistant commercial varieties in the world (Jones and Kapingu, 1982; Ngulu, 1982). As a follow-up to this achievement, a backcrossing and selection program was started to improve further on the level of resistance to Fusarium wilt in the commercial varieties.

The UK cotton varieties have the ability to compensate for early crop loss of fruiting points caused by either physiological stress or by *H. armigera* attack, provided soil moisture and nutrients are not limiting. Thus early sowing, preferably between the end of November and end of December, is strongly recommended. In seasons when *H. armigera* attack builds up early, the early sown cotton may lose its bottom crop, but can compensate later by producing a crop during the main rains in March–April. If the bollworm population builds up later, then the early sown crop would have set its main crop and will therefore escape damage. The sowing date and the compensatory ability of the UK varieties both contribute to minimizing the damage caused by *H. armigera* (Mackinlay and Geering, 1957; Brown, 1965).

To extend this integrated approach further, chemical spraying against the major insect pests was also evaluated and recommendations made for farmers (Reed, 1967, 1971, 1972; Kerridge *et al.*, 1969; Robertson, 1970). A fixed spraying regimen consisting of six sprays at 2-week intervals starting from first flower was recommended in 1968 to take care of *H. armigera* and the other minor pests during the flowering and fruiting period. This blanket recommendation is simple to implement but is not ideal as it ignores any variations in pest pressure that may occur during the season. In addition, it does not provide an environment conducive to the build-up of *H. armigera* natural enemies (Nyambo, 1990).

Due to the shortcomings associated with fixed spraying regimes, simple damage scouting thresholds for *H. armigera* were developed and recommended to farmers to further enhance the effectiveness of the bollworms' natural enemies and to optimize the benefits of chemical spraying (Nyambo, 1986, 1989). A pegboard was developed and fine-tuned in collaboration with farmers to make cotton scouting at farm level technically simple. Many farmers in WCGA are now scouting their crop before spraying (B. Nyambo, personal observations). Where farmers are doing effective scouting the number of sprays have been reduced from six to three, with yield advantage over the blanket recommendation (Anonymous, 1997a).

Seed dressing is used to reduce seedling diseases, such as *Rhizoctonia solani* (Ngulu, 1982). All commercial seeds sold to farmers are already dressed.

To control aphids, which attack the crop early in the season, farmers are advised not to spray their cotton crop early in the seasons to allow for the build-up of its natural enemies, which can give effective control. In addition, the heavy rain in March/April wipes out much of the population.

Cultural practices (early planting, early picking, clean weeding and close season) to control spiny bollworm, cotton stainers, *Calidea dregii*, pink bollworm, bacterial blight and cotton seed bug is also a strong component of the pest management strategies. The spiny bollworm, cotton stainers, *Calidea dregii* and the cotton seed bugs are late season pests (starting at boll formation stage) and hence early sowing and picking is recommended to avoid attack. The close season is also a strategy to further reduce carry-over of cotton stainers, the pink bollworm and bacterial blight since there is evidence to show that the pests can survive on cotton trash (Arnold and Arnold, 1969). A well-grown crop has good growth vigor and can also withstand a certain level of pest attack. Therefore, farmers are advised to apply manure and/or inorganic fertilizers to realize a good crop (Prentice, 1946; Peat and Prentice, 1949; Le Mare, 1954, 1959, 1972, 1974; Peat and Brown, 1960, 1962a,b).

Thus, the IPM practices in the WCGA are a combination of host-plant resistance to the major pests, cultural practices, use of synthetic pesticides, conservation of natural enemies and crop scouting.

It is also recognized that IPM strategies can be improved and/or changed in response to the development of new pest problems and/or a change in farming systems. Traditionally, cotton is grown as a monocrop. However, farmers have tended to intercrop it with other crops, particularly maize, sorghum and sunflower to optimize microeconomics at farm level. In response to this, research work is in progress to identify suitable plant arrangement and combinations to optimize the benefits of plant–plant interactions in pest management (Nyambo, 1988; Katua, Kolowa and Mkondo, personal communication).

The IPM strategies for the WCGA are a result of over 50 years of work involving step-by-step improvement of individual components. Host-plant resistance was recognized as the foundation for economic and viable plant protection practices at farm level from the inception of the research program and continues to form the backbone for all other strategies. All cotton varieties released to farmers have good tolerance and/or high resistance to the major pests. The significance of team work, involving plant breeders, entomologists, plant pathologists, agronomists, extensionists and fiber technologists was recognized as the key to the development of the cotton crop in Tanzania and was therefore emphasized and enforced. A supportive research policy ensured staff continuity and adequate funding until recently. Funding is now a major constraint and much of the work has been suspended. In addition, networking and free flow of information between the national, regional and international cotton research programs played a major role in the success of the Ukiriguru-based cotton research work. The information on mechanism of jassid resistance in cottons was discovered in South Africa (Parnell *et al.*, 1949). The sources of resistance to Fusarium wilt were obtained from West Africa and America (Jones and Kapingu, 1982). Some technology developed in neighboring countries was fine-tuned for local use to optimize resources (Mackinlay and Geering, 1957). Cotton scouting and the use of pegboard was developed in Malawi (Matthews and Tunstall, 1968) and modified for use in the WCGA (Nyambo, 1986).

Coconut

The coconut crop is a major source of food, shelter and household income for smallholders on the coastal belt of Tanzania and Zanzibar Islands. The crop provides fresh nuts, cooking oil and the surplus is sold for cash. Although not considered as one of the principal cash crops, coconut is exported as copra cake in limited quantities. During the 18th century coconut was an important export crop for Zanzibar. However, over the past 20–30 years, production has declined tremendously for a number of reasons including neglect of palms, mismanagement, old age of palms and pest problems. Consequently, production of copra, oil and coir declined for the Mainland and Zanzibar Islands. On the Mainland, copra production declined from 6597 t in 1973/74 to about 1600 t in 1980. Production for export has practically stopped. In Zanzibar, production followed the same trend. Copra export declined from 11,871 t in 1980 to 7360 t in a decade (Kabonge and Temu, 1997; Yusuf *et al.*, 1997).

The coreid bug, *Pseudotheraptus wayi* Brown (Heteroptera: Coreidae), and the African rhinoceros beetle, *Oryctes monoceros* Oliv (Coleoptera: Scarabaeidae), are the major insect pests of coconut in Tanzania. Other pests include mites and termites, which may cause serious damage to young nuts. The latter are very destructive on seed nuts in nurseries, on sprouts and on newly transplanted seedlings.

Pseudotheraptus wayi is by far the most damaging as it is responsible for premature nut fall which leads to extensive crop losses. The National Coconut Development Program (NCDP) has been developing feasible IPM measures to control this pest with emphasis on biological control by using its indigenous predator, the weaver ant, *Oecophylla longinoda* Latreille (Hymenoptera: Formicidae). Although insecticides could be used to give effective control of the bug, spraying is not a feasible solution because it is difficult to spray the tall palms.

The potential of the weaver ant was recognized in early 1950s as the only feasible control method against *P. wayi* (Way, 1951, 1953a,b, 1963; Vanderplank, 1958). Methods to enhance the effectiveness of *O. longinoda* were researched. Interplanting coconut fields with tree crops, e.g. citrus and cloves, that hosts the weaver ant, were recommended in the 1980s (Way, 1983, unpublished). Preservation of non-host tree plants such as wild custard apple as well as maintaining suitable ground vegetation were also suggested as strategies to enhance the efficacy of the weaver ant.

However, inimical ants, notably *Pheidole megacephala* Fabricius and the pugnacious ant *Anoplolepis custodiens* Smith, were observed to interfere seriously with successful biological control of *P. wayi* by preying on and inhibiting the establishment of *O. longinoda* in infested trees (Oswald and Rashid, 1992; Varela, 1992; Seguni, 1997). Thus, successful selective suppression of *P. megacephala* and *A. custodiens* was necessary to ensure effectiveness of the weaver ants. This was made possible by application of hydramethylene, a selective ant bait. Hydramethylene is applied at the rate of 6 g per tree at the base of the trees. This eliminates *P. megacephala* but not *A. custodiens*. *Anoplolepis custodiens* populations can be reduced by maintaining appropriate ground cover in coconut fields. Where hydramethylene is applied, there is fast colonization of host trees interplanted in coconut fields by aggressive colonies of *O. longinoda*, reduced populations of *P. wayi* and a fivefold increase of nut production within 24–36 months of establishment of the weaver ants (Varela, 1992; Seguni, 1997).

Another strategy used to enhance the effectiveness of the weaver ant is the use of artificial bridges. *Oryctes longinoda* is known to readily accept artificial aerial bridges to walk from one tree to another and so spreading to neighboring trees. This behavior has been used to enhance *O. longinoda* by creating aerial passages between trees in coconut plantations, thus bypassing *P. megacephala* and *A. custodiens* on the ground. Farmers use various types of rope material for the aerial bridges including discarded steel wires, manila ropes and locally available natural fibers. This method is simple and environmentally friendly and therefore likely to be sustainable (Z. Seguni, personal observations).

The rhinoceros beetle is very destructive to young palms. The only feasible management option recommended to farmers is physical removal or killing the beetle using a thin metal rod and field sanitation. Although hooking of the beetles is effective it is impractical for a farmer with many palms. However, farmers are advised to deal with a few palms at a time. Field sanitation involves removal of dead logs to eliminate breeding sites for the beetle (Kabonge and Temu, 1997).

The main disease affecting coconuts is the lethal disease associated with phytoplasmas. The disease occurs on the mainland in localized areas and kills palms of all ages. To date, the national research program has not been able to identify a viable solution (Kullaya *et al.*, 1997).

The IPM approaches so far developed by researchers in collaboration with farmers are only known by a few farmers, mostly because of lack of adequate funding for technology transfer. Luckily, pesticides are not used on the palms because it is not easy to spray the tall trees. However, quality production is hampered by a lack of technical know-how at farm level on how to manage the major pests.

Key Constraints to IPM

The development and practice of IPM in Tanzania has been constrained by a number of factors including:

1. Lack of an enabling national IPM policy. As presented above this has favored wide and injudicious use of pesticides.

2. The traditional top-down national research and extension policy, which ignored farmer participation in technology development and transfer. Consequently, technologies developed by researchers have not reached the end users and/or they were not appropriately formulated for use in different production systems.

3. Lack of coordinated multidisciplinary team and cropping systems approach at research and extension levels. Some pest management recommendations are sometimes antagonistic at farm level. For instance, the copper-base fungicides used for the control of CBD and CLR are not compatible with other insect pest management strategies. The copper-based fungicides are toxic to a wide range of potential indigenous natural enemies of leaf miners and hence the resurgence of the coffee leaf-miners in recent years.

4. Inadequate funding for IPM research and extension in all cropping systems. This is due partly to the poor national IPM policy. This has been the case in the coconut crop discussed above as well as in horticultural crops and cashew. In cashew for example, the insect pest pressure is escalating due to excessive use of sulfur to control powdery mildew (Anonymous, 2000). The recommended disease tolerant/resistant clones and cultural practices have not been promoted at farm level due to lack of funds

– a situation that applies to many other cropping systems/technologies (Nyambo, 2001, unpublished).

5. Frequent senior staff transfers due to poor policy and lack of priorities that interfered with program continuity in many crops and systems.

The case example of cotton has shown that development of IPM is a long-term endeavor requiring careful planning. It also shows the importance of teamwork, staff continuity, networking, free flow of information between the national, regional and international research programs, and adequate funding. Any new component in the system has to be careful evaluated to avoid/prevent possible adverse effects on the equilibrium of the system as happened in the case of coffee.

Thus, the success in cotton is fortuitous in the light of the existing national agricultural policy. A change of policy, reflected in the new Government Agricultural and Livestock Policy of 1997 and the National Environmental Policy of 1997 together with the new Plant Protection Legislation of 1997, should create a more favorable environment for the development and implementation of sustainable IPM in other production systems.

References

Ak'habuhaya, J. and Rusibamayila, C. (2000) Contamination of soils and the coffee plant by copper-based fungicides in Kilimanjaro Region; Northern Tanzania. *Tropical Pest Management Bulletin* 1(1), 20.

Anonymous (1997a) *Brief Outline on the Profile of the IPM project – Lake Zone.* Tanzania–German Project for Integrated Pest Management (IPM) Western Zone Tanzania, pp. 1–9.

Anonymous (1997b) *The Plant Protection Act, 1997.* The United Republic of Tanzania. The Government Printer, Dar-es-Salaam.

Anonymous (1997c) *The Agricultural and Livestock Policy of 1997.* Ministry of Agriculture and Co-operatives, Dar-es-Salaam, Jan 1997. The United Republic of Tanzania. The Government Printer, Dar-es-Salaam, pp. 40–41.

Anonymous (1997d) *The National Environmental Policy, Dec 1997,* sectoral policies. The United Republic of Tanzania. Vice President's Office Dar es Salaam. The Government Printer, Dar es Salaam, pp. 18–19.

Anonymous (2000) Annual cashew research report 2000/2001. Cashew Research Programme, ARI. *Naliendele Mtwara* 2, 2–14.

Arnold, H.M. (1957) Resistance to bacterial blight in Lake Province local cotton. *West African Research Conference,* Sumaru, Northern Nigeria, p. 159.

Arnold, H.M. (1963) The control of bacterial blight in rain grown cotton. 1. Breeding for resistance in African up-land varieties. *Journal of Agricultural Science* 60, 415.

Arnold, H.M. (1965) The control of bacterial blight in rain grown cotton. II. Some effects of infection on growth and yield. *Journal of Agricultural Science* 65, 29.

Arnold, H.M. and Arnold, K.M. (1969) Bacterial blight of cotton: trash borne infection. *Empire Cotton Growing Review* 38, 258.

Bardner, R. and Mcharo, E.Y. (1988) Confirmation of resistance of the coffee leaf miner, *Leucoptera meyricki* Gesq. (Lepidoptera: Lyonetiidae) to organophosphate insecticide spray in Tanzania. *Tropical Pest Management* 34, 52–54.

Bridge, J. (1984) *Coffee Nematode Survey of Tanzania.* Report on a visit to examine plant parasitic nematodes of coffee in Tanzania, February/March 1984. Tropical Plant Nematology Advisory and Research Unit, IPP, UK, 22 pp.

Brown, A.G.P. (1964) Field trials of three Fusarium wilt resistant cotton selections in Tanganyika. *Cotton Growing Review* 41, 194.

Brown, K.J. (1965) Response of three strains of cotton to flower removal. *Cotton Growing Review* 42, 279.

Bujulu, J. and Uronu, A. (1985) A revise of the coffee berry disease (CBD) chemical control in Northern Tanzania for the period 1980–84 seasons. *Tropical Pest Research Institute Miscellaneous Report* No. 1019.

Cross, J.E. (1963) Pathogenicity differences in Tanganyika populations of *Xanthomonas malvacearum. Empire Cotton Growing Review* 40, 125.

Cross, J.E. (1964a) Breeding for resistance to more pathogenic variant of *Xanthomonas malvacearum. Cotton Growing Review* 41, 48.

Cross, J.E. (1964b) Field differences in pathogenicity between Tanganyika populations of *Xanthomonas malvacearum. Cotton Growing Review* 41, 44.

Cross, J.E. and Hayward, A.C. (1964) Relationship between pathogenicity and phage type in *Xanthomonas malvacearum*. *Cotton Growing Review* 41, 49.

Cross, J.E. and Innes, N.L. (1963) Bacterial blight of cotton, effects of resistance genes. *Empire Cotton Growing Review* 4, 48.

Crowe, T.J. (1964) Coffee leaf miner in Kenya. II-Causes of outbreaks. *Kenya Coffee* 29, 223–231.

Ebbles, D.L. (1975) Fusarium wilt of cotton: a review with specific reference to Tanzania. *Cotton Growing Review* 52, 295–339.

Furtado, I. (1969) Effect of copper fungicide on the occurrence of the pathogenic form of *Colletotrichum coffeanum*. *Transactions of the British Mycological Society* 53(2), 325–328.

Hillocks, R.J. (1984) Production of cotton varieties with resistance to Fusarium wilt with special reference to Tanzania. *Tropical Pest Management* 30, 234–246.

Jones, G.B. and Kapingu, P. (1982) Breeding. In: Jones, G.B. and Nyambo, B.T. (eds) *Ukiriguru 50 Years of Research, 1932–1982: the Golden Jubilee*. Tanzania Agricultural Research Organization, Ukiriguru Research Institute, pp. 17–20.

Kabonge, M.D. and Temu, N.A. (1997) *The Coconut Industry in Tanzania. Country Report.* International Cashew and Coconut Conference, 1997, Dar es Salaam, Tanzania, pp. 2–6.

Kerridge, P., Kayumbo, H. and Reed, W. (1969) Insecticide use in cotton in the sucking pest areas of Western Tanzania. *East African Agricultural Journal* 35, 147.

Kullaya, I.K. (1983) A note on copper and coffee plants in Kilimanjaro Region. In: *TARO Coffee Research Programme Annual Report, 1983.* Lyamungu Agricultural Research Institute, Moshi, Tanzania, pp. 75–77.

Kullaya, A.K., Masumbuko, L., Mkumbu, K., Mpunambi, A. and Chipungahelo, G. (1997) *Coconut Breeding Programme for Tanzania.* International Cashew and Coconut Conference 1997, Dar es Salaam, Tanzania, pp. 367–374.

Le Mare, P.H. (1954) Tie-ridging as a means of soil and water conservation and yield improvement. *Proceedings Second International African Soil Conference* 44.

Le Mare, P.H. (1959) Soil fertility studies in three areas of Tanganyika. *Empire Journal of Experimental Agriculture* 27, 197.

Le Mare, P.H. (1972) A long term experiment on soil fertility and cotton yield in Tanzania. *Experimental Agriculture* 8, 28.

Le Mare, P.H. (1974) Responses of cotton to ammonium sulphate and superphosphate, and relationships between yield and some soil constituents in Tanzania. *Journal of Agricultural Science* 83, 47–56.

Mackinlay, K.S. and Geering, Q.A. (1957) Studies of crop loss following insect attack on cotton in East Africa. 1. Experiments in Uganda and Tanganyika. *Bulletin of Entomological Research* 48, 833.

Masaba, D.M. (1991) The role of saprophytic surface microflora in the development of coffee berry disease (*Colletotrichum kahawae*) in Kenya. PhD thesis, University of London.

Masaba, D.M., King'ori, P.N. and Opilo, V.L. (1993) Effectiveness of new fungicides against coffee berry disease (*Colletotrichum coffeanum*) in Kenya. *Kenya Coffee* 58(678), 1517–1524.

Matthews, G.A. and Tunstall, J.P. (1968) Scouting for pests and the timing and spray applications. *Cotton Growing Review* 45, 115–127.

Ngulu, F.S. (1982) Pathology. In: Jones, G.B. and Nyambo, B.T. (eds) *Ukiriguru 50 Years of Research, 1932–1982. The Golden Jubilee.* Tanzania Agricultural Research Organization, Ukiriguru Research Institute, pp. 17–20.

Notely, F.B. (1948) The Leucoptera leaf miner of coffee on Kilimanjaro. I – *Leucoptera coffeela*, Guer. *Bulletin of Entomolological Research* 39, 399–416.

Notely, F.B. (1956) The Leucoptera leaf miner of coffee on Kilimanjaro. II – *Leucoptera coffeina* Washb. *Bulletin of Entomological Research* 46, 899–912.

Nyambo, B.T. (1982) History and background information on the research centre and its substations. In: Jones, G.B. and Nyambo, B.T. (eds) *Ukiriguru 50 Years of Research, 1932–1982: the Golden Jubilee.* Tanzania Agricultural Research Organization, Ukiriguru Research Institute, pp. 12–14.

Nyambo, B.T. (1985) Cotton insect resistance studies in the Western Cotton Growing areas of Tanzania. *Insect Science and its Application* 6(3), 379–384.

Nyambo, B.T. (1986) Studies in the bollworm, *Heliothis armigera* Hubner, the key cotton pest in Tanzania, as a basis for improved integrated pest management. PhD thesis, University of London, UK.

Nyambo, B.T. (1988) Significance of host-plant phenology in the dynamics and pest incidence of the cotton bollworm, *Heliothis armigera* Hubner (Lepidoptera: Noctuidae) in Western Tanzania. *Crop Protection* 7, 161–167.

Nyambo, B.T. (1989) Use of scouting in the control of *Heliothis armigera* in the Western cotton

growing areas of Tanzania. *Crop Protection* 8, 310–317.
Nyambo, B.T. (1990) Effect of natural enemies on the cotton bollworm, *Heliothis armigera* Hubner (Lepidoptera: Noctuidae) in Western Tanzania. *Tropical Pest Management* 36, 50–58.
Nyambo, B., Murphy, S.T., Baker, P. and Waller, J. (1994) Biocontrol in coffee pest management. *Proceedings 3rd International Conference on Tropical Entomology, 30 October to 4 November, 1994*, Nairobi, Kenya.
Nyambo, B.T., Masaba, D.M. and Hakiza, G.J. (1996) Integrated pest management of coffee for small-scale farmers in East Africa: needs and limitations. *Integrated Pest Management Reviews* 1, 125–132.
Oswald, S. and Rashid, M. (1992) Integrated control of the coconut bug *Pseudotheraptus wayi* (Brown) in Zanzibar. *National Coconut Development Programme Technical Paper* 1/92, 104.
Parnell, F.R., King, H.E. and Ruston, D.E. (1949) Jassid resistance and hairiness of the cotton plant. *Bulletin of Entomological Research* 39, 539–575.
Peat, J.E. and Prentice, A.N. (1949) The maintenance of soil productivity in Sukumaland and adjacent areas. *East African Agricultural Journal* 12, 101.
Peat, J.E. and Brown, K.J. (1960) Effect of management on increasing crop yields in the Lake Province of Tanganyika. *East African Agricultural Journal* 26, 103.
Peat, J.E. and Brown, K.J. (1961) A record of cotton breeding for the Lake Province of Tanganyika. *Empire Journal of Experimental Agriculture* 29, 114.
Peat, J.E. and Brown, K.J. (1962a) The yield response of rain grown cotton at Ukiriguru in the Lake Province of Tanganyika. I: the use of organic manure, inorganic fertilisers and cottonseed ash. *Empire Journal of Experimental Agriculture* 30, 215.
Peat, J.E. and Brown, K.J. (1962b) The yield response of rain grown cotton at Ukiriguru in the Lake Province of Tanganyika. II: land use and other rotational treatments contrasted with the use of organic manure and inorganic fertilisers. *Empire Journal of Experimental Agriculture* 30, 305.
Perry, D.A. (1962) Fusarium wilt of cotton in the Lake Province of Tanganyika. *Empire Cotton Growing Review* 39, 14.
Prentice, A.N. (1946) Tie-ridging with special reference to semi-arid areas. *East African Agricultural Journal* 12, 101.
Reed, W. (1964) Problems posed by early sowing of cotton in Lake Region, Tanganyika. *Cotton Growing Review* 41, 255.
Reed, W. (1965) *Heliothis armigera* in Western Tanganyika. I. Biology, with special reference to the pupal stages. II. Ecology, and natural and chemical control. *Bulletin of Entomological Research* 56, 117–140.
Reed, W. (1967) Use of insecticides on cotton in Western Tanzania. *Journal of Agricultural and Veterinary Chemicals* 8, 8.
Reed, W. (1970) The ecology and control of *Earias* (Lepidoptera: Noctuidae) in Western Cotton Growing Areas of Tanzania. PhD thesis, University of Reading, UK.
Reed, W. (1971) A comparison of insecticide sprayed and unsprayed cotton at Ukiriguru, Western Tanzania. *Cotton Growing Review* 48, 200.
Reed, W. (1972) Use and abuse of unsprayed controls in spraying trials. *Cotton Growing Review* 49, 67.
Robertson, I.A.D. (1970) Trials of insecticides to control pests attacking cotton in Tanzania 1963–67. *Cotton Growing Review* 47, 112.
Seguni, Z.S.K. (1997) Biology and control of *Pheidole megacephala* (Hymenoptera: Formicidae, Myrmicinae) especially in relation to use of *Oecophylla longinoda* (Formicidae: Formicinae) for biological control of *Pseudotheraptus wayi* (Heteroptera: Coreidae) in Tanzanian coconut cropping systems. PhD thesis, Imperial College, London, UK.
Spence, J.R. (1967) The development of Ukiriguru cotton varieties. *Ukiriguru Research Notes* 10.
Vanderplank, F.L. (1958) Studies on the coconut pest *Pseudotheraptus wayi* Brown (Coreidae) in Zanzibar. I. Method of assessing the damage caused by the insect. *Bulletin of Entomological Research* 49, 559–584.
Varela, A.M. (1992) Role of *Oecophylla longinoda* (Formicidae) in control of *Pseudotheraptus wayi* (Coreidae) on coconuts in Tanzania. PhD Thesis, Imperial Colllege, London, UK.
Way, M. J. (1951) An insect pest of coconuts and its relationship to certain ant species. *Nature* 168, 302.
Way, M.J. (1953a) Studies on *Theraptus* sp. (Coreidae); the cause of the gumming disease of coconuts in East Africa. *Bulletin of Entomological Research* 44, 657–667.
Way, M.J. (1953b) The relationship between certain ant species with particular reference to biological control of the Coreid, *Theraptus*

sp. *Bulletin of Entomological Research* 44, 669–691.

Way, M.J. (1963) Mutualism between ants and honeydew-producing Homoptera. *Annual Review of Entomology* 8, 307–343.

Wickens, G.M. (1964a) Fusarium wilt of cotton: seed husk a potential means of dissemination. *Cotton Growing Review* 41, 23.

Wickens, G.M. (1964b) Methods for detection and selection of hereditable resistance to Fusarium wilt in cotton. *Cotton Growing Review* 41, 172.

Wood, C.M. and Ebbles, D.L. (1972) Host range and survival of *Fusarium oxysporum* f. sp. *vasinfectum* in North Western Tanzania. *Cotton Growing Review* 49, 79.

Yusuf, C.H., Hassan, H.A., Mbarouk, I.K., Khalifan, N.A. and Ali, S.H. (1997) Status of coconut industry in Zanzibar. *International Cashew & Coconut Conference*, Dar es Salaam, pp. 40–45.

Useful website

http://www.habari.co.tz/tpri

Chapter 14

Integration of Integrated Pest Management in Integrated Crop Management: Experiences from Malawi

Sieg Snapp[1] and Eli Minja[2]

[1]*Department of Horticulture, Michigan State University, East Lansing, Michigan, USA;*
[2]*International Crop Research Institute for the Semi-Arid Tropics, Nairobi, Kenya*

Introduction

IPM should be viewed as an integral part of ICM. To raise productivity of smallholder farmers in Malawi it is necessary to address soil quality and fertility management issues, along with IPM. This integrated crop management approach focuses on soil health as the foundation of plant health, and involves farmers and researchers in partnership to evaluate indigenous knowledge and promising new technologies. This integrated approach is illustrated through case studies from Malawi. Experiences presented include farmer-friendly strategies to reduce *Striga* infestation in maize (*Zea mays*) systems and to improve pest management in pigeonpea (*Cajanus cajan*).

IPM has brought biological-based decision-making aids to farmers around the globe, but the focus has primarily been on areas where cash crops are grown with relatively high pesticide input levels. Success stories of farmers adopting IPM approaches include sweet potato and vegetable management in Indonesia (van de Fliert and Braun, 1999) and campaigns to improve pesticide efficiency and reduce excess use in Vietnamese rice production (Heong, 1999). In general, IPM has been effective at using ecological principles to improve pest control while minimizing pesticide use and abuse in disrupted agroecosystems (Greathead, 1989). Reducing grower costs through efficient and effective use of pesticides, and replacing pesticides with biologically based interventions makes sense for cropping systems that are dependent on pesticide use, and is not necessarily applicable to subsistence production systems (Orr *et al.*, 2000a).

The Malawi context is characterized by limited use of pesticides and a lack of market integration (Bentley and Thiele, 1999). A case in point is the use of pesticides to control serious outbreaks of pod borer pests in pigeonpea, application was almost zero in a recent survey (Minja *et al.*, 1999). In Kenya, by contrast, pesticides were used on pigeonpeas by about one-third of the farmers surveyed. Malawi is one of the poorest countries in Africa (US$170 GNP/capita in 1995). Purchase of pesticides, and access to sprayers, is so limited in Malawi that even when it is an economically viable option, farmers may prioritize easy to apply inputs such as fertilizer. Further, small investments in

 Integrated Pest Management in the Global Arena (eds K.M. Maredia, D. Dakouo and D. Mota-Sanchez)

fertilizer frequently improve productivity greatly, as soil fertility is low (Kumwenda *et al.*, 1997). The lack of market integration in Malawi also limits the extent to which horticultural and high value crops can be grown, and thus influences what are appropriate IPM options. We suggest that the focus of IPM in Malawi must be broad – to include integrated crop management issues, such as replenishment of soil fertility and tradeoffs between investment in weeding and fertilizer. Orr *et al.* (1997, unpublished) put forth a similar hypothesis.

The productivity of smallholder farms is constrained by a unimodal precipitation tropical climate, and a heavily maize-based cropping system. Maize dominates 67–89% of the smallholder cropped land area, grown primarily for food security purposes. Many farmers are at the edge of survival, cropping less than 1 ha of land (Kanyama-Phiri *et al.*, 1998). There are many challenges to producing crops in a sub-humid tropical ecology, with low soil fertility and minimal resources. Farmers are pursuing a range of strategies in the face of rising economic pressure from the declining value of local currency and subsidy removal from inputs; further, farmers are aware of new opportunities from recent market liberalization trends. One approach is the diversification and intensification of maize-based systems into a wide range of crops, grown primarily to sell: e.g. tobacco, vegetables (tomatoes, leafy vegetables, pepper, onions, and green maize), legumes (pigeonpea, groundnut, *Phaseolus* bean), and root crops (sweet potato, cassava, and Irish potatoes) (Orr *et al.*, 2000b; Snapp *et al.*, 2002). However, market options are limited in part because Malawi is a land-locked country located at the bottom of the Great Rift Valley in southern Africa.

Researchers, farm advisors and change agents working in Malawi are finding that not only is it important to take an integrated approach, combining soil and pest management for holistic cropping systems, it is also crucial to look beyond assumed goals, such as maximizing productivity. Surveys have documented that farmer priorities encompass reducing risk, optimizing economic returns and returns to small investments of cash or labor (Rohrbach and Snapp, 1997). Particularly in very poor, resource limited environments, careful attention must be paid to understanding farmer priorities and beliefs. This can be addressed through farmer participatory approaches, surveys and whole farm budgets to improve understanding of the economic context, and farmer perceptions (Heong and Escalada, 1999; Snapp and Silim, 2002). Farmer participatory research has documented the need to include soil fertility in integrated pest management research. Soil fertility is frequently the number-one priority of smallholders in Malawi when households were asked to rank agricultural problems in surveys (Table 14.1). For specific crops, particularly among pest-vulnerable crops such as legume pulses tend to be, pests may be ranked high as major yield constraints (Table 14.2).

Beyond soil fertility, pest problems are identified as productivity constraints by farmers in some surveys, notably: *Striga* (*S. asiatica* L. Kuntze) and termites in maize; pod suckers, borers and *Fusarium* wilt in pigeonpea and other legumes; and sweet potato weevil (Riches *et al.*, 1993; Minja *et al.*, 1999; Ritchie *et al.*, 2000b). Sauerborn (1991) concluded that *Striga* is the largest, single biological constraint to food production in Africa, and pod-pests are the greatest constraint to productivity in pigeonpea (Minja *et al.*, 1999). Unfortunately, there are few easy answers for these serious pest problems. We present case studies here that explore integrated management of whole systems, including soil fertility, *Striga* issues in maize-based systems and pigeonpea pod pests. Through active partnerships we discuss how researchers can work with farmers' knowledge to help develop appropriate technologies to control these challenging pests. The importance of involving farmers as active partners in technology development and taking a whole-systems approach are key components of Malawi IPM experiences (Kanyama-Phiri *et al.*, 2000; Orr *et al.*, 2000a).

Table 14.1. Ranking of agricultural problems in terms of importance, where all farmers in three communities surveyed listed their three worst problems in Malawi (Snapp *et al.*, 2002).

	Chisepo		Dedza		Mangochi	
	MHH[a] (n = 100)	FHH (n = 19)	MHH (n = 42)	FHH (n = 48)	MHH (n = 87)	FHH (n = 33)
Ranking:						
Lack of fertilizer	1	1	1	1	1	1
Lack of seed		3	2	3	2	3
Lack of labor		2		2		2
Lack of food	3				3	4
Lack of cash	2					
Pests			3			

[a]Response reported separately for male headed households (MHH) and female headed households (FHH).

Table 14.2. Farmers' perceptions of important biotic constraints to pigeonpea production in Kenya, Malawi, Tanzania and Uganda (adapted from Minja *et al.*, 1999).

	Farmers concerned about a given constraint (%)	
Constraint	Southern eastern Africa	Malawi
Pod sucking bugs	83	88
Pod boring Lepidoptera	51	39
Flower (pollen) beetles	32	41
Termites	26	31
Bruchids	72	54
Fusarium wilt	42	72

Case Studies

Reducing *Striga* infestation of maize-based systems in Malawi

Biology of Striga

The major parasitic weed that infests maize in Malawi is *Striga asiatica*. *Striga* is effective at infesting maize fields in part because of its astoundingly prolific seed production (millions of seed produced per plant). Allowing even one *Striga* plant to produce seed can cause significant long-term damage to maize production. The seeds are long lasting, so it is difficult to deplete the seed bank. The problem is compounded by the erratic and heterogeneous nature of the parasitic weed: *Striga* infestation varies from nil to severe in the same field, and frequently varies from year to year as well (Riches *et al.*, 1993; Orr *et al.*, 2000a). This has negative implications as farmers are not sure how severe the infestation will be and what returns in increased productivity can be expected from controlling *Striga*. The one positive aspect of heterogeneous *Striga* distribution is that farmers note that the 'patchiness' of distribution makes it easier to concentrate on severely infested areas for localized, hand pulling to eradicate *Striga* (Riches *et al.*, 1993).

Scope of problem in Malawi

Striga infests about two-thirds of fields with depleted soil in southern and central Malawi – particularly where continuous maize has been grown for many decades, near urban areas such as Blantyre and Lilongwe (Kabambe and Mloza-Banda, 2000). *Striga* is heterogeneously distributed, causing losses of about 50% in a minority (~10%) of infected fields, and minor damage of 5–15% in other fields (Riches *et al.*, 1993). In some areas of central Malawi average maize yield loss is about 28% (Chanika *et al.*, 2000). Occasionally farmers in southern Africa will abandon to fallow severely *Striga* infested land (Riches *et al.*, 1993).

Indigenous knowledge of pest management

Farmers in Malawi appear to have some knowledge of the potential for severe negative impact from *Striga*, and that damage to maize occurs before the weed is visible when it emerges above the soil surface (Riches *et al.*, 1993). Farmers frequently associate *Striga* infestations with long-term, continuous maize production with insufficient soil fertility inputs. However, farmers interviewed by Riches *et al.* (1993) were not aware that *Striga* was a parasite on maize roots, that the attachment to the host plant is how the weed draws on nutrients and is the mechanism for its particularly competitive effect. Local knowledge associates evil witchcraft powers with this weed. In Malawi the name for *Striga* is witchweed (English) or *Kaufiti*, which means a witch or wizard in Chichewa, an important local language.

Potential control methods and resource constraints

The recommended control practice for *Striga* and all parasitic weeds is hand pulling and burning the plant, according to the Malawi Government Guide to Agricultural Practice (1995). Hand removal is a very arduous undertaking and in a labor-short agricultural system, such as Malawi smallholders face, other alternatives are urgently needed.

A long-term, low cost solution has been sought through breeding maize varieties that are resistant to *Striga*. This goal has remained elusive as no varieties have been identified that show any consistent degree of tolerance or resistance to *Striga* infestation (Kabambe and Mloza-Banda, 2000). However, some degree of escape may be possible as early maturing maize varieties frequently are damaged less by *Striga* parasitism, compared with longer season varieties. Another long-term biological solution would be to identify fungi that could act as mycoherbicides, as parasitic fungi that infect *Striga* occur in nature and could be exploited (Greathead, 1989).

Application of fertilizer and manure is recommended for *Striga* management in maize-based systems (Kabambe and Mloza-Banda, 2000). Soil fertility enhancement from addition of nitrogen-containing fertilizer, green manure biomass, animal manure or rotational systems can reduce the negative impact of *Striga* to almost zero. Urea and other nitrogen sources have been shown to limit germination and infection by *Striga*, so benefits of soil fertility inputs may be in part due to nitrogen release that reduces parasitic infestation (Mumera and Below, 1993). But the main benefit from adding fertilizer sources – organic and inorganic – appears to be growth enhancement that allows the crop to out-grow the negative effects of *Striga* by improving plant nutrition (Kabambe and Mloza-Banda, 2000).

Use of legume rotations and legume intercrops have been studied, both as nitrogen enhancing technologies and as trap crops. The role of a trap crop is to stimulate *Striga* seeds to germinate and subsequently die from lack of an appropriate host. A number of cowpea varieties, and other legume species – most notably Malawi's indigenous *Tephrosia vogelii* plant – can stimulate *Striga* to germinate and die (Chanika *et al.*, 2000). A maize/*Tephrosia* intercrop tends to enhance maize grain yield compared with sole cropped maize. This may be related to nitrogen contributions from the biomass, but a role for *Tephrosia* in limiting *Striga* infestation has also been hypothesized (Snapp *et al.*, 2000). Farmers have observed that *Striga* infestation of maize is often reduced when *Tephrosia* is intercropped, to a greater extent than either maize/pigeonpea intercrop or continuous maize. On-farm results support farmer observations, where *Striga* emergence was markedly reduced in a maize/*Tephrosia* intercrop (Table 14.3). Similar results were reported by Kabambe and Mloza-Banda (2000), where a *Tephrosia* intercrop with maize reduced *Striga* emergence by about 50% compared with emergence in continuous sole cropped maize. Malawi agronomists have followed up these observations with studies of *Tephrosia* as a *Striga* germination stimulator and found it to be the most effective of all tested trap crop plants (Chanika *et al.*, 2000).

Rotation of grain legumes such as soybean and groundnut with maize appears to help control *Striga* infestation, although not as effectively as a green manure legume

Table 14.3. Effect of legume intercrop on maize yield and *Striga* emergence in a central Malawi research station trial, 1999 where $n = 4$ (S. Snapp, unpublished data, 2000).

Cropping system	*Striga* emergence at tasseling (number/m²)	Grain yield (t/ha)
Continuous maize	92	1.9
Maize/pigeon pea intercrop	70	1.5
Maize/*Tephrosia vogelii* intercrop	37	1.9
Mean (LSD)	66 (10)	1.8 (0.4)
P	<0.01	<0.2

Table 14.4. Marginal return analysis of alternative cropping systems compared with unfertilized, continuous maize. Tested in central and southern Malawi on-farm trials where $n = 30$, 1997/98–1999/2000 (S. Snapp and J. Rusike, unpublished data, 2000).

Cropping system	Chisepo Return (%)	Mangochi Return (%)
Unfertilized maize	na	na
Maize/*Tephrosia vogelii* intercrop	49	39
Maize/pigeonpea intercrop	239	649
Groundnut + pigeonpea rotation with maize	220	184

na, not applicable.

rotation of maize/mucuna (Snapp, unpublished data). Rotation of maize with groundnut reduced *Striga* emergence to 4 plants/m² compared with 16 plants/m² in continuous maize (Kabambe and Mloza-Banda, 2000). *Sesbania sesban* relay intercrop and rotation systems have been shown to increase maize yields by 30–80% compared with continuous maize – which has been attributed to enhanced soil fertility and reduced *Striga* infestation (Kanyama-Phiri *et al.*, 1998; Phiri *et al.*, 1999; Sanchez, 1999).

Potential for adoption of integrated Striga *management*

An integrated approach is particularly useful to control *Striga* as there appears to be a cumulative positive benefit from combining practices (Kabambe, 1997). It is recommended that Malawi smallholders use manure in addition to inorganic fertilizer, plus hand pulling of parasitic weeds in heavily infested areas and rotating maize with legumes or intercropping legumes, to the extent practical. Farmer adoption of these control methods however will remain limited unless fertilizers are affordable and the policy context does not favor maize production over legume production (Snapp *et al.*, 2002).

The economics of adoption must be considered before green manure crops such as *Tephrosia* are considered for promotion as *Striga* controlling systems. Unfortunately the marginal returns analysis shown in Table 14.4 indicated that maize/*Tephrosia* intercrop was the poorest performing system in a comparison of legume integrated options with unfertilized, sole cropped maize. *Tephrosia* is not a very attractive option for farmers because it does not produce grain or another marketable product, compared with grain–legume systems. Fuel wood and a biopesticide from leaf extract are potential products from *Tephrosia*, but have no market value at present.

Farmer-friendly control strategies for pigeonpea pests

Biology of pigeonpea and associated seed pests

Pigeonpea is a short-lived perennial leguminous shrub that is grown for multiple products. The primary produce from pigeonpea is the seed, used as a vegetable green pea and as a dry grain, however farmers also use pigeonpea stems for fuel wood and the high quality residues as fodder to feed livestock and for soil fertility enhancement (Faris and Singh, 1990). The multiple uses and soil-nutrient building capacity of pigeonpea has fostered great interest in promoting this crop to resource-poor farmers (Ritchie *et al.*, 2000a; Snapp *et al.*, 2000). However, not all pigeonpeas are alike in terms of ability to fix nitrogen biologically, nor in yield potential. Pigeonpea genotypes vary greatly in growth habit, including annuals, biennials and perennials. Widely grown varieties of pigeonpea vary from short-lived, determinate, short-statured and high-yield potential types to longer-lived, indeterminate types that branch and are well adapted to intercropping systems with cereal crops such as maize (Snapp and Silim, 2002).

Pest protection measures are most urgently required for determinant varieties that have one reproductive period and are particularly vulnerable to pigeonpea flower and pod pests. Risk adverse farmers in a pest ridden environment – particularly humid regions – can grow indeterminate materials that produce multiple flushes of flowers and pods and thus avoid a one-time vulnerability. Climate and day-length sensitivity greatly influence growth habit of pigeonpea, and the suitability of varieties for different production systems (Silim *et al.*, 1994). Pod borer activity is influenced by climate as well, where damage to pigeonpea pods by *Helicoverpa armigera* is reported to be low during cool and dry months (Myaka, 1994).

Scope of problem in Malawi

The mean grain yield losses due to field insect pests on pigeonpea in farmers' fields in Malawi has been estimated at between 15 and 20% (Minja *et al.*, 1999; Minja, 2001). The economically important insect pests on pigeonpea include the pod-sucking bugs (Hemiptera), pod-boring caterpillars (Lepidoptera), pod-boring fly maggots (Diptera) and storage bruchids (Coleoptera). Of these, the damage levels pre-harvest are most severe from sucking bugs which generally account for about 60–70% of damaged seed. Pod borers can be severe, causing 5–35% losses and pod fly usually causes less than 2% damage (Table 14.5). Bruchids are highly variable and can cause up to 100% losses in stored grain, depending on storage conditions and period of storage (Silim-Nahdy, 1995).

Another major pest problem in pigeonpea is *Fusarium* wilt, which can cause substantial plant losses (up to 80%) depending on varietal susceptibility, spread of disease in soil, and soil types (Songa and King, 1994). Generally, pigeonpea suffers more from wilt in heavy and poorly drained (as in the drainage lines where vegetables and seed production are concentrated) than in light and well-drained soils. Fortunately, *Fusarium* wilt resistant varieties of pigeonpea are released and widely available in Malawi, such as ICP 9145 (Snapp and Silim, 2002). Use of wilt resistant genotype ICP 9145 and the newly released ICEAP 00040 has been widely promoted by development efforts and NGOs (Ritchie *et al.*, 2000b).

Indigenous knowledge of pest management

Farmers use their knowledge of insect behavior to advantage through avoidance techniques, such as growing long-duration pigeonpea varieties that mature during the dry season, to reduce damage from pests that are active and at high population densities during the wet season. The majority of farmers surveyed by Minja *et al.* (1999) also had traditional methods to control storage pests of pigeonpea grain, including wood ash mixed with grain. The survey highlighted, however, that farmer knowledge was lacking regarding which insect groups were pests and which their natural enemy

Table 14.5. Mean pigeonpea grain yield losses (%) and contribution (%) of each field pest group to the losses in Kenya, Malawi, Tanzania and Uganda (Minja *et al.*, 1999).

Country	Kenya	Malawi	Tanzania	Uganda
Mean grain yield loss (%)	22	15	14	16
Contribution (%) to losses by pest groups:				
Flower/pod borers	22	28	50	53
Pod-sucking bugs	52	69	47	30
Pod fly	26	3	3	17

(Minja *et al.*, 1999). Increased efforts to train lead farmers and extension staff on specific crop pests and associated parasitic and other natural enemies is urgently needed in Malawi. This was also a conclusion of the FSIPM project in southern Malawi (Orr *et al.*, 2000a). However, prioritization is important in an extremely resource-poor country and Orr and colleagues make the point that training farmers to use low-cost soil fertility improvements may be the first step in an integrated crop management approach, before knowledge of insect behavior is addressed.

Farmers in Malawi have developed a range of pigeonpea pest management control technologies, as documented by recent surveys (Minja *et al.*, 1999 and unpublished data). As mentioned earlier, farmers use selection of the appropriate genotypes to optimize environmental conditions and reduce pest activity, as much as is practical. For example, farmers in the drier parts of southern Malawi, near Mwanza and Zomba, use medium-duration cultivars of pigeonpea to avoid terminal drought and high pest populations. In contrast, farmers in wetter regions further south, such as Mulanje and Kyolo, grow long-duration cultivars that mature after the warm rainy period to escape high pest populations. Intercropping of pigeonpea and maize is used throughout southern and eastern Africa, partly to reduce labor requirements by intensifying crops, enhancing returns from labor and land invested (Table 14.4). However, intercropping may also reduce pest problems through enhancing natural enemy habitat. Post-harvest techniques to reduce damage include storing seed above the fireplace, mixing the seed with chillis, wood ash, and other herbs before storage and regularly sunning the stored food grain to keep away storage pests. Another strategy pursued by farmers is to sell off grain immediately after harvesting and process it quickly, to avoid storage losses (Minja *et al.*, 1999).

Research on control of pigeonpea pests

Promising pigeonpea cultivars with pest-resistant or avoidance properties are in the process of being rigorously evaluated throughout Malawi. On-farm testing has included comparison of high yielding medium-duration pigeonpea genotypes with respective local cultivars for field pest susceptibility. Over 100 farmers in Chiradzulu and Matapwata areas of southern Malawi were involved over several years (Ritchie *et al.*, 2000a). In addition, farmers experimented with maize intercropping with long-duration genotypes of pigeonpea. In terms of field pest damage, the local farmer control variety in the medium-duration genotypes was less damaged compared with the three improved genotypes. The importance of conducting evaluation on-farm was illustrated by ICP 6927. This variety was very promising in research station trials, but it turned out to be the most susceptible in these on-farm trials by Ritchie and colleagues. Among the long-duration genotypes, the improved variety ICP 9145 had the least damage. Pest damage at Matapwata was comparatively higher than at Chiradzulu and the difference was believed to be due to different pest loads resulting from climatic differences where Matapwata is wetter that Chiradzulu (Ritchie *et al.*, 2000b).

Assessment of improved pigeonpea varieties by client farmers and other stakeholders is essential before any adoption will occur (Snapp and Silim, 2001). Some 64 farmers in Chiradzulu and Matapwata areas of southern Malawi experimented further with four long-duration pigeonpea varieties identified in earlier research. The varieties included a local cultivar that is susceptible to *Fusarium* wilt, an improved ICRISAT line ICP 9145 that was selected in Kenya and released in Malawi in 1987 for its resistance to *Fusarium* wilt (Daudi, 1994), and two high yielding improved lines from the ICRISAT eastern and southern Africa pigeonpea improvement program – ICEAPs 00040 and 00053. ICEAP 00053 had desirable seed characteristics, including a large cream-colored seed, but was rejected by farmers because it was as susceptible to *Fusarium* wilt as the local cultivar. Farmers were particularly interested in ICEAP 00040 because it has large cream seeds and many seeds per pod (5–8) compared with ICP 9145 (4–5), which improved ease of harvest. In addition, the new variety is resistant to *Fusarium* wilt and pigeonpea processors (millers) preferred it for its seed size and color (Ritchie *et al.*, 2000a). The variety has been released for cultivation by farmers in southern Malawi.

Longer-term research continues on pest-resistance in pigeonpea. Genotype screening and selection for insect pest tolerance/resistance is on-going in Kenya (S. Silim, personal communication, 2001). In addition to the development of pest-resistant varieties research is targeting the production of plant-based insecticides that can be locally produced. On-station research from Malawi (unpublished data, E. Minja) and Uganda (Silim-Nahdy, 1995) has shown that dried *Tephrosia vogelii* leaf powder at 1 kg/100 kg pigeonpea grain can provide protection from storage bruchids when incorporated with the stored grain. Field work in Kenya and Uganda (Minja, 2001) indicates that mature *Tephrosia* leaf extract may be able to reduce field pest damage on pigeonpea satisfactorily, although the pest numbers may not be reduced substantially. The phytoinsecticide product from *Tephrosia* appears to be highly unstable and rapidly degraded under most conditions. This is desirable from the environmental view point, but it makes it difficult to use effectively in non-confined settings. Control of postharvest pests may be the only practical use for this bioinsecticide.

The need to identify biocontrol agents of pod sucking and pod boring bugs and develop practical pest management technologies has been widely acknowledged, however this is a long-term research and development project (Greathead, 1989). Preliminary results from an ICRISAT/ICIPE field trial test on a locally isolated *Helicoverpa* NPV gave promising results suggesting a new control measure may be possible to protect short-duration pigeonpea. This virus is, however, specific only to the bollworm. Laboratory bioassays with *Metarrhizium* spp. have also been conducted in Kenya and are showing the potential to control sucking bugs and a field trial has just been established at the University of Nairobi research farm (E. Minja, personal communication, 2001).

Potential for adoption of integrated management of pigeonpea pests

The market context must be considered carefully when assessing the potential of IPM technologies. Farmer investment in IPM and soil fertility inputs may only be economic for cash crops that provide sufficient return (Orr *et al.*, 2000b). Involvement of processors and traders, market representatives, and input suppliers in this process is very important (Jones *et al.*, 2000). The importance of market linkages and appropriate farming system economics is illustrated by attempts to promote pest-susceptible, and high yielding, cultivars of pigeonpea in Malawi. This is an example of how farmer–researcher partnerships can evaluate and find a niche for a new crop, the short duration and high yielding pigeonpea. In semi-arid areas of southern Malawi along the lakeshore cotton and tomato are the major cash crops. These crops are vulnerable to pest infestations, and generally require farmers to invest in pest control measures. Backpack sprays and moderately intense use of pesticides are common among cotton and fresh market tomato smallholder producers, in contrast to the lack of experience among the vast majority of Malawi smallholders (Minja *et al.*, 1999).

After unsuccessful attempts to promote pest-vulnerable short duration pigeonpeas in hundreds of on-farm trials across Malawi, a targeted approach was tried along the arid lakeshore in southern Malawi (R. Jones and S. Snapp, unpublished data, 1998). Twenty cotton farmers were provided with seed of new, high-yield potential varieties of pigeonpea, that were also short season and thus produced grain at the end of the rainy season when pod pest activity was high. Farmers were encouraged to intercrop the pigeonpea varieties with cotton, with the expected benefit of dual pest control on cotton and pigeonpea. The pigeonpea proved difficult to manage alongside cotton, however farmers experimented and developed a more successful intercrop of tomato and short-duration pigeonpea, along similar lines to the cotton/pigeonpea intercrop. However, these entrepreneurial pigeonpea farmers continually raised concerns about access to market. Over the last few years traders, millers and exporters have begun to consistently purchase large amounts of pigeonpea, and farmers are beginning to invest in pest control systems and higher yielding pigeonpea varieties (Jones *et al.*, 2000).

Conclusion

Improved understanding of integrated pest management and soil health can solve challenges and improve productivity on-farm, even under the severe resource constraints that face Malawi smallholders. However, we contend this requires systems-based research and extension that involves farmers at every step in the process. We present here two case studies that explore the challenges and opportunities for integrated crop management of *Striga* in maize systems and pests in pigeonpea. Progress has been achieved in the development of market and farmer acceptable pigeonpea varieties with improved resistance to some pests and *Fusarium* wilt. Farmers are experimenting with intercrops of pest-vulnerable pigeonpea varieties and other cash crops such as tomatoes, for dual purpose pesticide management.

Legume-based intensified systems show promise as means to reduce *Striga* infestation, including the use of *Tephrosia* and pigeonpea intercrops with maize. However, combined strategies of manure or fertilizer applied to legume–maize intercrops may be necessary. Finally we stress the requirement for market integration and economic returns that justify farmer investment in pest control, and building soils for improved productivity.

References

Bentley, J.W. and Thiele, G. (1999) Bibliography: farmer knowledge and management of crop disease. *Agriculture and Human Values* 16, 75–81.

Chanika, C.S.M., Abeyasekera, S., Ritchie, J.M., Riches, C.R., Mkandawire, C.B.K., Mputeni, H., Makina, D. and Daudi, A.T. (2000) On-farm trials of technologies for the management of *Striga asiatica* in Blantyre/Shire Highlands. In: Ritchie, J.M. (ed.) *Integrated Crop Management Research in Malawi: Developing Technologies with Farmers.* University of Greenwich, Natural Resources Institute, Chatham, UK, pp. 216–225.

Daudi, A.T. (1994) Plant protection research on pigeonpea in Malawi. In: Silim, S.N., Tuwafe, S. and Laxman Singh (eds) *Pigeonpea Improvement in Eastern and Southern Africa.* Annual research planning meeting 25–27 Oct 1993, Bulawayo, Zimbabwe. ICRISAT, Patancheru, Andhra Pradesh, India, pp. 63–64.

Faris, D.G. and Singh, U. (1990) Pigeonpea nutrition and products. In: Nene, Y.L., Hall, S.D. and Sheila, V.K. (eds) *The Pigeonpea.* ICRISAT and CAB International, Wallingford, UK, pp. 401–433.

Greathead, D.J. (1989) Present possibilities for biological control of insect pests and weeds in tropical Africa. In: Yaninek, J.S. and Herren, H.R. (eds) *Biological Control: a Sustainable Solution to Crop Pest Problems in Africa.* IITA, Ibadan, Nigeria, pp. 173–194.

Heong, K.L. (1999) From-research-to-farmer practice: a case study in rice pest management. In: Balasubramanian, V., Ladha, J. and Denning, G. (eds) *Resource Management in Rice Systems.* Kluwer Academic, The Netherlands, pp. 199–211.

Heong, K.L. and Escalada, M.M. (1999) Quantifying rice farmers' pest management decisions: beliefs and subjective norms in stem borer control. *Crop Protection* 18, 315–322.

Jones, R.B., Likoswe, A. and Freeman, H.A. (2000) Improving poor farmers' access to technologies and markets for pigeonpea in Malawi. In: Ritchie, J.M. (ed.) *Integrated Crop Management Research in Malawi: Developing Technologies with Farmers.* University of Greenwich, Natural Resources Institute, Chatham, UK, pp. 150–157.

Kabambe, V.H. (1997) Studies on some agronomic approaches in the control of witchweed (*Striga asiatica* L. Kuntze) in Maize (*Zea mays* L.) in Malawi. PhD thesis, University of Reading, UK.

Kabambe, V.H. and Mloza-Banda, H. (2000) Options for management of witch weeds in cereals for smallholder farmers in Malawi. In: Ritchie, J.M. (ed.) *Integrated Crop Management Research in Malawi: Developing Technologies with Farmers.* University of Greenwich, Natural Resources Institute, Chatham, UK, pp. 210–215.

Kanyama-Phiri, G.Y., Snapp, S.S. and Minae, S. (1998) Partnership with Malawian farmers to develop organic matter technologies. *Outlook on Agriculture* 27, 167–175.

Kanyama-Phiri, G.Y., Snapp, S.S., Kamanga, B. and Wellard, K. (2000) Towards integrated soil fertility management in Malawi: incorporating participatory approaches in agricultural research. *Managing Africa's Soils* No. 11. International Institute for Environment and Development, UK. www.iied.org/drylands

Kumwenda, J.D.T., Waddington, S.R., Snapp, S.S., Jones, R.B. and Blackie, M.J. (1997) Soil fertility management in Southern Africa. In: Byerlee, D. and Eicher, C.K. (eds) *Africa's Emerging Maize Revolution.* Lynne Publishers, Boulder, Colorado, pp. 153–172.

Malawi Ministry of Agriculture and Livestock Development (1995) *Guide to Agriculture Production.* Malawi Government Press, Lilongwe, Malawi, 135 pp.

Minja, E.M. (2001) Yield losses due to field pests and integrated pest management strategies for pigeonpea – a synthesis. In: Silim, S.N., Mergeai, G. and Kimani, P.M. (eds) *Status and Potential of Pigeonpea in Eastern and Southern Africa: Proceedings of a Regional Workshop, 12–15 Sept 2000, Nairobi, Kenya.* Gembloux Agricultural University, Gembloux, Belgium and ICRISAT, Patancheru, India, pp. 48–54.

Minja, E.M., Shanower, T.G., Songa, J.M., Ong'aro, J.M., Kawonga, W.T., Mviha, P.J., Myaka, F.A., Slumpa, S., Okurut-Akol, H. and Opiyo, C. (1999) Studies of pigeonpea insect pests and their management in Kenya, Malawi, Tanzania and Uganda. *African Crop Science Journal* 7, 59–69.

Mumera, L.M. and Below, F.E. (1993) Role of nitrogen in resistance to *Striga* parasitism of maize. *Crop Science* 33, 758–763.

Myaka, F.A. (1994) Pigeonpea agronomy research in Tanzania: Recent results and future plans. In: Silim, S.N., King, S.B. and Tuwafe, S. (eds) *Improvement of pigeonpea in Eastern and Southern Africa Annual Research and Planning Meeting. Southern and Eastern Africa.* ICRISAT, Nairobi, Kenya.

Orr, A., Mwale, B., Ritchie, J.M., Lawson-McDowall, J. and Chanika, C.S.M. (2000a) *Learning and Livelihoods: the Experience of the FSIPM Project in Southern Malawi.* Natural Resources Institute, Chatham, UK, 58 pp.

Orr, A., Mwale, B. and Saiti, D. (2000b) Markets and livelihoods: lessons from farming systems research in Blantyre/Shire Highlands. In: Ritchie, J.M. (ed.) *Integrated Crop Management Research in Malawi: Developing Technologies with Farmers.* University of Greenwich, Natural Resources Institute, Chatham, UK, pp. 279–291.

Phiri, R.H., Snapp, S.S. and Kanyama-Phiri, G.Y. (1999) Soil nitrate dynamics in relation to nitrogen source and landscape position in Malawi. *Agroforestry Systems* 47, 253–262.

Riches, C.R., Shaxson, L.J., Logan, J.W.M. and Munthali, D.C. (1993) Insect and parasitic weed problems in southern Malawi and the use of farmer knowledge in the design of control measures. *Agricultural Administration Research and Extension Network Paper* No. 42. Overseas Development Institute, London, UK, 17 pp.

Ritchie, J.M., Abeyasekera, S., Chanika, C.S.M., Ross, S.J., Mkandawire, C.B.K., Maulana, T.H., Shaba, E.R. and Milanzi, T.T.K. (2000a) Assessment of improved pigeonpea varieties under smallholder management. In: Ritchie, J.M. (ed.) I*ntegrated Crop Management Research in Malawi: Developing Technologies with Farmers.* University of Greenwich, Natural Resources Institute, Chatham, UK, pp. 180–189.

Ritchie, J.M., Polaszek, A., Abeyasekera, S., Minja, E. and Mviha, P. (2000b) Pod pests and yield losses in smallholder pigeonpea in Blantyre/ Shire highlands. In: Ritchie, J.M. (ed.) *Integrated Crop Management Research in Malawi: Developing Technologies with Farmers.* University of Greenwich, Natural Resources Institute, Chatham, UK, pp. 90–101.

Rohrbach, D.D. and Snapp, S.S. (1997) *Reconnaissance Survey of Cropping Systems and Soil Fertility Management Practices in Case Study Villages of Central and Southern Malawi.* ICRISAT, Lilongwe, Malawi, mimeo, 18 pp.

Sanchez, P.A. (1999) Improved fallows come of age in the tropics. *Agroforestry Systems* 47, 3–12.

Sauerborn, J. (1991) The economic importance of the phytoparasites *Orabanche* and *Striga.*

In: Ransom, J.K., Musselman, L.J., Worsham, A.D. and Parker, C. (eds) *Proceedings of the 5th International Symposium of Parasitic Weeds.* CIMMYT, Nairobi, Kenya, pp. 137–143.

Silim, S.N., Omanga, P.A., Johansen, C., Singh Laxman and Kimani, P.M. (1994) Use of the Kenya transect for selecting and targeting adapted cultivars to appropriate production systems. In: Silim, S.N., King, S.B. and Tuwafe, S. (eds) *Improvement of Pigeonpea in Eastern and Southern Africa Annual Research Planning Meeting Report.* ICRISAT, Patencheru, India, pp. 44–54.

Silim-Nahdy, M. (1995) Biotic and abiotic factors influencing the biology and distribution of common storage pests in pigeonpea. PhD thesis, University of Reading, UK, 260 pp.

Snapp, S.S. and Silim, S.N. (2002) Farmer preferences and legume intensification for low nutrient environments. *Plant and Soil* 245, 181–192.

Snapp, S.S., Kamanga, B.C.G. and Simtowe, F. (2000) Soil fertility options for Malawi smallholders: integrated legume systems. In: Ritchie, J.M. (ed.) *Integrated Crop Management Research in Malawi: Developing Technologies with Farmers.* University of Greenwich, Natural Resources Institute, Chatham, UK, pp. 246–255.

Snapp, S.S., Rohrbach, D.D., Simtowe, F. and Freeman, H.A. (2002) Sustainable soil management options for Malawi: can smallholder farmers grow more legumes. *Agriculture, Ecosystems and Environment* 91 (1–3), 159–174.

Songa, W. and King, S.B. (1994) Pigeonpea pathology research in Kenya: progress and future research strategies. In: Silim, S.N., Tuwafe, S. and Laxman Singh (eds) *Pigeonpea Improvement in Eastern and Southern Africa* – Annual research planning meeting 25–27 Oct 1993, Bulawayo, Zimbabwe. ICRISAT, Patancheru, India, pp. 45–50.

Van de Fliert, E. and Braun, A.R. (1999) *Farmer Field School for Integrated Crop Management of Sweetpotato: Field Guides and Technical Manual.* International Potato Center, Regional Office for East, Southeast Asia and the Pacific, PO Box 929, Bogor, Indonesia, in collaboration with: Mitra Tani, Yogyakarta, Indonesia; Research Institute for legume and tuber crops, Malang, Indonesia; and Users' perspective with agricultural research and development, c/o IRRI, Philippines, 365 pp.

Chapter 15

Integrated Pest Management in South Africa

D.S. Charleston[1], Rami Kfir[1], N.J. van Rensburg[1], B.N. Barnes[2], V. Hattingh[3], D.E. Conlong[4], D.Visser[5] and G.J. Prinsloo[6]

[1]*ARC–Plant Protection Research Institute, Pretoria, South Africa;*
[2]*ARC Infruitec-Nietvoorbij, Stellenbosch, South Africa;* [3]*Citrus Research International, Nelspruit, South Africa;* [4]*SASA Experiment Station, Mount Edgecombe, South Africa;*
[5]*ARC–Vegetable and Ornamental Plant Institute, Pretoria, South Africa;*
[6]*ARC–Small Grain Institute, Bethlehem, South Africa*

Introduction

The southernmost part of the African continent, known as the Republic of South Africa, runs latitudinally from 22°S to 35°S and longitudinally from 17°E to 33°E. South Africa borders the Atlantic and Indian Oceans and shares borders with Namibia, Botswana, Zimbabwe, Mozambique, Swaziland and Lesotho. It has nine provinces, a population of approximately 43 million and a surface area of just over 1.2 million km^2, of which approximately 85% is dedicated to agriculture and forestry (van Dyk, 2000). Agricultural land is mainly used for grazing, approximately 13% is dedicated to crop production, and a small portion is used for forestry (Anonymous, 1989). The total number of plant species in South Africa exceeds 20,000, the richest assembly in the world (Anonymous, 1989).

The most important factor limiting agricultural production is the availability of water. Only 10% of the country has an annual precipitation of more than 750 mm. The country is divided into three main rainfall regions. Approximately 86% of South Africa lies within the summer rainfall area, a small belt along the south coast receives rainfall throughout the year, and the southwest of the country has a Mediterranean climate with winter rainfall, and dry summers. Almost 50% of South Africa's water is used for agricultural purposes (Anonymous, 1989). Farming in South Africa involves virtually every type of animal, crop, fruit and vegetable production adapted to weather conditions ranging from temperate to subtropical.

Agricultural Policy

In the new political dispensation after 1994 the country was divided into nine provinces (Fig. 15.1), each with its own department of agriculture. The national department of agriculture remained as the central body overseeing national interest. In 1993 the Agricultural Research Council (ARC) was established in terms of the 'Agricultural Research Act 86 of 1990'. The ARC places an emphasis on agricultural research, while the agricultural extension

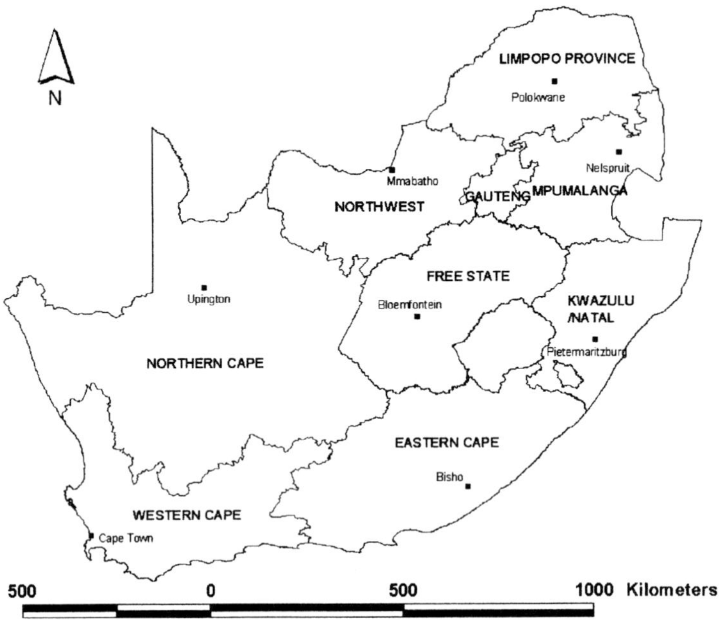

Fig. 15.1. The nine provinces of South Africa.

component remains with the department of agriculture.

Agriculture contributes about 3.4% of the total value of the South African economy, however it employs 13% of the economically active population (van Dyk, 2000). Therefore, agriculture in South Africa has a central role to play in building a strong economy, by reducing inequalities through increasing employment opportunities for the poor and by nurturing natural resources. The government has three major goals for policy reform: (i) to build an efficient and internationally competitive agricultural sector; (ii) to support the emergence of more diverse production with an increase in the number of successful smallholder farming enterprises; and (iii) to conserve agricultural natural resources and put in place policies and institutions for sustainable resource use. The role of the government is based on working as partners with others, including the private sector, farmer unions, and voluntary organizations (Simbi, 2001). Although there is still much to achieve, the past few years have seen a rapid change in the farming sector, allowing South Africa to move towards the future with confidence.

Agricultural Legislation

Much of the legislation protects the environment and promotes the use of integrated management strategies. However, there is no legislation specific to IPM, but there are various acts in place, which cover certain aspects of IPM. There are regulations governing the use of pesticides and herbicides, the importation and release of biological control agents, the improvement of plant species, the genetic modification of organisms, and the control and removal of problem plants. These regulations promote the use of IPM in agriculture in South Africa.

Pesticides and other agricultural chemicals represent a very large industry in South Africa. In terms of the law, pesticides have to be registered prior to sale. The use, sale and importation of fertilizers and chemicals in agriculture is under the control of the 'Fertilizers, Farm feeds, Agricultural Remedies and Stock Remedies Act 36 of 1947.' The Act provides for the appointment of a registrar, the registration of pest control operators, the registration of fertilizers, farm feeds, agricultural remedies (which include pesticides and herbicides), stock remedies and sterilizing plants. The Act regulates imports, sales, acquisitions, disposal and use of these products, and provides for the designation of technical advisors and analysts. Products must be registered with the registrar before they can be applied, and registration is only granted once the material has been proved to be effective for the purpose for which it is intended, and if it is in the public interest that registration be approved.

The main legislation involving biological control falls under the 'Agricultural Pests Act 36 of 1983.' The Act provides for 'measures by which agricultural pests may be prevented and combated; and for matters connected therewith'. It has particular relevance for the importation of natural enemies. The Act stipulates that live insects or other beneficial organisms may only be imported on the authority of a permit issued by the NDA, Directorate: Plant Health and Quality. Permits are only issued after consultation with the DEAT and experts in the appropriate fields. After the permit is obtained the organism must be kept in quarantine. The quarantine facility of the PPRI of the ARC is the only insect quarantine facility in South Africa. In quarantine the organism is reared for at least one generation to prevent the accidental introduction of unwanted organisms, and to ensure quality control. Once the organism has been declared clean and viable a release permit must be applied for. This is a long complicated process, which requires approval from both the NDA and also from DEAT. The release of the biological control agents is further controlled by the 'Environment Conservation Act 73 of 1989' which requires that an environmental impact assessment be carried out before the organism can be released; this often results in a delay, or may sometimes prevent, the release of biological control agents. This often impedes the implementation of IPM programs.

The 'Conservation of Agricultural Resources Act 43 of 1983' provides for 'control over the utilization of the natural agricultural resources of the Republic in order to promote the conservation of the soil, the water sources and the vegetation and the combating of weeds and invader plants; and the matters connected therewith'. This Act plays an important role in the classification and control of problem plants. The Act also lays down certain rules for the protection of biological control agents. In areas where biological control of weeds is effective, no additional control methods should be used that would destroy the biological control agents or make them less effective. Provision is made for certain areas to be set aside as biological control reserves, in these areas, no measures may be applied that would render the biological control agents ineffective. The Act also covers grazing and grazing capacity, soil conservation (i.e. the prevention of soil erosion) and conservation of water sources on agricultural land.

Plant cultivars are controlled by a variety of legislation, including: the 'Plant Improvement Act 53 of 1976', the 'Plant Breeders Rights Act of 1976' and the 'Seeds Act 28 of 1961'. These Acts require the registration of new varieties, of premises and of plant breeders. Varieties must also be evaluated and certified. Genetically modified organisms are controlled by the 'Genetically Modified Organisms Act 15 of 1997'. The Act provides for measures to promote the responsible development, production, use and application of genetically modified organisms, including the importation, production, release and distribution of these organisms.

Legislation therefore covers most aspects of IPM. Acts cover the importation and release of biological control agents, including biological control of problem plants. Pesticides, fertilizers and herbicides, as well as plant breeding and genetic modification of organisms are carefully monitored and controlled. Protection of agricultural resources including water and soil are provided for in various Acts. Therefore the legislation is already in place to promote IPM, however it is dispersed throughout the various Acts, and may be more effectively implemented if it were included in one Act that covers all aspects of IPM.

IPM Research and Practice in South Africa

History of IPM in South Africa

Before 1895 there were no full time agricultural entomologists in South Africa. Biologists were commissioned by the government of the Cape Colony to undertake entomological investigations, and S.D. Bairstow's research led to the publication of the first book on South African pests, published by E.A. Ormerod in 1889 (Annecke and Moran, 1982). C.P. Lounsbury was appointed in 1895 as the first 'Government Entomologist of the Colony of the Cape of Good Hope' (Anonymous, 1989). He became the first economic entomologist to take an interest in biological control. In the period until his retirement in 1927, economic entomology developed rapidly in South Africa (Annecke and Moran, 1982).

Biological control was initiated in the late 1800s, early 1900s. One of the first successful examples comes from the control of cottony cushion scale in citrus in 1892 (Lounsbury, 1897, unpublished). Other examples come from the control of invasive plants, initiated in 1913 with the release of cochineal insects, which resulted in virtual elimination of the invasive cactus *Opuntia vulgaris* (Hoffmann, 1991), which had become a major problem in agricultural areas. Biological control of insect pests in plantations was initiated in 1925 with control of the eucalyptus snout beetle (Anonymous, 1989). However, many pests could not be adequately controlled by action of their natural enemies alone. The challenge facing entomologists was to develop a financially attractive and ecologically sound program that integrated biological, insecticidal and other methods of control so that the greatest benefit could be derived from all components of the control program.

IPM represented a complete change in the philosophy of pest control, away from pest eradication towards pest management. Instead of a single control technique, emphasis was placed on the use of a combination of techniques aimed at providing cheap but long-term reliability with the minimum of harmful side effects (Dent, 1991). Successful IPM programs can produce many benefits including: lower production costs, reduced environmental pollution, reduced farmer and consumer risks, and ecological sustainability (Lim *et al.*, 1997). IPM is especially suited to subsistence farming in developing countries. Despite this, pesticides dominated attempts to control insect pests in most countries. Finally, the negative impact of these chemicals and the increasing pesticide resistance in a number of pest species resulted in the implementation of IPM in many crops. The earliest and most successful examples of IPM in South Africa come from citrus. IPM was initiated in citrus to control ants and thus the red scale infestations (Ulyett, 1938; Annecke, 1958; Bedford, 1968a,b). However, IPM suffered a setback in the late 1960s–early 1970s when intensive use of pesticides disrupted the system, resulting in the classical 'pesticide treadmill' (Hattingh and Tate, 1996a). Today, however, the reintroduction of IPM in most of the crops in South Africa has resulted in some of the most successful pest control programs. The government of South Africa established the PPRI in 1962 (see more details on ARC–PPRI in Appendix 15.1).

Successful Implementation of IPM in South Africa

IPM in citrus

Citrus is a major fruit crop worldwide with global production figures estimated at 90 million tonnes annually. In sub-Saharan Africa the majority of the commercial production of citrus occurs in the southern African countries of South Africa, Swaziland, Zimbabwe and Mozambique. Southern Africa produces approximately 1.5 million tons of citrus annually and exports approximately 900 million kg as fresh fruit. Although the southern African citrus industry is only the 13th largest citrus producing region globally, it is the third largest volume exporter of fresh citrus. Southern

African citrus is exported to over 60 countries, including some of the most discerning markets of the world and income from such exports accounts for 90% of the industry's income.

Before the late 1970s, the southern African citrus industry experienced an era of concentrated classical biological control effort, with great names such as Dr Eric Bedford demonstrating what phenomenal successes could be achieved with biological control. This started with the first example of the potential for classical biocontrol introductions in the form of the control of cottony cushion scale *Icerya purchasi* Maskell, effected through the introduction of the predator *Rodolia cardinalis* (Mulsant) in 1892 (Lounsbury, 1897, unpublished). Some other very successful classical biocontrol projects led to the complete control of the blackflies *Aleurocanthus woglumi* Ashby (Bedford and Thomas, 1965; Bedford, 1998) and *Aleurocanthus spiniferus* (Quaintance) (van den Berg and Greenland, 1997). This era of classical biocontrol introductions left the industry with a wide diversity of introduced biological control agents, which complement the rich diversity of indigenous biocontrol agents in the region.

The basic foundations for IPM were established in the southern African citrus industry at an early stage (Bedford, 1968a, 1979a,b; Annecke, 1969). Ant control has been a cornerstone to citrus IPM since it was promoted by Ulyett (1938), Annecke (1958) and Bedford (1968a,b). The conservation of biocontrol agents has also been an integral component of citrus IPM for many years since Searle (1963) demonstrated the negative impact of dust on the biocontrol fauna in citrus orchards. Likewise, the non-target effects of pesticides on biocontrol agents have received much attention over the years (Searle, 1961, 1965; Bedford *et al.*, 1992; Grout *et al.*, 1997; Hattingh *et al.*, 2003).

The late 1960s and early 1970s were marked by an increased usage of organophosphate pesticides, particularly in the northern production regions. In the mid-1970s the key pest, California red scale *Aonidiella aurantii* (Maskell), developed high levels of resistance to organophosphate pesticides in the northern production regions (Georgala, 1975). The crisis that ensued, necessitated the introduction of mineral spray oils as an alternative treatment, a strategy introduced to the industry by S.S. Kamburov and F. Honiball (Annecke and Moran, 1982). There were severe limitations on the times at which these products could be used, as well as quantities of product that could be applied before adverse effects on yield and fruit quality were experienced (Grout, 1993a). Reliance on spray oils for the control of red scale populations consequently necessitated the conservation of biocontrol populations. Likewise, controls applied for another key pest, citrus thrips, *Scirtothrips aurantii* Faure were also potentially disruptive to the populations of biocontrol agents for *A. aurantii* and other pests (Grout *et al.*, 1997). The situation became so critical in the northern production regions in the late 1970s that the development and implementation of IPM strategies became a prerequisite for continued economic production of citrus (Grout, 1993b; Hattingh, 1996b).

The pest control environment in the southern production regions was far simpler. The cold winters of the Mediterranean climate in the Western Cape resulted in slower development rates and populations were more synchronized than in the hotter northern regions (Grout *et al.*, 1989). This made control of the pest in these regions possible with far less reliance on pesticide usage. Likewise, the presence of an effective biocontrol agent for citrus thrips in the Eastern Cape, in the form of *Euseius addoensis addoensis* (McMurtry), reduced the need for reliance on pesticides (Grout and Richards, 1992).

In the far northern areas, including Zimbabwe, production of citrus commenced without initial heavy reliance on pesticides. This good start, together with the advantages of having a diversity of habitat structure maintained by virtue of plantations remaining relatively small and being surrounded by large tracts of indigenous vegetation (Magagula, 1998), has meant that the need for intensive chemical intervention has not arisen in these regions.

The forced adoption of IPM in the northern regions, and avoidance of a need to implement intensive chemical control practices in the far northern and southern region, ensured that IPM philosophies became an integral component of commercial citrus production in southern Africa throughout the 1980s. However, the early 1990s saw a dramatic increase in the number of thripicides introduced onto the citrus pest control market. Several of these new products proved to be highly detrimental to the biocontrol populations of other pests such as *A. aurantii* and mealybugs (Hattingh and Tate, 1996b). Organophosphate pesticides used for the control of *A. aurantii* also provided incidental control of mealybugs. The disruption caused by thripicides and the reduced use of organophosphate pesticides led to a dramatic increase in the pest status of mealybugs in the 1990s (Hattingh and Tate, 1996b).

Another negative impact on the status of IPM in the southern African citrus industry in the early 1990s came in the form of extensive use of the IGRs class of chemistry, as an alternative to spray oils for the control of *A. aurantii*. This class of chemistry was reported to have been successfully adopted into IPM practices on other crops (Ishaaya, 1990). Under the climatic conditions prevailing in Israel, these pesticides were also introduced into citrus IPM programs (Peleg, 1988). However, under southern African conditions the introduction of some of these pesticides proved to be highly detrimental to the stability of the pest–biocontrol population balance in many areas (Hattingh and Tate, 1995). The southern African citrus industry took a leading role in demonstrating the IPM incompatibility of this class of chemistry under certain conditions (Hattingh, 1996b; Grout, 1998).

The IGR experience also created an awareness in the local industry of the need to protect IPM systems developed over many years. There was a realization that a stable IPM system could be completely disrupted in one season by the use of a new pesticide that had not been adequately assessed for IPM compatibility under specific local conditions (Hattingh *et al.*, 2003). This led to the development of a standardized system for quantitatively evaluating the potential impact of new pesticides on indicator biocontrol agents of relevance to the local industry (Hattingh *et al.*, 2003). The need to introduce compulsory evaluation of these risks prior to commercialization of new products was recognized. The citrus industry developed standardized testing procedures, and in collaboration with the agricultural chemical industry, had such screening adopted into the registration process for new citrus plant protection products.

The development of effective scouting and monitoring systems for key pests provided a valuable basis for limiting chemical inputs (Reed and Rich, 1975; Newton and Mastro, 1989; Hofmeyr and Calitz, 1991; Grout *et al.*, 1998). The quantitative demonstration of the cost effectiveness of an IPM strategy (Hattingh, 1996a) highlighted the value of IPM systems to the growers. The development of a technique whereby plant systemic pesticides were applied to the stem of the tree instead of foliar sprays, made a great contribution towards enabling the realization of the potential of biocontrol agents in an IPM system (Buitendag and Bronkhorst, 1986; Buitendag and Naude, 1992). More recent developments in the strategy have entailed the development of a bio-intensive form of IPM, whereby key biocontrol agents are mass reared in insectaries and released into the orchards (Newton and Odendaal, 1990; Hattingh, 1997).

The lessons of the past led to adoption of mandatory procedures for screening the potential impact of new crop protection products and have enabled the development of systems for quantifying the level of IPM implementation by individual producers (Hattingh *et al.*, 2003). The consolidation of all these strategies into detailed pest management recommendations in the form of comprehensive Integrated Production Guidelines for Export Citrus (Grout *et al.*, 1998), made a major contribution towards the implementation of IPM practices on citrus farms. From the late 1990s to date, the southern African citrus industry has seen a reversion back to a stable IPM approach.

The high level of understanding of the intricacies of managing IPM systems that have developed in the industry over the past 30 years are paying off admirably for the industry. As the global fresh produce markets go into an over-supply status, the ability to differentiate product on the basis of socioeconomic and environmental considerations becomes ever more important. The southern African citrus industry consequently finds itself today continuing to enjoy the status of preferred supplier of fresh citrus to the world's most discerning and profitable markets.

IPM in deciduous fruit orchards

The South African deciduous fruit industry can justifiably claim to be one of the leaders in the application of IPM. This is partly because it exports to discerning international markets, which increasingly demand commodities produced under sustainable and ecologically compatible conditions. However, pest management practices have had to undergo considerable change especially during the past 50 years, as the extensive use of pesticides led to the development of pest resistance, and consumer concern for safety, health and the environment steadily increased.

Today, there is a greater realization than ever before that one of the main keys to sustainable deciduous fruit production is a reduction in synthetic pesticide usage, and, where possible, a return to more natural production methods. However, many crop protection scientists and practitioners believe that few IPM programs will develop without *some* pesticide input, particularly with high-value crops such as fruit where cosmetic perfection is still a requirement of most consumers. Nevertheless, in a successful IPM system pesticides are applied only when really needed, and using a product with the least detrimental impact on the orchard environment and in the lowest amounts that will accomplish the task.

A wide variety of IPM strategies are employed in South African deciduous fruit orchards. Effective monitoring systems are in place for many pests and diseases, allowing growers to make informed decisions on pest control interventions (Barnes, 1990). Pheromone-based mating disruption programs are effectively and economically used to control codling moth and oriental fruit moth, key pests on pome and stone fruit respectively (Barnes and Blomefield, 1997). Disruptive foliar sprays against fruit weevils on apples, nectarines and table grapes have been reduced by the use of trunk exclusion barriers (Barnes *et al.*, 1995). Trunk applications of certain systemic insecticides control root-infesting woolly apple aphid colonies through translocation. Phenology models are used to predict optimum periods for insecticide applications against oriental fruit moth and codling moth, resulting in reduced pesticide applications (Blomefield and Kleinhans, 1995; Blomefield and Barnes, 2000). Narrow-spectrum insect growth regulators and less-hazardous encapsulated pesticide formulations form part of some control and resistance management programs, particularly of codling moth (Blomefield, 1994, 1997). Low-volume bait sprays against fruit fly preclude the necessity for full-cover sprays which have a greater negative impact on natural enemies (Barnes, 1999). The use of the sterile insect technique against fruit flies has significantly reduced insecticide usage on table grapes, and other pests such as codling moth and false codling moth are being considered for control by this unique method (Barnes and Eyles, 2000). Mass-reared parasitoids are released in table grape vineyards to successfully control vine mealybug (Walton, 2000). Fluorescent lights placed around stone fruit orchards repel nocturnal invasions of fruit-piercing moths (Whitehead and Rust, 1972). In addition, sound orchard and vineyard management practices such as weed management, pruning management and orchard sanitation, help to reduce the pest infestation pressure (Blomefield, 1991; Riedl *et al.*, 1998).

As a result of these practices, increasing numbers of beneficial arthropod species survive in deciduous fruit orchards. One of the most striking examples has been the decrease in the status of phytophagous mites in pome fruit orchards. Until about 5 years ago, miticides were the most expensive item on a fruit grower's pesticide shopping list. Today, a complex of predators of phytophagous mites keep their populations below economic injury levels, and the use of miticides has radically decreased.

IPM in deciduous fruit orchards was given a considerable boost in the 1990s by the implementation of an integrated fruit production (IFP) initiative. Through this approach, fruit growers are guided towards a mature and responsible agricultural community by requiring them continuously to apply the latest technology in order to protect, conserve and improve the soil and the ecology, and to place the health and welfare of people above all else. The responsible use of IPM practices plays a major part in the IFP scheme. All registered agrochemicals are coded into four categories of acceptance according to their fate in soil and water, biodegradation, toxicological profile and ecotoxicology. Restrictions are placed on the use of certain pesticides. This system allows fruit growers to select spray programs having the least negative impact on the environment, significantly promoting IPM in deciduous fruit orchards.

IPM of cereal stem borers

Maize is the most important agricultural crop grown in South Africa. It is grown on 4 million ha with an annual production of 12 million tons. Sorghum is also an important crop, especially in the marginal grain producing areas, because of its drought resistant properties (van Rensburg and van Hamburg, 1975). It is grown on 300,000 ha with a yield of 0.5 million tons. The most important pests of cereal crops in South Africa are the indigenous African maize stem borer, *Busseola fusca* (Fuller) (Lepidoptera: Noctuidae), and the exotic spotted stem borer, *Chilo partellus* (Swinhoe) (Lepidoptera: Crambidae) (Skoroszewski and van Hamburg, 1987).

The profit margin for maize or sorghum grown in South Africa is relatively low due to high production costs; pest control alone may amount to 56% of the gross margin above cost for an average yield (van Hamburg, 1987). The timing of insecticidal application is crucial, as control measures are effective against young borer larvae only. Older larvae penetrate the stalks and are difficult to control with insecticides. Estimated yield losses from borer damage range between 10% and total loss (van Rensburg and Bate, 1987). In a study on the effect of stem borers on growth and yield of maize the economic threshold for modern pesticidal application was determined as 40% plants showing visible damage (Bate and van Rensburg, 1992).

Crop residues are important for carrying over diapausing stem borer larval populations from one growing season to the next (Kfir, 1988). Ploughing in order to bury maize stubble was an effective control measure used in South Africa to control *B. fusca* early in the past century (Mally, 1920). It was shown in South Africa that slashing maize and sorghum stubble destroyed 70% of *C. partellus* and *B. fusca* populations, and that ploughing and disking destroyed a further 24% of the pest population in sorghum and 19% in maize (Kfir *et al.*, 1989; Kfir, 1990a). Currently this system is not practiced in South Africa due to the advent of minimum tillage and the importance of winter grazing of maize to beef farmers (Kfir, 1992a). Manipulation of sowing dates is used to avoid severe borer infestations. In the Highveld region of South Africa, the second-generation population of *B. fusca* is larger and can cause more damage than the first generation (van Rensburg *et al.*, 1988). Farmers sow maize early in the growing season to avoid severe damage. In the lower elevations of South Africa, it is recommended that sorghum be planted after mid-October to avoid infestation from the first moth peak of *C. partellus* (van Hamburg, 1979).

Based on work on *B. fusca* by Revington (1987), pheromone traps for commercial use were developed in South Africa. Currently, omni-directional pheromone traps are commercially available and are widely used by farmers to determine the timing of insecticidal applications against *B. fusca*.

The accepted action threshold level for spraying against *B. fusca* is when the average catch from three traps per site exceeds two moths per week for 4 consecutive weeks.

There were early attempts in South Africa to develop maize cultivars resistant to *B. fusca* (Du Plessis and Lea, 1943; Walters, 1974). Several lines of maize with some resistance to first-generation (whorl-feeding) *B. fusca* larvae were identified (Barrow, 1989). Van den Berg *et al.* (1994) showed increased insecticide efficacy against stem borers in South Africa in grain sorghum lines showing some resistance. The resistance mechanism that is based on antibiosis and antixenosis apparently causes stress in *C. partellus* and *B. fusca*, making them more susceptible to the insecticides (van den Berg *et al.*, 1994).

Early studies on the natural enemies of *B. fusca* in South Africa by Mally (1920) and by Du Plessis and Lea (1943), identified the gregarious larval parasitoid, *Cotesia sesamiae* (Cameron) (Braconidae), as an important mortality factor. In more recent studies the parasitoids of *B. fusca* (Kfir, 1990b; Kfir and Bell, 1993) and *C. partellus* (Kfir, 1992b, 1995) were identified, as well as the predators (Watmough and Kfir, 1995) and the microbial pathogens (Hoekstra and Kfir, 1997). However, despite high mortality levels natural enemies were not able to prevent economically significant damage (Kfir and Bell, 1993). This factor and the inefficacy of insecticides to control stem borers in South Africa precipitated a biological control program using exotic parasitoids. Parasitoids were introduced from Pakistan, Mauritius, Indonesia, Trinidad, Brazil, Taiwan and the Philippines with emphasis on *Cotesia flavipes* (Cameron) from Pakistan (Skoroszewski and van Hamburg, 1987; Kfir, 1991, 1992a, 1994).

Several Bt-transgenic maize hybrids are commercially available in South Africa. During the 1998/99 growing season approximately 50,000 ha of irrigated land was planted with Bt-transgenic maize hybrids, all employing event Mon810 (van Rensburg, 2001). As the price of Bt-maize seed is still substantially higher than regular hybrids it is mainly planted in irrigated land. A price reduction will no doubt result in farmers planting Bt-maize also in rain-fed lands.

Pesticide applications are still the first option of control for cereal farmers in South Africa. Commercial farmers do not practice cultural control and the stubble, which harbors the hibernating stem borer larval populations stays in the fields in winter.

However, the introduction of synthetic sex pheromones, accurate economic threshold levels for chemical control, Bt-transgenic maize and resistant cultivars were important steps in the implementation of an IPM program in cereal production in South Africa.

IPM in sugarcane

George Morewood, in 1848, planted the first sugarcane in South Africa on the Compensation Flats, north of Durban in KwaZulu-Natal (KZN) (Osborne, 1964). From that small start with sugarcane imported from Mauritius (Osborne, 1964) has grown an industry, which, in 2001/02, has 426,597 ha under sugarcane (Anonymous, 2001). Rain-fed sugarcane is now grown along the east coast of South Africa, from just south of Umzimkulu in southern KZN to Mkuze in northern KZN, and inland of this area to just west of Pietermaritzburg. Irrigated sugarcane is grown in the Pongola area of KZN, and the Malelane/Komatipoort areas of Mpumalanga (Anonymous, 2001).

By 1950, southern African sugarcane had at least 33 species of indigenous insects feeding on it (Dick, 1950). Subsequent additions to this list are *Numicia viridis* Muir (Homoptera: Tropiduchidae) and an unidentified species of *Margarodes* (Homoptera: Margarodidae) (Leslie, 1986). *Perkinsiella saccharicida* Kirkaldy (Homoptera: Delphacidae) is the only exotic pest recorded (Carnegie, personal communication, cited in Conlong, 1994a). However, an exotic borer, *Chilo sacchariphagus* Bojer (Lepidoptera: Pyralidae), has recently been confirmed from Mozambiquan sugarcane, but has not yet been found in South Africa (Way and Turner, 1999).

Fortunately, very few of these insects have become major pests in the industry. During the 1880s it was reported that the Uba variety of sugarcane was little damaged by white ants or the 'borer' (Osborne, 1964). The identity of the borer was, however, unknown. In 1894 locust plagues appeared and devastated many cane fields around Durban (40% of sugarcane crop destroyed) (Osborne, 1964; Anonymous, 2001). This was the first report of identified insect damage in southern African sugarcane. Only two other insect species have caused serious damage to South African sugarcane, necessitating remedial control measures. These are *Numicia viridis* Muir (Hemiptera: Tropiduchidae) and *Eldana saccharina* Walker (Lepidoptera: Pyralidae), both of which are indigenous.

The leaf hopper, *N. viridis*, became of economic importance in 1962, when it was recorded on sugarcane in Swaziland and South Africa, particularly in inland irrigated fields (Carnegie, 1966). Most damage is caused by the developing nymphs, of which there are five instars. Biological studies showed that it had invaded sugarcane fields from adjacent wild grasses and sedges, which were its natural host plants (Carnegie, 1967, 1994). There is evidence that it did so unaccompanied by its two efficient egg parasitoids, *Ootetrastichus beatus* Perkins (Hymenoptera: Eulophidae) and *Oligosita* sp. (Hymenoptera: Trichogrammatidae), which control it well in wild grasses (Carnegie, 1975, 1980, 1994). Both parasitoids subsequently became common in sugarcane, and *N. viridis* ceased to be a continuing problem during the early 1970s. Occasional minor outbreaks now occur. Other parasitoids recorded from this insect include two Dryinidae, *Dryinus* sp. and *Lestrodryinus* sp. An epipyropid (*Epipyrops* sp.) attacked *N. viridis* when infestations became severe (Carnegie, 1975).

There are three well synchronized generations of *N. viridis* per year (Carnegie, 1969), and control measures can be timed appropriately. Mercaptothion (malathion) and endosulfan dusts, applied aerially give effective control in severe infestations (Carnegie, 1971, 1994). However, during normal years, the indigenous parasitoids maintain the populations of *N. viridis* below economic threshold levels, eliminating the need for pesticide applications.

The stem borer *Eldana saccharina* was first reported as a pest of South African sugarcane from the Umfolosi Flats of KZN in 1939 (Naude, 1940), where it attacked 20-month-old cane of the variety POJ 2725 (Dick, 1945). This variety was known to be relatively soft (Carnegie, 1974). The cane in question had a quota restriction imposed on it, which prevented it from being sent to the mill at the younger correct age (Naude, 1940). Subsequent knowledge has shown that older sugarcane is particularly vulnerable to *E. saccharina* (Carnegie, 1981).

Initially it was thought that *E. saccharina* was a recent introduction into southern Africa (Dick, 1945), but records from indigenous plants from various west and east African countries, and KZN soon revealed it was indigenous to Africa (Dick, 1945; Carnegie, 1974; Atkinson, 1980). It is thus apparent that this borer is a fairly recent invader of sugarcane in South Africa, being recorded as a localized pest on the Umfolosi Flats for the first time in 1939, 26 years after sugarcane was first planted there, and 91 years after this crop was first grown commercially in KZN (Conlong, 1994a). By 1954, however, it had disappeared, a number of reasons being given for this (Carnegie, 1974). The current infestation of southern African sugarcane began in 1970. Its spread through the southern African industry is summarized by Carnegie (1974) and Atkinson *et al.* (1981). Currently it is found in all parts of the industry.

Control of *E. saccharina* is multifaceted. Initial research advances and recommendations are summarized by Carnegie (1981). By having knowledge of its biology, early field management control measures were developed and recommended to growers. These include cutting the cane early, applying reduced nitrogen levels, good field hygiene, pre-trashing and selection of varieties (Carnegie, 1981). The variety selection program has developed well, and varieties such as N21 are showing resistance to this borer, and have been released to the industry to plant in areas where *E. saccharina* is a constant problem (Leslie and Keeping, 1997). A biological control program has been in operation since 1981. Conlong (1994b, 1997) reviews its progress and constraints since its inception. Recent research has concentrated on the relocation of indigenous parasitoids of *E. saccharina* from various habitats in other parts of Africa to South Africa (Conlong, 2000; Conlong and Mugalula, 2001), and habitat management practices to lure known indigenous local parasitoids out of the indigenous wetland habitats (Conlong, 1990) into sugarcane affected by *E. saccharina* (Conlong and Kasl, 2000, 2001). Pesticide research has concentrated on the application of various pesticides at periods of moth abundance, so as to target the dispersing neonate larvae before they move behind the protective dry leaf sheaths and into the stalk itself (Leslie, 1997, 2001; Leslie and Keeping, 1997).

In years of average and above average rainfall, *E. saccharina* is not a serious pest in sugarcane. However, in years of drought, and in plant stress situations it does become a serious problem. Farmers who have planted the more resistant cultivars of sugarcane, and who implement the published management control recommendations (Carnegie, 1981) to their fields are less affected by this borer than those who do not. It is thus clear that a combination of resistant varieties and management control options, when implemented do offer some measure of protection against *E. saccharina*. This measure of control will be enhanced by fully researched biological control and habitat management options, in addition to well-researched and judicious use of pesticides, only at the times when susceptible stages of the pest will be exposed to pesticide treatment.

IPM in cotton

The so-called American 'upland' cotton, *Gossypium hirsutum* is grown in South Africa on about 100,000 ha with an annual production of about 115,000 tons of seed cotton. The main factor limiting cultivation in South Africa is the high cost of production, and a major input is attributed to pest control. One spray against *Helicoverpa armigera* (Hubner) (Lepidoptera: Noctuidae), a key pest of cotton in South Africa, can amount to 11% of the profit of a dryland cotton farmer under normal circumstances (Anonymous, 1992).

Early in the past century damage by leafhoppers was neutralized by breeding more hairy cultivars of cotton from the imported base stock (Annecke and Moran, 1982). The principle was established that it is essential to destroy all remnants of the previous season's cotton crop along with any ratooning or volunteer plants before any new plantings were made in order to prevent the carry-over of pests from the previous season. With the advent of the cheap and broad-spectrum pesticides it became normal practice to spray mainly preventatively up to 15 times per season (van Hamburg and Guest, 1997). As a result, *H. armigera* populations became resistant to insecticides, which led to more frequent spraying and ever escalating costs (Witlock 1973; Anonymous, 1992). In addition a previously minor pest, red spider mite (*Tetranychus cinnabarinus* (Boisduval)), became a major pest because its natural enemies had been destroyed by the frequent pesticide treatments (Botha, 1990).

Since 1975, a spray program against *H. armigera* based on scouting for eggs was developed. Infested cotton was sprayed wherever an average of 0.5 or more eggs per plant was found. This led to a reduction in the average number of insecticidal applications from 15 when sprayed preventatively to eight when sprayed according to egg counts (van Hamburg and Kfir, 1982). Later it was shown that egg population is not the best indicator of the damaging larval population, owing to egg inviability and egg parasitism and predation (van Hamburg, 1981). A new scouting method based on larval counts was developed whereby only 2.3 sprays per season were needed if sprayed according to a larval index of eight larvae per 12 plants, without any decline in yield (Kfir and van Hamburg, 1983). This new system resulted in a 60% reduction of pest control costs.

A biological control program was initiated and several egg parasitoids, *Trichogramma* spp., and the larval parasitoid, *Cotesia kazak*, were introduced into South Africa. The importance of natural enemies has been widely recognized and to protect them restrictions were imposed on the use of certain insecticides harmful to natural enemies, including the prohibition of synthetic pyrethroids on cotton less than 12 weeks old.

Bt-transgenic cotton has been commercially available in South Africa since 1999. In general about 30% of cotton planted in South Africa is Bt cotton, but in some regions such as the Loskop Irrigation Scheme about 50% is Bt cotton. With a 'technology fee' imposed by the seed companies, the local price for Bt cotton seed is about four times the price of regular seed.

The introduction of scouting for insect pests, establishment of economic threshold levels, Bt-transgenic cotton, and imposing restrictions on the use of pesticides harmful to natural enemies brought about a great reduction in the number of pesticidal applications and important progress in implementing an IPM program in cotton production in South Africa.

IPM in potatoes

The potato tuber moth, *Phthorimaea operculella*, is the most important pest on potatoes in South Africa. It is responsible for losses of up to R40 million/annum. Tubers are attacked both in the field and in stores. Studies on the integrated control of the potato tuber moth have been done, mostly in the 1960s and 1970s (Broodryk, 1967; Zimmermann, 1967; Findlay, 1975). A new study is currently in progress by the Agricultural Research Council (ARC). Detailed studies on the biology of indigenous parasitoids have been done (Broodryk 1969; 1971). Before the mid-1960s when research began in South Africa on the biological control of tuber moth, there were a number of parasitoids which sometimes collectively attained high rates of parasitism (Annecke and Moran, 1982). They included the braconids, *Chelonus curvimaculatus* and *Orgilus parcus*, and the ichneumonids, *Diadegma molliplum* and *Temelucha picta*. Despite the fact that these parasitoids attained a percentage parasitism of 80% or higher during the season, losses were still very high (Annecke and Moran, 1982). From 1965 to 1969 two species of parasitoids, imported from South America, were released in South Africa (Watmough *et al.*, 1973). These two parasitoids, *Copidosoma koehleri* and *Apanteles subandinus*, replaced the indigenous parasitoids (Findlay, 1975). During 1978, 50% fewer 15 kg potato bags were rejected at the Pretoria market than would have been expected before the release of the two mentioned parasitoids (Annecke and Moran, 1982). In one field experiment done in 1994 where no insecticides were used, *C. koehleri* and *A. subandinus* were responsible for 86% parasitism at the end of the season, while all other parasitoids together could only reach 4.5% (D. Visser, in preparation). In the same fields where insecticides were used, total parasitism decreased to 23%. The precise effect and value of these imported parasitoids are extremely difficult to calculate. Parasitoids will only become valuable as a component of an IPM program when softer insecticides are more freely available. Parasitoids may be more advantageous to the small-scale farmer because they often do not have access to expensive chemical insecticides.

Potato tuber moth pheromones were registered for commercial use in South Africa in 1995. Since then some commercial farmers have started to use them to calculate thresholds for the timing of insecticide sprays. Most farmers, however, still start spraying insecticides for tuber moth control as soon as they see the first moth in their fields. Thresholds and parasitoids are therefore currently not part of the decision-making process of the majority of potato farmers.

A granulosis virus of the potato tuber moth was discovered and described in South Africa (Broodryk and Pretorius, 1974). Although no field tests have been performed yet, a current study has shown that the virus remains active at –20°C for at least 9 years (D. Visser, in preparation). Extensive tests to evaluate this virus as an IPM tool in stored potatoes are in progress.

The first trial with Bt-transgenic potatoes with tuber moth resistance in South Africa was conducted by First Potato Dynamics during 2000. Four potato cultivars were transformed by Vitality Biotechnologies in Israel and tested in the Ceres area of the Western Cape. Resistance was absolute, expressed in both the foliage and tubers. New trials are planned in the future, which would include transformations of all the major commercial potato cultivars planted in South Africa.

Other IPM components currently under investigation by the Agricultural Research Council for potato tuber moth control include: crude aqueous extracts of the syringa tree, *Melia azedarach*, as a bio-insecticide; the effect of all registered insecticides against tuber moth parasitoids; and the potential of pheromones as a mating disruption technique in potato stores.

IPM in wheat

Wheat is produced in South Africa in the winter rainfall regions of the Western Cape and the summer rainfall areas of the Free State Province (FSP), Northern Cape, North West and the Limpopo Province. The FSP is the largest production area, and is subject to considerable annual fluctuations due to variable annual rainfall patterns (Marasas *et al.*, 1997). Russian wheat aphid (RWA), *Diuraphis noxia* (Kurdujmov, 1913) (Homoptera: Aphididae) is the most devastating of several insect pests of wheat in South Africa (Prinsloo *et al.*, 1999). Since its detection in 1978 RWA spread rapidly through the country causing yield loss of up to 90% on susceptible cultivars when not treated with chemical insecticides (Aalbersberg, 1987).

The outbreak of RWA in South Africa and other countries probably resulted from the spread of an aggressive biotype of the aphid, which is controlled by stabilizing selection, causing it to become a sporadically conspicuous pest only in its native areas (Kovalev *et al.*, 1991). The absence of successful natural enemies in South Africa is probably a prime reason for population explosions occurring regularly in South Africa (Aalbersberg *et al.*, 1989).

Initially a chemical control program was developed to prevent high yield loss. Economic injury levels and thresholds were determined to encourage farmers to spray only once, however some farmers tended to spray insecticides routinely (Du Toit, 1986, 1990). Between 1980 and 1990 commercial farmers in the Free State sprayed up to four times annually. With the exception of two (imidacloprid and thiamethoxam), all insecticides registered for the control of RWA in South Africa are broad-spectrum systemic and contact organophosphates (LD_{50} 2–70 mg/kg) (Nel *et al.*, 1999) also killing natural enemies that attack RWA. The possible development of insecticide resistance could not be ignored especially when farmers were spraying RWA on a routine basis. As a cultural practice farmers were recommended to eradicate volunteer wheat and *Bromus* spp. grass as a method of preventing infestation early after crop emergence.

The Small Grain Institute of the ARC in South Africa (ARC–SGI) started an investigation into the development of a more sustainable integrated control program for RWA during the early 1980s. In the context of sustainable pest management for RWA, host plant resistance and biological control seemed the most suitable alternative control methods.

Several wheat lines resistant to RWA were identified (Du Toit, 1987; Harvey and Martin, 1990; Smith *et al.*, 1991) and a breeding program to incorporate resistance into good quality bread wheat cultivars was started. The first resistant cultivar Tugela-Dn was released during 1993. To date, the ARC–SGI and other seed companies have released to farmers in the FSP 15 cultivars containing different levels of plant resistance (ARC–Small Grain Institute, 2000). More than 70% of the wheat farmers in the eastern parts of the FSP now plant resistant cultivars with good results and insecticide treatment has decreased by 35.8% between 1990 and 1996 (Marasas *et al.*, 1997).

Although biological control of aphids seems to be limited to a few cases in the world (van Lenteren, 1991), RWA seems to be a pest that could be controlled through classical biological control as defined by DeBach (1979). This means that it is typically an insect that invaded a new area without its effective natural enemies and became a pest, and therefore could be controlled through the introduction of natural enemies from the countries of origin of the pest. Although several natural enemies including ladybirds and parasitoids attack RWA in South Africa, they are not able to protect the susceptible cultivars from damage (Aalbersberg *et al.*, 1988). Therefore the introduction of natural enemies was started during 1980. Between 1980 and 1994 six natural enemy species were introduced and released (*Adalia bipunctata*, *Hippodamia convergens*, *Coleomegilla maculata*, *Leucopis ninae*, *Aphidius maticariae*, *Aphelinus hordei*). Two species namely the ladybird *A. bipunctata* and the parasitoid *A. matricariae* have become established, although not very actively seen on aphid populations in wheat (G.J. Prinsloo, unpublished data). The parasitoid *Aphelinus hordei* spread rapidly and established in the mountainous northeastern parts of Lesotho about 200 km from where it was released. Here they were found parasitizing *D. noxia* on wheat during 1999 and 2000 (Prinsloo *et al.*, 2002).

The parasitoid *Aphelinus hordei* was tested in the field during 1998 and 1999 for compatibility with resistant cultivars (G.J. Prinsloo, unpublished data). RWA infestation was reduced by approximately 50% on both susceptible and resistant cultivars where *A. hordei* was released compared with the plots where *A. hordei* was absent. This parasitoid shows good potential as a biological control agent of *D. noxia* on both susceptible and resistant cultivars.

Recently the use of entomopathogenic fungi for the control of RWA and other wheat aphids was initiated. These pathogens could not only be used against aphids but could replace chemical insecticides, which are necessary to control other sporadic pests and therefore will enhance the sustainability of insect pest control in wheat in future.

The RWA control program has changed dramatically since the release of RWA resistant cultivars. As more farmers changed to these cultivars the use of organophosphorus sprays dropped dramatically. Where farmers are still planting susceptible cultivars, chemicals such as imidacloprid became more popular because small amounts are applied to the seed, and the cost of application is lower. The use of seed dressing is also more environmentally friendly, giving natural enemies a better chance of survival. The use of natural control methods, e.g. plant resistance, entomopathogenic fungi, predators, parasitoids and fungi will become more important in future.

IPM in forestry

The forestry industry in South Africa is more than 120 years old and is based almost exclusively on exotic tree species. The industry comprises about 1.5 million ha of large, fast-growing monocultures of *Pinus*, *Eucalyptus* and Australian *Acacia* (wattle) species. Such man-made monocultures are of course particularly susceptible to pest attack and South Africa was no exception. Through the years a number of serious insect pests arrived on the scene. These could be divided in two groups, exotic pests from the countries where the tree species originated from, and local insects that adapted to exploit the abundant food source provided by the large, healthy plantings of exotics (van Rensburg, 1984). The local pests are mainly species of Lepidoptera leaf feeders and subterranean fungus-growing termites, which cause losses during first rotation establishment of eucalyptus and wattle in grasslands.

The approach to pest control in South African forestry plantations was based on bio-intensive pest management from the beginning. Biological control played a major role, initially because modern pesticides or the means of application (aerial spraying) were not available (Tooke, 1953), but later, because of an appreciation of the ecologically sensitive nature of most of our forestry areas. Some of our major exotic pests were almost completely eliminated by the introduction of exotic natural enemies (Kirsten *et al.*, 2000). Others were dealt with through a combination of biological control and silvicultural procedures, while for some, we had to rely on chemical treatments. The latter group included the indigenous defoliators of pine trees and subterranean termites during eucalyptus establishment. For these pests our R & D efforts were focused on finding the safest formulations and application methods, both in terms of human safety and for the environment. The IPM approach followed in South Africa can best be described by looking at a few case studies.

The eucalyptus snout beetle, *Gonipterus scutellatus*, an Australian curculionid, appeared in South Africa early in the 20th century. Most eucalyptus species are attacked but some are more susceptible to beetle damage than others. Both the adults and the larvae feed on the growing tips, causing severe stunting or even die-back. By 1926 the eucalyptus industry was practically on its knees. This situation was rapidly turned around by the introduction of an egg parasitoid, *Anaphes nitens*, which quickly brought the pest under control in most areas (Tooke, 1953). The exception was for particularly susceptible eucalyptus species such as *Eucalyptus dunii*, *E. nitens*, *E. maideni*, *E. viminalis*, *E. globulus* and *E. punctata*, planted above 1200 m. Today, some of these susceptible species are still cultivated for oil production at high altitudes and here problems with the spring generation of the snout beetle persists. Chemical treatments are sometimes used but it is not generally prescribed because of the pesticide's interference with the egg parasitoid, which becomes very effective from early summer onwards.

Before its successful biological control, the black pine aphid, *Cinara cronartii*, was considered the most serious pest of pine trees in South Africa (van Rensburg, 1979). It is indigenous in the eastern USA, where it occurs mainly on galls of the fusiform rust fungus on pine trees. It was first observed in South Africa in 1974. It attacks all pine trees grown commercially in South Africa. The aphid extracts large amounts of sap from the host tree and the copious honeydew that it secretes gives heavily attacked trees a black appearance from the sooty mould fungal growth. When trees are under drought stress, heavy infestations may kill tops or even the whole tree (van Rensburg, 1981). Owing to the high reproductive rate of the aphid, the cost of aerial spraying and the ecologically sensitive nature of most forestry areas, chemical control was not considered feasible. A parasitic wasp, *Pauesia cinaravora*, was introduced in 1983 (Kfir *et al.*, 1985; van Rensburg, 1988) and has been effective in suppressing black pine aphid numbers.

The eucalyptus tortoise beetle, *Trachymela tincticollis* is endemic to southwestern Australia. It appeared in South Africa during the late 1970s. Both the adults and the larvae feed on the young leaves and shoots of eucalyptus, causing defoliation from the top downwards, severely stunting and even killing the host trees. The tortoise beetle, which threatened to become as serious a pest of eucalypts as the snout beetle had been, was brought firmly under control by an egg parasitoid, *Enoggera reticulata*, introduced from southwestern Australia in (1987) (Tribe, 2000; Tribe and Cillié, 2000). This happened in time to prevent the pest from spreading from the Cape to the main eucalyptus areas of the country.

The sirex woodwasp, *Sirex noctilio*, hails from Eurasia and North Africa and is perhaps the most notorious of all pests of pine trees in the world. Sirex appeared in Cape Town, South Africa, in the early 1990s at a time when the pine plantations were susceptible to sirex attack after several years of drought. Stressed trees are killed when the wasp injects mucus and a fungus (*Amylostereum areolatum*) into the wood during oviposition. The mucus causes the stomata to open and the tree to wilt, allowing the dry rot fungus to germinate and spread within the tree. The larvae, which feed on the fungus, make tunnels in the wood. A tree killed by sirex is a total loss (Tribe, 1995).

Chemical control of sirex is not feasible. Maintaining the vigour of plantation trees and biological control, which was successful against sirex in Australasia, are the best options. In South Africa, *Deladenus siricidicola*, a parasitic nematode, obtained under license from the CSIRO in Australia, was injected into sirex-infested trees at Tokai, Cape Town in 1996 and 1997. Three years later the parasitism rate of emerging sirex was over 90%. The nematode makes the females sterile and is spread when the female attempts to lay eggs (Kirsten *et al.*, 2000).

Two exotic species of bark beetles, from European/Asian origin, sporadically damage commercial pine stands in South Africa. The pine bark beetle, *Hylastes angustatus*, which appeared in 1938, is troublesome during the establishment of plantations. *Hylastes* breeds in the trunks and roots of pine. In clear-felled areas the abundance of such breeding spaces leads to enormous population build-ups. Newly planted seedlings are attacked and ring barked by female beetles that need the fresh cambium material for egg development (Webb, 1974). Before being banned, preventative chemical treatments with chlorinated hydrocarbons produced some results, but the applications had to be very well timed to be effective, which was often not the case. The best way to avoid *Hylastes* attack is to time plantings to avoid peak numbers of the beetle. Planting during good rains also ensures healthy seedlings able to survive attack.

The European bark beetle, *Orthotomicus erosus*, arrived in 1968. It attacks trees under stress such as during periods of severe drought or after fire. Large numbers of fan-shaped tunnel systems in the inner bark leads to eventual ring-barking. Timber quality is reduced by the introduction of blue-stain fungi into the wood. The beetle is kept under control by maintaining good forest hygiene, denying it opportunities for population build-ups. Good silvicultural practices aimed at maintaining maximum tree vigor are also essential. Beetle numbers can be monitored by using radiator-traps baited with the commercial pheromone, Pheroprax (Tribe, 1991).

Several indigenous Lepidoptera species have adapted to feeding on pine trees in South African plantations. The most important of these defoliators are the pine emperor moth, *Imbrasia cytheria* (Saturnidae), the pine brown tail moth, *Euproctis terminalis* (Lymantridae) and the Cape lappet moth, *Pachypasa capensis* (Lasiocampidae) (Webb, 1974). These moths, despite being heavily attacked by their natural enemies, sporadically become so abundant that the trees are totally defoliated, which results in slower growth and even death of trees. Three synthetic pyrethroids and one biological insecticide are registered for aerial application against the pine emperor moth but because of the presence of natural enemies, the use of the biological insecticide *Bacillus thuringiensis* var. *kurstaki*, is recommended. Silvicultural procedures to limit the effect of defoliation include forest hygiene and thinning the stands to reduce physiological stress levels in the trees (Kirsten *et al.*, 2000).

Eucalypts and wattle are attacked by termites, but pines are not. Transplants are ring-barked or whittled to a point just below the root collar. Once the canopy has closed termites are seldom a problem, and rarely attack second rotations. There are resistant eucalyptus selections suitable for planting in some areas. Chlordane and carbosulfan for application in the planting holes, are registered for use when needed (Atkinson, 1997).

IPM in forestry is implemented in most areas, and has had tremendous success. IPM practices in forestry mainly involve the use of biological control agents, silvicultural practices and correct timing of planting. Chemicals are only used when necessary. This combination of a variety of control methods provides a good example of IPM.

Constraints Facing IPM in South Africa

While the South African government is committed to the promotion of IPM there are still a number of constraints facing its implementation. There is no specific legislation in place governing IPM, and the various pieces of legislation that cover some of the aspects of IPM, are often too prohibitive. Insufficient extension workers and the lack of training and advice on IPM strategies hamper implementation. Farmers and growers are sometimes slow to adopt IPM due to technical, attitudinal, educational, management and economic barriers. As a result, it often takes a pest management crisis before a high level of IPM is widely accepted. However, as the public demands more change from agriculture, it must be prepared to support the research that can trigger these changes, and unfortunately the funding is often inadequate. More conventional methods, such as pesticide application, are often less expensive and easier to implement/apply than IPM technologies, and the new technology is more complicated than the older more conventional methods. Often by reducing pesticide input and relying on IPM the farmer/grower will initially have higher pest levels, and this is not acceptable to most. Finally, the AIDS pandemic may affect the manual labor available in rural communities, therefore less labor-intensive, but affordable plant protection practices must be put in place (van Dyk, 2000).

Final Conclusions

South African agriculture developed from modest beginnings. During the past it suffered many setbacks, which at times threatened to destroy agriculture, but through these problems, knowledge and experience were gained which allow the South African agricultural scientist to compete with leading experts throughout the world. As part of the African continent, South Africa has a significant advantage over other nations in respect of technical knowledge, as well as social and cultural information. Therefore, South Africa's strategic position at the southern tip of Africa enables it to make a big contribution towards agricultural development in general.

IPM has had a long and mostly successful history in South Africa and is currently practiced in many crops throughout the country. Current political climate emphasizes social upliftment and the support of small-scale rural farming communities. In light of this, IPM has many benefits to offer, by reducing the reliance on expensive chemical pesticides, which many small-scale rural farmers cannot afford and which are hazardous to their health. Food security is also vital in a country where an estimated 16 million people are living in poverty (Simbi, 2001). Emphasis is placed on food security at the household level, increasing the production on small-scale farms will improve the availability and nutritional value of food, particularly if IPM strategies are used to grow the crops. In commercial farming, IPM plays a major role in producing export crops, which must meet international standards. The increasing demand for agricultural commodities with minimal pesticide residues and consumer demand for 'ecologically friendly' products means that IPM technology will continue to be implemented and improved. Other countries on the African continent can benefit from the South African experiences and expertise that is available in South Africa.

Important Agricultural Websites

Sites of main research organizations

Agricultural Research Council: www.arc.agric.za
Citrus Research International: www.cri.co.za
South African Sugar Experimental Station: www.sasa.org.za/sasex/
Council for Scientific and Industrial Research (CSIR): www.csir.co.za
Entomological Society: www.up.ac.za/academic/entomological-society/entsoc.html

Links to museums and other entomology sites

www.nfi.org.za/inverts/Insectlinks.html

Government sites

National Department of Agriculture: www.nda.agric.za
Parliament of South Africa: www.parliament.gov.za
Department of Environmental Affairs and Tourism: www.environment.gov.za/
Department of arts, culture, science and technology: www.dacst.gov.za
Links to other government sites: www.polity.org.za/lists/govsites.html
Government policy document: www.nda.agric.za/docs/policy98.htm

Universities participating in IPM research

University of Pretoria: www.up.ac.za/academic/zoology/
Natal University: www.unp.ac.za/ UNPDepartments/zoo/zoodept.html
Rhodes University: www.ru.ac.za/academic/departments/zooento/
Stellenbosch University: www.sun.ac.za/agric/entomol
University of Cape Town: www.uct.ac.za/depts/zoology/
University of the Witwatersrand: www.wits.ac.za/apes/
University of the Free State: www.uovs.ac.za/faculties/nat/zent/index.htm

Sites with general South African agricultural news and information

www.agrelek.co.za
www.agri24.com

References

Aalbersberg, Y.K. (1987) Ecology of the wheat aphid *Diuraphis noxia* (Mordvilko) in the eastern Free State. MSc thesis, University of the Orange Free State, South Africa.

Aalbersberg, Y.K., Van der Westhuizen, M.C. and Hewitt, P.H. (1988) Natural enemies and their impact on *Diuraphis noxia* (Mordvilko) (Hemiptera: Aphididae) populations. *Bulletin of Entomological Research* 78, 111–120.

Aalbersberg, Y.K., Van der Westhuizen, M.C. and Hewitt, P.H. (1989) Japanese radish as a reservior for the natural enemies of the Russian wheat aphid *Diuraphis noxia* (Hemiptera: Aphididae). *Phytophylactica* 21, 241–245.

Annecke, D.P. (1958) Ants and citrus. *Citrus Grower* 296, 1–3.

Annecke, D.P. (1969) Recent developments in biological and integrated control of citrus pests in South Africa. *Proceedings of the First International Citrus Symposium* 2, 849–854.

Annecke, D.P. and Moran, V.C. (1982) *Insects and Mites of Cultivated Plants in South Africa.* Butterworth, Pretoria, 383 pp.

Anonymous (1989) *Agriculture in South Africa.* Chris van Rensburg Publications, Johannesburg, South Africa.

Anonymous (1992) Integrated pest management of cotton pests: a compromise between biological and chemical control. *Plant Protection News* 29, 6–7.

Anonymous (2001) *South African Sugar Industry Directory, 2001/2002.* External Affairs Division, South African Sugar Association, Mount Edgecombe.

ARC–Small Grain Institute (2000) *Guidelines for Wheat Production in the Summer Rainfall region*, pp. 54–57.

Atkinson, P.R. (1980) On the biology, distribution and natural host plants of *Eldana saccharina* Walker (Lepidoptera: Pyralidae). *Journal of the Entomological Society of Southern Africa* 43, 171–194.

Atkinson, P.R. (1997) Recommendations, practices and notes on the control of pests and diseases in commercial forestry in South Africa. *ICFR Bulletin Series* 1/9, 21 pp.

Atkinson, P.R., Carnegie, A.J.M. and Smaill, R.J. (1981) A history of outbreaks of *Eldana saccharina* Walker in Natal. *Proceedings of the South African Sugar Technologists Association* 55, 111–115.

Barnes, B.N. (1990) *FFTRI Monitoring Manual for Orchard Pests.* Fruit and Fruit Technology Research Institute, Stellenbosch, 21 pp.

Barnes, B.N. (1999) Fruit fly baiting in deciduous fruit – doing the right things right. *Deciduous Fruit Grower* 49, 5–7.

Barnes, B.N. and Blomefield, T.L. (1997) Goading growers towards mating disruption: the South African experience with *Grapholita molesta* and *Cydia pomonella* (Lepidoptera: Tortricidae). In: Witzgall, P. and Arn, H. (eds) *Technology Transfer in Mating Disruption.* IOBC/WPRS Bulletin 20(1), 45–56.

Barnes, B.N. and Eyles, D.K. (2000) Feasibility of eradicating *Ceratitis* spp. fruit flies from the Western Cape of South Africa by the sterile insect technique. In: Tan, K.H. (ed.) *Area-Wide Control of Fruit Flies and Other Insect Pests.* Penerbit Universiti Sains, Pulau Pinang, Malaysia, pp. 449–455.

Barnes, B.N., Calitz, F.J. and Knipe, M.C. (1995) Effective weevil control on apple trees with batting trunk barriers. *Deciduous Fruit Grower* 45, 376–378.

Barrow, M.R. (1989) Screening and breeding for resistance to *Busseola fusca*. Towards insect resistant maize for the 3rd world. *Proceedings International Symposium on Methodologies for Developing Host Plant Resistance to Maize Insects.* CIMMYT, Mexico, 9–14 March 1987, pp. 184–191.

Bate, R. and van Rensburg, J.B.J. (1992) Predictive estimation of maize yield loss caused by *Chilo partellus* (Swinhoe) (Lepidoptera: Pyralidae) in maize. *South African Journal of Plant and Soil* 9, 150–154.

Bedford, E.C.G. (1968a) Biological and chemical control of citrus pests in the Western Transvaal. An integrated spray programme. *South African Citrus Journal* 417, 9, 11, 13, 15, 17, 19, 21–28.

Bedford, E.C.G. (1968b) The biological control of red scale, *Aonidiella aurantii* (Maskell), on citrus in South Africa. *Journal of the Entomological Society of Southern Africa* 31, 1–15.

Bedford, E.C.G. (1979a) An integrated pest management programme applied to citrus in the Lowveld. *Citrus and Subtropical Fruit Journal* 549, 10–13.

Bedford, E.C.G. (1979b) Recommendations for the integrated control of the citrus pest complex in the Lowveld. *Citrus and Subtropical Fruit Journal* 550, 11–18.

Bedford, E.C.G. (1998) Citrus Blackfly *Aleurocanthus woglumi* Ashby. In: Bedford, E.C.G., van den Berg, M.A. and de Villiers, E.A. (eds) *Citrus Pests in the Republic of South Africa*, 2nd edn. Dynamic Ad, Nelspruit, South Africa, pp. 102–104.

Bedford, E.C.G. and Thomas, E.D. (1965) Biological control of the citrus blackfly, *Aleurocanthus woglumi* (Ashby) (Homoptera: Aleyrodidae) in South Africa. *Journal of the Entomological Society of Southern Africa* 28, 117–132.

Bedford, E.C.G., Vergoes-Houwens, G., Kok, I.B. and Vercueil, S.W. (1992) Impact of 46 pesticides on non-target pests under biological control in citrus orchards. Department of Agricultural Development, Republic of South Africa. *Technical Communication* 233, 112 pp.

Blomefield, T.L. (1991) Codling moth management, there's more to it than spraying. *Deciduous Fruit Grower* 41, 250–251.

Blomefield, T.L. (1994) Codling moth resistance: is it here, and how do we manage it? *Deciduous Fruit Grower* 44, 130–132.

Blomefield, T.L. (1997) Managing resistance of codling moth, *Cydia pomonella* (L), in South African pome fruit orchards. In: *Resistance 97 – Integrated Approach to Combating Resistance.* IARC-Rothamsted, Harpenden, UK, pp. 14–16.

Blomefield, T.L. and Barnes, B.N. (2000) Integrated control of oriental fruit moth, *Grapholita molesta*, on peaches using a spray-date prediction model. In: Müller, W., Polesny, F., Verheyden, C. and Webster, A.D. (eds) *Proceedings of the International Conference on Integrated Fruit Production, IOBC/WPRS Bulletin* 23(7), 161–168.

Blomefield, T.L. and Kleinhans, P.A. (1995) Codling moth management using a degree-day phenology model. *Proceedings of the 10th Congress of the Entomological Society of Southern Africa*, Grahamstown, 13 pp.

Botha, J.H. (1990) Natural enemies of the red spider mite in cotton. *Farming in South Africa, Cotton* G, 4, 1–2.

Broodryk, S.W. (1967) Bioecological studies on the potato tuber moth *Gnorimoschema operculella* (Zeller) and its hymenopterous parasitoids in South Africa. PhD thesis, University of Pretoria, Pretoria, South Africa.

Broodryk, S.W. (1969) The biology of *Orgilus parcus* (Turner) (Hymenoptera: Braconidae). *Journal of the Entomological Society of Southern Africa* 32(2), 243–257.

Broodryk, S.W. (1971) The biology of *Diadegma stellenboschense* (Cameron) (Hymenoptera: Ichneumonidae) a parasitoid of potato tuber moth. *Journal of the Entomological Society of Southern Africa* 34(2), 413–423.

Broodryk, S.W. and Pretorius, L.M. (1974) Occurrence in South Africa of a granulosis virus attacking potato tuber moth, *Phthorimaea operculella* (Zeller). *Journal of the Entomological Society of Southern Africa* 37, 125–128.

Buitendag, C.H. and Bronkhorst, G.J. (1986) Further aspects of trunk treatment of citrus with insecticides: phytotoxicity, side effects on incidental pests and development of application apparatus. *Citrus and Subtropical Fruit Journal* 604, 7–10.

Buitendag, C.H. and Naude, W. (1992) Insecticide stem treatments for the control of citrus pests. *Citrus Journal* 2(5), 36–39.

Carnegie, A.J.M. (1966) The progress of an untreated outbreak of *Numicia viridis* Muir. *Proceedings of the South African Sugar Technologists Association* 40, 319–326.

Carnegie, A.J.M. (1967) Field populations of *Numicia viridis* Muir. *Proceedings of the South African Sugar Technologists Association* 41, 178–180.

Carnegie, A.J.M. (1969) The development of populations of *Numicia viridis* Muir in sugarcane fields. *Proceedings of the South African Sugar Technologists Association* 43, 75–84.

Carnegie, A.J.M. (1971) Insecticide use against *Numicia viridis* Muir (Homoptera: Tropiduchidae). *Proceedings of the International Society of Sugar Cane Technologists* 14, 516–526.

Carnegie, A.J.M. (1974) A recrudescence of the borer *Eldana saccharina* Walker (Lepidoptera: Pyralidae). *Proceedings of the South African Sugar Technologists Association* 48, 107–110.

Carnegie, A.J.M. (1975) Natural enemies of *Numicia viridis* Muir (Homoptera: Tropiduchidae). *Proceedings of the Congress of the Entomological Society of Southern Africa* 1, 245–254.

Carnegie, A.J.M. (1980) Egg mortality of *Numicia viridis* Muir (Homoptera: Tropiduchidae) in sugarcane and in indigenous hosts. *Journal of the Entomological Society of Southern Africa* 43(2), 215–221.

Carnegie, A.J.M. (1981) Combating *Eldana saccharina* Walker: a progress report. *Proceedings of the South African Sugar Technologists Association* 55, 116–119.

Carnegie, A.J.M. (1994) Sugar cane sucking insects (Homoptera): Africa and close islands. In: Carnegie, A.J.M. and Conlong, D.E. (eds) *Biology, Pest Status and Control Measure Relationships of Sugar Cane Insect Pests. Proceedings of Second Sugar Cane Entomology Workshop of the International Society of Sugar Cane Technlogists.* Mount Edgecombe. South African Sugar Association Experiment Station, Mount Edgecombe, South Africa.

Conlong, D.E. (1990) A study of pest–parasitoid relationships in natural habitats, an aid towards the biological control of *Eldana saccharina* Walker (Lepidoptera: Pyralidae). *Proceedings of the South African Sugar Technologists Association* 64, 111–115.

Conlong, D.E. (1994a) Host–parasitoid interactions of *Eldana saccharina* (Lepidoptera: Pyralidae) in *Cyperus papyrus* L. PhD thesis, University of Natal, Pietermaritzburg, South Africa.

Conlong, D.E. (1994b) A review and perspectives for the biological control of the African sugarcane stalkborer *Eldana saccharina* Walker (Lepidoptera, Pyralidae). *Agriculture, Ecosystems and Environment* 48, 9–17.

Conlong, D.E. (1997) Biological control of *Eldana saccharina* Walker in South African sugarcane, constraints identified from 15 years of research. *Insect Science and its Application* 17, 69–78.

Conlong, D.E. (2000) Indigenous African parasitoids of *Eldana saccharina* (Lepidoptera, Pyralidae). *Proceedings of the South African Sugar Technologists Association* 74, 201–211.

Conlong, D.E. and Kasl, B. (2000) Stimulo-deterrent diversion to decrease infestation in sugarcane by *Eldana saccharina* (Lepidoptera: Pyralidae). *Proceedings of the South African Sugar Technologists Association* 74, 212–213.

Conlong, D.E. and Kasl, B. (2001) Stimulo-deterrent diversion: *Eldana saccharina* Walker (Lepidoptera: Pyralidae) and *Xanthopimpla stemmator* Thunberg (Hymenoptera: Ichneumonidae), preliminary results. *Proceedings of the South African Sugar Technologists Association* 75, 180–182.

Conlong, D.E. and Mugalula, A. (2001) *Eldana saccharina* (Lepidoptera: Pyralidae) and its parasitoids at Kinyara Sugar Works, Uganda. *Proceedings of the South African Sugar Technologists Association* 75, 183–185.

Dent, D. (1991) *Insect Pest Management.* CAB International, Wallingford, UK.

Dick, J. (1945) Some data on the biology of the sugarcane borer (*Eldana saccharina* Wlk.). *Proceedings of the South African Sugar Technologists Association* 19, 75–79.

Dick, J. (1950) Sugarcane entomology in Natal, South Africa. *Proceedings of the International Society of Sugar Cane Technologists* 7, 377–394.

DeBach, P. (1979) *Biological Control by Natural Enemies.* Cambridge University Press, Cambridge, UK.

Du Plessis, C. and Lea, H.A.F. (1943) The maize stalk borer, *Calamistis fusca* (Hmps). *Bulletin Department of Agriculture and Forestry, Union of South Africa* 238, 51 pp.

Du Toit, F. (1986) Economic thresholds for *Diuraphis noxia* (Hemiptera: Aphididae) on winter wheat in the eastern Orange Free State. *Phytophylactica* 18, 107–109.

Du Toit, F. (1987) Resistance in wheat (*Triticum aestivum*) to *Diuraphis noxia* (Hemiptera:

Aphididae). *Cereal Research Communications* 15, 175–179.
Du Toit, F. (1990) Field resistance in three bread wheat lines to the Russian wheat aphid *Diuraphis noxia* (Hemiptera: Aphididae). *Crop Protection* 9, 255–258.
Findlay, J.B.R. (1975) Biological and cultural control of the potato tuber moth in table potatoes. *Proceedings, First Congress of the Entomological Society of Southern Africa*, pp. 147–150.
Georgala, M.B. (1975) Possible resistance of the red scale *Aonidiella aurantii* (Mak.) to corrective spray treatments. *The Citrus and Subtropical Fruit Journal* 505, 5, 7, 9, 11, 13, 15, 17.
Grout, T.G. (1993a) Directions and precautions for the use of spray oils on citrus. *Citrus Journal* 3(4), 29–30.
Grout, T.G. (1993b) Restoring the balance in citrus pest control. *Shell Agriculture* 16, 7–9.
Grout, T.G. (1998) Silkworms as indicators of environmental contamination with diofenolan and fenoxycarb. *African Entomology* 6(1), 149–151.
Grout, T.G. and Richards, G.I. (1992) *Euseius addoensis addoensis*, an effective predator of citrus thrips, *Scirtothrips aurantii*, in the eastern Cape Province of South Africa. *Experimental and Applied Acarology* 15, 1–13.
Grout, T.G., du Toit, W.J., Hofmeyr, J.H. and Richards, G.I. (1989) California red scale (Homoptera: Diaspididae) phenology on citrus in South Africa. *Journal of Economic Entomology* 82(3), 793–798.
Grout, T.G., Richards, G.I. and Stephen, P.R. (1997) Further non-target effects of citrus pesticides on *Euseius addoensis* and *Euseius citri* (Acari: Phytoseiidae). *Experimental and Applied Acarology* 21, 171–177.
Grout, T.G., Hofmeyr, J.H., Hattingh, V., Buitendag, H.,Ware, A.B., Marais, J.L., le Roux, H.F., Schutte, G.C. and du Toit Pelser, P. (1998) *Integrated Production Guidelines for Export Citrus, Vol. III – Integrated Pest and Disease Management.* Outspan International.
Harvey, T.L. and Martin, T.L. (1990) Resistance to Russian wheat aphid, *Diuraphis noxia* in wheat, (*Triticum aestivum*). *Cereal Research Communications* 18, 127–129.
Hattingh, V. (1996a) Comparison between the cost of an IPM and a chemical approach to pest management. *Citrus Journal* 6(2), 20–28.
Hattingh, V. (1996b) The use of insect growth regulators – implications for IPM with citrus in southern Africa as an example. *Entomophaga* 41, 513–518.
Hattingh, V. (1997) Mass rearing for augmentation. *Fifth National Insect Rearing Workshop.* Outspan International, 34 pp.
Hattingh, V. and Tate, B.A. (1995) Effects of field-weathered residues of insect growth regulators on some Coccinellidae (Coleoptera) of economic importance as biocontrol agents. *Bulletin of Entomological Research* 85, 489–493.
Hattingh, V. and Tate, B.A. (1996a) The effects of insect growth regulator use on IPM in southern African citrus. *Proceedings of the International Society of Citriculture*, pp. 523–525.
Hattingh, V. and Tate, B.A. (1996b) The status of mealybugs on citrus in southern Africa. *Proceedings of the International Society of Citriculture*, pp. 560–563.
Hattingh, V., Ware, A.B. and Grout, T.G. (2003) *Proceedings of the International Society of Citriculture*, IX Congress 2000, Orlando, Florida, pp. 795–797.
Hoekstra, N. and Kfir, R. (1997) Microbial pathogens of the cereal stem borers *Busseola fusca* (Fuller) (Lepidoptera: Noctuidae) and *Chilo partellus* (Swinhoe) (Lepidoptera: Pyralidae) in South Africa. *African Entomology* 5, 161–163.
Hoffmann, J.H. (1991) Introduction. Special Issue: biological control of weeds in South Africa. *Agriculture, Ecosystems and Environment* 37, 1–3.
Hofmeyr, J.H. and Calitz, F.J. (1991) Field evaluation of synthetic sex attractants of the false codling moth, *Cryptophlebia leucotreta* (Lepidoptera: Tortricidae). *Phytophylactica* 23, 149–152.
Ishaaya, I. (1990) Benzophenyl ureas and other selective insect control agents – mechanism and application. In: Cassida, J.E. (ed.) *Pesticides and Alternatives: Innovative Chemical and Biological Approaches to Pest Control.* Elsevier, Amsterdam.
Kfir, R. (1988) Hibernation by the lepidopteran stalk borers, *Busseola fusca* and *Chilo partellus* on grain sorghum. *Entomologia Experimentalis et Applicata* 48, 31–36.
Kfir, R. (1990a) Prospects for cultural control of the stalk borers, *Chilo partellus* (Swinhoe) and *Busseola fusca* (Fuller), in summer grain crops in South Africa. *Journal of the Entomological Society of Southern Africa* 53, 41–47.
Kfir, R. (1990b) Parasites of the spotted stalk borer, *Chilo partellus* (Lepidoptera: Pyralidae) in South Africa. *Entomophaga* 35, 403–410.
Kfir, R. (1991) Selecting parasites for biological control of lepidopterous stalk borers in

summer grain crops in South Africa. *Redia* 74, 231–236.

Kfir, R. (1992a) Alternative, non-chemical control methods for the stalk borers, *Chilo partellus* (Swinhoe) and *Busseola fusca* (Fuller), in summer grain crops in South Africa. *Technical Communication Department of Agricultural Development, Republic of South Africa* 232, 99–103.

Kfir, R. (1992b) Seasonal abundance of the stem borer *Chilo partellus* (Lepidoptera: Pyralidae) and its parasites on summer grain crops. *Journal of Economic Entomology* 85, 518–529.

Kfir, R. (1994) Attempts at biological control of the stem borer, *Chilo partellus* (Swinhoe) (Lepidoptera: Pyralidae), in South Africa. *African Entomology* 2, 67–68.

Kfir, R. (1995) Parasitoids of the African stem borer, *Busseola fusca* (Lepidoptera: Noctuidae), in South Africa. *Bulletin of Entomological Research* 85, 369–377.

Kfir, R. and Bell, R. (1993) Intraseasonal changes in populations of the African maize stem borer, *Busseola fusca* (Fuller) (Lepidoptera: Noctuidae), in Natal, South Africa. *Journal of African Zoology* 107, 543–553.

Kfir, R. and van Hamburg, H. (1983) Further tests of threshold levels for control of cotton bollworms (mainly *Heliothis armiger*). *Journal of the Entomological Society of Southern Africa* 46, 49–58.

Kfir, R.A., Kirsten, J.F. and van Rensburg, N.J. (1985) Observations on the black pine aphid, *Cinara cronartii* (Homoptera, Aphididae), the insects associated with it and its introduced parasitoid, *Pauesia* sp. (Hymenoptera: Aphidiidae) in South Africa. *Environmental Entomology* 14, 597–601.

Kfir, R., van Hamburg, H. and Van Vuuren, R. (1989) Effect of stubble treatment on the post-diapause emergence of the stalk borer *Chilo partellus* (Swinhoe) (Lepidoptera: Pyralidae). *Crop Protection* 8, 289–292.

Kirsten, J.F., Tribe, G.D., van Rensburg, N.J. and Atkinson, P.R. (2000) Insect pests in South African forestry plantations. In: *South African Forestry Handbook*, 4th edn. Southern African Institute for Forestry, Pretoria, South Africa.

Kovalev, O.V., Poprawski, T.J., Stekolshchikov, A.V., Vereshchagina, A.B. and Gandrabur, S.A. (1991) *Diuraphis* Aizenberg (Hom. Aphididae) key to apterous viviparous females and review of Russian language literature on the natural history of *Diuraphis noxia* (Kurdjumov, 1913). *Journal of Applied Entomology* 112, 425–436.

Leslie, G.W. (1986) The arthropod predators of *Eldana saccharina* Walker (Lepidoptera: Pyralidae), their identification, and relative importance. MSc thesis, University of Natal, Pietermaritzburg, South Africa.

Leslie, G.W. (1997) Research strategies for the use of insecticides to control sugarcane pests in South Africa. *Proceedings of the South African Sugar Technologists Association* 71, 91–93.

Leslie, G.W. (2001) Field-scale insecticide trials against the sugarcane borer *Eldana saccharina* (Lepidoptera: Pyralidae) in South Africa: a preliminary assessment. *Proceedings of the International Society of Sugar Cane Technologists* 24, 395–397.

Leslie, G.W. and Keeping, M.G. (1997) Approaches to the control of the pyralid sugarcane borer *Eldana saccharina* Walker. *Proceedings of the International Society of Sugar Cane Technologists* 23, 561–565.

Lim, G.S., Sivapragasam, A. and Loke, W.H. (1997) Crucifer insect pest problems: trends issues and management strategies. In: Sivapragasam, A., Loke, W.H., Hussan, A.K. and Lim, G.S. (eds) *The Management of Diamondback Moth and Other Crucifer Pests. Proceedings of the Third International Workshop.* Malaysian Agricultural Research and Development Institute (MARDI), Kuala Lumpar, Malaysia, pp. 3–14.

Magagula, C.N.N. (1998) Impact of insect growth regulators on no-target species, with an emphasis on coccinellids on citrus, Swaziland. PhD thesis, University of Natal, Pietermaritzburg, South Africa.

Mally, C.W. (1920) The maize stalk borer, *Busseola fusca*, Fuller. *Bulletin, Department of Agriculture, Union of South Africa* 3, 111 pp.

Marasas, C., Anandajayasekeram, P., Tolmay, V., Martella, D., Purchase, J. and Prinsloo, G. (1997) *Socio-economic impact of the Russian Wheat Aphid Control Research Program.* Southern African Centre Botswana.

Naude, T.J. (1940) Work on the control of insect pests. Annual Report of the Division of Entomology. *Farming in South Africa,* 516 pp.

Nel, A., Krause, M., Ramautar, N. and Van Zyl, K. (1999) *A Guide for the Control of Plant Pests.* National Department of Agriculture, Republic of South Africa.

Newton, P.J. and Mastro, V.C. (1989) Field evaluation of commercially available traps and synthetic sex pheromone lures of the false codling moth, *Cryptophlebia leucotreta* (Meyr.) (Lepidoptera: Totricidae). *Tropical Pest Management* 35, 100–104.

Newton, P.J. and Odendaal, W.J. (1990) Commercial inundative release of *Trichogramatoidea cryptophlebiae* (Hym.: Trichogrammatidae) against *Crytophlebia leucotreta* (Lep.: Tortricidae) in citrus. *Entomophaga* 35, 545–556.

Osborne, R.F. (1964) *Valiant Harvest. The Founding of the South African Sugar Industry 1848–1926.* South African Sugar Association, Durban, South Africa.

Peleg, B.A. (1988) Effect of a new phenoxy juvenile hormone analog on California red scale (Homoptera: Coccidae) and the ectoparasite *Aphytis holoxanthus* DeBach (Hymenoptera: Aphelinidae). *Journal of Economic Entomology* 81, 88–92.

Prinsloo, G.J., Tolmay, V.L. and Hatting, J.L. (1999) *Field Guide for the Identification of Wheat Insects in South Africa.* ARC–Small Grain Institute, Bethlehem, South Africa.

Prinsloo, G., Chen, Y., Giles, K.L. and Greenstone, M.H. (2002) Release and recovery in South Africa of the exotic aphid parasitoid *Aphelinus hordei* Kurdjumov (Hymenoptera: Aphelinidae) verified by the polymerase chain reaction. *Biocontrol* 47, 127–136.

Reed, W. and Rich, J.R. (1975) A new survey technique for citrus thrips. *Journal of Economic Entomology* 68, 739–742.

Revington, J. (1987) Use of a synthetic sex pheromone in forecasting attacks on maize by maize stalkborer *Busseola fusca* (Fuller) (Lepidoptera Noctuidae). MSc thesis, University of the Witwatersrand, Johannesburg, South Africa.

Riedl, H., Blomefield, T.L. and Giliomee, J.H. (1998) A century of codling moth control in South Africa: II. Current and future status of codling moth management. *Journal of the Southern African Society for Horticultural Sciences* 8, 32–54.

Ripley, L.B. and Hepburn, G.A. (1934) Adhesives for cryolite suspensions. *Science Bulletin, Department of Agriculture, South Africa* 122, 12 pp.

Searle, C.M. (1961) Beneficial insects – insecticides and citrus. *Technical Communication* 12, Department of Agricultural Technical Services, Republic of South Africa, pp. 183–187.

Searle, C.M. (1963) An investigation into the problem of integrating biological and chemical control of citrus pests with special reference to the effect of insecticides on a parasite (*Pauridia peregrina* Timb.) and a predator (*Exochomus flavipes* Thunb.). MSc thesis, University of Pretoria, South Africa.

Searle, C.M. (1965) The susceptibility of *Pauridia peregrina* Timb. (Hymenoptera: Encyrtidae) to some pesticide formulations. *Journal of the Entomological Society of Southern Africa* 27, 239–250.

Simbi, T. (2001) www.nda.agric.za/docs/policy98.htm

Skoroszewski, R.W. and Van Hamburg, H. (1987) The release of *Apanteles flavipes* (Cameron) (Hymenoptera: Braconidae) against stalkborers of maize and grain-sorghum in South Africa. *Journal of the Entomological Society of Southern Africa* 50, 249–255.

Smith, C.M., Schotzko, D., Zemetra, R.S., Souza, E.J. and Schroeder-Teeter, S. (1991) Identification of Russian wheat aphid (Homoptera: Aphididae) resistance in wheat. *Journal of Economic Entomology* 84, 328–332.

Tooke, F.G.C. (1953) The eucalyptus snout beetle, *Gonipterus scutellatus* Gyll. A study of its ecology and control by biological means. *Entomology Memoir Department of Agricultural Technical Services, Union of South Africa*, 3, 95 pp.

Tribe, G.D. (1991) Phenology of three exotic pine bark beetle species (Coleoptera, Scolytidae) colonising *Pinus radiata* logs in the southwestern Cape Province. *South African Forestry Journal* 157, 27–31.

Tribe, G.D. (1995) The woodwasp, *Sirex noctilio* Fabricius (Hymenoptera: Siricidae), a pest of *Pinus* species, now established in South Africa. *African Entomology* 3(2), 215–217.

Tribe, G.D. (2000) Ecology, distribution and natural enemies of the eucalyptus-defoliating tortoise beetle, *Trachymela tincticollis* (Blackburn) (Chrysomelidae: Chrysomelini: Paropsina) in southwestern Australia, with reference to its biological control in South Africa. *African Entomology* 8(1), 23–45.

Tribe, G.D. and Cillié, J.J. (2000) Biological control of the eucalyptus-defoliating Australian tortoise beetle, *Trachymela tincticollis* (Blackburn) (Chrysomelidae: Chrysomelini: Paropsina) in South Africa by the egg parasitoid, *Enoggera reticulata* Naumann (Hymenoptera: Pteromalidae: Asaphinae). *African Entomology* 8(1), 15–22.

Ulyett, G.C. (1938) Ants and beneficial insects. *Farming in South Africa* 13, 144–120.

van den Berg, J., van Rensburg, G.D.J. and van der Westhuizen, M.C. (1994) Host plant resistance and chemical control of *Chilo partellus* (Swinhoe) and *Busseola fusca* (Fuller) in an integrated pest management system on grain sorghum. *Crop Protection* 13, 307–310.

van den Berg, M.A. and Greenland, J. (1997) Classical biological control of *Aleurocanthus spiniferus* (Hem.: Aleyrodidae), on citrus in southern Africa. *Entomophaga* 42(4), 459–465.

van Dyk, L.P. (2000) Status of plant protection in South Africa. *Proceedings of the Strategic Planning Workshop for the Plant Protection Sub-committee, 25–26 September (2000)*, pp. 28–32.

van Hamburg, H. (1979) The grain sorghum stalk borer, *Chilo partellus* (Swinhoe) (Lepidoptera: Pyralidae): seasonal changes in adult populations in grain sorghum in the Transvaal. *Journal of the Entomological Society of Southern Africa* 42, 1–9.

van Hamburg, H. (1987) A biological control approach to pest management on grain crops with special reference to the control of stalkborers. Department of Agriculture and Water Supply, Republic of South Africa. *Technical Communication* 212, 52–55.

van Hamburg, H. (1981) The inadequacy of egg counts as indicator of threshold levels for control of *Heliothis armigera* on cotton. *Journal of the Entomological Society of Southern Africa* 44, 289–295.

van Hamburg, H. and Guest, P.J. (1997) The impact of insecticides on beneficial arthropods in cotton agro-ecosystems in South Africa. *Archives of Environmental Contamination and Toxicology* 32, 63–68.

van Hamburg, H. and Kfir, R. (1982) Tests of threshold levels based on larval counts for chemical control of cotton bollworms. *Journal of the Entomological Society of Southern Africa* 45, 109–121.

van Lenteren, J.C. (1991) Biological control in a tritrophic system approach. In: Peters, D.C., Webster, J.A. and Chlouber, C.S. (eds) *Proceedings: Aphid–Plant Interactions: Populations to Molecules.* Stillwater, Oklahoma, USA, pp. 3–28.

van Rensburg, G.D.J. and Bate, R. (1987) Preliminary studies on the relative abundance and distribution of the stalk borers *Busseola fusca* and *Chilo partellus. Technical Communication.* Department of Agriculture and Water supply, Republic of South Africa, 212, 49–52.

van Rensburg, J.B.J. (2001) Larval mortality and injury paterns of the African stalk borer, *Busseola fusca* (Fuller) on various plant parts of Bt-transgenic maize. *South African Journal of Plant and Soil* 18, 62–68.

van Rensburg, J.B.J., Walters, M.C. and Giliomee, J.H. (1988) Plant population and cultivar effects on yield losses caused by the maize stalk borer, *Busseola fusca* (Lepidoptera: Noctuidae). *South African Journal of Plant and Soil* 5, 215–218.

van Rensburg, N.J. (1979) *Cinara cronartii* on the roots of pine trees (Homoptera: Aphididae). *Journal of the Entomological Society of Southern Africa* 42(1), 151–152.

van Rensburg, N.J. (1981) A technique for rearing the black pine aphid, *Cinara cronartii* T and P, and some features of its biology (Homoptera: Aphididae). *Journal of the Entomological Society of Southern Africa* 44, 367–379.

van Rensburg, N.J. (1984) Forest insect pest management with special reference to insect populations in fast growing plantations. *Conference Proceedings: Symposium on Site and Productivity in Fast Growing Plantations.* IUFRO, Pretoria and Pietermaritzburg, pp. 375–385.

van Rensburg, N.J. (1988) The biological control of the black pine aphid, *Cinara cronartii* T and P, in Southern Africa. *Abstracts of XV111 International Congress of Entomology*, Vancouver, Canada.

van Rensburg, N.J. and van Hamburg, H. (1975) Grain sorghum pests: an integrated control approach. *Proceedings 1st Congress Entomological Society of Southern Africa*, pp. 151–162.

Walters, M.C. (1974) Mechanisms of resistance to insects and application to maize stalk borer. *Technical Communication.* Department of Agricultural Technical Services, South Africa, 132, 83–87.

Walton, V.M. (2000) Rearing and releasing of a parasite for the control of grapevine mealybugs in South African vineyards. *Wynboer Technica*, pp. 100–101.

Watmough, R.H. and Kfir, R. (1995) Predation on pupae of *Helicoverpa armigera* Hbn. (Lep., Noctuidae) and its relation to stem borer numbers in summer grain crops. *Journal of Applied Entomology* 119, 679–688.

Watmough, R.H., Broodryk, S.W. and Annecke, D.P. (1973) The establishment of two imported parasitoids of potato tuber moth (*Phthorimaea operculella*) in South Africa. *Entomophaga* 18(3), 237–249.

Way, M.J. and Turner, P.E.T. (1999) The spotted sugarcane borer, *Chilo sacchariphagus* (Lepidoptera: Pyralidae: Crambinae), in Mozambique. *Proceedings of the South African Sugar Technologists Association* 73, 112–113.

Webb, D.v.V. (1974) Forest and timber entomology in the Republic of South Africa. *Entomology Memoir Department of Agricultural Technical Services, Republic South Africa*, 34, 21 pp.

Whitehead, V.B. and Rust, D.J. (1972) Control of the fruit-piercing moth *Serrodes partita* (Fabr.) (Lepidoptera: Noctuidae). *Phytophylactica* 4, 9–12.

Whitlock, V.H. (1973) Studies on insecticidal resistance in the bollworm *Heliothis armigera* Hubner (Lepidoptera: Noctuidae). *Phytophylactica* 5, 71–74.

Zimmermann, H.G. (1967) Biologiese studies op en die deoltreffendheid van kultuurmaatreels in die beheer van die aartappelmot. MSc (Agric) thesis, University of Pretoria, Pretoria, South Africa.

Appendix 15.1.

Plant Protection Research Institute (ARC–PPRI)

(website: www.arc.agric.za)

The ARC–PPRI was established in 1962 with the amalgamation of the Divisions of Entomology and Plant Pathology of the Department of Agriculture. These divisions had been in existence since Unification in 1910. In 1981 research on invasive weeds was formally added to the ARC–PPRI and during 2000 the Agricultural Biodiversity Information Unit joined the institute. This multidisciplinary institute follows a holistic approach to the pest, disease and alien invasive plant problems, in line with the principles of integrated pest management as defined in Agenda 21 of the Rio Convention.

ARC–PPRI is one of the Institutes of the ARC and currently employs 230 staff consisting of 68 researchers, 47 technicians and 115 support staff. The hub of the research activities is in Pretoria, with campuses at Roodeplaat, Rietondale, and Vredehuis. There are satellite units at Cedara, Uitenhage and Stellenbosch.

The ARC–PPRI provides expertise to agricultural and environmental concerns through research aimed at the promotion of economic and environmentally acceptable pest management strategies in support of sustainable land management in the subregion and many other African countries. To this end ARC–PPRI is a center of expertise on biosystematics, ecology and epidemiology of invertebrates, fungi, pathogenic and useful bacteria, viruses and the control of pests and invasive plants through optimization of pesticidal and biological control strategies in integrated management programs.

Vision of PPRI

ARC–PPRI's vision is to establish an agricultural production system in South Africa in which yields are maximized through sustainable integrated management strategies. The mechanisms involved in upsurges are therefore investigated in order to control harmful and invasive plants. Beneficial organisms such as biological control agents, nitrogen fixing bacteria and insect pollinators of plants will be cultured, nurtured and promoted for increased plant production.

Mission of PPRI

Through research and development, the ARC–PPRI aims to produce sound pest, disease and invasive plant management strategies and encourage the utilization of advantageous organisms to strengthen agricultural production at all levels.

Core competencies

ARC–PPRI is mandated to address plant protection issues that cut across commodities, affecting many crops and regions; thus the research impacts on all the provinces of South Africa and addresses the needs of many African countries. Research is directed at commercial, small-scale and resource-poor farmers to address current and anticipated threats.

1. **Biosystematic services** are provided for the benefit of researchers, agricultural industries and to governments to carry out their statutory obligations. To this end the institute is the custodian of the National Collections of Insects, Arachnids, Nematodes and Fungi. Biosystematic capacity in the region is promoted through participation in African initiatives such as those on arachnids, pollinators, fruit flies and SAFRINET (Southern African Network of BIONET International). Agricultural Biodiversity Information Systems are developed and maintained in keeping with government policy and international conventions.

2. **IPM of pests of crops, plantations and stored products is a central theme in much of the research of the institute and includes:**

- Classical biological control programs;
- Quarantining of imported organisms on behalf of government and industry;
- Cultural practices;

- Pesticide application and residue analyses;
- Monitoring of resistance to pesticides in pest populations;
- Bioprospecting to develop viable alternative control methods;
- Development of strategies to curb migrant pests in collaboration with neighboring countries and international institutions such as DFID, NRI and FAO.

3. Plant pathology research and services focus on fungi, bacteria and plant viruses and includes:

- Studies of disease epidemiology;
- Monitoring of disease resistance in plants in support of plant breeding programs;
- Development of diagnostic techniques;
- Diagnostic services;
- Indexing of virus diseases of banana and plantain on behalf of INIBAP;
- Development of disease-free seed schemes in collaboration with local industries and protocols of ISTA;
- Specialization in disease complexes such anthracnose and soil-borne diseases.

4. Research on weeds is directed at the development of integrated control strategies against alien invaders of rangeland, plantations, rivers and dams. This is in keeping with the objectives of the Working-for-Water Program of the Department of Water Affairs and Forestry, the Water Research Commission, the Landcare Initiatives of the National Department of Agriculture, the National Department of Environmental Affairs and Tourism and industries. Many of the weeds occur in other parts of Africa and regional collaboration, such as for the proposed African Water Hyacinth Initiative, gives substance to the President's 'African Renaissance'. Consultancies on weed management are provided to various African countries on behalf of CABI and FAO.

5. Expertise on beneficial organisms includes:

- Beekeeping for the benefit of commercial and resource-poor farmers for the production of honey and other bee products and the use of bees for pollination;
- Nitrogen-fixing rhizobia as effective substitutes for nitrogenous fertilizers;
- Mycorrhizal inoculants that promote nutrient uptake;
- Natural enemies for the control of pests, diseases and weeds.

6. Regional training courses for extensionists serving both the commercial and resource-poor sectors are undertaken independently and in collaboration with international partners.

Chapter 16

Integrated Pest Management in China

Zhen-Ying Wang[1], Kang-Lai He[1], Jian-Zhou Zhao[2] and Da-Rong Zhou[1]
[1]Institute of Plant Protection, Chinese Academy of Agricultural Sciences, Beijing, China; [2]Department of Entomology, Cornell University, New York Agricultural Experiment Station, Geneva, New York, USA

Brief History and Evolution of IPM in China

History of agriculture in China

Agriculture has played a central role in the history of China for more than 5000 years. Food production is important in China; reliable and productive agricultural systems are essential because of the large population. Crop pests, floods and drought have been the most serious natural disasters throughout the history of agricultural production in China, and the struggle with insect pests dates back for centuries. For example, the Oriental migratory locust, *Locust migratoria manilensis* (Meyen), has been known as a notorious insect pest for more than 13 centuries. The Emperor of the Tang Dynasty appointed the first full-time officers for locust control in AD 707. Since then and up to the year 1911, 536 serious outbreaks of the locust have been recorded. Early literature on pest control includes *The Etiquette of Zhou Dynasty* written in 240 BC, *Lushi Chunqiu* written in 239 BC, and *Qi Min Yao Shu* (an ancient agricultural encyclopedia) written in AD 528–549. Pest control through cultural practices such as cultivation, rotation, intercropping and irrigation were recommended in the *Nongzheng Quanshu* (*Encyclopedia of Agriculture*) published in 1639 (Chou, 1980). The red tree ant, *Oecophylla smaragdina* (Fabricius), was used to control citrus pests in AD 304, and is the first known example of using natural enemies against insect pests (Chen, 1962). It is still widely used in China today (Yang *et al.*, 1983).

Development of IPM in China

The development of IPM in China has been a gradual but continual process, which can be roughly divided into the following three periods.

Early attempts to develop IPM concepts (1950 to the early 1970s)

Efforts to develop integrated pest control tactics were initiated in the early 1950s (Ma, 1976). The goal was to integrate agricultural, chemical, biological, and physical control measures in order to increase the effectiveness of pest control and avoid the weaknesses of using each method alone. This concept is different from the modern concept of IPM, in that economic injury levels or threshold levels were not yet developed, and the use of broad-spectrum

 Integrated Pest Management in the Global Arena (eds K.M. Maredia, D. Dakouo and D. Mota-Sanchez)

insecticides was not especially discouraged (Li, 1990). The most notable achievement in pest control during this period was management of the locust. No large outbreaks of locusts have occurred since the program began. With the development of the national chemical industry during the 1950–1960s, DDT and BHC became widely used to control crop insect pests, but eventually these insecticides caused major problems such as environmental pollution, high residue, and pesticide resistance. These problems eventually led the public to become more concerned about insect pest management (Piao, 1999).

Acceptance of modern IPM concepts (1974–1982)

In 1974, the First Nationwide Conference on Integrated Pest Control for Crop Diseases and Insect Pests took place in Shaoguan City, Guangdong Province, under the supervision of the Chinese Ministry of Agriculture. This meeting became a milestone in the history of plant protection in China. During the conference, progress in plant protection research and implementation in China were thoroughly examined, emphasizing the problems with high residue, pest resistance, and pest resurgence. The highlight of the conference was the introduction of modern IPM concepts in China. During the latter part of the conference, a new basic principle for plant protection in China was suggested with the theme of 'Integrated Pest Control with Prevention First.' This theme was officially approved in 1975 by the Ministry of Agriculture as the guiding principle of national plant protection. At the time, the major IPM tactics included four methods: changing the habitats of pests and breaking the food chain or destruction of hibernating sites, using natural enemies, elimination of insects by attraction with physical or chemical factors, and rational application of insecticides. Cultural practices were considered the cornerstone of IPM systems in China (Ma, 1976; Qiu, 1976; Li, 1990).

Modern IPM in China (1983–present)

Since 1983, the Chinese government has funded National IPM Technique Research Projects as one of the State Key Research Programs in four successive State Five-year Plans. This marks the beginning of the era of modern IPM in China (Li, 1990; Guo, 1999). Approximately 1000 scientists from 50 research institutes, universities and extension agencies became engaged in IPM research and extension programs. The four major research areas included: IPM system for major crops (rice, wheat, maize, cotton, soybean, vegetables, fruit trees), biological control tactics, pesticide resistance management, and specific research on weed and rodent management. During 1983–1985, each one of the main pest groups (pathogens, insects, weeds and rodents) was regarded as a research target. During 1986–1990, the target was changed into each pest complex on a given crop. Further research on IPM systems during 1991–1995 demonstrated significant benefits for preventing outbreaks of several important pests, especially the cotton bollworm (*Helicoverpa armigera*) and the brown planthopper (*Nilaparvata lugens*). From 1996–2000, IPM systems appropriate for each of the major crops in the main regions of China were developed, evaluated and promoted on large scales.

A more comprehensive definition of the IPM concept is now in use in China:

> IPM is a scientific crop pest management system. It proceeds from the whole agro-ecosystem, and is based on the relationships between pests and environment, utilizing natural control factors and considering local conditions, for keeping pests under an economic injury level, and obtaining optimum economic, sociological and ecological benefits
>
> (Li, 1990)

The target of this strategy is not aimed at eliminating pests, but using various natural control factors to reduce the use of chemical pesticides. It is a long-term strategy that requires continuous development for agricultural production systems to meet the future needs of sustainable agriculture in China (Li, 1990).

Organizational Structure of the IPM System in China

Organization

Three agencies are involved in IPM research in China: the Ministry of Agriculture, Chinese Academy of Sciences and the university systems. The Institute of Plant Protection (IPP) and CAAS, and the National Agriculture Technology Extension and Service Center (NATESC), Ministry of Agriculture are the two organizers of IPM research and extension in China. The main tasks of IPP–CAAS are to organize and coordinate the national crop pests research program to avoid major pest outbreaks and to develop sustainable IPM techniques. The main tasks of NATESC are to lead the provincial Plant Protection Stations in order to extend IPM technology and to take part in IPM research activities. The key institutions involved in national IPM research are:

- **Institutes of the Chinese Academy of Agricultural Sciences.** IPP, Biological Control Institute, Cotton Research Institute, Institute of Vegetables and Flowers, Institute of Crop Germplasm Resources, China Rice Research Institute.
- **Institutes of the Chinese Academy of Sciences.** Institute of Zoology, Shanghai Institute of Entomology, Guangdong Entomological Institute.
- **Institutes of Plant Protection of Provincial Academy of Agricultural Sciences of the following provinces:** Beijing, Gansu, Guangdong, Hebei, Heilongjiang, Henan, Hubei, Hunan, Jiangsu, Jilin, Qinghai, Shaanxi, Shandong, Shanxi, Shanghai, Sichuan, Xinjiang, and Zhejiang.
- **Universities.** National universities such as China Agricultural University, Huanzhong Agricultural University, Nanjing Agricultural University, Northwest Sci-Tech University of Agriculture and Forestry, Zhejiang University, Zhongshan University, and many provincial agricultural universities.

IPM policy in China

The Ministry of Agriculture published the first National Safety Standard for Pesticide Application in 1980. Since 1980, 301 standards for 140 pesticides in 19 crops have been published. Since the first National Standards for residue limits of DDT and BHC in grains and vegetables were published in 1981, the maximum residue limits for 77 pesticides (including insecticides, fungicides, herbicides and growth regulators in rice, vegetables, orange, sugarcane, groundnut and vegetable oils) have been published (First Editorial Division, 1999). The production and use of DDT and BHC ended in China in 1983. On 22 November 1995, a regulation banning the use of high toxicity, high residue pesticides on vegetables grown close to urban areas was established. This included the commonly used insecticides methamidophos, parathion-methyl, methomyl, and phorate. The State Council issued the Pesticide Administration Regulation of China in 1997, which is the guiding principle for the registration, production, management and application of pesticides in China.

IPM Research and Practice in China

Pest monitoring and forecasting in IPM

A nationwide pest monitoring and forecasting system in China is used for all crop protection programs. The first forecasting station for rice stem borers was established in Zhejiang Province in the early 1950s. There have been 8, 68 and 1100 forecasting stations at provincial, city and county levels, respectively, in 13 rice planting provinces since the early 1970s (General Station of Plant Protection, 1988). Since 1983, 90 forecasting regional stations for rodent population surveys have been established (Zhao and Yuan, 1990). The county level forecasting stations report to the provincial forecasting center, and then to the General Forecasting Station of Crop Insect Pests and Plant Diseases of China (currently the Forecasting Department of Crop Insect Pests and Plant Diseases, NATESC of Ministry of Agriculture). The system provides information documenting the trends for major pests during the entire year in different regions over the last 40 years.

Data from the Crop Insect Pests and Plant Diseases Forecasting database were compiled by the General Forecasting Station of Crop Insect Pests and Plant Diseases in China and published in 1983. It documented the historical data of more than 100 major insect pests and plant diseases in different regions from 1964 to 1979, including the time of occurrence, climate factors, host, effect of natural enemies, insect occurrence in relation to migration patterns, losses due to insect density or disease severity, epidemic dynamics and the forecasting experiences (General Forecasting Station of Crop Pests of China, 1983). Extensive research on important migratory insect pests has been coordinated nationwide to increase the effectiveness and accuracy of forecasting.

The establishment of economic thresholds further increased the usefulness of the monitoring and forecasting activities. In the 1970s, economic thresholds for important insect pests in major crops were developed (Chiang, 1977). In order accurately to predict potential pest damage and decide whether control measures were warranted, more research on yield losses and economic thresholds for pests of rice, wheat, maize and cotton in different regions has been conducted. The economic thresholds for most pests have been raised to accommodate the compensatory ability of crops and the effects of natural enemies (Zeng *et al.*, 1988; Wen *et al.*, 1992). Some single and complex dynamic action thresholds of the main crop pests have been developed and used in IPM demonstration regions since the 1980s (Guo *et al.*, 1988, 1994; He *et al.*, 1991; Lu *et al.*, 1991; Ding *et al.*, 1994; Wang, R.Q. *et al.*, 1994).

Cultural practices

Cultural control practices are considered a major component of IPM in China. Cultural control of the Asian corn borer (*Ostrinia furnacalis*) by manipulating the ratio of areas planted to spring maize and other host crops has been an excellent example of the benefits of cultural controls. In the early 1970s, the Asian corn borer caused yield losses as high as 40–50% and reduced the quality of the harvestable grain. The major reason for this serious damage was the 4.3:1 ratio of the area planted to spring maize (and other host crops) versus summer maize. A large number of first generation egg masses were laid upon a relatively small area of summer maize and resulted in tremendous damage. After changing the cropping system into winter wheat rotated with summer maize, the area planted for the spring crop significantly decreased and the area planted for summer maize increased, resulting in a ratio of 1:40. A chemical control plan for spring maize further reduced the first generation progeny and resulted in excellent control (Zhou and He, 1995).

Multicropping patterns can influence the occurrence of pests and natural enemies. Growing rape and sorghum as trap crops in cotton fields significantly increased the predator density compared with cotton grown in a monoculture cotton field (Zhao *et al.*, 1991). Natural enemies easily transfer from winter wheat to cotton in cotton–wheat intercropping after the wheat is harvested (Nan *et al.*, 1987; Wang *et al.*, 1991). Such cultural practices are one of the common IPM components in cotton in the North China cotton belts.

Biological control

Biological Control in China (Bao and Gu, 1998) is an excellent reference book on research and application of biological control in China. Biological control has been accomplished using two methods: mass culture and release of natural enemies or microbial/antibiotic agents, and conservation and utilization of natural enemies. Studies on the major natural enemies of important pests on ten major crops were carried out from 1979 to 1982, recording the dynamic patterns of pests and natural enemies (Gao, 1983; Hou *et al.*, 1984; Zhao, 1984). There were 1303 species of natural enemies of rice pests, and 633, 434, 361 and 777 species in cotton, soybean, millet and vegetables, respectively (Piao, 1998).

Parasitoids of the genus *Trichogramma* have been mass cultured and released to control insect pests on a large scale in China. Twenty-four species of *Trichogramma* were reported, five of which are mass-produced: *T. dendrolimi*, *T. chilonis*, *T. ostriniae*, *T. evanesscens* and *T. japonicum*. Mechanized production and standardization of products of *T. dendrolimi* and *T. chilonis* with artificial host eggs has been successful (Liu *et al.*, 1980, 1996; Dai *et al.*, 1996). *T. dendrolimi* is the most extensively used parasitoid to control insect pests such as the Asian corn borer and sugarcane borers. This parasitoid is applied to an average of 400,000 ha each year, up to a maximum of 670,000 ha.

Microbial insecticides and agricultural antibiotics have been used widely to control insect pests and plant diseases. Formulations of Bt are commonly used to control Lepidoptera in cotton, maize, rice, and vegetables. Jinggangmycin is effective in controlling rice sheath blight (*Thanatephorus cucumeris*) and corn sheath blight (*Rhizoctonia solani*), the important diseases in rice and maize, respectively, in China (Zeng *et al.*, 1998). The area using Jinggangmycin to control rice sheath blight increased to 16 million ha in 1998 (Piao, 1998).

Conservation of natural enemies has gained attention in China. Studies have documented 227 species of parasitoids and predaceous insects of 27 families belonging to six orders as well as 90 species of spiders in rice fields of Guangxi Autonomous Region. In China 283, 204, 23 and 152 species of spiders have been recorded in rice, cotton, wheat fields and citrus orchards, respectively (Zhao, 1998). Many species of natural enemies of insect pests in cropland are conserved through planting trap crops, combined with *Trichogramma* releases or microbial insecticide applications, or rational application of chemical pesticides (Shi, 1996).

Pest-resistant crop varieties

Since the 1970s, much research has focused on plant resistance in four key crops: rice, wheat, maize and cotton. Rice cultivars resistant to *N. lugens*, *Sogatella furcifera*, *C. suppressalis* and *Cnaphalocrosis medinalis* were discovered, and some insect-resistant varieties, such as Zheli 1, Xiangzhong Xian 3, and Zhongxuan 13, have been developed (Lu and Gu, 1988). From 1973 to 1979, about 11,200 rice cultivars were tested in Guangxi rice blast nurseries for resistance, 23 of which showed highly to moderately stable resistance (Lay, 1981). From 1975 to 1983, a total of 1770 maize samples, including inbred, hybrids, and open-pollinated varieties were screened for resistance to the Asian corn borer. Of these, 47 were highly resistant to whorl infestation and an additional 86 showed moderate resistance (China National Corn Borer Research Group, 1983).

Since 1983, screening resistant varieties for the major plant diseases and insect pests has been the responsibility of the National IPM Technique Research Projects. More than 20,000 varieties, hybrids or lines of rice, wheat, cotton and maize have been screened. More than 2000 varieties have been verified to have some resistance, and over 40% of resistant germplasms have concurrent or multiple resistance to two or more pests (Guo, 1999).

Transgenic Bt cotton entered commercial use in China in 1998. Two different sources of Bt cotton expressing Cry1A insecticidal protein were used, one developed in the USA and the other developed in China by using the pollen tube pathway transformation method (Guo, 1995; Ni *et al.*, 1998). Acreage of Bt cotton in China increased from 80,000 ha in 1998 to over 0.3 million ha in 1999. In the northern China cotton belt, the total cotton acreage in Shandong and Hebei provinces in 2000 was over 0.6 million ha, 90% of which was Bt cotton (Zhao *et al.*, 2000). Bt cotton plants have resulted in a 60–80% decrease in the use of foliar insecticides (Xia *et al.*, 1999). Estimated economic benefits averaged US$250/ha in 1998–2000 (Jia *et al.*, 2001).

Successful Examples of IPM

In rice

The yellow stem borer, rice planthoppers and the rice leaf-folder are the most serious pests in Guangdong Province. Dasha Township in Guangdong is an experimental base for rice IPM established by the Institute of Entomology of Zhongshan University. It was the first township in China to implement IPM. IPM teams were formed at village and township levels and farmer schools were established. Farmers, village technicians, township extension experts, and university professors designed the IPM strategies jointly. The system includes using varieties with multiple pest resistance, cultural practices, sanitary treatments on seed, growing strong seedlings, planting in rational densities, fertilizer and water management, conservation and utilization of natural enemies, rational use of chemical pesticides and farmer training (Pu *et al.*, 1984; Zhang and Gu, 1998). By using these strategies, populations of rice insect pests in this township were lowest in the Zhaoqing City. Rice planthoppers were kept under control in 98% of the rice fields, and populations of other insect pests also declined due to natural enemies. Natural enemies in rice fields without chemical control kept populations of rice planthoppers lower than in those under chemical control. The long-term preservation and utilization of natural enemies in large rice fields can not only enhance the species and diversity of natural enemies, but also promote natural enemies and strengthen biological control of rice planthoppers (Zhang *et al.*, 1996). Demonstration and extension areas for the rice IPM system were 73,000 and 6,700,000 ha, respectively, in 1994–1995. The cost of pest control was reduced by 30%, and the net benefits increased by US$56/ha (He and Gu, 1996).

The brown planthopper migrates to the north in summer on the airflow of monsoons from the southwest, and back to the south in autumn on the northeastern wind (Cheng *et al.*, 1979). In the southern area, the major initial population comes from Southeast Asia outside the China mainland. A long-term forecasting method for the planthopper outbreak based on the insect origin was developed, and forecasting results were effective in predicting the actual population. Imidacloprid proved be an excellent insecticide in rice IPM, and was used for the control of planthoppers based on forecasting and field monitoring (He and Gu, 1996).

In wheat

Wheat stripe rust (*Puccinia striiformis*) is an important disease of wheat in China, with 31 physiological races of the rust found in the past 40 years (Wu *et al.*, 1993; Wang *et al.*, 1996). When a new race appears, some varieties of wheat resistant to other races may be susceptible to the new race. It is very important to breed wheat resistant to new races and to monitor for new races of wheat stripe rust. For example, a wheat stripe rust outbreak occurred on large areas in 1990. One of the key factors was a new, highly virulent race, CY29 (Wu *et al.*, 1993). Another factor was suitable weather for the disease. Researchers discovered the occurrence of this race and were able accurately to forecast that the stripe rust would be prevalent in 1990. In an IPM demonstration plot, the fungicide triadimefon was used at the proper time and efficient cultivation practices were used, keeping the disease under control. The average wheat yield was 40–70% higher than in non-demonstration areas (Guo, 1999).

The stripe rust pathogen cannot survive the summer in most of the wheat region in China where the mean temperature in mid-summer is over 20°C. It can only survive the summer in restricted regions at an altitude of

1450–1650 m. East and south Gansu and northwest Sichuan are the most important sources of rust infection (Wang *et al.*, 1988). IPM of wheat was implemented in the Longnan region (South of Gansu Province), a key source of wheat stripe rust infection. By analyzing the ecology of the wheat field system, a more complete IPM system was established for control of wheat pests, especially wheat stripe rust. The system changes the deployment of crops, making cultural techniques a foundation, then uses pest-resistant cultivars. By monitoring the population dynamics of local major wheat diseases and insect pests, and then applying highly effective and selective pesticides as auxiliary measurements, the program proved successful. In addition, the sustainable management of wheat diseases and insect pests in the Longnan region will not only delay the emergence of new races of the stripe rust, but also reduce the pathogen source and decrease the threat to major wheat production areas in central and eastern China (Wu *et al.*, 1999).

In maize

Maize is the most important crop in northeastern China. Major maize pests are the Asian corn borer, stalk rot, northern leaf blight and the corn head smut. Utilization of multiple resistant hybrids is considered the most fundamental component of maize IPM. As a result of IPM tactics with resistant hybrids, northern leaf blight is under control for most of the commercial maize hybrids. In the region where corn head smut is a major problem, use of resistant varieties is the first choice, and a seed-coating agent is also very effective at controlling the disease with efficacy over 80%. Maize stalk rot can be kept under control by using resistant varieties combined with increasing the potash fertilizer.

Asian corn borer

The Asian corn borer is the most important insect pest in maize and a number of IPM strategies have been established for its control. In early spring, piles of infested maize stalks are treated with *Beauveria bassiana* preparations to control the overwintering generation. The fungus can kill 82% of the overwintering larvae, significantly decreasing the number of egg masses in the maize field and reducing the percentage of infested plants. In addition, intermediate resistant hybrids were recommended before the introduction of highly resistant Bt maize.

A mass trapping system was also developed. During the emergence of overwintering moths, a light trap with a high intensity mercury-vapor lamp and a pool trap were spaced 150 m apart in a checkerboard pattern in a village where maize stalks were stored. From 1987 to 1989, 63% of the wild population was captured, leading to a reduction of 71.1% and 66.1% of egg masses and infested plants, respectively, in the adjacent area (Wang *et al.*, 1990). Based on the dispersal ability and migration possibility of the borer, the 'minimum effective area' for a population control program using a trapping system was determined to be not less than 50 km^2 surrounding each hibernation site (Wang, Z.Y., *et al.*, 1994).

Biological control with mass release of *Trichogramma* has been used against the Asian corn borer since the 1970s. The most effective release time was determined by monitoring the borer pupation rate. When the pupation rate of the overwintering generation reached 10%, the first release of *Trichogramma* was made 10 days later. A second release was made after 7 days. A total of 150,000–300,000 wasps/ha were released, which resulted in a parasitization rate of 65–85%. The cost in the areas where one generation of corn borer occurs was about US$1.6/ha, which was US$0.5/ha less than if chemical insecticides were used. In two-generation areas, additional *Trichogramma* releases were needed when the egg masses of the second generation were observed, leading to an average reduction of the Asian corn borer population of 46.3% to 73.6% in the autumn (Cong *et al.*, 2000; Liu *et al.*, 2000). In areas with continuous large-scale releases of *Trichogramma* for more than 10 years, the number of overwintering larvae per 100 stalks decreased from over 150 to fewer than 10. The mean parasitization rate of borer egg masses was 76% on 72,400 ha in 1988 in Yushu City of Jilin Province, compared with 12% in a non-released area (Wang *et al.*, 1998). In the case of an outbreak year, chemical insecticide granules and Bt formulations are applied when the maize is in the late whorl stage.

Key Constraints

One of the challenges for IPM implementation in China is the lack of a single organization in charge of IPM research and extension, although there is close collaboration between the different agencies. Also, most farmer families grow crops on a very small scale because of the rural reform policy of the early 1980s. It is difficult to reach millions of small farmers with an IPM extension program because of the many isolated farms. Further reform in agricultural education, research and extension systems will have significant influences on IPM in the future in China.

Important Websites and Publications in China

Information on IPM in China is usually in Chinese, although there are abstracts in English for the papers in the academic journals. Some important websites are: www.ipmchina.net www.nasesc.gov.cn www.weeds.net.cn Important publications include: *Acta Phytophylacica Sinica, Acta Entomologica Sinica, Scientia Agricultura Sinica, Acta Phytopathologica Sinica, Plant Protection, Chinese Journal of Biological Control, Natural Enemies of Insects, Plant Protection Technology and Extension, Entomological Knowledge, Journal of Weed Science, Pesticides*, and *Pesticide Science and Administration*. There are also journals published by agricultural universities and provincial/regional agricultural academies of different provinces with articles on IPM.

References

Bao, J.Z. and Gu, D.X. (eds) (1998) *Biological Control in China*. Shanxi Science and Technology Press, Taiyuan, Shanxi, 644 pp.

Chen, S.J. (1962) The earliest biological control method in the world – the liberation and breeding of the yellow citrus ant (*Oecophylla smaragdina* Fabr.) in citrus orchards and its significance in practice. *Acta Entomologica Sinica* 11, 401–408.

Cheng, S.N., Chen, J.C., SI, H., Yan, L.M., Chu, T.L., Wu, C.T., Chien, J.K. and Yan, C.S. (1979) Studies on the migrations of brown planthopper *Nilaparvata lugens* Stål. *Acta Entomologica Sinica* 22, 1–21.

Chiang, H.C. (1977) Pest management in the People's Republic of China. Monitoring and forecasting insect populations in rice, wheat, cotton and maize. *FAO Plant Protection Bulletin* 25, 1–8.

China National Corn Borer Research Group (1983) An evaluation of resistance of corn varieties to the Asian corn borer. *Plant Protection* 9(2), 41–42.

Chou, I. (1980) *A History of Chinese Entomology*. Entomotaxonomia Press, Wugong, Shaanxi, China, 213 pp.

Cong, B., Yang, C.C., Yang, S.W., Li, Y.Q., Wen, H., Liu, W., Yi, Y.Q., Fu, B. and Chen, G. (2000) Establishment, development and perspective of IPM of maize diseases and pest insects in Liaoning Province. *Journal of Shenyang Agricultural University* 31, 413–417.

Dai, K.J., Ma, Z.J., Pan, D.S., Xu, K.J. and Cao, A.H. (1996) Standardization on mechanized production of artificial host eggs of *Trichogramma* and technique of propagating wasp. In: Zhang, Z.L., Piao, Y.F. and Wu, J.W. (eds) *Proceedings of the National Symposium on IPM in China*. China Agricultural Scientech Press, Beijing, pp. 1138–1139.

Ding, H.J., Ni, H.X., Sun, J.R., Chen, J.L., Li, S.J., Ma, Q.X., Liu, H.Y., Zhang, Z.S. and Wang, J.L. (1994) Application of accumulated insect/disease/day value for determining the mixed damage and the dynamic economic threshold. In: Li, G.B. and Guo, Y.Y. (eds) *The Key Techniques of IPM on Cereal Crops and Cotton in China*. China Science and Technology Press, Beijing, pp. 267–275.

First Editorial Division, China Standard Press (1999) *Collection of National Standard for Pesticide Residue Limits in China*. China Standard Press, Beijing, 213 pp.

Gao, S.G. (1983) Protection and application of natural enemies in the integrated control of paddy insects. *Natural Enemies of Insects* 5, 49–53.

General Forecasting Station of Crop Pests of China (1983) *Data Collection for Crop Insect Pests and Plant Diseases Forecasting (1964 to 1979).* China Agriculture Press, 1719 pp.

General Station of Plant Protection of China (1988) Control of Diseases and Pests in the 36 years in China. In: General Station of Plant Protection of China (ed.) *The Developments in Integrated Control of Rice Diseases and Insect Pests in China.* Zhejiang Science and Technology Press, Hangzhou, pp. 1–8.

Guo, S.D. (1995) Engineering of insect-resistant plant with *Bacillus thuringiensis* crystal protein genes. *Scientia Agricultura Sinica* 28(5), 8–13

Guo, Y.Y. (1999) The measures and results in crop pest disaster reduction in China. In: China Association of Agricultural Sciences Societies (ed.) *Research Progress in Plant Protection and Plant Nutrition.* China Agriculture Press, Beijing, pp. 6–9.

Guo, Y.Y., Cao, Y.Z., Li, S.G., Wu, Y.Q., Liu, A.Z. and Hu, Y. (1988) The damage and dynamic action threshold of wheat aphids mixed population in wheat ear stage. *Plant Protection* 14(3), 2–7.

Guo, Y.Y., Dai, X.F., Liu, Q.X., Gao, Z.R., Li, Q.S. and Qiu, F. (1994) On the dynamic compound action thresholds of second generation cotton bollworm and spider mites damaged to cotton in Northern China cotton area. *Acta Phytopathologica Sinica* 21, 339–344.

He, M. and Gu, Z.Y. (1996) Development of rice IPM systems. In: Qiu, S.B. (ed.) *Advance in Plant Protection in China.* China Scien-Technology Press, Beijing.

He, M., Gao, J.C., He, Z.Q., He, S.L. and Mao, J.H. (1991) Studies on the mixed damage and economic threshold of sheath blight and stem borer to hybrid rice. *Acta Phytopathologica Sinica* 18, 241–242.

Hou, Z.D., Zhang, Y. and He, Z.X. (1984) Dominant species of the predators and their control effect on the *Ostrinia furnacalis* in corn field. *Natural Enemies of Insects* 6, 236–239.

Jia, S.R., Guo, S.D. and An, A.D. (eds) (2001) *Transgenic Cotton.* Science Press, Beijing, 281 pp.

Lay, S.H. (1981) On the screening of resistant rice cultivars (*Pyricularia oryzae*) in Guangxi. *Acta Phytopathologica Sinica* 11(2), 1–5.

Li, G.B. (1990) The concept and principle of IPM. In: Li, G.B., Zeng, S.M. and Li, Z.Q. (eds) *Integrated Management of Wheat Pests.* China Agricultural Scientech Press, Beijing, pp. 25–54.

Liu, Z.C., Wu, J.Q., Xie, W.D., Wu, Y.N., Wu, S.C., Liao, W.T. and Hu, L.B. (1980) Preliminary study on the mechanized production of *Trichogramma. Acta Phytophylacica Sinica* 7, 233–237.

Liu, Z.C., Liu, J.F., Yang, W.H., Li, D.S. and Wang, C.X. (1996) Research on technological process of *Trichogramma* produced with artificial host egg and quality standard. *Natural Enemies of Insects* 18(1), 23–25.

Liu, Z.C., Liu, J.F., Zhang, F., Li, D.S. and Feng, X.X. (2000) *Production and Field Application Techniques of* Trichogramma. Golden Shield Press, Beijing, 119 pp.

Lu, Z.Q. and Gu, Z.F. (1988) Progress and analysis for the insect-resistant breeding of rice. *Scientia Agricultura Sinica* 21(4), 1–8.

Lu, Z.Q., Shen, G.Q. and Han, Y.G. (1991) Injury equivalence system of leaf-mass consuming insect guild on rice and the multiple-species economic threshold. *Acta Entomologica Sinica* 34, 319–325.

Ma, S.J. (1976) On the integrated control of agricultural insect pests. *Acta Entomologica Sinica* 19, 129–141.

Nan, L.Z., Wang, D.A., Sun, X. and Li, X.Z. (1987) Study of protection and application of natural enemies of cotton insect pests. *Natural Enemies of Insects* 9(3), 125–129.

Ni, W.C., Zhang, Z. and Guo, S.D. (1998) Development of transgenic insect-resistant cotton plants. *Scientia Agricultura Sinica* 31(2), 8–13.

Piao, Y.F. (1998) *List of Natural Enemy Species of Major Crop Insect Pests in China.* China Agriculture Press, Beijing.

Piao, Y.F. (1999) Discussion on plant protection policy and the development of IPM in China. In: National Agriculture Technology Extension and Service Center (ed.) *Research Progress in Crop Pests Sustainable Management.* China Agriculture Press, Beijing, pp. 19–25.

Pu, Z.L., Gu, D.X., Zhou, H.H., Tang, J.Q., Zhang, Y.J. and Zhang, X.D. (1984) Integrated control of rice pest insects in Dasha Township, Sihui County, Guangdong Province. *Scientia Agricultura Sinica* 9(4), 73–80.

Qiu, S.B. (1976) Keep the key strategy – 'Putting prevention first and integrated control' in plant protection. *Scientia Agricultura Sinica* 17(1), 41–47.

Shi, K. (1996) Research and application on technique of protection the natural enemies to control wheat aphids by rape trap belts in wheat field. In: Zhang, Z.L., Piao, Y.F. and Wu, J.W. (eds) *Proceedings of the National Symposium on IPM in China.* China Agricultural Scientech Press, Beijing, pp. 448–452.

Wang, C.L., Zhang, J., Huo, S.T., Zhou, Y.N., Feng, J.G. and Liu, Z.C. (1998) Study, reproduction and utilization of *Trichogramma.* In: Bao, J.Z. and Gu, D.X. (eds) *Biological Control in China.* Shanxi Science and Technology Press, Taiyuan, Shanxi, pp. 67–123.

Wang, F.L., Wu, L.R., Xu, S.C., Jing, S.L., Jia, Q., Yuan, W. and Yang, J. (1996) The discovery and studies on new races CYR 30 and CYR 31 of wheat stripe rust in China. *Acta Phytophylacica Sinica* 23(1), 40–44.

Wang, K.N., Xie, S.X., Liu, X.K., Wu, L.R., Wang, J.X. and Chen, Y.L. (1988) Progress in studies on control of wheat stripe rust in China. *Scientia Agricultura Sinica* 21(2), 1–8.

Wang, R.Q., Zhang, S.F., Fan, X.X., Xiong, Y.Q., Zhou, B.L. and Gi, L. (1994) Studies on the complex damages by *Pectinophora gossypiella* (Saunders) and *Tetranychus cinnabarrinus* (Boisduval) and their dynamic economic threshold. *Acta Gossypii Sinica* 6(1), 53–56.

Wang, X.G., Liu, L.C., Ma, Y.H. and Cui, H.Y. (1991) The natural enemies in wheat field transfer to cotton field. *Natural Enemies of Insects* 13(2), 79–81.

Wang, Y.S., Zhang, Y., Yue, D.R. and Yang, G.H. (1990) A preliminary report on the application of lamp trap against Asian corn borer. *Plant Protection* 16(2), 32–33.

Wang, Z.Y., Zhou, D.R., Song, Y.Y., Li, B.X., Zhang, G.Y., Gao, S.L., Liu, Y., Zheng, L., Wang, Y.S., Xie, W.M., Li, W.D. and Pan, Y.C. (1994) Studies on behavior of dispersal and possibility of migration in adults overwintering generation Asian corn borer by using release-and recapture technique. *Acta Phytophylacica Sinica* 21(1), 25–31.

Wen, L.P., Wang, Z.Y., Ye, Z.H. and He, K.L. (1992) Yield losses and economic threshold of Asian corn borer, *Ostrinia furnacalis* (Guenee). *Scientia Agricultura Sinica* 25, 44–49.

Wu, L.R., Yang, H.A., Yuan, W.H. and Song, W.Z. (1993) On the physiologic specialization of stripe rust of wheat in China during 1985–1990. *Acta Phytopathologica Sinica* 23, 269–274.

Wu, L.R., Wan, A.M., Niu, Y.C., Jin, S.L. and Li, J.P. (1999) On the integrated pest (stripe rust mainly) management of wheat in Longnan region. In: China Association of Agricultural Sciences Societies (eds) *Research Progress in Plant Protection and Plant Nutrition.* China Agriculture Press, Beijing, p. 225.

Xia, J.Y., Cui, J., Ma, L., Dong, S. and Cui, X. (1999) The role of transgenic Bt cotton in integrated pest management. *Acta Gossypii Sinica* 11, 57–64.

Yang, P., Chen, S.P. and Zheng, J.H. (1983) Yellow citrus ant, *Oecophylla smaragdina* Fabricius, in Fujian Province. *Natural Enemies of Insects* 5, 59–60.

Zeng, G.R., Song, Y.X., Shen, Y.C., Xie, D.L., Lin, D.X., Kong, J. and Zhu, C.X. (1998) Research and application of agricultural antibiotics. In: Bao, J.Z. and Gu, D.X. (eds) *Biological Control in China.* Shanxi Science & Technology Press, Taiyuan, Shanxi, pp. 521–567.

Zeng, Z.H., Li, Y.C., He, L. and Li, D.Y. (1988) Strategy and set forming techniques for integrated control of rice insect pests and diseases. In: General Station of Plant Protection of China (ed.) *The Developments in Integrated Control of Rice Diseases and Insect Pests in China.* Zhejiang Science and Technology Press, Hangzhou, pp. 9–20.

Zhang, R.J. and Gu, D.X. (1998) Practice of integrated pest management in rice-based ecosystem in Guangdong Province. In: *Proceedings of Abstracts, International Conference of Integrated Pest Management (IPM), Theory and Practice, Developing Sustainable Agriculture.* Guangzhou, China, p. 57.

Zhang, W.Q., Zhang, G.R., Gu, D.X., Lin, Z.Y., Huang, Y.H., Xie, K.X., Yang, J. and Lin, X.D. (1996) The control effect of long-term preservation and utilization of natural enemies in paddy field against brown planthoppers. *Acta Phytophylacica Sinica* 23, 131–136.

Zhao, G.Z. and Yuan, K.L. (1990) The advance of ecological control and forecast of the damage by rodent in field. *Scientia Agricultura Sinica* 23(5), 53–59.

Zhao, J.Z. (1984) Species, distribution and the population fluctuation of predatory spiders in the cotton fields in China. *Natural Enemies of Insects* 6, 1–12.

Zhao, J.Z. (1998) Protection and utilization of farmland spiders. In: Bao, J.Z. and Gu, D.X. (eds) *Biological Control in China.* Shanxi Science and Technology Press, Taiyuan, Shanxi, pp. 251–273.

Zhao, J.Z., Yang, Q.H., Cheo, M.T., Ma, Z.S. and Li, B.L. (1991) The effects of attractive crops interplanted in cotton field on the protection and propagation of natural enemies of cotton insect pests. *Acta Phytophylacica Sinica* 18, 339–343.

Zhao, J.Z., Rui, C., Lu, M., Fan, X., Ru, L. and Meng, X. (2000) Monitoring and management of *Helicoverpa armigera* resistance to transgenic Bt cotton in Northern China. *Resistant Pest Management Newsletter* 11(1), 28–31.

Zhou, D.R. and He, K.L. (1995) *Asian Corn Borer and Its Integrated Management.* Golden Shield Press, Beijing, 102 pp.

Chapter 17

Integrated Pest Management in India

Amerika Singh[1], Saroj Singh[1] and S.N. Rao[2]

[1]National Centre for Integrated Pest Management, Pusa Campus, New Delhi, India;
[2]A.N.G. Ranga Agricultural University, Andhra Pradesh, India

Introduction

India's population is about 1 billion people. More than 70% of India's population lives in rural areas where the main occupation is agriculture. Agriculture engages around 66% of the total work force and contributes 25% to the gross domestic product of the country.

Insect pests and diseases are dynamic components of ecosystems. Changes in cropping pattern and other production inputs have resulted in major changes in pest complexes during the post Green-Revolution years. It is now well understood that the pests continue to be the major constraints in stabilizing crop yield and also in realizing the full potential of research findings. It has been estimated that on average 18% of the crop production worth more than rupees 292,400 million (US$8600 million) is lost annually due to pests (insect pests, plant pathogens, nematodes, rodents and weeds, etc. (Dhaliwal and Arora, 1996). The losses are likely to mount with increasing monocropping, fertilization, irrigation and other important features of intensive agriculture which may intensify further in coming years to produce more and more food and fiber to meet the growing demands of the increasing population. At least half of the total losses due to pests can be avoided by adopting the sustainable IPM, which is ecologically safe, cost-effective and farmer friendly.

It has been proved beyond doubt that IPM is a valid solution leading to sustainable production and food security. The greatest challenge is to do this without harming the environment and depleting the resource base for future generations. IPM is expanding in the Indian subcontinent also, but with less vigor. A rapid adoption of IPM is now called for, so that the goal to achieve long-term sustainable systems of crop protection and production can be achieved before it is too late. For this, education, an effective information dissemination system, motivation and mass production of the IPM program need to be developed.

Recent pest-related disasters in cotton (whitefly in 1983 and 1984, *Helicoverpa armigera* in 1987 and 1988) caused major public concern. Because of these disasters, cotton growers and extension workers are now more aware of problems associated with excessive insecticide use. Studying the underlying causes of pest outbreaks and modifying management systems to prevent them has become the focus of pest management. New technology, biotechnology and improved chemical pesticides are being developed to accelerate the adoption of IPM and increase its potential application.

(eds K.M. Maredia, D. Dakouo and D. Mota-Sanchez)

Losses Due to Pests and Emerging Pest Problems

The major constraints limiting higher productivity are the onslaught of insect pests and diseases, which are favored by intensive agriculture. It is estimated that herbivorous insects eat about 20% of the crop grown for human consumption. The development of resistance, secondary pest outbreaks and emergence of new pest problems had amounted to increased monetary losses. The annual crop losses due to insect pests and diseases in India vary up to 38% (insect pests and diseases 26%, weeds 10% and birds 1–2%). Reports indicate that the losses caused by the specific major pests may be higher. *Helicoverpa* in cotton causes yield losses of up to 20–25%. Raheja and Tiwari (1997) had reported that losses due to American boll worm alone may be around rupees 10,000 million annually while the losses due to insect pests and diseases in rice (18.6%) amounted to rupees 55,120 million (US$1102.4 million) (Table 17.1). The overall losses due to insect pests were estimated to be rupees 60 billion (US$1.2 billion) in 1983 (Rao Krishnamurthy and Murthy, 1983), rupees 200 billion (US$4 billion) in 1993 (Jayaraj and Regupathi, 1993) and rupees 290 billion (US$5.8 billion) in 1996 (Dhaliwal and Arora, 1996). In recent years, the leaf curl disease of cotton has been reported in the states of Rajasthan, Punjab and Haryana in virulent form and more than 100,000 ha of cotton have been infected with this disease in Rajasthan alone. Insect pests inflict direct losses and also act as vectors for transmission of several viral diseases, e.g. aphids are reported to transmit about 160 viruses.

Table 17.1. Estimated crop losses in important crops. (Source: Lal, 1996.)

Crop	Loss (%)	Rupees (million)
Rice	18.6	55,120 (US$1102.4 million)
Wheat	11.4	14,150 (US$283 million)
Jowar	10.0	1,732 (US$34.64 million)
Pulses	7.0	4,840 (US$96.8 million)
Oilseeds	25.0	41,800 (US$836 million)
Cotton	22.0	20,000 (US$400 million)
Sugarcane	15.0	13,360 (US$267.2 million)

New pest problems have appeared owing to intensive crop production technologies and changing cropping patterns. A change in the host scenario of the pest is noticed. The serpentine leaf miner (*Liriomyza trifolii*), spiraling white fly (*Aleurodicus disperses*), coffee white stem borer (*Xylotrechusm quadripes*) and mango borer (*Deonalis albizonalis*), cotton root rot (*Rhizoctonia solani*) and leaf curl virus disease on cotton are some of the new pest problems that need priority attention and warrant immediate management to prevent further spread.

Pesticide Use – Indian Scenario

Pesticides will remain a key means of intervention of most IPM strategies and their injudicious use represents the greatest threat to IPM, and yet has provided the catalyst for virtually all IPM programs (Singh and Dubey, 1996).

India is the second largest manufacturer of basic pesticides in Asia with 165 registered pesticides in the country and it accounts for less than 2.5% of the world markets in value terms. Consumption is also low in India at 288 g/ha compared with 12,000 g/ha in developed countries. Insecticides account for nearly 76% while herbicides account for only 10% of the pesticides usage in India.

The peculiar feature of this sector is that the use is skewed in favor of a few cash crops (Table 17.2).

Regional variation is also evident in pesticide use in India. The states of Andhra Pradesh, Karnataka and Gujarat account for 65% of the total pesticide use, with 33.6% in Andhra Pradesh alone. In the state of Tamil Nadu, pesticide use has decreased by more than 50% during the last 7 years. A decreasing trend is also evident in Andhra Pradesh and Karnataka. In contrast, pesticide use continues to rise rapidly in the states of Punjab and Rajasthan (Table 17.3).

Table 17.2. Pesticide use on major crops in India.

Major crops	Market share (%)
Cotton	40
Rice	14
Vegetables	8
Wheat	6
Pulses	5
Tea	5
Others	22

Table 17.3. Total pesticide use in different States. (Source: Directorate of Plant Protection Quarantine & Storage.)

State	Consumption (million t)
Uttar Pradesh	7459
Punjab	6972
Haryana	5025
Andhra Pradesh	4054
Gujarat	3646
Maharashtra	3614
West Bengal	3370
Karnatka	2484
Tamil Nadu	1685

Origin and Evolution of IPM in India

The government conducted pilot projects on IPM from 1975 to 1980. Pest surveillance activities begun in 1980 soon after India became part of the FAO Inter-country IPC Rice Program, and have now been extended to all major crops. Since then, Indian scientists have made many contributions to the development of IPM systems in India. Biological control, the development of resistant crop varieties, cultural controls, and the use of botanicals such as neem are all used in Indian agriculture. Crop diversity, intercropping, and the small size of most farms are the factors that favor the implementation of IPM. Indian scientists and extension workers are now well aware of problems that can result from improper use of pesticides, and the concept of an economic threshold justifying pesticide use is well recognized.

The Ministry of Agriculture has established 32 IPM centers under the Directorate of Plant Protection, Quarantine and Storage to act as catalysts and experiment stations for IPM programs. These centers educate state extension workers and farmers through training and demonstrations. But since these centers can reach only 5% of the crop area, it has been proposed by the Ministry of Agriculture to add 228 IPM centers to be established by state governments. Half of the cost would be borne by the central government. In the near future, there should be about 550 such centers in India.

The ICAR has established NCIPM in New Delhi to plan and coordinate the IPM research and development programs in collaboration with the SAUs and other ICAR institutions (Appendix 17.1). A project Directorate of Biological Control has been created to support biological control programs.

Easily adoptable and economically viable IPM strategies have been developed for the control of major pests in rice, cotton, pulses, oilseeds and sugarcane. Conservation of biological control has been especially successful, by either selective use of pesticides or their avoidance. Augmentative release of biological control agents has successfully controlled pyrilla and top borer of sugarcane, mealybug of coffee, and lepidopterous pests affecting cotton, tobacco, coconut and sugarcane. The development of mass-rearing technology for biotic agents such as *Trichogramma* spp., *Chrysoperla* spp. and NPVs of *Helicoverpa armigera* and *Spodoptera* spp. has been a major achievement.

IPM is also an integral component of crop improvement research. The various disciplines are incorporated in crop research institutes and the All India Coordinated Crop Improvement Projects of the ICAR.

National IPM Policy in India

In 1985, the Government of India adopted IPM as the official guiding principle of plant protection strategies in government-sponsored crop production programs. The Government of India is also a signatory to Agenda 21 of the United Nations Conference on Environment and Development, which promotes IPM to reduce the use of pesticides in agriculture. Overall, the Government of India has taken a number of initiatives for the promotion of IPM (Paroda, 1997). Notable initiatives taken by the Government of India for the promotion of IPM include:

1. Infrastructure development
- Enhanced budgetary provisions for promotion of IPM.
- Assistance to 30 states for establishing biocontrol laboratories.
2. Human resource development
- Organizing season-long training programs for IPM trainers.
- Establishing FFS to train agricultural extension officers and farmers.
3. Policy support
- Phasing out pesticide subsidies and diverting the savings to promotion of IPM programs.
- Emphasis on production and release of biological control agents.
- Phasing out, banning, or restricting the use of hazardous pesticides.

IPM Strategies Used by Farmers

Trap cropping

Cabbage

A successful example of IPM in cabbage is the use of trap crops for the control of diamondback moth. Growing Indian mustard in paired rows at the edge and after every 25 rows of cabbage attracts 80% of the diamondback moth and entire populations of leaf webber, stem borer, bugs and aphids. Control of remaining diamondback moth can be achieved with 4% neem seed kernels applied at the head-initiation stage of the crop. This can be repeated two or three times at 10–15 day intervals if necessary. This treatment is safe for *Cotesia plutellae*, a dominant natural enemy of diamondback moth. Irrigating in the evening can also help to control diamondback moth.

Tomatoes

Entomologists at the Indian Institute of Horticultural Research in Bangalore developed an IPM system for the management of tomato fruit borer, *Helicoverpa armigera*. Trap cropping using marigold after every eight rows of tomato plants attracts most of the ovipositing tomato fruit borers. Mechanically destroying the trap crop usually eliminates the pest population, but the use of conventional insecticides reduces its attractiveness. The residual pest populations on the tomato crop can be reduced through the application of nuclear polyhedrosis virus. However, this recommendation does not work in all situations. In temperate regions of the country, marigold does not attract *H. armigera*.

Maize

Sorghum, if interplanted with maize, can effectively trap *Chilo suppresalis* as it is a preferred host and does not require chemical treatment.

Host plant resistance

Host plant resistance has been used successfully in India. In a study on rice, higher mortality of leafhoppers and plant hoppers occurred on resistant varieties than on susceptible varieties (Heinrichs, 1994). Mortality of brown planthoppers reared on moderately resistant ASD7 or highly resistant Sinna Sivappu was higher than on a susceptible TN1 cultivar. The integration of host plant resistance and insecticides has a cumulative effect on *Nephotettix virescens*, the vector of rice tungro virus, and in at least one case, there was no tungro virus infection on a resistant cultivar IR28 without application of insecticides.

Biological and cultural controls

An IPM program based on a combination of biological and cultural controls has been developed for the management of rhinoceros beetle, *Oryctes rhinoceros*, a major pest of coconut palm. Two pathogens affect rhinoceros beetle: a baculovirus and a fungus, *Metarhizium anisopliae*. Release of baculovirus-infected beetles has been successfully used, but this approach occasionally fails due to development of resistance. Therefore, cultural control methods have been developed, such as planting legumes to cover potential breeding sites and leaving some dead standing palm to aid the spread of disease. In addition, beetle-breeding areas are treated with *M. anisopliae*. Virus-infected beetles are released in case the pest population increases (Pillai *et al.*, 1993).

Another successful example is mechanical control of cockchafer beetles (*Holotrichia* spp., *Adoretus* spp., *Schizonycha* spp., and *Anomala* spp.). These beetles are found throughout the country and have a wide host range. They are controlled by shaking host trees vigorously at night, and collecting them on a sheet of cloth. This collection method works best if started with the onset of pre-monsoon showers, when the beetles emerge, and continued for 5–6 days. This is the cheapest and the most effective method of control. Cultural methods also can be successful, by plowing frequently and exposing beetles to predators. Beetles are attracted by low intensity light and can be killed in kerosenized water. Many host plants such as *Moringa oleifera*, *Carissa caranda*, *Azadirachta indica*, and *Ziziphus mauriliana* can act as trap plants. The adults are attracted to heaps of manure and plant debris for egg laying, where they can be treated with pesticides. Other cultural methods include careful timing of planting. In sunflower, early sowing favors important predators such as the green lacewings, *Chrysopa* spp., and ladybird beetles. This method has been successful against green jassid, cabbage semi-looper and head bug. Similarly a number of microbials such as HaNPV, Bt, *Beauveria*, *Nomuraea*, *Verticillium*, *Aspergillus*, *Trichoderma*, *Gliocladium*, *Bacillus* and *Pseudomonas* have proved efficient, however, only a few could be used commercially. Among these the HaNPV, Bt and *Trichoderma* are becoming popular among farmers.

New ways of using pesticides

Pesticides will continue to be an important part of IPM programs, although currently, the availability of selective pesticides is very limited. To reduce exposure, broad-spectrum pesticides can be applied as seed treatments, granules, stem injection or leaf axial application to avoid injury to natural enemies. Sprays can be timed to avoid bee activity and adult natural enemies (Jayaraj *et al.*, 1994). For example, a single spray of systemic insecticide can easily control peach leaf curl aphid at the pink bud stage when natural enemies are scarce.

Botanical pest repellents

Plant extracts of some rice varieties are highly toxic to important rice pests such as *Chilo suppressalis*, *Nilaparvata lugens* and *Sogatella furcifera*, but safe to predators such as *Cyrtorhinus lividipennis* (Arora and Dhaliwal, 1994). Extracts of *Mentha arvensis*, *M. sylvestris* and *Adhatoda vasica* can help control ants on potato. *Mentha*, if grown with potatoes, has a repellent effect.

Examples of Successful IPM in India

Several organizations, including pesticide companies, have been actively involved in developing IPM techniques for adoption by farmers. Consequently, comprehensive IPM programs are developed for rice, cotton, sugarcane, pulses, oilseeds and vegetables.

Sugarcane

Two notable examples of successful biological control in India are control of the sugarcane top borer, *Scirpophaga excerptalia*, with an indigenous larval parasite, *Isotema javensis*, and the control of sugarcane pyrilla, *Pyrilla perpusella*, with *Epiricania melanoleuca*.

Inundative releases of *Trichogramma chilonis* and *T. japonicum* have been found to control borers effectively. The technology for mass-production of *Trichogramma* spp. and their field release is also available (ICAR, 1991).

Cotton

The IPM technology for rain-fed cotton was first formulated by NCIPM and tested in collaboration with cotton research station Nanded of Marathwada Agricultural University, Parbhani. Initially three modules, namely bio-intensive, biocontrol + intercrop, biocontrol + insecticide were evaluated in comparison to farmers' practices over 10 ha during 1997/98. The most promising module, namely bio-intensive, was taken up for large-scale validation and promotion in 200 ha during 1998/99 and was continued for 4 years. The main IPM interventions of this module were seed treatment with imidacloprid, scouting, placement of pheromone traps for monitoring, two releases of the egg parasitoid *Trichogramma chilonis*, one spray of HaNPV and two or three sprays of Neem Seed Kernal Extract (NSKE).

The IPM module resulted in substantial reduction of pesticide use and conservation of natural fauna. The IPM technology provided higher net returns and yields over the farmers' practices (non-IPM) were 1:1:76 and the increase in monetary gains were to the tune of 62.3%/ha.

Basmati rice

NCIPM initiated an IPM program in basmati rice in 1994 with a few acres of land area during *kharif* 2001. Later an entire village (Shikohpur) was taken for IPM validation in basmati rice, with a total of 400 acres of land under Pusa Basmati-1. With the negligible use of pesticide during the past 2 years, this year the natural enemy population was recorded in abundance and insect pest incidence was also found to be comparatively less. Some improved crop management practices helped in improving the plant vigor and stand, such as planting of two or three seedlings per hill, planting of 'Dhaincha' (*Sesbania* sp.) for green manure before transplanting of rice and judicious use of fertilizer with addition of potash at 40 kg/ha. Regular pest surveillance and monitoring along with natural enemies of insect pests helped in reducing the IPM interventions to a bare minimum, which included seed treatment with carbendazim at 2 g/kg of seed, and one release of parasitoids *Trichogramma japonicum* when the incidence of leaf folder was found to be on the increase. Pesticide interventions included the use of carbendazim against sheath blight in a few infected patches (total area less than 10 acres) and streptomycin against bacterial leaf blight in some fields, not exceeding an area of 5 acres in total. Spraying of monocrotophos was done in a few fields (in about 2 acres) against gundhi bug and pollen beetle *Chiloloba* sp.

Thus, from an average of five or six pesticide applications during earlier years, the farmers have come down to less than one spray. The farmers under the IPM program with negligible pesticide use harvested an average of 5740 kg/ha in contrast to non-IPM farmers who harvested 4560 kg/ha with four or five sprays of pesticides.

Rapeseed–mustard

The IPM program on rapeseed–mustard was initiated by NCIPM in 1995. A study with three treatments was undertaken: (i) IPM; (ii) chemical control measures; and (iii) farmers' usual practices. The major components of IPM treatments included: timely sowing of the crop (15–25 October), seed treatment with *Trichoderma viride* at 2 g/kg seed, judicious use of fertilizers, mechanical removal of aphid infested twigs at the initial stage of attack etc. This module was validated in village Bhora Khurd district Gurgaon, Haryana during 1995 to 2001 on large area of 100 acres. Due to timely sowing, i.e. 15–25 October, the crop completely escaped from the incidence of mustard aphid (*Lipaphis erysimin*) white rust (*Albugo candida*) and *Alternaria* leaf spot during all the years of trial.

The farmers were educated about the concept of IPM, identification of crop insect pests, benefits of regular monitoring of insect pests through 'FFS of IPM' regularly.

As a result of this, a higher yield average 2100 kg/ha was obtained as compared to farmer's practices (control) 1700 kg/ha. No chemical pesticide application was done in the IPM plots after 1995 (Singh and Kumar, 2001).

Chickpea

An eco-friendly IPM program in chickpea at four locations in three states was initiated by NCIPM. The main IPM components were seed treatment with *Trichoderma* + Vitavax (carboxin), chloropyrifos and *Rhizobium*, pheromone traps for pest monitoring and foliar spray of HNPV and NSKE and one spray of endosulfan (need based). In farmers' practice (FP) around three foliar sprays of insecticides were applied as mixture and or alone (parathion methyl, dichlorvos and monocrotophos). The average yield obtained in IPM trials was 2700 kg/ha as against 1500 kg/ha in FP fields.

Funding and Linkages of IPM Programs

National

Central government

The IPM program in India is funded by the Central, State and International organizations like UNDP. A provision of rupees 447.1 million (US$8.9 million) has been made by the Government of India during IX plan (1997–2002) for promotion of IPM. In addition, the Department of Agriculture and cooperation under various crop improvement programs are also giving substantial funds to various states for IPM training and demonstration (Rajak, 2001).

The Department of Biotechnology is also providing financial assistance to various SAUs and research centers for developing and producing biopesticides and biocontrol agents such as NPV, granulosis virus, *Trichogramma*, *Chrysopa* and *Trichoderma*. Presently, ten biopesticide production units are able to cover an area of 1 million ha/annum in ten crops. Under this program, biopesticide production units and plant protection clinical centers at regional research stations have also been established.

State governments

State governments have intensified their efforts to promote IPM through demonstrations and training for extension personnel and farmers. The Central Government, ICAR, and SAUs are extending technical assistance.

ICAR

ICAR Crop Institutes and SAUs have focused on developing IPM programs with emphasis on:

- breeding resistant varieties;
- developing improved agronomic practices; and
- identification of potential biocontrol agents.

Private sector/NGOs

A few private agencies have set up commercial insectaries for mass rearing and supply of egg parasites and pathogens to farmers. However, these are far from meeting the demand. With assistance from the Department of Biotechnology, a few NGOs have started Agricultural Development Centers. Private Plant Clinic Centers also help to promote IPM programs.

International

FAO–ICP program in rice

FAO–ICP has pioneered the IPM program in rice, which has been widely accepted by state extension workers. Based on the success of nine season-long training programs, many states have suggested extending the scope of the training to other crops, especially cotton and vegetables. The FAO-sponsored training programs have helped to build a core of IPM trainers.

UNDP–IPM project

Development and strengthening of IPM has been implemented in India since 1994 with an outlay of US$2.37 million mainly for consultancy (project personnel), development of master training programs, training (fellowship, study tour, international workshop, master training, national workshop), purchase of equipment, and impact assessment.

Research Efforts on IPM in India

The ICAR is a premier research organization at the national level with the NCIPM and a network of 13 crop-based institutes, which are engaged in IPM-related research. These institutes have developed IPM modules for rice, cotton, and some horticultural crops. In collaboration with SAUs, these institutes are also developing IPM strategies for other crops including pulses and oilseeds. The NCIPM emphasizes region-wide crop-based IPM programs (Appendix 17.1). Other institutions including the Department of Biotechnology and some NGOs are also developing and promoting IPM technology.

A network of IPM centers has been established throughout the country for organizing IPM field demonstrations and training of farmers and extension workers. Over 10,000 demonstrations have already been conducted and 9000 extension officers have been trained in IPM.

Infrastructure development

IPM is knowledge-based, so coordination between research institutes and farmers is essential. Lack of a reliable database has hampered progress of IPM programs. Technology development for IPM involves 29 SAUs, one Central Agricultural University and other National Institutes, Crop Research Institutes, National Research Centers and a network of All India Coordinated Crop Improvement Projects, besides traditional universities. Synthesis of IPM modules, their evaluation and socioeconomic impact analysis are being carried out at NCIPM.

Commercialization of biocontrol agents and biopesticides

Efforts are being made to use biodegradable and renewable organic materials in IPM whenever possible. The government of India is providing financial assistance for 30 state biological control laboratories to promote the use of biological control agents. The Registration Committee has decided to promote use of biopesticides such as Bt, and allow their commercialization to promote IPM. Requirements for registration of these pesticides have been simplified in order to boost their production and import.

Registration and quality control

About 139 commercial insecticides have been registered so far and 35 are widely used. The Government has already banned 11 insecticides and has restricted the use of several others. Historically, the government subsidized pesticides by as much as rupees 580 million. To promote IPM, this subsidy has been withdrawn, the money diverted to promote IPM (Rajak, 2001).

IPM publicity

To promote and implement IPM, links with national and international agencies, SAUs, NGOs and the private sector have to be strengthened. Joint efforts of the national and state Departments of Agriculture, ICAR and SAUs are needed to publicize IPM on a wide scale.

Major Constraints to IPM in India

A major limitation is the lack of trained personnel. Many farmers are not trained adequately in augmentative biological control, leading to misunderstanding of its potential efficacy. Logistical problems such as improper timing and delays in shipment can alter the effectiveness of natural enemies. Farmers often believe that natural enemies do not work well, and that low pest populations will cause losses. The use of biopesticides is limited due to moderate toxicity, slow action, host specificity and photo-instability as well as a higher cost. Many farmers are not yet aware of the proper usage and available suppliers of biocontrol agents and biopesticides.

A number of botanicals such as karanj, mahua, nuxvomica, custard apple, ipomoea, garlic and tobacco have been found to be effective against insect pests and diseases, however in absence of detailed scientific data, except for neem, most of them are localized to rural pockets.

Botanicals, particularly neem, have not found much favor with farmers. The necessity for repeated applications, low toxicity and persistence, cumbersome procedures of collection and extraction coupled with low yields have discouraged wide use of neem. IPM adoption is influenced by the cost versus efficacy of products, need for sophisticated information for decision making, ability to integrate new products and techniques into existing farm management practices and managerial skills. Strategies that are being used now may need to be modified to achieve the goal of wider adoption of IPM.

Conclusion

The most important aspect of the IPM program in India is the community approach. Both national and state research organizations, along with SAUs, have been actively involved in developing IPM technology for farmers. As a result, a comprehensive package of IPM practices has been developed for rice, cotton, mustard, chickpea, pigeon pea and sugarcane crops. The Indian Council of Agricultural Research and Department of Agricultural Research and Education of the Ministry of Agriculture, Government of India are fully committed to the development and promotion of IPM in the country as evident from the fact sheets of allocations and crop/pest priorities. It is the top priority mission of the ICAR and Government of India to provide safe and effective technologies to protect against unacceptable losses caused by weeds, diseases and insect pests. There is urgent need for decision support software to be developed so as to allow IPM practitioners to estimate cost/benefit for a variety of management inputs and examine profitability of a system. Genetic engineering to enhance the potential of LMOs also needs priority in order to ensure a clean environment and food security. The ICAR and SAUs are continuing to develop IPM programs for other crops such as vegetables, oilseeds and pulses. However, IPM efforts have so far remained restricted to the research activities of ICAR, the SAUs, and the Central IPM Centers of the Ministry of Agriculture. Even though some successful non-chemical methods for control of crop insect pests and diseases have been developed, the transfer of this knowledge to the farmers and extension officers has been relatively slow.

Ideally, the IPM approach seeks to understand the causes of pest outbreaks and modify the design and management system to prevent them. Coordinated efforts of research institutes and extension personnel will continue to educate the farming community on IPM practices. Active participation of the farmers, quick dissemination of the technology, area-wide approach and timely supply of inputs including quality biocontrol agents along with new technology such as precision farming, i.e. broad combination of hardware, software, information technology and new product technologies (biotechnology/bio-rational and new selective chemicals) will continue to increase the adoption of IPM.

References

Arora, R. and Dhaliwal, G.S. (1994) Botanical pesticides in insect pest management: ecological prespectives. In: Dhaliwal, G.S. and Kansal, B.D. (eds) *Management of Agricultural Pollution in India.* Commonwealth Publishers, New Delhi, pp. 213–245.

Dhaliwal, G.S. and Arora, R. (1996) *Principles of Insect Pest Management.* National Agricultural Technology Information Centre, Ludhiana, India, 374 pp.

Heinrichs, E.A. (1994) Development of multiple pest resistant crop cultivars. *Journal of Agriculture Entomology* 11, 325–353.

ICAR (1991) *Integrated Pest Management in Sugarcane with Special Emphasis on Biological Control.* Sugarcane Breeding Institute (Indian Council for Cotton Research), Coimbatore, India.

Jairaj, S. and Regupathi, A. (1993) Novel concept and future of pesticides in third world. In: Dhaliwal, G.S. and Singh, B. (eds) *Pesticides: Their Ecological Impact in Developing Countries.* Commonwealth Publishers, New Delhi, pp. 335–363.

Jayaraj, S., Ananthakrishnan, T.N. and Veeresh, H.K. (1994) *Biological Pest Control in India: Progress and Perspectives.* RGICS Project No. 2, Rajiv Ghandhi Institute of Contemporary Studies, New Delhi.

Lal, O.P. (ed.) (1996) *Recent Advances in Entomology.* APC Publications, New Delhi, 392 pp.

Paroda, R.S. (1997) Towards Food Secure India – a Policy Prospective. Paper presented at the India–IRRI Dialogue, New Delhi.

Pillai, G.B., Sathiamma, B. and Dangar, T.K. (1993) Integrated control of rhinoceros beetle. In: Nair, M.K., Khan, H., Gopalsundaram, H.P. and Bhaskara Rao, E.V.V. (eds) *Advances in Coconut Research and Development.* Oxford and IBH Publishing, New Delhi, pp. 455–463.

Raheja, A.K. and Tiwari, G.C. (1997) *Awareness Programme on Pesticides and Sustainable Agriculture.* ICAR Publication, 69 pp.

Rajak, R.L. (2001) Promotion of activities under IPM. *IPM Mitr* 1, 9–17.

Rao Krishnamurthy, B.H. and Murthy, K.S.R.K. (1983) Proceedings of National Seminar on Crop Losses due to Insect Pests. *Indian Journal of Entomology* (Special Issue), Vols 1 and 2, Hyderabad.

Singh, S. and Dubey, O.P. (1996) Integrated pest management beyond 2000. *IPM and Sustainable Agriculture – an Entomological Approach* 6, 29–32.

Singh, S. and Kumar, S. (2001) Evaluation of eco-friendly integrated pest management module on mustard crop. In: *National Symposium on Plant Protection Strategies for Sustainable Agri-Horticulture,* 12–13 Oct. 2001, held at S.K. University of Agricultural Sciences and Technology, R.S. Pura, Jammu, India.

Vaidya, K. (1992) Aurevedic pest management in Nepal. In: *Regenerative Agriculture Technology for the Hill Farmers of Napal.*

Appendix 17.1. National Center for Integrated Pest Management (NCIPM)

LBS Building, Pusa Campus, New Delhi – 110 012, India

The ICAR established the NCIPM in February 1988 to meet emerging plant protection needs in India. The activities of NCIPM extend across disciplines and agencies to partnerships with SAUs, government agencies, industries, NGOs and farmers. NCIPM plans and conducts IPM research and development programs to promote sustainable agriculture.

NCIPM is now making efforts to develop computer-based programs for storage and retrieval of information on IPM. Programs for developing and promoting environmentally sound IPM technologies for different crops are underway. The center is striving for effective cooperation with: All India Coordinated Crop Improvement Programs; Crop Research Institutes; State Agricultural Universities; Departments of Science and Technology, Environment and Biotechnology of Government of India; National Remote Sensing Agency; Indian Meteorological Department; National Informatics Center; Directorate of Plant Protection Quarantine and Storage; NGOs and industries.

Mandate of NCIPM

- To develop and promote IPM technologies for major crops so as to sustain higher crop yields with minimum ecological implications.
- To develop an information base on all aspects of pest management and to advise on related national priorities and pest management policies.
- To establish collaborative programs with other national and international institutes in the area of IPM.
- To extend technical consultancies.

Missions of NCIPM

- Development and promotion of biological and cultural control components in IPM.
- Synthesis and validation of region-specific IPM systems for rice, wheat, maize, cotton, pulses, oilseeds and vegetables.
- Creation of IPM database system and Pest Management Information Systems for major crops.
- Development of models for forewarning of key pests of major crops.

Significant achievements of NCIPM

- Rice IPM technology has been demonstrated successfully on 300 acres of farmers' fields.
- IPM in rain-fed cotton has been successfully demonstrated for 3 years, using the FFS approach. Biodiversity in cotton fields increased, and the program created some jobs. IPM adoption was high and the technology has spread to adjoining villages.
- Key pests of mustard (aphids and white rust) could be effectively controlled by timely sowing and use of bioagents (*Trichoderma viride*) as seed and soil treatments.
- IPM of chickpea as well as pigeonpea has been successfully demonstrated. Yields were substantially increased without the use of chemical pesticides. The use of locally available neem seeds as biopesticides is encouraged. Rural unemployed youths have been trained in production of HaNPV to meet local demands of a village.
- A forewarning system has been developed and validated for potato aphid.
- A rule to predict *Helicoverpa armiger* population in the Deccan region has been developed.

Appendix 17.2. IPM Technologies/Tools Available in India

Several technologies are available for the implementation of IPM, many of them refined through testing and actual demonstration. They aim to provide an ecologically sound pest management program with sustainable use of renewable natural resources.

Host habitat management

This involves modification of the environment to lower pest population densities. The strategy relies on enriching biodiversity of natural enemies in and around the cropping environment.

Botanicals

Plants are the richest source of renewable, natural insecticides. As many as 2121 plant species have been reported to possess pest control properties, 1005 species have insecticidal, 384 anti-feedants, 297 repellents, 27 attractants and 31 with growth inhibiting properties.

India has an estimated 18 million neem trees, with a potential to produce 0.7 t of fruit. On average, four neem trees planted on the border of a 1 ha field should yield enough neem seeds to protect the crop. Approximately 67 commercial neem-based formulations are currently available. However, necessity for repeated applications, low toxicity, rapid biodegradability, standard formulations and lack of standardized bioassay procedures have prevented widespread use.

Biopesticides

Currently, the production of biocontrol agents and biopesticides is not sufficient to cover large areas. However, the potential of these agents is continually improving. Microbial pesticides (Bt, fungi, and viruses) offer high potential for avoiding the development of resistance. These are more effective when used in combination with chemical insecticides. Entomopathogenic fungi and bacteria paralyze or kill their hosts by adversely affecting growth and development of host insects. *Beauveria bassiana* and *Metarhizium anisopliae* are now commercially available, but total consumption is only 15–20 t, or rupees 100–150 million (US$2–3 million) in a total pesticide market of rupees 25 billion (US$0.5 billion).

NPV is effective against *Helicoverpa armigera* and *Spodoptera litoralis*. The Department of Biotechnology has provided financial support to establish units for mass production of NPV at SAUs and ICAR institutes. Overall 14 central and 20 state biocontrol units have been established. Another group of biological control agents includes pheromones, which can be used for mass trapping, mating disruption and monitoring. Pheromones have been identified for more than 800 species of insects. However, host specificity, photodegradability, low persistence, timing of application, and difficulties in mass production need to be overcome for their use on a large scale.

Biotechnology

Host plant resistance is a unique approach to pest management, using molecular techniques to identify, qualify and monitor the genetic content of the pest population. The important contribution of biotechnology is the capacity to express pesticidal proteins within transgenic plants. Transgenic plants in tobacco, tomato, potato and cotton hold promise in the management of pests. Efforts are underway to produce Bt brinjal (aubergine), Bt cotton and transgenic mustard.

Novel compounds

Several novel compounds have been synthesized for management of pests and diseases. Avermectins and mylebimycins having antibiotic properties have been developed. Acetylene and furanocoumarins that act at novel points of insect neurotransmission hold promise.

Pesticide application technology

The equipment used for application of insecticides plays an important role in reducing the use of pesticides and ensuring proper coverage. Different types of nozzles and application equipment are now available to provide better coverage, improve the efficiency of the spray and reduce risk to the operator. Electrodyne sprayers, ultra low volume applicators, and controlled release droplet applicators are some recent introductions. Many problems resulting from the misuse of insecticides can be reduced by improving the application equipment available to farmers.

Information technology

Information technology is key to developing decision support systems to manage pests and pesticide resistance and integrate these with other crop management practices, such as irrigation and fertilizer regimes. Personal computers are now important tools for management of pests and diseases. Databases are available to help diagnose pest problems, for genetic cataloging and to provide a quarantine support system for pest risk analysis.

User friendly software packages on IPM are being developed. National Center for IPM, New Delhi, has released a first version of software on cotton IPM, which provides easy identification of pests and outlines management options. Models to forecast and monitor blast disease have been developed by Madras University, Guindy (Tamil Nadu).

Chapter 18
Integrated Pest Management in Indonesia: IPM by Farmers

Ida Pedanda Gde Nyoman Jelantik Oka
IRRI Liaison Office, Jalan Merdeka, Bogor, Indonesia

Country Profile

People vs. food

Indonesia is currently the fourth most populous nation in the world after the Peoples' Republic of China, India, and the USA. In the 1980s a national family planning program began, which reduced the population growth rate from 2.4% to 1.8%. During the years 1991 to 2000, the population growth rate was estimated at about 1.4%. By 2000, the population numbered approximately 206 million people. By the year 2025, the Indonesian population is predicted to number 265 million. Indonesia's population is expected to stabilize at approximately 353 million in the future (World Development Report, 1993).

The high population presents a difficult challenge for the country. Basic needs such as food, housing, clothing, health, education, and employment have become progressively more difficult to meet. The rice supply is an example of the urgent need for food in Indonesia. Rice is the staple food of the Indonesian diet, supplying nearly 60% of the total caloric intake of the average person, and even more for the poor. Indonesia has suffered chronic rice shortages since the days of Dutch colonialism. During the short period of Japanese occupation, major rice shortages occurred. Rice shortages have persisted during the country's independence. Two to three million tons of rice were imported from the global market each year to meet the demand, placing a drain on the nation's economy.

Attempts to boost rice production

In the early years of Indonesian independence, the Indonesian government attempted to boost rice production, but the increases could not keep up with the growing population. One reason for the rice shortage was the lack of high-yield technology. For example, the Indonesian-bred rice varieties (Bengawan, Dewi Tara, etc.) yield only 2.4 tons/ha under favorable conditions. They do not respond to chemical fertilizers, and need a relatively long 140 days to harvest.

In the late 1960s, President Soeharto's government addressed the rice problem. A comprehensive food production program was launched in 1969 with the following objectives:

- to achieve and maintain self-sufficiency in food production;
- to increase farmers' income;

- to provide job opportunities and alleviate poverty;
- to increase foreign earnings through exports of agricultural products;
- and to provide strong support for the heavily expanding industrial, business and service sectors.

To achieve these objectives, four programs were organized: rehabilitation, expansion, diversification, and intensification.

Rehabilitation, expansion, and diversification

The rehabilitation and expansion programs included rehabilitating outdated irrigation systems and neglected fields, and bringing new areas under cultivation.

The diversification program promoted the production of food crops other than rice, such as maize, groundnuts, soybean, cassava, potato, sweet potato and other tubers.

Intensification of food production

The intensification program covered both technical as well as socioeconomic aspects. It initiated large-scale planting of modern food crop varieties, increased use of chemical fertilizers (N, P, K and micronutrients), and the expansion of irrigation networks. Economic stimuli included improved price policies for rice, supporting farmers' cooperatives, expanding agricultural extension programs, and intensive use of pesticides.

Improvements in rice production

The IRRI and the FAO provided assistance with the development of the programs. The IRRI introduced two modern rice varieties, (MV) IR 5 and 8. Under good management practices, each variety yields 5–8 tons of rough rice per hectare, but both are highly susceptible to most rice pests. Therefore, heavy use of pesticides became necessary.

The intensification program contributed the most to increased rice production. Between 1951 and 1961, production increased by 3478 t, from 9336 to 12,084 t (Table 18.1). This was before the large-scale implementation of the intensification programs. From 1971 to 1981, rough rice production jumped from 20,058 to 32,774 t, a significant increase in only 10 years (Table 18.1). This showed that the intensification program was effective. Increases in rice production continued for several years.

For the first time in the history of the country, self-sufficiency in rice production was achieved during the 1983/84 season with an output of 38,134 t. This achievement encouraged policymakers to continue to work for self-sufficiency in rice production. To reach that goal, a yearly increase of rice yields of at least 2% was needed. Indonesia maintained this level until 1993/94. Rice production in that year was 48,181 t. The following year, rice production declined owing to extreme drought, compelling the government to resume importation of rice.

Economics of rice production in Indonesia

After the fall of Soeharto's government in 1997, significant changes took place. Rice production continued to decline, forcing the government to import 2–3 million tons of rice annually. The market price of imported rice fell to rupees 1500/kg, below the minimum profitable price for domestic rice of rupees 2500/kg. The globalization policies for Indonesia's economy meant

Table 18.1. Rice production in Indonesia. (Source: IRRI (1990); *Statistik Indonesia* (2001).)

Year	Rice production (tons)
1951	9,336
1961	12,084
1971	20,058
1981	32,774
1991	44,688
1996	51,102
1997	49,377
1998	49,237
1999	50,866
2000	51,899

that the government could not levy import duties for imported rice, a policy which threatened the economic viability of Indonesian rice production. The same was true for fruit and vegetables. Approximately 100 million people depend on agriculture as a source of income in Indonesia. The price policies and marketing strategies for harvested rice needed re-evaluation to make rice production profitable for farmers but still economically acceptable to consumers.

Alternative control methods by necessity

Increasing costs of production inputs such as chemical fertilizers and pesticides forced many farmers to search for inexpensive alternatives. For example, in parts of central Java and in North Bali, farmers used organic fertilizers exclusively (compost, dung, etc.). Many farmers' groups used botanical pesticides such as leaves of neem trees to prevent insect pests from infesting the rice crop. Finely chopped leaves of the neem tree were mixed with water, and the filtrate was sprayed on the plant. Farmers reported that this method does not kill insect pests, but acts as a repellent.

In the future, it is anticipated that botanicals and microbial pesticides will be increasingly in demand because of their effectiveness, relative safety to humans and the environment, and the low cost. For example, the microbial insecticide *Bacillus thuringiensis* is now extensively used to combat the diamondback moth, *Plutella xylostella*, a major insect pest of cabbage in Indonesia.

Pesticides and the Environment

Logistical problems lead to environmental threats

Various technical and socioeconomic problems surfaced during the early years of the intensification program, such as availability of production inputs (rice varieties, fertilizers, pesticides) in the rice-growing centers. Poor harvest technology and inadequate grain storage caused appreciable grain losses. Poor handling of production inputs made some of them useless. In particular, canned liquid pesticide formulations were sometimes left unused and stored inadequately, leading to corrosion and leakage into the environment.

Pesticide use before IPM

An unexpected problem with the intensified rice production programs was the emergence of various insect pests and diseases causing concern to both the government and farmers. At the time, pest problems were regarded as separate from the health of the rice ecosystem. The belief was that pests can and should be easily eliminated with regular applications of pesticides, at least three to four times during the season. If insect pests were still observed after regular insecticide applications, farmers were urged to apply more insecticides, even by increasing the application rate. Often, pesticides were sprayed prophylactically even if no pests were observed in the field.

In rice production centers, such as on the northern plain of West Java, rice fields were aerially blanketed with ultra low volume (ULV) formulations using small fixed-wing aircraft. Chemical insecticides were regarded as effective tools to maintain the health of the rice fields. This ad hoc philosophy of pest control led the government to subsidize pesticides for use in mass rice intensification programs by as much as 80%, costing from US$100–150 million/year.

Effects of pesticide use on human health and the environment

Because almost all pesticide formulations were broad spectrum and applied excessively they caused various environmental problems. They not only killed pests, but also beneficial organisms (such as parasitoids, predators, bees, earthworms, and birds), and fish in irrigated rice–fish ecosystems. Chemical insecticides polluted irrigation water, eventually reaching the rivers. Village people who used these waters for bathing, cooking and other needs were exposed to health hazards (Oka, 1997).

Human poisonings and pesticide-related deaths were also reported from the highly intensified rice production centers. In 1976, 450 insecticide poisoning cases were reported, with 26 deaths. Ten years later, 404 insecticide poisoning cases were reported and 32 deaths occurred (Mustamin, 1988). Most likely, many more cases went unreported because of poor communication systems in remote and isolated places.

Pesticide resistance: the brown and green planthoppers

The brown planthopper, *Nilaparvata lugens* Stahl, developed resistance to most chemical insecticides (Laba and Soeyitno, 1987; Sutrisno, 1987). The brown planthopper changed from a minor and occasional rice pest to the most feared insect pest in most Asian countries (Dyck and Thomas, 1979). During the 1976/77 rice season, the brown planthopper caused extensive damage to the rice crop, affecting at least 450,000 ha. Conservative yield loss estimates were approximately 364,500 tons of milled rice, enough to feed 3 million people for an entire year (Oka, 1979b).

The green leafhopper, *Nephotettix virescens* Distant (Tandiabang, 1986), also developed resistance to most chemical insecticides. *N. virescens* is an effective transmitter of the rice tungro virus. As the *N. virescens* population increased owing to developed resistance, the rice tungro virus spread rapidly in 1970 causing damage to up to 70,000 ha of rice fields in South Sulawesi. In the 1980s, the virus broke out in Bali, causing damage to at least 12,000 rice fields. Since then, the disease has become endemic in most of the rice centers throughout the country.

Some causes of pest problems in Indonesian rice production

Based on field observations and research, several factors have been identified that contributed to increasing pest problems in rice production centers. These include the following (Oka, 1997).

- Continuous and staggered planting in rice production allowed pests to build up continually, frequently reaching epidemic proportions. This happened with both insect pests and other pests such as field rats and the rice tungro virus.
- Excessive nitrogen fertilizer application tended to make the rice plant more susceptible to many pest species.
- Reduction of genetic diversity in rice fields by planting only one or two modern rice varieties over wide areas caused decreased stability of the rice ecosystem.
- Not all modern varieties possessed resistance to all pests, therefore, pest species to which the varieties are susceptible could multiply unchecked.
- Modern varieties with a small genetic base for resistance could trigger development of new biotypes/races of insect pests capable of breaking down the resistance of the rice varieties.

Fighting the brown planthopper with host plant resistance

This was the case with the brown planthopper. After the outbreaks of brown planthopper in the 1970s, the rice varieties IR 5 and 8 were replaced with IR 26, 28, and 30. These new varieties contained a monogenic dominant gene for resistance to brown planthopper, designated as *Bph 1*. These rice varieties were highly resistant to the existing population of brown planthopper, which dramatically reduced the planthopper population. However, after five or six rice seasons (2–3 years), these varieties became susceptible to the pest. To attempt to protect the rice crop, pesticides were applied more often, but the problem persisted. In many rice centers the brown planthopper population nearly doubled 2 or 3 weeks after heavy insecticide application.

It was suspected that the brown planthopper population had developed a new biotype or race capable of attacking resistant rice varieties. This was soon identified as Biotype 2. New rice varieties resistant to Biotype 2 (IR 32, 36, and 42) were requested from IRRI to control the spread of Biotype 2. Of these, IR 36 was most widely distributed, because it was most in demand for its high yield and field performance. Even with widespread planting, IR 36's resistance to the new brown planthopper biotype has remained stable, presumably it has a broader genetic base for resistance. The only drawback is that its taste is inferior to newer modern varieties with better cooking quality and a broad base for resistance to the brown planthopper biotypes, such as IR 56, 64, and 70.

An ecological approach to pesticide use

Heavy reliance on chemical pesticides often creates pest resistance, resurgence of resistant pests, secondary pest outbreaks and other environmental problems. A thorough understanding of the intricate interrelationships of the crop ecosystem should be the basis for sound control strategies. This philosophy is the basis of IPM (Oka, 1987).

It is most likely that in the future, chemical pesticides will still be widely used, with the potential for causing unwanted side effects. The first step to preventing adverse effects of pesticide use is education about their behavior in the environment. Farmers and other pesticide applicators are often unaware of the potential effects of pesticide use, and education at this level is essential. Figure 18.1 presents a model of the path of pesticides in the environment after application (Oka, 1995).

Pesticides applied as emulsifiable concentrate, dust, or ULV formulations mix with air before they reach the rice plant and the water in rice fields. In the air, pesticides may be blown by wind. Depending on the wind speed and direction, the pesticide particles may be deposited far from the source. Or, they may follow the air percolation and be photodecomposed by the sun's ultraviolet radiation. In the irrigation water, they might be biodegraded and diluted. After pesticides reach the plant, they often kill the pests but also beneficial organisms such as predators, parasitoids, the hyperparasitoids and other non-target organisms.

Biomagnification

Highly persistent pesticides, such as DDT and other organochlorines, do not degrade quickly in water or soil. Microplankton in rivers and oceans may absorb small amounts of pesticides while feeding. Zooplankton feeding on microplankton will have higher concentrations of pesticides in their bodies, depending how many microplankton they feed on. In turn, these microanimals will serve as food for small fish, and these small fish will be eaten by larger fish. Finally, humans may catch and eat the larger fish. The concentrations of pesticide increase at each level of the food chain. Persistent pesticides such as DDT are stored in fat and passed on at each level. Humans and other top-level consumers can accumulate the highest concentration of pesticide in their bodies.

In the soil, persistent pesticides may remain for years and be absorbed by plants and earthworms; in turn, these worms might be picked up by chickens and ducks. The eggs of these chickens are believed to contain some amount of pesticide. For example, the milk of nursing women was reported to contain traces of DDT in Yogyakarta (Untung, 1995, personal communication), probably because they regularly consumed the eggs of free-range chickens. This may be harmful to nursing infants, although the effects are not well studied.

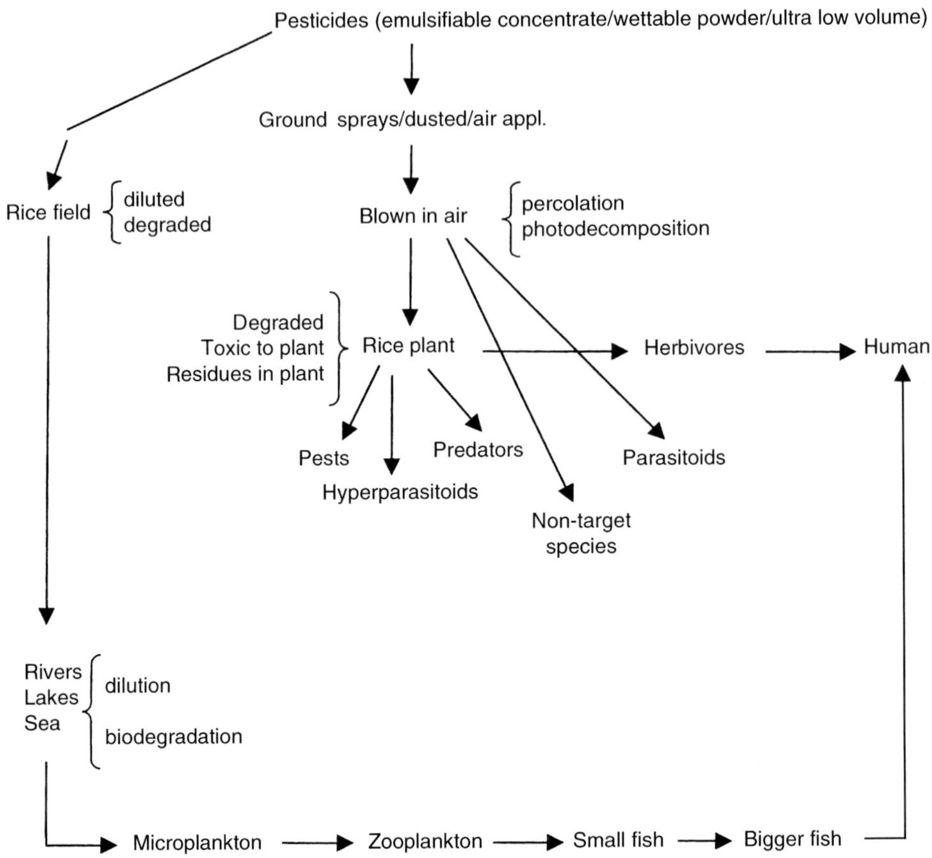

Fig. 18.1. Qualitative model of the behavior of pesticides in the environment, in particular broad-spectrum and persistent pesticides (Oka, 1995).

The ultimate cost of pesticide use: human life

Cases of death from DDT poisoning have been reported from Boyolali regency in central Java, where three out of 386 people died after eating locally prepared food, including 'limpung' and fried soybean cakes, in 1982 (Mustamin, 1988).

Worldwide human poisonings and deaths from pesticides are reported by WHO/UNEP (1989, in Pimentel *et al.*, 1992). An estimated 1 million people each year are poisoned by pesticides, including approximately 20,000 fatalities. Most deaths from pesticide poisoning occur in Third World countries. The greatest costs of pesticide use are outside the economic realm.

In light of these concerns, researchers in Indonesia sought to develop alternative methods to minimize crop pest problems, which would be environmentally sound, sustainable and economical. The IPM program was begun.

Introducing IPM as an Alternative Pest Control Strategy

Early obstacles to implementing IPM

Early on, the major challenge faced by the IPM program was convincing the government to adopt an IPM approach. Supporters of IPM lobbied policymakers to consider the new approach. Several meetings debating the issue were held, often involving heated discussions of the pros and cons. Some newspapers became interested in the IPM approach and covered the issue.

Finally, IPM was mentioned in official documents in the Third Five-Year Plan (1979–1984), but its implementation was still limited by the existing plant protection directorate within the Ministry of Agriculture. Technical aspects of IPM, such as cultural controls (synchronized planting, crop rotation, and sanitation), extensive use of resistant plant varieties, and an extensive pest monitoring system to encourage more judicious use of pesticides were endorsed and implemented on a wide scale.

However, the implementation of alternative pest control strategies had to face many obstacles. Pesticide subsidies remained in effect, which encouraged farmers to continue using the inexpensive pesticides. Massive advertising campaigns by pesticide companies also encouraged the use of pesticides. At the time, a systematic program for educating farmers about IPM and training them to use it did not yet exist. Extension workers were not yet trained in the IPM approach, so their pest control recommendations to farmers often did not change. Worse, the authorities were doubtful about the effectiveness of the new approach. Many were concerned that the new approach would not be able to control pest outbreaks. In addition, there were external pressures from pesticide manufacturers to promote certain pesticide formulations.

Political and economic crises spur introduction of IPM

A new IPM program to address these problems did not emerge until 1986, when a massive outbreak of brown planthopper occurred in Central Java, destroying 75,000 ha of rice fields. This threatened the self-sufficiency in rice production which had been achieved at great cost only a few years earlier. Rice shortages triggered price increases, which in turn caused widespread public unrest, heightened because of the importance of rice as a staple food for most Indonesians, especially the poor.

Resuming large-scale importation of rice would drain the country's economy and present an embarassing situation for the government. This greatly concerned the National Agency for Planning and Development (BAPPENAS). To solve the problem as quickly as possible, yet effectively and in the long term, the agency sought advice from the Department of Agriculture and leading universities. Their recommendation was to implement IPM strategies at the grassroots level, by training farmers to practice IPM in their fields. This was a major departure from the traditional method of telling farmers what to do, rather than teaching them the theory behind the new approach. This new program became known as 'IPM by Farmers'.

Recommendations of the IPM by Farmers program

Other recommendations included reducing pesticide use to a minimum, and banning the use of pesticide formulations which had caused pest resistance. Also, it was suggested that farmers should be educated through field training to implement IPM in their own rice fields. In addition, the number of field observers should be doubled from 1500 to 3000, and be thoroughly trained in IPM before they were permitted to train farmers.

Meanwhile, reports from several rice centers indicated that the brown planthopper epidemic had spread quickly to cover wide areas, in spite of farmers' efforts to intensify pesticide sprays. At the time, the Department of Agriculture was still convinced that the threat of brown planthopper was not serious enough and could be overcome by conventional control methods.

Despite these concerns, the recommendations of the scientists to the BAPPENAS were accepted and became a new policy. In 1986, a Presidential Decree was promulgated to support the IPM by Farmers approach. The three objectives of the Decree were as follows.

1. Develop manpower at the grassroots level, both farmers and field staff, to expand education and awareness of IPM.
2. Increase efficiency of input use, in particular of pesticides.
3. Improve environmental quality and prevent unwanted side effects on human health.

The Decree banned 57 broad-spectrum insecticides formerly approved for use in rice. Only the use of a few insecticides with a relatively narrow spectrum would be permitted, and their use would be based within the IPM program. Another important step was gradual withdrawal of the pesticide subsidies, from 75–80% early in 1986 to 40–45% in 1987. By January 1989, the subsidies were totally withdrawn. In addition, pesticide companies were challenged to change their orientation from a profit standpoint to showing more concern for the environment and human health (Oka, 1997).

These policies necessitated a change in the mentality and orientation of government officials. They were required by law to move from a pesticide-oriented pest control program to a comprehensive one, i.e. IPM, to be carried out by farmers in their own rice fields. This represented a shift from the 'top-down' approach to a 'bottom-up' philosophy of program implementation.

The government charged BAPPENAS with both planning and implementing the 'IPM by Farmers' program. This policy was controversial, since the mission of BAPPENAS was development planning, not program administration. The previous IPM program had been administered by the Department of Agriculture. This policy was meant temporarily to bypass the slow-moving bureaucracy of the Department. In addition, several Department of Agriculture officers with direct responsibility for crop protection were not yet ready to adopt the new IPM approach. When the program was established in 1994–1995, its management was transferred to the Department of Agriculture.

What is IPM by Farmers?

By definition, IPM is a comprehensive approach to pest control, utilizing compatible control tactics in one unified program (such as cultural controls, host plant resistance, biological control, pheromone disruption, and mechanical/chemical controls) to maintain pest populations below economically damaging proportions, in an environmentally sound and safe manner. In essence, it is an ecological approach to pest control. But in practice, this definition is not enough. Social, religious, and cultural practices must be taken into account when planning and executing the program. Otherwise, it is difficult to convince farmers to accept and practice IPM. For example, on the island of Bali, rat control through regular mass hunting by the farmers before planting is only undertaken after prayers and offerings are given at the temple. Participation in the ceremony would gain credibility for IPM trainers.

IPM practices employed in the IPM by Farmers program

Cultural control

Growing a healthy crop by using good seed, proper soil preparation, and good crop management practices such as balanced fertilizers, mulching, proper distance between plants, sanitation, and timely irrigation is recommended. A healthy crop is the first defense against pest attacks because it can withstand pests better than a poorly growing crop.

The customary practice of continuous or staggered planting throughout the year had been shown to cause a buildup of pests such as field rats, brown planthopper, and the rice tungro virus. Crop rotation was recommended to prevent this by rotating rice with other annual crops or fallowing between two rice seasons. Annual crops used in the rotation cycle should be non-hosts such as mungbean, soybeans, or sweet potatoes. One recommended rotation cycle included two rice plantings followed by another crop, then another rice crop followed by a non-rice crop, then three rice crops followed by a fallow season (Oka, 1979a,b). Crop rotation cycles varied depending on the availability of irrigation water and local customs (Oka, 1979a,b).

Biological control

Conserving natural enemies of insect pests, such as parasitoids, predators, and insect pathogens, is important for pest control. Natural enemy conservation is environmentally safe and relatively stable as a method of control. For example, the brown planthopper has about 80 species of parasitoids and predators (Chiu, 1979). Field observations and research show that spiders, especially the wolf spider, *Lycosa pseudoannulata*, play an important role in keeping brown planthopper populations at low levels.

Host plant resistance: the case of the brown planthopper

During the early years of the intensification program a number of pest species became more and more abundant, especially the brown planthopper. The IRRI provided modern varieties of rice resistant to this pest. Their resistance was due to a single gene, called *Bph-1*. Perhaps due to this narrow genetic base, the resistance broke down after only five or six cropping seasons. Over this short amount of time, the strong selection pressure exerted by the resistant varieties allowed the brown planthopper to develop a new biotype, called Biotype 2, capable of attacking the resistant rice varieties.

Another problem with the brown planthopper was the widespread application of broad-spectrum insecticides, which exerted another strong selection pressure on the population and encouraged the development of insecticide resistance in the brown planthopper. The resulting increase in the population of insecticide-resistant Biotype 2 planthoppers posed a serious threat. Since insecticides were no longer effective, the solution was to replace the susceptible rice varieties with new varieties that were resistant to Biotype 2. New varieties had been developed that possessed a monogenic recessive gene for resistance called *Bph-2*. Of course, these varieties might also become susceptible after the next few crop seasons as new biotypes developed. Replacing susceptible varieties with resistant ones over large areas was time-consuming. A quick and effective method for detecting the brown planthopper biotypes was needed.

A field method for detecting biotypes of brown planthopper

Detecting the different biotypes of brown planthopper in the field was difficult because the populations were usually mixed, consisting of several biotypes. New biotypes were usually confirmed after the pest had spread rapidly over wide areas. Another challenge was determining which rice varieties were planted in the area. These factors made it difficult to choose the appropriate rice variety to plant the next season.

A method for identifying the biotypes of brown planthopper was developed for use in the field. Two rice varieties susceptible to all brown planthopper biotypes are used as a standard for comparison. Two other varieties susceptible to Biotype 2, but resistant to Biotypes 3 and 1 are used, and two other varieties susceptible to Biotype 3 but resistant to Biotypes 2 and 1 are used. Seeds of the different varieties are glued between two sheets of absorbent tissue paper. The variety names are copied opposite each row. This method of seed preparation simplifies the work of field personnel and minimizes errors. About 1 cm of soil is placed in a plastic tray. Urea is mixed with the soil at the rate of 90 kg N/ha before the planting sheet is set on the soil surface. The plastic tray is then placed in an insect proof cage. The cages are placed under sunlight and watered as necessary.

Three or four days after seeding, the plants are infested with brown planthoppers collected from the surrounding area. About

100 to 150 adults are released into each box. Assuming that about half are females, they should produce about 2000 nymphs per box. Each tray contained about 500 seedlings. When nymphs are present in the area, plants are infested with second, third or fourth instar nymphs. Number of nymphs per plant is maintained at an average of four to five. Each of the infested seedlings is evaluated as soon as the universally susceptible plants die (according to IRRI Standard Evaluation System for Rice, 1980). The results are sent to the Directorate of Crop Protection in Jakarta and Central Research Institute of Agriculture in Bogor for analysis, and used to recommend the best rice varieties to plant the next season. This monitoring system is carried out once or twice a year.

Maps of brown planthopper biotype distributions are made to guide the field extension workers in recommending the correct rice varieties to plant in each area. This method also makes it possible to follow the biotype shifts and the distribution of the pest (Oka, 1978, 1980).

Mechanical/physical control

Mechanical methods are used to control field rats. In areas with synchronized planting and crop rotation, the farmers usually organize a hunt for field rats a few days before planting. Rat holes found in dykes and along the canals are plugged with mud, and a few are left open and filled with sulfur gas. Escaped rats are found with the aid of hunting dogs and killed. This method can reduce the initial rat population by almost half. If it is done regularly the rat population may remain low.

Pesticides

The Presidential Decree still allows the use of pesticides in the IPM program, but only those with high selectivity and little or no side effects on non-target animals and plants. For example, in rice–fish culture, selective pesticides are needed. However, such ideal pesticides are not available for every situation. Usually, their selectivity is limited to only a few groups of non-target species. Selective pesticides favoring natural enemies of pests and beneficial insects are most effective. For example, pyridaphenthion and tetrachlorvinphos are highly selective, favoring the wolf spider, *Lycosa pseudoannulata*, but toxic to their prey, the green leafhopper.

Molting inhibitors like buprofezin were reported to be highly effective against brown planthopper and safe for its natural enemies. This chemical has been widely used in the IPM program. It should be applied only once, at the time when the planthopper population is predominantly in the nymph stage. Factors such as time of application, formulation, methods of application, dosage, and biodegradability are important in increasing pesticide selectivity.

Pesticide formulations determine the method of application. Spray and dust formulations increase the risk of poisonings, because the fine particles can be inhaled, contaminate the skin and clothing and penetrate into the body. The advantage of the 'emulsifiable concentrate' formulations is that they are easy to transport and can be applied at any time. Dust formulations, on the other hand, are bulky and difficult to transport. Systemic pesticides as seed treatments are recommended to control downy mildew, *Sclerospora maydis* Butler, on maize. For example, only about 5 g of the fungicide Ridomil™ (metalaxyl) is needed for 1 kg of maize seed.

Spot treatments use much smaller amounts of pesticide and reduce the risk of contaminating the environment. For example, during the early stage of the brown planthopper outbreak, spot treatments on local populations with molting inhibitors proved to be effective. Rat control may be effectively carried out by the ecologically sound method of baiting with the anticoagulant rodenticide, Klerat RM™ (brodifacoum) (Oka, 1988). The use of pesticide mixtures should be discouraged because, in general, pests will develop simultaneous resistance to the chemicals (Metcalf, 1980).

A strong pesticide regulation program and its enforcement are critical to the success of the IPM program. Indonesia has developed a set of pesticide regulations,

Decree No. 7 1974, dealing with various aspects of pesticide management, such as registration, permits, safe handling and use, storage, transport, disposal, and sanctions to violators. The Pesticide Commission is chaired by the Director of Plant Protection within the Department of Agriculture. Its members are experts from research institutes and the Departments of Health, Labor, Trade and Environment. Although the formation of this committee is an important step, several barriers still remain to adequate enforcement. This is due to factors such as a lack of adequately trained personnel, inadequate facilities to check pesticide residues and quality control, and a lack of public awareness of the dangers of pesticides. Many farmers still see pesticides as effective medicines to heal their crops from pest damage, and banned pesticides are still available in many areas. Reporting of imports and exports of pesticides and of human poisonings and fatalities is still relatively weak (Oka, 1988).

The IPM by Farmers program included training the farmers to get them more acquainted with pesticide issues, particularly on their unwanted side effects to human health and the environment.

Organization of the National IPM by Farmers Program

Administration and funding

The National Agency for Planning and Development (BAPPENAS) took the lead to initiate the IPM by Farmers program, including program design, organization, and implementation in fields. The thrust of the program was to develop manpower capabilities at the grassroots level through intensive training in IPM. To do this, BAPPENAS established an organization consisting of three groups: (i) an advisory group; (ii) a group of directives; and (iii) a working group. The membership of these groups was chosen from the Departments of Agriculture, Domestic Affairs, Health, Environment, Bureau of Statistics, leading universities, FAO, and BAPPENAS.

The group of directives was charged with tackling policy issues, setting priorities, identifying problems, coordinating activities with foreign assistance, and identifying the right personnel to guarantee the success of the program. The working group assisted with working out the details of the directives and ensuring that the program ran according to plan. When problems emerged during the implementation of the program, alternative solutions were developed and offered to the group of directives.

The program was supported by USAID from 1989 to 1992. During the 1992/93 fiscal year, the program was funded by a World Bank loan. The program was intended to terminate in 1998, but was again funded by the World Bank loan. FAO provided technical assistance from the beginning, including program design, curriculum, training methodology, and field studies related to IPM, such as habitat studies and effects of pesticides on the environment and farmers' health.

The program scope

During the first year of the program (1989/90), it was limited to six provinces (West, Central, and East Java, Yogyakarta, North Sumatra and South Sulawesi). Those provinces provide approximately 70% of the rice production in Indonesia. Pest problems in the rice centers in those provinces were also very serious. After the first successful year, other provinces also wanted to be included in the program, but due to budget limitations and a shortage of field trainers, only six other provinces could be added (Aceh, Bali, West Sumatra, West Nusa Tenggara, North Sulawesi and Lampung). These provinces were second in rice production. The next year, the Minister of Agriculture requested that a similar program be initiated for high altitude vegetable crops (cabbage, potatoes and chili) and soybeans. These vegetables received 16–20 insecticide applications in one season, while soybean received at least 4–5 applications/season.

Developing the FFS program

The most critical part of the new IPM program was establishing a training program that would ensure accurate replication of IPM principles at each level, from extension agents to farmers, and result in farmers capable of teaching each other the new IPM methods. During the first year of the program, 22 senior field pest observers were selected from major provinces to receive intensive IPM training in Yogyakarta. The training facility was equipped with a laboratory for studying insects and diseases, and 2 ha of rice field to carry out field activities. First, participants underwent 4 months of field training, starting from soil preparation, seeding, transplanting, managing the crop, observing the occurrence and dynamics of pests, natural enemies, and their interactions. They also studied the effects of some insecticides to pests and their natural enemies and non-target species (i.e. frogs, fish), and performed field cage studies on the effect of the wolf spider on brown planthopper populations. Finally, they learned how to estimate rice yield after harvest. The next 4 months they were taught how to train farmers' groups in the rice centers. Then they went back to the facility in Yogyakarta to be trained in non-rice crop IPM, particularly soybean, for 4 months. Lastly they entered a socialization stage of IPM to farmers in the first six provinces. They were called Pemandu lapang (PL I) = (Field Leader I).

At about the same time as training the 22 Field Leaders, the program also trained another 90 senior field pest observers from various provinces in IPM for 2 weeks. After their training they were called Pemandu lapang II (PL II) = (Field Leader II). They were sent to the provinces to assist the Field Leader I in devising an IPM training program for the province. Local authorities helped provide field training facilities, including 2 ha of rice land. Large provinces with extensive rice acreage needed more than one facility, such as West, Central and East Java, which needed three, three and two facilities, respectively.

Each year each facility provided IPM training for 50–60 field pest observers by the Field Leader I and assisted by two Field Leader IIs, until all the field pest observers in the provinces were trained. The training was the same as that given in Yogyakarta: rice IPM, socialization, and non-rice IPM. After this, they were capable of training farmers in IPM for 12 weeks. They would stay with the farmers in their villages. Farmers located in the rice centers received highest priority.

Within the first 2 years of the program (1989–1991), 1000 field pest observers and 2000 field extension workers underwent intensive IPM training, along with 100,000 farmers. From 1992 to 1998, approximately 600,000 farmers, 3000 field pest observers and 6000 field extension workers were trained in IPM. During the last year of the program, a total of 800,000 farmers received IPM training. This number of farmers was thought sufficient to diffuse the knowledge of IPM to other farmers. For this a special program named Farmer to Farmer Training was devised.

Farmer to Farmer Training: the FFS program

Farmers' IPM training was designed to be a 'learning by doing' process (Dilts, 1985; Anonymous, 1989). Each participant was encouraged actively to observe pests and natural enemies in rice fields, carry out simple experiments, discuss their findings with fellow trainees, and draw their own conclusions. In this process, no difference was made between the trainers and the trainees. Each person contributed to teaching and learned along the way. The trainers acted more as facilitators than as traditional extension agents, to whom the farmers were expected to only listen carefully. The atmosphere was relaxed to encourage the participants to freely express their opinion. This participatory process of learning by doing was based on the assumption that the farmers already had many years of experience as rice farmers. They were familiar with farming rice and other food crops following traditional methods, so they were readily able to make comparisons among different management practices, including pest problems. This approach helped to sharpen their analytical skills, allowing them to arrive at more informed conclusions and act accordingly.

Participants in field schools were members of existing farmers' groups. Since one farmers' group included 75–100 farmers, the persons selected to participate in the IPM field school served as representatives and were responsible for teaching and passing on information to the rest of the group. All participants were active farmers, owners, tenants or sharecroppers.

During the field school program, the farmers were provided with brochures containing colored pictures of insect pests and their natural enemies, plant diseases, and weeds. These were designed to encourage analytical thinking, rather than simply providing descriptions. For example, a brochure would ask the farmer about the effects of factors such as irrigation, sunlight, soil, fertilizers, and pests on the crop. The trainers also encouraged the farmers to observe insects and discover their function in the ecosystem based on their observations.

Field training was conducted on 2 ha of rice land. The field was divided into two treatments: the rice crop in one hectare was managed following conventional methods, and rice in the other hectare was managed following IPM principles. One Field Leader managed four farmers' groups consisting of 25 farmers each. Each farmers' group was required to come to the field once a week for 12 weeks. On these field days, each group was divided into a smaller group of five farmers. After discussing the topic of the day, each small group went into the field, carefully observed plant growth, counted the tillers, and looked for pests and natural enemies. They took notes on what they found and recorded their observations.

After a period of observation in the field, each small group came together to discuss their findings. They drew a rice plant on paper and showed the pests and natural enemies they had observed. They also commented on the appearance of the crop, such as good, yellowish, stunted, or too weedy, and included their suggestions. Following this discussion, each group presented its findings to the rest of the farmers. Heated debates and disagreements among them were common, which were usually settled by the trainer.

Simple experiments were performed in the laboratory to demonstrate some of the ecological principles inherent in IPM. One simple experiment included caging a planthopper-infested rice plant with a wolf spider, and comparing it to a control with no wolf spider. In the cage with a wolf spider, the planthopper population declined sharply, while it increased in the other cage. The farmers were impressed to observe firsthand the effect of the wolf spider. Another experiment demonstrated the effect of an insecticide on two non-target species, fish and frogs. They put fish and frogs in each of two plastic jars, and added a few drops of insecticide to one jar. After an hour or so, they observed that those animals were dead, while those in the jar without insecticide remained healthy. This discovery demonstrated to the farmers firsthand that spraying the fields with insecticides could kill beneficial species.

Exercises in working as a team rather than individually were also included. The purpose was to show that working together as a group can produce better results than working alone, such as organizing the mass hunting of rats. At the close of the training, the participants were tested on IPM field skills. Following the test, they were awarded a certificate of achievement in a simple ceremony. The certificate was especially meaningful, since many of the farmers had not even received primary schooling.

Effects of the national IPM by Farmers program

Significant reduction of insecticide use

Field studies in 1991 and in 1993 revealed that insecticide applications were reduced by as much as 56.2% and 55.8% in some areas (Pincus, 1991; Anonymous, 1993). The same was true for IPM on high altitude vegetable crops. Surveys conducted in various cabbage centers in North Sumatra and Java revealed that the number of insecticide applications in IPM and non-IPM fields averaged 4.01 and 10.57 respectively. The number of fungicide applications averaged 0.07 and 2.74 respectively. For potatoes, the number of insecticide applications in IPM fields and non-IPM fields averaged 3.09 and 8.49 (Kusumah, 1994).

The amount of illegal insecticide use on rice in field studies before and after the IPM FFS training program was significantly reduced by 78% and 81% in two separate studies. This indicated that after the training was implemented, many farmers abandoned the illegal use of insecticides.

Careful scouting for pest:predator ratios

The IPM-trained farmers would apply insecticides only after they carefully investigated the presence of pests and predators in the field. If the two were in balance (for example, five planthoppers and two or three ladybugs or spiders) they decided not to spray, a marked departure from the traditional methods.

A follow-up program encouraged retention of IPM principles

To prevent farmers from reverting to older methods of pest control, a follow-up program was organized, called institutionalization of the program. Groups of farmers who had completed the IPM training were revisited to test them on what they had learned and survey what they were doing in the field.

The training methods developed for the IPM program are applicable to other programs

The organization of the IPM by Farmers program gained importance as an example of developing human resources at the grassroots level, a valuable asset for developing countries.

New philosophy of extension work

The old 'visit and training' methodology practiced by the extension workers was modified using the IPM by Farmers approach, to include a more interactive and hands-on teaching philosophy.

The FFS program has served as a model for other countries

Neighboring countries have become interested in Indonesia's rice IPM program. Policymakers and field officers from Sri Lanka, Bangladesh, India, Malaysia, Thailand, Vietnam and the Philippines have visited Indonesia to observe the field school program firsthand. Many countries have sent their field technicians to Indonesia to undergo a 3-week IPM training. Indonesian field staff have also been sent to other countries to conduct IPM training sessions.

Challenges and future needs

The IPM by Farmers program met with great success. By the end of the field school program, over 807,000 people had been trained in IPM. This included 800,000 farmers, more than 2000 field pest observers, 4000 field extension workers, 208 field extension leaders, 16 village heads, 95 subdistrict heads, 333 primary school teachers, 202 field laboratory workers, and more than 800 people from different branches of village organizations. The continuing education of farmers through farmer-to-farmer training has likely added many more farmers to these numbers.

But this achievement is just a beginning. Indonesia supports approximately 19 million farmers, creating a need for continuing outreach, education, and training. The IPM program can only be sustained through the dedicated efforts of both farmers and policymakers. Active extension programs are needed to support the farmers in training each other. In addition, government policy should continue to be informed by IPM principles.

References

Anonymous (1989) *Educare 'to Head Out' (Mengajak Keluar).* Materi Pelatihan Pengendalian Hama Terpadu, Bappenas.

Anonymous (1990) *World Rice Statistics.* International Rice Research Institute, the Philippines, 320 pp.

Anonymous (1993) *The Impact of IPM Training on Farmers' Behavior: a Summary of Results From the Second Field School Cycle.* IPM National Program Monitoring and Evaluation Team, August 1993, National Intergrated Pest Management, Jakarta, 30 pp., with appendices.

Anonymous (1993–94) *World Rice Statistics.* International Rice Research Institute, the Philippines, 260 pp.

Chiu, Shui-chen (1979) Biological control of the brown planthopper. In: *Brown Planthopper Threat to Rice Production in Asia.* IRRI, the Philippines, pp. 335–355.

Dilts, R. (1985) Training: reschooling society. *Prisma* 38, 79–90.

Dyck, V.A. and Thomas, B. (1979) The brown planthopper problem. In: *Brown Planthopper Threat to Rice Production in Asia.* IRRI, the Philippines, pp. 13–17.

IRRI (1980) *Stand Evaluation Systems for Rice.* Los Baños, the Philippines, 44 pp.

IRRI (1990) *World Rice Statistics.* 320 pp.

Kusumah, E. (1994) Dampak ekonomi penerapan konsep PHT pada petani sayuran dataran tinggi. *Lokakarya Dampak Sosial Program PHT Pusat Penelitian Sosial Ekonomi Pertanian*, Bogor, 7–9 Maret, 10 pp.

Laba, I.W. and dan Soejitno, J. (1987) Resistensi pada wereng coklat (*N. lugens* Stal)> Edisi khusus no.1 Wereng coklat. Balai Penelitian Tanaman Pangan, Bogor, pp. 69–76.

Metcalf, R.L. (1980) Changing role of pesticides in crop protection. *Annual Review of Entomology* 25, 219–256.

Mustamin, M. (1988) Hazards due to the use of pesticides in Indonesia: data collected and surveys. *Proc. SEA Pesticide Management and Integrated Pest Management Workshop.* 23–27 Feb., Thailand, pp. 301–309.

Oka, I.N. (1978) Quick method for identifying brown planthopper biotypes in the field. *International Rice Research Newsletter* 3(6), 11–12, IRRI, PO Box 933, the Philippines.

Oka, I. (1979a) Cultural control of the brown planthopper. In: *Brown Planthopper, Threat to Rice Production in Asia.* IRRI, the Philippines, pp. 359–369.

Oka, I. (1979b) Feeding populations of people versus populations of insects: the example of Indonesia and the rice brown planthopper. In: Commendahl, T. (ed.) *Proceedings Symposium IXth International Congress of Plant Protection. Vol. I. Fundamental aspects.* UAS, Washington, DC, Aug. pp. 23–26.

Oka, I. (1980) Brown planthopper survey technique. Rice improvement in China and other countries. IRRI and China Academy of Agricultural Science, pp. 287–291.

Oka, I. (1987) Pest management in Southeast Asia, problems and prospects. *Proceedings International Congress of Plant Protection.* Manila, the Philippines, 5–9 October, pp. 7–10.

Oka, I. (1988) Future needs for pesticide management in Southeast Asia. *Proceedings Pesticide Management and Integrated Pest Management in SEA Workshop*, 23–27 Feb., pp. 1–11.

Oka, I. (1995) *Pengendalian Hama Terpadu dan Implementasinya di Indonesia.* Gadjah Mada University Press, Bulak Sumur, Yogyakarta, Indonesia, 255 pp.

Oka, I. (1997) Integrated crop pest management with farmer participation in Indonesia. In: *Reasons for Hope. Instructive Experiences in Rural Development.* Kumarian Press, pp. 184–199.

Pimentel, D., Aquay, H., Biltonen, M., Rice, P., Silva, M., Nelson, J., Lipner, V., Giordano, S., Horowitz, A. and D'Amore, M. (1992) Environmental and economic costs of pesticide use. *BioScience* 42(10), 750–760.

Pincus, J. (1991) Farmers Field School survey: impact of IPM training on farmers' pest control behavior. *Integrated Pest Management National Program*, 36 pp., with figures and tables.

Statistik Indonesia (2001) BPS, Jakarta, 608 pp.

Sutrisno (1987) Resistensi wereng coklat, *N. lugens* (Stal), terhadap insektisida di Indonesia. Edit. Khusus no. 1 wereng coklat. Balai Penelitian Tanaman Pangan, Bogor, pp. 55–68.

Tandiabang, J. (1986) Tingkat resistensi wereng daun hijau, *Nephotettix virescens* Distant terhadap insektisida pada empat tempat di Sulawesi Selatan. *Bull. Pen. Pert. Maros* 1(1), 4.

World Development Report (1993) *Investing in Health,* Table 26, *Population Growth and Projections.* The Bank, Oxford University Press, 328 pp.

Chapter 19

Integrated Pest Management in the Philippines

Pio A. Javier[1], Maria Lourdes Q. Sison[2] and Reynaldo V. Ebora[2]
[1]National Crop Protection Center, University of the Philippines, Los Baños College, Laguna, the Philippines; [2]National Institute of Molecular Biology and Biotechnology, University of the Philippines, Los Baños College, Laguna, the Philippines

History of IPM in the Philippines

Early IPM efforts

IPM in the Philippines in the early 1940s began with individual farmers who planted pest or disease resistant crops, practiced crop rotation and intercropping and used botanical repellents such as tobacco and papaya extracts, tubli (*Derris* sp.) and kakawati (*Glyricidia sepium*), or used biological control agents (*Trichogramma*) to control pests (Baltazar, 1963). *Trichogramma japonicum* Ashmead was introduced in the late 1940s for the control of yellow stemborer (*Scirpophaga incertulas* Wlk.) in rice (Sumangil *et al.*, 1991). Similarly, microbial control agents like Bt, baculoviruses, *Metarrhizium anisopliae* and other entomopathogens were also used for insect control.

In the early 1970s, a nationwide rice shortage prompted the government to intensify rice production with the 'Masagana 99 Rice Program' (M-99). Under this program, the package of technology included pesticides as a part of production loan. The production guide recommended the calendar application of pesticides from six to nine times per cropping season. The heavy use of pesticides led to pest outbreaks. During the M-99 program, a cost reduction technology known as IPM was pilot tested in strategic rice production areas in the Philippines by the International Rice Research Institute (IRRI), state colleges and universities and the Bureau of Plant Industry Crop Production Division (Callo, 1990). Economic demonstration plots tested old and new management techniques that would be economically feasible.

In 1986, President Corazon Aquino issued a policy declaring IPM the core of crop protection policy (Adalla, 1986). This directed agricultural programs to incorporate development, dissemination and transfer of IPM technology to important agricultural crops such as rice, maize, vegetables, banana, sugarcane and root crops. The Ministry of Agriculture and Food created multisector IPM technical working groups, the Research and Development Committee and the Training and Extension Committee to support the country's new agricultural policy.

The emphasis on IPM was based on its ability to provide desired economic, as well as social and environmental benefits. Farmers' profit was stressed over gross production. It was strongly regionalized with high dependence on various sectors (government, private and non-government

 Integrated Pest Management in the Global Arena (eds K.M. Maredia, D. Dakouo and D. Mota-Sanchez)

organizations) to solve problems. Both farmers and trainers learned by practical experience in rice fields over an entire season. Through IPM adoption the country was able to save more than US$3.5 million in rice production (Davide *et al.*, 1990).

In 1993, President Fidel V. Ramos created the Philippine National IPM Program (Dar, 1994). The program was named KASAKALIKASAN (acronym for *Kasaganaan ng Sakahan at Kalikasan – Prosperity of the Farm and Nature*) and established IPM as the standard approach to crop production. Based on techniques developed in the Indonesian IPM program, the program trains farmers to understand the crop–field agroecosystem interactions that affect plant growth. Farmers are trained to make informed decisions in crop management. Together with field technicians and workers from other organizations, farmers learn IPM skills in FFS. Conservation biological control is the program's ecological foundation.

Organizational structure

The Department of Agriculture defined the structure of the National IPM Program (Fig. 19.1). The Central Program Management Organization oversees activities such as research, training and extension, communication, policy formulation and advocacy, and supports local IPM programs. The National Management Committee sets policies and operating guidelines for the National IPM Program. It also provides funds in support of locally based IPM programs. Under the National Management Committee, the KASAKALIKASAN became an integral part of the Department of Agriculture system.

The Program Working Group supervises all conceptual, technical and operational aspects of KASAKALIKASAN, directs program operations and approves annual IPM budgets. It includes representatives of 15 government organizations, agencies and state universities (Table 19.1).

The Program Working Group also collaborates with international agencies and donor institutions such as the AVRDC, IIBC and the FAO through the Intercountry Program for IPM in Rice and Vegetables. Other international collaborators are also included (Table 19.2).

The National Program Office oversees implementation of KASAKALIKASAN on a day to day basis. The Regional Program Committee coordinates programs in all regions of the country. State colleges and universities located in different regions of the country are also involved in regional IPM research and development.

The Provincial Program Committee oversees the IPM field-related tasks performed by the training team in each province. The IPM training team consists of two full-time IPM-trained field workers who conduct the FFS. The Department of Agrarian Reform and the Department of Education, Culture and Sports also collaborate with the IPM program.

Pesticide policy

The FPA is the government agency concerned with pesticide regulation and safety in the Philippines. Created in 1977 by President Ferdinand Marcos, the FPA controls the importation, manufacture, formulation, registration, distribution, sale, transport, storage, labeling, use and disposal of pesticides. The FPA has codified its pesticide registration requirements in accord with the international standards set by FAO and WHO (Bayani, 1998). A Pesticide Technical Advisory Committee evaluates and reviews the toxicology, efficacy, environmental fate and residue data, and recommends registration of pesticides to the FPA.

In addition to registration of new pesticide products, the FPA is also mandated to restrict or ban the use of hazardous pesticides or pesticide formulations, issue licenses to pesticide handlers, educate the public on safe and judicious use of pesticides, and protect the public against the potential hazards of pesticide residues on food (FPA, 1985).

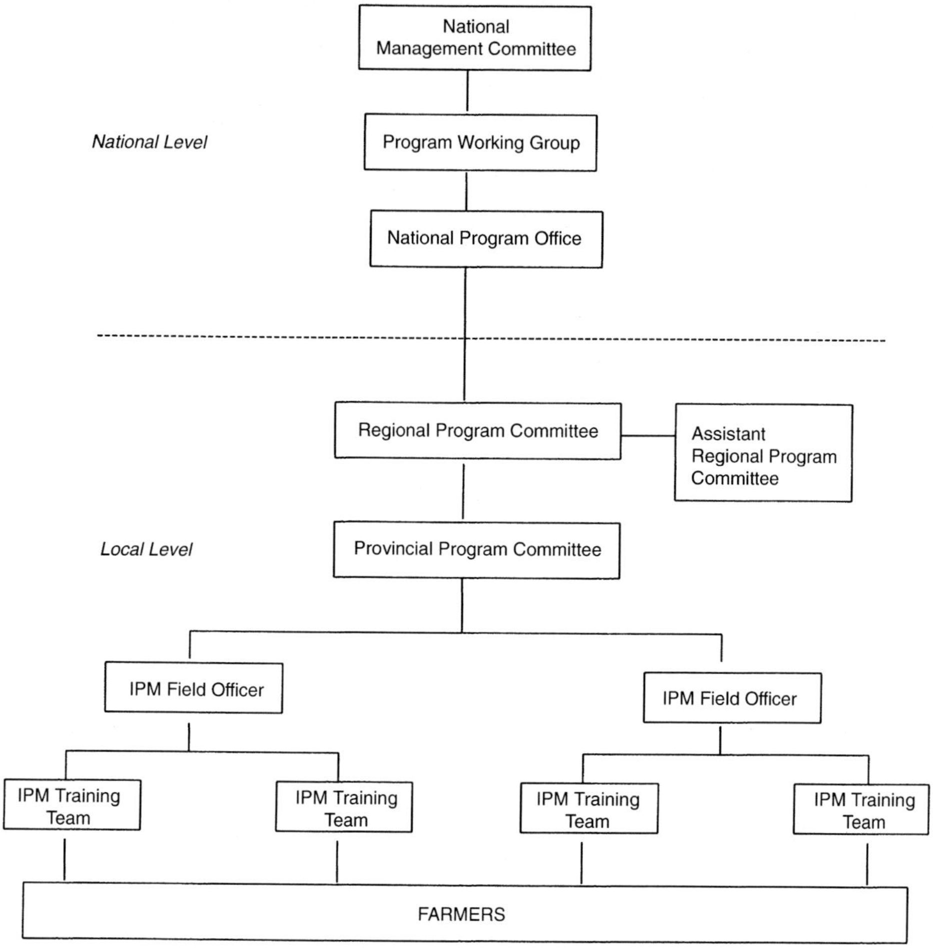

Fig. 19.1. Organizational structure of national IPM program (KASAKALIKASAN) in the Philippines. (Source: Medina and Callo, 1999.)

In 1991, the Secretary of Agriculture created a Pesticide Policy Task Force to provide recommendations on ways to improve pesticide policy. Analysis by the task force revealed a need to integrate an improved pesticide regulatory system with the broader goals of pest management (Versteeg, 1992). To this end, health and environmental concerns are now integrated by encouraging the reduced use of synthetic pesticides. The development of alternative pest management strategies complements the reduction in pesticide use. Farmer training programs that focus on the safe use of pesticides are based on IPM principles.

The 2000 revision of pesticide policies incorporated new developments in pesticide legislation, policies and guidelines to safeguard human health and the environment. Two new topics were included: policy guidelines on the fast developing group of biorational pesticides, and the principles of product stewardship and responsible care (FPA, 2000). Biorational pesticides include semiochemicals (pheromones and similar substances), biochemical plant control agents (phytohormones and enzymes) and microbials (naturally occurring or improved bacteria, fungi, protozoa and viruses). The principle of product stewardship and

Table 19.1. Composition of the Philippine IPM Program Working Group. (Source: Department of Agriculture Special Order No. 70 Series of 1997.)

Designation	Position/Agency
Chairman	Undersecretary of Agriculture for Field Operations
Vice Chairman	Assistant Secretary of Agriculture for Irrigation, Soils, Research and Training
Members	Director, Bureau of Agricultural Research
	Director, Agricultural Training Institute
	Administrator, Fertilizer and Pesticide Authority
	Executive Director, National Agricultural and Fishery Council
	Director, Bureau of Plant Industry
	Director, Bureau of Soil and Water Management
	Director, National Crop Protection Center
	Director, Philippine Rice Research Institute
	President, Benguet State University
	President, University of Southern Mindanao
	Executive Director, Philippine Council for Agriculture, Forestry and National Resources Research and Development
	National Program Officer, KASAKALIKASAN
	Deputy Administrator, Philippine Coconut Authority

Table 19.2. International collaborators for the Philippine IPM Program. (Source: Medina and Callo, 1999.)

Agency	Nature of Collaboration
UNDP/FAO/CARE International Rice IPM Trainor Exchange Program	Technical assistance in training of trainers provided by KASAKALIKASAN rice IPM specialists to Bangladesh, Cambodia, India (from 1995 to date)
IIBC DA-IIBC-ADB Technical Assistance Grant for Vegetable IPM in the Highlands	Technical assistance in the development of vegetable IPM in Benguet and Mt Province (1995–1996)
Centro International de la Papa (CIP)UPWARD Lowlands Potato IPM Program	Training of potato farmer – trainers and LGU extension workers and conduct of potato FFS in Bukidnon (1996 to date)
SEAMEO SEARCA ASEAN IPM Knowledge Network	Networking of IPM knowledge capital that can be reused and shared by National IPM Program in the ASEAN Region
FAO Inter country Program in Rice and Vegetables IPM	Provide capability-building expertise in education and training, equipment for national office and paying for participation of senior government staff in international meetings and workshops

responsible care requires pesticide manufacturing companies to ensure human health and safety of pesticide workers, dealers and end-users.

Research and Extension Focus

IPM in rice: a FFS approach

The Philippine IPM program in rice employs a farmer training program similar to the Indonesian FFS approach. In 1993, 13 rice farmers from Pila, Laguna (Luzon) participated in an experiment to evaluate whether early season insecticide use was necessary. Each farmer allotted a small portion of his land to a treatment of no insecticide in the first 40 days after transplanting. After five planting seasons, all 13 farmers were convinced of the importance of IPM (Cadiz and Suva, 1995). From June 1993 to December 1994, the NCPC conducted a farmer training program in Infanta and General Nakar, Quezon

(Luzon), consisting of three phases, with each phase extending for one cropping season or about 5 months (Medina *et al.*, 1995).

Phase I: experiential learning

Many natural enemies have been shown to regulate insect pest populations in rice (Shepard *et al.*, 1987). These include predators, parasitoids, spiders and microorganisms. In Phase I, farmers were taught to recognize natural enemies and incorporate them into sustainable pest management strategies. The training was conducted in the farmers' own fields, focusing on crop protection and pest dynamics.

Phase II: application

The current IPM program for rice relies on conservation of natural enemies and need-based insecticide applications. Phase II allows the participating farmers to apply what they have learned in Phase I and enables them to develop their own site-specific crop production and pest management strategies.

Phase III: replication

In Phase III, the farmers learn to teach others to use the IPM strategies they have learned in Phases I and II. Trained farmer participants introduce IPM concepts to farmers in other areas. Farmers are encouraged to redesign their farm environment to favor natural control of the ecosystem. This learning process is facilitated by the NCPC scientists, while the farmers choose the technology they will use based on the results of their experiments (Medina *et al.*, 1995).

An evaluation of the program in January 1995 showed that after the IPM training (two rice cropping seasons), the percentage of pesticide users dropped to almost none. The percentage of inorganic fertilizer users also declined, from about 80% to 10%. All the participants shifted from the conventional distance of planting (less than 40 × 10 cm) to the recommended 40 × 10 cm distance. Of the training participants, 83% have already shared the knowledge they gained from the training. Based on the problem solving and decision-making skills learned in IPM training, the farmers were able to develop their own approach and move toward self-reliance in pest management (Seminiano *et al.*, 1996).

In addition, an IPM Collaborative Research Support Program supporting innovative IPM research in rice–vegetable systems in Asia has been operating in the Philippines for the past 4 years. It is funded by the United States Agency for International Development with collaborative efforts of scientists at the Philippine Rice Research Institute, University of the Philippines at Los Baños, the Central Luzon State University, the Visayas State College of Agriculture, IRRI, the Asian Vegetable Research and Development Center, Pennsylvania State University, Ohio State University, and Virginia Tech. Results of research activities during the past year have been very encouraging (Gapud *et al.*, 1999).

IPM in maize

Next to rice, maize is the second most researched crop in the Philippines. Research is primarily focused on the Asian corn borer, *Ostrinia furnacalis* (Guenee), the most destructive insect pest of maize in Asia. Asian corn borer can reduce maize yield by as much as 20–80% (Gabriel, 1971; Sanchez, 1971; Morallo-Rejesus *et al.*, 1982a,b). Basic research on its biology, ecology, yield loss, and control tactics had already been conducted before the National IPM Program. Studies such as these are prerequisites for the development of rational and sound pest management strategies.

The major management strategy employed against Asian corn borer is field releases of the egg parasitoid *Trichogramma evanescens* West; an especially effective control because it prevents the corn borer from reaching the destructive larval stage. Successful control of Asian corn borer hinges on the area-wide use of *Trichogramma* by maize growers. *Trichogramma* are made available to farmers at rearing laboratories located throughout the major maize growing regions. Releases of *Trichogramma* are based on threshold levels determined by monitoring corn borer egg masses starting 20–25 days after planting. If there are 3–5 egg masses/100 plants, 70–100 *Trichogramma* cards (containing 1500–2000 parasitoids per card) are released per hectare. If the percentage of egg mass parasitism is less than 20%, field releases of *Trichogramma* are continued for 2–3 times at weekly intervals. If 30% of maize plants show symptoms of corn borer damage, then granular/systemic insecticides are applied directly into the whorl or the plants are sprayed with Bt.

Detasseling is also used as a mechanical control method. Immediately after tassel emergence or before pollen shedding, 75% of the tassels (three rows are removed for every four rows) are removed. Detasseling contributes to a 40–50% reduction in the corn borer population (Morallo-Rejesus and Javier, 1985).

The FFS program has contributed to the success of IPM in maize. Several research projects at the University of the Philippines are addressing the effectiveness of the FFS approach. One study showed that IPM was highly effective in controlling Asian corn borer in six regions of the Philippines (Javier *et al.*, 2001). Other biological control agents are also known to affect Asian corn borer. Earwigs, *Euborellia annulata* (Fab) and flower bugs, *Orius tantillus* (Motschulsky) can be effective natural enemies and can be conserved in the field, helping to maintain the corn borer population below damaging levels (Javier and Morallo-Rejesus, 2000).

Other pests of maize

The most important disease of maize, downy mildew, is controlled with a combination of the use of resistant varieties, cultural control and seed treatment with Ridomil™ (metataxyl–m+ mancozeb) (Exconde and Molina, 1978). For weed management, a combination of cultural and chemical control is used. To control rodents, sustained baiting with anti-coagulant rodenticides has been effective in reducing losses (Bato, 1990).

IPM in cabbage

The diamondback moth, *Plutella xylostella* (L.), is the most destructive insect pest of crucifers such as cabbage, broccoli, pechay, cauliflower, radish, kale and mustard. The larvae feed on the foliage in all stages of plant growth and are capable of destroying the crop if left uncontrolled. In a 1992 study in Atok, Benguet, a crucifer-growing area in Luzon, all of the farmers interviewed depended on the use of chemical pesticides for controlling diamondback moth (Cardona, 1992), spraying an average of 12 to as many as 32 times per season. Mixtures of two or more insecticides and high dosages were applied frequently, often on a calendar schedule irrespective of the diamondback moth population. In some cases, insecticides were applied until harvest to preserve the cosmetic value of the crop (Magallona, 1986). The indiscriminate use of synthetic insecticides resulted in the development of resistance to many insecticides. Other problems included the rising cost of insecticides, elimination of natural enemies and pollinators, toxicity hazards, and contamination of soil, water and food chains.

An IPM program for diamondback moth was developed that proved successful in reducing the population below ETLs. The major components of the program included field releases of parasitoids and the use of microbial insecticides. Economic thresholds were defined for diamondback moth at two stages of the crop: from seedling up to the mid-vegetative stage, the ETL is two larvae per plant, and from the pre-heading stage until harvest, the ETL is five larvae per plant. The endoparasitoids *Cotesia plutellae* (Kurdj.) and *Diadegma semiclausum* (Hellen), imported from Taiwan in collaboration with the Asian Vegetable Research and Development Center, are capable of regulating the moth population. *Cotesia* attacks the second instar larva of diamondback moth. It is more effective at elevations less than 800 m above sea level, with temperature above 25°C. *Diadegma* attacks all instars of diamondback moth, especially the second instar. *Diadegma* is more suited to elevations higher than 800 m above sea level, at temperatures below 24°C.

Tangible benefits of the IPM program for diamondback moth include a 75% increase in yield and income, a 41% reduction in production costs, an 86% reduction in pesticide application, and fewer health hazards posed by pesticide residues (Morallo-Rejesus *et al.*, 2000).

Impact of KASAKALIKASAN on Agricultural Production

The impact of FFS

In 1997, the KASAKALIKASAN Program was evaluated by the SEAMEO SEARCA (Medina *et al.*, 1998). One of the objectives was to determine the impact of FFS on rice, maize and cabbage farmer-participants. The results showed that a majority of the farmers applied most of the IPM principles they learned during the field schools. Farmers stated that they practiced sound cultural management in their crops, such as land preparation, and water, nutrient and pest management. In addition, farmers indicated that they practiced agroecosystem analysis in assessing the status of their crops and in making management decisions, especially for insect pest management. The National IPM Program was able to improve farmer pest management practices in several areas: greater reliance on biological and cultural control, reduced insecticide use, less frequent insecticide application, and greater use of less toxic insecticides (Medina *et al.*, 1998).

Economic impacts

The economic impact of the National IPM Program was also evident. Rice yield increased by an average of 0.5 t/ha, maize yield increased by an average of 0.8 t/ha, and cabbage yield increased from an average of 11.9 t/ha to 13.3 t/ha (wet season) and 14.3 t/ha to 15 t/ha (dry season). In addition, most farmers sustained lower production costs after attending the FFS (due to lower inputs for maize, rice and cabbage farmers, respectively). Economic returns due to the FFS increased significantly. The net income of farmers on per hectare basis for the different crops increased as follows:

- Rice – US$401 (wet season) and US$438 (dry season);
- Maize – US$198 (wet season) and US$269 (dry season);
- Cabbage – US$716 (wet season) and US$2598 (dry season).

Benefits of increased genetic diversity in rice

With effective pest management, new rice varieties can be used longer, and the possibility of bringing back older varieties is increased because resistance to pests is not as critical. This is an important contribution to genetic diversity in rice. In addition to the great diversity of the rice varieties, rice paddies harbor a tremendous diversity of both pests and beneficial organisms. Many rice growing areas are in diverse landscapes, with rice paddies interspersed among forests, other crops and grassy dikes (Mew and Cohen, 1995).

The most outstanding achievement of the KASAKALIKASAN Program is a shift in the paradigm of agricultural extension. Through the FFS approach, farmers in rice, maize and cabbage have been taught to make informed decisions. The higher yield and profits obtained by FFS farmers have substantially improved their economic conditions. The risks of pesticides to humans and the environment have been substantially reduced. Another outstanding achievement of the KASAKALIKASAN Program is the creation of a group of skilled, highly motivated farmers working together with local government units to mobilize resources for the national IPM program (Medina *et al.*, 1998).

Constraints and Challenges

Supply and demand issues

The use of biological control agents is the most important component of IPM for several crops including rice, maize, and vegetables. However, in some cases the availability of biological control agents is limited. In maize, for example, the supply of *Trichogramma* cards can be a major bottleneck in the biological control of Asian corn borer. If the government or other agencies can not supply sufficient quantities of *Trichogramma* cards at the necessary time, farmers are compelled to use pesticides. When this occurs, natural enemy establishment is limited. Therefore, there is a need to coordinate the mass production of biological control agents to ensure the sustainability of the IPM program.

Emerging secondary pests

In some areas, secondary pest outbreaks have become a concern. In crucifers, diamondback moth has been effectively reduced to non-damaging levels because of the release of highly effective parasitoids. However, these parasitoids are highly specific to second instar diamondback moth larvae, which can result in outbreaks of secondary pests such as the cabbage moth (*Crocidolomia binatolis* Zeller), common cutworm (*Spodoptera litura* Fab.) and other species. While some of these are still considered minor pests, research is needed (i.e. life history studies, crop loss assessment, identification and evaluation of associated natural enemies, etc.) so that if they become major pests, management strategies will be at hand.

Challenges for the farmer using IPM

Farmers generally find it difficult to differentiate between pests and natural enemies. In the rice agroecosystem, for instance, there is a complex of pests and natural enemies that cannot be differentiated by most farmers because they are too small for field identification. Also, the complex monitoring system used by farmers, AESA, is time-consuming. AESA requires farmers to regularly visit the field on a predetermined schedule and sample the pest population or count damaged plants, then calculate the infestation levels. Although this is necessary for IPM, practicing farmers have little time to spare. Of course, a farmer who practices IPM should know the level of pest damage in order to implement control strategies and avoid yield loss.

ETLs are a basic part of the definition of IPM. Recently, however, the ETL concept has been challenged, because a fixed injury threshold may not be valid when several pests affect a crop, and the response of a crop to injury may depend on its physiology (Rubia *et al.*, 1995). New and more practical ETLs must be developed to meet the changing needs of farmers.

Financial concerns

Insufficient financial support is a major constraint to the successful implementation of IPM. The budget for IPM research and development is chronically low. In general, the policy on R&D has been fragmented and under financed (http://www.da.gov.ph). However, despite this limitation, the implementation of IPM in the Philippines has been generally successful due to hard work and determination of the government.

Philippine IPM Information Websites

Information on IPM practices and experiences (published or unpublished) in the Philippines are accessible from the following websites:

www.da.gov.ph – for information about KASAKALIKASAN

www.uplb.edu.ph/institutes.html – for information about NCPC-UPLB

www.pcarrd.dost.gov.ph – for information about pest management programs http://asean-ipm.searca.org – ASEAN IPM Knowledge Network.

The ASEAN IPM Knowledge Network, based in the Philippines, includes information from the Association of Southeast Asian Nations (ASEAN) member countries (ASEAN IPM, 2000). It is a knowledge watershed of all current IPM technologies, reference libraries and case studies of proven IPM concepts and best management practices. For IPM-Philippines, about 250 references are listed in the ASEAN IPM Center.

References

Adalla, C.B. (1986) Integrated pest management as a national crop protection policy – are we ready for it? 2nd Diamond Jubilee Professional Chair Lecture in Entomology, University of the Philippines College of Agriculture, Los Baños, Laguna, Philippines, 17 December, 1986, 21 pp.

ASEAN IPM Knowledge Network (2000) *Gateway to Farmer Empowerment.* SEAMEO Regional Center for Graduate Study and Research in Agriculture (SEARCA), College, Los Baños, Laguna, Philippines, 8 pp.

Baltazar, C.R. (1963) Import and export of biological control agents in the Philippines (1952–1960). *Philippine Journal of Agriculture* 28(1–2), 1–30.

Bato, S.M. (1990) Integrated pest management research and development status on corn in the Philippines. In: *Proceedings of the National Conference on Integrated Pest Management in Rice, Corn and Selected Major Crops.* National Crop Protection Center, University of the Philippines Los Baños, College, Laguna, Philippines, pp. 15–20.

Bayani, E.M. (1998) Fertilizers and pesticides: aids to growth. In: Dy, R.T. (ed.) *The Food and Agriculture Centennial Book 100 Years of Philippine Agriculture.* University of Asia and the Pacific, Manila, Philippines, pp. 120–123.

Cadiz, M.C.H. and Suva, M.M. (1995) *Success Stories of IPM Farmer-cooperators: Participatory Research in IPM Implementation.* Institute of Development Communication, University of the Philippines Los Baños, College, Laguna, Philippines, 27 pp.

Callo, D.P. Jr (1990) The Department of Agriculture's training program in the extension of integrated pest management technology. In: *Proceedings of the National Conference and Workshop on IPM in Rice, Corn and Selected Major Crops.* 1–3 March, 1990. National Crop Protection Center, University of the Philippines Los Baños, College of Agriculture, College, Laguna, Philippines, pp. 135–141.

Cardona, E.V. (1992) Field releases of *Diadegma semiclausum* in Benguet (highland), Philippines. Paper presented at the AVNET Workshop, Lembang, Indonesia, 21–27 September 1992, 24 pp.

Dar, W.D. (1994) The KASAKALIKASAN: the Philippines' national IPM effort. In: *Technotrends* April–September 8 (2 & 3), 11–14. Bureau of Agricultural Research, Quezon City, Philippines.

Davide, R.G., Litsinger, J.A., Paller, E.C. and Rola, A.C. (1990) Evolving concepts, principles and problems of IPM. Paper presented at the *21st Conference of the Pest Control Council of the Philippines.* Bacolod City, Philippines, 7–10 May 1990.

Department of Agriculture Special Order No. 70 Series of 1997 (1997) *Strengthening the Central Program Management Organization of KASAKALIKASAN.* Quezon City, Philippines.

Exconde, O.R. and Molina, A.B. Jr (1978) Note: Ridomil (Ciba-Geigy) a seed dressing fungicide for the control of Philippine downy mildew. *Crop Science Journal* 3, 60–64.

FPA-Fertilizer and Pesticide Authority (1985) *Pesticide Regulatory Policies and Implementing Guidelines and Procedures.* Manila, Philippines, 30 October 1985, 304pp.

FPA-Fertilizer and Pesticide Authority (2000) *Pesticide Regulatory Policies and Implementing Guidelines* 2nd edn. Manila, Philippines, August 2000, 337 pp.

Gabriel, B.P. (1971) Insect pests of field corn in the Philippines. *Technical Bulletin No. 26.* College of Agriculture University of the Philippines Los Baños, College, Laguna, Philippines, 60 pp.

Gapud, V.P., Obien, S.R., Norton, G.W. and Rajotte, E. (1999) Integrated Pest Management Collaborative Research Support Program (IPM CRSP). *Philippine Rice Technical Bulletin* 4, 35–40.

Javier, P.A., Gonzales, P.G., Rosales, A.M., Tividad, J., Labios, R.V., Tamisin, L.L. Jr, Ocampo, E.P., Miciano, D.M. and Yadao, L.A. (2001) On-farm verification of developed IPM technologies against the Asian corn borer, *Ostrinia furnacalis* (Guenee) in San Jose, Occidental, Mindoro. Paper presented during the *32nd Annual Scientific Conference of the Pest Management Council of the Philippines*, 2–5 May 2001, Pili, Camarines Sur, Philippines, 12 pp.

Javier, P.A. and Morallo-Rejesus, B. (2000) Conservation methods for the natural enemies of Asian corn borer, *Ostrinia furnacalis* (Guenee), *Annual Report of DA-BAR Funded Project.* University of the Philippines, College of Agriculture, College, Laguna, Philippines, 15 pp.

Magallona, E.D. (1986) Development of diamondback moth management in the Philippines. In: *Diamondback Moth Management.* Asian Vegetable Research and Development Center, Taiwan, pp. 157–171.

Medina, J.R. and Callo, D.P. Jr (eds) (1999) *Empowering Farmers: the Philippine National Integrated Pest Management Program*, 2nd edn. SEAMEO Regional Center for Graduate Study and Research in Agriculture (SEARCA), College, Laguna, Philippines, 146 pp.

Medina, J.R., Magsino, G.L., Javier, P.A., Peralta, G.A., Hautea, R.A., Bariuan, J.V. and Zorilla, R.A. (1995) Sustainable agriculture program for resource-poor communities. *Proceedings of the 26th Annual Conference of the Pest Management Council of the Philippines.* La Trinidad, Benguet, Philippines, pp. 35–49.

Medina, J.R., Catelo, S.P., Callo, D.P. Jr and Calumpang, S.M.F. (1998) Implementation of IPM: the Philippine experience. *RIS Biotechnology Review*, pp. 72–87.

Morallo-Rejesus, B. and Javier, P.A. (1985) Detasseling technique for controlling corn borer, *Ostrinia furnicalis* (Guenee). *Philippine Entomologist* 6(3), 287–306.

Morallo-Rejesus, B., Javier, P.A., Gabriel, B.P., Costales, C.C. and Baniqued, F.C. Sr (1982a) The corn borer population in relation to sowing date and stage of corn plant at Villaverde, Nueva Vizcaya. *Annual Report.* National Crop Protection Center, University of the Philippines Los Baños, College, Laguna, Philippines, p. 5.

Morallo-Rejesus, B., Zabate, P., Javier, P.A. and Robles, R.P. (1982b) The corn borer population in relation to sowing date and stage of corn plant in Malaybalay, Bukidnon. *Annual Report.* National Crop Protection Center, University of the Philippines Los Baños, College, Laguna, Philippines, p. 6.

Morallo-Rejesus, B., Sayaboc, A.S. and Malabanan, J.M. (2000) Guide to control diamondback moth. *PCARRD Farm Primer No. 4-A/2000.* Philippine Council for Agriculture and Natural Resources Research and Development, Los Baños, Laguna, Philippines, 15 pp.

Mew, T.W. and Cohen, M.B. (1995) Future research priorities in international plant protection – focus on rice. *Agriculture and Rural Development*, pp. 24–27.

Rubia, E.G., Santiago, G.F., Barroga, C., Heong, K.L., Justo, H.D., Gapud, V.P., Benigno, E.A., Teng, P.S., Fernandez, R. and Barroga, R.F. (1995) Improving regional decisions in rice pest management. *Proceedings of a National Workshop*, 18–20 October 1995, Philippine Rice Research Institute, Muñoz, Nueva Ecija, Philippines, 93 pp.

Sanchez, F.F. (1971) The economics of corn borer control in the Philippines. *Proceedings of the 7th Inter-Asian Corn Improvement Workshop.* University of the Philippines College of Agriculture, College, Laguna, Philippines, 1–16 January 1971, 272 pp.

Seminiano, S.C., Valencia, G.Z., Lalican, N.M. and Bariuan, J.V. (1996) Evaluation of technology dissemination for integrated pest management towards sustainable agriculture in irrigated lowland rice. *The Philippine Agriculturist* 79(1), 81–95.

Shepard, B.M., Barrion, A.T. and Litsinger, J.A. (1987). *Friends of the Rice Farmers – Helpful Insects, Spiders and Pathogens.* International Rice Research Institute, Los Baños, Laguna, Philippines, 13 pp.

Sumangil, J.P., Dancel, A.J. and Davide, R.G. (1991) National IPM in the Philippines: a country report. *Conference on IPM in the Asia Pacific Region.* 23–27 September 1991. Kuala Lumpur, Malaysia, 55 pp.

Versteeg, H. (1992) *Recommendation on Ways to Improve Pesticide Policy in the Philippines.* College, Laguna, Philippines and Halifax, Nova Scotia, Environment and Resource Management Project (ERMP), 30 pp.

Chapter 20
Integrated Pest Management in the USA

Larry Olsen[1], Frank Zalom[2] and Perry Adkisson[3]
[1]*Department of Entomology, Michigan State University, East Lansing, Michigan, USA;*
[2]*Department of Entomology, University of California, Davis, California, USA; and*
[3]*Department of Entomology, Texas A&M University, College Station, Texas, USA*

Introduction

IPM has over a 40-year history in the USA. This chapter provides a glimpse of that history, current USDA programs and funding and case examples of IPM programs in California and Michigan. The states provide a brief history of their program structure, management, funding, implementation networks, successful programs and interactions with each other. A discussion of the 2001 completed Government Accounting Office review of IPM summarizes their findings and recommendations.

Historical Perspective on IPM in the USA

IPM arose owing to the ecological problems caused in the late 1950s and 1960s by the unilateral use of insecticides for controlling several major crop and orchard pest insects. By this time the limitations of pesticides, including the development of pesticide-resistant pest strains, destruction of natural enemies leading to key pest resurgence and secondary pest outbreaks, direct hazards to humans and other non-target organisms, pesticide residues in food, water and air; and their overall adverse impact on the environment had become well known (Smith, 1969). The crop and environmental disasters caused by the misuse and overuse of pesticides led several entomologists to a different and less chemically dependent approach to crop protection. These entomologists believed that more effective, profitable, sustainable and environmentally safer methods of suppressing crop pests could be achieved through integrated management systems combining cultural, biological and chemical control techniques in an ecologically sound context.

Although IPM was born in mid-century as a consequence of pesticide-induced disasters, it was conceived much earlier by the philosophies and concepts of a number of early entomologists. Foremost among those who advocated an ecological approach to controlling agricultural arthropod pests were S.A. Forbes (Illinois), L.O. Howard (USDA), C.W. Woodworth, A.E. Michelbarger and Harry Smith (California), and Dwight Isely (Arizona). These men not only had a great impact on the insect control practices of their era, they also had great influence later through their students (e.g. Ray Smith, H.T. Reynolds, Robert van den Bosch, P.L. Adkisson, and many others) on the development of future IPM programs. That their legacy has long survived them becomes apparent if one reads papers such as one by Hoskins, Borden and Michelbarger (1939) which states:

> biological and chemical control are considered as supplementary to one another or as the two edges of the same sword which is raised for the protection of man's food from his insect rivals . . . it is undoubtedly true that nature's own balance provides the major part of the protection that is required for the successful pursuit of agriculture . . . insecticides should be directed only against specific pests which are known to need chemical control . . . (use of insecticides may be decreased by) determination of the time at which most effective control of the pest may be secured and the least injury done to its natural enemies.

These sentences present the major concepts for modern integrated control.

Also, several early entomologists developed ecologically oriented multi-tactic approaches for managing insect pests which had all the components of an IPM system. Notable among these is the multi-tactic strategy developed in Texas in the early 1900s for control of the boll weevil, *Anthonomus grandis* Boh. on cotton by Fred Malley, W.D. Hunter, W.E. Herids, B.R. Coad and others. This system continued to provide the basic foundation for cotton IPM in Texas (Bottrell *et al.*, 1972). Isely also pioneered modern crop pest control in the 1920s by utilizing systematic field inspection, economic thresholds, trap crops and insecticides to control the boll weevil in cotton (Newsom, 1974). The term 'integrated control' was first used in entomological literature by Michelbarger and Bacon (1952) in their research report on the chemical control of walnut insects and spider mites in California. The first discussion of the concept of integrated control was presented in a paper by R.F. Smith and W.W. Allen (1954). This was followed by more comprehensive presentations in their seminal publication 'The Integrated Control Concept' in the October 1959 issue of *Hilgardia* (Stern *et al.*, 1959). They presented the first detailed concepts for integrated control, defined economic injury levels and economic thresholds and established the principles for their use. Smith and van den Bosch (1967) further elaborated the integrated control philosophy and expanded its concepts in a second paper 'Integrated control' published in 1967. These were the key papers in establishing the integrated control concept.

The concept of integrated control was first based on an approach that applied ecological principles in utilizing biological and chemical control methods against insect pests. It was subsequently broadened to include all control tactics (Smith and Reynolds, 1966). The idea of 'managing' insect pest populations was proposed by Geier and Clark (1961), and the term 'pest management' was advocated by Geier (1966) in preference to integrated control. From these presentations, the term integrated pest management (IPM) has evolved and now has been broadened to include the management of all classes (arthropods, pathogens, weeds and nematodes) of crop pests. For detailed discussions see Smith *et al.* (1976) and Frisbie and Adkisson (1985).

An early accepted definition of IPM is the one coined by the FAO/UNEP Panel of Experts on Integrated Pest Control in Agriculture which defines integrated control as:

> a pest management system that, in the context of the associated environment and the population dynamics of the pest species, utilizes all suitable techniques and methods in as compatible a manner as possible and maintains the pest population at levels below those causing economic injury. Integrated control achieves this ideal by harmonizing techniques in an organized way by making control practices compatible and by blending them into a multifaceted, flexible, evolving system.
>
> (FAO, 1966)

The most accepted definition today is:

> Integrated Pest Management (IPM) is a sustainable approach to managing pests by combing biological, cultural, physical and chemical tools in a way that minimizes economic, health and environmental risks.
>
> (Anonymous, 1994)

IPM is probably one of few scientific terms to be redefined by a US President. In a 1979 environmental message to Congress,

President Jimmy Carter described IPM as 'a systems approach to reduce pest damage to tolerable levels through a variety of techniques, including predators and parasites, genetically resistant hosts, natural environmental modifications and when necessary and appropriate, chemical pesticides'. IPM is now the most widely accepted term used with reference to integrated control (Smith *et al.*, 1976; Frisbie and Adkisson, 1985).

The IPM movement

Although the conceptual basis for IPM was developed in the late 1950s and early 1960s, the philosophy was accepted and integrated control programs for major crops were developed in the late 1960s and 1970s. The IPM movement began in the 1960s when a number of major pests developed such a high level of resistance to chlorinated hydrocarbon and carbamate insecticides that they could no longer be controlled. This was particularly acute in cotton where production was in a crisis in the Lower Rio Grande Valley of Texas, the Imperial Valley of California, Egypt, The Sudan, Peru and Colombia (Adkisson, 1969, 1973; Smith *et al.*, 1976). At the same time there was an increasing concern and awareness that DDT and other insecticides were having a disastrous effect on many species of birds and fish. Also, many farm workers and pesticide applicators were being exposed to dangerous levels of pesticides. These concerns were amplified in the general public by Rachel Carson (1962) in her famous book *Silent Spring* and led to the banning of most uses of DDT in the USA by the EPA in 1972.

Several very successful early IPM systems were developed for a number of crops including cotton, lucerne, certain fruits, sorghum and groundnuts. These early successes provided the impetus and encouragement needed to convince a few key government officials of the wisdom for funding a large national IPM project. Out of this the Huffaker project was born.

Concurrently with the above activities, Smith in the early 1970s obtained funding from USAID for the University of California Pest Management project which later evolved into a 12 university-USDA CICP. This group sponsored workshops, conferences and intensive training programs on IPM, pesticide management, safety and pesticide residue analysis.

The Huffaker Project

In the early 1970s, the US Environmental Protection Agency was under increasing public pressure to ban most uses of DDT, especially on major crops such as cotton. This pressure was counterbalanced by farm organizations who feared that a DDT ban would be followed by bans on other pesticides making crop protection more difficult and expensive.

The Nixon Administration, caught in a dilemma, sought ways to satisfy both groups and eventually made a decision to couple the ban on DDT with expanded research on alternative methods of crop insect control. One of the top administrators of the NSF US International Biological Programs approached Carl Huffaker about developing a research proposal in biological control as an alternative to broad-spectrum toxic chemicals for insect pest control. Soon thereafter, with the support of Vice President Gerald Ford of Michigan, a Presidential initiative was issued directing NSF, EPA and USDA to develop and fund a major national research initiative focused on alternative methods for controlling crop insect pests (Huffacker Project).

The research phase of the Huffaker Project began in March 1972 and continued until 1978, with federal funding of approximately US$14 million. The research centered on five major crops and pine forests. The crops and their sub-project directors were: cotton (P.L. Adkisson), lucerne (E.J. Armbrust), pome and stone fruits (B. Croft), soybeans (L.D. Newsom), citrus fruits (L. Riehl) and pine (W.E. Waters). Multidisciplinary teams of agronomists, ecologists, economists, entomologists, plant pathologists and systems analysts, were formed at 18 major land-grant universities. The goals of the subprojects depended on the crops. For example, larger amounts of insecticides were used in cotton and fruit so their goal was to reduce use without adversely affecting yields. In contrast, the use of insecticides on lucerne, soybeans and pine forests was relatively low and their goal was to keep it that way.

The Huffaker Project was an application of applied ecology using a holistic systems approach to agroecosystem management designed to maintain pest numbers below crop damaging levels. Integrated strategies were developed based on accurate, up-to-date information of the crop ecosystem, including its pests, their parasites and predators, the weather, crop plant growth and development and how all of these influence one another.

The project was the first agricultural research project in the USA to use computer technology and system analysis intensively to build models of crop ecosystems. These models developed by A. Gutierrez, W. Ruesink, D. Hayes, J. Stamic, L. Brown, G. Curry, P. Sharpe, A. Hartstack, etc., provided researchers, extension specialists and farmers with reliable information on controlling insect pests and other aspects of crop management. This part of the project had a significant impact on future interactions between Michigan and California, as evidenced by systems science publications by Bird and Thomason (1980) and Caswell *et al.* (1986).

The Huffaker Project made great progress in clearly identifying research gaps and furnished leads to developing a better system for conducting agricultural research (Huffaker and Messenger, 1976). During the project much improved, but only partial or preliminary systems of integrated insect control were developed for lucerne, apples, cotton, citrus and soybeans. The Project demonstrated that the arthropod pests of these crops can be controlled with much less use of insecticides, while maintaining as good or higher yields. Farmers who used IPM increased their profits and reduced their use of energy resources, with less damage to the environment and less hazard to people when compared with those whose methods were completely dependent on chemicals (Sprott *et al.*, 1976; Kogan, 1977). IPM systems also helped to avoid the long-term problems of pesticide-resistance (Armbrust, 1978; Georghiou and Taylor, 1986).

On its termination in 1978, the Huffaker Project improved integrated insect control programs that lessen the dependence on chemical insecticides, especially on cotton and apples. Several computer simulation plant–insect growth models and insect forecasting models were developed and system science was recognized as a good way for organizing agricultural research. Also, the benefit of economic analysis for crop protection was demonstrated.

The Adkisson Project

Concurrent with the Huffacker Project, five states were provided with Adkisson Extension funds to complement the research funds. These were 3-year 'pilot projects' to deliver the research results that were renewed for an additional 3 years as 'implementation projects'. The goals were to refine, test and evaluate available technology to limit the use of pesticides while maintaining crop yield and quality. These demonstration projects were structured so that participating farmers would help pay the cost of scouts during the first 3 years of the demonstration project, then assume full costs (Fitzner, 2002). The ultimate result of the Adkisson project was Extension formula funding beginning in 1978 to all states and territories to implement educational IPM programs. A much more detailed discussion of the history and development of these projects and IPM policy development is discussed in Fitzner, 2002.

National IPM Initiative of the Cooperative Extension Service

The Adkisson Extension IPM programs were expanded in the USA and by 1979 included all 50 states and three protectorates, covering 45 commodities (Blair and Edwards, 1979). These federally funded programs emphasized field inspection to monitor pest densities and advised the application of pesticides only when economically damaging pest populations were present. They later included educational and technical assistance to producers on varietal selection, cultural practices, biological control, economic evaluations of crop management practices, pesticide-resistance management and pesticide safety.

In addition to the federally funded program several states developed large IPM programs for commodities of major importance which were supported by a cadre of well-trained pest management specialists who worked directly with producer groups. As of 1982, 42 states had developed Extension IPM education programs. Currently, all states have IPM education programs. The two most successful technology transfer programs were established in Texas and California. Both states made large appropriations in support of these programs. Other notable programs were developed in Michigan and New York.

Success in implementing IPM programs, as is demonstrated in Texas, California, New York and Michigan, depends on two important elements. First, farmers must be shown that new practices will be more profitable than the old ones being displaced before they will be implemented. Second, farmers need technical assistance and encouragement from extension specialists and others while gaining knowledge and confidence in their ability to make pest management decisions based on new practices.

Impact of IPM

IPM has had major impacts on science and agricultural production. The science of IPM evolved to a state where Pimentel published a 2271 page (three volume) *Handbook of Agricultural Pest Management in Agriculture* in 1991. Scientifically, IPM research has expanded knowledge of basic ecological and physiological principles governing pest population dynamics, pest behavior, and crop–pest interactions. It has pioneered the use of systems research to improve agricultural production and environmental quality.

IPM has reshaped crop protection philosophies here and abroad. Across the world, farmers are more receptive to using ecologically sound crop protection methods. More emphasis is being placed on the use of pest resistant varieties, cultural practices, field monitoring and preservation of natural control agents and less on the unilateral use of chemical insecticides.

A comprehensive evaluation of university-led IPM programs revealed that the adoption of IPM methods resulted in lower pesticide use for seven out of eight commodities, production costs decreased or were unchanged in four out of five commodities, yield increased for six out of seven crops and risks of crop loss decreased in all three cases for which it was evaluated (Norton and Mullen, 1994). This report studied 61 economic evaluations of IPM programs to reach their conclusions. Another report estimated that the use of IPM strategies saves US agricultural producers more than US$500 million/year due to reductions in pesticide use and better management practices (Rajotte *et al.*, 1987).

The future of IPM

The greatest impact of IPM is still in the future and will likely occur in the developing and less developed countries. IPM has been designated by the United Nations agencies, The World Bank and Allied Banks and most donor countries as the preferred option for crop protection. IPM technology is gaining wider acceptance in developing countries where simple IPM systems are being used by producers of rice, cotton, sorghum, soybeans, groundnuts, maize, pulses, cassava and certain vegetables. IPM is a technology that is still being developed and perfected and its application is being directed towards all pest species – insects, diseases, weeds and nematodes – of a crop. Results produced for funds invested have been excellent and show the need for continued investment in developing and implementing new IPM technologies, especially those needed to counter or manage pesticide resistance. For those involved in the IPM movement, it is gratifying to see that at last it has become the most common option for managing crop pests.

Organizational Structure, Evaluation and Funding of IPM Programs in the USA

USDA IPM organizational structure

Within USDA, a National IPM Coordinating Committee was established to provide interagency communication and guidance on policies, programs, and budgets. In 1988, the committee published a brief summary of the current state of IPM, and sponsored a national IPM Workshop held in the spring of 1989 (Bird, 1989). The committee now gives staff from eight agencies an opportunity to develop policies and strengthen coordination of programs that form the IPM Initiative.

Evaluation of IPM

Grower adoption

The prevailing method to evaluate IPM programs measures grower adoption along a continuum as stated in the USDA report 'Adoption of Integrated Pest Management in the United States' (Vandeman *et al.*, 1994). Charles Benbrook (1996) supported and refined this concept in his book *Pest Management at the Crossroads*. This concept recognizes that IPM in any production system has a number of tactics, and farmers adopt them based on numerous factors. The continuum recognized that the more tactics adopted, the higher level of IPM practiced. USDA (1998a) developed a 'working definition of what growers must do to be considered IPM practitioners. Where appropriate, each site should have in place a management plan for Prevention, Advoidance, Monitoring and Suppression of pest populations (the PAMS approach)'. In order to qualify as IPM practitioners, growers should be utilizing tactics in three or more of the PAMS components. USDA's National Agricultural Statistics Service used this definition in a survey to measure adoption of IPM in the 1997–2000 growing seasons. In 2000 (Table 20.1) they found that IPM practices had been adopted at some level on approximately 70% of crop land, just less than the stated goal (USDA–NASS, 2001).

Eisley *et al.* (2001) developed a more specific definition and point system on 20 commodities to define the continuum. Growers were urged to accumulate 80% of the available points to be considered a high level IPM grower. This can be used as an educational tool and as a measure of IPM adoption by an industry.

Producer workshops

The benefits of IPM are well understood, but the adoption rate of IPM has not reached the level anticipated. To understand the constraints to adoption of IPM and moving away from chemically based pest control and to define ways to overcome them, a series of workshops were held around the USA in 1992 and 1993. USDA and EPA brought together about 40 producers at five sites to identify factors constraining adoption of IPM (Sorensen, 1993). Some of the major impediments to adoption were:

1. Lack of incentives to use IPM or a perception that the economic benefits of IPM programs did not justify the increased demands on management.

Table 20.1. Percentage of acres under IPM practices, crop year 2000 (USGAO, 2001).

Crop	USDA IPM estimate
Cotton	86
Fruits and nuts	62
Vegetables	86
Soybeans	78
Maize	76
Barley	71
Wheat	65
All other crops and pasture	63
Lucerne hay	40

2. Differing agendas and conflicting messages from governmental agencies.
3. Loss of funding for applied research.
4. Lack of funding for IPM education and research programs.
5. Slowness of the EPA pesticide registration system.
6. Lack of availability of pesticides, thereby limiting control options for growers and reducing the flexibility that IPM programs need to respond quickly to pest outbreaks.
7. Commodity programs that discourage crop rotation which is a key IPM tactic.
8. The need to market IPM more effectively.
9. Problems with Delaney Clause restrictions which may limit pesticide availability.
10. Lack of understanding by producers about the total production system approach to IPM.

IPM Initiative

One outcome of the regional Producer Workshops was a renewed commitment to IPM. In September 1993, USDA, EPA and FDA announced to Congress an IPM Initiative to develop tools needed to adopt IPM programs, and set a national goal to have some IPM methods implemented on 75% of the US crop acreage by the year 2000 (USDA, 1994). The Initiative was designed to develop new technologies to implement biologically based pest management to reduce reliance on broad-spectrum pesticides. The plan offered a description of how the Departments would manage the Initiative, establish a process for setting IPM priorities locally, link research and educational efforts to meet those priorities, and coordinate effort across agencies. The IPM Initiative recognized that successful achievement of its goals is dependent on producer involvement in priority setting of research and education activities, and improved agency and producer cooperation.

Another outcome to the Producer Workshops was that the USDA Office of Pest Management Policy was formed in September 1997 to 'improve USDA's ability to address the FQPA by improving integration and coordination of pest management and pesticide data programs, and by strengthening communication with the existing network of grower organizations and crop specialists at land-grant universities' (USDA, 1998b). The activities coordinated by OPMP should help increase USDA's responsiveness to pest management needs and interface with EPA, FDA, growers and interest groups. The office, in cooperation with CSREES, is now developing a document called the IPM Roadmap that proposes a detailed course of action for the next 5 years to improve coordination and integration of IPM efforts in the USA. The final draft of the IPM Roadmap was be presented at the Fourth National IPM Symposium/Workshop in Indianapolis, Indiana on 8–10 April, 2003.

Progress reporting under the IPM Initiative

As a result of the IPM Initiative, USDA developed a process to measure progress that is mandated under the Government Performance and Results Act (Public Law 103-62). States provide to the federal government a 4-year Plan of Work called Performance Planning and Reporting System indicating state areas of emphasis, indicators for the programs they deliver, target numbers of audiences they will reach, and how they will measure the impacts. After each year of the plan, states report on their progress at the website (www.pprs.info).

GAO report evaluates the IPM Initiative

The US General Accounting Office (USGAO) conducted an in-depth audit of the IPM Initiative announced in 1994. This joint USDA and EPA goal stated that 'the agricultural producers would implement IPM practices on 75 percent of the nation's crop acreage by the year 2000'. Their report *Agricultural Pesticides: Management Improvements Needed to Further Promote Integrated Pest Management* (USGAO, 2001) released 17August 2001 and found at http://www.gao.gov/gao-01–815, said that 'although 70% of the acreage has implemented some level of IPM, the implementation rate is a misleading indicator of the progress made toward an original purpose of IPM – reducing chemical pesticide use'. The report also said that 'federal efforts to support IPM adoption suffer from shortcomings in leadership, coordination, and management'. More detailed evaluations stated:

> The IPM initiative is missing several management elements identified in the Government Performance and Results Act that are essential for successful implementation of any federal effort. Specifically, no one is effectively in charge of federal IPM efforts; coordination of IPM efforts is lacking among federal agencies and with the private sector; the intended results of these efforts have not been clearly articulated or prioritized; and the methods for measuring IPM's environmental and economic results have not been developed. Until these shortcomings are effectively addressed, the full range of potential benefits that IPM can yield for producers, the public, and the environment is unlikely to be realized.

Recommendations to improve IPM adoption in the USA included:

> 1. Establish effective department-wide leadership, coordination, and management for federally funded IPM efforts.
> 2. Clearly articulate and prioritize the results the department wants to achieve from its IPM efforts, focus IPM efforts and resources on those results, and set measurable goals for achieving those results.
> 3. Develop a method of measuring the progress of federally funded IPM activities toward the stated goals of the IPM initiative.
> 4. If the Secretary of Agriculture determines that reducing the risks of pesticides to human health and the environment is an intended result of the IPM initiative, we also recommend that the Secretary collaborate with EPA to focus IPM research, outreach, and implementation on the pest management strategies that offer the greatest potential to reduce the risks associated with agricultural pesticides.

The significance of the GAO report to the national and state IPM programs is that the goals need to be clarified and how to measure progress toward those goals needs to be determined. Once this is determined and implemented, then IPM programs can respond to GAO and promote the positive impacts of the results.

Funding

USDA-CSREES is responsible for providing coordination and leadership to land-grant universities where many research programs are conducted and extension programs delivered. CSREES provides funding to state land-grant universities through formula funds and competitive grant programs.

Extension

CSREES continues to provide formula funds to states based on the dollar value of pesticide sales. States and territories receive a total of US$10.75 million annually for IPM extension programs. The funds are used to link the state programs with the federal partner, and to develop and help coordinate IPM research and extension programs. They are leveraged with funds from other agencies, state budgets, commodity organizations and other private funds nearly 4 to 1 to manage state and local programs.

Research

IPM has been fortunate to receive new federal funding the last several years for competitive grants to conduct research and educational programs to respond to pest management issues that have arisen due to the implementation of the 1996 FQPA. FQPA requires the EPA to re-evaluate the use of pesticides beginning with the most risky organophosphate and carbamate insecticides and B2 carcinogens. The criteria incorporate risk from aggregate sources of exposure including both dietary and non-dietary; accumulate exposures for all types of pesticides with a common mode of action; and incorporate up to an additional 10× safety factor to protect children's health. The regulatory decisions made under FQPA greatly restrict or eliminate many pesticide uses that growers have relied on in the past to manage their pests. Below are some of the new funding sources administered by CSREES to help US agriculture adjust to changes resulting from FQPA decisions.

- **RAMP** – The Risk Avoidance and Mitigation Program is funded at US$4.9 million/year. This program funds longer term proposals that show multidiscipline and multistate or regional cooperation that minimize pesticide residues of concern on foods, in drinking water and in the environment. Funded projects involve research, education and extension and cooperation between the public and private sectors. Projects are funded up to US$625,000/year for up to 4 years.
- **CAR** – The Crops At Risk program supports intermediate-term pest management research and extension efforts with at-risk crops, usually minor crops or major crops with minor pesticide uses. Its goal is to develop new multiple tactic IPM strategies to assist in the transition period for agriculture. It is competitively funded at US$1.5 million/year and projects up to US$200,000 are typically for 2–4 years in duration.
- **MBT** – The Methyl Bromide Transitions program supports the discovery and implementation of practical pest management alternatives for commodities affected by the phase-out of methyl bromide. These projects focus on short-to intermediate-term solutions. A total of US$2.5 million is available annually with each project funded for up to 2 years.
- **PMAP** – Pest Management Alternatives Program currently provides US$1.6 million annually for a national competition that supports projects that help farmers respond to the environmental and regulatory issues resulting from EPA decisions. The aim is to fund effective alternative pest management tactic projects that are already developed and need demonstration to farmers to have them quickly adopted.
- **RIPM** – The Regional Integrated Pest Management program has been in existence since 1980. This program allows regional prioritization and peer review panel into the annual competitive research and education programs to ensure relevancy to local problem solving and addressing new or emerging pest management threats. Federal project funds total US$2.4 million annually.
- **Regional Pest Management Centers** – CSREES established four regional centers in 1999 to provide leadership and coordination of both urban and agricultural IPM issues. The centers have Advisory Committees composed of stakeholders who assist in priority setting and increasing communication between the numerous agencies and groups conducting IPM programs. The centers integrate research, extension, and education programs; promote multistate activities within regions; and bring together multidisciplinary teams for problem solving. The centers are funded competitively for 3–5 years with US$4.5 million total. These centers provide to states competitive funds to establish an information network that responds to USDA and EPA questions related to pest management and pesticide issues. The state Project Leaders also develop information to aid in the FQPA reassessment decisions.

Case Studies of IPM Programs in the USA

California case study

Historical development

California has a rich history in biological control dating back to the introduction of the vedalia beetle, *Rodolia cardinalis*, into southern California citrus orchards from Australia in 1888. Within a year, this predator controlled the cottony-cushion scale, *Icerya purchasi*, an 'invader' from Australia which was devastating the citrus crop. This was the first successful introduction of a predaceous insect into any country to control a pest species. Soon thereafter, the California Horticultural Department established a state insectary for research and release of biological control organisms. In the 1920s, this insectary was transferred to the University of California's Citrus Experiment Station (now University of California Riverside), and a second University of California insectary and biological control unit was established at University of California Berkeley in 1945. Research on biological control in particular helped to establish the ecological basis for pest control that provided a groundwork for what was to become IPM in California.

The development of synthetic pest control chemicals during World War II, in combination with improvements in application technology, increased the potential for pest control dramatically. The rapid expansion of California's agriculture and growers' ability to produce a variety of high value fruits and vegetables made pesticides an important and economical way of reducing production risks and increasing yields. Entomologists in California and elsewhere soon began to recognize problems associated with reliance on applications of these new synthetic pesticides to control insect pests including secondary pest outbreaks, pest resurgence, and pest resistance to the pesticides. University of California entomologists addressing problems associated with pesticide applications were among the first to use the terms 'integrated control' and 'integrated pest management' in scientific literature. Michelbacher and Bacon (1952) called for 'integrated control' in their article to describe the judicious use of chemicals to control the codling moth, *Cydia pomonella*, in walnut orchards so as to preserve the walnut aphid, *Chromaphis juglandicola*, parasite, *Trioxys pallidus*. The walnut aphid, formerly under good biological control by the imported parasitoid, had become an important pest as a result of being disrupted by applications of broad-spectrum insecticides. Stern *et al.* (1959) used 'integrated pest management' in a groundbreaking article in which they described the concept of managing insect pests below economic thresholds by using a variety of control approaches including biological and chemical controls.

Although the IPM research base gradually expanded in the 1960s, IPM was not widely utilized in practice. A few pioneering crop consultants actively practiced IPM for many years, but their employment by growers was not common. To gain wider acceptance, growers needed to understand the benefits of pest and beneficial monitoring, and the results of practicing a more ecological approach to managing pests.

The need for professional crop consultants, formally trained in the pest management sciences, who could regularly meet with and advise growers in IPM strategies was officially recognized when the state of California initiated licensing of pest control advisers (PCAs) in 1971. In order to use a pesticide in California, growers were required to obtain a written recommendation from a state-licensed PCA. This action institutionalized pesticide-use decisions. It also created the challenge of providing these individuals with suitable tools for monitoring pest populations, and with reliable, science-based IPM information. By law, anyone (except for some public sector representatives specified by law) who offers a grower a recommendation to use a pesticide or any other pest control method or device must be licensed by the CDPR as a PCA.

University of California IPM Program

An Environmental Assessment Team was established within California's Department of Food and Agriculture by then-Governor Jerry Brown in the mid-1970s to determine how to reduce the impacts of pesticides in California's agricultural system. It identified a need to implement existing IPM practices that were not being fully utilized, to support the newly formed PCA licensing program by developing training materials and practices, and by supporting the development of new IPM systems for California's major agricultural crops. In 1979, with strong support from University of California Vice President for Agriculture and Natural Resources, Dr J.B. Kendrick, Jr, the California

State Legislature established the UCIPM project in part to address these identified needs. The Legislature specified that in addition to sponsoring short-term (5 years or less) research on eight commodities (lucerne, almonds, citrus, cotton, grapes, rice, tomatoes, and walnut), other activities should include implementing IPM locally through Cooperative Extension (CE) IPM Farm Advisors, writing a series of IPM manuals, and developing a computer network for delivering pesticide registration information, pest control guidelines, and predictive models. Its mission was to reduce the pesticide load in the environment, increase the predictability and thereby effectiveness of pest control techniques; develop pest control programs that are economically, environmentally and socially acceptable; marshal agencies and disciplines into IPM programs; and increase the utilization of natural pest controls. Among its primary external clientele were the state-licensed PCAs, and its initial efforts primarily targeted those individuals. The program had no regulatory function, and was never intended to become the only focus for IPM research and extension within the University of California. Its intent was to facilitate research and education that might not otherwise occur. Because it was not linked to a disciplinary department, it supported activities across disciplines. Because it was a statewide program supporting both research and extension activities, it was designed to foster linkages between these activities across theUniversity and between the University, government agencies and the private sector. Dr I.J. Thomason, a Professor of Nematology at UC-Riverside, was appointed as the first Director of the UC Statewide IPM Project. Under Thomason's leadership, including a 1980 sabbatical leave to study Systems Science at Michigan State University, a catalytic and long-term two-state IPM relationship was initiated between California and Michigan[1]. The UCIPM Program became an integral part of the university's IPM efforts, and its activities and scope have expanded considerably over the years.

In 1987, the UCIPM commodity-focused Workgroups, which set priorities for the research program, were restructured to emphasize broader areas of IPM research, including biological controls, cultural controls, commodity pest interactions, decision support, and systems applications. Much of the commodity-focused research involved more fundamental studies about when and under what condition organisms become pests. Researchers emphasized the development of simulation models to help them understand the relationships between crops and pests. With the new Workgroup structure, research emphasis shifted more to biological and cultural alternatives to pesticides, and better understanding of pest biology. These focus area Workgroups have been reviewed using a facilitated process at 5-year intervals by ad hoc research advisory committees consisting of stakeholders including commodity boards and organic growers, consultants, agencies, environmental and consumer groups and agrochemical and biotechnology industry groups. In its first 20 years, over 200 individuals served on UCIPM research Workgroups. These Workgroups recommended funding of 374 research grants (e.g. Grieshop and Pence, 1990; Klonsky and Shouse, 2000). These grants targeted 45 different crops or sites, but about 15% of projects dealt with general techniques. Most research projects (69%) had two or more researchers as principal investigators, and 25% involved principal investigators from different institutions. About 19% involved researchers from two different agricultural disciplines. The UCIPM research projects resulted in over 1000 publications, 324 of which were accepted in peer-reviewed journals at the time of the Klonsky (a Michigan State University IPM Program graduate in Agricultural Economics) and Shouse (2000) study.

[1] As continuing evidence of the interactions between Michigan and California, Dr Kendrick was the keynote speaker for a Michigan Governors Conference on Pesticides, Benefits, Risks and Alternatives held at Michigan State University on 4 December 1987. Highlights of this landmark assessment of IPM are included in The Integrated Pest Management Experience (Bird, 1989).

The UCIPM Project employs a group of CE Area IPM Advisors who are located in important agricultural regions of California. Their goal is to adapt and implement research-based IPM practices in the field working with and through county CE Farm Advisors and licensed PCAs. They are closely linked with the UCIPM research program and its publications and computer units. This close link to the field serves to maintain IPM implementation as a major program goal.

The UCIPM Publications unit lead by Dr Mary Louise Flint produces books, manuals and guides on pest management, and content for the UCIPM World Wide Web (http://www.ipm.ucdavis.edu). During its initial years UCIPM publications emphasized the creation and expansion of a series of IPM manuals for major California commodities. The manuals were produced through consultation and review of research and extension staff, and quickly received national recognition for their quality, style, excellent color photographs, and clear and concise presentation of IPM information for field use. Many of the manuals are in their third or fourth editions. A variety of other IPM publications are now produced which include the *University of California Pest Management Guidelines*, a series of *Pest Notes* on urban and garden pest problems for homeowners, and books intended for homeowners and for those taking the state PCA licensing examination. The *UC Pest Management Guidelines*, written by a UCIPM Project staff writer, are companion publications to the IPM manuals and contain peer-reviewed suggestions from research and extension staff for various pest specific controls including pesticides, alternatives to pesticides, sampling and monitoring methods, treatment thresholds, and organically-acceptable approaches where they are known. The major distribution method is via the UCIPM website, but camera-ready copies are also sent to county Cooperative Extension offices for reprinting and distribution.

Computers and their application to crop and pest management has been an important focus for the UCIPM Program since its inception (Zalom and Strand, 1990). Originally known as the IMPACT system, the IPM Computer System was initially created to support research and extension staff statewide with various tools for improving their productivity, and as a vehicle to transfer research models to county-based CE staff. The system was the first readily available computing resource for many California extension staff. It included common statistical tools and data entry utilities, weather data, and several types of models. A degree–day program made it simple for many people to apply phenology models for many pest species for the first time. The IMPACT system evolved with available technology from its initial configuration of terminals linked directly to a central computer, to a dial-up central database and processing system accessible by users with accounts from virtually any terminal or microcomputer. The user base also expanded from campus-based researchers affiliated with the UCIPM Project and CE Farm Advisors in 11 counties, to all University of California agricultural scientists and over 500 agency employees and private individuals.

Today, the primary focus is delivery via the UCIPM website which is accessible by anyone on the Internet. Website usage has increased dramatically with the user base expanding from about 7000/month in 1997 to over 350,000/month in 2002. Primary features of the UCIPM website are the Pest Management Guidelines and Pest Notes databases (illustrated with over 5000 color images), weather database, various degree–day and management models, PestCast disease forecasting, reports of UCIPM-sponsored research grants, and databases of pesticide registrations and total pesticide use reporting summaries developed in cooperation with the CDPR. Weather data are important to drive phenology and various other types of pest and crop models. The IPM Project's weather database is the largest online collection of quality-controlled current and historic weather data available for California, and is linked to the predictive tools on the system. The database includes over 400 stations in agriculturally important areas with historic weather data available for many stations dating from 1951. Over 70 of the stations are accessed electronically from automated weather stations. Although the weather database was developed primarily for research, it has become important to government agencies and pest management consultants as well.

Implementation of a number of monitoring methods and phenology models has been a documented IPM success. For example, a 1987 survey of California PCAs (Flint and Klonsky, 1989) indicated a substantial number used pheromone traps or bait traps, degree–day calculations to predict insect phenology, and University of California guidelines to make treatment decisions for various pests. The survey indicated that 93.8% and 87.1% of PCAs working in pome fruit crops and walnuts, respectively, used pheromone or bait traps, degree–day calculations, and UC guidelines for codling moth. Similarly, 71.5% and 66.1% of almond and stone fruit PCAs used these practices. A national survey of tree fruit and nut growers by the USDA Economic Research Service (Vandeman *et al.*, 1994) indicated that the percentage of growers using IPM and employing professional scouting exceeded 50% in many California crops including grapes, oranges and pears.

Crop consultants

The large number of consultants in California can be attributed to the state's regulatory and enforcement system which licenses and regulates their activities. PCA licensing has had a far-reaching effect on implementation of IPM, and for most crops a higher proportion of acres are scouted in California than elsewhere in the USA. In spite of this, it has been suggested (e.g. Wearing, 1988) that the licensing of pest control advisors in California has not reached its potential for reducing pesticide use in practice because the licensing program does not distinguish private consultants from the majority who are employed by farm supply dealers or other chemical retailers.

CDPR

California has the most comprehensive state-level pesticide regulatory program in the USA. It is administered by the CDPR, and is almost half the size of the US EPA's pesticide programs. The challenge to CDPR is to enforce the pesticide laws and regulations of the most populous of US states without significantly increasing production costs or reducing efficacy of pest control methods for the state's US$25 billion agricultural industry. Along with licensing of PCAs and certifying pesticide applicators, the agency has a state pesticide registration process, a use reporting process, and a food pesticide residue monitoring program. For example, growers must file a notice of intent with the county Agricultural Commissioner before a pesticide is applied, and a use report after the application. Since 1990, pesticide use reports have been required for all agricultural pesticide applications and commercial applications in urban situations. It seems likely that California pesticide regulations create an incentive for growers to use consultants to assist them in pest management decision making and use reporting.

CDPR actively promotes the philosophy of IPM through its Innovators Program which recognizes organizations that are trying to practice IPM. This is a competitive research grants program called Pest Management Alliance Program which directs funding of up to US$100,000/year to commodity-board led partnerships of university, industry, agency and non-governmental organizations. This government agency model has led to the formation of a number of other stakeholder alliances formed through increasingly available grant support from Federal and state sources, and from private foundations.

Grower organizations and food industry groups

California produces over 250 commercial crops, and over 40 of these are represented by commodity boards or marketing orders which support research and extension activities of university and USDA scientists. Through their grower research advisory committees, these commodity organizations help to identify important pest management issues for their growers. They often play an important part in updating their growers on IPM-related research, and in dealing with pesticide regulatory issues. Several of these commodity boards have been quite proactive in encouraging IPM research and implementation, particularly in recent years. They can be especially effective in facilitating discussion and setting priorities with growers and researchers, advisers, regulators and others interested in IPM.

One particularly notable grower group with regard to IPM is the Lodi-Woodbridge Winegrape Commission which was created by a grower referendum in 1991 (Klonsky *et al.*, 1998). A significant commission activity has been its establishment of an IPM program which conducts on-farm research and demonstration projects in cooperation with university scientists and growers, holds regular educational meetings, produces an IPM newsletter and other publications. The commission employs a PhD level IPM Coordinator to supervise these activities.

Perhaps the very best example of a grower-initiated IPM program in California is the Fillmore Citrus Protective District which was established in 1922 (Graebner *et al.*, 1984) as a community effort to control the California red scale, *Aonidiella aurantii*. Area growers created the district hoping to gain economically by acting as a collective rather than as individuals in the purchasing and application of pest controls. Both chemical and biological controls are used by the District to control the red scale, and the scope has expanded to target a number of other citrus pests as well. The District employs pest managers who monitor area citrus orchards and implement biological control and other IPM practices. The District owns insectaries for rearing required biological control agents.

Campbell's Soup Company and the Sun-Maid Growers cooperative have made significant progress in promoting IPM to their growers through well organized demonstration and educational programs led by IPM coordinators employed by each organization (e.g. Bolkan and Reinert, 1994). This model has proven very successful for implementing IPM practices on products intended for processing. Historically, concerns for meeting high quality standards for branded products created a situation where the buyer had little tolerance for damage and consequently growers were afraid of taking the chance that any damage might occur in their fields or vineyards. While both the company and the cooperative maintain high quality standards, good communication between the processor and grower have led to greater IPM adoption by growers of these crops and reduced use of pesticides.

Status of California IPM

California has a well-developed infrastructure for effective development and implementation of IPM systems. The IPM philosphy has become generally embraced, a decidedly positive shift over the past 20 years. Unfortunately, in practice most growers and their advisers use IPM tactics which typically target one or a limited number of related pest species, and chemicals remain a primary means for pest control. This means that although the level of IPM adoption of a few practices might be high, the biological intensity of IPM being practiced remains fairly low for many crops. Technical, financial, institutional and social challenges remain which inhibit further movement of California growers to higher levels of integration or toward more biologically intensive IPM approaches. As a description of the current state of IPM, a prognosis and continuing evidence of the two-state interaction between Michigan and California, Norris, Caswell-Chen (Michigan native and MSU graduate) and Kogan published a textbook entitled, *Concepts in Integrated Pest Management* (2003).

Michigan case study

Historical development

Michigan State University (MSU) has a 30-year tradition of developing innovative pest management programs and information. Its strength has been in the fruit area as Michigan is a leading state in fruit crop production, and has a significant number of insect and disease pests. MSU was very successful in obtaining Hoffacker and Adkisson project funds to implement IPM research and Extension programs in the early 1970s (Jones and Croft, 1981). The research projects were notable in developing practical approaches to manage pests, and MSU was exceptionally strong in the principles of system science and modeling for predicting pest occurrences and outbreaks (Haynes *et al.*, 1973; Welch *et al.*, 1978; Jones *et al.*, 1980). IPM experts around the USA received their training at MSU. Researchers developed integrated mite control in the early 1970s (Croft, 1975) and it is still implemented in nearly every apple orchard in the state. Plant pathologists established apple scab monitoring networks that monitor spore release and weather parameters to predict disease infection periods (Sutton and Jones, 1976). Resistance management programs for insects and diseases were established early (Jones, 1981), and are still the major topics in educational programs with farmers and pest managers. A comprehensive overview of the first 20 years of the MSU IPM program was presented by Bird *et al.* (1990), and the process of IPM described by Bird and Berney in *Michigan Field Crop Ecology* (2000). Research and extension specialists developed educational materials to aid pest monitoring and decision making including *Common Tree Fruit Pests* by Howitt, 1993, *Diseases of Tree Fruits in the East* by Jones and Sutton, 1996, and *A Pocket Guide for IPM Scouting in Michigan Apples* by Epstein and Gut, 2000.

The apple scouting program was strong and influenced growers all over the state. It has also had a significant impact on the overall evolution of the pest disciplines. For example the role of nematodes in IPM was described by Bird (1987), and by Duncan and Noling (1998). The book by Barker *et al.* (1998) also included a phytopathological chapter on concomitant pathogen interactions by Abawi and Chen (1998). In the mid-1970s growers paid for the MSU trained and supervised scouts. Federal funds paid for Pest Management Field Assistants (PMFA) who were mostly non-thesis Masters of Science students who worked as interns for District Horticulture Extension agents in the growing season. They provided supervision for scouts but also had their own insect pheromone trap lines and disease and weather monitoring stations. These individuals were invaluable assistants to the District Horticulture agents and advisors to growers. They conducted applied research and had demonstration plots. They maintained code-a-phone 3 minute prerecorded telephone messages to update growers in pest management recommendations, harvest timing and storage conditions for their respective commodities. Many of these individuals have progressed into leadership positions in the University and private industry. This highly successful program is being partially replicated today through a Professional Masters in IPM at MSU where students have seasonal internships similar to the PMFA while completing a non-thesis Masters of Science.

PROJECT GREEEN In more recent years, Project GREEEN (**G**enerating **R**esearch and **E**ducation to meet **E**conomic and **E**nvironmental **N**eeds) is a unique aspect of the Michigan case example that has major implications for the MSU IPM program. Project GREEEN is a state-funded initiative to MSU of over US$6 million annually with about half of the funds going to competitive projects every year. Its mission is to boost the state's economy by expanding Michigan's plant-based agriculture and processing systems through research and educational programs while protecting and preserving the quality of the environment and the safety of our food supply.

Project GREEEN provides significant funds for ongoing programs and staff, and for competitive basic and applied research, value added, extension and education activities. For the IPM program, the funds pay the salary and benefits for the IPM Coordinator, four district ICM agents, half-time for two Crop Integrators and operating funds for each of these individuals. The project makes Michigan unique in the USA by generating funds to assist growers solve plant agriculture problems. It is through grower and commodity leadership foresight that this initiative was developed and funded by the legislature, and they are still receiving its benefits.

Michigan State University IPM program

The IPM program at MSU is well recognized within the College of Agriculture and Natural Resources departments and across the state. This is a result of continual federal and state funding, and a commitment to the goals of the IPM program which are to:

- research methods to prevent pest problems;
- teach IPM practitioners;
- convince pest managers to adopt IPM practices;
- create and distribute useful IPM information;
- measure the level of IPM adoption; and
- secure adequate resources to solve new problems and enhance support services.

The staff are dedicated to these goals, and we are continually developing programs and materials, conducting training sessions and seeking funds to meet them.

The current IPM program continues to stress basic and applied research and demonstrations along with delivering programs that emphasize biological and meteorological monitoring. It produces print, Internet, and video resources to aid in commodity specific pest management decisions. Recent projects have included pocket-size scouting guides (Epstein and Gut, 2000; Brown-Rytlewski, 2002) and annual reports on nursery research/extension activities at MSU (Nursery and Landscape Research Projects and Educational Programs). The staff have cooperated with the sustainable agriculture program at MSU to develop several publications that include IPM within ecologically based cropping systems (*Michigan Field Crop Pest Ecology and Management* by Mutch *et al.* (2000a); *Fruit Crop Ecology and Management* edited by J. Landis *et al.* (2002); and *Ecologically-Based Farming Systems* edited by Harwood *et al.* (2003)). The programs website (www.msue.msu.edu/ipm) includes newsletters, useful links to commodity-specific resources and information for the general public on topics like the Asian multi-colored lady beetle and West Nile virus. The program meets urban and homeowner needs by producing materials such as *What's Bugging You?* by Ellis and Landis (1999), and has a cooperative program on IPM in schools with the Michigan Department of Agriculture.

IPM PROGRAM OFFICE/STAFF The MSU IPM program and its affiliates have staff on campus and located throughout the state in key areas where the commodities for which they have pest management responsibilities are grown. Figure 20.1 shows an organizational chart. Photos of these people and their position responsibilities are found at the website under Publications-IPM Report, Spring 2002.

The IPM Coordinator has a full-time faculty rank which has reporting lines to the departments of Entomology and Plant Pathology, MSU Extension and Michigan Agricultural Experiment Station. The Coordinator manages on-campus staff and their activities, and serves as a liaison that links MSU Extension field staff and campus faculty with growers, consultants, public agencies and private organizations. His overall responsibility is to coordinate the development and implementation of IPM programs, but he has reporting and mentoring responsibilities as well.

The Communications/Publications Specialist and Assistant IPM Coordinator is responsible for publications and other communication resources produced by the IPM Program. The specialist is the editor and program coordinator of the Crop Advisory Team (CAT) Alert newsletters and has responsibilities for producing educational resources both in print and on the Internet, including management of the Program's website. As assistant coordinator, some administrative responsibilities involving organization, reporting, and public relations are shared with the IPM Coordinator.

The Publications/Internet Specialist assists with print publishing and promotional materials for the IPM program. This

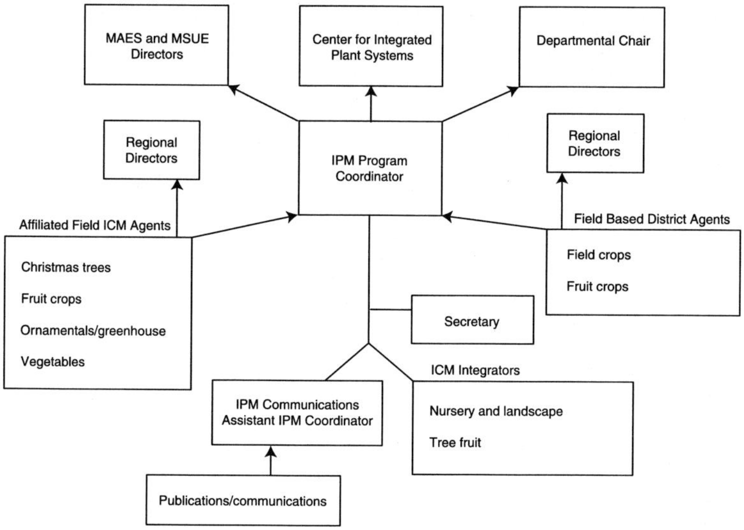

Fig. 20.1. Table of organization of the Michigan State University IPM Program.

position is a joint appointment with the Pesticide Safety Education Program, and provides publication design assistance for the program.

Crop Integrators are full-time, non-tenured staff who take a broad ICM perspective to help address critical crop management needs that enhance adoption of ICM and IPM strategies. They are paid half by MSU Project GREEEN funds (see below) and half by their industry. They have an industry advisory committee and university mentoring committee to direct and prioritize their efforts. Their performance is jointly evaluated by the IPM Coordinator and the industry Executive Director with input from their advisory committees. Currently there are two integrators, one working in tree fruit crops and the other in nursery and landscape crops. Additional integrators may be added in the future for vegetables, field crops, specialty field crops, turfgrass and small fruit if these industries obtain funds to match GREEEN funds for their salaries. Their specific duties are to:

- coordinate multi-disciplinary teams that enhance linkages between industry and the university;
- identify key program needs;
- prioritize programs to address industry identified problems;
- write grants to fund programs;
- analyze data and evaluate programs;
- communicate the benefits of new IPM technology to industry;
- produce resource materials;
- share program results.

The IPM program also has staff located around the state key agricultural areas. The District Field Crops IPM Agent is located at the W.K. Kellogg Biological Station where his area of specialty is cover crops and carbon sequestration. He teams with local agents to coordinate field crop IPM activities in southwest Michigan as well as statewide demonstrations and educational opportunities for sustainable agriculture and organic production. The District Fruit IPM Agent is based at the Northwest Michigan Horticultural Research Station in Traverse City. He coordinates similar activities in the northwest region of the state along with some statewide training sessions.

The MSU IPM Program is affiliated with and cooperates with activities of Integrated Crop Management agents funded through Project GREEEN. These District agents have crop-specific assignments and are located at county extension offices. They include agents assigned to work with vegetables, fruit, floriculture/greenhouse, and Christmas trees. They coordinate with growers, agents and campus specialists to conduct integrated crop management projects on their respective crops (Fig. 20.1).

INFORMATION SHARING AND COOPERATION THROUGH AREA OF EXPERTISE (AOE) TEAMS AOE teams were first created at MSU by the Director of Extension in 1994 (Leholm *et al.*, 1999). These teams are modeled after self-directed teams in private industry. At MSU they have these common features:

- largely self-directed and link extension, research and stakeholders;
- share self-developed vision, mission and goals, are mutually and individually accountable for performance, and are directly engaged with clientele;
- given responsibility, authority and resources to design and conduct programs;
- are co-chaired by on- and off-campus extension staff members and rotate on alternate years;
- members are both from campus and off-campus;
- teams are composed of agents, specialists, researchers and industry in some cases.

One of their objectives is to identify IPM issues and concerns with the clients they serve. The teams then design research and extension activities to address these concerns. IPM staff serve on all the plant AOE teams to assist in the coordination of program efforts. There are seven plant-based AOE teams that are actively conducting IPM research and education. They are the field crops, fruit crops, forestry, forages, landscape/ornamentals, vegetable, and Christmas tree teams.

These teams are not expensive to form, and the limited funds provided to them by extension has provided significant new funds to MSU to carry out research and education programs that solve industry-identified problems. The AOE team concept has worked well at MSU and has been adopted at other US land-grant universities and can be adopted internationally to enhance cooperation among universities, agencies and commodity groups working to solve agricultural problems.

CAT ALERTS PROVIDES IPM INFORMATION AND STRENGTHENS TEAMWORK Crop Advisory Team Alerts are newsletters published during the growing season to provide current advice on pest management for insects, nematodes, weeds and plant diseases. There are five separate editions of the newsletter: fruit, vegetable, field crop, landscape/turfgrass/Christmas tree, and greenhouse. Most editions have 18–20 issues. Each alert has a team of extension faculty and field staff that combine their observations and expertise to provide timely articles based on current conditions. During a morning conference call, the agricultural meteorologist discusses weather forecasts and data, county-based extension agents provide regional reports, and specialists offer specific pest and crop management recommendations. Articles are written, edited, and formatted with graphics added and posted on the Internet at http://www.msue.msu.edu/ipm/aboutcat/htm Current season and searchable archives issues are available online. The same day as the conference call the newsletter is printed and mailed to those who purchase a print subscription. This newsletter is an excellent method for providing current and long-term pest management recommendations to clients throughout the state and in the Great Lakes region. It provides a framework where unusual crop and pest developments can be quickly addressed and information delivered to those who need it.

COMMODITY SCOUTING INFORMATION The Michigan IPM program develops and delivers scouting information in several production systems including commercial fruit; field crop production; turfgrass for lawns, golf courses and athletics fields; Christmas tree production; commercial vegetable production; nursery plant and landscape maintenance; and home and urban pest management. This website (http://www.msue.msu.edu/ipm/scoutingIPM.htm) provides links to cropping systems and then specific pests and how to monitor for them. A few pests have action thresholds and management practices. All sites have links to additional sources of information including historical and current issues of the MSU Crop Advisory Team Alert newsletters, other information within MSU and at other universities.

Successful programs

FOOD SAFETY MSU and Michigan Department of Agriculture with commodity groups realized a public concern of pesticide residues in food and a potential loss of pesticide uses due to the implementation of the FQPA. Studies were initiated in 1998 to analyze fresh and processed produce for residues. The study compared samples taken fresh from the field with those taken after the produce had gone through processing. After 3 years of sampling, it was discovered that processing reduced the amount of pesticides in processed foods in almost every case. This data has been provided to EPA to aid in the reassessment and re-registration of specific pesticide uses. To view those reports, visit the following website: http://www.cips.msu.edu/share/MIFQPAresidue1999.pdf

MAIPMIP The Michigan Apple IPM Implementation Project is an excellent example of private foundation, university, processor, private consultant and grower cooperation to increase the adoption of IPM. The apple industry faces many challenges in pest management with loss of many pesticides due to the FQPA, pest resistance and secondary pests becoming damaging again, marketing and labor issues, and negative public opinion of pesticide residues and *E. coli* in apple cider. The goal of this project was the wide-scale implementation of an economically viable and environmentally sound pest management and production system that significantly reduced reliance on broad-spectrum pesticides and reduced the potential for residues on both raw and processed products. This program began in 1999 with 47 growers participating on 877 acres and increased by 2001 to 106 growers with 8300 acres or 20% of the state acreage. On those acres growers reduced their use of organophosphate insecticides compared with their conventionally treated blocks by 49% in year 1, 25% in year 2 and 30% in year 3 by reducing their reliance on these products and increasing the use of mating disruption for lepidopteran control and pest models for better timing of applications. Pest damage was reduced and the percent of marketable fruit was increased. The project also created a very effective industry network, conducted numerous workshops, improved overall pest management skills of participating growers and consultants, trained seven new consultants, and produced several new educational fact sheets and manuals. The final report is found at http://www.cips.msu.edu/maipmip/ It can be used as an example of cooperation to solve grower identified problems.

AOE–FIELD CROP TEAM DEMONSTRATIONS The AOE–Field Crops Team has for several years written an *On-Farm Research and Demonstration* (Mutch *et al.*, 2000b) report edited by the District IPM agent. These booklets report on the results of on-farm IPM field trials conducted across Michigan. These trials are large plots, replicated, taken to yield when appropriate and statistically analyzed. The team is striving to assist those involved in Michigan crop production with current, research-based information that is agronomically sound, profitable and environmentally responsible.

IPM INSTRUCTION The MSU IPM program has conducted IPM schools for agents, industry, chemical company representatives and growers for many years. The annual fruit school has over 120 participants and has in-state and out-of-state speakers to provide new information. Classroom instruction in entomology, plant pathology, crop and soil sciences and horticulture provides IPM information. The Professional Masters in IPM curriculum includes traditional technical classes and capstone seminars. Students in this non-thesis degree program also have an internship with industry to complete their broad-based experiential learning. A description of the degree program can be found at http://www.ent.msu.edu/dept/docs/ipm-ms.htm IPM staff have assisted with the International

IPM Short Course since 1995, providing speakers and field site visits for the international participants.

Summary

The MSU IPM program has strengths in the areas of new information development and delivery through CATS Alerts, other publications and the Web; and planning and delivery through AOE teams, extension agents and private consultants. The current state of the philosophy, principles and practices of the MSU IPM program are expressed in the 2002 publication entitled *Fruit Pest Ecology and Management* (Landis *et al.*, 2002), published as the third in a series of books designed for use by educators and farmers in their progress towards an economically viable, socially acceptable and environmentally sound agriculture for both current and future generations.

Acknowledgment

The authors wish to thank the following persons for reviewing this chapter: Dr Mike Fitzner, National Program Leader for IPM, USDA, CSREES; Dr Arlen Leholm, past Director of Extension, Michigan State University; Dr George Bird, Professor, Entomology, Michigan State University, and Ms Joy Landis, Assistant IPM Coordinator, Michigan State University.

References

Abawi, G.S. and Chen, J. (1998) Concomitant pathogen and pest interactions. In: Barker, K.R., Pederson, G.A. and Windham, G.L. (eds) *Plant Nematode Interactions*. Agronomy Series No. 36. American Society of Agronomy, Madison, Wisconsin, pp. 135–158.

Adkisson, P.L. (1969) How insects damage crops. In: *How Crops Grow – a Century Later*. Connecticut Agricultural Experiment Station Bulletin (New Haven) No. 708.

Adkisson, P.L. (1973) The principles, strategies and tactics of pest population regulation and control in major crop ecosystems: the cotton system. In: Geier, P.W. (ed.) *Studies in Population Management*, Vol. 1. Memoirs of the Ecological Society of Australia, Canberra, Australia, pp. 274–282.

Anonymous (1994) *Integrated Pest Management Practices on 1991 Fruits and Nuts*. RTD Updates: Pest Management. USDA:ERS, 8 pp.

Armbrust, E.J. (1978) Alfalfa weevil pest management system for alfalfa. In: Smith, E.H. and Pemitel, D. (eds) *Pest Control Strategies*. Academic Press, New York, pp. 85–99.

Barker, K.R., Pederson, G.A. and Windham, G.L. (eds) (1998) *Plant Nematode Interactions*. Agronomy Series No. 36. American Society of Agronomy, Madison, Wisconsin, 771 pp.

Benbrook, C.M. (1996) *Pest Management at the Crossroads*. Consumers Union, Yonkers, New York.

Bird, G. (1987) Role of nematology in IPM. In: Veech, J. and Dickenson, D. (eds) *Vistas in Nematology*. Society of Nemotologists. Painter Publishing, DeLoin Springs, Florida, pp. 114–121.

Bird, G.W. (1989) The integrated pest management experience. Reform and innovation of science and education: Planning for the 1990 Farm Bill. In: *101st Congress. Senate Committee Print 101–61*. US Senate Committee on Agriculture, Nutrition, and Forestry. pp. 31–41.

Bird, G.W. and Berney, M.F. (2000) Pest ecology and management. In: Cavigelli, M.A., Deming, S.R., Probyn, L.K. and Harwood, R.R. (eds) *Michigan Field Crop Ecology*. Michigan State University Extension Bulletin, E-2646, 86 pp.

Bird, G.W. and Thomason, I.J. (1980) Integrated pest management: the role of nematology. *Bioscience* 30, 670–674.

Bird, G.W., Edens, T., Drummond, F. and Groden, E. (1990) Design of pest management systems for sustainable agriculture. In: Francis, C.A., Flora, C.B. and King, L.D. (eds) *Sustainable Agriculture in Temperate Zones*. John Wiley & Sons, New York, pp. 55–95.

Blair, B.D. and Edwards, C.R. (1979) *The Development of Integrated Pest Management Programs in the United States*. Science and Education Administration, Extension. USDA, Washington, DC.

Bolkan, H. and Reinert, W. (1994) Developing and implementing IPM strategies to assist farmers: an industry approach. *Plant Disease* 75, 545–550.

Bottrell, D.G., Rummel, D.R. and Adkisson, P.L. (1972) Spread of the boll weevil into the high plains of Texas. *Environmental Entomology* 1, 136–140.

Brown-Rytlewski, D. (2002) *A Pocket IPM Scouting Guide for Woody Landscape Plants.* Michigan State University Extension Bulletin. E-2839, 116 pp.

Carson, R. (1962) *Silent Spring.* Houghton Mifflin, Boston, Massachusetts.

Caswell, E.P., MacGuidwin, A.E., Milne, K., Nelsen, C.E., Thomason, I.J. and Bird, G.W. (1986) A simulation model of *Heterodera schactii* infecting *Beta vulgaris. Journal of Nematology* 18, 512–519.

Croft, B.A. (1975) *Integrated Control of Apple Mites.* Michigan State University Extension Bulletin E-825, 12 pp.

Duncan, L. and Noling, J.W. (1998) Agricultural Sustainability and Nematode Integrated Pest Management. In Barker, K.R., Pederson, G.A. and Windham, G.L. (eds) *Plant Nematode Interactions.* Agronomy Series No. 36. American Society of Agronomy, Madison, Wisconsin, pp. 251–288.

Ellis, T.L. and Landis, J.N. (1999) *What's Bugging You? How to Deal With Insects Around Your Home.* Michigan State University Extension Bulletin E-2649.

Epstein, D. and Gut, L. (2000) *A Pocket Guide for IPM Scouting in Michigan Apples.* Michigan State University Extension Bulletin E-2720.

Eisley, B., Gastier, T. and Jasinski, J. (2001) *Integrated Pest Management Program: Field Crop, Fruit and Vegetable IPM Definitions.* The Ohio State University, IPM Program.

Fitzner, M. (2002) Three decades of federal integrated pest management policy. In: Wheeler, W. (ed.) *Pesticides in Agriculture and the Environment.* Marcel Dekker, New York.

Flint, M.L. and Klonsky, K. (1989) IPM information delivery to pest control advisors. *California Agriculture* 43(1), 18–20.

[FAO] Food and Agriculture Organization of the United Nations (1966) *Symposium on Integrated Pest Control.* 11–15 October. Rome. Proceedings, Part 1. FAO, Rome, 91 pp.

Frisbie, R.E. and Adkisson, P.L. (eds) (1985) *Integrated Pest Management on Major Agricultural Systems.* Texas Agricultural Experiment Station, College Station, Texas, MP-1616.

Geier, P.W. (1966) Management of insect pests. *Annual Review of Entomology* 11, 471–490.

Geier, P.W. and Clark, L.R. (1961) An ecological approach to pest control. In: *Proceedings: Technical Meeting of the International Union for Conservation of Nature and Natural Resources. 8th, 1960.* Warsaw, Poland, pp. 10–18.

Georghiou, G.P. and Taylor, C.E. (1986) Factors influencing the evolution of resistance. In: *Pesticide Resistance Strategies and Tactics for Management.* National Research Council. National Academic Press, Washington, DC, pp. 157–169.

Graebner, L., Moreno, D.S. and Baritelle, J.L. (1984) The Fillmore Citrus Protective District: a success story in integrated pest management. *Bulletin – Entomological Society of America* 30(4), 27–33.

Grieshop, J.I. and Pence, R.A. (1990) Research results: Statewide IPM's first 10 years. *California Agriculture* 44(5), 24–26.

Harwood, R.R., Mutch, D., Smalley, S., Renner, K., Fettig, K. and Olsen, L. (eds) (2003) *Ecologically-based Farming Systems.* Michigan State University Extension Bulletin E-2786 (in press).

Haynes, D.L., Brandenburg, R.K. and Fisher, P.D. (1973) Environmental monitoring network for pest management systems. *Environmental Entomology* 2, 889–899.

Hoskins, W.M., Borden, A.D. and Michelbacher, A.E. (1939) Recommendations for a more discriminating use of insecticides. *Proceedings of the sixth Pacific Science Congress of the Pacific Science Association* 5, 119–23.

Howitt, A.H. (1993) *Common Tree Fruit Pests.* Michigan State University Extension Bulletin NCR-63.

Huffaker, C.B. and Messenger, P.S. (eds) (1976) *Theory and Practice of Biological Control.* Academic Press, Orlando, Florida.

Jones, A.L. (1981) Fungicide resistance: Past experiences with benomyl and dodine and future concerns with sterol inhibitors. *Plant Disease* 65, 990–992.

Jones, A.L. and Croft, B.A. (1981) Apple Pest Management Research in Michigan. *Plant Disease* 65, 223–229.

Jones, A.L. and Sutton, T.B. (1996) *Diseases of Tree Fruits in the East.* Michigan State University Extension Bulletin NCR-45.

Jones, A.L., Lillevik, S.L., Fisher, P.D. and Stebbins, T.C. (1980) A microcomputer-based instrument to predict primary apple scab infection periods. *Plant Disease* 64, 69–72.

Kogan, M. (1977) The role of chemical factors in insect/plant relationships. *Proceedings 15th International Congress of Entomology*, Washington, 1976, pp. 211–227.

Klonsky, K. and Shouse, B. (2000) IPM research profiled: 10 year trends. *California Agriculture* 54(6), 20–21.

Klonsky, K., Zalom, F., Chandler, M., Ohmart, C., Elmore, C. and Tourte, L. (1998) The Lodi-Woodbridge Winegrape Commission: a framework for implementing IPM. *Integrated Pest Management Reviews* 3, 243–255.

Landis, J.N., Sanchez, J.E., Bird, G.W., Edson, C.E., Isaacs, R., Lehnert, R.H., Schilder, A.M. and Swinton, S.M. (eds) (2002) *Fruit Crop Ecology and Management.* Michigan State University Extension Bulletin E-2759.

Leholm, A., Hamm, L., Suvedi, M., Gray, J.I. and Poston, F. (1999) Area of expertise teams: the Michigan approach to applied research and extension. *Journal of Extension* 37(3). Eugene, Oregon.

Michelbarger, A.E. and Bacon, O.G. (1952) Walnut insect control in northern California. *Journal of Economic Entomology* 45, 1020–1027.

Mutch, D.R., Cavigelli, M.A., Deming, M.A., Frost, L.A. and Probyn, L.K. (2000a) *Michigan Field Crop Pest Ecology and Management.* Michigan State University Extension Bulletin E-2740.

Mutch, D.R., Kells, J.J. and Poindexter, S. (2000b) *AOE-Field Crops Team: On-farm Research and Demonstration – Annual Report.* IPM Program, Michigan State University.

Newsom, L.D. (1974) Pest management: history, current status and future progress. In: Maxwell, F.G. and Harris, F.A. (eds) *Proceedings of the Summer Institute on Biological Control of Plant Insects and Diseases.* Mississippi State University Press, pp. 1–18.

Norris, R.F., Caswell-Chen, E.P. and Kogan, M. (2003) *Concepts in Integrated Pest Management.* Prentice Hall, New Jersey, 586 pp.

Norton, G.W. and Mullen, J. (1994) *Economic Evaluation of Integrated Pest Management Programs: a Literature Review.* Virginia Polytechnic Institute and State University, Blacksburg, Virginia, Extension Publication 448-120.

Pimentel, D. (1991) *Handbook of Pest Management in Agriculture*, 2nd edn, vol. 1–3. CRC Press, Boca Raton, Florida, 2271 pp.

Rajotte, E.G., Norton, G.W., Kazmierczak, R.F., Lambur, R.F. and Allen, W.A. (1987) *The National Evaluation of Extension's Integrated Pest Management (IPM) Programs.* Virginia Polytechnic and State University, Blacksburg, Virginia.

Smith, R.F. (1969) The new and the old in pest control. In: *Proceedings of the Accadimi Nazionale dei Lineci (Rome)*, 1968. 366(128), 21–30.

Smith, R.F. and Allen, W.W. (1954) Insect control and the balance of nature. *Scientific American* 190(6), 38–42.

Smith, R.F. and Reynolds, H.T. (1966) Principles, definitions and scope of integrated pest control. *Proceedings FAO Symposium in Integrated Pest Control* 1, 11–17.

Smith, R.F. and van den Bosch, R. (1967) Integrated control. In: Kilgore, W.W. and Doutt, R.L. (eds) *Pest Control: Biological, Physical and Selected Chemical Methods.* Academic Press, New York, pp. 295–340.

Smith, R.F., Apple, J.L. and Bottrell, D.G. (1976) Origins of integrated pest management concepts for agricultural crops. In: Apple, J.L. and Smith, R.F. (eds) *Integrated Pest Management.* Plenum, New York, pp. 1–18.

Sprott, J.M., Lacewell, R.D., Niles, G.A., Walker, J.K. and Gannaway, J.R. (1976) *Agronomic, Economic, Energy, and Environmental Implications of Short-Season, Narrow-Row Cotton Production.* Texas Agricultural Experiment Station, MP-1250C.

Sorensen, A.A. (1993) *Regional Producer Workshops: Constraints to the Adoption of Integrated Pest Management.* National Foundation for IPM Education, Austin, Texas.

Stern, V.M., Smith, R.F., van den Bosch, R. and Hagen, K.S. (1959) The integrated control concept. *Hilgardia* 29, 81–101.

Sutton, T.B. and Jones, A.L. (1976) Evaluation of four spore traps for monitoring discharges of ascospores of *Venturia inaequalis. Phytopathology* 66, 453–456.

USDA (1994) *USDA's Integrated Pest Management (IPM) Initiative.* Office of Communications, News Backgrounder, Release No. 0942.94. Washington, DC.

USDA (1998a) *Determining the Practice of Integrated Pest Management: a Working Definition for the year 2000 goal.* USDA Integrated Pest Management Committee, October, 1 page.

USDA (1998b) *Report on Integrated Pest Management and Related Activities Supported by the USDA.* Submitted to the House Agricultural, Rural Development, Food and Drug Administration, and Related Agencies Appropriation Subcommittee, 18 May 1998.

USDA, National Agricultural Statistics Service (2001) *Pest Management Practices: 2000 Summary.* Sc Cr 1(01).

[USGAO] United States General Accounting Office (2001) Report to the Chairman, Subcommittee on Research, Nutrition, and

General Legislation, Committee on Agriculture, Nutrition and Forestry, US Senate: *Agricultural Pesticides: Management Improvements Needed to Further Promote Integrated Pest Management.* GAO 01-815, August 2001, 36 pp.
Vandeman, A., Fernandez-Cornejo, J., Jans, S. and Lin, B.H. (1994) Adoption of integrated pest management in U.S. agriculture. *Agricultural Information Bulletin* No. 707, United States Department of Agriculture Economic Research Service, Washington, DC.
Wearing, C.H. (1988) Evaluating the IPM implementation process. *Annual Review of Entomology* 33, 17–38.
Welch, S.M., Croft B.A., Bunner, J.F. and Michels, M.F. (1978) PETE: An extension phenology modeling system for management of a multi-species pest complex. *Environmental Entomology* 7, 482–494.
Zalom, F.G. and Strand, J.F. (1990) Expectations for computer decision aids in IPM. *AI Applications in Natural Resource Management* 4, 53–58.

Chapter 21
Integrated Pest Management in Mexico

David Mota-Sanchez[1,2], Francisco Santos Gonzalez[2], Benito Alvarado-Rodriguez[3], Ovidio Diaz-Gomez[4], Hiram Bravo Mojica[1], Guillermo Santiago Martinez[5] and Rafael Bujanos[6]

[1]Colegio de Postgraduados en Ciencias Agricolas, Montecillo, México; [2]Department of Entomology, Michigan State University, East Lansing, USA; [3]Independent Consultant, Sinaloa, México; [4]Universidad Autonoma de San Luis, México; [5]Direccion General de Sanidad Vegetal, México City; [6]Instituto Nacional de Investigaciones Agropecuarias y Forestales (INIFAP), Guanajuato, México

Introduction

The major Mexican agricultural exports are coffee, cotton, sugar, fresh vegetables and tropical fruits, for a total value of US$5.6 billion/year. Imported agricultural commodities include maize, sorghum, beans, oilseeds, wheat, barley, and fresh and dried fruits, totaling US$2.3 billion (Agri-Food Trade Service, 1998). Although 23% of Mexico's population of 100 million is rural, agriculture makes up a relatively small share of the country's economy. In 1990, agriculture comprised 6.2% of the Gross Domestic Product. By 2000 this had lessened to 5% (Ruiz-Funes, 2001). Total Mexican territory is about 1.96 million km^2, from this area 21.8 million ha are arable. About 4.8 million ha of the arable land is irrigated (INEGI, 2002).

History of IPM in Mexico

Mexico has a long history of proactive pest management. In 1854, the first pest control training courses were taught at the National School of Agriculture and Veterinary (now Universidad Autonoma Chapingo). Forty years later, the Mexican government created a Commission of Plant Protection to assist farmers with pest management issues (Rodriguez, 2000). In 1924, the first law related to pest control was created following a devastating outbreak of the Central American locust, *Schistocerca piceifrons* Walker. The Commission of Plant Protection published the regulations of plant protection. Further, the Commission signed an agreement with the USDA to control the pink bollworm, *Pectinophora gossypiella* (Saunders), an introduced pest from Egypt, and to study Mexican fruit flies (Rodríguez, 2000). As early as 1927 and 1928, the first pest quarantine laws were passed in Mexico. In the 1930s, several parasitoids were successfully introduced to control sugarcane borers and the citrus blackfly, *Aleurocanthus woglumi* Ashby (Rodríguez, 2000).

In 1935, the Department of Plant Protection was founded at the National School of Agriculture in Chapingo. In 1959, the first

 Integrated Pest Management in the Global Arena (eds K.M. Maredia, D. Dakouo and D. Mota-Sanchez)

graduate program in agriculture was offered at the National School of Agriculture. Graduate research programs cover a broad range of topics in plant protection. In addition, the Graduate Program of the ITESM and the University Agraria Antononio Narro offer similar programs, contributing to scientific and technological developments in Mexican agriculture. In the past 50 years, several societies related to plant protection were founded including entomology, plant pathology, and weed sciences. In 1991, the Mexican Society of Biological Control was created, indicating the growing interest in developing biological control methods.

In the early 1960s, the first successful cotton IPM program was developed in northern Mexico. Integrated management of major cotton pests such as pink bollworm, bollworm, and cotton boll weevil included mandatory planting and harvest dates, destruction of cotton residues after harvest, scouting of pest populations and natural enemies, release of *Trichogramma*, and careful use of insecticides. These measures notably reduced pest populations and damage. This successful demonstration of IPM stimulated the creation of several facilities for mass-rearing biological control agents. About 64 Mexican companies now mass-rear more than 55 species of beneficial organisms for biological control (Rodriguez del Bosque and Arredondo, 1999). The top ten biological control agents sold in Mexico are listed in Table 21.1.

Table 21.1. List of the top biological control agents produced and sold in Mexico.

Natural enemies	Number of companies
Trichogramma pretiosum	33
Chrysoperla carnea	25
Beauveria bassiana	17
Metarrizium anisoplie	10
Paecilomyces fumosoroseus	7
Muscidifurax raptor	7
Chrysoperla rufilabris	5
Trichogramma exiguum	4
Spalangia endius	4
Encarsia formosa	3

Organizational Structure of IPM in Mexico

The SENASICA, under the Secretary of Agriculture, is the main body for all aspects related to pest control in the country. The SENASICA includes General Direction of Plant Protection (DGSV), General Direction of Animal Health (DGSA), and the General Direction for Animal and Plant Health Inspection (DGIF). The SENASICA provides support to comply with all international agreements, including the NAFTA (Trujillo, 1998).

The DGSV develops pest control regulations including: (i) certification of specialists in plant protection and diagnostics; (ii) phytosanitary programs for the prevention and control of pests; (iii) regulation of pesticide dealers and applicators; and (iv) certification of diagnostic laboratories. More than 1100 specialists have been accredited as national agents. Areas covered by the specialists include the fruit fly program, seed certification, IPM in vegetables, quarantine treatments, verification of free fruit pests in the country of origin, and IPM in avocados (Carreón, 1998).

IPM Policy in Mexico

The second chapter of the Federal Law of Plant Protection has as objectives to maintain product quality, to reduce cropping damage, and to be competitive at national and international levels (Reyes, 1994). A common approach to pest control in Mexico is designing programs that can be carried out at regional and national levels. Organized campaigns for the control of karnal bunt, coffee borer, coconut palm lethal yellowing, white fly, fruit flies, cotton pests, and the Central American locust have all been developed from specific regulations imposed by state and federal governments. The primary purpose of these programs is to prevent the further spread of pests. Resistant host plant varieties and biological control techniques are often used as strategies.

When an agricultural pest warrants a nationwide or regional campaign, the DGSV sets policies for a control program. The first step in this process is the detection and evaluation of the pest problem. Second, the control program is put in place with the cooperation of growers. Throughout the process, the organization, coordination and supervision of the effort is under the administration of the DGSV. The execution of programs includes training of specialists and growers, and coordination with growers, technicians, government and non-governmental agencies (Cárdenas, 1994). In 2000, about 1000 local, regional and state committees provided support in plant protection. These committees plan, organize, finance and evaluate the programs. In addition, committees acquire equipment and other supplies, and assign resources for research, diagnostic laboratories, pesticide residue detection, and production of biological control agents. The National Farm Program, together with state and local governments, supports grower organizations in Plant Protection, one of 26 programs it manages. This program also receives financial support from The World Bank and the FAO.

Most Mexican growers believe that IPM reduces costs of production and that it is vital to plant protection programs against polyphagous pests including the silverleaf whitefly, *Bemisia argentifolii* Bellows & Perring. In northwest Mexico, mandatory planting and harvesting dates, postharvest sanitation, and host-free period, together with the use of selective insecticides are important factors to control whiteflies (Ellsworth and Martinez-Carrillo, 2001). One of the constraints of IPM implementation is the lack of communication among the agencies.

Research and Extension Focus

Most Mexican universities now have IPM curricula at both the undergraduate and graduate levels. Each of the 31 states in Mexico has at least one university with an agronomy department. In addition, The National Institute of Research in Forestry, Agriculture and Animal Sciences (INIFAP) has established more than 81 experiment stations throughout Mexico. These stations are responsible for research and technology transfer. The research programs carried out in the areas of entomology and plant pathology are IPM oriented.

Successful IPM Case Studies

Impact on international trade

In recent years, IPM has become even more important as trade regulations have begun to restrict the amounts of pesticide residue or insects that may be present on produce exported to the USA and Canada. In order to maintain the extensive trade in fresh fruits and vegetables, these commodities must comply with strict regulations that are difficult to meet with conventional pest control methods. In many cases, IPM has become the only viable option for growers intending to export their products. In this section we will focus on three unique cases of IPM in Mexico representing principal agricultural commodities.

IPM in tomatoes

In the state of Sinaloa, tomatoes, including both processing and fresh-market tomatoes, are annually grown on more than 50,000 ha with a total value of US$1.1 billion. Insects feeding on tomatoes are the principal pests. In the 1980s, the tomato pinworm, *Keiferia lycopersicella* (Walsingham), emerged as the most important insect pest of tomatoes in Mexico. Other important pests included the tomato fruitworm, *Helicoverpa zea* (Boddie), the tobacco budworm, *Helicoverpa virescens* F., and the beet armyworm, *Spodoptera exigua* (Hübner). At that time, the percentage of damaged tomatoes ranged from 25% to 100%, despite frequent use of broad-spectrum insecticides such as methamidophos, methomyl, permethrin and fenvalerate (Alvarado-Rodriguez and

Rivera-Rubio, 1990). During this period, the tomato industry in Mexico faced serious problems due to new quality standards for tomatoes and tomato paste exported to the USA. High levels of insect fragments and pesticide residues in tomato paste threatened the entire processing industry. In the late 1980s, Campbell's Sinalopasta and other tomato industry companies initiated an IPM program for commercially grown tomatoes used for processing (Alvarado-Rodriguez, 1988).

Fresh-market tomato growers also faced many of the same problems related to the excessive use of broad-spectrum insecticides. These problems included the resurgence of insect pests, outbreaks of secondary pests, insecticide resistance, adverse effects on natural enemies, and high pesticide residue levels causing the denial of export to the USA, as well as the risk of effects on human health and the environment (Oatman and Kennedy, 1976; FDA, 1979; Johnson *et al.*, 1980a,b; Oatman *et al.*, 1983; Trumble, 1985; Brewer *et al.*, 1990; Brewer and Trumble, 1991; Trumble and Alvarado-Rodriguez, 1993).

Fresh-market tomato growers in Sinaloa supported research to develop an IPM program for fresh-market tomatoes. The objective of the IPM program was to develop efficient and economical pest management alternatives, comply with quality standards for export established by the USA, and minimize adverse effects on the environment and human health (Trumble and Alvarado-Rodriguez, 1993).

The tomato pinworm

A successful management program for tomato pinworm includes careful scouting, cultural control, biological control, mating disruption, and the use of selective insecticides. The tomato pinworm has developed resistance to most commercial insecticides. Selective application of control measures is important, as infestations tend to start along field edges. Therefore, control measures need to be directed to edges where infestations are just beginning. Scouting is done primarily with pheromone traps from planting to harvesting. Monitoring of eggs is done by sampling leaves from below the inflorescence. The larval population is assessed by inspecting complete plants. Cultural control methods include plowing under crop residues promptly after harvesting, cleaning drainage ditches and irrigation canals where alternate hosts grow, and establishing a tomato-free period during summer or winter to break the cycle of tomato pinworm reproduction.

Conservation biological control is also important. The parasitoid wasp, *Trichogramma pretosium* is the only native egg parasitoid of tomato pinworm, and levels of parasitism can reach up to 35% in late plantings. Other parasitoids include the braconids *Apanteles scuttelaris* and *Pseudoapanteles dignus* that attack the larval stage. The synthetic pheromones ((E) 4 Tridecen-1 and 1-Acetate and (Z)-4 Tridecen-1 and 1-Acetate) (Consep Inc., Bend, Oregon) have been used as an effective mating disruption technique. Due to the overuse of insecticide applications, the tomato pinworm has developed resistance to conventional insecticides. However, selective insecticides including abamectin, spinosad and modified abamectin are still effective on tomato pinworm larvae. Combined use of pheromones, biological control and selective insecticides have reduced damage and number of insecticide applications.

The tomato fruitworm

The tomato fruitworm (*H. zea* and *H. virescens*) is a widely distributed pest on tomatoes, cotton and chickpeas. Damage from tomato fruitworm can reach up to 15% in tomato fruit. Control with a single method is not usually effective, so the most effective control is achieved with a combination of tactics. IPM of tomato fruitworm combines pest monitoring, biological control, and use of biopesticides. Egg density is monitored by sampling leaves located below the inflorescences, while fruit damage is assessed by inspecting fruits in the center of the field.

The tomato fruitworm has been effectively managed with mass releases of *Trichogramma* wasps, when natural parasitism is lower than 60%. Naturally occurring parasitism rates range from 35% to 96%, increasing as the season progresses. In the state of Sinaloa, natural parasitism by *T. pretiosum* on eggs of tomato fruitworm commonly occurs at high levels such as 90% on both processing and fresh-market tomatoes. Naturally occurring parasitism by the braconids *Cotesia marginiventris* and *Meteorus laphygmae* ranges from 35% to 40% on tomato fruitworm larvae, but parasitism does not necessarily mean a reduction in pest damage, since parasitized larvae feed on fruit, and cocoons of parasitoids can reduce fruit quality. Biopesticides (Bt) are effective and do not interfere with the biological control. In contrast, applications of conventional insecticides reduced natural parasitism. However, sprays around edges of tomato fields reduced their impact in natural enemies (Alvarado-Rodriguez, 1988).

The beet armyworm

Spodoptera exigua is also an important pest of tomatoes, reaching damage levels in fruit of 25%. In addition to tomatoes, this insect has several host plants including wheat, sorghum, soybeans, lucerne and weeds. Larvae usually feed on foliage, but they may also damage fruit. The tolerance level for direct feeding damage in fruit is 4%. IPM of beet armyworm is based on careful monitoring of pest populations and damage, biological control, and the use of Bt. Monitoring of *Spodoptera* eggs involves inspection of the whole plant for presence of egg clusters and larvae. Data are gathered from plants in the center of the field. The critical monitoring time for beet armyworm is at the fruiting stage.

Beet armyworm has several natural enemies that can have a significant effect on pest populations. Parasitoids and an NPV attack *S. exigua* larvae (Alvarado-Rodriguez, 1988). NPV is the major mortality factor affecting *S. exigua* larvae, with multiple epizootics occurring at different times during the season. Parasitism rates naturally occurring in the field are lower early in the season and increase as the season progresses, reaching up to 60%. These mortality factors acting in concert are generally able to reduce populations below the economic injury level. However, the role of these natural enemies at fruiting stage is of partial value because even parasitized larvae feed on fruit, and parasitoid cocoons can reduce fruit quality. Although an economic threshold for Bt application has not yet been developed, the recommended level is an average of 0.25 young larvae per plant at any time after the beginning of the fruiting stage until harvest.

Secondary pests

Important secondary pests in tomatoes include the vegetable leafminer, *Liriomyza sativae*, the western flower thrips, *Frankliniella occidentalis* (Pergande), the green peach aphid, *Myzus persicae* and whiteflies, *Bemisia tabaci* and *Bemisia argentifolii*. Sprays with abamectin reduce damage by leafminers, and soaps control whiteflies and thrips. Entomopathogenic fungi, e.g. *Paecelomyces fumosoroseus* reduce both species of whiteflies, and *Verticillium lecanii* is very effective against several species of aphids.

Implementation of tomato IPM

The Mexican tomato industry has widely adopted the IPM program (Alvarado-Rodriguez, 1988; Trumble and Alvarado-Rodriguez, 1993). The major motivation is compliance with the standards for export to the USA. These standards were achieved during the first year of the IPM program. From 1988 to 1991, the amount of active ingredient applied to tomato fields, the percentage of insect damage and control costs were all noticeably reduced (Table 21.2). IPM reduced control costs by 45% and the quantity of i.a./ha by more than 30%. In addition, insect damage in IPM-managed fields was reduced by as much as 76%. IPM adoption by fresh-market tomato growers has been steadily increasing.

Table 21.2. Comparison of the IPM program and conventional insecticides in commercially grown tomatoes at Campbell's-Sinalopasta in Sinaloa, Mexico.

Growing season	Control strategy	Area (ha)	Active ingredient (kg/ha)	Overall insect damage (%)	Cost/ha (US$)
1988–1989	Conventional insecticides	2300	2.45	8.5	350.00
1989–1990	IPM	3000	0.99	2.2	96.95
1990–1991	IPM	1800	0.77	2.7	75.00
2000–2001	IPM	1200	0.55	2.4	55.00

IPM in crucifers: the diamondback moth

Crucifer production in Mexico has grown significantly during the last 25 years. More than 30,000 ha of broccoli and cauliflower are grown in the state of Guanajuato. Of these crucifers 95% are grown for export by the frozen food industry in the states of Guanajuato and Queretaro. Currently, 50% of the broccoli consumed in the USA is imported from Mexico.

The diamondback moth, *Plutella xylostella* (L.), is a major insect pest of crucifers, along with *Trichoplusia ni* (Hubner) and *Brevicoryne brassicae* (L.). Reasons for this include the expanding acreage of crucifer production, year-round crop production, and the destruction of natural enemies by broad-spectrum insecticides. In 1986, control of the diamondback moth was very critical and it was necessary to apply up to 15 insecticide sprays per season, as compared to four in 2000 after establishment of IPM.

Economic importance

In 1986, the diamondback moth became a severe problem, although it had been present at low levels since 1970 (Domínguez and Carrillo, 1976; Laborde, 1992). As a result, a regional plan for IPM in crucifers was organized in 1987. Results of this program were not immediately effective because a zero tolerance policy for diamondback moth larvae or pupae in crucifers exported to the USA led to a 16% rejection rate of shipments in 1989. Economic losses during the period from 1988 to 1992 were estimated at US$3 billion (Martinez-Castillo *et al.*, 2002). In response, the processing industry in the Bajio created the Frozen Food Packers Association and a Technical Committee composed of members representing several industry and research institutes. These organizations shared research and field observations, identified and supported key research projects, organized technical and professional training seminars, supported technical extension publications, and generally promoted IPM of broccoli and cauliflower (Laborde, 1992).

Coordination at the regional level for education and training of extension agents and growers was vital to the success of the IPM program. The education program included pest identification, pest scouting, use of economic thresholds, cultural control methods, microbial control and use of selective pesticides. Pests were identified and monitored by farmers' consultants; in addition, the crop's maturity and health, weather conditions, and population level of beneficial organisms were monitored. Sampling methods included the inspection of plants and the use of pheromone traps, color traps, and light traps (Vargas, 1993; Hoy, 1999). Pheromone traps indicated both the local population level and also revealed the change in patterns of the population over time (McCully and Salas, 1992).

In the processing industry, pests are classified as contaminants of harvested product and secondaries. This first term is used to define the presence of immature stages on the harvested products. Among the first are: *P. xylostella*, *T. ni* and *B. brassicae*; the second are: *Artogeia rapae* (L.), *Leptophobia aripa* (Boisduval), *Spodoptera exigua* (Hübner), *Copitarsia consueta* (Walker), *Myzus persicae* (Sulzer) and *Lipaphis erysimi* (Kaltenbach), different species of *Diabrotica* and *Murgantia histrionica* (Hahn) (Bujanos *et al.*, 1995).

Action thresholds and crop phenology

The crucifer plant is divided into three developmental stages: seedling, reproductive stage, and harvest. At the seedling stage, control is recommended when the density of larvae is over 0.5 larvae/plant. At the second and third stages, control is needed at over 0.2 larvae/plant. The application of these thresholds reduced the number of insecticide applications per season. As a result of this reduction in pesticide use, the risks of worker poisoning and high pesticide residue levels were significantly reduced. In addition, populations of natural enemies increased (Díaz-Gomez and Jasso, 1989; Bujanos *et al.*, 1995).

Control strategies

Effective cultural control methods included plowing to eliminate crop residue, and rotation with non-host crops. In addition, careful inspection of nursery plants for diamondback moth eggs and larvae helped to prevent accidental introduction of diamondback moth into the field (Bujanos *et al.*, 1993). Another tactic included a 'host-free season' in which neither crucifers nor other hosts of diamondback moth were planted. This reduced pest densities for the following seasons and slowed the potential development of pesticide resistance.

The diamondback moth is ranked third in resistance development to insecticides in the world (Mota-Sanchez *et al.*, 2002). Due to the ability of diamondback moth to adapt to several insecticides, and the necessity of use effective insecticides, a common practice in the Bajio valley region is monitoring the resistance of *P. xylostella* to insecticides. This procedure is by the use of discriminating concentrations. The current data not only established a resistance/susceptibility database that includes several pests and insecticides, but have been also used for quick diagnosis of resistance or any shift to tolerance (Díaz-Gomez *et al.*, 2000).

Biological control

A key factor for the conservation of parasitoids and predators relies on the use of formulations of Bt at the beginning of the season and the reduction of broad-spectrum pesticide use. To avoid resistance, Bt is recommended for use on only one generation per crop season and should be rotated with groups of insecticides with other modes of action including spinosad, emamectin benzoate, and indoxacarb (Shelton *et al.*, 1999; Díaz-Gomez *et al.*, 2000). Biological control also includes the identification and assessment of native parasitoid species and the introduction of effective exotic species. In the Bajio region, *Diadegma insulare*, an introduced species, parasitoid of the last instar of diamondback moth, has been found at high levels (57% in San Luis Potosi and 10–30% in Guanajuato) (Martinez-Martinez *et al.*, 1996). Other introduced species of parasitoids include *Cotesia plutella* and various species of *Trichogramma*. Natural enemies of *Trichoplusia ni* larvae such as *Voria ruralis*, *Copidosoma truncatellum*, *Microplitis* sp., and *Hyposoter* sp. have also been found (Bujanos, 2000), while *Diaretiella rapae* and *Aphidius testaceipes* are specific to *Brevicoryne brassicae* in the region. In addition, an important group of predators including *Hippodamia convergens*, *Chrysoperla* spp., *Allograpta* sp. and *Orius* spp. contributed to suppression of pest populations.

Chemical control

Ultimately, the justification for heavy insecticide use in recent decades has been consumer demand for high quality and flawless products. At first, the new synthetic insecticides dramatically reduced pest damage and yield loss. However, over time the frequent use of broad-spectrum insecticides has become less effective and has led to a host of other problems. In some cases, however, the use of chemical insecticides is still necessary for growers to meet market demands, but guidelines for more judicious use of insecticides have been developed. Recommendations for the use of chemical insecticides now include:

1. Analysis of insecticide use and evaluation of effectiveness;
2. Monitoring for insecticide resistance and determining the mechanisms of resistance (Díaz-Gomez *et al.*, 1994);
3. Use of economic thresholds;
4. Eliminating the use of pesticide mixtures;
5. Restricting the use of pyrethroids to the end of the season;
6. Stricter laws regarding registration and use of pesticides and verification of the MLOs established by EPA; and
7. Use of diversified tactics to reduce chemical control (Bujanos *et al.*, 1993, 1995; Díaz-Gomez and Rodriguez, 2000).

IPM in crucifers has been fully or partially adopted by almost all Mexican growers in the Bajio region. IPM has proved to be critical for compliance with strict international standards for pest damage and pesticide residues. The IPM program for diamondback moth in Mexico is one of the most successful examples of crucifer IPM in the world (Talekar and Shelton, 1993). This successful program has allowed sustainable production of broccoli and cauliflower and reduced economic and ecological costs.

IPM of fruit flies

Fruit flies (both native and introduced) are the most important pest in the production of tropical fruits. Mexico is the third producer and the first exporter of mangoes in the world (158,000 ha and 207,000 tons, respectively) (SAGARPA, 2002 unpublished) and provides 5% of the world citrus production. Guava, papaya, and mamey are also important fruits in Mexico. In 1975, the Mediterranean fruit fly, *Ceratitis capitata* Widemann was introduced to Mexico from Guatemala. This sparked a major multinational effort to eradicate the flies. A sterile insect release program coordinated by the Mexican government, the USDA, and Guatemala built the largest facility in the world for mass production of the Mediterranean fruit fly in 1978. By 1980, 1000 million sterile flies were produced weekly. Although this program was largely successful, migrant flies are still detected sporadically at the Mexican–Guatemalan border.

However, outside of the Mediterranean fruit fly eradication project, native fruit flies continued to cause serious damage. Important resources to study the ecology of fruit flies, identification, population dynamics, and strategies of control were allocated. In 1982, a national campaign against these native fruit flies was established (Gutierrez *et al.*, 1992). In 1992, the national campaign against fruit flies was revived. Agreements were established between the federal and state governments and farmers to control, suppress or eradicate four species of economic importance: the mango fruit fly, *Anastrepha obliqua* (Macquart), the Mexican fruit fly, *Anastrepha ludens* (Loew), the sapote fruit fly, *Anastrepha serpentina* (Widemann) and the guava fruit fly, *Anastrepha striata* Shiner. These species caused an estimated US$710 million in damage per year. Additional losses were caused by increased costs of control and loss of international markets due to strict quarantine restrictions

Key points to implement areawide fruit fly IPM

Areawide fruit fly IPM involved activities that should be realized continuously and permanently. Some important points are:

1. Delimitation of the areas, taking into consideration the fruit system production, ecology, number of species of fruit flies and number of hectares under fruit fly IPM.
2. Technology for the control, suppression, and eradication of fruit flies.
3. Determining of the technical, economical and operational feasibility of the program in the areas.
4. Legislation of the IPM of fruit flies.
5. Tight collaboration between all participants in the IPM, particularly the farmers with the federal, state and local government.

Legislation and participation of producers

In 1995, Mexico passed legislation establishing mandatory procedures for plant protection and the national campaign against fruit flies. The objective was to create certified and protected pest-free orchards, at least temporarily (SAGAR, 1999).

Participating growers registered their orchards and made a coordinated regional effort to apply IPM strategies, including both commercial and non-commercial areas (back yard fruit trees with alternate hosts nearby). In addition, they contracted with certified plant protection specialists to supervise and monitor the progress of each grower. Each year, a regional IPM plan was defined, listing the goals, methodology, responsibilities, and financial funds for each participant. Evaluation and follow-up was also part of the plan. Both growers and specialists participated in IPM workshops organized by the Secretary of Agriculture and various universities. Extension workers actively promoted IPM through radio, newspapers, and bulletins (SAGAR, 1999).

Management strategies in fruit fly IPM programs

IPM of fruit flies relies on early detection and identification. Correct identification of the species is important to design management programs. Reduction of the fly population includes strategies such as cultural control, application of selective baits, the production and release of parasitoids and sterile fruit flies, and strict limitations on fruit movement out of infested areas. Adult fruit flies are monitored using glass McPhail traps baited with hydrolyzed protein at a density of one to five traps per hectare depending on the species. Fruit sampling is complementary to the trapping and is useful for the detection of larvae. Fruit sampling starts as soon as the orchards and areas outside of the orchards (fruit trees in yards of houses or other hosts in non-commercial areas) have fruits big enough to be infested by fruit flies. One to five samples of 1 kg of commercial fruits, or wild hosts are collected in places where adults of fruit flies were caught in McPhail traps. The decision to apply control measures is based on the results of trapping. The number of flies per trap per day (FTD) is used as an index to determine the level of infestation. This index is applied at several spatial scales: per orchard, county, state, per fruit fly species and the sexual proportion. Control is applied when the FTD is equal or more than 0.0100. However, IPM is applied to an area that is free of fruit flies as soon as the first fruit fly is caught. An emergency plan to eradicate the pest is then established quickly.

In 1992, the Mexican government built facilities to produce sterile fruit flies and parasitoids as a means of control and eradication of fruit flies. Every year, about 153 million *A. ludens* and 21 million parasitoids are produced each week. Since 1994, *A. ludens* and the parasitoid *Diachamimorpha longicaudata* (Ashmead) have been released in both infested areas and non-commercial areas (organic production areas and urban areas where chemical control is not applied) to suppress or eradicate fruit flies (Reyes *et al.*, 2000). Additional control methods have been used, including the destruction of fallen or leftover fruit and tillage to eliminate weeds and pupae in the soil. Trap cropping is also used by planting especially attractive fruit varieties. Chemical control with a mixture of malathion, hydrolyzed protein and water is used as a selective bait. This bait is applied with either a ground sprayer or aerially, leaving some areas untreated to reduce development of resistance (Gutierrez *et al.*, 1992). Research is currently being done on environmentally friendly compounds such as Spinosad as a substitute for malathion.

IPM of fruit flies is now mandatory for farmers, brokers, transport, exports and packaging industries in Mexico. In addition, compliance with the 1998 quarantine regulations governing the movement of fruit from infested areas is required. Infested areas are defined according to the results of the grower's application for orchard registration. The area-wide results are used to determine the category of each region. There are three categories (SAGAR, 1999):

1. An infested area, or with an FTD index higher than 0.0100 at any period of the year. It also applies to areas with no IPM record.
2. Low infestation, or an FTD index equal to or lower than 0.0100 for at least 6 months. In addition, this area must be under IPM.
3. Completely free of infestation – the FTD index must be equal to zero during the past 12 months under IPM. Movement and marketing of fruit from this category has many advantages over fruit produced in the two other categories.

In 1998, the SAGARPA declared three states (Baja California, Baja California Sur, Chihuahua and Sonora) free of the species *A. ludens*, *A. obliqua* and *A. serpentina*. It also categorized the states of Sinaloa (center and south), Nuevo Leon, and Tamaulipas (north and center) as low infestation. In 2001, the state of Coahuila and five municipalities in the state of Sinaloa were pronounced free of fruit flies, and the states of Durango and Aguascalientes were declared a low-infestation area. In 2002, 48 municipalities in the state of Zacatecas were also declared as fruit fly low-infestation areas. As a result of this program, insecticide use was reduced by as much as 35,000 kg of a.i./year. For additional protection, a contingency plan is ready in case of an outbreak of fruit flies.

Elimination of trade barriers is one of the most economically important benefits of IPM. The international recognition of areas declared free of fruit flies is a successful example. In 1999, the USDA recognized certain area in the State of Baja California Sur, Sonora and Chihuahua as demonstrably free of fruit fly infestation (USDA, 1999). The total area of these counties was 35,000 ha, with fruit such as oranges, grapefruit, tangerines, apples, peaches and persimmon. In 1999, a trade agreement allowed Mexico to export oranges from Baja California Sur and Sonora to the USA without quarantine treatments. This is an important and tangible cost-saving benefit. Fruit from a low infestation area can also be exported, but must undergo a fumigation or hot water quarantine treatment and IPM orchard certification.

Conclusions

IPM has been tremendously important to tomato, crucifer and fruit production in Mexico. Significant reductions in insecticide use, pesticide residues and enviromental contamination have been important benefits of IPM. The economic impact of IPM is most clearly demonstrated by the acceptance of these commodities on the international market.

Mexico has many research institutes and universities that are actively involved in research and training activities in IPM. Regulation in plant protection has made possible close links among growers and federal, state and local governments. Scientific societies in entomology, biological control and plant pathology, specialists in plant protection, graduate colleges and industry work together to set goals for sustainable crops and fruit production. However, Mexico is a country of contrasts where 50 million people live in poverty including poor farmers who cannot afford to implement IPM. Some government programs have been dedicated to improve conditions of poor farmers.

References

Agri-Food Trade Service (1998) Mexico-Agri-Food Export market assessment report. http://atn-riae.agr.ca/public/htmldocs/e0098.htm#1

Alvarado-Rodríguez, B. (1988) Manejo Integrado de Plagas en Tomate Industrial en Sinaloa. In: Ramirez, D.J.M. and Alvarado-Rodríguez, B. (eds) *Taller sobre el Manejo Integrado de Plagas en Tomate*. CAADES, Culiacán, Sinaloa, México, pp. 33–41.

Alvarado-Rodriguez, B. and Rivera-Rubio, E. (1990) Tomato pinworm (Lepidopera: Gelechiidae) an increasing pest on tomatoes in Sinaloa, Mexico. *Florida Entomologist* 73, 677–680.

Brewer, N.M. and Trumble, J.T. (1991) Classifying resistance severity in field populations: Sampling inspection plans for an insecticide resistance-monitoring program. *Journal of Economic Entomology* 84, 379–389.

Brewer, N.M., Trumble, J.T., Alvarado-Rodriguez, B. and Chaney, W. (1990) Beet armyworm,

(Lepidoptera: Noctuidae) adult and larval susceptibility to fenvalerate permethin, and methomyl in managed habitat and selection experiments. *Journal of Economic Entomology* 83, 2136–2146.

Bujanos, M.R. (2000) Manejo integrado de plagas en cruciferas. In: Bautista, M.N., Suarez, V.A.D. and Morales, G.O. (eds) *Temas Selectos en Fitosanidad y Producción de Hortalizas.* Instituto de Fitosanidad, C.P., Montecillo, Estado de Mexico, pp. 47–61.

Bujanos, M.R., Marin, J.A., Galvan, C.F. and Byerly, M.K.F. (1993) Manejo integrado de la palomilla dorso de diamante (*Plutella xylostella* (L.) Lepidoptera: Yponomeutidae) en el Bajio México. *INIFAP Asociación de Procesadores de Frutas y Vegetales en General A.C. Publicacion Especial* No. 4, 36 pp.

Bujanos, M.R., Marin, J.A., Galvan, C.F. and Byerly, M.K.F. (1995) Plagas de los Cultivos de Cruciferas en el Bajio. *INIFAP Asociación de Procesadores de Frutas y Vegetales en General A.C. Publicacion Especial* No. 5, 25 pp.

Cárdenas, M.J. (1994) Campañas fitosanitarias. In: CONACOFI (ed.) *Memoria de la Primera Asamblea Anual del Consejo Nacional Consultivo Fitosanitario.* CONACOFI-SARH. Colegio de Postgraduados, Montecillo, Mexico, 98 pp.

Carreón Z.M.A. (1998) Comentarios y recomendaciones al programa de trabajo del Consejo Nacional Consultivo Fitosanitario. In: CONACOFI (ed.) *Memoria de la Cuarta Reunión anual de CONACOFI.* CONASAG, Secretaria de Agricultura, Ganaderia y Desarrollo Rural, Mexico, DF.

Díaz-Gomez, O. and Jasso, P.I. (1989) Evaluación de insecticidas para el control de *Plutella xylostella* (L:) (Lepidoptera: Yponomeutidae) en col en la región de Pozos S.L.P. In: *Memorias del XXIV Congreso Nacional de Entomologia.* Sociedad Mexicana de Entomologia. Oaxtepec, Morelos, México, SME, pp. 363–364.

Díaz-Gomez, O. and Rodriguez, J.C. (2000) Resistencia a insecticidas: conceptos y procedimientos generales para su manejo. In: Bautista, M.N., Suarez, V.A.D. and Morales, G.O. (eds) *Temas Selectos en Fitosanidad y Producción de Hortalizas.* Instituto de Fitosanidad, C.P., Montecillo, Estado de Mexico, pp. 91–100.

Díaz-Gomez, O., Lagunes-Tejeda, A., Sanchez, A.H. and Alatorre, R.R. (1994) Susceptibility of *Plutella xylostella* (L.) to microbial insecticides. *Southwestern Entomology* 19, 403–408.

Díaz-Gomez, O., Rodriguez, J.C., Shelton, A.M., Lagunes-Tejeda, A. and Bujanos, M.R. (2000) Susceptibility of *Plutella xylostella* (L.) (Lepidoptera: Plutellidae) populations in Mexico to commercial formulations of *Bacillus thuringiensis. Journal of Economic Entomology* 93, 963–970.

Domínguez, R.Y. and Carrillo, S.J.L. (1976) *Lista de Insectos en la Colección Entomologica del INIA.* 2°. Suplemento INIA. SAG. México. Folleto Miscelaneo 29, 245 pp.

Ellsworth, P.C. and Martinez-Carrillo, J.L. (2001) IPM for *Bemisia tabaci*: a case study from North America. *Crop Protection* 20, 853–869.

FDA (1979) Results of FDA surveillance of Mexican produce for pesticide residues (compliance program 7305.008). Bureau of Foods and Office of Association Committee for Regulation Affairs, FDA Administration, Washington, DC.

Gutierrez, S.J., Reyes, J.F., Villasenor, A.C., Enkerlin, W.H. and Perez, A.R. (1992) *Manual para el Control de Moscas de la Fruta.* Secretaria de Agricultura y Recursos Hidraulicos. Serie Sanidad Vegetal, Mexico, DF, 34 pp.

Hoy, C.W. (1999) Making the connections between pest management and agroecosystem management. In: *Proceedings of Integrated Pest Management of Cole Crops International Workshop.* SAGAR, INIFAP, CESAVEG, SDAyR, Asociacion de Procesadores de Frutas y Vegetales en General, Guanajuato, Mexico, 57 pp.

INEGI (2002) Anuario estadistico de los Estados Unidos Mexicanos. http://www.gob.mx/difusion/espanol/fnuevo.html

Johnson, M.W., Oatman, E.R. and Wyman, S.A. (1980a) Effect of insecticides on populations of the vegetable leafminer and associated parasites on summer pole tomatoes. *Journal of Economic Entomology* 73, 61–66.

Johnson, M.W., Oatman, E.R. and Wyman, S.A. (1980b) Effect of insecticides on populations of the vegetable leafminer and associated parasites on fall pole tomatoes. *Journal of Economic Entomology* 73, 67–71.

Laborde, C.J.A. (1992) Palomilla dorso de diamante en el Bajio, control mediante un programa integral regional. In: Anaya, R.S., Bautista, N.M. and Dominguez, B. (eds) *Manejo Fitosanitario de las Hortalizas en Mexico.* CP-SARH. Chapingo, Mexico, 412 pp.

Martínez-Martínez, L., Díaz-Gómez, O. and Benrey, B. (1996) Parasitism on *Plutella xylostella* (L.) on wild and cultivated crucifers. In: *VI Congreso Internacional de Manejo Integrado de Plagas and V Taller Lationoamericano sobre Moscas blancas y Geminivirus.* Acapulco Gro. México. SAGAR, UACH. Departamento de Parasitología, p. 34.

Martinez-Castillo, M., Leyva, J.L., Cibrian-Tovar, J. and Bujanos, M.R. (2002) Parasitoid diversity and impact on populations of the diamondback moth *Plutella xylostella* (L.) on Brassica crops in central Mexico. *Biocontrol* 47, 23–31.

Mota-Sanchez, D., Bills, S.P. and Whalon, M.E. (2002) Arthropod resistance to pesticides: Status and overview. In: Wheeler, W. (ed.) *Pesticides in Agriculture and the Environment.* Marcel Dekker, Gainesville, Florida, pp. 241–272.

McCully, J.E. and Salas, A.M.D. (1992) Seasonal variation in populations of the principal insects causing contamination in processing broccoli and cauliflower en central Mexico. In: Talekar, N.S. (ed.) Diamondback moth and other crucifer pests. *Proceedings of the Second International Workshop.* AVRDC, Tainan, Taiwan, 603 pp.

Oatman, E.R. and Kennedy, G.C. (1976) Methomyl induced outbreak by *Liriomyza sativa* on tomato. *Journal of Economic Entomology* 69, 667–668.

Oatman, E.R., Platner, G.R., Wyman, J.A., van Steenwyk, R.A., Johnson, M.W. and Browing, H.W. (1983) Parasitization of lepidopterous pests on fresh market tomatoes in southern California. *Journal of Economic Entomology* 76, 452–455.

Reyes, F.J., Santiago, M.G. and Hernández, M.P. (2000) The Mexican fruit fly eradication programme. In: Tan, K.H. (ed.) *Area-Wide Control of Fruit Flies and others Insect Pests.* Penerbit Universiti Sains Malaysia, Penang, pp. 377–380.

Reyes, P.H. (1994) Ley federal de sanidad vegetal y su aplicación. In: CONACOFI (ed.) *Primera Asamblea anual del Consejo Nacional Consultivo Fitosanitario.* CONACOFI-SARH. Colegio de Postgraduados en Ciencias Agrícolas, Montecillo, Mexico, 98 pp.

Rodriguez del Bosque, L.A. and Arredondo, B.H.C. (1999) Quien es quien en Control Biologico en Mexico. INIFAP-CIRNE. Campo Experimental Rio Bravo. *Folleto Tecnico* Numero 23. Tamaulipas, Mexico, 147 pp.

Rodríguez, V.J. (2000) *Historia de la Fitosanidad en México, siglo XX.* Universidad Autónoma Chapingo, Departamento de Parasitología Agrícola, Chapingo, Mexico, 180 pp.

Ruiz-Funes, M.M. (2001) Perspectivas economicas, agropecuarias y politicas. *Claridades Agropecuarias* 93, 1–16.

SAGAR (1999) Norma Oficial Mexicana NOM-023-FITO-1995, por la que se establece la Campaña nacional contra Moscas de la Fruta. *Diario Oficial de la Federación.* 11 de febrero de 1999. Talleres de la Nacion, México, DF, pp. 10–26.

Shelton, A.M., Sances, F.V., Hawley, J., Tang, J.D., Boune, M., Jungers, D. and Farias, J. (1999) Susceptibility of standard (Ambush, Lannate and Javelin) and novel (Success, Proclaim and Alert) insecticides in California populations of diamondback moth. In: *Proceedings of Integrated Pest Management of Cole Crops International Workshop.* SAGAR, INIFAP, CESAVEG, SDAyR, Asoc. de Proc. de Frutas y Vegetales en General, 57 pp.

Talekar, N.S. and Shelton, A.M. (1993) Biology, ecology and management of the diamondback moth. *Annual Review of Entomology* 38, 275–301.

Trumble, J.T. (1985) Integrated Pest Management of *Liriomyza trifolii*: influence of avermectin, cyromazine, and methomyl on leafminer ecology in celery. *Agriculture Ecosystems and Environment* 12, 181–188.

Trumble, J.T. and Alvarado-Rodriguez, B. (1993) Development and economic evaluation of an IPM Program for fresh market tomato production in Mexico. *Agriculture Ecosystems and Environment* 43, 267–284.

Trujillo, A.J. (1998) Programa anual de la Dirección General de Sanidad Vegetal. In: CONACOFI (ed.) *Memoria de la Cuarta Asamblea anual de CONACOFI.* CONACOFI-CONASAG, SAGAR, México, DF, 339 pp.

USDA (1999) Importation of fruit and vegetables. Animal and plant health. Final rule, 7CFR part 319. Docket No. 97-197-3; Vol. 64. No. 12. Washington, DC, pp. 2993–2995.

Vargas, L.A. (1993) *Monitoreo de Plagas en el Cultivo de Brócoli y Coliflor.* Birds Eye de Mexico SA de CV, Guanajuato, Mexico. (Bol. Inform. No. 1. 11 pp.)

Chapter 22

Integrated Pest Management in Brazil

C.B. Hoffmann-Campo[1], L.J. Oliveira[1], F. Moscardi[1], D.L. Gazzoni[1], B.S. Corrêa-Ferreira[1], I.A. Lorini[2], M. Borges[3], A.R. Panizzi[1], D.R. Sosa-Gomez[1] and I.C. Corso[1]

[1]*Embrapa Soja, Londrina, PR, Brazil;* [2]*Embrapa Trigo, Passo Fundo, RS, Brazil;* [3]*Embrapa Recursos Geneticos e Biotechologia, Brazilia, DF*

History and Evolution of Integrated Pest Management in Brazil

Following World War II, a myriad of new chemical products became available. In Brazil, as in the rest of the world, newly available chemical pesticides were regarded as a dramatic and effective solution for pest control, and were widely used after the war (Kogan and Turnipseed, 1987). Eventually, negative effects of indiscriminate pesticide usage became apparent, such as pesticide resistance, pest resurgence, worker poisoning, and ecological imbalances.

Brazilian scientists started paying attention to problems caused by the indiscriminate use of pesticides (Crocomo, 1990). In addition, research was carried out on sampling methods on pests and natural enemies, use of threshold levels, and the correct timing for insecticide application (Parra, 2000). Highly successful IPM programs were developed for several crops including sugarcane, tomato, wheat, and soybean. IPM programs in Brazil are based on careful scouting for pests and natural enemies, and using biological control methods when control is warranted. In soybean, a combination of several control tactics is used (Parra, 2000).

Wheat

Until 1977, control of wheat aphids in Brazil relied heavily on insecticides, with an average of three insecticide applications per year in the state of Rio Grande do Sul (Salvadori, 1990; Parra, 2000). At the time, the major pest species included *Schizaphis graminum*, *Metopolophium dirhodum* and *Sitobion avenae*. These European and Asian species were highly invasive in Brazil (Parra, 2000). In 1978, the Wheat Pest Management program from Embrapa (Brazilian Agricultural Research Corporation) (http://www.embrapa.br/english.units/centers/wheat.htm) began a classical biological control program for wheat aphids with support from the FAO and the University of California (Gassen and Tambasco, 1983; Salvadori, 1990).

Fourteen hymenopteran parasitoids from Europe, the Middle East and Chile and two coccinellid predators from the USA and Israel were mass-produced in Brazil and released (Gassen and Tambasco, 1983). The parasitoids *Aphidius rhopalosiphi*, *A. uzbekistanicus* and *Praon volucre* became established and effectively reduced the aphid population, largely exceeding the

 Integrated Pest Management in the Global Arena (eds K.M. Maredia, D. Dakouo and D. Mota-Sanchez)

goal of 10–15% parasitism (Salvadori, 1990). In the state of Rio Grande do Sul, the number of insecticide applications was reduced from three on the entire area to one on less than 5% of the cultivated area (Gassen and Tambasco, 1983). Other tactics of wheat aphid IPM include the careful use of selective insecticides based on action threshold levels. Maintaining crop diversity and avoiding the use of fire to destroy crop residues can also help to conserve natural enemies (Salvadori, 1990).

Stored grain

In 1999–2000, Embrapa Wheat established an IPM program for stored grain on a growers' cooperative in the state of Paraná. After the use of IPM treatments, the grain showed no noticeable insect damage. This successful IPM management program saved US$10/ton/year for the growers' cooperative. According to Lorini (1999, 2000), methods of the stored grain IPM program included:

- training storage facilities personnel;
- inspection of the entire storage facility;
- thorough cleaning of the equipment and premises followed by application of a persistent insecticide in empty bins;
- correct identification of stored grain pests to ensure appropriate insecticide use;
- avoiding the development of pest resistance to chemical pesticides;
- predicting the level of potential pest damage;
- protecting stored grain with an appropriate insecticide after drying and cleaning;
- fumigating infested grain with phosphine for at least 5 days;
- continually monitoring grain in storage silos for pest infestation.

The first step, training personnel and providing reference materials for follow-up, was a critical part of the IPM program. Embrapa Wheat held 14 training sessions for 437 stored grain operators. An IPM program manual, an illustrated list of key pests and a guide to IPM procedures were distributed. The program also included a new website providing information on IPM practices in stored grains (http://www.cnpt.embrapa.br/armazena/). The success of the stored grain IPM program led to a national policy by the Ministry of Agriculture of Brazil is planned to be issued, overspreading the Stored-Grains IPM in the country.

Sugarcane

In the 1970s, high infestation levels (10% or more) of the sugarcane borer *Diatraea saccharalis* caused losses of US$100 million in the state of São Paulo alone. The parasitoid *Cotesia flavipes* (Braconidae) was imported from Trinidad and Tobago to control *D. saccharalis* in the southeast of Brazil (Parra, 2000). Laboratories were established to rear the parasitoids. From 1981 to 1996, this biological control program reduced pest infestation levels by 0.4%/year (Almeida *et al.*, 1997). Currently, there are 60 rearing facilities in Brazil, and as many as 2 billion parasitoids are released on 360,000 ha each year. Parasitoids are released (at the rate of 6000/ha) when the sugarcane-borer population reaches 3% to 4% infestation. Currently, the average sugarcane-borer infestation is around 2%.

In the northeast of Brazil, the spittlebug *Mahanarva posticata* is the key pest of sugarcane. A fungus, *Metarhizium anisopliae*, is now used as a substitute for chemical insecticides (Parra, 2000). The fungus is well established in the region, with infection levels ranging from 70% to 80%. Before using this fungus to control *M. posticata*, 17% of the saccharose content was lost, reducing the value of the crop (Parra, 2000).

Tomatoes

The pyralid *Tuta absoluta* is the major pest of tomatoes in the São Francisco River Valley, the most important irrigation area in Brazil. In 1989, *T. absoluta* caused 50%

reduction in tomato yield (Haji, 1997). Several methods are used in tomato IPM to *T. absoluta*. Cultural practices (tillage, fertilization, irrigation, a concentrated sowing period, destruction of crop residues, crop rotation), microbial control (application of *Bacillus thuringiensis*), legislative action, pest monitoring with sexual pheromones, and cleaning storage boxes and transportation vehicles are all components of the tomato IPM program (Haji, 1992).

IPM technology using the parasitoid *Trichogramma pretiosum* was developed in Colombia and adapted by Embrapa Semi-Arido for use in Petrolina and Pernambuco (Garcia Roa, 1989; Haji, 1997). Following a program of ten *T. pretiosum* releases per season a significant reduction in fruit damage was achieved. This parasitoid release program is one of the largest tomato IPM programs in the world (used on about 1500 ha) (Parra, 2000). Despite this success, the program has faced challenges in recent years from a virus vectored by thrips and the occurrence of *Bemisia tabaci*, biotype B (synonym *B. argentifolii*) requiring a large number of insecticide applications, thus preventing systematic releases of the parasitoid.

IPM Policy

Brazilian government policy on IPM is not collected under one major piece of legislation, but falls under several programs that collectively involve aspects of IPM. For example, official agricultural development programs always have an IPM component. Cotton growers from the central region of the country and tropical orchard farmers must follow official IPM guidelines in order to be eligible for credit lines and tax reduction advantages. Several government IPM programs also involve important quarantine pests, such as the star-fruit fly (*Bactrocera carambolae*) on the northern border of the country. For the registration of pesticides, stringent testing protocols must be followed, including assessment of environmental impact and effects on non-target organisms. Special attention is given to possible adverse effects on biological control agents (parasitoids, predators and entomopathogens).

Researchers at universities and research institutes also play an important role by continually updating and improving IPM techniques. New IPM recommendations are issued annually for each crop, with an emphasis on achieving adequate pest control. Official crop protection institutes maintain permanent programs of agricultural health training and education for technical staff.

International Cooperation in IPM

The NARS, responsible for the development and implementation of agricultural research in Brazil, is coordinated by Embrapa. In 1998, Embrapa became part of an international collaborative effort (called Labex) with the US ARS on five major areas of mutual interest, i.e. Precision Farming, Soil Resource Management, Integrated Disease and Pest Management, Integrated Control of Animal Diseases, and Intellectual Property Rights and Biotechnology (http://www.ars.usda.gov/is/AR/archive/may00/brazil0500.htm). The ARS–EMBRAPA/Labex program began in the USA in 1998, focusing on integrated pest and disease management (http://www.embrapa.br/labex). The mission of this joint commitment is to maximize food production, provide healthy food for the consumer, maintain minimum production costs and conserve non-renewable natural resources.

Case Study: Soybean IPM

Before the soybean IPM program, soybean insect pests were controlled exclusively by chemical insecticides, often on a preventive or calendar basis. On average, five insecticide applications were done per year, varying from three to ten per soybean season (Gazzoni, 1994). Broad-spectrum insecticides such as endrin, DDT, toxaphene, parathion-methyl and mixtures of these were commonly used. The problems resulting from this high rate of insecticide use led to the development of a diverse soybean IPM program.

History of soybean IPM in Brazil

During the 1974/75 season, an IPM pilot program was conducted in cooperation with Clemson University, University of Illinois, IAPAR, National Soybean Research Center (Embrapa Soybean) and FECOTRIGO, an agricultural cooperative located at Cruz Alta, RS. The IPM program was tested on nine farms in two major soybean-producing areas (Kogan *et al.*, 1977), using information from similar soybean IPM programs in the USA (for review see Moscardi, 1993; Gazzoni, 1994).

The pilot program was designed in paired plots so IPM control strategies and conventional methods could be compared. IPM plots were monitored weekly for pest populations, incidence of the entomopathogenic fungus *Nomuraea rileyi*, and native natural enemies of the major soybean defoliator, velvetbean caterpillar (*Anticarsia gemmatalis*) and stinkbugs. Defoliation level and stage of plant development were assessed weekly to dictate pest control decisions. Action threshold levels were determined based on insect pest density, defoliation level and stage of soybean development. When necessary, minimum effective rates of selected insecticides were used. At the end of the season, insecticide sprayings in the IPM plots were reduced by 78% compared with conventional fields, without a reduction in soybean yield (Kogan *et al.*, 1977).

In the following year, IPM training programs were conducted in several states. Particular focus was placed on technology transfer and implementing IPM at the growers' level. Demonstration fields were coordinated by Embrapa Soybean in partnership with official extension services and growers' cooperatives. Extension workers responsible for training farmers were instructed by Embrapa researchers and received audiovisual materials describing the program, as well as pictures to facilitate the identification of pests, natural enemies and the type of damage caused by each pest. In the first 4 years, 250 lectures were presented to about 100,000 participants. During the program, 70,000 ground cloths and 500,000 recording sheets were distributed to the growers to help them with scouting activities.

In 1978, a national awareness campaign involving mass media was conducted to encourage farmers to use IPM strategies (Gazzoni and Oliveira, 1984). Researchers and extension workers gave talks and wrote articles promoting IPM. Interviews describing successful IPM experiences were given by leading growers and broadcast on television, radio or in newspapers. Current pest population levels, determined by field scouting, were transmitted by radio and television to growers in the regions covered by the program. As a result of this successful IPM campaign, the accumulated production costs of soybean were reduced in 25 years to about US$3 billion and the number of insecticide applications dropped to less than two per season (Gazzoni, 1994).

The soybean IPM program is continually being updated as the research focus changes to meet consumer demands. An extensive survey in several regions of Brazil described the population dynamics of major soybean insects and their natural enemies (Corrêa *et al.*, 1977). Updated publications with recommendations for pest management and general information about IPM in Brazil are released as soon as new information becomes available (Panizzi *et al.*, 1977b; Gazzoni *et al.*, 1981).

The fauna associated with soybean has increased from ten species at the end of the 1960s to 60 species in 1997 (Panizzi and Corrêa-Ferreira, 1997). Currently 37 species of insects and other organisms are considered primary, secondary or sporadic soybean pests (Table 22.1). A few of them can be considered major pests of soybean in certain regions, e.g. *Anticarsia gemmatalis* and the stinkbug complex (*Euschistus heros*, *Nezara viridula* and *Piezodorus guildinii*).

Action thresholds

In the initial phase of the program, action thresholds for lepidopterous defoliators

Table 22.1. Pests of soybean, part of the plant attacked and its relative importance (Hoffmann-Campo *et al.*, 2000).

Pest	Part of plant attacked[a]	Importance
Anticarsia gemmatalis	Le	Main pest
Epinotia aporema	Le, Lb, Po	Secondary, some importance in restricted areas
Omiodes indicatus	Le	Secondary, usually occurring at the maturation, when defoliation is no longer important
Pseudoplusia includens	Le	Secondary
Rachiplusia nu	Le	Secondary
Cerotoma sp.	Le(A), No(L)	Secondary, in soybean areas, preceded by beans
Diabrotica speciosa	Le(A), No(L)	Secondary, in soybean areas, preceded by autumn season maize
Aracanthus mourei	Le, Pe	Secondary, occurring at the beginning of soybean development
Maecolaspis calcarifera	Le	Secondary
Megascelis sp.	Le	Secondary
Chalcodermus sp.	Le	Secondary, regionally important pest
Bemisia argentifolii	Le	Secondary, with high potential of damage
Grasshoppers	Le	Sporadic
Thrips	Nl	Secondary, important in restricted areas as vectors of the virus that causes bud blight disease
Nezara viridula	Po, Se	Main pest
Piezodorus guildinii	Po, Se	Main pest
Euschistus heros	Po, Se	Main pest
Dichelops spp.	Po, Se	Secondary
Edessa meditabunda	Po, Se	Secondary
Thyanta perditor	Po, Se	Secondary
Acrosternum sp.	Po	Secondary
Etiella zinckenella	Po	Secondary, with some importance in restrict areas
Spodoptera latifascia	Po	Sporadic
Spodoptera eridania	Po	Sporadic
Maruca testulalis	Po	Sporadic
Sternechus subsignatus	St	Regionally important pest
Elasmopalpus lignosellus	St	Sporadic, usually important in seasons with extended dry period, in the initial phase of the crop
Myochrous armatus	Ha	Sporadic
Blapstinus sp.	Sd, Ha	Sporadic
Miriapods	Sd, Sp	Secondary, important in no-till areas
Snails and slugs	Sd, Co, Yl	Secondary, important in no-till areas
Phyllophaga spp. (white grubs)	Ro	Regionally important pest
Scaptocoris castanea	Ro	Secondary, important in no-till areas
Scales	Ro	Secondary, important in no-till areas

Lb, leaf bud; Co, cotyledons; Yl, young leaves; Le, leaves; St, stems; No, nodules; Pe, petioles; Sd, seedlings; Sp, small plant; Ro, roots; Se, seeds; Po, pods. (A), adult, (L), larvae.

and stinkbugs were the same as practiced in the USA (Kogan *et al.*, 1977). Studies in Brazil confirmed and/or refined the action thresholds for these pests (Panizzi *et al.*, 1977a; Gazzoni *et al.*, 1981; Villas-Boas *et al.*, 1990). Thresholds (Fig. 22.1) were also established for the shoot and axil borer *Epinotia aporema* (Gazzoni and Oliveira, 1979) and the curculionid gall maker stem borer *Sternechus subsignatus* (Hoffmann-Campo *et al.*, 1990).

Sampling

Sampling of defoliators such as caterpillars, stinkbugs and beetles is done with the

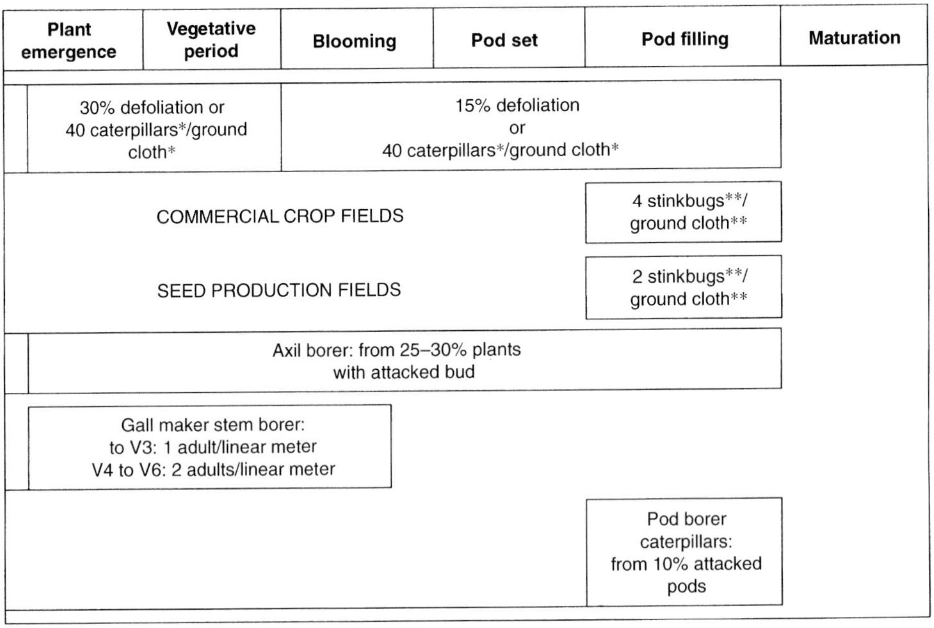

* > 1.5 cm
** > 0.5 cm

Fig. 22.1. Injury threshold levels for key pests of soybean.

shake–cloth technique (Boyer and Dumas, 1969). For assessment of other insects such as *E. aporema*, pod borers, leaf rollers and *S. subsignatus*, the examination of randomly selected plants is the standard method. Soil samples are recommended to estimate the population of root insects such as white grubs (*Phyllophaga cuyabana*) and burrowing bugs (*Scaptocoris castanea* and *Atarsocoris brachiariae*). Soil sampling is also used to estimate the potential population of *S. subsignatus* for the next season, helping farmers to decide if crop rotation is necessary (Silva, 1996).

Biological control

Natural enemies are important components of the soybean IPM program (Corrêa *et al.*, 1977; Corrêa-Ferreira, 1980; Gazzoni *et al.*, 1981). Seven species of predators (carabids, geocorids and nabids), eight species of parasitoids (microhymenopterans and tachinids) and six species of entomopathogens (virus and fungi) are the most common natural enemies of soybean pests (Hoffmann-Campo *et al.*, 2000). The entomopathogen, *N. rileyi* is responsible, especially in rainy seasons, for maintaining *A. gemmatalis* and Plusiinae populations under control, and virtually eliminate the requirement of insecticide use to control these pests. Other natural enemies are produced for release in the field, such as the nuclear polyhedrosis virus (*Ag*NPV) of *A. gemmatalis* (Moscardi, 1999) and the parasitoid of stinkbugs eggs, *Trissolcus basalis* (Corrêa-Ferreira and Moscardi, 1995).

Cultural practices

Cultural changes in cropping systems affect the dynamics of soybean pests (Kogan and Turnipssed, 1987). Some cultural practices, such as manipulating planting dates, modifying tillage systems, crop rotation and mulch management can be used to reduce the pest population. Early planting dates are currently being used as a powerful strategy to avoid synchrony between high pest populations and the most susceptible stage of the soybean plant. Early-maturing soybean cultivars that can escape stinkbug damage are preferred by growers in the south of Brazil, lowering costs and reducing the need for insecticide application (Panizzi, 1985). However, after these short-cycle cultivars are harvested, the stinkbug population can migrate to late-maturing fields and reach high levels, requiring frequent insecticide applications. Crop rotation also has some disadvantages, primarily by the difficulty of persuading growers to use it specially when soybean prices are high; in such years, tend to plant soybean over large areas. The low cost of some insecticides also limits the use of cultural practices; sometimes growers prefer to risk insect populations build up and later apply low cost insecticides.

Chemical control

Chemical control is an important tool in IPM systems. However, broad-spectrum insecticides can have an adverse effect on natural enemies and result in pest resurgence after frequent application (Oliveira *et al.*, 1988; Silva *et al.*, 1988). With the implementation of IPM tactics in Brazil, the number of insecticide applications decreased, and the next step was to seek more selective products. In 1988, selectivity of insecticides became a major criterion for recommendation by regional entomology committees. These committees are responsible for annual revision and recommendation of insecticides for soybean IPM. One major result of this action was a significant reduction in the number of insecticides that were recommended for velvetbean caterpillar (*A. gemmatalis*).

Empirical field observations had suggested that stinkbugs were attracted by sweat (Corso and Gazzoni, 1998). Sodium chloride was the most likely sweat composing substance to exert such effect. However, tests indicated that salt was not attractive in itself, but affected stinkbug behavior by causing them to remain still on treated plants, resulting in longer exposure time to the insecticide (Corso and Gazzoni, 1998). Adding salt allows a reduction in dosage without a loss of effectiveness. Some insecticides are currently recommended at half of their former rates mixed with 0.5% sodium chloride. Growers are successfully adopting this technique to control stinkbugs (Moscardi and Sosa-Gómez, 1996).

Velvetbean caterpillar (*Anticarsia gemmatalis*)

Control of A. gemmatalis *with NPV*

*Ag*NPV is a naturally occurring nuclear polyhedrosis virus of *A. gemmatalis*. In Brazil, this virus was first detected in 1972, and was found in several areas by 1976 (Allen and Knell, 1977; Corso *et al.*, 1977; Gatti *et al.*, 1977). But in spite of its widespread range, the natural incidence of *A. gemmatalis* baculovirus was low (less than 10%). The potential of *Ag*NPV to control *A. gemmatalis* was demonstrated in field experiments (Moscardi *et al.*, 1981). Basic studies at Embrapa Soybean focused on the development of a microbial insecticide to control *A. gemmatalis* (Moscardi, 1993). From 1980 to 1982, a pilot program for *Ag*NPV was tested on 21 soybean farms, in partnership with extension services and growers' cooperatives (Moscardi and Corrêa-Ferreira, 1985). Three plots (virus-treated, insecticide-treated and untreated control), measuring 1 ha each, were compared at each site. *Ag*NPV was applied at the rate of 1.5×10^{11} polyhedral inclusion bodies per ha, when *A. gemmatalis* were at an early instar (length <1.5 cm) and the population was under 20 larvae/m. One *Ag*NPV application successfully controlled *A. gemmatalis* in each trial, but on the insecticide-treated plots an average of 1.3 insecticide applications were necessary. Yield in the virus-treated and insecticide-treated plots was similar, but yield in untreated plots was significantly reduced (Moscardi and Corrêa-Ferreira, 1985).

Production of Ag*NPV*

In 1982, 2000 ha were treated with NPV produced by Embrapa Soybean, following the methodology described by Moscardi (1989). Extension workers collected and distributed frozen NPV-killed larvae to growers. The next year, other institutions and growers' cooperatives began mass production of NPV. *Ag*NPV was used widely, reaching a total area of 200,000 ha by 1984 (Moscardi, 1993). Early on, most NPV were produced under laboratory conditions by private industries, but this practice was discontinued because of the high cost of labor, equipment, and diet (Moscardi *et al.*, 1997).

Until 1985, a preparation of filtered macerated *Ag*NPV-killed larvae was used (Moscardi and Corrêa-Ferreira, 1985). In 1986, a kaolin-based wettable powder was developed (Moscardi, 1989). Virus-killed larvae were collected and stored at –4 to –8°C until processed and formulated as a wettable powder. Embrapa Soybean provided quality control by comparing the amount of viral polyhedra per gram and biological activity on *A. gemmatalis* larvae in laboratory tests (Moscardi and Sosa-Gómez, 1992, 1996; Sosa-Gómez and Moscardi, 1996). Currently, the production of virus in the field is the only method used commercially (Moscardi, 1999). Growers also collect and freeze NPV-killed larvae, so they can be used in the following season and eliminate the expense of purchasing formulated virus. When fields are highly infested with *A. gemmatalis*, this provides a ready source of inoculum for mass production. From 1991 to 1999, virus production was enough to treat from 650,000 to 1,750,000 ha. An average of 1.8 kg of *Ag*NPV-killed larvae/person/day was collected. In the 1996/97 growing season, approximately 35 t of virus-killed larvae were collected, enough to treat 1.75 million ha with *Ag*NPV. As technology continued to improve, the use of *Ag*NPV to control *A. gemmatalis* increased rapidly, reaching 1.4 million ha in the 1997/98 season (Moscardi, 1999). Since the beginning of the program, over 18 million ha have been virus-treated, saving more than US$150 million in production costs and reducing insecticide application by a volume of over 20 million l (Moscardi and Sosa-Gómez, 1996). The cost of formulated *Ag*NPV is currently US$0.40/ha, at a retail cost of US$0.70–1.00, lower than that of chemical insecticides.

In 1991, the formulated *Ag*NPV started to be produced and commercialized by five private Brazilian companies under contract with Embrapa Soybean (Moscardi and Sosa-Gómez, 1992).

Limitations of Ag*NPV use*

Despite the advantages of using NPV to control *A. gemmatalis*, some factors can limit its use by growers. First, demand for NPV is potentially higher than the available supply. The disadvantage of field collection of *Ag*NPV is its reliance on an adequate pest population. If the abundance of the pest population is low (due to factors such as weather), the production of *Ag*NPV can be seriously affected (Moscardi *et al.*, 1997). From 1990 to 1998 an average of 10% of the soybean-cultivated area (1.2–1.4 million ha) in Brazil was treated with *Ag*NPV (Moscardi *et al.*, 1999). Currently, around 1.6 million ha/year are treated with this biological insecticide. Potentially, approximately 4.0 million ha could use *Ag*NPV treatment. To meet such a demand, production would have to be increased fourfold without significantly increasing the cost of the final product (Moscardi, 1999). Large-scale laboratory production of *Ag*NPV is currently being implemented. Recent results indicate that the cost of laboratory-produced virus is competitive with the cost of many chemical insecticides.

Second, growers must know how to use the virus correctly, and understand its delayed mode of action. Lack of information is the primary cause of unsuccessful *Ag*NPV application. For example, most growers are unprepared for the careful monitoring of *A. gemmatalis* phenology necessary to ensure effective application of NPV (Moscardi, 1989; Moscardi and Sosa-Gómez, 1996). In 20 years of virus usage, grower education by research and extension services has been the single most important factor in increasing *Ag*NPV use in Brazil. In regions where the extension program is weak, the use of virus is negligible (Moscardi, 1999).

One characteristic of *Ag*NPV is the longer time-to-death (average 7 days) compared with most chemical insecticides. This is a hurdle for growers accustomed to fast-acting chemical insecticides. In regions with lower mean temperature, the time-to-death of NPV-treated larvae can be even longer, prompting growers to apply insecticide after 7–8 days. Formulations of virus with enhanced viral activity and speed of action are being investigated (Moscardi, 1999). The addition of a very low concentration of additional substances, such as boric acid and optical brighteners of the stilbene group, can increase viral activity and speed up larval death (Shapiro, 1995; Morales *et al.*, 1997, 2001).

Potential for resistance to AgNPV

The possibility of resistance developing to *Ag*NPV has been raised as a potential cause of virus failure in some regions. However, *A. gemmatalis* populations collected in several regions of the country and submitted to 3–8 years of *Ag*NPV application have shown high susceptibility (Abot *et al.*, 1996). But under some laboratory conditions with high selection pressure, *A. gemmatalis* is capable of developing resistance to *Ag*NPV (Abot *et al.*, 1996). Such resistance can be rapidly lost (in four generations) when resistant insects are crossed with susceptible wild-types (Abot, 1993), suggesting that immigration of susceptible insects to virus-treated areas may act as a mechanism to decrease the risk of resistance development by field populations. In addition, stilbenes can in some cases prevent the resistance of *A. gemmatalis* to its baculovirus (Fuxa and Richter, 1998; Moscardi, 1998; Morales *et al.*, 2001).

Reducing insecticide dosage with AgNPV

Most insecticides can be mixed with *Ag*NPV at rates up to one-quarter to one-sixth below the usual recommended dose (Moscardi, 1999). When a sublethal dosage of insecticide is used, the virus can attack the remaining insects and serve as source of inoculum for the next generation. This strategy can be effective when *A. gemmatalis* larvae have exceeded the threshold level used for *Ag*NPV application alone (Moscardi and Sosa-Gómez, 1992).

Stinkbugs

Biological control with egg parasitoids

Thirty-two species of stinkbug pests are associated with soybean in Brazil (Panizzi and Slansky, 1985). Among them, *N. viridula*, *P. guildinii* and *E. heros* represent 98% of the total stinkbug population (Corrêa-Ferreira and Moseardi, 1996). Over 20 species of stinkbug egg parasitoids have been reported by Corrêa-Ferreira (1991). The most important are *T. basalis*, that preferentially attacks *N. viridula* eggs, and *Telenomus podisi*, that prefers eggs of *E. heros* and *P. guildinii*. However, these parasitoids tend to occur in high populations when the stinkbugs have already reached threshold levels. For this reason, a program for laboratory production and inoculative release was developed at Embrapa Soybean (Corrêa-Ferreira, 1993). Before IPM implementation, only a few species of stinkbug egg parasitoids were present in the fields (Panizzi and Slansky, 1985). After the IPM program, 90% (*N. viridula*), 65% (*P. guildinii*), and 78% (*E. heros*) of stinkbug eggs were killed by egg parasitoids (Corrêa-Ferreira and Moscardi, 1995).

After 4 years of field experiments, a pilot program was developed in collaboration with the extension service of Paraná (Emater-PR) using *Trissolcus basalis* (Corrêa-Ferreira and Moscardi, 1996). To protect the early-season population of *T. basalis*, only biological agents such as *Ag*NPV, Bt, or highly selective insecticides such as IGRs were used to control *A. gemmatalis* and other early season insects. When stinkbugs began to colonize the field, or no later than the end of flowering, 5000 parasitoid eggs were released per ha. The egg parasitoids successfully lowered and maintained the stinkbug population below the threshold level. The quality of seeds, as evaluated by the tetrazolium test, was similar in parasitoid release areas and IPM plots (Corrêa- Ferreira, 1993). Following the successful results of the pilot program, the technology was implemented at the farm level in the 1993/94 soybean season. In 1994/95, a modified program was implemented in micro river basins (Corrêa-Ferreira *et al.*, 2000).

Production of Trissolcus basalis

Trissolcus basalis is multiplied in the laboratory on *N. viridula* eggs produced all over the year at Embrapa Soybean facilities under laboratory and glasshouse conditions (Corrêa-Ferreira, 1993). Egg masses frozen or stored in liquid nitrogen remain viable for up to 1 year (Corrêa-Ferreira and Oliveira, 1998). About 1500 parasitized eggs are glued to cardboard and covered with nylon screen to protect them from predation. These cards are sent to farmers for release along the edges of soybean fields. The cards are hung on soybean plants at the beginning of the flowering period when stinkbugs are just starting to colonize the field. An average of 5000 parasitoids/ha are recommended. Some limitations to the use of *T. basalis* include parasitoid availability and the amount of labor involved. Another constraint is the reluctance of many growers to adopt *T. basalis* in lieu of more conventional control methods.

Host plant resistance

Host plant resistance is a highly desirable IPM tactic, as its adoption does not require users to adopt complex changes in their routine. Soybean cultivars resistant to stinkbugs have been developed in Brazil (Rosseto, 1989; Hoffmann-Campo *et al.*, 1996). The soybean cultivar 'IAC-100' released by the breeding program of the Instituto Agronômico de Campinas (São Paulo State) presents double resistance to stinkbugs and defoliators, as well as good agronomic traits (Rosseto, 1989). At Embrapa Soybean, the breeding program has been incorporating resistant genes from identified genotypes (PI 171451, PI 227687, PI 229358 and PI 274454) into advanced genotypes presenting good agronomic traits and high yields.

Cultural control

Euschistus heros overwinters as diapausing adults under leaf litter (Panizzi, 1997). A cultural strategy that has been successful is to incorporate this leaf litter, or to apply insecticides on the overwintering sites (Corrêa-Ferreira and Panizzi, 1999).

The gall maker stem borer (*Sternechus subsignatus*) and white grubs

The gall maker stem borer, *S. subsignatus*, rose from a minor to an important pest as soybean fields expanded to new areas and cultivation systems changed. The adult *S. subsignatus* typically inhabits the lower third of soybean plants, protected by leaves and out of reach of insecticide sprayings (Silva *et al.*, 1998). Larvae develop inside a stem gall (Hoffmann-Campo *et al.*, 1991). White grubs are root feeders and are also difficult to reach with insecticides. Before the IPM program, more than five applications per month of broad-spectrum insecticides were used against the gall maker stem borer. These applications were done early in the season and often resulted in increased pest problems. Recently, seed treatment with systemic insecticides has been successfully used, preserving natural enemies. A border strip of 30–50 m width sown with treated seeds has maintained the pest population below threshold levels. Currently, insecticide applications have been reduced to only one per season.

Cultural control methods

The population of *S. subsignatus* usually becomes apparent in November, reaching a peak after mid-December (Hoffmann-Campo *et al.*, 1991). Early planting of soybean (mid to late October) can prevent the peak population of *S. subsignatus* from coinciding with the most vulnerable plant stages. Crop rotation is also an effective cultural control strategy, especially for oligophagous and long-cycle insects such as *S. subsignatus* and white grubs. Populations of *S. subsignatus* have been adequately controlled by rotation with non-legumes such as maize, sunflower, or cotton (Hoffmann-Campo *et al.*, 1999). During crop rotation, one-third to one-half of the previous soybean area is planted with other crops. This strategy is essential for *S. subsignatus* control, and particularly efficient when it is used along with soybean seed treated in the border strip. Cotton, *Crotalaria juncea* and *C. spectabilis* are options for crop rotation in areas infested with white grubs, especially *P. cuyabana*, since these plants affect insect biology (Oliveira *et al.*, 1997). The timing of tillage can also be used as a strategy to aid in white grub control (Oliveira *et al.*, 2000). Proper timing of tillage can be an effective component of IPM for soil dwelling insects, although further studies are needed to recommend its use for other species.

Soybean IPM in micro river basins

The soybean IPM program was modified slightly for implementation in environmentally sensitive river basins. The pilot program took place in the Rio do Campo basin. Laboratory rearing facilities were established in the river basin by farmers' associations in cooperation with county and state organizations. Strategies used in the river basin IPM program included: (i) weekly pest monitoring; (ii) using action thresholds to dictate control decisions; (iii) use of *Ag*NPV or other highly selective products against *A. gemmatalis*; (iv) releasing *T. basalis* throughout the river basin for adequate dispersion and control of stinkbugs; and (v) application of insecticides only in border rows or at a reduced dosage mixed with sodium chloride to control stinkbugs (Corso and Gazzoni, 1998).

A dramatic change in pest control practices occurred in the Rio do Campo basin after 4 years of the IPM program (Corrêa-Ferreira *et al.*, 2000). The average number of insecticide applications in IPM-treated fields dropped by 56%, from 2.80 to 1.23/season. In non-participating fields, the average number of insecticide applications ranged from 2.09 to 2.74 (Corrêa-Ferreira *et al.*, 2000). The use of microbial insecticides also increased. In the 1997/98 season, 62% of the river basin was treated with *Ag*NPV, and large areas outside of river basins started to be treated with *Ag*NPV as well (L. Morales, unpublished data).

When pest populations were high, *Ag*NPV was often applied together with reduced dosages of insecticides, especially Bt and IGRs (Moscardi, 1999). A significant increase in the use of more selective insecticides was a successful hallmark of the river basin IPM program. Before the IPM program, 97% of insecticides used were broad-spectrum and highly toxic. After 4 years of the river basin IPM program, microbial insecticides and IGRs accounted for about 75% of insecticide use.

Conclusions

Most IPM programs in Brazil utilize the production and release of biological control agents and strongly emphasize the reduction of broad-spectrum insecticide use. New IPM programs for several crops are being developed, with an emphasis on conservation and augmentative biological control, cultural practices and host plant resistance. IPM programs under development include vegetables (tomato, cole crops), fruit crops (citrus, apples, etc.), maize, cotton, coffee, and forests, as well as the control of widespread polyphagous pests such as the white fly *B. tabaci*. Continuous improvement of these programs is necessary to achieve wider use of IPM tactics against major pests, and to develop appropriate control tactics against newly emerging pests. Considerable improvements are expected in the methods of production and release of indigenous entomopathogens, parasitoids, and predators. Some systems are exploring classical biological control, a method with wide potential for application in several cropping systems. Organic farms are a growing sector in Brazil, with an increasing demand for pest control methods that can be used on organic crops.

Farmers are increasingly using safer and more selective insecticides, such as the biologicals, the IGRs and nicotinoids. Most likely this trend will continue, as private companies release new products, and since the public is demanding more high quality food and fewer chemical pesticides used in food production. In addition, the policies for registration and use of insecticides in the country have become more stringent.

There is no legislation prohibiting OGM use in Brazil. According to the Brazilian legislation, presently in force, the Biosafety Commission (Comissão Ténica Nacional de Biossseguranç (CTNBio)) has the authority to decide upon cultivation and commercialization of OGM crops. Meanwhile, a consortium of NGOs argued on the Brazilian Court that the environmental impact data provided for the Roundup Ready Soybean was not enough and consistent to assure the absence of negative impact of the technology on the environment. Once the matter is *sub judice*, no OGM can be cultivated (except for research purposes) or commercialized in Brazil. Presently research on the application of genetically modified crops to IPM programs is underway. Several aspects of genetically modified crops are being evaluated, especially their impact on natural enemies, non-target insects, and other arthropods that feed on these crops, as well as the possibility of pest resistance. Thus, if genetically modified crops are approved, it is likely that IPM programs will need to be re-evaluated and modified to account for this change.

Although continuous development and improvement of IPM programs in Brazil is important, improved technology transfer and outreach to growers is fundamental. IPM tactics must be made widely available to farmers through research institutions, official and private (farmer's cooperatives) extension services, and private companies. It is only by educating farmers on the importance and benefits of using IPM tactics for pest control that IPM programs can have a broader impact on agriculture in Brazil.

References

Abot, A.R. (1993) Avaliação da resistência de *Anticarsia gemmatalis* Hübner 1818 (Lepidoptera: Noctuidae) ao seu vírus de poliedrose nuclear, *Baculovirus anticarsia.* Dissertação de Mestrado. Universidade Federal do Paraná, Curitiba, Brazil.

Abot, A.R., Moscardi, F., Fuxa, J.R., Sosa-Gómez, D.R. and Richter, A.R. (1996) Development of resistance by *Anticarsia gemmatalis* from Brazil and the United States to a nuclear polyhedrosis virus under laboratory selection pressure. *Biological Control* 7, 126–130.

Allen, G.E. and Knell, J.D. (1977) A nuclear polyhedrosis virus of *Anticarsia gemmatalis*: I. Ultrastructure, replication, and pathogenicity. *Florida Entomologist* 60, 233–240.

Almeida, L.C., Arrigoni, E.B. and Rodrigues Filho, J.P. (1997) Modelo de análise econômica para avaliação do controle biológico da cana-de-açucar, *Diatraea saccharalis.* In: *Anais do VII Seminário Cooperçucar de Tecnologia Agronômica.* Fealq, Piracicaba, pp. 95–104.

Boyer, N.P. and Dumas, B.A. (1969) Plant-shaking methods for soybean insect survey in Arkansas. *Survey Methods for Economic Insects,* USDA–ARS 81-31, pp. 92–94.

Corrêa, B.S., Panizzi, A.R., Newman, G.G. and Turnipseed, S.G. (1977) Distribuição geográfica e abundância estacional dos principais insetos-pragas da soja e seus predadores. *Anais da Sociedade Entomologica do Brasil* 6, 40–50.

Corrêa-Ferreira, B.S. (1980) Controle biológico de pragas da soja. In: Ramiro, Z.A., Grazia, J. and Lara, F.M. (eds) *Anais do XI Congresso Brasileiro de Entomologia,* Campinas, SP, Fundaçao Cargill, pp. 277–301.

Corrêa-Ferreira, B.S. (1991) Parasitóides de ovos de percevejos: incidência natural, biologia e efeito sobre a população de percevejos da soja. PhD Thesis, Universidade Federal do Paraná, Curitiba, Brazil.

Corrêa-Ferreira, B.S. (1993) *Utilização do Parasitóide de Ovos* Trissolcus basalis *(Wollaston) no Controle de Percevejos da Soja.* EMBRAPA-CNPSo, (EMBRAPA-CNPSo. Circular Técnica, ll), Londrina, PR, Brazil.

Corrêa-Ferreira, B.S. and Moscardi, F. (1995) Seasonal occurrence and host spectrum of egg parasitoids associated with soybean stink bugs. *Biological Control* 5, 96–201.

Corrêa-Ferreira, B.S. and Moscardi, F. (1996) Biological control of stink bugs by inoculative releases of *Trissolcus basalis. Entomologia Experimentalis et Applicata* 79, 1–7.

Corrêa-Ferreira, B.S. and Oliveira, M.C.N. (1998) Viability of *Nezara viridula* (L.) eggs for parasitism by *Trissolcus basalis* (Woll.) under different storage techniques in liquid nitrogen. *Anais da Sociedade Entomológica do Brasil* 27, 101–108.

Corrêa-Ferreira, B.S. and Panizzi, A.R. (1999) *Percevejos da Soja e seu Manejo.* EMBRAPA-CNPSo, (EMBRAPA-CNPSo Circular Tecnica, 24), Londrina, PR, Brazil.

Corrêa-Ferreira, B.S., Domit, L.A., Morales, L. and Guimaraes, R.C. (2000) Integrated soybean pest management in micro river basins in Brazil. *Integrated Pest Management Reviews* 5, 75–80.

Corso, I. and Gazzoni, D.L. (1998) Sodium chloride: an insecticide enhancer for controlling pentatomids in soybeans. *Pesquisa Agropecuária Brasileira* 33, 1563–1571.

Corso, I.C., Gazzoni, D.L., de Oliveira, E.B. and Gatti, I.M. (1977) Ocorrência de poliedrose nuclear em *Anticarsia gemmatalis* Hübner, 1818 na região sul do Brasil (Nota prévia). *Anais da Sociedade Entomológica do Brasil* 6(2), 312–314.

Crocomo, W.B. (1990) O que é o manejo. In: Crocomo, W.B. (ed.) *Manejo Integrado de Pragas.* Editora Unesp, São Paulo, pp. 9–34.

Fuxa, J.R. and Richter, A.R. (1998) Repeated reversion of resistance to nucleopolyhedrovirus by *Anticarsia gemmatalis. Journal of Invertebrate Pathology* 71, 159–164.

Gassen, D.N. and Tambasco, F.J. (1983) Controle biológico dos pulgões do trigo no Brasil. *Informe Agropecuário* 9(104), 49–51.

Garcia Roa, F. (1989) *Plagas del tomate y su manejo.* ICA, Palmira, Colombia, 19 pp.

Gatti, I.M., Silva, D.M. and Corso, I.C. (1977) Polyhedrosis occurrrence in caterpillars of *Anticarsia gemmatalis* Hübner, 1818 in South of Brazil. *IRCS Medical Science: Cell and Membrane Biology; Environmental Biology and Medicine; Experimental Animals, Microbiology, Parasitology and Infections Diseases* 5, 136.

Gazzoni, D.L. (1994) *Manejo de Pragas da Soja: Uma Abordagem Histórica.* MAARA/ EMBRAPA-CNPSo. Brasília: EMBRAPA-SPI. (EMBRAPA-CNPSo. Documentos, 78). Londrina, PR, Brazil.

Gazzoni, D.L. and Oliveira, E.B. de (1979) Distribuição estacional de *Epinotia aporema* e seu efeito sobre o rendimento e seus componentes, e características agronômicas da soja cv. UFV-1, semeada em diversas épocas. *Anais do I Seminário Nacional de Pesquisa de Soja* 2, 94–105.

Gazzoni, D.L. and Oliveira, E.B. (1984) Soybean insect pest management in Brazil: I. Research effort; II Programme implementation. In: Matteson, P.C. (ed.) *Proceedings of International Workshop on Integrated Pest Control of Grain Legumes.* EMBRAPA, Goiania, Brazil, pp. 312–325.

Gazzoni, D.L., Oliveira, E.B., Corso, I.C., Corrêa-Ferreira, B.S., Villlas Boas, G.L., Moscardi, F. and Panizzi, A.R. (1981) *Manejo de Pragas da Soja.* (CNPSO-EMBRAPA, Circular Técnica 5) Londrina, PR, Brazil.

Haji, F.N.P. (1997) Controle biológico da traça do tomateiro com *Trichogramma* no Nordeste do Brasil. In: Parra, J.R.P. and Zucchi, R.A. (eds) *Trichogramma e o Controle Biológico Aplicado.* FEALQ, Piracicaba, pp. 319–324.

Haji, F.N.P. (1992) Manejo de pragas do tomateiro no submédio São Francisco. In: Fernades, O.A., Corrêa, A.B. and Bortoli, S.A. (eds) *Manejo Integrado de Pragas e Nematóides.* FUNESP/UNESO, Jaboticabal, pp. 341–352.

Hoffmann-Campo, C.B., Oliveira, E.B. de and Mazzarin, R.M. (1990) Níveis de infestação de *Sternechus subsignatus* Boheman, 1836: Influência nos rendimentos e características agronômicas da soja. *Pesquisa Agropecuária Brasileira* 25, 221–227.

Hoffmann-Campo, C.B., Parra, J.R.P. and Mazzarin, R.M. (1991) Ciclo biológico, comportamento e distribuição estacional de *Sternechus subsignatus* Boheman, 1836 (Col.: Curculionidae) em soja, no norte do Paraná. *Revista Brasileira de Biologia* 51, 615–621.

Hoffmann-Campo, C.B., Gazzoni, D.L. and Toledo, J.F.F. de (1996) Comparação de métodos para a avaliação de de linhagens de soja (*Glycine max*) resistentes a percevejos (Hemiptera: Pentatomidae). *Pesquisa Agraopecuária Brasileira* 31, 305–316.

Hoffmann-Campo, C.B., Silva, M.T.B. da and Oliveira, L.J. (1999) Aspectos biológicos e manejo integrado de *Sternechus subsignatus* na cultura da soja. Embrapa-Soja/Cruz Alta: FUNDACEP_FECOTRIGO, (Embrapa-Soja. Circular Técnica, 22), Londrina, PR, Brazil.

Hoffmann-Campo, C.B., Moscardi, F., Corrêa-Ferreira, B.S., Oliveira, L.J., Sosa-Gómez, D.R., Panizzi, A.R., Corso, I.C., Gazzoni, D.L. and Oliveira, E.B. de (2000) *Pragas da Soja no Brasil e seu Manejo Integrado.* (Embrapa Soja. Circular Técnica, 30), Londrina, PR, Brazil.

Kogan, M. and Turnipseed, S.G. (1987) Ecology and management of soybean arthropods. *Annual Review of Entomology* 32, 507–538.

Kogan, M., Turnipseed, S.G., Shepard, M., Oliveira, E.B. and Borgo, A. (1977) Pilot insect pest management program for soybean in Southern Brazil. *Journal of Economic Entomology* 70, 659–663.

Lorini, I. (1999) Pragas de grãos de cereais armazenados. *Embrapa Trigo* (Comunicado Técnico, 17). Passo Fundo, RS, Brazil.

Lorini, I. (2000) Manejo integrado de pragas de grãos armazenados. *Embrapa Trigo* (Comunicado Técnico, 17). Passo Fundo, RS, Brazil.

Morales, L., Moscardi, F., Sosa-Gómez, D.R., Paro, F.E. and Soldorio, I.L. (1997) Enhanced activity of *Anticarsia gemmatalis* Hüb. (Lepidoptera: Noctuidae) nuclear polyhedrosis virus by boric acid in the laboratory. *Anais da Sociedade Entomologica do Brasil* 26, 115–120.

Morales, L., Moscardi, F., Sosa-Gómez, D.R., Paro, F.E. and Soldorio, I.L. (2001) Fluorescent brighteners improve *Anticarsia gemmatalis* (Lepidoptera: Noctuidae) nucleopoplyhedrovirus (AgMNPV) activity on Ag MNPV-susceptible and resistant strains of the insect. *Biological Control* 20, 247–253.

Moscardi, F. (1989) Use of viruses for pest control in Brazil: the case of the nuclear polyhedrosis virus of the soybean caterpillar, *Anticarsia gemmatalis. Memórias do Instituto Oswaldo Cruz* 84, 51–56.

Moscardi, F. (1993) Soybean integrated pest management in Brazil. *FAO Plant Protection Bulletin* 41, 91–100.

Moscardi, F. (1998) Resistance of *Anticarsia gemmatalis* to its nucleopolyhedrosis virus (AgNPV) and evaluation of substances that enhance viral activity. In: *Simpósio De Controle Biológico*, Fundação Oswaldo Cruz, Rio de Janeiro, Brazil, pp. 421–425.

Moscardi, F. (1999) Assessment of the application of Baculoviruses for control of Lepidoptera. *Annual Review of Entomology* 44, 257–258.

Moscardi, F. and Corrêa-Ferreira, B.S. (1985) Biological control of soybean caterpillars. *World Soybean Research Conference* 3, 703–711.

Moscardi, F. and Sosa-Gómez, D.R. (1992) Use of viruses against soybean caterpillars in Brazil. In: Copping, L.G., Green, M.B. and Rees, R.T. (eds) *Pest Management in Soybean.* Elsevier Applied Science, London, pp. 98–109.

Moscardi, F. and Sosa-Gómez, D.R. (1996) Soybean in Brazil. In: Persley, G.J. (ed.) *Biotechnology and Integrated Pest Management.* CAB International, Wallingford, UK, pp. 98–112.

Moscardi, F., Allen, G.E. and Greene, G.L. (1981) Control of the velvetbean caterpillar by nuclear polyhedrosis virus and insecticides and impact of treatments on the natural incidence of the entomopathogenic fungus *Nomuraea rileyi. Journal of Economic Entomology* 74, 480–485.

Moscardi, F., Leite, L.G. and Zamataro, C.E. (1997) Production of nuclear polyhedrosis virus of *Anticarsia gemmatalis* Hübner (Lepidoptera: Noctuidae): Effect of virus dosage, host density and age. *Anais da Sociedade Entomologica do Brasil* 26, 121–132.

Moscardi, F., Sosa-Gómez, D.R. and Corrêa-Ferreira, B.S. (1999) Soybean IPM in Brazil, with emphasis on biological control tactics. *World Soybean Research Conference VI, Proceedings*, Chicago, Illinois, pp. 331–339.

Oliveira, E.B. de, Gazzoni, D.L., Corso, I.C., Villas Boas, G.L. and Hoffmann-Campo, C.B. (1988) *Pesquisa com inseticidas em soja: sumário dos resultados alcançados entre 1975 a 1987.* (CNPSo-EMBRAPA, Documentos 30) Londrina, PR, Brazil.

Oliveira, L.J., Garcia, M.A., Hoffmann-Campo, C.B., Farias, J.R.B., Sosa-Gómez, D.R. and Corso, I.C. (1997) *Coró-da-soja* Phyllophaga cuyabana (Embrapa-CNPSo. Circular Técnica, 20), Londrina, PR, Brazil.

Oliveira, L.J., Hoffmann-Campo, C.B. and Garcia, M.A. (2000) Effect of soil management on the white grub population and damage in soybean. *Pesquisa Agropecuária Brasileira* 35(5), 887–894.

Panizzi, A.R. (1985) Dynamics of phytophagous pentatomids associated with soybean in Brazil. In: *World Soybean Research Conference III.* Westview Press, Boulder, Colorado, pp. 674–680.

Panizzi, A.R. (1997) Wild hosts of pentatomids: ecological significance and role in their pest status on crop. *Annual Review of Entomology* 42, 99–122.

Panizzi, A.R. and Corrêa-Ferreira, B.S. (1997) Dynamics in the insect fauna adaptation to soybean in the tropics. *Trends in Entomology* 1, 71–88.

Panizzi, A.R. and Slansky, F. Jr (1985) Review of phytophagous pentatomids (Hemiptera: Pentatomidae) associated with soybean in Americas. *Florida Entomologist* 68, 184–203.

Panizzi A.R., Corrêa-Ferreira, B.S., Newman, G.G. and Turnipseed, S.G. (1977a) Efeito de insecticidas na população das principais pragas da soja. *Anais da Sociedade Entomologica do Brasil* 6, 264–275.

Panizzi A.R., Corrêa-Ferreira, B.S., Gazzoni, D.L., Oliveira, E.B., Newman, G.G. and Turnipseed, S.G. (1977b) *Insetos da soja no Brasil.* (CNPSO-EMBRAPA, Boletim Técnico 1), Londrina, PR, Brazil.

Parra, J.R. (2000) Manejo integrado de pragas. In: Paterniani, E. (ed.) *Agricultura Brasileira e a Pesquisa Agropecuária.* Embrapa Comunicação para Transferência de Tecnologia, Brasilia, pp. 92–105.

Rosseto, C.J. (1989) Breeding for resistance to stink bugs. In: Pascale, A.J. (ed.) *Actas Proceedings of IV World Soybean Research Conference.* Asociación Argentina de la Soja, Buenos Aires, pp. 2046–2060.

Salvadori, J.R. (1990) Manejo integrado das pragas do trigo no Brasil: situação ataul e perspectivas. In: Fernades, O.J., Corrêa, A.C.B. and Bortoli, S.A. (eds) *Manejo integrado de pragas e nematóides.* Unesp, Jaboticabal 1, pp. 237–253.

Shapiro, M. (1995) Radiation protection and activity enhancement of viruses. In: Hall, F.R. and Barry, J.W. (eds) *Biorational Pest Control Agents: Formulation and Delivery.* American Chemical Society, Washington, DC, pp. 153–164.

Silva, M.T.B. da (1996) Influência da rotação de culturas na infestação de *Sternechus subsignatus* Boheman (Coleoptera: Curculionidae) em plantio direto. *Ciência Rural* 26(1), 1–5.

Silva, M.T.B. da, Corso, I.C., Belarmino, L.C., Link, D., Tonet, G.L., Gomez, S.A. and Santos, B. (1988) Avaliação de inseticidas sobre predadores das pragas da soja, em dez anos agrícolas, no Brasil. *Trigo e Soja* 96, 3–16.

Silva, M.T.B. da, Neto, N. and Hoffmann-Campo, C.B. (1998) Distribution of eggs, larvae, and adults of *Sternechus subsignatus* Boheman on soybean plants under a no-till system. *Anais da Sociedade Entomologica do Brasil* 27(4), 513–518.

Sosa-Gómez, D.R. and Moscardi, F. (1996) Producción de virus patógenos de ácaros y insectos. In: Lecuona, R.E. (ed.) *Microrganismos Patógenos Empleados en el Microbiano de Insectos Control Plaga.* Talleres Gráficos Mariano Mas., Buenos Aires, pp. 223–236.

Villas-Boas, G.L., Gazzoni, D.L., Oliveira, M.C.N. de, Costa, N.P., Roessing, A.C., França Neto, J.B. and Henning, A.A. (1990) *Efeito de Diferentes Populações de Percevejos sobre os Rendimentos e seus Componentes, Características Agronômicas e Qualidade de Soja.* (EMBRAPA-CNPSo, Boletim de Pesquisa 1), Londrina, PR, Brazil.

Chapter 23

Integrated Pest Management in Peru

María Palacios Lazo[1], Alfonso Lizárraga Travaglini[2], Ricardo Velásquez Ochoa[3], Enrique Carranza Hernández[4] and Isaías Segovia[5]

[1]Centro Internacional de la Papa, Lima, Perú; [2]Red de Acción en Alternativas al Uso de Agroquímicos and Universidad Nacional Federico Villarreal, Lima, Perú; [3]Instituto Nacional de Investigación Agraria, Lima, Perú; [4]Servicio Nacional de Sanidad Agraria, Lima, Perú; [5]Universidad Nacional Agraria La Molina, Lima, Perú

History and Evolution of IPM in Peru

Peru was the site of one of the first large-scale IPM programs. In the early 1950s, the German–Peruvian entomologist Johannes Wille developed the concept that agricultural entomology was a branch of applied ecology, and recommended that insecticides be used only as a last resort (Wille, 1952). IPM in Peru began in the mid-1950s in response to problems caused by the use of organochlorines on crops such as cotton, citrus, olives and sugarcane (Risco, 1954, 1960; Herrera, 1961; Beingolea *et al.*, 1969; Beingolea and Salazar, 1970).

Cotton

An IPM program for cotton in the Cañete valley began in the mid-1950s after a period of intensive use of organochlorines (DDT, BHC, toxaphene and aldrin) from 1949 to 1956. Problems resulting from the overuse of insecticides were the development of insecticide resistance, pest resurgence, and appearance of new pests due to the elimination of natural enemies. In 1955–1956, cotton production collapsed, and the Farmers' Association of the Cañete Valley organized an IPM program with technical support from the government. This program banned the use of synthetic insecticides and initiated a program for biological restoration of the valley (Herrera, 1961). The biological restoration program was facilitated by the fact that cotton, its insect pests and the complex of natural enemies are all native to Peru. The IPM program was soon expanded to all coastal valleys of the country. In each valley, farmers' associations were promoted and plant protection services were established.

Citrus

By 1961, insect pests in citrus had become a major problem due to the use of organophosphate insecticides. Initially, alternative strategies for managing citrus pests included the use of selective insecticides and the release of natural enemies (Beingolea, 1961). Since citrus and many of its insect and mite pests had been

 Integrated Pest Management in the Global Arena (eds K.M. Maredia, D. Dakouo and D. Mota-Sanchez)

introduced into the country, the lack of effective natural enemies was understandable. Several beneficial insects were introduced successfully, as demonstrated in the rehabilitation of an 8-year-old citrus orchard in the Chincha valley (Central Coast of Peru) that had been severely infested by pests after frequent applications of insecticides (Beingolea *et al.*, 1969).

Olives

Biological control of olive pests began in 1937 with the first introductions of natural enemies of the black scale, *Saissetia oleae* (Wille, 1952). This system, with some fluctuations, was maintained until 1954. In that year, major outbreaks of the black scale occurred in some olive-producing areas, induced by the intensive use of insecticides such as parathion and azinphos-methyl. Following these outbreaks, an integrated management system was established for olives. Integrated management of olive pests was based on the action of predators and cultural measures to increase mortality and improve the effectiveness of parasitoids. An effective monitoring system, mass-rearing of natural enemies and the use of high-pressure sprayers were all part of the integrated program, among other practices (Beingolea, 1993).

Sugarcane

Significant losses from the sugar cane borer, *Diatraea saccharalis* in 1949 induced farmers to try new measures for the management of this pest. In 1951, the parasitoid, *Lixophaga diatraea* was introduced into the country. However, this species was not successfully established. Conversely, the mass release of a native parasitic fly *Paratheresia claripalpis* reduced by 83% the damage in sugarcane as a result of high levels of parasitism (88%) (Risco, 1954, 1960). During the Agrarian Reform in the 1970s, large sugarcane farms became cooperatives and the pest management program was significantly impaired due to new government policies (Pollack, 1994).

IPM education

In 1971, graduate programs (MSc level) in entomology and plant pathology were initiated at the National Agrarian University 'La Molina'. Important concepts of IPM were taught, for instance, Cisneros (1980) defined IPM as integrating insect pest, disease, and weed management, and emphasizing the inclusion of two or more pest control techniques based on economic damage level. Pest control techniques should be ecologically and economically sound, minimizing undesirable effects.

Policies and legal issues related to IPM

As early as 1909, Peru passed a law related to Plant Protection. This law was modified and expanded in 1949. In 1976, a newly created Ministry of Food issued a Plant Sanitation Regulation for importation and exportation of plant products. In 1993, a plant sanitary certificate from the country of origin became a requirement for the introduction of plant products. A law issued in 1997 promotes IPM as the major policy in agriculture. Several other policies also affect IPM implementation directly or indirectly.

In recent years, the Government of Peru has reinitiated technical assistance to farmers through special programs that included the extension of IPM. These programs include PRONAMACHCS (a national program for the management of soils and watersheds), and SENASA (the national service for plant and animal health).

Organizations Involved in IPM in Peru

Three types of institutions develop IPM research and extension in Peru: public institutions and universities, farmers' associations, and private organizations (Table 23.1).

Table 23.1. Institutions that do research, extension and set regulations for IPM techniques in Peru.

		Methods of control								
	Institutions	Cultural	Biological	Behavioral	Use of varieties	Plant breeding	Physical or mechanical	Chemical or botanical	Legal	IPM
Research	INIA	X	X		X			X		
	University						X	X		
	CIP	X	X	X	X	X		X		X
Farmers as IPM clients	Cotton	X	X	X	X		X	X	X	X
	Citrus	X	X	X			X	X		X
	Potatoes	X	X	X	X	X	X	X	X	X
Institutions involved in IPM extension	INIA	X	X		X	X	X		X	X
	University	X	X	X	X		X	X	X	X
	CIP	X	X	X	X	X	X	X	X	X
	NGOs	X	X	X	X		X	X	X	X
	SENASA	X	X	X	X		X	X	X	X
Regulatory Institution	SENASA	X	X			X		X	X	X
Institutions providing IPM inputs	SENASA		X							X
	Industry		X							
	CIP		X	X		X				X
	NGOs		X	X				X		X
	INIA		X		X					X

Public institutions

Research institutes

The first experiment station was established in La Molina in 1946, with several departments including entomology and plant pathology (Vilchez, 2000). In 1981, a National Institute for Agricultural Research and Promotion was created with a research program in entomology. Later on, this program has since been superseded by an INIA with four major research programs, one of which is the National Research Program for IPM.

In 1989, research began on the use of hydrothermal treatments of export mangos for the immature stages of the Mediterranean fruit fly, *Ceratitis capitata*. As a result, the USA approved the importation of mangos from Peru in 1991. Research on the control of the fruit flies *Ceratitis capitata* and *Anastrepha fraterculus* has continued with techniques such as mass releases of sterile flies, use of traps for monitoring, biological control, cultural control, and chemical control. Other research programs have included testing the efficacy of sticky yellow traps for leafminer flies and whiteflies, and the identification of several entomopathogenic fungi including *Neozygites*, *Verticillium lecanii*, *Pandora neoaphidis*, *Entomophthora planchoryana*, *Conidiobolus*, *Erynia* spp., *Erynia dipterigena*, Zoop Tera *phalloides* in crops such as coffee, citrus, tomato, potato, cucumbers and beans.

The biological control work has been conducted in two geographic areas: in the mountains and in the coast area. In the mountains, research and extension activities were carried out on the utilization of entomopathogens including *Beauveria bassiana* and *B. brongniartii* for the control of the Andean potato weevil; Baculovirus for the control of potato tuber moths; rearing and release of *Campoletis* sp. for control of larvae of Lepidoptera; *Copidosoma koehleri* for the control of tuber moth; and *Pachycrepoideus* spp. for fruit flies. In the coast area, research was concentrated on the use of entomopathogenic fungi; *Verticillium lecanii* and *Paecilomyces farinosus* for the control of whiteflies (*Bemisia tabaci* and *Aleurodicus cocois*) in cotton, cowpea, cucumber, melon, watermelon, and mango. *Metarhizium anisopliae* and *Beauveria bassiana* were used for the control of the diamondback moth, *Plutella xylostella*, and the antagonistic fungus, *Gliocladium roseum* for the control of the strawberry gray rot, *Botrytis cinerea*.

Universities

Research conducted at the universities focused on the development of IPM components. Sixteen agronomy faculties throughout the country are involved in basic and applied research related to IPM (Arroyo, 1988, 1989). These projects are most often related to thesis research required to obtain undergraduate degrees. For instance, in 1995, 75% of thesis research projects were related to the chemical control of insect pests at the Agrarian University at La Molina (Lizárraga *et al.*, 1995). However, in the past 5 years, research on biological control, host plant resistance, and other non-chemical measures has been given more importance.

Farmers' associations

Research in agriculture was initiated by the farmers belonging to the National Agrarian Society. In 1926, farmers of the Cañete valley founded an experiment station designed to increase productivity in export crops such as cotton and sugarcane. The pest resistance in cotton inspired research to explain the factors associated with increased pest populations and to develop new methods of control. This work established a foundation for IPM in Peru.

In the 1970s, due to change in the government policy, the Agrarian Reform truncated this unusual system of 'farmers promoting research for the control of pests'. Currently, only large agricultural enterprises with adequate economic resources can provide facilities for research related to the development of IPM.

International programs and local NGOs

FAO and NGOs

In 2000, a special IPM project known as 'Integrated Pest Management in Major Food Crops of Peru' was implemented. This project has been sponsored by FAO and run by several governmental and non-governmental organizations, including SENASA, INIA, PRONAMACHS, UNALM, Catholic Agency For Overseas Aid and Development, CARE (Network of relief and development organizations) and RAAA. The major objective of this program is to improve the quality of life of small farmers in Peru, by increasing their income, reducing pesticide exposure and promoting sustainable production. The program was patterned after the highly successful FFS approach. As a result of this training, many farmers, especially potato and cotton producers, will be able to implement IPM in their fields.

The CIP

In 1971, the CIP was created to generate improvements in potato production. CIP has contributed to the development of potato IPM, particularly in the management of nematodes, fungi and insects. Initially, research emphasized the development of resistant plants. From 1978 until 1990, CIP's entomological research was focused on the management of three key pests: the Andean potato weevil, *Premnotrypes* spp., the potato tuber moth, *Phthorimaea operculella* and the leafminer fly, *Liriomyza huidobrensis* (Raman and Palacios, 1983; Ramau, 1984, 1987, 1988a,b).

In 1988, CIP organized the first IPM Pilot Unit, and within 2 years Pilot Units were established in several areas of the country. Pilot Units demonstrate the use of IPM strategies in farmers' communities, and help train farmers, extension workers and professionals involved in potato cultivation. Since 1998, new research findings have been incorporated into the system to improve IPM implementation. In 1999 CIP initiated a study of alternatives for the management of potato late blight and the possibility of using the FFS approach, sponsored by FAO, as a training method.

Until 1990, CIP's research findings were transferred to the National Potato Program of the INIAA for on-site testing and demonstration of management strategies. Since 1992, these functions have been passed to other organizations, primarily NGOs. NGOs working in rural areas have partially replaced the role of the Peruvian Government in technology transfer among farmers. Several organizations are involved in these activities. In general, NGOs are not involved in research, except for those that are working on specific projects in collaboration with universities or other research institutions such as CIP.

To raise awareness of IPM programs and management strategies, CIP produces a variety of materials (bulletins, videos, posters, and portfolios). Major components of these management programs include cultural practices, use of sex pheromones, colored traps, shelter and food traps, and the introduction and protection of natural enemies (Cisneros *et al.*, 1995).

Successful Cases of IPM

Peru has a long history of successful IPM programs. The most well-known is the integrated management of cotton pests, begun in the mid-1950s and still in practice today. More recently, integrated management of potato pests has been implemented with the support of the CIP.

Integrated management of cotton pests

Cotton has been grown in Peru for over 5000 years. On the northern coast, the varieties Pima and del Cerro are grown. On the central and southern coasts, the most common variety is Tangüis, which is resistant to wilting disease. In the north and central jungle, a native white to red colored cotton (algodón áspero) is cultivated in small areas. Cotton in the coast is irrigated (surface irrigation), whereas cotton in the jungle is grown under rainfall conditions.

In recent years, cotton production has varied between 145,000 t and 268,000 t. The area under cultivation has also varied, from 73,000 ha to 137,000 ha (Perú Acorde, 2000). Currently, most cotton is grown on small farms of 1–3 ha. Only a few producers grow 50–500 ha.

Major cotton pests

Major pests of cotton in Peru are the cotton stainer, *Dysdercus peruvianus*, the cotton leaf perforator, *Bucculatrix thurberiella*, the Peruvian bollweevil, *Anthonomus vestitus* and the Indian pink bollworm, *Pectinophora gossypiella*. Other pests include mites, *Eriophyes gossypii*, armyworms, the cotton plant crown weevil, *Euthinobotrus gossypii*, the cotton aphid, *Aphis gossypii*, and other insects that feed on leaves, squares, flowers and bolls (*Anomis texana, Alabama argillacea, Mescinia peruella, Pococera atramentalis, Heliothis virescens*).

Control methods

Pest management in cotton is primarily applied against insects, since disease and weed problems are minimal. Control methods used in the integrated management of cotton pests vary, but emphasis is placed on cultural, legal and biological control. Management strategies for the cotton pests in Peru had been reviewed by Herrera (1998) and González (2000) (Table 23.2).

Cotton IPM is based on the knowledge of plant phenology, use of biological control agents, correct timing, planting deadlines and managing of the irrigation. This program has allowed cotton to be grown without the use of pesticides (i.e. organic production). IPM for organic cotton production also uses measures such as a fallow period, use of local varieties, certified seed, and adequate use of irrigation and fertilization (Morán *et al.*, 1999). Other treatments include the use of Bt, rotenone, natural oils, pheromones, copper sulfate, sulfur, and releases of *Trichogramma*.

Scouting is essential for decision making and for using preventive measures. In cotton agroecosystems, parasitoids, predators and entomopathogens are important pest mortality factors. It would be impossible to cultivate cotton without the regulating action of beneficial insects (Herrera, 1998). The implementation of an IPM program in cotton including the release of *Trichogramma* spp. and use of pheromone traps for the Indian pink bollworm reduced by 70% the use of pesticides on more than 500 ha of small farms in the Ica valley (Castro *et al.*, 1997).

In recent years, the populations of the silver leaf whitefly, *Bemisia argentifolii*, has increased due to climatic changes linked to the El Niño phenomenon. Fortunately, the whitefly population was reduced by an epizootic due to two fungi, *Paecilomyces fumosoroseus* and *P. farinosus*.

In organic cotton in the Cañete Valley, the use of good cultural practices together with sprays of Bt, rotenone, oils, pheromones, sulfur and release of *Trichogramma* wasps reduced cost of production by 50% (Van Elzakker, 1999).

Integrated pest management of potato pests

Geographical distribution of the potato crop

The potato was first domesticated near Lake Titicaca (between Peru and Bolivia). It is a staple food for about 8 million Peruvians and a source of income for farmers. The potato is grown from sea level to altitudes higher than 4200 m. The potato growing area varies from 200,000 to 320,000 ha. About 80% of this area is located in the higher sierra (above 3000 m), 15% at medium altitude (500–3000 m) and 5% at the coast (0–500 m).

HIGH ALTITUDES Small-scale farmers at higher altitudes have the lowest yields (3–4 t/ha). The production technology is largely traditional, but some farmers are beginning to use modern techniques. At high altitude, potato production occurs during the rainy season (September to June). The major pest is a complex of Andean potato weevil species (*Premnotrypes latithorax, P. suturicallus* and *P. vorax*).

Table 23.2. Components of the Peruvian model of IPM for cotton pests (adapted from González, 2000).

Pest		
Common name	Scientific name	Control measure
Armyworms	*Agrotis ypsilon* Roll *Prodenia eridania* Cramer *Prodenia ochrea* Hampson *Spodoptera frugiperda* Sm	Poisoned baits Heavy irrigation Minimum tillage Light traps
Whiteflies	*Bemisia tabaci* *B. argentifolii*	Entomopathogens Irrigation management Oil and rotenone Not planting near infested fields Organic fertilizers Not planting hybrid cotton
Cotton crown weevil	*Eutinobothrus gossypii* Pier.	Avoiding ratoon cotton (SENASA supervision) Domestic quarantine Light traps
Aphids	*Aphis gossypii* Glov.	Protection of natural enemies: predators and parasitoids
Peruvian weevil	*Anthonomus vestitus* Bohm.	Avoiding ratoon cotton Deadline for crop residue destruction Avoiding excessive foliage Destruction of pest-hosting weeds Topping (Piura); goat feeding (Pisco) Deadlines for planting Picking infested squares and placing them in cages to recover parasitoids
Leafworms	*Anomis texana* Riley *Alabama argillacea* (Hub)	Use *Bacillus thuringiensis* Protect natural enemies: predators Release *Trichogramma* spp. Light traps
Small bollworm	*Mescinia peruella* Schauss	Protect natural enemies Picking of dried flowers Light traps
Boll-end worm	*Pococera atramentalis*	Protect natural enemies Picking of dried flowers Avoid maize fruiting at the time of cotton fruiting Light traps
Bollworm	*Heliothis virescens* (Fab.)	Protect natural enemies Apply *Bacillus thuringiensis* in terminals Release *Trichogramma* spp. Irrigation management Light traps
Pink bollworm	*Pectinophora gossypiella* S.	Release of *Trichogramma bactrae* Light traps

continued

Table 23.2. *Continued.*

Pest		
Common name	Scientific name	Control measure
Cotton leaf perforator	*Bucculatrix thurberiella* B.	Protect natural enemies Organic fertilization
Cotton stainer	*Dysdercus peruvianus* Guer (remaining populations)	Frequent hand picking Destroy host plants Comply with 'clean field' regulations Poisonous baits with crushed cotton seed Light traps
	(migratory populations)	Destroy focal infestation in the upper valley Comply with 'clean field' regulations Destroy host plants (guava, loquat, aubergine, tomato, etc.)

MEDIUM ALTITUDES (INTER-ANDEAN VALLEYS) In the inter-Andean valleys, potato yields are highest (50–60 t/ha). Irrigated potato production occurs from July to February. The potato tuber moth complex (*Symmetrischema tangolias* and *Phthorimaea operculella*) is the predominant pest.

LOW ALTITUDES (THE COAST) Yields on the coast average 25 t/ha. Potato production is irrigated and uses modern technology, including high chemical inputs (Ewell *et al.*, 1990; Egúsquiza, 2000). The most important pest is the leafminer fly, *Liriomyza huidobrensis*. At the coast and at the mountains, the most important disease is the late blight, *Phytophthora infestans*.

Potato pests

ANDEAN POTATO WEEVIL The Andean potato weevil is endemic to the high areas of the Andean region (Peru, Bolivia, Ecuador, Colombia and Venezuela). The larva is the most damaging stage of this pest. When infestations are high, losses of more than 50% have been reported (Raman, 1984; Alcázar and Cisneros, 1997). In areas of intensive potato production where insecticides such as carbofuran, parathion, aldicarb and methamidophos are used, damage can reach 20–30% (Alcázar and Cisneros, 1997).

In Peru, the Andean potato weevil has only one generation per year. Adult weevils feed on leaves. The female lays eggs on wheat, barley or other plant debris. Larvae tunnel inside the tubers and then pupate in the soil. The adult has two phases: a diapausing phase in the soil and an active phase on the crop. The diapausing phase lasts about 4 months. Adults start emerging from the soil after the first rains and live from 4 to 5 months.

POTATO TUBER MOTH The tuber moth, *Phthorimaea operculella*, is an important pest in warm areas of the world where potato is cultivated. In Peru, this pest occurs at a wide range of altitudes. During the last decade, populations of another tuber moth species, *Symmetrischema tangolias*, have increased significantly at altitudes between 2500 and 4000 m (Palacios and Cisneros, 1997). Both species damage tubers in the field and in storage. Tuber damage is around 30% in the field and above 50% in storage. The larvae also damage the stems and leaves. Damage caused by *P. operculella* has no significant effect on yield, but tunnels produced by *S. tangolias* in potato stems can reduce yield depending on the potato variety. The populations of both species can increase significantly under dry and warm conditions. Farmers use toxic chemicals against this pest such as parathion, aldrin, foxim, malathion, methamidophos, propoxur and deltamethrin (Ewell *et al.*, 1990; Palacios and Cisneros, 1997).

The duration of the potato tuber moth life cycle varies with environmental conditions. At high and medium altitudes, the life cycle takes 2–4 months, with three to five generations per year. At lower altitudes, the life cycle is shorter and six to ten generations may occur in a year.

THE LEAFMINER FLY The leafminer fly, *Liriomyza huidobrensis*, has become a pest particularly in the Cañete valley, where the crop is grown intensively outside of its native range. This pest damages the leaves by larvae feeding or female oviposition. Larvae feed on the parenchyma and make serpentine tunnels. Mined leaves dry out and photosynthesis and yields are affected. Yield loss due to this pest is around 30–40%. To control this pest, farmers in the Cañete valley typically make 8–13 insecticide applications. This intensive use of chemicals has caused the development of insecticide resistance to carbamates, organophosphates and pyrethroids.

SECONDARY PESTS Secondary pests including the budmidge, *Prodiplosis* sp., the white mite, *Poliphagotarsonemus latus*, and whiteflies have been observed in recent years in several crops, including potato (Mujica and Cisneros, 1997). The whitefly is a polyphagous pest, with four to five generations per year, that infests a great number of cultivated and ornamental plants and weeds, which favors the presence of the pest the whole year.

Integrated management

Potato IPM is based on the knowledge of biology and pest behavior, seasonal occurrence, spatial distribution and plant phenology. The principal strategies rely on cultural, behavioral, and biological control methods. These methods have to be applied preventively to avoid economic damage both in the field and storage, pest migration from the field to the storage area, multiplication of the pest in plant residues, volunteer potatoes and alternate hosts (Table 23.3). In the design of IPM, several IPM strategies are available for farmers according to their needs (Cisneros, 1995; Alcázar and Cisneros, 1997; Mujica and Cisneros, 1997; Palacios and Cisneros, 1997; Cisneros *et al.*, 2001).

IPM technology transfer to farmers: Pilot Units

The CIP has defined phases of development for IPM programs, from initial evaluation to application by farmers in the field (Cisneros *et al.*, 1995). IPM training in Pilot Units is designed to first identify farmer knowledge gaps in relation to pests and control methods, so that training is focused on filling these gaps and reinforcing prior knowledge (Ortiz *et al.*, 1997). The implementation of IPM in Pilot Units and its extension by CIP and collaborating NGOs had resulted in the training of 37,702 farmers covering 15,098 ha, which corresponds to about 6% of the potato growing area (Alcázar, Palacios and Ortiz, personal communication). IPM in Pilot Units has resulted in a significant reduction of key potato pest damage and a reduction in the use of insecticides (Cisneros *et al.*, 1998). Currently, the IPM strategies developed by CIP to manage potato pests are being expanded to all the country by various institutions. These IPM programs have been used as models in the Andean region (Bolivia, Ecuador, Colombia and Venezuela) and the Caribbean region (Dominican Republic).

Final Comments

The successful Peruvian cotton IPM program, begun in the 1950s, has now been extended to various other countries. Currently, IPM in export crops such as cotton, citrus, sugarcane, mango and asparagus has improved marginal profits for Peruvian producers. In crops for domestic consumption such as potato, IPM has improved the food supply for the Andean population. In addition, it has reduced the risk of pesticide exposure, pesticide residues in food and in the environment. Potato IPM has also socially impacted the resource-poor farmers on Peruvian mountains. Many of these mountain communities are now practicing IPM strategies adapted to local conditions.

Table 23.3. IPM strategies for key pests of potato in Peru.

Andean potato weevil	Potato tuber moth	Leaf miner fly
Population reduction in the field	Crop protection: planting – harvest	Population reduction in the field
Early planting	Good plowing	Good quality seed
Healthy seed	Planting timely	Yellow sticky traps
Weevil hand picking	Good coverage of seed	Appropriate irrigation
Destroy volunteer plants	High hilling	Increase natural enemies
Harvest timely	Pheromone traps	Use selective insecticides
	Frequent irrigation	Destroy harvest residues
	Use selective insecticides	
Interruption of weevil migration	Protection of harvested tubers	Interruption of fly migration
Plant barriers	Harvest timely	Avoid neighboring fly-host crops
Chemical barrier	Tuber sorting	
Perimeter trenches	Cover harvested tubers	
Bait traps	Destroy harvest residues	
Use sheets at harvest		
Store in diffuse light		
Reduction of wintering population	Protection of stored tubers	
Plowing soil where tubers piled up at harvest	Cleaning and disinfestations of stores	
Winter plowing of harvested field	Use baculovirus	
	Use repellent plants	
	Store in diffuse light	
	Check stored tubers periodically	

Several organizations, both public and private, participate in the development and transfer of IPM strategies in Peru. Communication and coordination between these groups is occasionally limited. The lack of farmer organizations also limits a rapid IPM implementation, leaving pesticides as a major management strategy. In Peru, IPM is a model that despite some social and economic constraints has evolved to offer several pest management alternatives.

References

Alcázar, J. and Cisneros, F. (1977) Integrated management for andean potato weevils in pilot units. *International Potato Center Program Report 1995–96*. Lima, Peru, pp. 169–176.

Arroyo, B.O. (1988) *Políticas y Estrategias de Generación Tecnologías para el Desarrollo Agrícola y Rural del Perú*. Instituto Nacional de Investigación Agraria, Lima, Peru, 15 pp.

Arroyo, B.O. (1989) *Generación y Transferencia de Tecnología Agropecuaria en América Latina*. Serie Técnica. Instituto Nacional de Investigación Agraria, Lima, Peru, 44 pp.

Beingolea, G.O. (1961) El valle de Palpa como ejemplo de las posibilidades de integrar medidas de control químico y biológico en las plagas de los árboles de cítricos. *Revista Peruana de Entomología* 4(1), 1–4.

Beingolea, G.O. (1993) Situación actual del control integrado de plagas en el Peru. Presentado en *la Reunión sobre Código Internacional de Conducta (FAO/ONU)*. Ministerio de Agricultura, Lima, Peru.

Beingolea, G.O. and Salazar, T.J. (1970) Experiencias en el control integrado de las plagas del olivo. *Revista Peruana de Entomología* 13(1), 45–63.

Beingolea, G.O., Salazar, T.J. and Murat, I. (1969) La rehabilitación de un huerto de cítricos como ejemplo de la factibilidad de aplicar sistemas de control integrado de las plagas de los cítricos en el Peru. *Revista Peruana de Entomología* 12(1), 3–45.

Castro, Z.J., Loayza, C.F., Castro, M.T., Meza, P.M., Peña, V.L. and Molinari, N.E. (1997) *Control*

Integrado de Plagas y Producción de Controladores Biológicos en el Valle de Ica y el Callejón de Huaylas. CEDEP/RAAA, Lima, Peru, 149 pp.

Cisneros, F. (1980) *Principios del Control de las Plagas Agrícolas*, 1st edn. Centro Internacional de la Papa, Lima, Peru, 189 pp.

Cisneros, F. (1995) *Control de Plagas Agrícolas*, 2nd edn. Centro Internacional de la Papa, Lima, Peru, 313 pp.

Cisneros, F., Alcazar, J., Palacios, M. and Ortiz, O. (1995) Una estrategia para el desarrollo e implementación del manejo integrado de plagas. *CIP Circular* 21, 1–7.

Cisneros, F., Alcázar, J., Palacios, M. and Mujica, N. (1998) *Implementación de Programas de Manejo Integrado de Plagas del Cultivo Papa en Áreas Específicas de la Región Andina.* Informe Técnico Final. Proyecto Especial Banco Interamericano de Desarrollo – Centro Internacional de la Papa, Lima, Peru, 88 pp.

Cisneros, F., Alcázar, J., Palacios, M. and Mujica, N. (2001) Implementación de programas de manejo integrado del Gorgojo de los Andes, Polilla Guatemalteca y Mosca Minadora. In: *Taller Manejo Integrado de Plagas (MIP) en el Cultivo de la Papa.* Centro Internacional de la Papa, Lima, Peru.

Egúsquiza, B.R. (2000) *La Papa Producción, Transformación y Comercialización.* Universidad Nacional Agraria La Molina/ADEX – USAID, Lima, Peru, 192 pp.

Ewell, P., Fano, H., Raman, K.V., Alcazar, J., Palacios, M. and Carhuamaca, J. (1990) *Farmer Management of Potato Insect Pest in Peru.* International Potato Center, Lima, Peru, 77 pp.

González, B.J. (2000) *Manual de Evaluación y Control de Insectos y Ácaros del Algodonero.* Boletín Técnico No. 1, 3rd edn. FUNDEAL, Lima, Peru, 80 pp.

Herrera, A.J. (1961) Problemas entomológicos en el cultivo de los algodones Tangüis y Pima en el Perú. Medidas de control y su organización. *Revista Peruana de Entomología* 4(1), 58–66.

Herrera, A.J. (1998) Control biológico de plagas en el algodonero. In: Lizárraga, A., Barreto, U. and Hollands, J. (eds) *Nuevos Aportes del Control Biológico en la Agricultura Sostenible.* RAAA, Lima, Peru, 397 pp.

Lizárraga, T.A., Reyes, M., Ayala, L., Murrugurra, A., Montoro, I. and Hollands, J. (1995) Situación de los bioplaguicidas en el Perú. In: Sabillóu, A. and Bustamante, M. (eds) *Primer Taller Latinoamericano sobre Bioplaguicidas.* El Zamorano, Honduras, 34 pp.

Morán, C., Ugás, R., Lizárraga, T.A. and Gomero, L. (1999) Organic cotton in the Cañete Valley of the Peruvian coast. In: Mayers, D. and Stolton, S. (eds) *Organic cotton 'From Field to Final Product'.* Intermediate Technology Publications, London, 267 pp.

Mujica, N. and Cisneros, F. (1997) Developing IPM components for leafminer fly in the Cañete Valley. *International Potato Center, Program Report 1995–96.* CIP Lima, Peru, pp. 177–183.

Ortiz, O., Alcázar, J. and Palacios, M. (1997) La enseñanza del manejo integrado de plagas en el cultivo de la papa: la experiencia del CIP en la zona Andina del Perú. *Revista Latinoamericana de la Papa* 9/10, 1–22.

Palacios, M. and Cisneros, F. (1997) Integrated management for the potato tuber moth in pilot units in the Andean region and the Dominican Republic. *International Potato Center, Program Report 1995–96.* CIP, Lima, Peru, pp. 162–168.

Peru Acorde (2000) *Estudio Económico Productivo del Peru*, 2nd edn. Perú Acorde, Lima, Peru, 104 pp.

Pollack, M. (1994) *Manual de Plagas de la Caña de Azúcar.* Red de Acción en Alternativas al uso de Agroquímicos, Lima, Peru, 79 pp.

Raman, K.V. and Palacios, M. (1983) Control of major potato insect pests in Peru. *Annals of Plant Resistance Newsletter* 9, 69–72.

Raman, K.V. (1984) Evaluation of a synthetic pheromone funnel trap for potato tuberworm moths (Lepidoptera: Gelechiidae). *Environmental Entomology* 13, 61–64.

Raman, K.V., Booth, R.H. and Palacios, M. (1987) Control of potato tuber moth *Phthorimaea operculella* (Zeller) in rustic potato stores. *Tropical Science* 27, 175–194.

Raman, K.V. (1988a) Control of potato tuber moth with sex pheromones in Peru. *Agricultural Ecosystems Environment* 21, 85–99.

Raman, K.V. (1988b) Integrated insect pest management for potatoes in developing countries. *CIP Circular* 16, 1–8.

Risco, S.H. (1954) *La Mosca Indígena* Paratheresia claripalpis *W. en el Control Biológico de* Diatraea saccharalis *Fabr. en el Perú.* Sociedad Nacional Agraria-Comité de Productores de Azúcar. Lima, Peru, 5 pp.

Risco, S.H. (1960) La situación actual de los barrenadores de la caña de azúcar del género *Diatraea* y otros taladradores en el Perú, Panamá y Ecuador. *Revista Peruana de Entomología* 3(1), 6–10.

Van Elzakker, B. (1999) Comparing the costs of organic and conventional cotton. In: Mayers,

D. and Stolton, S. (ed.) *Organic cotton 'From Field to Final Product'*. Intermediate Technology Publications, London, pp. 86–110.
Vilchez, B.J.P. (2000) *Investigación y Transferencia de Tecnología en el INIA y Estrategias para su Cambio.* INIA, Lima, Peru, 22 pp.
Wille, J.E. (1952) *Entomología Agrícola del Perú*, 2nd edn. Estación Experimental Agrícola La Molina, Lima, Peru, 543 pp.

Chapter 24
Integrated Pest Management in Argentina

Dora Carmona[1], Marcelo Huarte[1], Graciela Arias[2], Alicia López[1], Ana M. Vincini[1], Héctor A. Alvarez Castillo[1], Pablo Manetti[1], Silvia Capezio[1], Eliseo Chávez[1], Mónica Torres[1], Juan Eyherabide[1], Jorge Mantecón[1], Liliana Cichón[3] and Darío Fernández[3]

[1]FCA, UNMdP-INTA Balcarce, Buenos Aires, Argentina; [2]INTA Saenz Peña, Chaco, Argentina; [3]INTA Alto Valle, Río Negro, Argentina

Country Profile

History of IPM in Argentina

Since the mid-20th century, Argentina has followed worldwide agroindustrial trends by using pesticides for pest control. The 'Green Revolution' with the use of high input techniques has been widely adopted in Argentina. New effective broad-spectrum pesticides were often applied upon the first detection of a pest, rather than on a pest density/economic threshold basis. As often results from this practice, populations of beneficial organisms were eliminated. This situation led to the resurgence of pest outbreaks, and increased the frequency of agrochemical treatments. Gradually, pest managers realized that pest outbreaks are an ecological problem, and solutions to pest problems should be sought in developing pest management techniques. Therefore, IPM was initiated as an attempt to reduce the sole reliance on pesticides and to make better use of natural resources.

In Argentina, farmers were introduced to IPM strategies in the 1970s. Government institutions like INTA and national universities started developing research projects in alternative pest control strategies. A Latin American meeting on IPM organized by INTA and sponsored by FAO was held in 1978. This meeting was very important for exchanging information and launching the Argentinean National IPM projects. In the past few years, Argentina has experienced major changes in many areas including cropping systems, fertilization, plant protection, biotechnology, machinery use and irrigation.

Pesticide regulations

In 1995, a new pesticide registration system by IASCAV (Argentinean Institute of Animal and Vegetal Health and Quality), resolution (17/95), was established in Argentina. This required toxicological, environmental and food security testing certificates for pesticide registration. In 1997, a pilot program on 'Safe pesticide use and disposal of pesticide containers' was developed by CASAFE and carried out in the orchard production area of Alto Valle Rio Negro.

Current IPM education in Argentina

Extension and technology transfer focus on the rational use of pesticides (proper rate and timing). The adoption of IPM strategies is demonstrated and recommended to growers. IPM programs are currently adopted in different areas of the country, the cotton IPM program being the oldest program in Argentina. In addition, IPM programs in soybeans, potatoes, and orchard crops, among others, have been developed and researchers are continuously working on their improvement.

IPM Case Studies

Cotton IPM in Argentina

National Research Organizations involved in cotton

INTA: Agricultural Experimental Stations of Saenz Peña, Reconquista, Las Breñas, Santiago del Estero.

Cotton production in Argentina

The cotton-growing area of Argentina is large, comprising most of the northeastern area of the country including the provinces of Chaco, Formosa, Corrientes and Santa Fé. The IPM Cotton Program was introduced and implemented in Argentina in the 1970s by Jorge Barral's group at the INTA Agricultural Experimental Station of Saenz Peña, in the province of Chaco (Barral and Zago, 1983). This program was developed using both local experience and foreign expertise. As time progressed, this program evolved to accommodate new production methods and new generations of pesticides. The IPM program in cotton consists of conservation of natural enemies, management of insecticide resistance, cultural techniques and the use of insecticides.

The use of chemical seed treatment or granulated insecticides applied to the soil at seeding time is a recommended practice. Cotton production occurs in dry land and under irrigation. The cotton plant undergoes three defined stages of growth, which are affected by different pests (Table 24.1).

Strategies used in cotton IPM

CONSERVATION OF BENEFICIAL INSECTS An essential component of pest management in cotton is the preservation of beneficial insects. The conservation of natural enemy populations is important as they regulate pest populations and reduce the number of pesticide applications. Recommended control strategies prevent pests from reaching damage thresholds and allow the establishment of beneficial insects in the field.

In the first stage of the crop, insecticides are recommended for use only if the pest population has reached the damage thresholds and the existing beneficial insect fauna is too low to regulate the pest population. Farmers are aware of the need for monitoring the field to determine the density of the insect populations, and to make control decisions. Although farmers understand that this practice is important, it is not widely used.

PREVENTION OF INSECTICIDE RESISTANCE Insecticide rotation is recommended to delay insecticide resistance.

CULTURAL PRACTICES The adjustment of planting date and the elimination of mulching stubble after harvest has decreased the population of pink bollworm *Platiedra gossypiella* by 90%. In addition, elimination of weeds in the crop, which are alternative pest hosts, and the use of short-cycle cotton varieties are important pest management strategies.

Eradication of the cotton boll weevil

The cotton boll weevil, *Anthonomus grandis*, is a potential threat to Argentinian cotton production. Since 1994, the counties of Pilcomayo and Pilaga in the province of Formosa have been an eradication zone for the boll weevil. So far, it is considered successful. This area is the only part of the world where the cotton boll weevil was detected at its point of introduction and is

Table 24.1. Cotton insect pests in Argentina.

Cotton stages					
Initial From planting to the beginning of boll formation Duration: 50 to 60 days		Intermediate From the floral stage through the beginning of fiber lignification Duration: approx. 60 days		Final From boll maturation until harvest time	
Principal pests	Secondary pests	Principal pests	Secondary pests	Principal pests	Secondary pests
Cotton aphid (*Aphis gossypii*)[a]	Cutworms (*Agrostis ypsilon*)[b]	Leafworm (*Alabama argillacea*)	Flower thrips (*Caliothrips brasiliensis*)	Pink bollworm (*P. gossypiella*)[e]	Leafworm (*A. argillacea*)
Thrips (*Frankliniella paucispinosa*)[a]	Wireworms (*Pyrophrus* spp.)[b]	Stink horcias (*Horcias nobilellus*)	Stink bug (*Gargaphia torresi*)		Bollworm group (*H. gelotopoeon, H. virescens, S. frugiperda, S. latisfacia*)
	Two spotted mite (*Tetranychus telarius*)[a]	Bollworm group (*Heliothis virescens, Helicoverpa gelotopoeon, Spodoptera frugiperda, S. latisfacia*)	Tinctorial bug (*Dysdercus chaquensis*)		
	'Broca worm' (*Eutinobothrus brasiliensis*)[b]		Whitefly (*Bemisia tabaci*)[d]		
	Pink bollworm (*Platiedra gossypiella*)[c]				

[a]In humid conditions aphids appear; in dry seasons, thrips are prevalent. In some circumstances both pests are observed. [b]Occur depending on environmental conditions. [c]Presence is directly related to the cultural management of the previous season. [d]Occur in irrigated areas. [e]High populations do not occur at this stage, if planting dates and mulching stubble are well managed.

still confined to the same area. This introduction of cotton boll weevil in Argentina changed the management practices used from an IPM perspective to an eradication attempt. Currently, the presence of large populations of this pest in cropping areas of the neighboring countries close to the eradication zone is not controlled. This may pose a potential threat to cotton in Argentina.

Cotton IPM technology transfer

Early in the development of the IPM program, different extension activities for farmers and professional technicians were offered, such as insect identification and monitoring training, including classroom training and field days in experimental plots. However, few growers put the IPM knowledge into practice.

In 1995, a new technology transfer program was organized to teach IPM philosophy. This new program had the advantage of beginning while a serious problem was occurring in cotton: lack of control of the leafworm, *Alabama argillacea* by pyrethroids, and farmers suffering severe economic problems. Under these circumstances, the Cotton IPM Program reappeared. This new training program was more effective. The training was 80% practice and it was taught during the cotton crop season. As a result of this IPM training, farmers understood that adequate insecticide use at the proper timing and at the correct dose reduces costs of production and provides more efficient crop management. The best outcome of the new teaching strategy was the establishment of technically and methodologically competent professionals to act as new trainers. Today, this program extends to four provinces of cotton production in the northeastern region of the country (Chaco, Formosa, Corrientes and Santa Fé) although it is implemented at different levels. For IPM programs, professionals rely on extension materials (bulletins, videos, audio records, etc.) which are used for training in other regions.

Currently, 500 insect diagnosticians are registered in the province of Chaco. As new personnel are trained, other provinces are covered as well. Demonstration plots have also helped cotton growers to learn more about IPM techniques. These experimental plots are generally established in fields where new technologies for IPM have been applied. In addition, two types of training courses are offered: the first to clarify doubts about the methodologies, and the second to introduce IPM philosophy.

IPM in potatoes

National Research Organizations involved in potatoes

- INTA: Balcarce, Rama Caída, Córdoba
- Instituto de fisiología vegetal, Castelar Centro de Investigaciones en Ciencias Veterinarias.
- INGEBI (Institute of Genetic and Molecular Biology, Buenos Aires).
- National Universities: Mar del Plata, Córdoba, Cuyo (Mendoza), La Plata.
- Provincial Ministries: Buenos Aires.

Potato production and use in Argentina

The total potato production area in Argentina covers approximately 100,000 ha over different cropping seasons: winter (5%), spring (27%), summer (47%), and autumn (21%). Yield is highest for the summer crops (30 tons/ha) and lowest for the autumn crops (20 tons/ha). The primary production area for the summer crop is located in the southeast of the province of Buenos Aires. Because of the different cropping seasons, freshly harvested potatoes are available throughout the year.

A rotation scheme (about 6 years) with pastures, wheat and maize is used as a predecessor to the potato crop. Potatoes are irrigated and usually produced on large farms (110 ha). Often in these large potato fields, a number of rows are sown with maize that is later used to cover the potato clamps.

Until recently, the major potato market was fresh potatoes with an annual consumption of 50 kg per capita, but in the past few years, the processing industry has been rapidly developing. Cold storage is primarily used by the seed and processing industries.

Varieties and seed used in potato production in Argentina

The most common variety grown in Argentina is Spunta. Other varieties used include Kennebec, Araucana, Huincul, Pampeana, Frital and Ballenera. Argentina has developed its own seed industry and has an official breeding program. The main emphasis in seed breeding was originally placed on virus diseases, but other quality characteristics, such as physiological age and soil seed-borne diseases, have been considered in recent years.

The pre-basic seed comes from laboratories that perform *in vitro* multiplication. In certain cases, the pre-basic seed from the laboratories can be multiplied in the field only for a limited number of generations because the infection pressure of PVY is relatively high. There are specialized laboratories that test seed samples for viruses. The ELISA test is used on sprouted tubers previously treated with Rindite® to break dormancy. Only 10% of the seeds used are certified. However, farmers also send non-certified seed samples to the laboratories for virus testing. Therefore, all the seed planted has passed some post-control test as well as field inspections.

Major constraints to the production and utilization of potatoes

The Argentine Potato IPM Program has focused on major diseases and pests of the crop. A series of studies are carried out on PLRV, PVY, PVX, late blight, *Phytophthora infestans* (Mont.) De Bary, stem canker or black scurf, *Rhizoctonia solani* Khun, nematodes, aphids, grubs, weeds, and quality factors (dry matter and reducing sugars, texture of final products).

IPM strategies have been developed and communicated to potato growers from the potato research group of FCA (College of Agricultural Science)–INTA Balcarce (southeast of Buenos Aires Province).

Virus management

Seed varieties resistant to PLRV, PVX and PVY have been successfully developed, leading to a reduction in insecticide applications against aphids and also extending seed viability. Both diploid and tetraploid material has been bred (Huarte *et al.*, 1986, 1990a; Huarte, 1989; Bofu *et al.*, 1996). Rapid multiplication techniques have also facilitated the multiplication of susceptible cultivars, by increasing the quantities of clean initial stocks.

The green peach aphid, *Myzus persicae* (Sulzer), is the most efficient vector of PLRV. Virus dissemination occurs exclusively in the crop, during the first flight of females that feed on infected plants and spread the virus to healthy plants. Systemic insecticides are used to control the aphids, and infected plants are discarded. During the second summer flight of aphids, the females arrive in crops carrying a high percentage of virus infection. To determine the moment of the first green peach aphid arrival, two water yellow traps (Moëricke) are placed in the crop and checked every day. When the first specimen of *Myzus* is detected in the traps, the potato foliage is destroyed with a dessicant (during the 8–10 following days) to cut down the virus pathway from the leaves to the tubers, avoiding virus infection.

Late blight management

In Argentina, the late blight, *Phytophthora infestans* De Bary, is the most serious fungal disease in potato producing areas. Under favorable weather conditions, periods of moderate temperatures, high humidity, and rain, the disease can cause high economic losses. During the last decade, in the southeast of Buenos Aires Province, the disease has produced losses up to 50% of the yield in commercial fields. The most important effect of late blight is the loss of commercial quality of tubers (Mantecón, 1998a, 2000a,b). Infected tubers must not be stored, because the disease remains latent under low temperature and reactivates when the tubers are taken out of cold storage, and become a primary inoculum that will infect the new crop when planted.

Late blight management strategies consider the implementation of genetic, cultural and chemical measures. Although genetic sources of disease resistance exist, at this time most of the cultivars used in Argentina are susceptible. As a result, the available resistance level is not enough to reduce the use of fungicides significantly. Recent research shows that genetic knowledge of QTLs governing late blight resistance in a *Solanum chacoense* population will render excellent tools for marker assisted breeding (Micheletto *et al.*, 1999, 2000) in a background of absence of major gene effects.

USE OF FUNGICIDES IN IPM The Argentine IPM Program has developed strategies for the control of late blight including chemical control and the use of resistant cultivars. Pampeana INTA is the most important late blight resistant variety that has achieved high yields without any fungicide spray for late blight control, although one or two applications of mancozeb can increase its yield by up to 30% because Pampeana INTA is highly susceptible to early blight. For susceptible cultivars, better formulations and lower doses of mancozeb at slightly higher frequencies have rendered good results (Mantecón, 1993, 1998b). A 30–80% reduction in the use of fungicides is likely to be achieved as well as an increased productivity and food safety in existing potato growing areas (Mantecón, 2000b). Also, as the fungus has turned out to be resistant to metalaxyl, continuous monitoring of the fungus population is needed to address its control properly (Mantecón *et al.*, 1995).

The preventive application of fungicides throughout the cycle of the crop is a common practice. The weather conditions and the phenology of the crop determine the frequency of applications (Mantecón, 1996). The intervals between applications are from 7 to 14 days, for non-systemic (contact) or systemic fungicides, respectively, and tend to be shorter when weather conditions are favorable for the development of the disease and the crop is at the tuberization stage, because the intervals of applications are too large for disease control efficiency (Mantecón, 1998b, 2000a).

The use of systemic fungicides for late blight control is important during the first stages of the crop (up to 80–100 days). However, systemic fungicides can also be used in the final stages of the crop mainly for early blight control. To avoid the possible selection of resistant strains of the pathogen, the use of fungicides in a continuous and systematic way is not recommended. There is no technical reason to support the widely accepted concept that mixing non-systemic and protective fungicides in the same application increases the efficiency of late blight control and diminishes pathogen resistance. The rotation of non-systemic fungicides during the crop cycle does not have any effect on the selection of resistant strains of *P. infestans* (Mantecón, 1998a).

CULTURAL MANAGEMENT TECHNIQUES Late blight is a pathogen inhabiting the soil and fresh tissue in the crop. Therefore, cultural control by reduction of secondary sources of infection (infected plants, volunteer plants, and discarded tubers) is practiced. In addition, the height of the planting 'ridges' is an important factor to reduce the disease incidence in tubers, because it is an important barrier of zoospore transfer through the soil.

Stem canker or black scurf management

The fungus causing black scurf, *Rhizoctonia solani* Khun., is present in all soils due to its wide host range and survival in plant debris as sclerotia (dormant bodies) for long periods. In Argentina, this fungus is present in most potato areas and the damage is more severe in the regions where crop rotations are shorter. It causes considerable damage in emerging crops planted in cold and wet soils. Hence, black scurf is a disease frequently found in early-planted crops in the southeastern region of the province of Buenos Aires. The disease can be present at several stages of the crop causing irregular emergence, diminution of plant stand, stem and root canker, rolling and purple colored leaves, weak growth and consequent yield reductions. In seed production, the damage thresholds should not be greater than 5%, although the absence of the disease is preferred.

Rhizoctonia solani is pathogenic in different crops and wild plants, which makes control difficult using crop rotations. Only very long rotations with tolerant crops, like cereals, reduce the inoculum levels in the soil. The use of certified potato seed is a recommended practice to avoid the introduction of the pathogen in the crop. Other strategies, such as avoiding the use of very susceptible cultivars and deep planting in cold, wet, or easily flooded soils to obtain fast emergence of the crop, are common practices to reduce the incidence of the disease. Fungicides can be applied to the tuber seed before planting when using infected seed, or applied to the soil at planting when 'pathogen free' seed is used. Chemical control reduces the symptoms of the disease by 70% in plants and 55% in tubers (Mantecón and Manetti, 2000).

Nematode management

In Argentina, *Meloidogyne* spp. and *Nacobbus aberrans* are the major plant-parasitic nematodes found in potato producing areas. The following *Meloidogyne* species are found in different provinces: *M. hapla*, *M. incognita* and *M. chitwoodi* in Buenos Aires; *M. arenaria*, *M. hapla*, *M. incognita* and *M. javanica* in Mendoza and Catamarca; *M. incognita* and *M. javanica* in Tucumán; and *M. arenaria* and *M. incognita* in Río Negro (Vega and Galmarini, 1970; Costilla, 1973; González de Ojeda *et al.*, 1978; Chaves and Torres, 1993, 2001).

Nacobbus aberrans is found in a restricted area of Tucumán and Catamarca provinces (Costilla *et al.*, 1978; Doucet *et al.*, 1986) and on potato tubers in Buenos Aires,

Río Negro and Santa Fé Provinces (Chaves and Torres, 1993, 2001). *Pratylenchus scribneri* was first found parasitizing potato tubers in a restricted area of Río Negro (Chaves and Torres, 2001). The potato cyst nematode, *Globodera rostochiensis*, has been found in different areas of the high mountain in Jujuy associated with wild *Solanum* species, but not in commercial potato producing areas (Chaves, 1993).

TOLERANCE LEVELS FOR NEMATODE INFECTION Currently, there are no data on yield losses caused by *Meloidogyne* spp. and *N. aberrans* on potato crops. However, in potatoes used in the processing industry, where quality is very important, yield losses of 5–10% due to galls in tubers parasitized by *Meloidogyne* have been reported by farmers in the south-east of Buenos Aires Province. The presence of *Nacobbus aberrans* in tubers does not cause severe damage or a decrease in potato quality. However, both *Meloidogyne* spp. and *N. aberrans* render the commercialization of seed potatoes more difficult. In 1983, the INASE established nematode tolerance levels for seed potatoes produced and imported into the country.

EXPERIMENTS WITH CHEMICAL CONTROLS Chemical control trials were conducted from 1983 to 1987 in the seed potato production area of Malargüe, Mendoza, to obtain nematode-free seed potatoes. The systemic nematicides aldicarb, fenamiphos, carbofuran and ethoprop (ethoprophos), applied to the soil for *Meloidogyne* spp. control, in doses up to 6 kg a.i./ha at planting or at hilling or in both times, were not effective in reducing the tuber infestation or increasing yields. The efficiency of different nematicides to control *Meloidogyne* in the tubers was also studied. An immersion of whole tubers in a solution of 300 ml of fenamiphos 40% in 100 l of water for 5 min produced a significant decrease in the infestation of the seed tubers. Even though the treatments did not have any phytotoxic effect on whole tubers, in cut tubers the sprouting was affected. The soil fumigants D-D and fenamiphos caused a significant decrease in juvenile stages of *Meloidogyne* in the soil and increased tuber yields. Costilla and Basco (1984) conducted trials on chemical control of *N. aberrans* on potato tubers and determined that 100% of the nematodes in tubers were eliminated by applying ethoprop and fenamiphos, so they can be recommended for nematode control.

CULTURAL CONTROL BY ADJUSTING PLANTING DATES A relationship between planting dates and tubers' infestation by *Meloidogyne* has been reported. In the Malargüe seed producing area, the worst damage occurred when potatoes were planted at the highest soil temperature (December–January), but decreased in the previous and following months. Due to the high cost of nematicides, farmers are more interested in planting potatoes in fields free or with low densities of *Meloidogyne*. In areas where this is not possible, potato commercialization depends on chemical treatment.

PREPLANTING SURVEY OF NEMATODE INFESTATION LEVELS Official laboratories usually carry out soil analysis to estimate the density of juvenile stages of *Meloidogine* prior to planting, as a strategy to prevent tuber infestation. This technique has been tested and adopted by farmers in Buenos Aires and Mendoza (Chaves and Torres, 1993, 2001). Field data indicate that it is possible to use preventive strategies to avoid potato infestation by *Meloidogyne* spp. and *N. aberrans*. In some tuber seed production areas, there are public and private laboratories for nematode diagnosis. To prevent nematode dispersion, the seed is analyzed through standard methods to estimate the percentage of infested tubers. Control focusing on prevention is a very useful method to avoid unnecessary nematicide use and environmental contamination.

Although some resistant genotypes have been reported in other countries, little effort has been made towards developing resistant varieties in Argentina (Huarte *et al.*, 1990b).

Insect pest management

The shining green beetle *Maecolaspis bridarollii* and the blond beetle *Cyclocephala signaticollis* (Coleoptera: Chrysomelidae and Scarabaeidae) are the most important insect pests of potatoes in the southeast of Buenos Aires Province (Alvarez Castillo *et al.*, 1993). Both of these species are univoltine. After many years of unsuccessful attempts to control these pests with insecticides, studies on the population dynamics, management and cultivar preferences were investigated.

The shining green beetle feeds on the leaves of the potato crop during the day beginning in mid-December (López *et al.*, 1993). The blond beetles, which do not feed on potato leaves, appear in November (spring) and peak in mid-December (Carmona *et al.*, 1994; Mondino *et al.*, 1997). Larvae of both species (white grubs) reach the last and most aggressive instar in March, coincident with the maturity of tubers and harvest time. Since the larvae feed on the tubers, they cause economic loss and decrease the commercial quality and value of potatoes.

CULTURAL CONTROLS The first factor to consider when using cultural control is to choose an appropriate seed potato and adjust the planting time. The potato varieties Spunta, Kennebec and the clone B.86.525.1 are tolerant to white grubs (Manetti *et al.*, 1996). Late planting causes late harvest and longer exposure of tubers to white grubs. Choosing the optimum harvest time is an important strategy to avoid major tuber damage. On the other hand, delaying harvest time will result in longer exposure of the tubers to grubs.

SCOUTING IN POTATO IPM From October to January, screen and light traps are used to determine the beginning of adult activity of *M. bridarolli* and *C. signaticollis*, respectively. Soil sampling to estimate the egg and larval populations must also be done after planting, in December and January. Since birds feed on the white grubs, plowing the field during the day is recommended to increase the natural control. Predators of the larval stage have been found regulating larvae and adults of *C. signaticollis* populations: *Scotobius miliaris* Billb. (Coleoptera: Trogidae) and *Cardigenus laticollis* Scolter (Coleoptera: Tenebrionidae); endoparasitic flies (Diptera: Tachinidae); 'killer or hunter flies' (Diptera: Asilidae), ectoparasites, and big and small wasps (Hymenoptera: Scolidae and Tiphiidae respectively) (Lopez *et al.*, unpublished data).

INSECTICIDE USE Soil insecticides (chlorpyrifos, imidacloprid, teflutrina and fipronil) to control grubs are applied once before planting (Manetti *et al.*, 1994). Pyrethroids are applied to the foliage to decrease the number of adults of *Maecolaspis* and the number of larvae at harvest time. Crop rotation and cultivar preference have been important factors in designing better use of insecticides.

Annual weed management

In the southeast of Buenos Aires Province, soils have a high organic matter content (5–6.5%), are slightly acid (pH 6.5), and have a low phosphorus content. Annual weeds (as well as some perennials) grow actively in these soils and compete successfully with potatoes, reducing crop yields between 32% and 81%, depending on the year and species of weeds (Eyherabide, 1995a).

The main species of annual broadleaved weeds in the area are: *Amaranthus quitensis*, *Brassica campestris*, *Chenopodium album*, *Datura ferox*, *Galinsoga parviflora*, *Polygonum aviculare*, *Polygonum convolvulus*, *Portulaca oleracea*, *Raphanus sativus*, *Stellaria media*, *Tagetes minuta*, *Xanthium spinosum*; and grasses *Digitaria sanguinalis*, *Echinochloa crusgalli* and *Setaria* spp. Perennial weeds, such as *Cynodon dactylon*, *Solanum sisymbriifolium*, *Cyperus esculentus*, *Sorghum halepense* and *Convolvulus arvensis* are present in the area, but the control of these species requires special programs (Eyherabide, 1995a).

Since mechanization began (late 1940s) and up to the early 1980s, potato growers planted in furrows spaced 65 cm apart, placing tubers 5–6 cm deep and controlling annual weeds by mechanical cultivation alone. Farmers believed that because of the high organic matter content, these soils easily suffered compaction after several months without cultivation. The methods used to control weeds were: (i) hilling after planting; (ii) passing a light harrow at pre-emergence; (iii) one (sometimes two) post-emergence light harrows; (iv) slight hilling; followed by (v) final hilling. This method resulted in cultivating the planted area as many as five times, and damage of the leaves by harrowing. In addition, roots were cut by hilling, favoring the formation of clods and soil compaction, and destroying the soil structure.

When farmers began to introduce integrated potato harvesters, shallow planting and more distance between furrows became necessary to make mechanical harvesting more feasible. However, problems with control of annual weed through 'traditional' mechanical methods appeared.

MECHANICAL WEED CONTROL The weeds research team (Integrated Unit, FCA–INTA) together with the potato growers identified the main problems associated with weed control when adapting the new technology of planting potatoes. According to potato growers, the traditional method of weed control became impossible when using this method of planting. The main reason was that the tuber seeds were close to the soil surface and pulled out by harrows, so that the plant emerged sooner, leaving less time to perform pre-emergence weed control cultivation. Therefore, alternative mechanical methods for weed control were imported from Europe, but were not very well accepted by potato growers. Some of these methods were too aggressive, and weeds were not successfully controlled, because weed seeds were germinating in small and stable clods, especially when soil was wet.

HERBICIDE USE Researchers identified the herbicides used in other countries and adapted the use of those chemicals to the needs of the southeast of Buenos Aires Province. Studies were carried out to determine herbicide selectivity, efficacy, and proper application timing for each cultivar. The critical period of weed–crop competition and the competitive species were investigated (Eyherabide *et al.*, 1983a,b, 1984; Manetti and Eyherabide, 1989; Eyherabide *et al.*, 1989; Eyherabide, 1995a,b,c).

The following conclusions were drawn:

1. Incorporating herbicides into the soil at pre-planting was not the best time for spraying, because after hilling, weeds germinate from the low area between rows.
2. If pre-emergent herbicides were properly chosen and sprayed at pre-emergence of the crop, the crop could be maintained free of weeds without further mechanical labor.
3. The presence of soil clods and soil compaction was minimized when less post-planting cultivation was performed.
4. Cultivars had different tolerances to metribuzin sprayed at postemergence. This is the most widely used herbicide for broadleaf weed control.
5. The cultivar tolerance to herbicides affecting the passing of electrons from photosystem II to photosystem I could be determined by low-cost laboratory tests (Manetti and Eyherabide, 1989; Eyherabide, 1995c).

Almost 100% of potatoes are grown either applying herbicide after hilling or at pre-emergence and then hilling at post-emergence.

IPM in apple and pears

National Research Organizations involved in IPM

INTA Alto Valle Río Negro, Río Negro Province.

Orchard crop production in Argentina

The High Valley of the Río Negro, in the province of Río Negro, is the major area for pome fruit production in Argentina. This region is located in the northern part of the Patagonia region (39°01′S, 67°40′W, 200 m altitude). The mean annual temperature is 15°C, and the mean winter and summer temperatures are 5.5°C and 22.2°C, respectively. The annual precipitation is lower than 200 mm and is concentrated mainly in the winter. The production area covers approximately 60,000 ha under irrigation; 33,600 ha are dedicated to apple orchards and 16,500 ha to pear orchards. The total production of these two orchard species reaches 1,300,000 tons. Although less than 10% of the production is under IPM certification (integrated or organic fruit production), most of the farmers apply IPM technologies. Various microorganisms, mites and insects affect the production, and many natural enemies are present in the same system (Table 24.2).

Diseases

Because of the semiarid characteristics of the area, there is low or no incidence of fungal diseases. Apple and pear scab (*Venturia inaequalis* and *V. pirina*) does not require specific controls throughout the productive orchard season. On the other hand, oidium (*Podosphaera leucotricha*) is restricted to susceptible apple cultivars like Granny Smith, Braeburn, Fuji and Gala.

Insects

The key pest in apple and pear orchards is the codling moth, *Cydia pomonella* (Table 24.2). In some years, this species completes three generations, and when the weather is exceptionally warm, it may complete a partial fourth generation. One of the critical problems in *Cydia pomonella* control is the duration of the first flight, which can continue for almost 3 months. This results in an overlap of the first and second generations, and makes control more difficult. The risk of pest attack for late crops is about 160 days.

Other pests present in the area include *Proxenus rionegresnsis*, *Agrotis* sp., *Cydia molesta*, and *Archips* sp., which occur in small populations in orchards under conventional control. In orchards under organic production, or with the use of mating disruption techniques applied to the key pest, they may become more prevalent. The bagworm moth, *Oiketicus platensis* is an occasional pest, since it is usually a pest of the windbreak *Populus* sp. surrounding the fruit orchard. The larvae of this univoltine species appear in November and are transported by the wind to the fruit trees. They feed on vegetative parts (Cichón *et al.*, 1996). Although there are some predators and parasitoids that attack this group of pests, biological control options are limited for bagworm moths. The pear psylla, *Cacopsyla pyricola* is another important pest that has resurged in recent years after an extensive period of very low incidence. In orchards under organic production, an increase of *Caliroa* sp. populations in pear trees and *Edwardsiana crataegi* in apple trees has been observed.

Other pests such as scale insects including *Quadraspidiotus perniciosus*, *Lepidosaphes ulmi*, *Lecanium* sp. and *Pseudococcus* sp. occur with abundant populations of the parasitoid *Aphytis* sp. *Aphytis* attacks primarily *Quadraspidiotus perniciosus* and *Lepidosaphes ulmi*. Unfortunately, the action of natural enemies does not maintain the pest populations below the economic threshold.

The primary aphid pests are the black aphid, *Aphis gossypii* in pears and the woolly apple aphid, *Eriosoma lanigerum* in apples. Many predators and parasitoids help to control aphids, including *Aphidoletes aphidimyza*, syrphids, coccinellids and microhymenopteran species. Among microhymenopterans, *Aphelinus mali* provide excellent control of the woolly apple aphid.

The European red mite, *Panonychus ulmi* is the major phytophagous mite that attacks apple and pear trees. The mite complex competes with the common red mite, *Tetranychus urticae* and with other less important mites like the brown mite, *Bryobia rubrioculus* and the flat scarlet mite, *Cenopalpus pulcher*. The eriophyid mites, *Epitrimerus pyri* and *Eriophes pyri* in pear orchards, occur in cycles and can cause major damage if not scouted and controlled in time. Although the mite *Aculus schlechtendali* can be present in great populations in apple trees, it does not cause significant damage.

Table 24.2. Pests and natural enemies occurring in apple and pear orchards in the Black River High Valley, Argentina.

Key pest	Codling moth (*Cydia pomonella*)
Secondary pests	
Direct damage	Bagworm moth (*Oiketicus platensis*)
	San José scale (*Quadraspidiotus perniciosus*)
	Oystershell scale (*Lepidosaphes ulmi*)
	Pear rust mite (*Epitrimerus pyri*)
	Pear leaf blister mite (*Eryophyes pyri*)
	Pear psylla (*Cacopsyla pyricola*)
	Oriental fruit moth (*Cydia molesta*)
	Leafrollers (varied spp.)
	Pear slug (*Caliroa* sp.)
	Thrips (*Frankliniella ocidentalis*)
	Oidium (*Podosphaera leucotricha*)
Indirect damage	European red mite (*Panonychus ulmi*)
	Two-spotted mite (*Tetranychus urticae*)
	Fruit brown mite (*Bryobia rubrioculus*)
	Flat scarlet mite (*Cenopalpus pulcher*)
	Apple rust mite (*Aculus schlechtendali*)
	Woolly apple aphid (*Eriosoma lanigerum*)
	Pear root woolly aphid (*Eriosoma lanuginosum*)
	Black aphid (*Aphis gossypii*)
	Green peach aphid (*Myzus persicae*)
	Apple leafhopper (*Edwardsiana crataegi*)
	Pear blight (*Pseudomonas syringae*)
	Mealybug (*Pseudococus marítimus*)
	Brown peach scale (*Lecanium* sp.)
Natural enemies	
Predators	Convergent lady beetle (*Hippodamia convergens*)
	Lady beetle (*Eriopes connexa*)
	San José lady beetle (*Cycloneda sanguinea*)
	Two-spotted lady beetle (*Adalia bipunctata*)
	Lady beetle (*Coccidophilus* sp.)
	Spider mite destroyer (*Stehtorus punctum*)
	Minute pirate bug (*Orius insiduosus*)
	Damsel bug (*Nabis* sp.)
	Bigeyed bug (*Geocoris pallipis*)
	Hover flies (*Syrphidae*, other spp.)
	Gall midges (*Aphidoletes aphidimiza*)
	Lacewings (*Crysoperla* spp., *Hemerobius* spp., *Sinpherobius* spp. and *Micromus* spp.)
	Mites (*Neoseiulus californicus*)
	(*Agistemus mendozensis*)
	(*Zetzellia mali*)
	(*Pyemotes ventricissus*)
Parasitoids	Afelino (*Aphelinus mali*)
	Clear wasp (*Aphytis longiclave*)

IPM management strategies

MATING DISRUPTION OF CODLING MOTH The introduction of the mating disruption (sexual confusion) technique made possible the implementation of organic and integrated fruit production programs in the Río Negro region. Farmers began to use mating disruption on a 40-ha experimental plot in 1991. Currently, the area under the mating

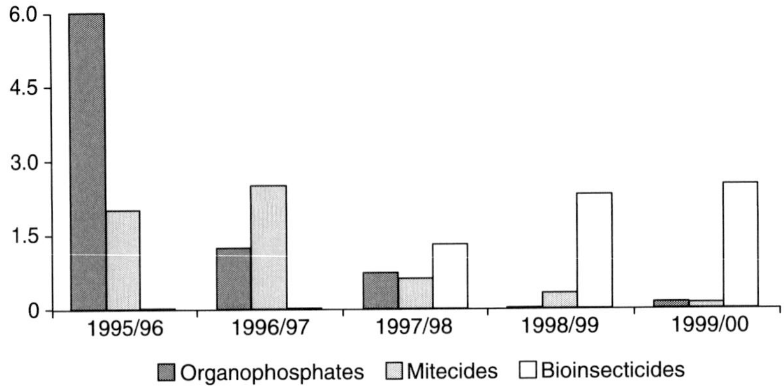

Fig. 24.1. Number of seasonal spraying of organophosphates for the codling moth, *Cydia pomonella*, miticides and biological insecticides, after 5 years of use of the confusion sexual technique in apples.

disruption technique is 6000 ha. Some of the obstacles to extensive adoption of this technique have been the high cost of materials and labor for application and scouting.

The outbreak of new pests (*Archips* sp.) or a change of status in other pests such as *C. molesta* and *E. crataegi*, is an important factor to monitor while implementing these programs. Nevertheless, the long-term advantage of the sexual confusion technique has been a substantial reduction in the number of applications of organophosphates and broad-spectrum insecticides (Fig. 24.1).

BIOLOGICAL CONTROL OF MITES The primary mite predators found in orchard crops include the flat mite, *Neoseiulus californicus*, the spheric acarus, *Mesoseiulus longipes*, the reticulated mite, *Zetzellia mali*, the spider mite destroyer, *Stethorus punctum* and the ladybird beetles, *Coccidophilus* sp. After a few years of use, selective control techniques such as mating disruption of codling moth together with biological control of phytophagous mites have significant reduced the number of pesticide applications (Fig. 24.1).

The use of IPM techniques including biopesticides, insect growth regulators, mating disruption techniques, and rational pesticide use has become the solution for pest control in Argentina.

References

Alvarez Castillo, H.A., López, A.N., Vincini, A.M., Carmona, D. and Manetti, P.L. (1993) Relevamiento de los insectos de suelo en el cultivo de papa del sudeste bonaerense. SAGyP (Secretaría de Agricultura, Ganadería y Pesca), CERBAS (Centro Regional Buenos Aires Sur), INTA (Instituto Nacional de Tecnología Agropecuaria), Balcarce. *Boletín técnico No. 118*, 18 pp.

Barral, J.M. and Zago, L.B. (1983) IPM Program of insects and mites of Cotton. *Extension Bulletin No. 71*. INTA (Instituto Nacional de Tecnología Agropecuaria), Saenz Peña Chaco, Argentina, 20 pp.

Bofu, S., Weiming, T., Jimin, W., Chunlin, W., Zhengui, Y., Shengwu, W. and Huarte, M. (1996) Economic Impact of CIP-24 in China. In: Walker, T.S. and Crissman, C.C. (eds) *Case Studies of the Economic Impact of CIP-Related Technologies*. International Potato Center, Lima, Perú, pp. 31–50.

Carmona, D., Vincini, A.M., López, A.N., Alvarez Castillo, H.A. and Manetti, P.L. (1994) Cambios estacionales en la comunidad de 'insectos del suelo' en el cultivo de papa en el sudeste bonaerense. SAGyP (Secretaría de Agricultura, Ganadería y Pesca), CERBAS (Centro Regional Buenos Aires Sur), INTA (Instituto Nacional de Tecnología Agropecuaria), Balcarce. *Boletín Técnico No. 126*, 15 pp.

Chaves, E. (1993) Revisión de los nematodes formadores de quistes de la República Argentina. CERBAS-INTA, *Boletín Técnico No. 113*, 1–11.

Chaves, E. and Torres, M. (1993) Nematodes parásitos de la papa del sudeste bonaerense. CERBAS-INTA, *Boletín Técnico No. 115*, 1–21.

Chaves, E. and Torres, M. (2001) Nematodos parásitos de la papa en regiones productoras de papa semilla en la Argentina. *Revista de la Facultad de Agronomía* 21, 245–259.

Cichón, L., Di Masi, S., Fernandez, D., Magdalena, J., Rial, E. and Rossini, M. (1996) *Guía Ilustrada para el Monitoreo de Plagas y Enfermedades en Frutales de Pepita*, 2nd edn. INTA EEA, Alto Valle, Argentina, 72 pp.

Costilla, M.A. (1973) Nematodes. Estación Experimental Agrícola de Tucumán, *Publicación Miscelánea* 50, 181–183.

Costilla, M.A. and Basco, H.J. (1984) Control químico del falso nematode del nudo *Nacobbus aberrans* (Thorne, 1935) Thorne and Allen, 1944, en tubérculos de papa. *Revista Industrial y Agrícola de Tucumán* 61, 39–45.

Costilla, M.A., González de Ojeda, S. and Hasselrot de Gómez, T. (1978) El falso nemátode del nudo, *Nacobbus aberrans* (Thorne, 1935) Thorne and Allen, 1944 (Nematoda: Nacobbidae) en cultivo de papa de Tucumán. *III Jornadas Fitosanitarias Argentinas.* Universidad Nacional de Tucumán, San Miguel de Tucumán, pp. 323–340.

Doucet, M.E., Gardenal, N., Costilla, M.A. and Vattuone, E. (1986) Caracterización de poblaciones del género *Nacobbus* (Nematoda: Tylenchidae) en la República Argentina. In: *VI Jornadas Fitosanitarias Argentinas. Tomo I, Mesa de Nematología.* Universidad Nacional del Comahue, Alto Valle del Río Negro, pp. 309–310.

Eyherabide, J.J. (1995a) *Problemática y Control de Malezas en Papa.* Unidad Integrada Balcarce (ed.), Balcarce, Argentina, 110 pp.

Eyherabide, J.J. (1995b) Evaluation of pre-emergence application of herbicides on weed control in potatoes. *Annals of Applied Biology* 126, 62–63.

Eyherabide, J.J. (1995c) Comparación entre el método de evaluación en el campo y el de flotación de discos de hojas para determinar la susceptibilidad de cultivares de papa (*Solanum tuberosum* L) al herbicida metribuzín. *Revista de la Facultad de Agronomía de La Plata* 71(1), 31–41.

Eyherabide, J.J., Bedmar, F. and Leaden, M.I. (1983a) Comportamiento de algunos herbicidas en el control de malezas en papas de variedad Huinkul y Spunta. *Malezas* 11(3), 185–195.

Eyherabide, J.J., Leaden, M.I., Bedmar, F. and Rohatscht, P. (1983b) Control de malezas y efectos en el rendimiento de algunos herbicidas aplicados en papas cvs. Huinkul y Spunta. *Malezas* 12(III), 21–28.

Eyherabide, J.J., Bedmar, F., Leaden, M.I. and Rohatsch, P. (1984) Evaluación de tratamientos con herbicidas para control de malezas en papa cvs Huinkul y Spunta. *Malezas* 6, 24–32.

Eyherabide, J.J., Manetti, P.L., Bedmar, F. and Leaden, M.I. (1989) Evaluation of herbicides applied at different times for weed control in potatoes and their phytotoxicity to the crop. *Annals of Applied Biology* 118, 84–85.

González de Ojeda, S., Costilla, M.A. and Hasselrot de Gómez, T. (1978) Nemátodes identificados en cultivos de papa de la provincia de Tucumán. *Revista Industrial y Agrícola de Tucumán* 55, 65–69.

Huarte, M.A. (1989) Situación actual de la producción de semilla de papa en la Argentina. In: Hidalgo, O. and Rincón, H. (eds) *Avances en la Producción de Tubérculo-semilla de Papa en los Países del Cono Sur, Memorias de las Reuniones.* Centro Internacional de la papa Canoinhas, Brasil, pp. 15–21.

Huarte, M.A., Mendiburu, A.O., Monti, M.C. and Butzonitch, I.P. (1986) Serrana INTA: a widely adapted virus resistant variety from Argentina. *American Potato Journal* 63, 695–699.

Huarte, M.A., Butzonitch, I.P. and Mendiburu, A.O. (1990a) Mejoramiento para resistencia a virus en el programa Argentino de papa. In: Hidalgo, O.A. and Rincón, H. (eds) *Avances en el Mejoramiento Genético de la Papa en los Países del Cono Sur.* CIP, Lima, Peru, pp. 85–94.

Huarte, M.A., Capezio, S. and Chaves, E. (1990b) Resistencia genética a *Meloidogyne* spp en papa. In: Hidalgo, O.A. and Rincón, H. (eds) *Avances en el Mejoramiento Genético de la Papa en los Países del Cono Sur.* CIP, Lima, Peru, pp. 191–194.

López, A.N., Alvarez Castillo, H.A., Carmona, D., Manetti, P.L. and Vincini, A.M. (1993) Aspectos morfológicos y biológicos de *Cyclocephala signaticollis* Burm. (Coleoptera: Scarabaeidae). *Boletín Técnico No. 123*, 18 pp.

Manetti, P. and Eyherabide, J.J. (1989) Ajuste de un método rápido para detección de susceptibilidad de cultivares de papa (*Solanum tuberosum* spp. *tuberosum* al herbicida metribuzin. *Malezas* 16(1), 85–89.

Manetti, P.L., Alvarez Castillo, H.A., Carmona, D. and López, A.N. (1994) Effects of chemical control of soil insects on potato tuber quality. *Annals of Applied Biology* 124, 12–13.

Manetti, P.L., Alvarez Castillo, H.A., Carmona, D., López, A.N., Vincini, A.M. and Huarte, M. (1996) Response of potato cultivars to natural infection of soil insects. Test of agrochemicals and cultivars. *Annals of Applied Biology* 128 (Suppl.), 82–83.

Mantecón, J.D. (1993) Performance of sticker addition to Dithane DG (mancozeb) for potato late blight control. *Fungicide and Nematicide Test* 49, 129.

Mantecón, J.D. (1996) Effectiveness of several fungicide foliar sprays on potato late blight. *Fungicide and Nematicide Test* 52, 155.

Mantecón, J.D. (1998a) Potential yield losses caused by late blight in Argentina during the last decade. *Fungicide and Nematicide Test* 53, 202–204.

Mantecón, J.D. (1998b) Minimum dose of mancozeb that offer effective control of potato late blight in Argentinean conditions. *Fungicide and Nematicide Test* 53, 200–201.

Mantecón, J.D. (2000a) Influence of acylalanine fungicide application interval on potato late blight control. *Fungicide and Nematicide Test* 55, 221.

Mantecón, J.D. (2000b) Management of potato late blight with several mancozeb formulations. *Fungicide and Nematicide Test* 55, 217–218.

Mantecón, J.D. and Manetti, P. (2000) Chemical strategies for controlling black scurf and soil insects and their influence on yield and quality of potato tubers. *Fungicide and Nematicide Test* 55, 219–220.

Mantecón. J., Van Damme, M., Zamudio, N., Lobo, R. and Huarte, M. (1995) Fungicide resistance in *P. infestans* from Argentina. *Resistant Pest Management* 8 (2).

Micheletto, S., Andreoni, M. and Huarte, M. (1999) Vertical resistance to late blight in wild potato species from Argentina. *Euphytica* 110(2), 133–138.

Micheletto, S., Boland, R. and Huarte, M. (2000) Argentine wild diploid *Solanum* species as sources of quantitative late blight resistance. *Theoretical and Applied Genetics* 101, 902–906.

Mondino, E.A., López, A.N., Alvarez Castillo, H.A. and Carmona, D. (1997) Ciclo de vida de *Cyclocephala signaticollis* Burm. (Coleoptera: Scarabaeidae) y su relación con algunos factores ambientales. *ELYTRON* 11, 145–156.

Vega, E. and Galmarini, H.R. (1970) Reconocimiento de los nematodes que parasitan los cultivos hortícolas de los Departamentos de San Carlos y Tunuyán, Mendoza, Argentina. *Revista de Investigación y Desarrollo Agropecuario* 272, 17–41.

Chapter 25

Integrated Pest Management in Greenhouses: Experiences in European Countries[1]

Joop C. van Lenteren

Laboratory of Entomology, Wageningen University, Wageningen, The Netherlands

Introduction

Biological control and IPM are reliable crop protection methods and are economically profitable endeavors for growers of greenhouse crops. The fast evaluation and introduction of a number of natural enemies in situations where chemical control was either insufficient, impossible or undesired, has taught growers and crop protection specialists that biological control, within IPM programs, is a powerful option in pest control (van Lenteren, 1995; Albajes *et al.*, 1999).

The total world area covered by greenhouses is about 300,000 ha, 50,000 ha of which are covered with glass, and 250,000 ha with plastic (e.g. Albajes *et al.*, 1999; Parrella *et al.*, 1999). Vegetables are produced on 195,000 ha and ornamentals on 105,000 ha. Developments in biological control in this cropping system have been unexpectedly fast and illustrate the great potential of alternatives to chemical methods. Greenhouses offer an excellent opportunity to grow high quality products in large quantities on a small surface area. For example, in The Netherlands only 0.5% of the area in use for agriculture is covered with glasshouses. On this small area of 10,000 ha, about 20% of the total value of agricultural production is realized.

Few specialists in biological control anticipated being able to employ natural enemies in greenhouses, because growing vegetables and ornamentals in this protected situation is very expensive and pest damage is not tolerated. This means that the usually well-trained, intelligent greenhouse growers will not run the risk of any damage from insects, just because of ideological reasons such as reduced environmental side effects compared with chemical control. If chemical control works better, they will certainly use it. In tomatoes, for example, pest control represents less than 2% of the total overall cost of production, so costs are not a limiting factor for chemical control (van Lenteren, 1995). The same situation occurs in ornamentals where the cost of pest control using chemicals (including material and application) is usually less then 1% of the overall cost of producing the crop (Parrella *et al.*, 1999). Yet despite the serious constraint that chemical control is comparatively simple and inexpensive, adoption of biological control has been remarkably quick in greenhouses first in northwestern Europe (van Lenteren and Woets, 1988), and later in other greenhouse areas (Parrella *et al.*, 1999). The growers now clearly see the specific advantages of biological control in greenhouses (see below for special section on this

[1] Reprinted from publication *Crop Protection*, Vol. 19, Joop van Lenteren (2000) A greenhouse without pesticides: fact or fantasy? pp. 375–384. Copyright (2003), with permission from Elsevier Science.

topic). Examples of commercially applied programs for biological and integrated control of pests and diseases in vegetables and ornamentals can be found in van Lenteren (1995) and Albajes *et al.* (1999). The main reason for use of biological control methods in the 1960s was the occurrence of resistance to pesticides in several key pests in greenhouses. Nowadays, other important stimuli include demands by policy makers for a reduction in usage of pesticides, and consumers requiring production of residue-free food and flowers.

In this chapter, the current situation concerning pest and disease control in greenhouses is summarized and new research leading to pesticide-free production of greenhouse crops is described.

The Greenhouse Environment

In temperate zones, differences between greenhouse and field environments may partly explain the success of biological control in greenhouses. Greenhouses are relatively isolated units, particularly during the cold season. Before the start of a cropping period, usually during the winter, the greenhouse can be cleansed of pest organisms and subsequently kept pest free for several months. Later in the season, isolation prevents massive immigration of pest organisms. Furthermore, a limited number of pest species occur in greenhouses, partly because of isolation and partly because not all pests specific to a certain crop have been imported into countries with greenhouses. Many greenhouse pests cannot survive in the field in winter or develop very slowly. This makes biological control easier because the natural enemies of only a few pest species have to be introduced. In addition, cultivars resistant to a number of diseases (viruses and fungi) had been developed already for the most important vegetable crops. As a result, chemical control of fungi – which may lead to high mortalities of natural enemies used for pest control – does not have to be applied frequently. During the past 20 years growing crops on inert media instead of in the soil, has considerably decreased soil diseases and nematode problems.

Another factor easing implementation of biological control in protected crops is that cultural measures and pest management programs can be organized for each separate greenhouse unit. Interference with pest management in neighboring greenhouses is limited. The influence of pesticide drift on natural enemies, which is a common problem in field crops, does not play a very important role here.

On the other hand, pest control is complicated by the virtually year-round culture of crops and by continuous heating during cold periods. These conditions provide excellent opportunities for the survival and development of a pest or disease once it has invaded the greenhouse. Some organisms that normally show diapause when they occur in the field, e.g. spider mites, have adapted to the greenhouse climate by no longer reacting to diapause-inducing factors (W. Helle, personal communication). As a result, rates of population growth are often much higher than in the field. These complications do not, however, create specific problems for biological control. The greenhouse climate is managed within certain ranges, and this makes prediction of the population development of pest and natural enemies easier and more reliable than in field situations (van Roermund *et al.*, 1997). The time of introduction of natural enemies, the number and spacing of releases can be fine-tuned, resulting in season-long economic control.

In warm climates, the situation is more complicated. In the Mediterranean for example, greenhouse frames are often constructed of wood, which harbors pests and diseases and is very difficult to clean. Growers in the sub-tropics and tropics are often less specialized than those of temperate zones, grow a diversity of crops on one holding, while at the same time some of the crops may also be present in the field. Most crops are grown in the soil, which can lead to nematode and fungal problems. Often, little attention is paid to farm hygiene. Climate control is limited to opening and closing of the greenhouse, the use of shade screens or whitewashing of the plastic. The mild climate outside enables pests to develop year-round and pest pressure is, therefore, very high. Ventilation leads to continuous migration of organisms in and out of the greenhouse. The pest and disease spectrum is much broader here (Albajes *et al.*, 1999). On the other hand, numerous natural enemies which occur in the field can invade the greenhouse and exert natural control free of charge (van Lenteren *et al.*, 1992).

Integrated Pest and Disease Management in Greenhouses

IPM is used on a large scale in all main vegetable crops. In The Netherlands for example, more than 90% of all tomatoes, cucumbers and sweet peppers are produced under IPM. Worldwide 5% of the greenhouse area is under IPM, and there is potential for increase to about 20% of the area in the coming 10 years.

A good example of an IPM program is the one for tomato used in Europe. It involves ten natural enemies and several other control methods like host-plant resistance, climate control and cultural control (Table 25.1). When tomatoes are grown in soil, soil sterilization by steaming is used shortly before planting the main crop to eliminate soilborne diseases such as Tomato Mosaic Virus (TMV), *Fusarium*, *Verticillium* and pests such as *Lacanobia oleracea* (tomato moth) and three *Liriomyza* spp. (leafminers). Previously, cultivars lacking TMV resistance were inoculated as young plants with a mild strain of the TMV virus to make them less susceptible. Now,

Table 25.1. Integrated Pest and Disease Mangement program as applied in tomato in Europe.

Pests and diseases	Method used to prevent or control pest/disease
Pests	
Whiteflies (*Bemisia tabaci, Trialeurodes vaporariorum*)	parasitoids: *Encarsia, Eretmocerus* predators: *Macrolophus* pathogens: *Verticillium, Paecilomyces, Aschersonia*
Spider mite (*Tetranychus urticae*)	predator: *Phytoseiulus*
Leafminers (*Liriomyza bryoniae, L. trifolii* & *L. huidobrensis*)	parasitoids: *Dacnusa, Diglyphus* and *Opius* and natural control
Lepidoptera (e.g. *Chrysodeixis chalcites, Lacanobia oleracea, Spodoptera littoralis*)	parasitoids: *Trichogramma* pathogens: *Bacillus thuringiensis*
Aphids (e.g. *Myzus persicae, Aphis gossypii, Macrosiphum euphorbiae*)	parasitoids: *Aphidius, Aphelinus* predators: *Aphidoletes* and natural control
Nematodes (e.g. *Meloidogyne* spp.)	resistant and tolerant cultivars, soil-less culture
Diseases	
Gray mold (*Botrytis cinerea*)	climate management, mechanical control and selective fungicides
Leaf mold (*Fulvia = Cladosporium*)	resistant cultivars, climate management
Mildew (*Oidium lycopersicon*)	selective fungicides
Fusarium wilt (*Fusarium oxysporum lycopersici*)	resistant cultivars, soil-less culture
Fusarium root rot (*Fusarium oxysporum radicis-lycopersici*)	resistant cultivars, soil-less culture, hygiene
Verticillium wilt (*Verticillium dahliae*)	pathogen-free seed, tolerant cultivars, climate control, soil-less culture
Bacterial canker (*Clavibacter michiganesis*)	pathogen-free seed, soil-less culture
Several viral diseases	resistant cultivars, soil-less culture, hygiene, weed management, vector control
Pollination	Bumble bees or bees

Natural control: natural enemies spontaneously immigrating into the greenhouse and controlling a pest.

TMV-resistant cultivars are available. Furthermore, many tomato cultivars in Europe are resistant to *Cladosporium* and *Fusarium*. Some cultivars are also tolerant to *Verticillium* and root-knot nematodes (van Lenteren and Woets, 1988). Problems with soil diseases can also be strongly reduced by growing the crop in inert media, which has become common practice in western Europe. In tomatoes, therefore, only foliage pests and *Botrytis cinerea* require direct control measures. The few pest organisms that 'overwinter' in greenhouses and survive soil sterilization are the greenhouse red spider mite (*T. urticae*) and the tomato looper (*Chrysodeixis chalcites*). Transferring young plants free of the other pest organisms into the greenhouse is important to prevent early pest development. For 20 years, the bulk of greenhouse tomatoes have been grown on rockwool systems, which makes soil sterilization redundant. With the cessation of soil sterilization more organisms, such as *Liriomyza* spp. and their natural enemies, and *Lacanobia oleracea* 'overwinter' in greenhouses. A recent development which gave a strong stimulus to the application of biological control is the use of bumblebees for pollination (van Lenteren, 1995).

IPM programs for cucumber, sweet pepper and aubergine are somewhat more complicated that the one for tomato, mainly because of a richer pest and disease spectrum. Detailed examples of IPM programs for vegetables used in different parts of the world are presented in Albajes *et al.* (1999). Until 1980 biological and integrated control of pests was almost exclusively applied in tomato and cucumber, which are by far the largest vegetable crops. Today IPM is being used in other important vegetable crops such as sweetpepper, aubergine, melon, strawberries and even in leaf vegetables like lettuce (Albajes *et al.*, 1999; van Lenteren, 1999).

Development of IPM for ornamentals is more complicated than for vegetables. The first problem is that many different species and cultivars of ornamentals are grown. In western Europe for example, more than 100 species of cut flowers and 300 species of potted plants are cultivated, and for several ornamentals more than 100 cultivars are produced. Other problems for implementation of IPM in ornamentals are that: (i) more pesticides are available than for vegetables; (ii) the whole plant is marketed, instead of only the fruits, so no leaf damage is allowed; and (iii) a zero-tolerance is applied to export material. But, since the 1990s the use of biological control has grown steadily in cut flowers (gerberas, orchids, roses and chrysanthemums) and pot plants (poinsettia) (van Lenteren, 1999; Parrella *et al.*, 1999). In gerberas the developments have been particularly fast and natural enemies were used on 78% of the Dutch gerbera area in 1998 (W. Ravensberg, personal communication). Biological control was applied on more than 10% (600 ha) of the total greenhouse area planted with flowers and ornamentals in 1998 in The Netherlands. Commercially used IPM programs for ornamental crops are presented in Parrella *et al.* (1999) for chrysanthemums, in van Lenteren (1995) for gerbera, and for various ornamentals in Gullino and Wardlow (1999). Worldwide, it is estimated that about 1000 ha of ornamentals are under IPM.

Production of natural enemies for pest control in greenhouses

At present greenhouse pests are managed through biological control on some 15,000 ha compared with 200 ha under biological control in 1970 (van Lenteren, 1995, 2000). In 1968, when commercial biological control in greenhouses started in Europe, two small commercial producers were active. Today Europe has 26 natural enemy producers including the world's three largest, and there are about 65 producers worldwide. These three largest companies serve more than 75% of the greenhouse biological control market. Of the circa 100 biological control agents that are marketed today, about 30 make up 90% of the total sales. Very limited information was available about prices of commercially produced organisms, but recently data for the North American market (Cranshaw *et al.*, 1996) and European market (van Lenteren *et al.*,

1997) became available. It appears that many more species of biological control agents are available in Europe than in North America or in other areas with a greenhouse industry such as Latin America (e.g. Bueno, 1999; de Vis, 1999), Japan (Yano, 1999), Australia (Goodwin and Steiner, 1996) and New Zealand (Martin *et al.*, 1996). This is caused by the much larger European greenhouse industry and a longer history of research in greenhouse biological control in Europe.

Although on-farm production of natural enemies is possible, most growers purchase them from commercial suppliers. Many of the mass production companies are, understandably, reluctant to provide information on many aspects of mass production. Our experience is that many of the natural enemies produced for biological control in protected cultivation are reared on their natural hosts (the pests) and host plants. Rearing on purely artificial media (without organic additives) is very rare, primarily because this technology is insufficiently developed for mass production and because this way of production may lead to poor performance of natural enemies when exposed to their target hosts (van Lenteren, 1993). Rearing conditions should be as similar as possible to the conditions under which the natural enemies will have to function in commercial greenhouses (van Lenteren and Woets, 1988).

Mass production of natural enemies has seen a very fast development during the past three decades. The numbers produced have greatly increased (up to 50 million individuals per week), the spectrum of species available has widened dramatically (from two in 1970 to almost 100 nowadays), and mass production methods clearly have evolved (Bolckmans, 1999; van Lenteren and Tommasini, 1999). Developments in the areas of mass production, quality control, storage, and shipment and release of natural enemies have decreased production costs and led to better product quality, but much more can be done. Innovations in long-term storage (e.g. through diapause), shipment and release methods may lead to a further increase in natural enemy quality with a concurrent reduction in costs of biological control, thereby making it easier and more economical to apply. In addition to developments in biological control with arthropod natural enemies, we also see currently other types of beneficial organisms used for pest control in greenhouses. Snakes from America are used for control of rats and mice, reptiles from Indonesia for control of thrips and scale insects, and tropical birds (*Alcippe brunnea* (Gould)) for control of Lepidoptera (van der Linden, 1999).

Companies starting the production of natural enemies usually have little knowledge about the obstacles and complications related to mass rearing. They are even more ignorant about the development and application of quality control. A special point of concern is the lack of knowledge about the sources of variability of natural enemy behavior and methods to prevent genetic deterioration of natural enemies. Mass-rearing of natural enemies often takes place in small companies with little know-how and understanding of conditions influencing performance, which may result in natural enemies of bad quality and failures of biological control programs. The few large companies employ entomologists who develop quality-control tests, but methods differ widely and are not always adequate. And even when the natural enemies leave the insectary in top condition it does not mean that they are in top shape when released in the greenhouse. Shipment and handling by the producers, distributors and growers may result in deterioration of the biological control agents. This makes robust quality-control programs a necessity (van Lenteren and Tommasini, 1999).

In the 1990s scientists and commercial producers of biological control agents started to work on development and standardization of quality control methods. Quality-control procedures for natural enemies have been developed for the 20 most important species of natural enemies that are commercially applied in greenhouses. Quality-control criteria relate to product control and are based on laboratory measurements, which are often easy to carry out. The criteria will soon be complemented with flight tests and field performance tests (van Lenteren, 2003).

Consumer demands for pesticide-free food also stimulate the use of biological control.

Specific advantages of biological pest control in greenhouses

Why do greenhouse growers use biological control? There are, of course, the general advantages of biological control such as reduced exposure of producer and applier to toxic pesticides, the lack of residues on the marketed product and the extremely low risk of environmental pollution. These, however, are not of particular concern for the grower. More important are the specific reasons that make growers working in greenhouses prefer biological control:

1. With biological control there are no phytotoxic effects on young plants, and premature abortion of flowers and fruit does not occur.
2. Release of natural enemies takes less time and is more pleasant than applying chemicals in humid and warm greenhouses.
3. Release of natural enemies usually occurs shortly after the planting period when the grower has sufficient time to check for successful development of natural enemies; thereafter the system is reliable for months with only occasional checks; chemical control requires continuous attention.
4. Chemical control of some of the key pests is difficult or impossible because of pesticide resistance.
5. With biological control there is no safety period between application and harvesting fruit; with chemical control one has to wait several days before harvesting is allowed again.
6. Biological control is permanent: once a good natural enemy, always a good natural enemy.
7. Biological control is appreciated by the general public.

Costs of biological control are similar to those of chemical control, and this, in combination with points 1, 2 and 5, makes it an attractive pest management approach.

Integrated management of diseases: a research priority

IPM was limited mainly to the control of insects until a few years ago. In Europe, disease problems are considerable, particularly in tomatoes, cucumbers and cut flowers (Gullino, 1992). Some fungicides can be integrated with the use of natural enemies, but as problems of fungicide resistance are strongly increasing, fewer 'relatively safe' fungicides remain available. The lack of biological control agents of diseases is of major concern. Although use of fungicides remains substantial for foliar pathogens, disease management is now evolving towards strategies relying on the use of resistant cultivars and manipulation of the environment. During the past decade several initiatives have led to research in non-chemical control, such as the effect of soil solarization on nematodes and fungi, and the potential use of antagonistic leaf fungi (Albajes *et al.*, 1999).

Disease suppressive soils, i.e. soils with antagonistic and antibiotic organisms reducing populations of disease-causing organisms, could provide another good opportunity for control of soil-borne diseases, but has not advanced to practical use as yet. Organisms isolated from suppressive soils can already be used against some soil-borne pathogens, like e.g. the antagonist bacterium *Streptomyces griseoviridis* (Mycostop), which is registered in at least seven European countries for use against *Fusarium oxysporum* wilt and *Pythium ultimum* seedling blight both in flowers and in cucurbits. Its mode of action is based on antibiotic effects. The microbe colonizes the rhizosphere before the pathogens, and secretes antibiotic substances, which inhibit the growth of fungal pathogens (Lahdenperä *et al.*, 1990). It can be applied as seed dressing or in the soil.

In the future, use of antagonistic *Fusarium* spp. and fluorescent pseudomonads active against *Fusarium oxysporum* will permit biological control of *Fusarium* wilts (Albajes *et al.*, 1999). Also the use of *Trichoderma* spp. as seed dressing or soil treatment will provide control of damping off (*Pythium* spp.) and root rot (*Phytophthera* spp., *Rhizoctonia solani*) (Gullino, 1992). In tomatoes, crown and root rot (*Fusarium oxysporum*) can be controlled with *Trichoderma harzianum* (van Steekelenburg, 1992).

Foliar diseases can often be reduced by proper climate regulation, and the use of computers to control temperature, light, humidity, water, ventilation, carbon dioxide and nutrition has resulted in improved disease management. Manipulation of the interactions of temperature and humidity is the most important but also the most costly to achieve, and it is easier in modern glasshouses than in plastic houses or tunnels that are mainly used in the subtropics, resulting in disease-prone situations in the plastic structures. Therefore, growers rely heavily on fungicides in such situations. Serious negative effects of fungicides on natural enemies of insects and widespread resistance of foliar pathogens to fungicides demands for alternatives (Gullino, 1992). As yet only one fungal biological control agent for foliar pathogens is registered, *Trichoderma harzianum*, for control of *Botrytis cinerea* in strawberry, and as soil fungicide for control of *Fusarium*, *Pythium* and *Rhizoctonia* spp.

Important recent successes in disease control concern biological control of gray molds (*Botrytis cinerea*) and powdery mildews (*Sphaeroteca fuliginea*, *Oidium* spp.) in cucumber, tomato and several ornamentals (Elad *et al.*, 1996; Dik *et al.*, 1999). *Botrytis cinerea* is a pathogen occurring in many fruit, vegetable and ornamental crops. *Botrytis cinerea* infections can be reduced by pre-inoculation of the leaves with yeasts, filamentous fungi and bacteria. These biological control agents compete with *B. cinerea* for nutrients and possibly they induce host-plant resistance against *B. cinerea* (Elad *et al.*, 1996). From 16 isolates of yeasts, filamentous fungi and bacteria, *Trichoderma harzianum* and *Aureobasidium pullulans* performed best in controlling *Botrytis* on tomato and cucumber in greenhouses. Control results were similar or even better than with currently used fungicides (Dik and Elad, 1999).

Also, biological control of *Botrytis* in cyclamen is now possible. Treatments of leaves with a conidial suspension of the competing saprophytic fungi *Ulocladium atrum* and *Gliocladum roseum* under commercial growing conditions were as effective as the standard chemical fungicide program (Koehl *et al.*, 1998). This work was continued in roses, where it was shown also that *U. atrum* had a strong reducing effect on *Botrytis*. The antagonist *U. atrum* was found to be insensitive to most fungicides used in conventional disease control. So this biological control agent can be integrated easily in programs where chemical fungicides are used for control of other fungi.

Most vegetables and ornamental plants grown in greenhouses suffer from powdery mildews. Several microorganisms have been found effective as biological control agent against various powdery mildew fungi. Their modes of action differ from *Botrytis* biological control agents: mildew is killed by hyperparasitism of the microorganism. Literature suggests the following hyperparasitic fungi to be potential candidates for biological control: *Ampelomyces quisqualis* and *Verticillium lecanii* for control of powdery mildews in cucurbits, and *Sporothrix flocculosa* for control of powdery mildews on roses and cucumber (Elad *et al.*, 1996). Testing at a semi-commercial scale in greenhouses showed that *Sporothrix flocculosa* was the only hyperparasite giving sufficient control of cucumber powdery mildew (Dik *et al.*, 1998).

Other research on disease control in closed-culture systems has shown that inoculation of artificial substrates with certain antagonistic microorganisms resulted in control of *Pythium* species (Postma *et al.*, 1996).

For control of the white molds *Sclerotinia sclerotiorum* and *S. minor* in glasshouse lettuce and in field crops, *Coniothyrium minitans* is available and registered in Europe (Whipps and Budge, 1992).

Nowadays 15 microbial products are registered and used for pest and disease control in greenhouse vegetables and ornamentals in Europe: five bacterial and fungal products for control of fungi, seven bacterial and fungal products for control of insects and three baculoviruses for control of insects (Oogst, 1999). Another three bacterial and fungal products for control of fungi are in the last phase of the registration procedure.

New research for durable control of greenhouse pests

Worldwide, a continuous search and evaluation of natural enemies (parasitoids, predators and pathogens) of insect and mite pests takes place, either to improve control of current pests or to develop control of new pests (Albajes *et al.*, 1999; van Lenteren, 1999, 2000). The most serious pest problems are currently caused by thrips species (Lewis, 1997), by *Bemisia* whiteflies (Parrella *et al.*, 1999) and by several species of aphids (Rabasse and van Steenis, 1999).

To control thrips, growers are either forced to use intensive application of broad-spectrum chemical pesticides that upset commercially successful greenhouse IPM programs, or to use biological control. Chemical control of thrips often proves to be extremely difficult and expensive. Although a large variety of predators (anthocorids, mirids, thrips and mites), entomopathogenic fungi, thrips attacking nematodes and parasitoids are known (Lewis, 1997), cheap and/or effective biological control programs for thrips are still rare. *Orius* and *Amblyseius* spp. provide adequate control of thrips in greenhouse crops like sweet pepper and cucumber worldwide, while performance in floriculture was less satisfactory until recently (van Lenteren and Loomans, 1998). Pathogenic fungi might be useful as additional control agents. Parasitoids, though the only specific natural enemies of thrips, have not shown much potential for control to date. An interesting new development for improving biological control of thrips is the open rearing system of *Amblyseius degenerans* on potted, pollen-bearing *Ricinus communis* plants (Ramakers and Voet, 1996). These 'banker plants' can be put in the greenhouse (e.g. sweet pepper or roses) to establish early colonies of the predator in a crop that does not yet have pollen, or even in plant propagation houses, where biological control is also applied nowadays.

The emergence of a new whitefly pest, the sweet potato whitefly *Bemisia argentifolii*, has complicated IPM programs in America, the Mediterranean and other parts of the world, although problems in northern Europe are fewer than initially expected after accidental importation of this pest in the 1980s (Drost *et al.*, 1998). A worldwide search for new natural enemies and pathogens of *Bemisia* is in progress (e.g. Gerling and Mayer, 1996; Hoddle *et al.*, 1998; Drost *et al.*, 1999; Hoelmer and Kirk, 1999; Meekes *et al.*, 2000). As a result, *Bemisia* can be controlled currently by introducing a mix of *Encarsia formosa* and *Eretmocerus eremicus* (northern Europe and northern America) or *Eretmocerus mundus* (Mediterranean). The predator *Macrolophus caliginosus* is generally added to the parasitoids, as this mix of two parasitoids and one predator results in better control over a long period.

Aphids have always created complications in IPM, as their populations can develop so quickly that introduction of natural enemies is often too late (Rabasse and van Steenis, 1999). Many genera are represented in greenhouses, each demanding a specific set of natural enemies for proper biological control. Despite numerous studies of aphidophagous insects (parasitoids and predators) and pathogenic fungi, only a few species have shown potential in greenhouses on a large scale because few natural enemies have the potential to match the reproductive and developmental rates of aphids (van Steenis, 1995a). The best method to prevent aphid populations escaping from control is to bring natural enemies into the greenhouse even before aphids have been discovered. This can be done in an elegant, yet very effective way by introducing open rearing units (also called 'banker plants') into the greenhouse that consist of wheat plants with wheat aphids (which cannot live on the greenhouse crop) and predators or parasitoids (van Steenis, 1995b).

Recently, a strongly increased activity in the area of resistance breeding to pests and diseases for greenhouse crops took place; about 30% of all breeding activities of important greenhouse breeding companies is now spent on resistance breeding (Cuartero *et al.*, 1999; A. Poolman, personal communication). In addition, partial resistance can often (but not always) be used in combination with biological control to obtain a sufficient control result. Further, plant breeders and biological control researchers have joined forces to develop plant cultivars, which help natural enemies to perform better. An example is the research on cucumber cultivars, where lines with fewer hairs were selected which resulted in a higher search efficiency of the parasitoid *Encarsia formosa* and more parasitized whiteflies (van Lenteren *et al.*, 1995). Effects of hairiness of gerbera cultivars on biological control of whitefly with *Encarsia formosa*, as well as on biological control of the spider mite *Tetranychus uriticae* with the predatory mite *Phytoseiulus persimilis* have also been evaluated recently (Sütterlin and van Lenteren, 1997; Krips *et al.*, 1999).

Another area where plant breeders and biological control workers can mutually benefit is in that of chemical communication between plant, herbivores (pests) and natural enemies. It appears that several crops start to produce volatile chemicals after being attacked by a pest insect or mite (Dicke, 1999). These chemicals are used by natural enemies to detect infested plants. Cultivars of the same plant species show large variation in the amount of volatiles produced after attack. Selection and use of plant cultivars that produce higher amounts of natural enemy attracting volatiles may improve biological control.

Modeling plant–herbivore–natural-enemy relationships has always played a role in the process of selecting and improving the efficacy of releases of natural enemies, but often biologically unrealistic simplifications were part of these models which strongly limited their predictive value. Recently a model was developed which is unique in that it is individual based and simulates the local searching and parasitization behavior of individual parasitoids (*Encarsia formosa*) in a whitefly-infested crop. The model includes stochasticity and spatial structure based on location coordinates of plants and leaves. This model comprises several submodels for: (i) the parasitoid's foraging behavior; (ii) the whitefly and parasitoid population development; (iii) the spatial distribution of whitefly and parasitoid within and between plants in the crop; and for (iv) leaf production. With the model, temporal and spatial dynamics of pest and natural enemy can be simulated. The model will help: (i) to explain why the parasitoid *E. formosa* can control whiteflies on some crops and not on others in large commercial greenhouses; (ii) to improve introduction schemes of parasitoids for crops where control was difficult; and (iii) to predict effects of changes in cropping practices (e.g. greenhouse climate, choice of cultivars) on the reliability of biological control; and finally (iv) to develop criteria for the selection of natural enemies (van Roermund *et al.*, 1997; van Lenteren and van Roermund, 1999). This model is in the process of being adopted to be able to simulate other plant–pest–natural-enemy relationships. Other, more simple models were developed to understand pest–natural-enemy dynamics and/or to adapt introduction schemes of natural enemies, e.g. Heinz *et al.* (1993) for biological control of leafminers, and Janssen and Sabelis (1992) for control of spider mites.

Several expert systems, or decision-support systems, are being developed for pest diagnosis and integrated control. An important factor favoring the use of such systems in the greenhouse industry is the fact that this sector is technologically highly advanced with a widespread use of computerized control of environmental conditions (Parrella *et al.*, 1999). Recently developed expert systems in this field have included pest and diseases diagnosis, integrated management in specific crops, information on natural enemy release programs and data on side effects of pesticides on natural enemies (Shipp and Clarke, 1999). This type of decision-support system helps growers manage increasingly complex production systems. The continuous updating of decision-support software packages, which is essential for their correct function, is still problematic. Online services concerning biological control of greenhouse pests and the effects of pesticides on natural enemies as provided by producers of natural enemies (e.g. www.koppert.nl/ information in English) can be updated easier and quicker. The information in this section does not summarize all the new developments in greenhouse IPM, but aims to show the creativity and innovativeness of this field of horticulture.

Future Prospects

The tremendous success achieved with biological control in greenhouses has set a very high standard that is difficult for other segments in agriculture to match (Parrella *et al.*, 1999). This success has occurred primarily as a result of outstanding cooperation between research, extension, growers and producers of natural enemies, often within the framework of IOBC (see e.g. van Lenteren, 1999). Several current trends will lead to a strong increase in the use of biological and integrated control of pests and diseases in greenhouses. First, fewer new insecticides are becoming available because of skyrocketing costs for development and registration, particularly for the relatively small greenhouse market. Second, pests continue to develop resistance to any type of pesticides, a problem particularly prevalent in greenhouses, where intensive management and repeated pesticide applications exert strong selective pressure on pest organisms. Third, there is a strong demand from the general public (and in an increasing number of countries also from governments) to reduce the use of pesticides. Finally, in order to escape from the 'pesticide treadmill', more sustainable forms of pest and disease control will have to be developed (Lewis *et al.*, 1997).

Because of the desire to reduce pesticide use, the future role of biological and integrated control is expected to increase strongly. This is aided by the extensive demonstration of its positive role and because many new natural enemy species still await discovery. Cost–benefit analyses show that biological control is the most cost-effective control method (Bellows and Fisher, 1999). With improved methods for evaluation of beneficial insects, an increased insight into the functioning of natural enemies, and more efficient mass production methods, the cost effectiveness of biological control may even be increased. Together with other control methods such as mechanical and physical control, control with semiochemicals, and host-plant resistance, new IPM programs will be developed. During the first decade of this century a greenhouse without conventional chemical pesticides could become a fact!

References

Albajes, R., Gullino, M.L., van Lenteren, J.C. and Elad, Y. (eds) (1999) *Integrated Pest and Disease Management in Greenhouse Crops.* Kluwer Publishers, Dordrecht.

Bellows, T.S. and Fisher, T.W. (eds) (1999) *Handbook of Biological Control.* Academic Press, New York.

Bolckmans, K.J.F. (1999) Commercial aspects of biological pest control. In: Albajes, R., Gullino, M.L., van Lenteren, J.C. and Elad, Y. (eds) *Integrated Pest and Disease Management in Greenhouse Crops.* Kluwer Publishers, Dordrecht, pp. 310–318.

Bueno, V.H.P. (1999) Protected cultivation and research on biological control of pests in greenhouses in Brazil. In: van Lenteren, J.C. (ed.) Proceedings of the IOBC/WPRS working group 'Integrated Control in Glasshouses'. *Bulletin IOBC/WPRS* 22(1), 21–24.

Cranshaw, W., Sclar, D.C. and Cooper, D. (1996) A review of 1994 pricing and marketing by suppliers of organisms for biological control of arthropods in the United States. *Biological Control* 6, 291–296.

Cuartero, J., Laterrot, H. and van Lenteren, J.C. (1999) Host-plant resistance to pathogens and arthropod pests. In: Albajes, R., Gullino, M.L., van Lenteren, J.C. and Elad, Y.

Integrated Pest and Disease Management in Greenhouse Crops. Kluwer Publishers, Dordrecht, pp. 124–138.

de Vis, R., Fuentes, L.E. and van Lenteren, J.C. (1999) Development of biological control of *Trialeurodes vaporariorum* with *Encarsia formosa* and *Amitus fuscipennis* on greenhouse tomatoes in Colombia. In: van Lenteren, J.C. (ed.) Proceedings of the IOBC/WPRS working group 'Integrated Control in Glasshouses'. *Bulletin IOBC/WPRS* 22(1), 270–277.

Dicke, M. (1999) Direct and indirect effects of plants on beneficial organisms. In: Ruberson, J.R. (ed.) *Handbook of Pest Management.* Marcel Dekker, New York, pp. 105–153.

Dik, A.J. and Elad, Y. (1999) Comparison of antagonists of *Botrytis cinerea* in greenhouse-grown cucumber and tomato under different climatic conditions. *European Journal of Plant Pathology* 105, 123–137.

Dik, A.J., Verhaar, M.A. and Bélanger, R.R. (1998) Comparison of three biological control agents against cucumber powdery mildew (*Spaerotheca fuliginea*) in semi-commercial-scale glasshouse trials. *European Journal of Plant Pathology* 104, 413–423.

Dik, A.J., Koning, G. and Koehl, J. (1999) Evaluation of microbial antagonists for biological control of *Botrytis cinerea* stem infection in cucumber and tomato. *European Journal of Plant Pathology* 105, 115–122.

Drost, Y.C., van Lenteren, J.C. and van Roermund, H.J.W. (1998) Life-history parameters of different biotypes of *Bemisia tabaci* (Hemiptera: Aleyrodidae) in relation to temperature and host plant: a selective review. *Bulletin of Entomological Research* 88, 219–229.

Drost, Y.C., Qiu, Y.T., Posthuma-Doodeman, C.J.A.M. and van Lenteren, J.C. (1999) Life-history and oviposition behavior of *Amitus bennetti* a parasitoid of *Bemisia argentifolii. Entomologia Experimentalis et Applicata* 90, 183–180.

Elad, Y., Malathrakis, N.E. and Dik, A.J. (1996) Biological control of *Botrytis*-incited diseases and powdery mildews in greenhouse crops. *Crop Protection* 15, 229–240.

Gerling, D. and Mayer, R.T. (eds) (1996) *Bemisia 1995: Taxonomy, Biology, Damage, Control and Management.* Intercept, Andover, UK.

Goodwin, S. and Steiner, M.Y. (1996) Survey of Australian natural enemies for control of thrips. In: van Lenteren, J.C. (ed.) Proceedings of the IOBC/WPRS working group 'Integrated Control in Glasshouses'. *Bulletin IOBC/WPRS* 19(1), 47–50.

Gullino, M.L. (1992) Integrated control of diseases in closed systems in the sub-tropics. *Journal of Pesticide Science* 36, 335–340.

Gullino, M.L. and Wardlow, L.R. (1999) Ornamentals. In: Albajes, R., Gullino, M.L., van Lenteren, J.C. and Elad, Y. (eds) *Integrated Pest and Disease Management in Greenhouse Crops.* Kluwer Publishers, Dordrecht, pp. 486–506.

Heinz, K.M., Nunney, L. and Parrella, M.P. (1993) Toward predictable biological control of *Liriomyza trifolii* (Diptera: Agromyzidae) infecting greenhouse cut chrysanthemums. *Environmental Entomology* 22, 1217–1233.

Hoddle, M.S., Driesche, R.G., Sanderson, J.P. and Minkenberg, O.P.J.M. (1998) Biological control of *Bemisia argentifolii* (Homoptera: Aleyrodidae) on poinsettia with inundative releases of *Eretmocerus eremicus* (Hymenoptera: Aphelinidae): do release rates affect parasitism? *Bulletin of Entomological Research* 88, 47–58.

Hoelmer, K.A. and Kirk, A.A. (1999) An overview of natural enemy explorations and evaluations for *Bemisia* in the U.S. In: van Lenteren, J.C. (ed.) Proceedings of the IOBC/WPRS working group 'Integrated Control in Glasshouses'. *Bulletin IOBC/WPRS* 22(1), 109–112.

Janssen, A. and Sabelis, M.W. (1992) Phytoseiid life-histories, local predator-prey dynamics, and strategies for control of tetranychid mites. *Experimental and Applied Acarology* 14, 233–250.

Koehl, J., Gerlach, M., de Haas, B.H. and Krijger, M.C. (1998) Biological control of *Botrytis cinerea* in cyclamen with *Ulocladium atrum* and *Gliocladium roseum* under commercial growing conditions. *Phytopathology* 88, 568–575.

Krips, O.E., Kleijn, P.W., Willems, P.E.L., Gols, R. and Dicke, M. (1999) Leaf hairs influence searching efficiency and predation rate of the predatory mite *Phytoseiulus persimilis* (Acari: Phytoseiidae). *Experimental and Appllied Acarology* 23, 119–131.

Lahdenperä, M.L., Simon, E. and Uoti, J. (1990) Mycostop – a novel biofungicide based on *Streptomyces* bacteria. In: Beemster, A.B.R. (ed.) Biotic *Interactions and Soil-borne Diseases.* Elsevier, Amsterdam, pp. 258–263.

Lewis, T. (ed.) (1997) *Thrips as Crop Pests.* CAB International, Wallingford, UK.

Lewis, W.J., van Lenteren, J.C., Phatak, S.C. and Tumlinson, J.H. (1997) A total systems approach to sustainable pest management. *Proceedings of the National Academy of Sciences, Washington* 94, 12243–12248.

Martin, N.A., Workman, P.J. and Marias, T. (1996) IPM for greenhouse crops in New Zealand: progress, problems and prospects. In: van Lenteren, J.C. (ed.) Proceedings of the IOBC/WPRS working group 'Integrated Control in Glasshouses'. *Bulletin IOBC/WPRS* 19(1), 99–102.

Meekes, E.T.M., van Voorst, S., Joosten, N.N., Fransen, J.J. and van Lenteren, J.C. (2000) Persistence of the fungal whitefly pathogen *Aschersonia aleyrodis*, on three different plant species. *Mycological Research* 104, 1234–1240.

Oogst (1999) Overview of commercial microbial crop protection agents for pest and disease control in Europe. *Plant Production and Crop Protection Journal for Dutch Growers* [in Dutch] 3 December, 1999.

Parrella, M.P., Stengard Hansen, L. and van Lenteren, J.C. (1999) Glasshouse environments. In: Bellows, T.S. and Fisher, T.W. (eds) *Handbook of Biological Control*. Academic Press, New York, pp. 819–839.

Postma, J., Willemsen-de Klein, M.J.E.I.M. and Hoogland, A.E. (1996) Biocontrol of *Pythium aphanidermatum* in closed culture systems. *IOBC/WPRS Bulletin* 19(6), 42–46.

Rabasse, J.M. and van Steenis, M.J. (1999) Biological control of aphids. In: Albajes, R., Gullino, M.L., van Lenteren, J.C. and Elad, Y. (eds) *Integrated Pest and Disease Management in Greenhouse Crops*. Kluwer Publishers, Dordrecht, pp. 235–243.

Ramakers, P.M.J. and Voet, S.J.P. (1996) Introduction of *Amblyseius degenerans* for thrips control in sweet peppers with potted castor beans as banker plants. In: van Lenteren, J.C. (ed.) Proceedings of the IOBC/WPRS working group 'Integrated Control in Glasshouses'. *IOBC/WPRS Bulletin* 19(1), 127–130.

Shipp, J.L. and Clarke, N.D. (1999) Decision tools for integrated pest management. In: Albajes, R., Gullino, M.L., van Lenteren, J.C. and Elad, Y. (eds) *Integrated Pest and Disease Management in Greenhouse Crops*. Kluwer Publishers, Dordrecht, pp. 168–182.

Sütterlin, S. and van Lenteren, J.C. (1997) Influence of hairiness of *Gerbera jamesonii* leaves on the searching efficiency of the parasitoid *Encarsia formosa*. *Biological Control* 9, 157–165.

van der Linden, A. (1999) Insectivorous birds for biological control of pests in glasshouses. In: van Lenteren, J.C. (ed.) Proceedings of the IOBC/WPRS working group 'Integrated Control in Glasshouses'. *Bulletin IOBC/WPRS* 22(1), 149–152.

van Lenteren, J.C. (1993) Parasites and predators play a paramount role in pest management. In: Lumsden, R.D. and Vaughn, J.L. (eds) *Pest Management: Biologically Based Technologies*. American Chemical Society, Washington, DC, pp. 68–81.

van Lenteren, J.C. (1995) Integrated Pest Management in protected crops. In: Dent, D. (ed.) *Integrated Pest Management*. Chapman and Hall, London, pp. 311–343.

van Lenteren, J.C. (ed.) (1999) Proceedings of the IOBC/WPRS working group 'Integrated Control in Glasshouses'. *Bulletin IOBC/WPRS* 22(1), 294 pp.

van Lenteren, J.C. (2000) Measures of success in biological control of arthropods by augmentation of natural enemies. In: Gurr. G. and Wratten, S. (eds) *Measures of Success in Biological Control*. Kluwer Academic Publishers, Dordrecht, pp. 77–103.

van Lenteren, J.C. (ed.) (2003) Quality Control and Production of Biological Control Agents: Theory and Testing Procedures. CAB International, Wallingford, UK.

van Lenteren, J.C. and Loomans, A.J.M. (1998) Is there a natural enemy good enough for biological control of thrips? *Proceedings 1998 Brighton Crop Protection Conference – Pests and Diseases* 2. British Crop Protection Council, Farnham, UK, pp. 401–408.

van Lenteren, J.C. and van Roermund, H.J.W. (1999) Why is the parasitoid *Encarsia formosa* so successful in controlling whiteflies? In: Hawkins, B.A. and Cornell, H.V. (eds) *Theoretical Approaches to Biological Control*. Cambridge University Press, pp. 116–130.

van Lenteren, J.C. and Tommasini, M.G. (1999) Mass production, storage, shipment and quality control of natural enemies. In: Albajes, R., Gullino, M.L., van Lenteren, J.C. and Elad, Y. (eds) *Integrated Pest and Disease Management in Greenhouse Crops*. Kluwer Publishers, Dordrecht, pp. 276–294.

van Lenteren, J.C. and Woets, J. (1988) Biological and integrated pest control in greenhouses. *Annual Review of Entomology* 33, 239–269.

van Lenteren, J.C., Benuzzi, M., Nicoli, G. and Maini, S. (1992) Biological control in protected crops in Europe. In: van Lenteren, J.C., Minks, A.K. and de Ponti, O.M.B. (eds) *Biological Control and Integrated Crop Protection: Towards Environmentally Safer Agriculture*. Pudoc, Wageningen, pp. 77–89.

van Lenteren, J.C., Li Zhao Hua and Kamerman, J.W. (1995) The parasite–host relationship between *Encarsia formosa* (Hymenoptera:

Aphelinidae) and *Trialeurodes vaporariorum* (Homoptera: Aleyrodidae). XXVI. Leaf hairs reduce the capacity of *Encarsia* to control greenhouse whitefly on cucumber. *Journal of Applied Entomology* 119, 553–559.

van Lenteren, J.C., Roskam, M.M. and Timmer, R. (1997) Commercial mass production and pricing of organisms for biological control of pests in Europe. *Biological Control* 10, 143–149.

van Roermund, H.J.W., van Lenteren, J.C. and Rabbinge, R. (1997) Biological control of greenhouse whitefly with the parasitoid *Encarsia formosa* on tomato: an individual-based simulation approach. *Biological Control* 9, 25–47.

van Steekelenburg, N.A.M. (1992) Novel approaches to integrated pest and disease control in glasshouse vegetables in the Netherlands. *Journal of Pesticide Science* 36, 359–362.

van Steenis, M. (1995a) Evaluation of four species of aphelinid parasitoids for biological control of *Aphis gossypii*. *Entomologia Experimentalis et Applicata* 75, 151–157.

van Steenis, M. (1995b) Evaluation and application of parasitoids for biological control of *Aphis gossypii* in glasshouse cucumber crops. PhD thesis, Wageningen University, The Netherlands, 217 pp.

Whipps, J.M. and Budge, S.P. (1992) Biological control of *Sclerotinia sclerotiorum* in glasshouse crops. *Proceedings 1992 Brighton Crop Protection Conference – Pests and Diseases* 1, British Crop Protecton Council, Farnham, UK, pp. 127–132.

Yano, E. (1999) Recent advances in the study of biocontrol with indigenous natural enemies in Japan. In: van Lenteren, J.C. (ed.) Proceedings of the IOBC/WPRS working group 'Integrated Control in Glasshouses'. *Bulletin IOBC/WPRS* 22(1), 291–294.

Chapter 26

Integrated Pest Management in the Mediterranean Region: the Case of Catalonia, Spain

R. Albajes[1], M.J. Sarasúa[1], J. Avilla[1], J. Arnó[2] and R. Gabarra[2]
[1]Universitat de Lleida, Centre UdL-IRTA, Lleida, Catalonia, Spain;
[2]IRTA, Centre de Cabrils, Cabrils, Catalonia, Spain

Introduction

Catalonia is located in northeast Spain (Fig. 26.1). Its climate and agriculture are typically Mediterranean, with hot, dry summers and mild winters. The economy is mainly devoted to services and industry, and the primary sector represents only about a 1.5% of the total gross Catalan product. Agriculture covered 1,140,480 ha of the Catalan soil in 1998 and produced a gross income of 1178 million Euros in 1998 (Anonymous, 2000), 85% of which came from six main commodities: pome and stone fruits (29.5%), summer and winter cereals (15%), vegetables (15%), grapes (10%), olives (8.5%), and cut flowers and ornamental plants (7%).

There are no reliable surveys to estimate crop losses due to arthropod pests, plant pathogens and weeds in Catalonia, but it can be assumed that the figures are similar to

Fig. 26.1. Location of Catalonia.

those of other Mediterranean countries. In spite of the relatively abundant scientific activity in entomology in the last century in this country, it has mainly focused on insect taxonomy and very few studies have dealt with insect ecology and insect pest control. Some early attempts to introduce biological control into agriculture in the mid-20th century were interrupted by the Spanish Civil War and were not subsequently continued. Since the late 1970s several real programs of research, development and technology transfer involving Integrated Pest Management have been undertaken in Catalonia.

The use of chemical pesticides showed sustained growth in Catalonia in the 1970s and 1980s, and though it fluctuated in the 1990s it still showed a slight increase overall (Fig. 26.2). If it is taken into account that modern pesticides are more expensive than those used earlier, it may be hypothesized that the use of pesticides in Catalonia has became stable in the last 10 years.

IPM Policy in Catalonia

Registration of pesticides

The commercialization and use of pesticides in Catalonia were regulated only by Spanish laws until the mid-1990s. Since then the European Commission (EC) of the European Union (EU) has been responsible for registering new pesticide active ingredients and is gradually dealing with the re-registration of old pesticides. A provisional positive list of active ingredients that can be commercialized in the whole EU was approved and published by the EC in the early 1990s. A definitive positive list in which far fewer active ingredients will be registered for commercialization in the whole EU should be ready by 2003. It is expected that only between one-third and one-half of the pesticides registered in most of Western Europe at the end of the 20th century will be left in the new positive list; the rest will be banned owing to toxicological and environmental concerns and lack of economic interest by manufacturers. The imminent disappearance of many insecticides from the market is pushing European governments to stimulate and support the research, extension and technology transfer of non-chemical methods for controlling insect pests and diseases.

Integrated production

The Integrated Production Guidelines for Catalonia were drawn up by the regional government (Generalitat de Catalunya). Information on the way the system works and the approved guidelines can be found on its Internet homepage http://www.gencat.es/darp/pi.htm The IOBC/WPRS (http://www.iobc-wprs.org) concept and definitions of Integrated Production (IP) are accepted as the conceptual framework in

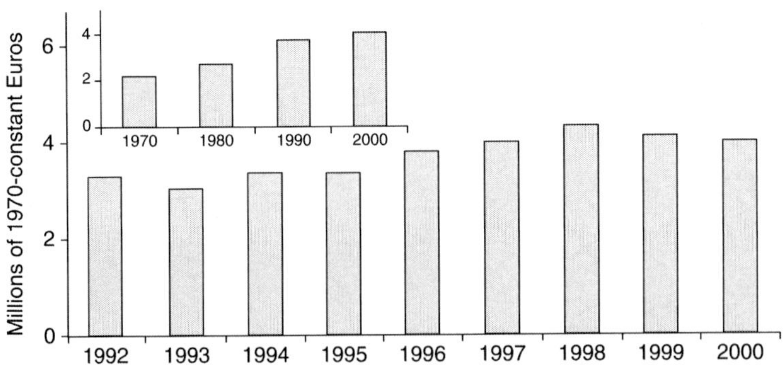

Fig. 26.2. Volume of sales of chemical pesticides in Catalonia in the last 30 years (estimated from AEPLA, 2000).

which the guidelines are to be developed (http://www.iobc.ch/). So far 18 guidelines covering the most important crops of Catalonia have been approved, including tomatoes, apples and pears. An external inspection system was applied for the first time in 2001 and included farm visits, field book inspections and residue analyses. The inspection system was improved in 2002, and included an initial inspection visit to all the farms included in the list of IP growers. Integrated Production in Catalonia was subsidized in 2001 for the first time, and consequently the hectarage under IP in Catalonia has dramatically increased, reaching 38,000 ha in many different crops: pome fruits, stone fruits, olives and nuts are the most important in terms of hectarage.

Research, education, extension, and technology transfer involving IPM

The development of research, extension, and education involving IPM in Catalonia has been linked to the general political evolution of Spain. In general terms, positive advances in scientific research and technology transfer in the last century have mainly taken place under the auspices of the Generalitat of Catalonia. Research is mainly concentrated in universities – in particular the University of Lleida in Western Catalonia, but also in Barcelona and Girona – and the Institute for Food and Agricultural Research and Technology in Cabrils (north of Barcelona). Major programs in greenhouse and outdoor vegetables, field crops, and pear and apple orchards are being developed.

Extension in IPM is the main concern of the Plant Protection Service, which belongs to the regional Ministry of Agriculture. Among other tasks, it diagnoses pest and disease problems, develops and implements warning systems, makes recommendations about the most efficient control methods, and is responsible for the tutelage of pest control advisers (PCAs). Only large farms have their own PCA; small growers tend to associate to create Growers' Associations for Plant Protection (ADVs), which are subsidized by the Spanish and Catalan governments to engage a PCA. This kind of association has been one of the keys to the faster development and implementation of IPM in recent years in some areas and crops. About 181,000 ha (a total of about 37,500 growers) of Catalan farmland are under the tutelage of the 115 PCAs who are particularly trained to implement and innovate IPM in the field. On this land olives (for which the main task is cooperative control of olive fly), cereals, rice, fruit (mostly pears, apples and peaches) and vines are the most important crops, accounting for 85% of the total area covered by ADVs in Catalonia (Fig. 26.3). A saving of between 40% and 85% of the chemical treatments that were applied before growers associated to engage the PCAs has been made.

Implementation of IPM in Catalonia: Case Studies

Although significant advances in the implementation of IPM have been made in several crops, the cases of tomatoes, apples and pears are especially relevant because complete and successful programs are now available thanks to several years of R&D, extension and tutelage of ADVs. Also, the programs are periodically updated as a result of advances in research on new

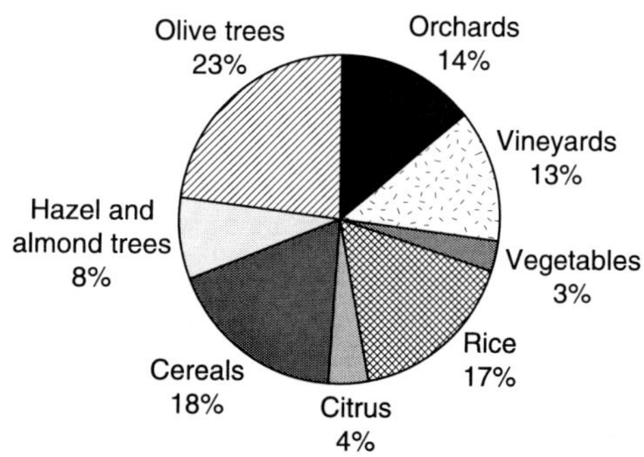

Fig. 26.3. Percentages of the total area (181,000 ha) of Catalan farmland under the tutelage of grower's associations for plant protection (ADVs) that are devoted to each of the indicated crops.

control techniques and the introduction into Catalonia of new pests. Tomatoes, apples and pears are among the crops that cover a relatively large area in comparison with other European countries (Table 26.1). Catalonia is also the main apple and pear producing area in Spain.

Tomatoes

Tomatoes were grown on 2900 ha of land in Catalonia in 1998 (Table 26.1), part in greenhouses and the rest in outdoor conditions. Protected tomato cultivation takes place in two different cycles, in spring (February to July or August) and in autumn (early August to November–December). Outdoor tomatoes are transplanted from March to June and harvested in October–November depending on the climatic conditions.

Greenhouses are built with wood or metal frames and covered with plastic film. Since most of them are unheated, daily temperatures can vary greatly, ranging from almost 0°C overnight to more than 25°C during the day in winter, and reaching temperatures of over 30°C during the warm season to autumn. The greenhouses have side and roof openings and are usually open from spring in order to improve ventilation. Pests can reproduce all year round inside and outside the greenhouses with a continuous movement of pest populations, between old and young crops, looking for the most suitable microclimate (Alomar *et al.*, 1989).

Table 26.1. Hectarage in Catalonia, Spain and the European Union devoted to the three crops for which an IPM program has been implemented in Catalonia, as discussed in this chapter.

	Area (1000 ha)		
Crop	Catalonia	Spain	European Union
Tomato	2.9[a]	55[b]	176[b]
Apple	17[a]	49[b]	307[b]
Pear	18[a]	43[b]	136[b]

[a]1998 data. Source: http://www.gencat.es/darp/estadist.htm
[b]Anonymous, 1999.

IPM is used in greenhouse tomato crops in many countries. In northeast Spain, IPM programs are based on inoculative and conservative biological control of the main pests and some secondary pests, the use of selective pesticides for the remaining pests (Table 26.2), and the use of fungicides with low toxicity on natural enemies for disease control. In 2001, these programs were applied in about 175 ha outdoors and 52 ha of greenhouse tomato (J. Ariño, M. Martí and M. Pagès, personal communication).

The most important pests affecting tomato crops are polyphagous, but their importance varies for protected or open field conditions. Whiteflies are a major pest in both of them. *Trialeurodes vaporariorum* is the predominant species in Catalonia, whereas *Bemisia tabaci* is not widely distributed in the region. The latter can be found, especially at the end of summer, in some particularly warm areas where tomatoes coexist with ornamental crops. In this chapter, the name *B. tabaci* will be used to refer to all biotypes/species.

Helicoverpa armigera is another major pest for outdoor tomatoes, causing very high yield losses (up to 30% of fruits in heavy infested fields), especially in fields transplanted around June. The second major pest in greenhouses, after whitefly, are leafminers. Aphids and the tomato russet mite may cause major economic damage but their incidence is very variable according to the climatic area and the year.

A number of foliar and soil-borne diseases affect greenhouse and outdoor tomatoes in the area. Gray mold in greenhouses, and powdery mildew in both greenhouses and open fields, are the main aerial diseases in tomato crops. Late blight causes severe damage but its incidence is very variable according to the climatic area and the year. The use of tomato varieties resistant to verticillium wilt, *Fusarium oxysporum* and the nematode *Meloidogyne* spp. is recommended (Gabarra and Besri, 1999). Viral diseases are an increasing threat to Mediterranean vegetable crops. Usually growers use varieties resistant to the TSWV transmitted by *Frankliniella occidentalis*, the most important virus affecting tomato in the area. *Bemisia tabaci* transmits the TYLCV, which causes also major economic losses. TYLCV was recorded for the first time in a small area north of Barcelona in summer 2000 (SSV, 2001). During 2001 no spread of the virus was observed.

Integrated arthropod pest management in both greenhouse and outdoor tomato is based on biological control. Additionally certain cultural practices – aimed at decreasing the biotic potential of the pest or favoring the action of natural enemies – may complement and enhance the efficacy of biological control.

Biological control in tomatoes is practiced with inoculative releases and conservation of native natural enemies (Table 26.2). From 1989 until 1998, control of *T. vaporariorum* was successfully achieved

Table 26.2. Control methods, decision thresholds and rates used for insect and mite control in IPM programs for tomato crops.

Insects & mites	Control agent/ Technique or method	Strategy for BC[a] C	Strategy for BC[a] I	Rate/action threshold and general remarks	Tomato crop[b]
Whiteflies					
Trialeurodes	*Macrolophus caliginosus*	X	X	1.5 Mc m^2, in 2 releases	GH, OF
vaporariorum (Tv)	(Mc)		(Only	In GH: 1 Tv/plant in margins	
T. vaporariorum &	*Dicyphus tamaninii* (Dt)	X	GH)	Monitor whiteflies and mirid bugs.	
Bemisia tabaci	*Encarsia pergandiella*	X		Management of Dt according to a	
	Eretmocerus mundus	X		decision chart	
				E. mundus is specific of *B. tabaci*	
Lepidoptera					
Helicoverpa	*Bacillus thuringiensis*			Labeled rates. For Ha control	GH, OF
armigera (Ha)	*Trichogramma*	X		use high units formulation +	
Chrysodeixis chalcites	*evanescens*			pinolene + sugar	
Autographa gamma	*Telenomus ullyetii*	X		Action threshold for Ha: 1 egg	
	M. caliginosus	X		or larvae/14 plants	
	D. tamaninii	X		Action threshold for loopers:	
				2 young larvae/plant	
Leafminers					
Liriomyza trifolii	*Diglyphus isaea* (Di)	X	X	0.2–0.4 Di m^2, 2–3 releases	GH
L. bryoniae	*Dacnusa sibirica*		(Only	Presence of first mines in plants,	
L. huidobrensis			GH)	check for less than 25% natural	
				parasitism	
Aphids					
Macrosiphum	Pirimicarb			Labeled rates	GH
euphorbiae	*Aphidoletes aphidimyza*	X		Presence of first foci. Treat	
Myzus persicae	*Aphelinus abdominalis*	X		foci if not widespread in the	
	Aphidius spp.	X		greenhouse.	
	Praon spp.	X		When aphids or damage are	
	M. caliginosus	X		first seen, check for natural	
	D. tamaninii	X		enemies	
Mites					
Aculops lycopersici	Specific acaricides,			Labeled rates	GH, OF
	compatible with			Treat foci if not widespread	
	biocontrol agents			in the greenhouse.	
				Treat at first sign of damaged	
				plants	

[a]Shows if the strategy is conservative (C) or inoculative (I).
[b]Shows if the method is applied in greenhouse (GH) or open field (OF) production.

by seasonal inoculative releases of the parasitoid *Encarsia formosa* (Albajes *et al.*, 1994). However, in the 1990s naturalized populations of the autoparasitoid *Encarsia pergandiella* spontaneously colonized the greenhouses and interfered with *E. formosa*. As a result, at the end of the spring crop *E. pergandiella* pupae were found in 90% of the greenhouses in which *E. formosa* had been released, and whitefly parasitism was lower than 40%. Moreover, 76% of adults that emerged from black pupae were *E. pergandiella* males (Gabarra *et al.*, 1999). The parasitism levels reached since the establishment of *E. pergandiella* were much lower than those found by Albajes *et al.* (1994) before the parasitoid spread over the whole area, and do not effectively control the whitefly. As a result, *E. formosa* is no longer used for whitefly control.

Macrolophus caliginosus and *Dicyphus tamaninii* are mirid bugs that spontaneously colonize greenhouses and open field tomatoes when no broad-spectrum insecticides are released. Castañé *et al.* (2000) show that their action is complementary to that of the parasitoid *E. formosa* and may help to control greenhouse whitefly. They are also known to predate on *B. tabaci* (Barnadas *et al.*, 1998). Since *M. caliginosus* is mass reared and sold by many companies, inoculative releases of this mirid bug have been performed to control whitefly in spring tomato greenhouses. In these crops, natural colonization often occurs too late or in too low numbers to provide acceptable control. However, some problems in the establishment of the predator have been observed and further studies must be done in order to improve it. Arnó *et al.* (2000) have shown that native *M. caliginosus* can be maintained in unheated greenhouses during winter in banker plants and enhance early colonization of the tomato crop.

In outdoor crops, a conservation strategy has been used since the late 1980s. The IPM program for field tomato crops was based on the use of naturally occurring populations of *M. caliginosus* and *D. tamaninii*. Because the latter species can cause some damage to tomato fruits, a decision chart was developed by Alomar and Albajes (1996) to advise growers of the need to spray either to avoid whitefly damage or to minimize the risk of injury by *D. tamaninii*.

Macrolophus caliginosus and *D. tamaninii* are the most common mirid bug species on tomato crops. In the late 1980s *D. tamaninii* was the predominant species but the relative abundance of the two species has changed recently. In 1990, 80% of mirid bugs found on tomato crops were *D. tamaninii* (Alomar *et al.*, 1991). In contrast, this species represented 26% of the mirid populations in tomato greenhouse in 1993/94 and just 10% in a survey done in open fields in 1999 (Castañé *et al.*, 2000).

Many natural enemies of *H. armigera* are present in our area. Mirid bugs can prey on lepidopteran eggs and young larvae, and in field surveys we have seen that predation can reach 80% (Gabarra *et al.*, 1996). Egg parasitism may reach 50% in *H. armigera*, and *Trichogramma* spp. and *Telenomus* spp. are the most abundant genera (Gabarra *et al.*, 2000). Despite this natural enemy complex, *H. armigera* is still a severe pest that needs sprays of Bt. A very low action threshold has been established for this pest, and intensive sampling is necessary to localize the *H. armigera* eggs and young larvae – the stages that can be managed with Bt sprays (Arnó *et al.*, 1994).

Natural populations of leafminer parasitoids are abundant all year round and often natural parasitism controls leafminers in the crop (Albajes *et al.*, 1994). Augmentative *D. isaea* releases are used only when natural parasitism is low, especially in spring tomato under greenhouse conditions.

Biological control of aphids with inoculative releases of natural enemies is not used in most greenhouses. However, in the IPM tomato crops aphids do not normally reach economic thresholds due to the presence of indigenous populations of natural enemies (Alomar *et al.*, 1997).

The application of the described IPM programs in tomatoes has led to a great reduction in pesticide use. Averaged over several years, the number of insecticide sprays was less than one per season and fungicides were reduced by 80% (Albajes *et al.*, 1994). In outdoor tomatoes, insecticide applications decreased by 78% and fungicide applications by 60% (Arnó *et al.*, 1996). Safer use of bumble-bees – an increasingly common technique for pollination in tomatoes as it improves crop yield and quality – was another important achievement of the application of IPM, and reciprocally it led to greater demand for IPM systems in greenhouse tomatoes as most insecticides are not compatible with pollinators.

Use of polyphagous predators, originally managed for whitefly control, reduces the incidence of other pests such as leafminers, caterpillars, aphids and mites since polyphagous predators also feed on those prey. In addition, native natural enemies are better adapted to local conditions and the agroecosystems resulting from conservation strategies are more stable than those produced with exotic beneficial insects. An additional advantage of using generalist predators is their lower cost – or even no cost at all when the system is managed to have the right number of beneficial insects at the right time.

A few weak points in the actual IPM systems for tomatoes may be identified. There are very few selective pesticides to control pests that cannot be managed with non-chemical methods. For example, only one selective insecticide is used for aphid control in protected tomatoes. Also, new compounds in the pesticide market create difficulties in the application of biological control, because effects their side on natural enemies are usually not evaluated before these pesticides replace older ones.

Apples

Apple and pear orchards share several insect pests whose importance varies between the two plant species and also among varieties within the same species, but they also have some specific pests (Table 26.3).

The hectarage for apple cultivation in Catalonia has remained steady at around 17,000 ha since about 1996. This hectarage accounts for 35% of the total Spanish hectarage for apple production and 5.5% of the total EU hectarage (Table 26.3). In terms of production, more than 400,000 t are harvested/year.

The IPM program for apples in Catalonia (Torà *et al.*, 1995) is based on the biological control of European red mite (ERM) (*Panonychus ulmi*). Several sprayings with acaricides against ERM were used in the early 1980s, but they were not always able to keep its populations under control. A fauna study carried out in non- or less-sprayed apple orchards demonstrated the importance of *Amblyseius andersoni* (Acari: Phytoseiidae) as a control agent. Under non-disturbed conditions, natural populations of *A. andersoni* are able to keep ERM populations well below the economic injury level. The change from a chemical-based mite control to a biologically based one takes 2–3 years. A decision chart for predicting whether successful biological control will take place has been developed, and it takes into account the time of the growing season and the populations of the pest and the predator, sampled by means of a presence–absence method (Avilla *et al.*, 1992). As *A. andersoni* is quite sensitive to conventional insecticides, mainly pyrethroids, selective methods must be used against other pests.

The codling moth *Cydia pomonella* (Lepidoptera: Tortricidae) is a key pest at present in most of the area. Its populations are monitored with pheromone traps (several hundreds of traps are placed each season), and a degree–day phenology model is used at the beginning of the season to time insecticide spraying. Chemical control with broad-spectrum insecticides (mainly organophosphates) is still the most commonly used method for controlling its populations. The use of these chemicals in IPM programs is restricted to a maximum number of applications. Several IGR insecticides (chitin synthesis inhibitors, juvenile hormone analogs and agonists of the molting hormone) are also registered for codling moth control. Codling moth resistance to insecticides is a big concern and an insecticide resistance management program is recommended. Mating disruption was not registered until very recently, and was used

Table 26.3. Main arthropod pests in apple and pear orchards, sampling techniques and control measures.

Scientific name	Common name	Host	Sampling system	Main control measure
Panonychus ulmi (Prostigmata Tetranychidae)	European Red Mite	Apple/pear	Visual presence–absence sampling of leaves	Biological control by natural populations of *Amblyseis andersoni*.
Cydia pomonella (Lep. Tortricidae)	Codling moth	Apple/pear	Pheromone traps. Visual counts of injured fruits	Chemical control (OPs, IGRs). Mating disruption.
Pandemis heparana *Adoxophyes orana* (Lep. Tortricidae)	Leafrollers	Apple/pear	Pheromone traps	Chemical control (IGR, organophosphates).
Quadraspidiotus perniciosus (Hom. Diaspididae)	San José scale	Apple/pear	Visual sampling in winter on wood, and during season on fruits	Chemical control (broad spectrum insecticides, IGR).
Dysaphis plantaginea (Hom. Aphididae)	Rosy apple aphid	Apple	Visual sampling	Chemical control (imidacloprid, selective aphicides).
Eriosoma lanigerum (Hom. Aphididae)	Woolly apple aphid	Apple	Visual sampling	Chemical control. Biological control by *Aphelinus mali*.
Phyllonorycter spp. (Lep. Gracillariidae) *Leucoptera* spp. (Lep. Lyonetiidae)	Leafminers	Apple/pear	Pheromone traps	Biological control by natural populations of natural enemies. Chemical control.
Zeuzera pyrina. *Cossus cossus* (Lep. Cossidae) *Synanthedon myopaeformis* (Lep. Sesiidae)	Wood borers	Apple/pear	Pheromone traps	Chemical control. Mass trapping with pheromone traps. Mating disruption.
Ceratitis capitata (Dipt. Tephritidae)	Mediterranean fruit fly	Apple/pear	Pheromone traps. Visual counts of injured fruits	Chemical control. Mass trapping with attractive traps (experimental).
Cacopsylla pyri (Hom. Psyllidae)	Pear psylla	Pear	Beating-trays sampling. Visual sampling	Chemical control. Biological control by natural populations of natural enemies. Cultural control.
Dasyneura pyri (Dipt. Cecidomyiidae)	Pear leafcurling midge	Pear minor pests	Visual	These pests are not common and are not a consideration in the IPM program.
Hoplocampa brevis (Hym. Tenthredinidae)	Pear fruit sawfly		Visual. White traps	
Stephanitis pyri (Het. Tingidae)	Pear lace-bug		Visual	
Janus compressus (Hym. Cephidae)	Pear shoot sawfly		Visual	

on about 100 ha in 2001. The system works very well when the appropriate conditions are met, and an increase in the hectarage under mating disruption is expected in the near future (Bosch *et al.*, 1998).

San José scale (SJS), *Quadraspidiotus perniciosus* Comstock (Homoptera: Diaspididae), is an important pest due to its zero tolerance level for exportation. Several natural enemies are well established (*Encarsia perniciosi*, *Aphytis* spp.), but their action is not sufficient to maintain SJS populations under control. Winter (lime sulfur) and prebloom treatments are used in IPM programs to avoid the use of insecticides when the individuals of *A. andersoni* are active.

The rest of the species listed in Table 26.3 have a variable importance, depending on specific areas or years. Several species of aphids are present. Chemical control is used against rosy aphid, *Dysaphis plantaginea* (pre-bloom sprayings) and woolly aphid, *Eriosoma lanigerum*. The fauna of natural enemies is important in IPM orchards and is composed of general predators (Chrysopidae: Coccinellidae) as well as specific parasitoids (*Aphelinus mali*). Leafrollers and leafminers have a wide fauna of parasitoids, which are able to control their populations, especially for the latter. When insecticides are necessary, leafroller control is usually achieved with IGR (juvenile hormone analogs) use. Woodborers may have an increasing importance under IPM programs when broad-spectrum insecticides are replaced by selective control measures. Mass trapping and mating disruption can be used in these cases. Mass trapping has proved to be efficient for the control of apple clearwing, *Synanthedon myopaeformis*, and, to a far lesser extent, that of the leopard moth, *Zeuzera pyrina*. Recent field trials have shown a very good control of leopard moth with mating disruption (Avilla and Bosch, 2001). The Mediterranean fruit fly, *Ceratitis capitata*, has had an increasing importance in the last 3–4 years, mainly due to the mild winter temperatures, which have decreased the mortality of winter medfly populations. This may be the main constraint of IPM programs for apples, as selective insecticides are not available and mass trapping with attractive traps is only at an experimental stage. The other constraint is SJS control; if it is necessary to spray against it during the growing season, there is no selective chemical registered for that use.

Pears

Spain is the second producer of pears in the EU, accounting for 31% of the hectarage (Table 26.1). Some 41% of the Spanish hectarage is located in Lleida province in western Catalonia.

Pears have many common pests with apples (Table 26.3), so they can be managed with the same methods with slight differences. For example, the economic thresholds for the codling moth are higher in pears. Among the specific pests, the pear sucker, *Cacopsylla pyri*, is actually the most important pest and it has been the main target of research devoted to developing and implementing an IPM program in pears based on its biological control. Other specific pests are normally minor pests (Table 26.3) that cause serious damage only under special circumstances. *Dasyneura pyri* and *Janus compressus* are particularly harmful in nurseries and young orchards. *Hoplocampa brevis* may seriously affect pear yield only in years when concurrently there is low flowering and/or fruit setting, adult flight coincides with flowering, and the population in the previous fall has been high. Otherwise, damage is reduced to biological thinning. *Stephanitis pyri* is only an occasional problem in untreated orchards.

Pear psylla started to be a problem on the most vigorous varieties during the 1970s and on all varieties by the 1980s. Several factors have contributed to psylla population outbreaks. The hectarage of pear orchards increased by 50% from the 1970s to the 1980s, and by 100% from the 1970s to the 1990s. This means a big increase in food resources and, moreover, the dominant variety has vigorous vegetative growth, which is therefore a favorable host plant trait for psylla development and reproduction. In the 1970s pyrethroids were also introduced and used profusely in orchards. The broad-spectrum activity of pyrethroids reduced the natural enemies of psylla and hence contributed to the insect population outbreaks. Reducing initial spring populations by treating overwintering psylla adults with pyrethroids and DNOC, and thus preventing egglaying, was the main strategy used in pear orchards during the late 1980s and most of the 1990s.

The research we have been conducting for the last 12 years has allowed us to identify a complex of natural enemies that under favorable conditions can prevent psylla population outbreaks (Artigues *et al.*, 1996; Sarasua *et al.*, 1999). The most interesting natural enemies identified are those that start their activity early in the season and remain in the orchard even if the psylla population is low, as is the case of Miridae, whose population at the end of winter is composed mainly of young nymphs that cannot leave the orchard, or the generalist predators belonging to the *Orius* genus or the Dermaptera that can attack prey other than psylla. Also, the parasitoids appear early and act on the first generation. On the other hand, *Anthocoris nemoralis*, the most commonly reported natural enemy of psylla in Europe, overwinters in the orchard only if the psylla population in the previous autumn has been very high. Even if it overwinters in the orchard, it migrates in early spring if the psylla populations are low and only goes back when pest populations are high and have already reached economic thresholds. As these natural enemies mostly overwinter on pear trees, winter sprays in pear orchards are extremely harmful for predators and parasitoids – indirectly by depriving them of food and hence causing nymph starvation or forcing adults to migrate, or directly by causing their mortality. Eliminating winter sprays and avoiding excessive vegetative growth in relation to production of pear trees are therefore two key actions for enhancing biological control of pear psylla by native natural enemies, which is the cornerstone of current IPM in pear orchards in the area.

Growers in the region are showing increasing interest in adopting IPM systems in pear orchards, partially because of current and forthcoming limitations of pesticide use (e.g. DNOC has been recently banned by the EC in the whole EU), partially to meet requirements for IP labels, and also because of the increasing difficulty in controlling psylla efficiently. However, in spite of the great advances that have been made in the implementation of IPM, many other aspects are still poorly developed. Economic thresholds of psylla should be adapted to the real risk of damage in each of the main pear varieties grown in the region. Furthermore, suppression of winter sprays may cause an increase in SJS, for which use of lime sulfur late in winter is recommended, when necessary, instead of summer oil and organophosphates, which are more harmful for natural enemies. However, the economic threshold for SJS is the mere presence and good control is really difficult. A more selective control of SJS and also of codling moth remain two unsolved problems in IPM in pear orchards. Additionally, a better knowledge of key factors in the ecology of native natural enemies may improve their management for controlling psylla and other pear pests. With regard to the latter, provision of artificial refuges for predators in winter, supplementing food resources in early spring to keep natural enemies on orchards even at low pest densities, and managing the diversity and composition of flora around pear orchards are some of the possibilities considered in our research into improving the IPM program.

Achievements and Main Constraints for a Wider Application of IPM in Catalonia

To summarize the main achievements and constraints in the application of IPM in Catalonia in the last few decades, it is important to recall two characteristics of this control strategy: that IPM systems are site-specific – they do and must vary by crop, cropping system and geographical area – and that in the development of IPM systems there is a continuum of activities and efforts from basic through applied research to field development and finally implementation by farmers or other uses. The schedule used by Gliss (1992) to review the activities needed for the implementation and adoption of IPM may allow us to discuss the progress made in the last few years and to identify the main constraints for a wider application of IPM in our area. The activities reviewed are research (basic, applied and field-oriented), extension, education, training, regulation, policy, and economic aspects. Of course, there is no discrete separation among all the categories but there are – and probably must be more – overlapping activities among two or more elements.

Research

There has been a significant increase in the last 30 years in the number of entomologists on the staff of Catalan institutions devoted to research: from seven in the late 1970s to about 25 in early 2000. Half of the current staff do field-oriented research in IPM, slightly less than a quarter are involved in more basic research – mainly insect physiology and behavior – and slightly more than a quarter deal mainly with insect taxonomy. There is an obvious gap in the research on insect ecology focusing on agroecology that is being filled by researchers working in IPM. Clearly, it is impossible to design management systems for our agroecosystems without a profound knowledge of how they function. The management of native natural enemies that has been described above in tomato crops and pear and apple orchards exemplifies how one can take advantage of the research into the ecology of agroecosystems.

Field-oriented research tends to produce results and hence publications at slower rates than more basic and laboratory research. In addition, field-research tends to be more expensive and more demanding in manpower. Consequently, young scientists are not stimulated to undertake such work without endangering their professional careers. Research projects in Catalonia may be funded by Catalan, Spanish and European administrations from public funds. All three R&D programs have the development of IPM systems as one of their priorities and, in fact, most of the research work carried out on IPM in Catalonia has been funded under their auspices. However, projects are generally funded for a maximum duration of 3–4 years to the detriment of long-term projects. Facilities like experimental fields or demonstration plots are somewhat lacking and only in the last few years have some experimental stations been created. Unfortunately, experimental stations are mostly supported by grower's organizations that prefer, once more, to give priority to short-term research.

Extension

Two services in particular have dealt with extension in Catalonia: the Extension Service and the Plant Protection Service. The former perished some years ago under mountains of bureaucracy; the latter is trying to survive budget cuts and lack of personnel. Many regular funds and resources are frequently devoted to special campaigns that have more impact on public opinion than the silent daily work, like the recent program for the prevention of fire blight of rosaceous crops. A worse consequence of the severe budget restrictions of the Catalan administration in the last few years has been the lack of young graduate recruitment leading to a decrease in the innovation potential. The most stimulating and profitable activity of the Plant Protection Service is the above-mentioned tutelage of pest control advisers of ADVs, a successful form of IPM technology transfer.

Education

European public opinion – particularly in northern countries – is very sensitive to the environmental impact of production technologies in agriculture. Public alarm about recent health problems of food (BSE, dioxins in chickens) and distrust of some recent scientific progress (e.g. transgenic crops) by Europeans have led many consumers to prefer 'natural' food or 'naturally-produced' food. Labels such as 'organic food' or 'integrated production' have proliferated in Europe under the auspices of regional governments or – in fewer cases – of scientific organizations like the IOBC. Retailers are also sensitive to this increasing demand in Europe for 'green' food and they promote agreements with farmers to impose clean production methods. In general, the perception that IPM is safer and healthier than chemical-based control is increasing among European consumers and also among farmers.

Training

The higher degree in agronomy, which in Catalonia is only taught at Lleida University, includes several courses on IPM and related matters. Great progress has been made in this area in Catalonia in the last 25 years. However, as in many other fields of scientific and technological knowledge, IPM needs to be constantly updated and curricula for training lecturers and researchers should be developed. Pest control advisers, who usually have a medium-level degree, and personnel in the private sector are in demand for periodic formal and less formal training programs in IPM tools and decision making.

Regulation

Regulations affecting IPM implementation and adoption in Catalonia, as in the rest of Europe, mainly deal with the registration of pesticides and non-chemical products. As stated above, most pesticide active ingredients are expected to disappear from the market in the coming 2 years. It is hoped that this will have a positive influence on the adoption of IPM, as several arthropod pests will not have an 'ad hoc' pesticide, but it may pose a problem because of the lack of selective insecticides for controlling secondary and occasional pests which will be treated with chemicals of a broader spectrum.

Several biological products, such as pheromones, need thick dossiers for registration, similar to those required for conventional chemical pesticides. This – and also their excessive prices – is seriously limiting the faster adoption of synthetic pheromones.

The register of macrobial (predators and parasitoids) and microbial insecticides is expected to change in Europe in the near future. Biological control has traditionally been considered as an environmentally safe technology. Recently, however, some scientists have expressed their concern about trading insect natural enemies between different geographical areas, an activity that may endanger native species. A positive list of natural enemies that can be commercialized in the whole of Europe with no special restrictions because they are native to the Continent or have been released for many years with no apparent side effects is being developed by the European and Mediterranean Plant Protection Organization; the other natural enemies will need careful screening procedures to prove that their release involves no risks for native fauna. Clearer registration rules for microbials, including those produced from genetic engineering processes, are also needed in Europe.

Growers in southern Europe, where there are only a few natural enemy producers, often complain about the quality of natural enemies that are commercialized. Quality control guidelines have been developed by researchers of private firms and public institutions and are nowadays available for the most commonly used natural enemies. These guidelines may be the basis for defining the official procedures of quality control and quality standards required for commercialization in Europe.

Catalonia is very active in the production and trade of ornamental plants and other commodities that are shipped quickly around the world. In addition, it is a pathway for millions of tourists and travelers. It seems difficult to prevent the entrance of exotic arthropod pests despite the establishment of quarantine (on a European scale) to regulate the movement of pest-risk items.

Policy

National and international R&D programs have supported research on IPM. As mentioned above, the development of IPM programs to reduce the impact of pesticides on the environment and human health has been one of the priorities of Catalan, Spanish, and European research agencies. Less support has been given to extension: validation and demonstration plots, training programs for field technicians and growers, and recruitment of young personnel are among the actions that could activate the extension of IPM in agriculture. Without doubt the support and subsidizing of growers' associations for plant protection (ADVs) have been key actions for speeding up the adoption of IPM systems in Catalan agriculture.

The organization of a network to coordinate all the components dealing with research, development, extension, technology transfer, and application of IPM would probably at least provide better communication and perhaps greater efficiency in the implementation of IPM in Catalan agriculture. Note that now there are at least eight public and several private institutions involved in this process.

Economic aspects

Though pesticides have increasingly evident negative impacts on the environment, human health and the technical efficiency of agriculture, they are still easy to apply and relatively cheap. These are two key advantages of conventional chemical control in comparison with innovative IPM, and they should therefore be two dominant objectives of new pest control programs. It is highly unlikely that IPM will be implemented on a large scale in the Mediterranean to replace chemically based methodologies if growers are only stimulated by environmental considerations and consumer health. This may be a factor – undoubtedly an important one as the proliferation of integrated production labels shows – but it is probably not the most decisive one, as Wearing (1988) points out in his review of incentives for the adoption of IPM systems in the world. We must implement IPM systems that are economically profitable and technically feasible. We can expect from authorities strict regulations of pesticide use but not its disappearance.

Lack of patents for biological products has often been put forward to explain the lack of incentives for private firms to become involved in the research, implementation and commercialization of IPM techniques. Some symptoms show that this situation is changing: more and more private funds are being invested in several fields of typical IPM tools and methods such as pheromones, biological control, monitoring techniques, devices for selective pest control, and insect-resistant crops. Additionally, selectivity is a major goal of the development of new insecticide active ingredients instead of the old objectives of broad-spectrum activity. Figure 26.4 shows a container that a Catalan grower's association uses to sell

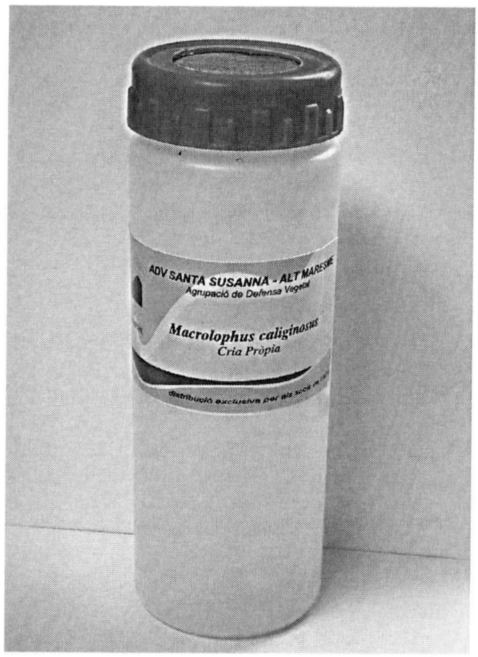

Fig. 26.4. A Catalan growers' association produces natural enemies for its own consumption. *Macrolophus caliginosus* is distributed in this bottle for use in greenhouses.

the self-produced predator *M. caliginosus* among its members. Without doubt, in the last few years some things have changed in the direction of IPM adoption.

Acknowledgments

We thank J. Ariño, M. Marti and M. Pagès, ADVs pest control advisers, for supplying information on acreage under IPM programs for tomatoes.

References

AEPLA (2000) Asociación Empresarial para la Protección de las Plantas. *Memoria 2000.* Madrid.

Albajes, R., Gabarra, R., Castañé, C., Alomar, O., Arnó, J., Riudavets, J., Ariño, J., Bellavista, J., Martí, M., Moliner, J. and Ramirez, M. (1994) Implementation of an IPM program for spring tomatoes in Mediterranean greenhouses. *IOBC/WPRS Bulletin* 17(5), 14–21.

Alomar, O. and Albajes, R. (1996) Greenhouse whitefly (Homoptera: Aleyrodidae) predation and tomato fruit injury by the zoophytophagous predator *Dicyphus tamaninii* (Heteroptera: Miridae). In: Alomar, O. and Wiedenmann, R.N. (eds) *Zoophytophagous heteroptera: implications for life history and integrated pest management.* Entomological Society of America, Lanham, Michigan, pp. 155–177.

Alomar, O., Castañé, C., Gabarra, R., Bordas, E., Adillon, J. and Albajes, R. (1989) Cultural practices for IPM in protected crops in Catalonia. In: Cavalloro, R. and Pelerents, C. (eds) *Integrated Pest Management in Protected Vegetable Crops.* A.A. Balkema, Rotterdam, The Netherlands, pp. 347–354.

Alomar, O., Gabarra, R. and Castañé, C. (1997) The aphid parasitoid *Aphelinus abdominalis* (Hym.: Aphelinidae) for biological control of *Macrosiphum euphorbiae* on tomatoes grown in unheated plastic greenhouses. *IOBC/WPRS Bulletin* 20(4), 203–206.

Alomar, O., Castañé, C., Gabarra, R., Arnó, J., Ariñó, J. and Albajes, R. (1991) Conservation of native mirid buds for biological control in protected and outdoor tomato crops. *IOBC/WPRS Bulletin* 14(5), 33–42.

Anonymous (1999) Eurostat, 2001. *Statistical yearbook 2000.* European Commission, Luxembourg.

Anonymous (2000) *Anuari Estadistic de Catalunya.* Generalitat de Catalunya, Barcelona, pp. 271 and 285.

Arnó, J., Ariño, J., Martí, M. and Tió, M. (1994) Seguimiento de las poblaciones de *Helicoverpa armigera* (Hübner) (Lepidoptera: Noctuidae) en cultivo de tomate. *Boletin de Sanidad Vegetol Plagas* 20, 251–260.

Arnó, J., Martí, M., Ariño, J., Ramírez, M. and Gabarra, R. (1996) El control integrat de plagues en tomàquet d'aire lliure. *Catalunya Rural i Agrària* 28, 29–32.

Arnó, J., Ariño, J., Español, R., Martí, M. and Alomar, O. (2000) Conservation of *Macrolophus caliginosus* Wagner (Het. Miridae) in commercial greenhouses during tomato crop-free periods. *IOBC/WPRS Bulletin* 23(1), 241–246.

Artigues, M., Avilla, J., Jauset, A.M. and Sarasúa, M.J. (1996) Predators of *Cacopsylla pyri* in NE Spain. Heteroptera: Anthocoridae and Miridae. *IOBC/WPRS Bulletin* 19(4), 231–235.

Avilla, J. and Bosch, D. (2001) Mass trapping and mating disruption for the control of leopard moth and apple clearwing moth. *Proceedings of the European Apple Symposium Biological and Alternative Protection in Apple: Orchards and Storage.* Centre Technique Interprofessionnel des Fruits et Légumes, Bordeaux (France).

Avilla, J., Bosch, D., Sarasúa, M.J. and Costa-Comelles, J. (1992) Biological control of *Panonychus ulmi* in apple orchards in Lleida (NE of Spain). *Acta Horticulturae* 347, 267–272.

Barnadas, I., Gabarra, R. and Albajes, R. (1998) Predatory capacity of two mirid bugs preying on *Bemisia tabaci. Entomologia Experimentalis et Applicata* 86, 215–219.

Bosch, D., Burballa, A., Sarasúa, M.J. and Avilla, J. (1998) Control de carpocapsa (*Cydia pomonella*) mediante confusión sexual y fenoxycarb. *Fruticultura Profesional* 99, 52–62.

Castañé, C., Alomar, O., Goula, M. and Gabarra, R. (2000) Natural populations of *Macrolophus caliginosus* and *Dicyphus tamaninii* in the control of the greenhouses whitefly in tomato crops. *IOBC/WPRS Bulletin* 23(1), 221–224.

Gabarra, R. and Besri, M. (1999) Tomatoes. In: Albajes, R., Gullino, M.L., van Lenteren, J.C. and Elad, Y. (eds) *Integrated Pest and*

Disease Management in Greenhouse Crops. Kluwer Academic Publishers, Dordrecht, The Netherlands, pp. 420–434.

Gabarra, R., Riudavets, J., Arnó, J., Castañé, C. and Albajes, R. (1996) Natural enemies associated to Lepidoptera pests in IPM tomato fields. Abstract. *IOBC/WPRS Bulletin* 19(8), 207 pp.

Gabarra, R., Arnó, J., Alomar, O. and Albajes, R. (1999) Naturally occurring populations of *Encarsia pergandiella* (Hymenoptera: Aphelinidae) in tomato greenhouses. *IOBC/WPRS Bulletin* 22(1), 85–88.

Gabarra, R., Arnó, J., Castañé, C., Izquierdo, J., Alomar, O., Riudavets, J. and Albajes, R. (2000) Fauna útil trobada en els cultius d'horta de Catalunya. In: Ticó, J. (ed.) *Enemics Naturals de Plagues en Diferents Cultius d'Horta.* Dossiers Agraris 6, pp. 83–103.

Gliss, E.H. (1992) Constraints to the implementation and adoption of IPM. In: Zalom, F.C. and Fry, W.E. (eds) *Food, Crop Pests, and the Environment.* APS Press, St Paul, Minnesota, pp. 167–177.

Sarasúa, M.J., Avilla, J., Artigues, M. and Jauset, A.M. (1999) Estrategia de control de la psylla del peral *Cacopsylla pyri* (L.), para el desarrollo de un programa de control integrado en peral. *Phytoma-España* 114, 86–89.

[SSV] Servei de Sanitat Vegetal de la Generalitat de Catalunya (2001) Incidencia de las plagas y enfermedades en las Comunidades Autónomas en el 2000. Cataluña. *Phytoma-España* 127, 20–28.

Torà, R., Sió, J., Sarasúa, M.J. and Avilla, J. (1995) Control integrado de plagas en huertos de manzano y de peral en Cataluña. *Fruticultura Profesional* 70, 36–51.

Wearing, C.H. (1988) Evaluating the IPM implementation process. *Annual Review of Entomology* 33, 17–38.

Chapter 27

Integrated Plant Protection Management in Russia

V.A. Zakharenko[1] and A.L. Il'ichev[2]

[1]*Russian Academy of Agricultural Sciences, Moscow, Russia;* [2]*Central Institute for Agrochemical Services, Moscow, Russia*

History of IPM in the Former USSR and Russia

Agriculture in Russia functions under adverse climatic and economic conditions. Despite a rich natural resource base, agricultural production in Russia has historically been far below its potential. Russia occupies nearly a seventh of the earth's land area (over 1.7 billion ha). Of this area 220 million ha (13%) are devoted to agriculture, and of that, 60% is considered arable. Most of the arable land is subject to significant limitations, such as inadequate rainfall, extensive salinity or moisture, limited growing season or difficult terrain. Only about 2 million ha of a vast belt of black soils have adequate rainfall and growing conditions.

In Russia, IPM is known as IPPM. IPPM is a complex system, using technical, cultural, physical and biological methods to counter a wide range of pests. Careful monitoring of pests and diseases, determination of economic thresholds and selection of the most appropriate control methods work to achieve cost-effective and sustainable IPPM systems. The development of IPPM in Russia and the USSR has a long history.

Biological control was pioneered in Russia by scientists such as I.I. Mechnikov (1879), who discovered a fungus affecting larvae of flour beetles, and Krasil'shik (1886), who produced green muscardine to control flour beetles. In 1929, the All-Union Scientific Research Institute of Plant Protection was formed, with the support of the First President of the All-Union Academy of Agricultural Sciences, N.I. Vavilov. Vavilov studied host-plant immunity and described the co-evolution of pests and host plants (Vavilov, 1935). Several biological control methods were developed at that time. In 1938, the official concept of 'control systems in plant protection' was developed, which consolidated several plant protection methods (Shegolev *et al.*, 1938) and established the fundamentals for IPPM.

In the 1940s the introduction of synthetic chemical pesticides changed the face of pest management in the USSR. But by the mid-1960s, it became apparent that broad-spectrum chemical treatments often led to the elimination of beneficial natural enemies. Eventually, high amounts of conventional chemical applications created problems such as resistance and environmental contamination (Fedorov and Yablokov, 1999). Biological control methods started to improve in the mid-1960s. The All-Union Scientific Research Institute of Biological Plant Protection was formed in Kishinev in 1969. By the end of the 1970s, biological control methods were well established. Enough biological pesticides and beneficial species were produced to treat an area of 23.8 million ha, a significant achievement at the time.

The IPPM system in the USSR included a widespread network of 40 All-Union plant protection systems and more than 100 regional plant protection systems. These systems provided IPM strategies for a wide diversity of agricultural crops (Chooraev *et al.*, 1981; Zakharenko *et al.*, 1985). Economic thresholds were developed for several insect pests, weeds and pathogens (Tansky, 1988; Zakharenko, 1979, 1990). A scientific basis for IPPM was outlined in the book *Integrated Plant Protection* in 1981 (Fadeev and Novojilov, 1981). Regional IPPM programs were developed in Ukraine by M.P. Lesovoy (1989, 1990), in Bielorussia by V.F. Samersov (1988) and in Uzbekistan by S.N. Alimukhamedov (1985).

Several Russian scientists contributed to the development of IPPM: N.N. Mel'nikov (1974) (pesticide chemistry), K.V. Novozhilov (1980) (insecticide application), N.M. Golyshin (1993) (fungicides), I.I. Gunar, and M.Ja. Beresovsky (1952) (herbicides). Research investigations of many Russian scientists have contributed to preparation and introduction of the rules for application of pesticides in plant protection (Fadeev and Novojilov, 1981).

After the collapse of the USSR, plant protection systems continued to develop under the auspices of the Russian Academy of Agricultural Sciences. Several applied scientific research institutes are involved in IPM research and adapting IPPM systems to changing economic conditions in Russia. In 1994, a description of IPM-based management strategies was published in the booklet: *The economical and organizational management of phytosanitary conditions of the agrocenosis* (Anonymous, 1994), very often used as a guidelines.

IPPM is designed to promote sustainable agriculture, minimize waste, and use energy efficiently. Basic tenets of IPPM include careful monitoring (of insect pests, diseases, and beneficial organisms), preventive measures (crop rotation, crop husbandry and hygiene, fertilization, irrigation, intercropping, proper harvesting and storage), and control measures (use of pest resistant and herbicide tolerant plant varieties, and biological, chemical, and cultural control methods).

Organization of the IPPM System in the Former USSR and Russia

The collapse of the USSR in 1991 brought about major changes in the structure of agriculture in Russia at every level. Russia has enacted several economic reforms intended to facilitate private sector participation in agriculture. In 1990, almost 99% of arable land was owned by collective and state-owned farms. By 2000, collective and state-owned farms controlled only 21% of the agricultural land. Most of the remaining arable land was owned and operated by 21,990 co-operatives and joint venture companies, and 6.9% of farmland was privately held in 261,000 farms. Some 6% was owned by 40 million private landowners, including collective orchards and vegetable farms. The former collective and state farms produced primarily grain, sunflower seed, sugarbeet, fiber flax and soybean. Private owners grew mostly potatoes, vegetables and fruits.

Besides the changes in ownership, the composition and distribution of cropland has also changed significantly. From 1991 to 2000, the area of cultivated agricultural cropland declined from 115.1 million ha to 88.3 million ha. Land area devoted to grain decreased from 61.8 million ha to 46.6 million ha, and to forage crops from 44 million ha to 30 million ha. Meanwhile, the area under industrial crops (fiber flax, sunflower, sugarbeet, etc.) has increased from 5.6 million ha to 7.5 million ha, and potatoes and vegetables have also increased from 4.1 million ha to 4.2 million ha. In general, crop yield has declined, with the exception of potatoes. One major reason for the decline in crop yield is the reduction in use of inputs such as fertilizers (from 11 million tons to 1.2 million tons) and pesticides, due to economic changes (Russian Agriculture, 2000).

Organization of the IPPM program

The Department of Plant Protection of the Russian Academy of Agricultural Sciences oversees research and development of IPPM in Russia. Local Plant Protection Stations are spread throughout the country. Plant Protection Stations work together with Diagnostic and Forecast Laboratories, which are directly involved in pest monitoring and threshold determination. The Department of Plant Protection includes the following research organizations, which employ a total of 491 scientists:

- All-Russian Scientific Research Institute of Plant Protection [VIZR], (St Petersburg–Pushkin);
- All-Russian Scientific Research Institute of Phytopathology [VNIIF], (Bolshie Viazemi, Moscow region);
- All-Russian Scientific Research Institute of Biological Plant Protection [VNIIBZR], (Krasnodar);
- The Regional Far East Scientific Research Institute of Plant Protection (Primorye Territory).

Research efforts are coordinated among more than 50 different departments, laboratories, scientific institutes and universities. The focus of IPPM research includes crop health and monitoring, host plant resistance, developing biological and chemical controls, equipment design, cost effectiveness and program development. IPPM research is usually carried out according to 5-year plans. The current 5-year plan (2001–2005) includes four major foci:

- monitoring of beneficial and harmful organisms and forecasting potential pest outbreaks;
- development of disease- and pest-resistant plant varieties, biological control products, and the design of genetically diverse agricultural ecosystems;
- testing new chemical products and engineering their application;
- development of principles for agricultural and ecological systems designed with optimum phytosanitary, ecological and toxicological characteristics.

Important crop pests in Russia

Table 27.1 illustrates the pest complexes of several important agricultural crops in Russia. The pest complexes were identified by several monitoring efforts by 'walk' through crops, pheromone trapping, sample identification, diagnostics, etc. These pests are the focus of IPPM research.

Weeds are also important pest species on agricultural crops. More than 120 weed species are frequently observed. The top ten weeds for each of six major crops are listed in Table 27.2.

Potential crop losses due to damage by pests, diseases and weeds are substantial (Zakharenko, 1975). The annual potential loss as a result of damage by pests, diseases and weeds was calculated (data from Zakharenko, 1975. Crop loss estimated using the method of Ouerke *et al.*, 1994) as follows: on grain crops –42.3%, fiber flax –42.8%, sugarbeet –48%, sunflowers and soybean –42.8%, potatoes –51.2%, vegetables –58.6%, fruit and berry crops –60.4% and forage crops –38.8%. The quantitative data on the significance and effect of pest damage to the yield of crops are stated as crop loss. Crop loss was calculated with respect to the data of weighted average percentage (%) of yield loss, yield and area of crop for the period of the 1996–2000 years and presented in Table 27.3.

The average loss was about 105 million t, expressed as grain equivalent, on the annual average yield for the 1996–2000 years (Zakharenko, 1997).

Table 27.1. Major insect pests and diseases of some important agricultural crops in Russia.

Field crops	Grain crops	Sugarbeet	Fiber flax	Sunflower	Potato
Locusts (Acrididae)	Sun bug (*Eurygaster integriceps*)	Sugar beet weevils (*Bothynoderes punctiventris*, *Tanymecus palliatus*)	Flax fleas (*Psylliodes* spp.)	Weevils (*Psalidium* spp., *Tanymecus* spp.)	Colorado potato beetle (*Leptinotarsa decemlineata*)
Cutworms (*Agrotis* spp.)	Ground beetle (*Zabrus tenebrioides*)	Leaf beetles (*Chaetocnema* spp.)	Flax thrips (*Thrips lini linorius*)	Aphid (*Brachycaudus helichrysi*)	Twenty-eight-spotted potato ladybird (*Epilachna vigintisexpunctata*, or *E. vigintioclomaculata*)
Beet webworm (*Pyraustra sticticalis*)	Leaf beetle (*Lema melanopus*)	Wood moths (*Cassidae* spp.)	Flax leafroller (*Cochylis epilinana*)	Sunflower pyralid (*Homoeosoma nebulellum*)	Late blight (*Phytophthora infestans*)
Click beetles (*Elateridae*)	Aphids (*Aphididae* spp.)	Bean aphid (*Aphis fabae*)	Crane flies (*Tipula paludosa*)	Lygus bug (*Lygus pratensis*)	Macrosporium leaf spot (*Macrosporium solani*)
Gophers (*Citellus* spp.)	Grain cutworm (*Apamea anceps*)	Seedling rot (*Pythium* spp.)	Anthracnose (*Colletotrichum lini*)	Bean aphid (*Aphis fabae*)	Bacterial ring rot (*Corynebacterium sepedonicum*)
Mice (*Microtinae* spp.)	Loose smut (*Ustilago tritici*)	Fusarium wilt (*Fusarium* spp.)	Ascochyta stem blight (*Ascochita linicola*)	Gray mold (*Botrytis cinnerea*)	Bacterial slimy soft rot (*Erwinia caratovora*)
	Stinking smut (*Tilletia tritici*)	Powdery mildew (*Erysiphe communis*)	Flax rust (*Melampsora lini*)	White rot (*Sclerotinia libertiana*)	Virus diseases
	Powdery mildew (*Erysiphe graminis*)	Cercospora leaf spot (*Cercospora beticola*)	Browning and stem break (*Polyspora lini*)	Charcoal rot (*Sclerotium bataticola*)	
	Rust (*Puccinia* spp.)	Virus diseases	Bacteriosis (*Bacillus macerans*)	Rust (*Puccinia helianthii*)	
	Fusarium wilt			Phomopsis rot	
	Septoria leaf spot				

Table 27.2. The top ten noxious weeds in major crops. (Data from Zakharenko and Zakharenko, 2001.)

Winter grain	Summer grain	Maize	Sugarbeet	Sunflower	Potatoes
Field bindweed (*Convolvulus arvensis*)	Sow thistle (*Sonchus arvensis*)	Field bindweed (*Convolvulus arvensis*)	Fat hen (*Chenopodium album*)	Field bindweed (*Convolvulus arvensis*)	Fat hen (*Chenopodium album*)
Creeping thistle (*Cirsium arvensis*)	Creeping thistle (*Cirsium arvensis*)	Sow thistle (*Sonchus arvensis*)	Pigweed (*Amaranthus* spp.)	Sow thistle (*Sonchus arvensis*)	Sow thistle (*Sonchus arvensis*)
Sow thistle (*Sonchus arvensis*)	Field bindweed (*Convolvulus arvensis*)	Pigweed (*Amaranthus* spp.)	Winter-cress (*Barbarea vulgaris*)	Creeping thistle (*Cirsium arvensis*)	Creeping thistle (*Cirsium arvensis*)
Fat hen (*Chenopodium album*)	Spring wild oat (*Avena fatua*)	Fat hen (*Chenopodium album*)	Creeping thistle (*Cirsium arvensis*)	Pigweed (*Amaranthus* spp.)	Common couch (*Agropyron repens*)
Winter-cress (*Barbarea vulgaris*)	Cockspur (*Echinochloa crus-galli*)	Winter-cress (*Barbarea vulgaris*)	Sow thistle (*Sonchus arvensis*)	Fat hen (*Chenopodium album*)	Wild radish (*Raphanus raphanistrum*)
Pigweed (*Amaranthus* spp.)	Green bristle grass (*Setaria* spp.)	Green bristle grass (*Setaria* spp.)	Field bindweed (*Convolvulus arvensis*)	Winter-cress (*Barbarea vulgaris*)	Common chickweed (*Stellaria media*)
Mayweed (*Matricaria inodora*)	Fat hen (*Chenopodium album*)	Cockspur (*Echinochloa crus-galli*)	Cockspur (*Echinochloa crus-galli*)	Green bristle grass (*Setaria* spp.)	Hemp nettle (*Galeopsis* spp.)
Green bristle grass (*Setaria* spp.)	Pigweed (*Amaranthus* spp.)	Spring wild oat (*Avena fatua*)	Green bristle grass (*Setaria* spp.)	Spring wild oat (*Avena fatua*)	Pigweed (*Amaranthus* spp.)
Wild oat (*Avena fatua*)	Winter-cress (*Barbarea vulgaris*)	Johnson-grass (*Sorghum halepense*)	Spring wild oat (*Avena fatua*)	Cockspur (*Echinochloa crus-galli*)	Cockspur (*Echinochloa crus-galli*)
Cockspur (*Echinochloa crus-galli*)	Scentless mayweed (*Matricaria inodora*)	Common couch (*Agropyron repens*)	Common couch (*Agropyron repens*)	Common couch (*Agropyron repens*)	Cleavers (*Galium aparine*)

Table 27.3. The quantitative crop loss due to damage done by pests, diseases and weeds for the period of the years 1996–2000.

Crops	Area (000 ha)	Yield (t/ha)	Yield losses		
			Rate (%)	Thousand t	Thousand t in grain equivalent
Grain crops	49,982.2	1.30	42.3	27,751	27,751
Fiber flax	116.4	0.28	42.8	16	399
Sugarbeet	920.6	15.22	48	6,371	3,231
Sunflower	4,558.8	0.74	42.8	1,426	3,322
Potatoes	3,306.4	10.26	51.2	17,627	31,376
Vegetables	773.8	14.73	58.6	6,678	17,363
Fruit and berries	1,006	3.19	60.4	1,938	9,692
Forage crops	31,831.2	2.00	38.8	24,791	11,609
Total:	93,215				104,743

IPPM and Pesticide Use Policy in the Former USSR and Russia

The current list of 391 registered pesticides (based on 227 active ingredients) was authorized in 1998. The list includes: 97 insecticides (49 pyrethroids and 17 organophosphate products), 70 fungicides (18 azoles, eight benzimidazoles, five dithiocarbamates and five copper-based formulations); and 139 herbicides (19 sulfonylurea, three phenoxy acetic acid, ariloxyphenoxypropionic acid, thiocarbamatic acid, and 12 organophosphates).

Search for better chemical pesticides

Chemical control is still an important part of plant protection systems. Research seeking improved, safer and effective chemical pesticides is conducted at the research institutes of the Department of Plant Protection of the Russian Academy of Agricultural Sciences. The search is more effective when the use of dangerous substances influencing synthesis of amino acid and photosynthetic processes, absent in human and animals, is minimized. These are products such as pheromones, inductors of protective reactions of plant to pest damage (immunocitoifit, analogs chitosan, etc.), sulfonylurea herbicides, imidasolinon, herbicide mixtures (cowboy, kross, kronos) and complex products including mixture of fungicide, insecticide and herbicide (koprangs), which are active in doses of grams per hectare.

Biological control

The Department of Plant Protection of the Russian Academy of Agricultural Sciences is actively developing biological control methods and agents. The All-Russian Scientific Research Institute of Plant Protection (VIZR) has created a bank of microorganisms, including more than 500 cultures of bacteria and *Actinomycetes*, 120 fungi, 22 insect pathogenic nematodes and *Microsporidia*, and 15 insect viruses. The bank is used for biological research and development of microbiological products for plant protection.

There are 60 biological control products developed for agriculture in Russia, including 26 formulations for insect pests and 19 products based on Bt: six varieties of Bt (*Bt.* var. *dendrolimus*, *Bt.* var. *galleria*, *Bt.* var. *insektus*, *Bt.* var. *kurstaki*, *Bt.* var. *tenebrionis*, *Bt.* var. *thuringiensis*), two *Beauveria bassiana*, three *Verticillium lacanii*, one nematode (*Steinernema carpocapsa*), and microbiological products based on *Bacillus subtilis*, *Penicillium vermiculatum*, *Pseudomonas syringae*, *P. fluorescens*, *Streptomices griceoviridis*, *S. lavendula*, *S. falleus*, *Trichoderma lignorum* and virus products.

The Department established a regional system for manufacture of microbiological products that includes 75 biological laboratories, 13 biological factories, 41 regional plant protection stations, 159 laboratories in hothouse facilities, and 56 small private enterprises and co-operative organizations. The laboratories receive virulent strains of microorganisms directly from the producers.

In 1996, the laboratories produced about 350 t of Bactorodenced with the active strain of *Salmonella eneritidis* var. *Issatschenko*; 150 t of Trichodermin with *Trichoderma* spp.; 280 t of Zhizoplant based on *Pseudomonas* spp.; 19 t of Lepidoced with culture of *Bt* var. *kurstaki*; 15 t of Boverin based on *Beauveria bassiana*; 11 t of Agat with *Pseudomonas fluorescens*; and 10 t of Bitoxibacillin with *Bt* var. *thuringiensis*.

Currently, the following biological products may be effective for protection:

- of grain crops from root rot – agat 25, rhizoplant, phitolavin;
- of technical crops from beet webworm (*Pyraustra sticticalis*), cutworms (*Agrotis* spp.) – baxin, bitoxibacillin, bicol, lepidoced, dendrobacillin;
- of potatoes from Colorado potato beetle – bitoxibacillin, bicol, dicimid, colorado, from *Phytophthora* infection – rhizoplan, agat;

- of vegetable crops from *bacterial wilt* – rhizoplan, phitolavin, from wireworm – trichodermin, from *Lepidoptera* insects – astur, baxin, lepidoced, dendrobacillin, bitoxibacillin, bicol, homelin, *Trichogramma* spp.;
- of fruit crops from *Podosphaera leucotricha* and *Venturia* spp. – bactofit, from *Ervinia amylovora* – pentafag.

The following biological products which are not considered pesticides are used to control diseases in greenhouses such as:

- root rot – bactofit, trichodermin, agat 25, rhizoplan;
- powdery mildew – bactofit;
- angular leaf spot – pentafag;
- bacteriosys – phitolavin;
- white fly – verticillin, boverin, *Encarsia*;
- spider mite – *Phytoseiulus*, bicicol, bitocsibacillin;
- tobacco thrips – *Amblyseius*, boverin;
- aphids – *Chrisopa*, *Aphidoletes aphidimyza*, *Cycloneda limbifera*, *Aphidius matricariae* (Bondarenko, 1986; Zakharenko, 2000a).

Advantages of biological control

Economically, biological control methods have been demonstrated to be efficient when used for vegetables and a wide range of fruit and berry crops. Biological methods are also advantageous for companies producing baby and health food, and when crops are grown near large urban centers, in water reserve and sanitary zones, and near areas with radioactive pollution. Biological control methods are used on 1.5–2 million ha (1–2 % of all arable land).

Management of pest resistance using transgenic crops

Field experiments done by VIZR, VNIIF, VNIIBZR (1998–2000) demonstrated considerable potential for transgenic plants to control pest insects such as Colorado potato beetle (*Leptinotarsa decemlineata*) and corn borers (*Ostrinia* spp.). Some herbicide-tolerant transgenic crops have been developed (maize, sugarbeets, soybean) (Zakharenko, 2000b). The genetic make-up of transgenic plants has been described in *Bio-Pesticide Manual* (Copping, 1998).

ULV application

Ultra low volume (ULV) pesticide formulations have several benefits, including cost effectiveness and reducing pesticide pressure to the environment. There is a need for modern, efficient spray equipment capable of complying with new stringent environmental requirements for pesticide application. Application equipment is being designed under the conversion program of the Defence Ministry involving the Institutes of the Plant Protection Department and an ex-military producer. A prototype sprayer is able to ionize droplets during the working cycle, reducing spray loss and enhancing the efficiency of application.

Pesticide production has declined in recent years, in part due to national economic reforms. Pesticide production declined from 215,600 t in 1986–1990 to 17,000 t in 1991–1997, and as a result, pesticide use has declined to 29,600 t and agricultural land treated with pesticides has decreased from 76.9 to 28.4 million ha (Table 27.4). Average pesticide use per hectare in Russia was estimated at 0.1–0.16 kg of formulated product (124,537,000 ha), but at the same time the world use of pesticides in 1996 equalled 1.6 kg of active ingredient per hectare of arable land at the sales level (Calderoni, 1997).

The same decrease happened with pesticide use on major agricultural crops in Russia (Table 27.5). For example, in 1998 only 47% of total grown area of cereals has been treated with pesticides, where herbicides were applied on 11,783,000 ha. Maize for grain had been treated only with herbicides and crops grown for grain legume had been treated only against insects with insecticides (Table 27.5).

The effectiveness of the actual crop protection practice was calculated for the 1991–1995 years as the percentage of prevented losses by chemical treatments. The amounts of losses prevented by the use of pesticides was determined on the basis of the data of crop and area treated with pesticides and the amount of additional yield per hectare of areas treated with pesticides (Table 27.6).

Crop losses prevented by the use of chemical and biological products for crop protection were 14.2 million tons of grain units (Zakharenko, 1998). Considering the enormous potential of plant protection in preventing yield losses, the possible future

Table 27.4. Average annual use of agricultural pesticides (t) in Russia during 1986–2000.

	1986–1990	1991–1995	1996–2000
General pesticide use, formulated product, t	215,566	51,710	29,625
Herbicide application/thousand ha	32,442	16,273	16,007
Insecticide application/thousand ha	23,352	12,049	9,273
Fungicide application/thousand ha	13,155	5,829	2,924
Total pesticide application/thousand ha	68,949	34,151	28,400

Table 27.5. Pesticide use on major agricultural crops in Russia in 1998.

Crops	Areas grown ('000s ha)	Total areas treated ('000s ha)	Areas by product type ('000s ha)			Pesticide usage (%)
			Herbicide	Insecticide	Fungicide	
Cereals	29,821	13,988	11,783	981	1,234	47
Rice	146	27		13	14	19
Grain legume (peas and other)	1,784	311		311		17
Maize, grain	791	633	633			80
Foot root crops	3,265	152	100	52		5
Rapeseed	276	213		213		77
Sugarbeet	810	1,014	802	132	80	125
Fiber flax	106	140	116	22	1.5	132
Sunflower	4,167	208	190	15	3	5
Soybean	487			7		
Potatoes	3,265	2,251	101	1,730	420	69
Vegetables	745	356	123	193	40	48
Fruit and berry	919	656		383	273	0.71
Grape	80	442		114	328	5.53

Table 27.6. The losses prevented by chemical treatments in Russia during 1991–2000.

Crops	Treated area ('000 ha)				Additional yield	
	Herbicide	Insecticide	Fungicide	Biological control	('000 t)	To potential losses (%)
Grain crops	11,203	3,726	3,275	913	3,127	10.3
Fiber flax	285	78	10	30	2.21	34.9
Sugarbeet	1,330	441	145	365	1,447	32.6
Sunflower	751	143	130	61	428	9
Potatoes	229	1,878	1,213	87	4,206	13.9
Vegetables	250	324	229	571	3,074	17.9
Fruit and berries	15	1,272	1,105	151	1,378	21
Forage crops	1,511	2,035	126	50	332	3

trends over the period of 1999–2005 should include enhanced pesticide use and other control measures in IPPM.

Research and Extension Focus in IPPM

Wide-scale adoption of IPPM practices has been encouraged by both government and non-government organizations. The Department of Plant Protection network has made monitoring data available for pest, disease and weed infestations. Plant protection systems have been introduced on state, collective and private agriculture enterprises throughout the country.

The IPPM program is a complex and continually improving system. Four institutions in Russia share responsibility for education, training, technology development and plant protection research: the Russian Academy of Agricultural Sciences, the Ministry of Agriculture, the Ministry of Education, and the Ministry of Science and Technology Policy. Education in IPPM practices includes secondary vocational technical schools, agricultural colleges or specialized secondary education institutions, and higher education institutions. There are also a number of academies and universities in Russia offering higher level agricultural education such as Moscow Agricultural Academy and St Petersburg State University. The majority of the agricultural research institutes are under the auspices of the Russian Academy of Agricultural Science and supervised by the Ministry of Agriculture.

The state plant protection service in Russia organizes education and training programs for farmers in subjects such as sampling methods, economic and action thresholds, chemical, natural and biological products, critical stages for pest control, and application techniques. In 1999, the state service organized 1996 seminars, 21 exhibitions, 1891 TV and radio interviews, published 5886 articles, and gave 29,853 lectures and 258,001 consultations.

In 1993, the Russian Ministry of Agriculture began to improve advisory services focused on the re-structuring of Russian agriculture. In 2000, a network of 53 regional advisory agencies was established with 37 extension services. These agencies provide a wide range of necessary services, such as consultations in pest, weed and disease control and crop monitoring. The agencies were provided with recommended retail price lists to make the service affordable for all customers (Goats, 2000).

The state plant protection service has organized support services for 27,000 state and co-operative agricultural companies, 279,000 private farms and 42,000,000 primary agricultural producers on an area of 208.4 million ha, including 126 million ha of arable land and 90.9 million ha of cultivated land. Cultivated land consisted of 50.8 million ha of grain crops, 4.1 million ha of sunflower, 0.82 million ha of sugarbeet, 0.48 million ha of soybean, 0.11 million ha of fiber flax, 3.3 million ha of potatoes, 0.74 million ha of vegetable crops, 30 million ha of fodder cultures and 1 million ha of fruit, berry and grapes.

Case Studies of IPPM in the former USSR and Russia

Reduction in herbicide use

From 1975 to 1993, the researchers from Moscow Agricultural Academy conducted long-term field trials of crop rotation with different herbicide application programs at the experimental farm in the Moscow region. The trials involved rotating crops such as winter wheat, potatoes, barley, vetch, and oat mixture in control (conventional herbicide application program) and experimental (IPPM with reduced herbicide application program) treatments. Control blocks were treated with 1.96 kg/ha of herbicides annually from 1975 to 1993. IPPM blocks were treated with herbicides at 0.9 kg/ha annually, reducing chemical inputs by 54.1%. The average crop yield in the experimental blocks was 5.08 t/ha, 1.2% higher than the yield in the control block of 4.57 t/ha (Zakharenko and

Zakharenko, 1995). Similar results were observed during long-term trials at the state farm enterprises for grain cultivation (Rtischeva *et al.*, 1994). The results of these are shown in Table 27.7.

IPPM strategies in orchards

Insect growth regulators

New improved recommendations for protection of apple orchards in Krasnodar region against Lepidopteran pests were developed by 'Slavajanskaja' Plant Protection Station of VIZR. The insect growth regulator Insegar was recommended, as it has low toxicity for mammals and does not affect beneficial insects. Insegar was effective against insects that have developed resistance to organophosphates and synthetic pyrethroids. Insegar applications on 1230 ha of apple orchards in Orchard Gigant Ltd in the Krasnodar region decreased the number of chemical sprays by half compared with organophosphate and pyrethroid-based spray programs. Crop yield during these trials reached 23,000 kg/ha.

Biological control in apples

IPPM in apples includes several biological control methods (Sazonov, 1988). *Bacillus thuringiensis*, the terpene-based product Biostat, an animal-based product Hitozan, entomophagous *Habrobracon* spp. and *Elasmidae*, and pheromones are all used. The research centre 'Kuban' conducted field trials in the Krasnodar region. Results showed that biological control methods were at least as effective as 12 applications of conventional chemicals. The biological methods were also half as expensive.

Table 27.7. Efficiency of wheat production under IPPM with reduced herbicide application program at state owned and collectives farm enterprises in Russia during 1987–1993.

Farm name and location	Additional yield (t/ha)	Reduction in pesticide use (%)
'Gigant', Rostov	1.06	66.7
'Elizavetinscoje', Saratov	1.03	57.7
'Promcor', Voronezh	0.33	0.1
'Katchevskoje', Novosibirsk	0.27	75

Area-wide mapping and mass-trapping of major pests using sex pheromone traps on vegetable fields

Pheromone-based IPM strategies have been used for many years in the former USSR. Insect sex pheromones were probably the most widespread and certainly the most widely documented IPM tools in the former USSR and Russia (Lebedeva *et al.*, 1984). Insect sex pheromones have been used in IPM: for discovering species of insects in natural ecosystems and agricultural environments for taxonomic and plant protection investigations (Il'ichev *et al.*, 1981; Il'ichev, 1987); for detection (early warning) and monitoring of pests, threshold determination (timing treatments and sampling methods) and for density estimation (risk assessments, effects of control measures) (Lebedeva *et al.*, 1984); for forecasting of population density, trends and dispersion (Il'ichev *et al.*, 1989); for mapping of infestation distribution (hot spots) and risk assessment (Il'ichev, 1991); for application of sex pheromone barriers to localize hot spots (Il'ichev, 1991); for mass trapping (Boorov and Sazonov, 1987; Il'ichev and Zakharenko, 1991); and for mating disruption application (Sazonov, 1988; Zakharenko and Il'ichev, 1991).

This case study describes area-wide monitoring, mapping and mass-trapping of major vegetable pests in 220 ha of the Issyk-Kuhl Lake region of Kirgizia and in 110 ha of the Crimea region of Ukraine during 1987–1990 (Il'ichev, 2000).

Pheromone traps (Attracon AA) with the sticky base Pestifix (Flora Ltd, Tartu, Estonia) and sex pheromone dispensers for the noctuid moths *Agrotis segetum* Schiff., *Agrotis exclamationis* L., *Amathes c-nigrum* L., *Autographa gamma* L., *Mamestra brassicae* L., *Scotogramma trifolii* Hbn., were used for monitoring and control. The sex pheromone traps were distributed as follows: one trap/3–5 ha for each species for monitoring of initial pest population, one trap/ha for detailed mapping and identification of hot spots, and four traps/ha for mass-trapping in hot spots. Daily averages of 0.8 *A. segetum*, 1.7 *A. exclamationis*, 0.6 *Am. c-nigrum*, 0.8 *S. trifolii* per trap were recorded during mapping (880 traps in 220 ha) of tomato fields in the Issyk-Kuhl Lake region. These levels of infestation were below the recommended economic threshold and consequently, regular insecticide applications were postponed.

Daily averages of 4.8 *A. segetum*, 6.2 *A. exclamationis*, 2.6 *Au. gamma*, 6.4 *M. brassicae* per trap were found in tomato and cabbage fields in Crimea (440 traps in 110 ha). These levels were above the recommended economic threshold and 'hot spots' of each species were identified. Mass-trapping in hot spots with four traps/ha for each pest reduced the infestations and avoided insecticide applications. Detailed mapping of species distribution and movement with pheromone traps indicated that the cutworms *A. segetum* and *A. exclamationis* were concentrated on the edges of the vegetable fields, but armyworms including *M. brassicae*, *S. trifolii* and *Au. gamma* were distributed randomly and concentrated in hot spots throughout the field. In this study, area-wide application of mapping and mass-trapping reduced insecticide application during four consecutive seasons, therefore benefiting the local environment.

Important Websites, Publications and Reports on IPM in Russia

http://www.rsl.ru/ – home page of the Russian State Library

http://www.mosinfo.com.ru – home page of the East View Publications (periodicals)

http://home.eastview.com/epubs.shtml – home page of the East View Publications (periodicals) where Russian Scientific Periodical Journals can be found.

http://www.gpntb.ru/ – home page of the State Public Scientific and Technical Library

http://www.cnshb.ru/csal/general.htm – home page of the Central Scientific Agricultural Library

http://ben.irex.ru/ – home page of the Library for Natural Sciences of Russian Academy of Sciences (English version)

http://www.cnshb.ru/csal/izdat/cx_bl_e.htm – The journal *Selskokhozyaistvennaya biologiya*

http://www.cnshb.ru/csal/izdat/cx_bl_e.htm – The journal *Agrokhimicheskii vestnik*

http://www.cnshb.ru/csal/izdat/dokl_ak_e.htm – The Journal *The Report of Russian Academy Agricultural Sciences* (*Doklady Rossel'khozakademiya*)

http://www.integrum.ru/eng/ – The home page of the information agency 'Integrum-Tekhno'. They collaborate in using IRS 'Artefakt' in libraries, the system that allows retrieval from full-text documents with regard for peculiarities of the Russian and English languages.

http://www.cnshb.ru/aw/nii/nii_list.htm – home page of Agricultural Scientific Research Institutes in Russia (AgroWeb in Russia)

http://www.aris.ru/GALLERY/ROS/NAUCH/NII/RASHN/1.html – home page of the Russian Agricultural Academy

http://www.aris.ru/GALLERY/ROS/NAUCH/NII/RASHN/perehen0.html – home page and list of all Scientific Research Institutes belonging to the Russian Agricultural Academy

http://www.cnshb.ru/csal/indengl.htm – home page of the Central Scientific Agricultural Library

http://www.ras.ru/ – home page of the Russian Academy of Sciences

References

Alimukhamedov, S.N. (1985) [Improvement of integrated plant protection.] *Zashchita Rastenii, Moscow* 9, 2–4.

Anonymous (1994) [*The Economical and Organizational Management of Phytosanitary Conditions of the Agrocenosis.*] Rosselkhozakademia, Moscow, 38 pp.

Bondarenko, N.V. (1986) [*Biological Plant Protection.*] Kolos, Moscow, 280 pp.

Boorov, V.N. and Sazonov, A.P. (1987) [*Biological Active Products for Plant Protection.*] Agropromizdat, Moscow, 199 pp.

Calderoni, P. (1997) Pesticide Industry, Overview. *Chemical Economics Handbook* – SRI

(Salient Research Institute) Industry International, SRI International, USA, Zurich, Tokyo.

Chooraev, I.A., Zakcharenko, V.A. and Goncharov, N.P. (1981) [*Organization and Economy of Plant Protection – Integrated Plant Protection.*] Kolos, Moscow, 113 pp.

Copping, L.G. (ed.) (1998) *The Bio-Pesticide Manual*, 1st edn. BCPC, Farnham, 323 pp.

Goats, V.V. (2000) [*Development of Regional Information Advisory Service by Example of Nizniy Novgorod region.*] FGNY, Rosinformagrotech, Moscow, 384 pp.

Golyshin, N.M. (1993) [*Fungicides.*] Kolos, Moscow, 316 pp.

Gunar, I.I. and Beresovsky, M.Ja. (1952) [*Chemical Method to Control Weeds.*] Sel'khozgiz, Moscow, 200 pp.

Fadeev, Y.N. and Novojilov, K.V. (1981) [*Integrated Plant Protection.*] Kolos, Moscow, 334 pp.

Fedorov L.A. and Yablokov, A.V. (1999) *Pesticides: a Toxic Blow to the Biosphere and Humans.* Nauka Publishing House, Moscow, 462 pp.

Il'ichev, A.L. (1987) [*The Sex Pheromones of Harmful Insects. Part 1 (Clothes-moths and Tortricid moths).*] NIITEKHIM, Moscow, USSR, ISSN 0203-7955, 59 pp.

Il'ichev, A.L. (1991) [Use the pheromones traps for the mapping of vegetable crop pests centres and creation of the pheromone barriers.] *Zashchita Rastenii, Moscow* 5, 42–43.

Il'ichev, A.L. (2000) Areawide mapping of major pests abundance with the sex pheromone traps and mass-trapping in hot spots on vegetables in the former Soviet Union. In: *Abstracts of the XXI-th International Congress of Entomology*, Foz do Iguassu, Brazil, 20–26 August, 2000, Vol. 1, 171.

Il'ichev, A.L. and Zakharenko, V.A. (1991) [Use of sex pheromones in horticulture.] *Chemizatcia Selskogo Khozaistva, Moscow* 5, 54–57.

Il'ichev, A.L., Kondrat'ev, Iu.A. and Roslavtceva, S.A. (1981) [*Sex attractants of Lepidoptera and their Use in Plant Protection.*] NIITEKHIM, Moscow, ISSN 0203–7955, 35 pp.

Il'ichev, A.L., Binkin, V.A. and Dvusherstov, M.G. (1989) [*Methodical Instructions by Use Sex Pheromones of* Agrotis segetum *and* Agrotis exclamationis *for prognosis and diagnostics.*] CINAO, Moscow, 8 pp.

Krasil'shik, I.M. (1886) [The reasons of pest disappearance and fungus manufacture for pests control.] In: *VI Regional Entomological congress in Odessa.* 11 pp.

Lebedeva, K.V., Miniailo, V.A. and Piatnova, Y.B. (1984) [*Insect Pheromones.*] Nayka, Moscow, 267 pp.

Lesovoy, M.P. (ed.) (1989) [*Environmental Plant Protection.*] MP, Kiev, 168 pp.

Lesovoy, M.P. (ed.) (1990) [*Plant protection of intense agriculture in Ukraine Soviet Socialist Republic.*] MP, Kiev, 141 pp.

Mechnikov, I.I. (1879) [*Flour beetles. Diseases of flour beetle larva. Investigations done by I.I. Mechnikov.*] Frantsov Printery, Odessa, 32 pp.

Mel'nikov, N.N. (1974) [*Pesticide Chemistry.*] Khimia, Moscow, 495 pp.

Novozhilov, K.V. (1980) [*Plant Protection Chemicals.*] Khimia, Moscow, 268 pp.

Ouerke, E.-C., Dehne, H.-W., Schonbeck, F. and Weber, A. (1994) *Crop Production and Crop Protection.* Elsevier Science, Amsterdam, 808 pp.

Rtischeva, I.A., Chenkin, A.F., Goncharov, N.P. and Zakharenko, A.V. (1994) [*The economical and organisational management of phytosanitary conditions of the agrocenosis.*] Rosselkhozakademia, Moscow, 38 pp.

Russian Agriculture (2000) Department of Economy, Ministry of Agriculture, Moscow, Russija, FGNY *'Rosinformagrotekh'*, 48 pp.

Samersov, V.F. (1988) [*Integrated protection of grain crops against pest insect.*] Yuradjay, Minsk, 207 pp.

Sazonov, A.P. (ed.) (1988) [*Insect pheromones and development of their use in practice.*] VIZR, Leningrad, 133 pp.

Shegolev, V.N., Znamenskyi, A.V. and Bei-Bienko, G.Ja. (1938) [*Insects – pests of field crops.*] Sel'khozgiz, Moscow–Leningrad, 538 pp.

Tansky, V.A. (1988) [*Biological base of insect harmfulness.*] Agropromizdat, Moscow, 180 pp.

Vavilov, N.I. (1935) [*Science about plant immunity to infections and diseases.*] Sel'khozgiz, Moscow–Leningrad, 100 pp.

Zakharenko, V.A. (1975) Economy and perspective of pesticide application in conjunction with intensification of agriculture. In: *Abstracts of the Eighth International Plant Protection Congress, Moscow, USSR, 1975*, Vol. 2, 1–6.

Zakharenko, V.A. (1979) [*Method of Economic Threshold Estimation for Weeds and Expediency of Herbicide Application.*] Vaskhnil, Vniiash, Moscow, 38 pp.

Zakharenko, V.A. (1990) [*Herbicides.*] Agropromizdat, Moscow, 239 pp.

Zakharenko, V.A. (1997) [Tendency of changes in yield losses of agricultural crops due to pest damage under the economy reformation conditions in Russia.] *Agrokhimiia, Moscow, Nauka* 3, 65–75.

Zakharenko, V.A. (1998) [Economy and effectiveness of chemical plant protection under the economy reformation conditions in Russia.] *Agrokhimiia, Moscow, Nauka* 11, 84–92.

Zakharenko, V.A. (2000a) [The state of biological plant protection, promising direction of its fundamental and applied priority research within the system for phytosanitary optimisation of crop production]. In: *Urgent Problems of Biological Plant Protection.* The Russian Academy of Agricultural Sciences, Pushchino, Moscow, pp. 11–26.

Zakharenko, V.A. (2000b) [Selection of transgenic plants and their use in plant protection.] *Selskokhozyaistvennaya Biologiya* 3, 30–49.

Zakharenko, V.A. and Il'ichev, A.L. (1991) [Planning work volumes and pheromone material requirements for forecasting and plant protection.] *Zashchita Rastenii* 9, 44–47.

Zakcharenko, V.A., Chenkin, A.F. and Cherkasov, V.A. (1985) [*Plant Protection Guide.*] Agropromizdat, Moscow, 415 pp.

Zakharenko, V.A. and Zakharenko, A.V. (1995) [Estimation of an economic efficiency of the integrated protection of plants (with an example of protection crops from weed plants).] *Bulletin IOBC /EPRS, Moscow* 31, 10–16.

Zakharenko, V.A. and Zakharenko, A.V. (2001) [Phytosanitary situation of agroecosystems and prospects optimization its weediness in Russia.] In: *Novel Approaches to Weed Control Using New Classes of Herbicides and Transgenic Plants Resistant to Herbicides.* Bioengineering, Russian Academy of Sciences Information Centre, Moscow, pp. 15–25.

Chapter 28
Integrated Pest Management in Australia

D.G. Williams and A.L. Il'ichev
Institute of Sustainable Irrigated Agriculture, Tatura, Victoria, Australia

Introduction

The history of IPM in Australia began with biological control. Australia has been actively involved in biological control of pests and weeds since the early 1900s (Wilson, 1960). IPM in Australia followed similar trends to other countries with the development of damage thresholds and monitoring systems that allowed reliable estimates of when pest populations approached action thresholds. Eventually, the discovery of pesticide resistant strains of predators and parasitoids allowed integration of biological control with chemical control methods. IPM systems in Australia are most advanced in high-value crops such as pome and stone fruits (Williams, 2000a), cotton (Ives *et al.*, 1984), wine grapes (Madge *et al.*, 1993; Glenn *et al.*, 1998), and citrus (Smith and Papacek, 1993). The pome fruit industry has played a leading role in the development of IPM with national guidelines for integrated fruit production (IFP) in apples (Williams, 2000b) and the funding of a large-scale collaborative project to facilitate the adoption of integrated pest management. Other industries such as tomatoes, crucifers, sweetcorn, greenhouse crops and potatoes are rapidly adopting IPM.

Registration of Agricultural and Veterinary Chemicals

Since 1995, the Commonwealth (national) government has been responsible for evaluation, registration, review and control of agricultural and veterinary (AgVet) chemicals to the point of retail sale. The States and Territories are responsible for the control of use of AgVet chemicals, including licensing of pest control operators and aerial spraying, and they administer separate legislation for this purpose. Prior to 1995, the states, territories and national government performed these functions independently.

Criteria for registration include no unacceptable risk to humans, the environment or agricultural exports, and an accurate description of the product and its efficacy for the recommended uses. Several government bodies assess different aspects of the pesticide registration process. Environment Australia undertakes assessment of the environmental hazards associated with use of AgVet chemicals. The Department of Health and Family Services reviews the potential to affect human health. The National Occupational Health and Safety Commission reviews risk to those handling, applying, or otherwise exposed to the chemicals in the workplace. If the chemical poses a risk of poisoning if improperly used, the National Drugs and Poisons Schedule Committee conducts a review. Preliminary Maximum Residue Limits are suggested as part of the registration process. The Australia New Zealand Food Authority is responsible for the final assessment and approval of the residue limits.

In 1993, the Commonwealth established an independent statutory body, the National Registration Authority, and codified its regulatory powers in the Agricultural and Veterinary Chemicals Code (AgVet Code). Further details are available at http://www.affa.gov.au/nra/legislat.html The National Registration Authority is also responsible for periodic review of existing AgVet chemicals. Recent review of a number of organophosphate and organochlorine pesticides has led to the restriction or withdrawal of several pesticide products from the market. These products were either an occupational health and safety risk or a risk to export of other commodities by accidental contamination (drift or animal consumption of treated feed). Changes in pesticide registration status can have major effects on the implementation of IPM, especially if alternative products are not available. Often, alternative products are either not as effective or have toxic effects on natural enemies that are resistant to other pesticides.

Regulation of biological control agents

The need for caution when importing biological control agents is best illustrated by the great cane toad *Bufo marinus* (L.), imported to control cane beetles in Australian sugarcane fields (Wilson, 1960). The toad has a voracious appetite for native insects, frogs, small birds and rodents, and has glands that exude highly toxic secretions.

Today, permits are required before any biological control agent can be introduced into Australia. Permits are obtained through the Australian Quarantine and Inspection Service. There has been considerable discussion over the last decade regarding the appropriate protocols for testing host specificity and other criteria for potential biological control agents (Cullen, 1992, 1993; Field, 1993; Sands, 1993, 1998; Heard and van Klinken, 1998).

Policy Support for IPM in Australia

In 1987 the World Commission on Environment and Development released the Brundtland Report. This document clearly identified that economic growth patterns at the time could not be sustained without major changes in attitudes and actions. In 1989 the Australian Government responded with the release of a public discussion paper on a proposal for a NSESD. After extensive consultation and negotiation between key interest groups within Australia the NSESD was endorsed by Heads of Government in 1992 and established the framework for cooperative decision making in government and the promotion of ecologically sustainable development throughout Australia. Working groups were established in nine areas, including agriculture. The Agriculture Working Group reported that the development of integrated policies and programs for natural resource management were essential, and should promote community self-reliance. Farmers were encouraged to integrate property management plans with regional land management approaches and good business practices. The strategy also sought to reduce and manage the impact of pest species, and to ensure that AgVet chemicals were managed safely.

One of the five objectives in the Agriculture section of the NSESD stated that the national government will encourage research in IPM and decision-support systems, and continue to support integrated approaches to pest management through a combination of biological, chemical and cultural control measures. Further details of the NSESD can be found at http://www.environment.gov.au/psg/igu/nsesd/

National agricultural policies are jointly administered by the ARMCANZ and the

SCARM. ARMCANZ includes leaders responsible for agriculture and resource management from the state, territorial, and national governments of Australia and New Zealand. SCARM represents government agencies responsible for agriculture, soil, water and rural adjustment policy. Animal and plant health issues are overseen by the Australian Department of Agriculture, Fisheries and Forestry. A Plant Health Committee consisting of representatives of relevant SCARM member agencies advises SCARM on the use of biological control and integrated pest management for pest and disease control.

Agricultural entomology

IPM has been a major focus for agricultural entomologists in Australia for decades. In 1969, The Australian Entomological Society held the first of a series of Applied Entomological Research Conferences to discuss recent developments in applied entomology, particularly the management of insect pests and environmental impacts of current management practices. Conferences are held about every 5 years. The proceedings of the 1992 conference were published in the book *Pest Control and Sustainable Agriculture*, a valuable reference on Australian IPM in the early 1990s (Corey *et al.*, 1993). The 1998 conference proceedings were also published (Zalucki *et al.*, 1998).

Industry has responded to consumer pressure to reduce the detrimental effects of AgVet chemicals. Most producer groups support research and development in IPM. Chemical companies are assessing the effects of new chemicals on existing IPM systems. In 1991 the Australian Apple and Pear Growers Association took the unprecedented step of signing a Pesticides Charter with the Australian Consumers Association. The Charter recognized that pesticides vary in their toxicity, effect on the environment, and compatibility with biological control agents and other beneficial species (Anonymous, 1991).

In 1995, Penrose *et al.* (1995) developed PestDecide©, a decision support system that balanced desirable and undesirable characteristics of pesticides used on apples, to assist apple growers in meeting the requirements of the Pesticides Charter. The growers association also commissioned an analysis of their investment in IPM research and development. The report found that the investment in IPM research had caused significant changes in industry practices and increased the rate of adoption of IPM, resulting in a rate of return of at least 12% (AGTRANS Research, 1999).

Commercial suppliers of biological control agents in Australia

The early releases of biological control agents in Australia were government sponsored. Once the potential for commercial production of biological control agents was recognized, several independent companies were established. Production of large numbers of biological control agents requires considerable investment in rearing facilities, techniques, quarantine and quality control, information and marketing. The companies rearing commercial quantities in Australia formed the Australasian Biological Control Association to ensure that high standards are maintained.

Case Studies of IPM in Australia

Kitching and Jones (1981) provided an excellent synopsis of IPM case histories across a range of ecosystems in Australia. Their book contains detailed ecological studies on skeleton weed (Groves and Cullen, 1981), kangaroos (Cunningham, 1981), crown of thorns starfish (Potts, 1981), aphids (Maelzer, 1981), codling moth (Geier, 1981), lightbrown apple moth (Geier and Briese, 1981), mosquitoes (Kay *et al.*, 1981), Australian bushfly (Hughes, 1981), sheep blowfly (Kitching, 1981), cabbage butterfly (Jones, 1981), and sirex wasp (Taylor, 1981). Two other case studies are included here: IPM in pome and stone fruits, and cotton.

Development of IPM in pome and stone fruits in the state of Victoria

The state of Victoria is a major producer of fruit, nuts and berries in Australia. Government policy in Victoria encourages sustainable production and adoption of IPM. The Department of Natural Resources and Environment (NRE) in Victoria is responsible for policy, research and extension services to agriculture. NRE has had a long history of integrating biological control into production systems. In the early 1900s, the parasitoid wasp *Aphelinus mali* was released to help control woolly aphid. *Aphelinus mali* is still active today, and research projects conducted over the last 20 years to reduce pesticide usage have enhanced its potential. For example, a large national collaborative multi-disciplinary approach led by NRE to integrate the management of weevils, woolly aphid and powdery mildew improved grower knowledge and understanding of how poor choice of pesticides impacts on the biological control of woolly aphid (Williams, 2000c).

In the 1930s, scientists at the Tatura Horticultural Research Station (now the Institute of Sustainable Irrigated Agriculture) showed that traps containing fermenting brown sugar solution could be combined with daily temperature data (converted to the developmental units of Shelford (1927)) to improve the timing of sprays against codling moth in apple and pear orchards (Miller, 1943). However, the technology was not widely adopted for a number of reasons. Traps were messy and difficult to use, mathematical calculations had to be made each day, suitable weather stations and communication systems were not available, and new pesticides (organochlorines and organophosphates) became available with such efficacy and persistence that accurate timing was not important (Williams, 1984).

This situation changed in the 1960s when the side effects of broad-spectrum pesticides became known. Also during the 1960s, entomologists at Burnley and Tatura collaborated with their interstate and CSIRO colleagues in a major ecological study in apple orchards (Geier *et al.*, 1969; Lloyd *et al.*, 1970). This work set the framework for orchard pest management research over the next 40 years and considerable effort has been expended since then to develop viable alternatives for pest management.

Codling moth granulosis virus

Codling moth granulosis virus was cultured, converted into sprayable formulations and tested against local codling moth populations (Morris, 1972). Despite some success with this method overseas, the Australian trials were not able to obtain economical control. More recently, commercial preparations of the virus from France did not control Australian populations of codling moth even at 100 times the recommended dose (Dingey and Williams, 1995, unpublished report).

Predatory mites

In the 1970s, the two-spotted mite *Tetranychus urticae* (Koch) became a major problem in orchards, in part because spraying to control pests such as codling moth *Cydia pomonella* (Linnaeus), oriental fruit moth *Grapholita molesta* (Busck), and lightbrown apple moth *Epiphyas postvittana* (Walker) decimated populations of predatory beetles and phytoseiid mites (Readshaw, 1971, 1975a). An Australian strain of the phytoseiid *Typhlodromus occidentalis* Nesbitt resistant to parathion existed in Victoria, but its effectiveness was limited by susceptibility to azinphos-methyl (Field, 1974, 1976). A North American strain resistant to both parathion and azinphos-methyl was introduced into Australia in 1972 and mass-released between 1976–1978 (Readshaw, 1975b; Webster and Field, 1977; Field and Webster, 1978; Field *et al.*, 1979). A carbaryl-resistant strain was released in 1983. To date, the predatory mites have been released on over 4500 ha of orchards and have been extremely successful (Williams, 2000a). They are now available from commercial suppliers.

Pest phenology modeling

Pheromone traps were developed to monitor populations of codling moth, oriental fruit moth and, later, lightbrown apple moth. This development along with the advent of computers and electronic weather stations made possible the prediction of codling moth populations. The PETE (Pest Extension Timing Estimator) model (Welch *et al.*, 1978) was obtained from Michigan State University and used during the 1982/83 season, but a simpler model was developed for the 1983/84 season (Williams, 1984). The system was expanded to include all three moth pests and has been widely used in Victoria, largely as a result of the highly successful Cropwatch service operated by NRE (Williams and McDonald, 1991; Williams and Pullman, 1995). Mating disruption for codling moth often requires intervention with insecticides if the population is large, and the predictive model has been used successfully to schedule such interventions (Vickers *et al.*, 1998).

Pheromone-mediated mating disruption

Mating disruption for orchard pests considerably reduced pesticide usage (Brown and Il'ichev, 1999), but also caused unexpected side effects. Reduction in pesticide usage in pome fruit as a result of mating disruption for codling moth created an opportunity for oriental fruit moth to build up in pome fruit blocks, especially those located adjacent to stone fruit. Oriental fruit moth had not been recorded in pome fruit in Victoria up to that time. It now appears that where pome and stone fruit blocks are adjacent, the oriental fruit moth can move between the blocks. Mated females from pome fruit blocks can migrate into stone fruit blocks and cause significant damage (Il'ichev, 1997).

Area-wide mating disruption

To counter this, Il'ichev *et al.* (1998) developed an area-wide mating disruption project for oriental fruit moth. Many growers in northern Victoria had successfully used mating disruption in peach and nectarine blocks for more than 10 years. However, some growers were reporting an increase in oriental fruit moth damage to shoot tips and fruit. The most severe damage was typically found at the edge of peach blocks treated with mating disruption adjacent to pear blocks treated with insecticides. Peach shoot tips and fruit attracted mated females from adjacent pear blocks, where they had developed a high population under ineffective insecticide treatments. The migration of mated females from pears under insecticide treatment to adjacent peaches treated with mating disruption resulted in damage at the edge of the peach blocks. This pattern of damage is known as an 'edge effect'. Experiments in 1996–1999 investigated whether applying mating disruption to all orchards in a given area (area-wide) would improve the effectiveness of mating disruption in hot spots and edges. This approach was expected to be more reliable and cost effective than combining mating disruption and insecticide treatments.

Experimental methods – area-wide mating disruption

In 1997/98, 800 ha on 18 orchards in Cobram, northern Victoria, were chosen to conduct the project. The area included 550 ha of peaches and nectarines, which had been treated with mating disruption in the previous season. The other 250 ha included pears, apples, plums and apricots that had not been previously treated with mating disruption. For the area-wide mating disruption test, every host tree in this area was treated with oriental fruit moth (OFM) pheromone at the recommended rate of 1000 dispensers of 'Isomate OFM Plus' per hectare. Traps were placed around the area to monitor the presence of oriental fruit moth in adjacent areas. Within the treatment area, traps were placed in each block of each fruit variety, on average one trap/4 ha. Detailed shoot tip and fruit damage assessments were made before the first color picking and at the time of harvest. Monitoring data from the first oriental fruit moth flight revealed two distinct hot spots. One hot spot was eliminated and all locations with edge damage effects were also successfully controlled by the end of the 1997/98 season.

In 1998/99, the experiment was expanded to over 1100 ha on 40 orchards. 'Isomate OFM Rosso' dispensers at the rate of 500 dispensers/ha replaced 'Isomate OFM Plus'. One season of this treatment controlled a localized population (5–10 OFM/trap/week). Two consecutive seasons of area-wide mating disruption were able to control a higher population level (up to 80 OFM/trap/week). Fruit damage in the worst hot spot was reduced from 25% to 15% in the first year and fell to almost zero in the second year. Overall, area-wide mating disruption was effective in reducing the oriental fruit moth population in hot spots, migration of mated females, and edge effects (Il'ichev *et al.*, 1999a,b).

The project stimulated a community approach to pest management. All participating growers received written reports of oriental fruit moth numbers in their fields weekly and were regularly informed of the population on the whole experimental area. In the first year, growers reduced the use of insecticides against oriental fruit moth by half, and in the second year, most growers did not apply insecticides against oriental fruit moth at all.

Solving an unexpected problem

The reduction in pesticide use resulted in *Carpophilus* beetles becoming a problem. This often caused growers to apply pesticides and question the value of area-wide mating disruption. To maintain the benefits of the area-wide mating disruption program, an effective method of controlling *Carpophilus* became essential. In response, a program using aggregation pheromone and co-attractants was developed (Hossain *et al.*, 2000a).

The level of *Carpophilus* infestations varies from year to year. Prior to harvest, *Carpophilus* beetles are rarely seen in orchards because of their small size and cryptic behavior, although trap data indicates the beetles are active in orchards from September onwards. Prior to the mating disruption program, control of *Carpophilus* in stone fruit was based on the use of broad-spectrum insecticides applied near harvest time. This control was often unsatisfactory and difficult due to the required preharvest interval for insecticide application.

Identification and synthesis of male-produced aggregation pheromones of *Carpophilus* and subsequent field trials have identified the potential of these pheromones for management (James *et al.*, 1996; Hossain *et al.*, 2000b). Mass-trapping the beetle has proved effective and contributed to the understanding of *Carpophilus* population dynamics and species composition. Field trials demonstrated that pheromone stations are able successfully to trap and kill a large number of beetles (about 500–9000 beetles/station/week) outside the orchard and protect ripening stone fruit.

IPM of plant diseases

IPM has also been applied to the development of control programs for plant diseases in pome and stone fruits. The epidemiology of diseases including apple scab *Venturia inequalis*, pear scab *Venturia pirina*, and powdery mildew *Podosphaera leucotricha* has been studied extensively. This has resulted in improvements such as disease forecasting systems (Villalta *et al.*, 2002), 'softer' pesticides (Washington *et al.*, 1998a) and a better understanding of the susceptibility of apple varieties to both scab and powdery mildew (Washington *et al.*, 1998b).

Non-target impacts

The impact of pesticides on biological control agents and other non-target organisms should also be elucidated to aid in the implementation of control programs. To assist in this endeavor, NRE provides a commercial bioassay service. Advances in spray application technology have resulted in increased popularity of the more efficient low-volume spray application technique (Cole *et al.*, 1998). Some effects of this on non-target organisms have been quantified and work has started on improving the habitat value of orchard floors for predators and parasitoids (Cole and Laukart, 2000).

Flexibility of IPM systems and integrated fruit production

Changes in pest management practices often result in a shift in the composition of pest populations, indicating a need for flexibility in any IPM program. Pest management has traditionally been addressed as a stand-alone issue, but experience with the development of IPM for the fruit industry has demonstrated that many factors including economics and soil, water and fertilizer management have considerable impact on the effectiveness of pest management. Pome fruit growers in Australia have recognized this and are funding a national system for IFP. The IFP system developed for pome fruit in Australia incorporates whole-farm planning, site-specific selection of scion/rootstock combinations, IPM, irrigation and nutrition, crop management, quality assurance, food safety, and occupational health and safety (Williams, 2000a,b). IFP takes a broad approach to pest management decision-making by encouraging integration and understanding of the interactions occurring in the orchard and their impacts on crop quality.

Whole-farm planning requires consideration of remnant native vegetation and siting of buildings, tracks, dams, windbreaks and other works to minimize environmental impacts. Retention of remnant native vegetation and re-vegetation of marginal areas is important for several reasons. Besides the ecological value, retaining native vegetation provides a valuable service by acting as a reservoir for natural enemies. It may also provide a haven for native pests such as fruit bats, parrots and leafrollers, but proper planning can reduce the likelihood of unwelcome pest species. Selection of native species for windbreaks and spray buffers should consider their potential as a host for pests and their capacity to act as refugia for predatory mites and parasitoids (Williams, 2000a).

A major motivation for the IFP program is the requirement of European and UK markets for crops grown in ways that are safe to the environment, consumers, and farm workers. Currently, growers can be audited to confirm that they meet these expectations, so careful and accurate record keeping is indispensable. Australian growers who supply export markets have adopted Quality Assurance systems that use a HACCP to identify and control food safety issues. Domestic supermarket chains also request suppliers to have HACCP certification. The Australian IFP system is designed to complement the HACCP based system so that only one audit is required.

A preliminary study to determine how Australian growers were performing against the IFP guidelines was conducted in 1999 (Williams, 2000b). Growers in all production areas were surveyed (Table 28.1). Slightly more growers selected pesticides on the basis of compatibility with predators (86%), than efficacy against the target pest (84%), suggesting that they are prepared to balance pest control with the value of maintaining a predator population. Nearly

Table 28.1. Responses to survey of pest and disease management practices. Figures are the percentage of the total number of responders who claimed to be using the practice.

Management practice	Percentage
Use orchard sanitation and cultural controls to avoid pest and disease build-up	88
Monitor pest and disease levels in individual blocks within the orchard	88
Select pesticides on basis of compatibility with predators and parasitoids	86
Select pesticides on basis of efficacy against target pest	84
Base spraying decisions on results of monitoring	92
Use a consultant for pest management	57
Own staff are trained and conduct pest and disease monitoring	45

all (92%) based spraying decisions on the results of monitoring, while 57% used a consultant for their pest management. Consultants are not always able to attend every few days to monitor, and growers are encouraged to have their staff trained in pest scouting. Overall 45% had their own staff trained to monitor pest populations.

The high level of monitoring and documentation required by a formal, audited system is encouraging growers and their staff to take a greater interest in the ecology of their orchards. Practical training programs conducted by NRE help growers to develop a greater understanding, sense of ownership, and confidence in the biological systems they are managing.

IPM in cotton

Cotton is produced in the northern states of Queensland, Western Australia, Northern Territory, and northern New South Wales. It is attacked by a wide range of pests including cotton bollworm (*Helicoverpa armigera* Hübner), native budworm (*H. punctigera* Wallengren), green mirids (*Creontiades dilutus* Stål), and two-spotted mites (*Tetranychus urticae* Koch). Severe outbreaks of these pests in recent seasons (1997/98) have resulted in very high costs (Aus$800–1000/ha) of insecticide usage. This has focused the industry on the need to reduce reliance on synthetic insecticides. Growers, consultants, researchers and extension officers, the Australian Cotton Cooperative Research Center, and the Cotton Research and Development Corporation are collaborating in an effort to develop IPM guidelines for cotton (Mensah and Wilson, 2000). The Cooperative Research Center comprises participants from CSIRO, New South Wales Agriculture, Department of Primary Industries Queensland, NT Department of Primary Industries and Fisheries, Agriculture Western Australia, University of Sydney, University of New England, Cotton Research and Development Corporation, Cotton Seed Distributors, Queensland Cotton, and Western Agricultural Industries.

Australian cotton production utilizes many IPM strategies such as augmentation of beneficial insect populations, host plant resistance, selective insecticides, incorporating the compensatory capacity of the plant, cultural control, and sampling systems and thresholds. Despite this, the most common tool for pest control is still application of chemical pesticides; conventional cotton crops receive about 8–15 sprays/season (Fitt, 2000).

The Cooperative Research Center conducts programs related to cotton breeding, crop agronomy, weed control, soil structure, plant diseases, and information technology as well as entomology. Pest resistance, the ever-increasing cost of pesticides, the need to reduce environmental pollution, and the essential task of maintaining profitability are driving the development of management approaches that optimize the ability of pest managers to deal with multiple problems (Kauter, 2001). Formal integration of the many pest management strategies in cotton systems is now underway (Anonymous, 1999) and should result in the cotton equivalent of IFP. However, the integration of various best management practices is complex, since many management practices are contradictory. For example, the cultivation of soil to kill insecticide-resistant moth pupae conflicts with the minimum tillage system used to maintain soil structure.

Bt cotton

The use of Bt cotton (tradename INGARD®) can potentially reduce insecticide usage against *Helicoverpa* by 50–70% (Fitt, 2000). The adoption of Bt cotton in Australia has been relatively gradual (Fitt and Wilson, 2000). The sustainable use of transgenic cotton depends on managing the risk of resistance developing to the engineered toxins. Growers of Bt cotton must adhere to an Insect Management Plan as part of the required INGARD® Grower Agreement. The Insect Management Plan requires each grower to plant a refuge crop capable of producing sufficient moths to dominate any survivors of Bt crops and keep resistance at low levels. Also, Bt crops must be planted in narrow time windows designed to minimize pest pressure, and all volunteer cotton plants must be removed from all fields being planted with INGARD® and from fallows after INGARD® (Holloway *et al.*, 2000).

Fitt (2000) calls for Bt cotton to be viewed as part of an IPM system that incorporates a broad range of other tactics. IPM systems for future cotton production will likely be more complex than pesticide-based systems, and transgenic cotton alone will not be a sustainable technology if left alone. Intelligent use of new technologies requires a thorough understanding of the ecology of the crop system.

Area-wide programs

The IPM Guidelines for Australian Cotton recognize that to be most effective, IPM requires a strategic approach involving year-long planning deployed in a coordinated way through district or regional area-wide strategies (Mensah and Wilson, 2000). Cotton growers in Australia have used area-wide strategies to manage insecticide-resistant populations of *Helicoverpa armigera* (Forrester *et al.*, 1993). Area-wide strategies include the coordinated use of trap crops to concentrate moths where the larvae and pupae can be controlled by cultivation (Mensah and Wilson, 2000). Region-specific strategies have also been devised for both conventional and transgenic cotton (Holloway *et al.*, 2000).

Important Websites, Publications and Reports on IPM in Australia

Government research organizations in Australia

http://www.nre.vic.gov.au/ – Home page of the Department of Natural Resources and Environment including research institutions and projects throughout Victoria.

http://www.agric.nsw.gov.au/ – Home page of the New South Wales Agriculture including research institutions and projects throughout NSW.

http://www.dpi.qid.gov.au/ – Home page of the Department of Primary Industries including research institutions and projects throughout Queensland.

http://www.agric.wa.gov.au/ – Home page of the Western Australia Agriculture including research institutions and projects throughout Western Australia.

http://www.dpif/ – Home page of the Northern Territory Department of Primary Industries and Fisheries with Agriculture.

http://www.agric.sa.gov.au/ – Home page of the South Australia Agriculture including research institutions and projects throughout South Australia.

http://www.dpiwe.tas.gov.au/ – Home page of the Department of Primary Industries, Water and Environment in Tasmania.

http://www.affa.gov.au/ – Home page of Agriculture, Fisheries and Forestry Australia.

http://www.ento.csiro.au/ – CSIRO Division of Entomology (Australia).

http://www.horticulture.com.au/ – Home page of the Horticultural Research and Development Corporation (now Horticulture Australia Limited) with project descriptions.

Universities and associated IPM research centres in Australia

http://www.uws.edu.au/ – University of Western Sydney. Horticulture and integrated pest management.

http://www.cpitt.uq.edu.au/ – Cooperative Research Centre for Tropical Pest Management, University of Queensland, Australia. Research Topics: Modeling, Insect Identities and Behavior, Insect/ Plant Interactions, Field Analysis and Application, Decision Analysis and Implementation, Computer Assisted Learning and Decision Support, Biocontrol.

http://www.tpp.uq.edu.au/ – Cooperative Research Centre for Tropical Plant Pathology: University of Queensland, Australia.

http://cotton.pi.csiro.au/ – Home page of the Australian Cotton Cooperative Research Centre

http://www.une.edu.au/ – Insect Pest Management at University of New England: this document introduces the activities and interests of the Insect Pest Management Group within the Department of Agronomy and Soil Science, at the University of New England, Armidale, New South Wales, Australia.

http://www.waite.adelaide.edu.au/ – Department of Crop Protection, Waite Campus, University of Adelaide. Research Groups: Entomology, Nematology, Plant–Microbe Interactions, Virology, Weed Science.

http://www.unisearch.com.au/html/cerit.html – Centre for Entomological Research & Insecticide Technology (CERIT), University of New South Wales, Sydney, Australia. CERIT Home Page: outlines research and teaching activities of the Centre together with industry services provided, pest species available for commercial testing and latest developments.

http://groucho.ucc.usyd.edu.au:9000/public/RMAS6905/ – Web page for on-line IPM unit of study conducted by the Orange campus of the University of Sydney. Designed to give worldwide access to IPM education.

Agricultural universities

http://www.jcu.edu.au/school/tbiol/Botany/ – James Cook University (Queensland): Department of Botany and Tropical Agriculture.

http://www.latrobe.edu.au/ – La Trobe University School of Agriculture (Victoria): School of Agriculture.

http://www.landfood.unimelb.edu.au/ – The University of Melbourne: this university offers programs in Agriculture, Forestry & Horticulture.

http://ansc.une.edu.au/ – University of New England (New South Wales).

http://www.uq.edu.au/entomology/home.html – University of Queensland: School of Entomology.

Private IPM companies (other useful sites associated with IPM)

http://www.biocontrol.com.au/ – Biocontrol Ltd is an Australian company based in southern Queensland and has been in operation since 1981. Biocontrol Ltd specializes in development of effective soft option products for pest control, particularly those based on insect behavior modifying chemicals. Their first three products for the Australian market are formulations of sex pheromones for key pests of horticulture.

http://www.goodbugs.org.au/-Australasian – Australasian Biological Control Association.

http://www.goodbugs.org.au/home.html – Bio Resources – site includes IPM in sweetcorn and macadamias. It also details biocontrol agents available in Australia as well as general information on using Biological Control Agents in IPM programs. (Bugs for Bugs – a specialist citrus IPM site. IPM Technologies Ltd. Bio-protection. Greennem – Suppliers of parasitic nematodes.)

http://www.nre.vic.gov.au/ – Crop Health Services, Knoxfield, Victoria.

http://www.nre.vic.gov.au/isia/plantprotection/projects – Area-wide mating disruption projects in orchards in Victoria, Australia.

IPM related literature in Australia

http://www.farmonline.com.au/ – Home page of the Farmonline – a network of agricultural news from around Australia from leading rural newspapers and magazines. Select your publication for all the latest news plus an extensive list of properties for sale in rural Australia, job vacancies and a comprehensive list of classified advertisements. Farmonline also features a calendar of events, rural bookshop and an extensive database of rural businesses, trades and services.

References

AGTRANS Research (1999) AP980667: Evaluation of HRDC/AAPGA Investment in R&D for the Apple and Pear Industry with Emphasis on IPM, 1999. *Horticultural Research and Development Corporation Final Report Series.* (Project AP98064)

Anonymous (1991) Pesticides charter. In: *Towards a National Food Policy.* National Working Group on Food Policy. Australian Consumers Association, Sydney, p. 21

Anonymous (1999) New emphasis on cotton farming systems. Cotton CRC website http://www.mv.pi.csiro.au/publicat/articles/

Brown, D.J. and Il'ichev, A.L. (1999) The potential for the removal of organophosphate insecticides from stone-fruit production in the Goulburn Valley, Australia. *Acta Horticulturae* 525, 85–91.

Cole, P. and Laukart, N. (2000) Concentrate spraying in Pome fruit: preserving the beneficials.

Horticultural Research and Development Corporation Final Report Series.
Cole, P., Riches, D. and French, H. (1998) Improved spray application in apples and pears. *Horticultural Research and Development Corporation Final Report Series.*
Corey, S., Dall, D. and Milne, W. (eds) (1993) *Pest Control and Sustainable Agriculture.* CSIRO Publishing, Melbourne.
Cullen, J.M. (ed.) (1992) *Report of a joint SCA and CONCOM workshop on Environment Conservation and the Introduction of Biological Control Agents, Canberra 1990.*
Cullen, J.M. (1993) Opportunities and challenges in biological control. In: Corey, S., Dall, D. and Milne, W. (eds) *Pest Control and Sustainable Agriculture.* CSIRO Publishing, Melbourne, pp. 44–50.
Cunningham, D.K. (1981) Arid-zone kangaroos – pest or resource? In: Kitching, R.L. and Jones, R.E. (eds) *The Ecology of Pests: Some Australian Case Histories.* CSIRO, Melbourne, pp. 19–54.
Dingey, A. and Williams, D. (1995) Field assessment of Carpovirusine in Southern Australia. Confidential report for Calliope, France.
Field, R.P. (1974) Occurrence of an Australia strain of *Typhlodromus occidentalis* (Acarina: Phytoseiidae) tolerant to parathion. *Journal of the Australian Entomological Society* 13, 255–256.
Field, R.P. (1976) Integrated pest control in Victorian peach orchards: the role of *Typhlodromus occidentalis* Nesbitt (Acarina: Phytoseiidae). *Australian Journal of Zoology* 24, 565–572.
Field, R.P. (1993) Guidelines for reviewing applications for importation and release of biological control agents of weeds. In: Corey, S., Dall, D. and Milne, W. (eds) (1993) *Pest Control and Sustainable Agriculture.* CSIRO Publishing, Melbourne, pp. 193–195.
Field, R.P. and Webster, W.J. (1978) Predators keep down twospotted mites – but watch your spray. *Victorian Horticulture Digest* 74, 17–20.
Field, R.P., Webster, W.J. and Morris, D.S. (1979) Mass rearing *Typhlodromus occidentalis* Nesbitt (Acarina: Phytoseiidae) for release in orchards. *Journal of the Australian Entomological Society* 18, 213–215.
Fitt, G.P. (2000) An Australian approach to IPM in cotton: integrating new technologies to minimise insecticide dependence. *Crop Protection* 19, 793–800.
Fitt, G.P. and Wilson, L.J. (2000) Genetic engineering in IPM: case study – Bt plants. In: Kennedy, G.G. and Sutton, T.B. (eds) *Emerging Technologies in Integrated Pest Management: Concepts, Research and Implementation.* APS Press, St Paul, Minnesota, pp. 108–125.
Forrester, N.W., Cahill, M., Bird, L.J. and Layland, J.K. (1993) Management of pyrethroid and endosulfan resistance in *Helicoverpa armigera* (Lepidoptera: Noctuidae) in Australia. *Bulletin of Entomological Research* (Suppl. 1), 132
Geier, P.W. (1981) The codling moth, *Cydia pomonella* (L.): profile of a key pest. In: Kitching, R.L. and Jones, R.E. (eds) *The Ecology of Pests: Some Australian Case Histories.* CSIRO, Melbourne, pp. 109–130.
Geier, P.W. and Briese, D.T. (1981) The light-brown apple moth, *Epiphyas postvittana* (Walker): a native leafroller fostered by European settlement. In: Kitching, R.L. and Jones, R.E. (eds) *The Ecology of Pests: Some Australian Case Histories.* CSIRO, Melbourne, pp. 131–156.
Geier, P.W., Harris, W.B., Hillman, T.J., Hudson, N.M., Jones, E.L., Lloyd, N.C., Lower, H.F., Morris, D.S. and Webster, W.J. (1969) A cooperative programme of research into the management of pest insects in pome fruit orchards of South-eastern Australia. *Progress Report for 1966–67.* CSIRO, Melbourne.
Glenn, D., Sward, R., Braybrook, D., Aitken, D., Hayes, P. and Smith, M. (1998) Improving adoption of IPM in Viticulture: from Research to Practice. In: Zalucki, M.P., Drew, R.A.I. and White, G.G. (eds) (1998) *Pest Management – Future Challenges. Proceedings of the Sixth Australasian Applied Entomological Research Conference, Brisbane, Australia,* Vol. 1, pp. 57–68.
Groves, R.H. and Cullen, J.M. (1981) *Chondrilla juncea*: ecological control of a weed. In: Kitching, R.L. and Jones, R.E. (eds) *The Ecology of Pests: Some Australian Case Histories.* CSIRO, Melbourne, pp. 7–18.
Heard, T.A. and van Klinken, R.D. (1998) An analysis of test designs for host range determination of insects for biological control of weeds. In: Zalucki, M.P., Drew, R.A.I. and White, G.G. (eds) *Pest Management – Future Challenges. Proceedings of the Sixth Australasian Applied Entomological Research Conference, Brisbane, Australia,* Vol. 1, pp. 539–546.
Holloway, J., Gunning, R., Tucker, G., Pyke, B., Wilson, L. and Fitt, G. (2000) Insecticide resistance management strategy for conventional cotton 2000–2001. In: Shaw, A.J.

Cotton Pest Management Guide 2000/2001. NSW Agriculture, pp. 27–33.

Hossain, M.S., Mansfield, C.M., Jerie, P.H. and Il'ichev, A.L. (2000a) Mass trapping of *Carpophilus* spp. (Coleoptera: Nitidulidae) using aggregation pheromone: a possible strategy for suppressing *Carpophilus* population in stone fruit orchards. *Abstracts of the 31st Annual General Meeting and Scientific Conference of the Australian Entomological Society, Darwin, 25–30 June 2000*, p. 16.

Hossain, M., Mansfield, C. and Brown, C. (2000b) Control of *Carpophilus* beetle by using aggregation pheromone. *Australian Fresh Stone Fruit Quarterly* 2(3), 13–14.

Hughes, R.D. (1981) The Australian bushfly: a climate-dominated nuisance pest of man. In: Kitching, R.L. and Jones, R.E. (eds) *The Ecology of Pests: Some Australian Case Histories.* CSIRO, Melbourne, pp. 177–192.

Il'ichev, A.L. (1997) Problems with mating disruption of oriental fruit moth *Grapholita molesta* (Busck). *Abstracts of the 28th Annual General Meeting and Scientific Conference of the Australian Entomological Society, Melbourne, Victoria*, p. 37.

Il'ichev, A.L., Jerie, P.H. and Hossain, M.S. (1998) Wide area mating disruption of oriental fruit moth *Grapholita molesta* Busck. (Lepidoptera: Tortricidae) in Victoria. In: Zalucki, M.P., Drew, R.A.I. and White, G.G. (eds) *Pest Management – Future Challenges. Proceedings of the Sixth Australasian Applied Entomological Research Conference, Brisbane, Australia.* Vol. 1, pp. 348–355.

Il'ichev, A.L., Hossain, M.S. and Jerie, P.H. (1999a) Application of wide area mating disruption for control of oriental fruit moth *Grapholita molesta* Busck. (Lepidoptera; Tortricidae) migration in Victoria, Australia. *IOBC/WPRS Bulletin* 22(9), 95–103.

Il'ichev, A.L., Hossain, M.S. and Jerie, P.H. (1999b) Migration of oriental fruit moth *Grapholita molesta* Busck. (Lepidoptera: Tortricidae) under wide area mating disruption in Victoria, Australia. *IOBC/WPRS Bulletin* 22(11), 53–62.

Ives, P.M., Wilson, L.T., Cull, P.O., Palmer, W.A., Hagwood, C., Thomson, J.J., Hearn, B. and Wilson, A.G.L. (1984) Field use of SIRATAC: an Australian computer based pest management system for cotton. *Protection Ecology* 6, 1–21.

James, D.G., Bartelt, R.J. and Moore, C.J. (1996) Mass-trapping of *Carpophilus* spp. (Coleoptera: Nitidulidae) in stone fruit orchards using synthetic aggregation pheromones and a co-attractant: development of a strategy for population suppression. *Journal of Chemical Ecology* 22(8), 1541–1556.

Jones, R.E. (1981) The cabbage butterfly, *Pieris rapae* L.: a just sense of how not to fly. In: Kitching, R.L. and Jones, R.E. (eds) *The Ecology of Pests: Some Australian Case Histories.* CSIRO, Melbourne, pp. 217–230.

Kay, B.H., Sinclair, P. and Marks, E.N. (1981) Mosquitoes: their inter-relationship with man. In: Kitching, R.L. and Jones, R.E. (eds) *The Ecology of Pests: Some Australian Case Histories.* CSIRO, Melbourne, pp. 157–176.

Kauter, G. (2001) IPM – Sticking to the challenge. Cotton CRC website http://www.mv.pi.csiro.au/publicat/conf/coconf00/areawide/07/07.htm

Kitching, R.L. (1981) The sheep blowfly: a resource-limited, specialist species. In: Kitching, R.L. and Jones, R.E. (eds) *The Ecology of Pests: Some Australian Case Histories.* CSIRO, Melbourne, pp. 193–216.

Kitching, R.L. and Jones, R.E. (eds) (1981) *The Ecology of Pests: Some Australian Case Histories.* CSIRO, Melbourne, 254 pp.

Lloyd, N.C., Jones, E.L., Morris, D.S., Webster, W.J., Harris, W.B., Lower, H.F., Hudson, N.M. and Geier, P.W. (1970) Managing apple pests: a new perspective. *Journal of the Australian Institute of Agricultural Science* 36, 251–258.

Maelzer, D.A. (1981) Aphids – introduced pests of man's crops. In: Kitching, R.L. and Jones, R.E. (eds) *The Ecology of Pests: Some Australian Case Histories.* CSIRO, Melbourne, pp. 89–108.

Madge, D.G., Buchanan, G.A. and Stirrat, S. (1993) Practical application of Integrated Pest Management in Vineyards in Sunraysia. In: Corey, S., Dall, D. and Milne, W. (eds) *Pest Control and Sustainable Agriculture.* CSIRO, Melbourne, pp. 102–104.

Mensah, R. and Wilson, L. (2000) IPM guidelines for Australian cotton. In: Shaw, A.J. (2000) *Cotton Pest Management Guide 2000/2001.* NSW Agriculture, pp. 8–26.

Miller, L.W. (1943) Codling moth in Williams' pears: investigations in the Goulburn Valley district 1936–41. *Journal of Agriculture, Victoria* 41, 39–46, 141–148.

Morris, D.S. (1972) A cooperative programme of research into the management of pest insects in pome fruit orchards of South-eastern Australia: 1. Evaluation of a nuclear granulosis virus for control of codling moth. *Abstracts of the 14th International Congress of Entomology, 1984 Canberra*, p. 238.

Penrose, L.J., Thwaite, W.G. and Maguire, M. (1995) AP430: a rating index as a basis for decision making on pesticide use reduction and IPM accreditation. *Horticultural Research and Development Corporation final report series.* Sydney, Australia.

Potts, D.C. (1981) Crown of Thorns starfish – Man-induced pest or natural phenomenon? In: Kitching, R.L. and Jones, R.E. (eds) *The Ecology of Pests: Some Australian Case Histories.* CSIRO, Melbourne, pp. 55–88.

Readshaw, J.L. (1971) An ecological approach to the control of mites in Australian orchards. *Journal of the Australian Institute of Agricultural Science* 37(3), 226–230.

Readshaw, J.L. (1975a) The ecology of tetranychid mites in Australian orchards. *Journal of Applied Ecology* 12, 473–495.

Readshaw, J.L. (1975b) Biological control of orchard mites in Australia with an insecticide-resistant predator. *Journal of the Australian Institute of Agricultural Science* 41, 213–214.

Sands, D.P.A. (1993) Effects of confinement on parasitoid/host interactions: interpretation and assessment for biological control of arthropod pests. In: Corey, S., Dall, D. and Milne, W. (eds) *Pest Control and Sustainable Agriculture.* CSIRO, Melbourne, pp. 196–199.

Sands, D.P.A. (1998) Guidelines for testing host specificity of agents for biological control of arthropod pests. In: Zalucki, M.P., Drew, R.A.I. and White, G.G. (eds) *Pest Management – Future Challenges. Proceedings of the Sixth Australasian Applied Entomological Research Conference, Brisbane, Australia.* Vol. 1, pp. 556–560.

Shelford, V.E. (1927) An experimental investigation of the relations of the codling moth to weather and climate. *Bulletin of the Illinois Natural History Survey* 16, 311–440.

Smith, D. and Papacek, D.F. (1993) Integrated pest management in Queensland citrus – recent developments. In: Corey, S., Dall, D. and Milne, W. (eds) *Pest Control and Sustainable Agriculture.* CSIRO, Melbourne, pp. 112–115.

Taylor, K.L. (1981) The Sirex woodwasp: ecology and control of an introduced forest insect. In: Kitching, R.L. and Jones, R.E. (eds) *The Ecology of Pests: Some Australian Case Histories.* CSIRO, Melbourne, pp. 231–250.

Vickers, R.A., Thwaite, W.G., Williams, D.G. and Nicholas, A. (1998) Control of codling moth in small plots by mating disruption: alone and with limited insecticide. *Entomologia Experimentalis et Applicata* 86(3), 229–239.

Villalta, O., Washington, W.S., Kita, N. and Bardon, D. (2002) The use of weather and ascospore data for the forecasting of apple and pear scab in Victoria, Australia. *Australasian Plant Pathology* 31(3), 205–215.

Washington, W., Villalta, O. and Appleby, M. (1998a) Control of pear scab with hydrated lime alone or in schedules with other fungicide sprays. *Crop Protection* 17, 569–580.

Washington, W., Villalta, O., Ingram, J. and Bardon, D. (1998b) Susceptibility of apple cultivars to apple scab and powdery mildew in Victoria, Australia. *Australian Journal of Experimental Agriculture* 38 (6), 625–629.

Webster, W.J. and Field, R.P. (1977) Mass release of predatory mites in Goulburn Valley orchards. *Victorian Horticulture Digest* 72, 1–4.

Welch, S.M., Croft, B.A., Brunner, J.F. and Michels, M.F. (1978) PETE: an extension phenology modeling system for management of multi-species pest complex. *Environmental Entomology* 7, 482–494.

Wilson, F. (1960) A review of the biological control of Insects and Weeds in Australia and Australian New Guinea. *CAB Technical Communication No. 1.*

Williams, D. (1984) Development of computerized pest management in Victorian orchards. In: Bailey, P. and Swincer, D. (eds) *Proceedings of the 4th Australian Applied Entomological Research Conference, September 1984,* pp. 146–149.

Williams, D. (2000a) Integrated fruit production as a means of improving biological control of pests of pome and stone fruit orchards in Australia. *Proceedings of the International Symposium: Biological Control for Crop Protection, 24–25 February 2000, Korea,* pp. 203–214.

Williams, D. (2000b) National guidelines for Integrated Fruit Production in Australia: apples. *Horticultural Research and Development Corporation final report series.*

Williams, D. (2000c) WWAPM: Integrated management of weevils, woolly aphid and powdery mildew. *Horticultural Research and Development Corporation final report series.*

Williams, D.G. and McDonald, G. (1991) Computer models and pest management in Victoria. *Plant Protection Quarterly* 6(1), 45–47.

Williams, D. and Pullman, K. (1995) Cropwatch: an integrated pest management service in Victoria. *Apple Integrated Pest Management Forum, Tasmania 1995*, pp. 1–4.

Zalucki, M.P., Drew, R.A.I. and White, G.G. (eds) (1998) Pest Management – Future Challenges. *Proceedings of the Sixth Australasian Applied Entomological Research Conference, Brisbane, Australia.* Volumes 1 and 2. ISBN 1864990511.

Chapter 29
Integrated Pest Management in New Zealand Horticulture

D.M. Suckling[1], C. McKenna[2] and J.T.S. Walker[3]
HortResearch, [1]Lincoln, Canterbury, New Zealand; [2]Te Puke, Bay of Plenty, New Zealand; [3]Havelock North, Hawke's Bay, New Zealand

History and Evolution of IPM in New Zealand

New Zealand is an island nation located in the South Pacific. The horticultural production areas of the country span a wide range of subtropical to temperate latitudes (Fig. 29.1, 35° to 45° South). The New Zealand horticultural industry is export-driven and highly focused on exports to distant markets, due to the low population (3.8 million people) and remote geographical position of

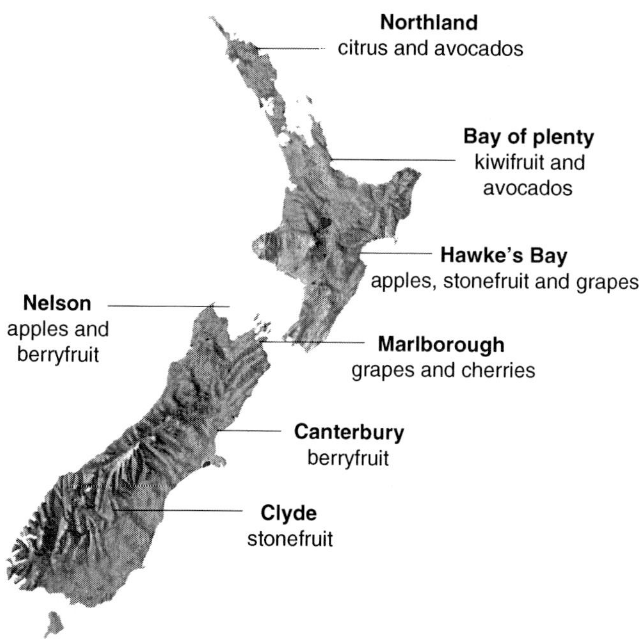

Fig. 29.1. The main horticultural areas of New Zealand.

the country. Hence the economic sustainability of the major horticultural industries depends on international trade.

Important subtropical crops include kiwifruit, avocados and citrus, while further south the important temperate crops include apples and pears, grapes and stonefruit. The wide range of crops represents a diversity of pest management problems, as might be expected. However, a unifying theme across these crops is quarantine issues that impact upon market access. Export crops must meet international sanitary and phytosanitary standards, as well as the customer demands for safe food that have been produced using environmentally benign production systems (Christie, 1993). Hence the problem facing the New Zealand horticultural industry is complex, involving production of food that meets stringent quality standards, and which is free from pests or pesticides.

IPM has a long research history in New Zealand horticulture, but the availability of inexpensive broad-spectrum insecticides (organophosphates, carbamates and pyrethroids) had a major adverse impact on the development and implementation of IPM from the early 1960s until the 1980s (Wearing *et al.*, 1982). During the 1980s, there was interest by horticultural industries and growers in reducing overall pesticide usage but any reduction was over-shadowed by an increasing need to eliminate the risk of quarantine-actionable pests in export crops. The first serious attempts to reduce pesticide use developed in the early 1990s with the emergence of increasingly important market signals (Wearing, 1993), problems experienced with pesticide resistance (Suckling, 1996), and significant environmental legislation, such as the Resource Management Act (1991).

Despite the development of alternative strategies to reduce pesticide use in the 1960s and early 1970s, e.g. 'integrated control' (Collyer and van Geldermalsen, 1975; Wearing and Thomas, 1978) and 'supervised control' (Wearing *et al.*, 1980), these approaches never gained acceptance with growers over routine calendar applications. The development of integrated mite control programs in the late 1970s resulted in the first serious attempts by growers to implement a strategy to reduce pesticide use (Wearing *et al.*, 1978; Martin *et al.*, 1984). Some progress was achieved with insect pest management, e.g. the 'window' program (Shaw *et al.*, 1993), but significant progress was not achieved until the development of the 'KiwiGreen' (Steven *et al.*, 1994; McKenna *et al.*, 1995), 'Integrated Fruit Production' (Batchelor *et al.*, 1997) and 'SummerGreen' (McLaren *et al.*, 1999) programs.

By 2000, all of the major horticultural industries in New Zealand had developed or had begun to develop IPM programs, and to use a named program as a marketing platform (Table 29.1). In some cases, the systems are well advanced.

A considerable amount of information on pest ecology is typically required to permit the withdrawal of broad-spectrum insecticides, without causing major crop and export losses. Management of leafrollers, scale insects, mealybugs, thrips, and other horticultural pests in the absence of broad-spectrum insecticides requires new information. Research has targeted the factors that affect the pest status of insects including: (i) the development of pest sampling methods and ways of measuring changing pest status; (ii) population ecology and regulating factors; (iii) pest movement, host finding mechanisms, including plant resistance; and (iv) pest and natural enemy interactions underlying the efficacy of biological control. More details of two specific IPM programs are best considered in the context of each crop (apples and kiwifruit), below.

Table 29.1. New Zealand IPM systems being developed in selected horticultural crops.

Crop	IPM program name
Apples	Integrated Fruit Production
Kiwifruit	KiwiGreen
Grapes	Integrated Wine Production
Stonefruit	SummerGreen
Avocados	AvoGreen

Organizational Structure of IPM Systems

In New Zealand, the former government research and extension agencies have gone through significant organizational change since the late 1980s when the government withdrew from direct involvement in IPM research or extension. The Ministry of Agriculture and Fisheries (now Forestry) has undergone significant reduction to focus on primary sector policy development, regulatory functions and biosecurity. The New Zealand government department that was primarily responsible for IPM research in horticultural crops (the Department of Scientific and Industrial Research, or DSIR) was disbanded in 1992, and replaced with CRIs, that operate under both the Companies Act (1996) and the CRI Act (1992). The CRIs have boards of directors and operate with commercial objectives. Typically, researchers compete for 1 or 2 year research contracts that are obtained with funding bodies including government agencies and private companies. Most of the research relevant to IPM in horticulture is done under this framework by staff of the Horticulture and Food Research Institute of New Zealand Ltd (www.hortresearch.co.nz). Accountability is typically very high from both private sector and government contracts, including presentation of grower seminars, popular articles, grower manuals and published papers. Current government policy is to seek 'outcomes', and the focus is towards evidence of industry change and environmental improvement as a result of research. Private sector research by individuals or small companies also exists, but tends to focus on pesticide efficacy and registration. However, private consultants, industry groups, grower groups or individual growers play an important role in information dissemination. There is relatively little university research on IPM.

IPM policy and infrastructure

Key strengths in the New Zealand horticultural industry that have aided the development of IPM have been the industry structure, the pesticide registration policy and strong, crop-based, IPM research teams. A cohesive industry structure based around regulated marketing at a national level has served to assist with, and encourage, adoption of IPM. Coordination of applied research on projects such as the development of pest monitoring systems, more selective insecticides, biocontrol, with related underpinning research on pest biology has been possible for many years under a complex framework of 'user-pays' and 'public-good' values.

However, each of these key factors is undergoing rapid change. Deregulation of the apple industry, for example, is fragmenting the industry focus on research and development, and weakening research capability. In addition, the Hazardous Substances and New Organisms Act (1997) is providing a much more rigorous and risk-averse framework for the introduction of new pesticides and/or biological agents in New Zealand. Government agencies that determine the direction of science funding support have also signaled a marked shift away from primary sector research towards research that underpins the development of a knowledge-based economy. The implications of changes in these structural factors on IPM development and implementation remain uncertain.

From research to extension to grower uptake

A number of approaches to technology transfer have been used in New Zealand. Farmer empowerment is currently achieved by requiring groups of farmers (or companies) to understand and access limited government funding for business development on a competitive basis. Various schemes are designed to support and promote uptake of a research and development culture across all sectors of the economy (not just agriculture). Farmers must show business benefits for the proposal, and in the case of IPM and other initiatives aiming at more sustainable production systems, the proposals typically take account of market signals for a preference towards 'green' technologies. In other cases, limited central government funding has also been made available for 'focus orchards', which form the basis of practical and regional demonstrations of new techniques.

All of the recent and successful IPM programs crops have involved the development of manuals on how to use the system. The manuals have been developed by research and extension personnel for use by growers and consultants, and have been funded directly from the industries involved. The central tenet is for 'continuous improvement', requiring updates and refinements. In addition, bulletin boards, e-mail to scientists, and other electronic media have supported grower uptake with information technology (e.g. www.hortnet.co.nz and www.hortnet.co.nz/key/pipfruit.htm). Publications of scientific papers from the New Zealand Plant Protection Society are also available online (1994–), at www.hortnet.co.nz/publications/nzpps (see references).

Integrated Fruit Production (IFP) of New Zealand Apples

The New Zealand apple industry is predominantly export-focused, and pests in New Zealand apple orchards are mostly cosmopolitan species (Table 29.2). These introduced pests vary in importance between regions, but include only a subset of the fauna associated with *Malus* sp. elsewhere. This places New Zealand in a situation of competitive advantage for developing more sustainable pest management systems with reduced reliance on broad-spectrum insecticides. However, the application of quarantine restrictions on pests in export crops imposes constraints that significantly reduce this advantage for certain markets. In some cases, such barriers to market access are based on incorrect taxonomy (Charles *et al.*, 2000).

These arthropods were controlled with the use of broad-spectrum organophosphate insecticides and various acaricides from the early 1960s onwards. This system was successful at meeting quarantine requirements

Table 29.2. Pests of apple crops in New Zealand in approximately descending order of economic importance, taking into account market access, yield, and fruit quality issues.

Insect species		Production pest	Quarantine pest
Primary pests			
Leafrollers	*Epiphyas postvittana*	✔	✔
	Planotortrix octo	✔	✔
	Planotortrix excessana	✔	✔
	Ctenopseustis obliquana	✔	✔
	Ctenopseustis herana	✔	✔
Codling moth	*Cydia pomonella*	✔	✔
San José scale	*Quadraspidiotus perniciosus*	✘	✔
Oystershell scale	*Quadraspidiotus ostreaeformis*	✘	✔
Mussel shell scale	*Lepidosaphes ulmi*	✘	✔
Secondary pests			
Longtailed mealybug	*Pseudococcus longispinus*	✘	✔
Obscure mealybug	*Pseudococcus viburni*	✘	✔
Citrophilus mealybug	*Pseudococcus calceolariae*	✘	✔
Apple leafcurling midge	*Dasineura mali*	✘	✔
Woolly apple aphid	*Eriosoma lanigerum*	✔	✘
European red mite	*Panonychus ulmi*	✔	✔
Two-spotted spider mite	*Tetranychus urticae*	✔	✔
Froggatt's apple leafhopper	*Edwardsiana crataegi*	✔	✘

of overseas markets, including those with a nil tolerance of pests. However, use of broad-spectrum pesticides was recognized long ago as undesirable (Collyer and van Geldermalsen, 1975), and research continued on a range of IPM tactics. An acceleration of the search for alternatives resulted from problems with insecticide and miticide resistance (Suckling, 1996), and more importantly, recognition of the potential for negative trade implications due to mounting consumer concerns in key markets (Christie, 1993). The search for insect pest management systems with minimal environmental and human health impact gained momentum in the 1980s and 1990s (e.g. Wearing *et al.*, 1993). Solutions that have been adopted for individual pests include a decision-support model for European red mite *Panonychus ulmi* (Koch), biological and chemical control (e.g. Hayes *et al.*, 1993), and pheromone traps to reduce insecticide applications against leafrollers (Shaw *et al.*, 1993; Bradley *et al.*, 1998; Clare *et al.*, 2000).

Over the last 5 years, a very large change in pest management practice has followed the introduction of insect growth regulators for leafroller and codling moth (Walker *et al.*, 1991, 1997a; Batchelor *et al.*, 1997; Bradley *et al.*, 1998). The attributes of insect growth regulators enabled the development of selective pest management and increased the focus on integrating natural enemies within the apple IPM program. This development has been an integral part of the apple industry's IFP program that requires pest monitoring and justification of all pesticide use, backed up by auditing and certification systems. Very significant reductions in the use of broad-spectrum insecticides have been achieved after several years of development and implementation of IFP (Fig. 29.2). For example, from 1997 to 2000, there was a fully documented 72% reduction in organophosphate use in New Zealand apple orchards, with a 90% reduction in use of azinphos-methyl use (Fig. 29.2). By the 2000/01 season, this is equivalent to an overall 90% reduction in organophosphate use, with the industry-wide adoption of the IFP program by growers occurring in the 2001 season (Fig. 29.3).

Enthusiasm for the IFP pest management goal (i.e. the elimination of broad-spectrum pesticides) in the New Zealand apple industry has been widely supported by growers because it also met an underpinning desire by many growers to move away from the use of highly toxic pesticides. Now, enhanced awareness of the changing

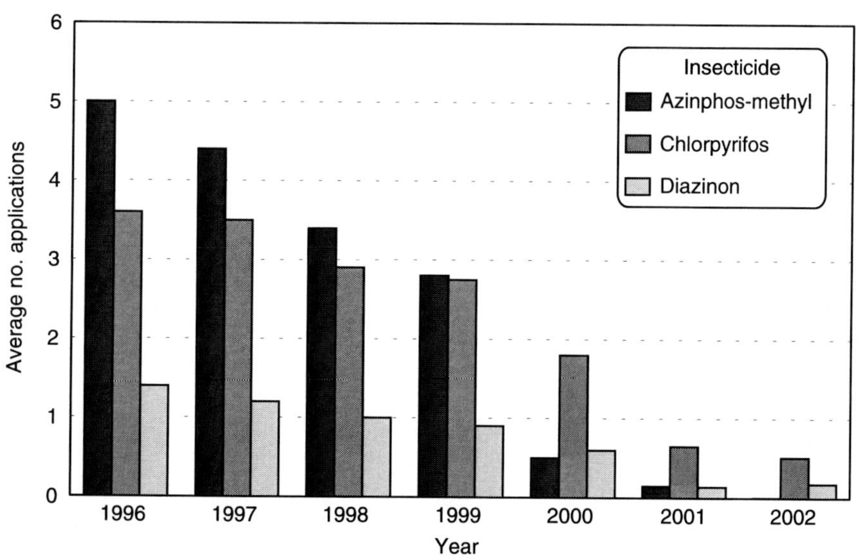

Fig. 29.2. Reduction in organophosphate insecticide use in New Zealand apples, following the adoption of integrated fruit production.

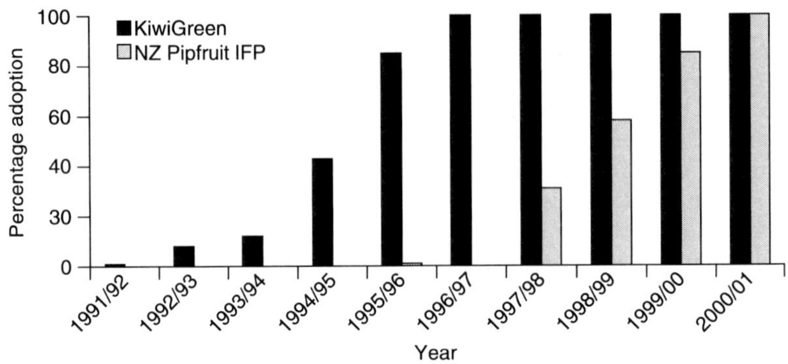

Fig. 29.3. Rate of adoption by New Zealand growers of KiwiGreen and integrated fruit production for apples.

status of key pests, due to the increased role of biological control, has encouraged the development of organic production that will in 2002 account for almost 10% of New Zealand's export apple crop.

IPM and Sustainability

Sustainability of the new IPM program is reliant on the use of strategies to minimize the likelihood of resistance developing to new selective insecticidal products. An insecticide resistance management strategy is being implemented based on minimal intervention, insecticidal class rotation (Lo *et al.*, 2000) and the enhanced role of biological control. Recognition of the need for more sustainable production systems, enshrined in New Zealand's Resource Management Act (1991), has highlighted the need to develop the basis for measuring sustainability. Practical measures of this widely discussed concept (Wearing, 1997; Suckling *et al.*, 1999) are elusive and likely to contain multiple elements. For example, the agrochemical inputs can be compared between alternative production systems using a pesticide rating system (Walker *et al.*, 1997b). Measures of the ecological impact of management practices on pests and non-target organisms are also likely to be important components of any definition of sustainability (Wearing, 1997; Suckling *et al.*, 1998). These measures can be used to compare specific management practices (Burnip *et al.*, 1998). Comparative assessments of ecological impact must be accompanied by an evaluation of economic sustainability. While IPM is an essential part of the process of continuous improvement, a holistic view demands the integration of IPM with other components of the production system, to minimize the impact of any required intervention.

KiwiGreen – IPM for New Zealand Kiwifruit Crops

KiwiGreen is an example of the successful development and implementation of an IPM program across an entire fruit industry. It is an IPM program driven by commercial need, and which reflects the restraints of producing a high quality export crop that is free from quarantine pests and pesticide residues. KiwiGreen consists of a documented and audited program of pest control measures that can only be applied in response to a demonstrable need.

Pre-KiwiGreen

When kiwifruit were first grown commercially in New Zealand in the late 1960s, there were few pests. As the area planted expanded to 10,000 ha and plantings matured, pests became a significant problem. Spraying of organophosphate insecticides became increasingly common, and by 1980 kiwifruit growers were following a schedule that recommended a broad-spectrum spray every 3–4 weeks from pre-flowering in November until harvest in May.

The key pests of kiwifruit are two species of leafrollers and three species of armoured scale insects (Table 29.3) (Berry, 1989; Steven, 1990). A number of secondary pests can also cause sporadic problems on kiwifruit (Table 29.3) (Steven, 1990), but generally pest control practices aim to protect the crop from leafroller feeding damage and to ensure the fruit is free from insects at harvest. Calendar spraying was a simple and successful means of achieving this, but researchers recognized early on that such pest control practices were unsustainable, disruptive to the predator–parasite complex, had potential to lead to insecticide resistance, and carried unacceptable environmental risks.

In response to these threats, a research program specific to kiwifruit pests was set up in the early 1980s with the aim of reducing the reliance on organophosphate sprays and broadening the range of pest management tools available. Initial studies focused on developing an understanding of the biology of the key pest species. It was determined that the majority of leafroller damage to kiwifruit occurs in the 8 weeks immediately after fruit set and was mostly due to a single endemic species, *Ctenopseustis obliquana* (Stevens *et al.*, 1995). After this time, the risk of leafroller damage was shown to be low, but late-season infestations of a second endemic species, *Cnephasia jactatana*, were recognized as possible in some orchards. Concurrently, the number, timing and distribution of armoured scale generations in kiwifruit were determined (Greaves *et al.*, 1994; Blank *et al.*, 1996, 1997). This information allowed researchers to identify the periods when control measures would be critical, and conversely, the periods when insecticides could potentially be omitted from the spray schedule (Stevens *et al.*, 1993; McKenna, 1998a). However, pest pressure can vary markedly among kiwifruit blocks, and so it was considered important that growers were provided with a means of determining whether omission of a spray would be likely to put the crop at risk of pest damage. Using knowledge accumulated from the pest biology studies, systems for monitoring scale and leafroller populations in kiwifruit were developed and action thresholds identified (Steven *et al.*, 1991; Blank *et al.*, 1994; Stevens *et al.*, 1997; McKenna, 1998b). These monitoring systems are now an integral part of the KiwiGreen program and form the basis of the decision-making process on whether or not a spray is required in a kiwifruit block.

Table 29.3. Common pests of kiwifruit and their status as a production or quarantine problem.

Insect species		Production pest	Quarantine pest
Primary pests			
Leafrollers	*Ctenopseustis obliquana*	✔	✔
	Cnephasia jactatana	✔	✔
Armoured scales	*Hemiberlesia rapax*	✘	✔
	Hemiberlesia lataniae	✘	✔
	Aspidiotus nerii	✘	✔
Secondary pests			
Passionvine hopper	*Scolypopa australis*	✔	✘
Fuller's rose weevil	*Asynonychus cervinus*	✘	✔
Collembola	*Xenylla maritima*	✘	✔
Thrips	*Heliothrips haemorrhoidalis*	✘	✔
	Thrips obscuratus	✘	✔
	Nesothrips propinquus	✘	✔
Orabatid mites	*Irgella bullager* and others	✘	✔
Two-spotted spider mite	*Tetranychus urticae*	✘	✔
Fungal feeding beetles	*Aridius* sp., *Orthoperus* sp.	✘	✔

A second key component of the kiwifruit research program was targeted at identifying alternative, environmentally benign insecticides that would enable the production of residue-free fruit. Field studies showed products containing Bt could provide excellent control of leafrollers (McKenna *et al.*, 1995; McKenna, 1998a; Stevens and McKenna, 1999), but an alternative was also needed for armoured scale control. Mineral oil was one such option, but trials in the 1970s had shown it to be phytotoxic to the crop (Sale, 1972). These problems have since been overcome with the development of more highly refined mineral oil products in the late 1980s, and the subsequent identification of factors that influence the occurrence of mineral oil damage to kiwifruit (McKenna and Steven, 1993; McKenna *et al.*, 1997).

The impetus for change

While an extensive amount of knowledge on the biology of kiwifruit pests and potential control options was accumulated over the 1980s and early 1990s, it was not until 1992 that it was first brought together in a package that could be used by the industry. The detection of spray residues on New Zealand kiwifruit, although well under the acceptable European Union guideline (Bull, 1993), was essentially being used as a trade barrier to some European markets. The NZKMB[1] responded in 1991 by requesting researchers to devise a pest management strategy that would enable the production of fruit with no detectable residues. This required the various components and outcomes of the kiwifruit research to be placed within a program that the industry and growers could readily adopt. The program, 'KiwiGreen', was launched a year later. In year one (1992), only 1% of the national crop was produced using KiwiGreen, and in year 2 this reached 8%. Thereafter grower uptake of the technology was unprecedented, and 6 years after its inception the total export crop was being produced using KiwiGreen (Fig. 29.3).

Implementation of KiwiGreen

The key factor contributing to the rapid adoption and expansion of KiwiGreen within the industry was a highly successful implementation process. This process was based on three key elements: the writing of a manual, the establishment of a pest monitoring infrastructure and the transfer of the technology to growers.

The KiwiGreen manual, written by a group of HortResearch and NZKMB personnel, contains information on the identification and biology of kiwifruit pests, detailed instructions on the procedures for monitoring and recording pest levels, and recommendations for pest control when threshold levels are exceeded (e.g. McKenna *et al.*, 1995).

Monitoring kiwifruit pests requires expertise and laboratory facilities with microscopes and so the establishment of a pest monitoring infrastructure was essential. In the first 2 years of KiwiGreen, NZKMB staff, whom had been trained by the researchers, supplied this service to growers at no cost. However, as the program expanded across the industry (Fig. 29.3), the NZKMB undertook licensing and training of pest monitoring facilities. These facilities were generally set up by kiwifruit packhouses as an extension of their service to their grower clients.

Transfer and adoption of the technology by growers was another critical step in the implementation process. In the early years of KiwiGreen, two field operators were employed by NZKMB. These field operators were primarily responsible for providing technical support to the growers, and acted as the link between researchers and the wider industry. In subsequent years this responsibility was transferred to technical personnel employed by the packhouse/pest monitoring centers. The technique of using

[1] Now Zespri International Limited, the sole marketer of New Zealand kiwifruit.

field operators was hugely successful and has since been replicated in other technology adoption processes.

KiwiGreen benefits

The benefits of KiwiGreen are primarily to the environment, market access and consumer acceptability. Environmental benefits arise from both the reduced number of sprays and the use of more environmentally benign sprays. Over the last decade the number of broad-spectrum sprays being applied has decreased from an average of eight per annum in 1980 to just three per annum in 2000. This equates to a reduction of 100 t of pesticide per annum. This has had a positive impact on orchard biodiversity and has created opportunities for greater use of biological controls (Thomson *et al.*, 1996; Thomson, 1997). In regions where kiwifruit orchards and urban settlements have to coexist, this reduction in insecticide use has also alleviated the potential for conflict between the two communities.

The benefits for market access are twofold. New Zealand kiwifruit typically has residue levels that are less than 5% of the maximum permitted residue levels in destination markets, but perhaps of greater importance are the consumer acceptability benefits. Key customers are now demanding evidence of food safety and environmental integrity in the production of food; without the KiwiGreen program it is unlikely this demand could be met.

The costs of KiwiGreen versus calendar spraying are similar but KiwiGreen has resulted in a shift in spending on pest control. The additional costs of pest monitoring and the use of environmentally benign sprays have been offset by a reduced number of sprays being applied. KiwiGreen has also created significant employment opportunities in the rural regions with the formation of the PMC. These centers have become an important point in the technology transfer chain, and most now have their own dedicated technology transfer person whose primary responsibility is information dissemination to grower clients. KiwiGreen has instigated an upskilling of growers and the industry in general, and the technology is now being cited as a key reason behind the increase in organic production (now 6% of the total kiwifruit production).

Future of KiwiGreen

KiwiGreen is a dynamic program that is continually evolving. While the focus remains on seeking a more diverse range of sustainable pest control options for managing both primary and secondary pests, the entire production process is now being considered. This includes all chemical inputs, as well as broader environmental and production issues such as canopy, ground cover and waste management. The demands of key customers for safe fruit that has been produced using environmentally sound practices are likely to become even more stringent in the future. This will ensure that programs such as KiwiGreen will continue to be developed and enhanced.

Use of industry agrochemical data in the development of IFP

There have been a variety of approaches taken towards assessing farmer or grower inputs, with the aim of reducing unsustainable pest management inputs (e.g. Reus, 1993). In several New Zealand fruit industries, the accumulation and analysis of agrochemical use information has been a very powerful tool for benchmarking current practice, setting goals, and measuring progress towards them over time (Manktelow *et al.*, 2000, 2001).

Key Constraints and the Future of IPM

Many of the major constraints to the adoption of IPM in New Zealand horticulture relate to the technical difficulties of cost-effective pest management and market access. The costs of technical solutions to some pest, disease and weed management problems are significant, but often grower perception of higher costs, financial risks and complexity also reduce adoption. Perception problems probably equal the significance of genuine technical problems (Wearing, 1988). There are also the opposing demands of export markets that present a difficult conundrum for producers and exporters. While the customers demand 'no pesticides', the foreign regulators demand 'no pests'. The Montreal Protocol will reduce the availability of methyl bromide, which has played an important role in achieving market access for many countries. Alternative, more sustainable technologies to achieve postharvest disinfestation are needed to meet the market demand. This is particularly the case because pest incidence rises as broad-spectrum pesticide use declines.

Trend analysis indicates that continuous improvements in pest management are needed to meet changing market needs. The widening debate about the benefits and risks of genetic modification in pest management highlights how rapidly markets can change on issues of food safety and food security. Furthermore, changing pest pressure from new biological invasions, changing land use, changes in area-wide pest management practices, and other factors will ensure a dynamic future for IPM.

References

Batchelor, T.A., Walker, J.T.S., Manktelow, D.W.L., Park, N.M. and Johnson, S.R. (1997) New Zeland integrated fruit production for pipfruit: charting a new course. *Proceedings 50th New Zealand Plant Protection Conference*, pp. 14–19. www.hortnet.co.nz/publications/nzpps/

Berry, J.A., Morales, C.F., Hill, M.G., Lofroth, B.J. and Allan, D.J. (1989) The incidence of three Diaspid scales on kiwifruit in New Zealand. *Proceedings 42nd New Zealand Weed and Pest Control Conference*, pp. 182–187.

Blank, R.H., Gill, G.S.C. and Olsen, M.H. (1994) Relationship between armoured scale infestations on leaves and fruit. *Proceedings 47th New Zealand Plant Protection Conference*, pp. 304–309.

Blank, R.H., Gill, G.S.C. and Upsdell, M.P. (1996) Greedy scale, *Hemiberlesia rapax* (Hemiptera: Diaspididae), phenology on kiwifruit leaves and wood. *New Zealand Journal of Crop and Horticutural Science* 24, 239–248.

Blank, R.H., Gill, G.S.C. and Dow, B.W. (1997) Determining armoured scale distribution within kiwifruit blocks. *Proceedings 50th New Zealand Plant Protection Conference*, pp. 293–297. www.hortnet.co.nz/publications/nzpps/

Bradley, S.J., Walker, J.T.S., Wearing, C.H., Shaw, P.W. and Hodson, A.J. (1998) The use of pheromone traps for leafroller action thresholds in pipfruit. *Proceedings 51st New Zealand Plant Protection Conference*, pp. 173–178. www.hortnet.co.nz/publications/nzpps/

Bull, P. (1993) An exporters view of plant protection. In: Suckling, D.M. and Popay, A. (eds) *Plant Protection: Costs, Benefits and Trade Implications.* New Zealand Plant Protection Society, Christchurch, pp. 155–161.

Burnip, G.M., Gibb, A.R. and Suckling, D.M. (1998) Target and non-target impacts from diazinon in a Canterbury apple orchard. *Proceeding 51st New Zealand Plant Protection Conference*, pp. 143–147. www.hortnet.co.nz/publications/nzpps/

Charles, J.G., Froud, K.J. and Henderson, R.C. (2000) Morphological variation and mating compatibility within the mealybugs *Pseudococcus calceolariae* and *P. similans* (Hemiptera: Pseudococcidae), and a new synonymy. *Systematic Entomology* 25, 285–294.

Christie, R. (1993) Plant protection and international trade. In: Suckling, D.M. and Popay, A. (eds) *Plant Protection: Costs, Benefits and Trade Implications.* New Zealand Plant Protection Society, Christchurch, pp. 3–7.

Clare, G., Suckling, D.M., Bradley, S.J., Walker, J.T.S., Shaw, P.W., Daly, J.M., McLaren, G.F. and Wearing, C.H. (2000) Pheromone trap colour determines catch of non-target insects. *New Zealand Plant Protection* 53, 216–220.

Collyer, E. and van Geldermalsen, M. (1975) Integrated control of apple pests in New Zealand 1. Outline of experiment and general results. *New Zealand Journal of Zoology* 2, 101–134.

Greaves, A.J., Davys, J.W., Dow, B.W., Tomkins, A.R., Thomson, C. and Wilson, D.J. (1994) Seasonal temperatures and the phenology of greedy scale populations (Hemiptera: Diaspididae) on kiwifruit vines in New Zealand. *New Zealand Journal of Crop and Horticultural Science* 22, 7–16.

Hayes, A.J., Walker, J.T.S., Shaw, P.W. and White, V. (1993) A decision model for miticide use in apple orchards. *Proceedings 46th New Zealand Plant Protection Conference*, pp. 162–165.

Lo, P., Walker, J.T.S. and Suckling, D.M. (2000) Insecticide resistance management of leafrollers (Lepidoptera: Tortricidae). *New Zealand Plant Protection* 53, 163–167. www.hortnet.co.nz/publications/nzpps/

Manktelow, D.W.L., Walker, J.T.S., Beresford, R.M., Hodson, A.J. and Laurenson, M.R. (2000) Ten years in the evolution of horticultural decision support systems in New Zealand. *British Crop Protection Council Conference Proceedings, Pests and Diseases*, pp. 1271–1278.

Manktelow, D.W.L., Walker, J.T.S., Hodson, A.J. and Suckling, D.M. (2001) Use of industry agrochemical use data in the development of integrated fruit production programs in the New Zealand horticultural industry. *Bulletin IOBC/WPRS* 24(5), 51–56.

Martin, N.A., Workman, P.J., Burgess, E.P. and Wearing, C.H. (1984) Integrated pest control in greenhouse crops. *Proceedings 37th New Zealand Weed and Pest Control Conference*, pp. 253–256.

McKenna, C.E. (1998a) The effectiveness of different insecticide programmes for kiwifruit. *Proceedings 51st New Zealand Plant Protection Conference*, pp. 184–188.

McKenna, C.E. (1998b) Pest monitoring in kiwifruit crops. *New Zealand Kiwifruit Journal* 125, 12.

McKenna, C.E. and Steven, D. (1993) Phytotoxicity to kiwifruit of oil sprayed after flowering. *Proceedings 46th New Zealand Plant Protection Conference*, pp. 75–79.

McKenna, C.E., Stevens, P.S. and Steven, D. (1995) A new *Bacillus thuringiensis* product for use on kiwifruit. *Proceedings 48th New Zealand Plant Protection Conference*, pp. 135–138.

McKenna, C.E., Stevens, P.S. and Steven, D. (1997) Phytotoxicity to kiwifruit from sprays of mineral oil. *Acta Horticulturae* 444(2), 779–784.

McLaren, G.F., Grandison, G., Wood, G.A., Tate, G. and Horner, I. (1999) *Summerfruit in NZ: Management of Pests and Diseases.* University of Otago Press, 136 pp.

Reus, J.A.W.A. (1993) An environmental yardstick for pesticides. *Proceedings 46th New Zealand Weed and Pest Control Conference*, pp. 91–96.

Sale, P.R. (1972) Oil and Chinese gooseberries do not agree. *The Orchardist of New Zealand* 44, 97–99.

Shaw, P.W., Cruickshank, V. and Suckling, D.M. (1993) Commercialisation of pheromone trapping in Nelson. *Proceedings 46th New Zealand Plant Protection Conference*, pp. 135–140.

Steven, D. (1990) Entomology in kiwifruit. In: Warrington, I.J. and Weston, G.C. (eds) *Kiwifruit Science and Management.* Ray Richards Publisher in association with the N.Z. Society for Horticultural Science, pp. 362–412.

Steven, D., Blank, R.H. and Tomkins, A.R. (1991) Monitoring scale to reduce sprays in kiwifruit. *New Zealand Agricultural Science* 26, 4.

Steven, D., Tomkins, A.R., Blank, R.H. and Charles, J.G. (1994) A first-stage integrated pest management system for kiwifruit. *Proceedings Brighton Crop Protection Conference* Vol 1, pp. 135–142.

Stevens, P.S. and McKenna, C.E. (1999) Factors affecting the efficacy of *Bacillus thuringiensis* against *Cnephasia jactatana* in kiwifruit. *Proceedings 52nd New Zealand Plant Protection Conference*, pp. 89–93. www.hortnet.co.nz/publications/nzpps/

Stevens, P.S., Steven, D., Malcolm, C. and Holland, P.T. (1993) Reducing pesticide residues on kiwifruit. *Proceedings 46th New Zealand Plant Protection Conference*, pp. 62–66.

Stevens, P.S., McKenna, C.E. and Steven, D. (1995) The timing of lepidopteran damage to kiwifruit. *Proceedings 48th New Zealand Plant Protection Conference*, pp. 130–134.

Stevens, P.S., McKenna, C.E., Blank, R., Tomkins, A. and Steven, D. (1997) Comparison of armoured scale thresholds in kiwifruit. *Proceedings 50th New Zealand Plant Protection Conference*, pp. 288–292. www.hortnet.co.nz/publications/nzpps/

Stewart, T.M., Norton, G.A., Mumford, J.D. and Fenemore, P.G. (1993) Pest and disease decision support for Hawke's Bay apple growers – a survey. *Proceedings 46th New Zealand Plant Protection Conference*, pp. 152–161.

Suckling, D.M. (1996) The status of insecticide and miticide resistance in New Zealand. In: Bourdôt, G.W. and Suckling, D.M. (eds) *Pesticide Resistance: Prevention and Management.* New Zealand Plant Protection Society, Christchurch, pp. 49–58.

Suckling, D.M., Walker, J.T.S., Wearing, C.H., Burnip, G.M. and Gibb, A.R. (1998) Measures of sustainability in New Zealand apple orchards: investigating biodiversity in managed ecosystems. *British Crop Protection Council Conference, Pests & Diseases*, pp. 637–642.

Suckling, D.M., Walker J.T.S. and Wearing, C.H. (1999) Ecological impact of three pest management systems in New Zealand apple orchards. *Agriculture Ecosystems and Environment* 73, 129–140.

Thomson, C. (1997) The body invaders or scale parasitoids. *New Zealand Kiwifruit Journal* 119, 15–16.

Thomson, C., Tomkins, A.R. and Wilson, D.J. (1996) Effect of insecticides on immature and mature stages of *Encarsia citrina*, an armoured scale parasitoid. *Proceedings 49th New Zealand Plant Protection Conference*, pp. 1–5. www.hortnet.co.nz/publications/nzpps/

Walker, J.T.S., Baynon, G.T. and White, V. (1991) Insect control on apples with RH-5992 a novel insect growth regulator. *Proceedings 44th New Zealand Weed and Pest Control Conference*, pp. 66–69.

Walker, J.T.S., Hodson, A.J., Wearing, C.H., Bradley, S.J., Shaw, P.W., Tomkins, A.R., Burnip, G.M., Stiefel, H.E. and Batchelor, T.A. (1997a) Integrated fruit production for New Zealand pipfruit: evaluation of pest management in a pilot programme. *Proceedings 50th New Zealand Plant Protection Conference*, pp. 258–263. www.hortnet.co.nz/publications/nzpps/

Walker, J.T.S., Hodson, A.J., Batchelor, T.A., Manktelow, D.W. and Tomkins, A.R. (1997b) A pesticide rating system for monitoring agrichemical inputs in New Zealand horticulture. *Proceedings 50th New Zealand Plant Protection Conference*, pp. 529–534. www.hortnet.co.nz/publications/nzpps/

Walker, J.T.S., Manktelow, D.W.L., Wearing, C.H., Lo, P.L. and Suckling, D.M. (2001) Development of integrated fruit production programmes in the New Zealand horticultural industry. *Bulletin IOBC/WPRS* 24(5) 39–44.

Wearing, C.H. (1988) Evaluating the IPM implementation process. *Annual Review of Entomology* 33, 17–38.

Wearing, C.H. (1993) IPM and the consumer: the challenge and opportunity of the 1990s. In: Corey, S.A., Dall, D.J. and Milne, W.M. (eds) *Pest Control and Sustainable Agriculture.* CSIRO, Australia, pp. 21–27.

Wearing, C.H. (1997) Indicators of sustainable pest management in orchard production systems. *Proceedings 50th New Zealand Plant Protection Conference*, pp. 506–513. www.hortnet.co.nz/publications/nzpps/

Wearing, C.H. and Thomas, W.P. (1978) Integrated control of apple pests in New Zealand. 13. Selective insect control using diflubenzuron and *Bacillus thuringiensis. Proceedings 31st New Zealand Weed and Pest Control Conference*, pp. 221 – 228.

Wearing, C.H., Collyer, E., Thomas, W.P. and Cook, C. (1978) Commercial implementation of integrated control of European red mite in Nelson. *Proceedings 31st New Zealand Weed and Pest Control Conference*, pp. 214 – 220.

Wearing, C.H., York, M. and O'Kane, P. (1980) Supervised spraying of lepidopterous pests of kiwifruit in the Bay of Plenty. *Proceedings 33rd New Zealand Weed and Pest Control Conference*, pp. 100–104.

Wearing, C.H., Thomas, W.P. and Collyer, E. (1982) The decision to implement IPM: prerequisites and consequences. In: Cameron, P.J., Wearing, C.H. and Kain, W.M. (eds) *Proceedings of Australasian Workshop on Development and Implementation of IPM, Mt Albert Research Centre, Auckland, New Zealand, 20–22 July, 1982*, pp. 198–210.

Wearing, C.H., Walker, J.T.S. and Suckling, D.M. (1993) Development of IPM for New Zealand apple production. Proceedings 2nd IOBC/ISHS Symposium on Integrated Fruit Production. *Acta Horticulturae* 347, 277–284.

Chapter 30
FAO Integrated Pest Management Programs: Experiences of Participatory IPM in West Africa

Pieter Stemerding[1] and Souleymane Nacro[2]
[1]Global IPM Facility, FAO/AGPP, Rome, Italy; [2]INERA, Bobo-Dioulasso, Burkina Faso

Introduction

The FAO has been actively providing technical assistance and building capacity in plant protection. The FAO Plant Protection Service addresses international aspects of plant protection and closely cooperates with regional and national plant protection organizations and programs. The Plant Protection Service is hosting the Global IPM Facility which supports IPM initiatives worldwide, with a particular focus on Africa, Latin America and Central Asia. This chapter highlights some of the early experiences in Asia before providing some insights on IPM development in West Africa, where FAO involvement in IPM goes back to the early 1990s.

The Global IPM Facility: Promoting IPM Since 1997

The need to establish the Global IPM Facility first emerged when the UNCED Agenda 21 assigned a central role to IPM in agricultural programs and policies in 1992. In 1993, the FAO inter-country program on IPM rice in Asia organized a Global IPM meeting to enable interested policymakers from other regions to familiarize themselves with its successful IPM approach. This triggered many requests for assistance in setting up farmer-centered IPM pilot activities, particularly from African countries. Consequently, FAO, UNEP, UNDP and The World Bank established a task force, which outlined the Facility's functions and helped raise the necessary funding. The Facility was formally established in 1995 with FAO as the host organization. It became fully operational in 1997, after adequate funding had been secured from the Facility's co-sponsors and the Governments of The Netherlands, Switzerland and Norway.

The mandate of the Global IPM Facility is to assist interested Governments and NGOs to initiate, develop and expand IPM programs that aim to reduce pesticide use and associated negative impact on health and environment, while increasing production and profits through improved crop and pest management.

Effective IPM programs include both capacity building and policy development. Capacity building involves: training of facilitators for participatory IPM education; science and technology development aimed at farmer needs; development of sustainable

 Integrated Pest Management in the Global Arena
(eds K.M. Maredia, D. Dakouo and D. Mota-Sanchez)

funding mechanisms and networking. Policy development focuses on national and international policies that affect pest management, including international conventions and standards[1]. These two components are often mutually reinforcing.

The Facility currently supports IPM initiatives in Asia[2], Africa, Latin America, Middle East, Central Asia and Eastern Europe.

IPM and FFS

In a nutshell, the IPM[3] approach aims at helping farmers to take independent, well-informed crop production and management decisions. IPM is primarily brought to farmers through participatory, season-long farmer field schools, in which farmers and extensionists come together to study their fields. Rather than receiving instructions from outsiders, field schools help farmers to uncover and strengthen their own local knowledge.

IPM is a participatory approach to crop production and protection based on ecosystem management and aimed at maintaining a natural equilibrium. As such it reduces the risk of damage by pests. IPM helps farmers to enhance their understanding of the agro-ecosystem and develop capacities to take well-informed, independent decisions on how to manage their crops more efficiently, in a more sustainable manner.

As said, a key tool to support farmers in understanding and applying IPM principles is the season-long FFS. A field school is a discovery-based learning process in which farmers themselves design and carry out field experiments to find solutions to field problems and challenges such as diseases and pests, soil degradation and nutrient management. Field schools have four core principles:

1. Grow a healthy soil and crop.
2. Observe the field regularly.
3. Conserve natural enemies.
4. Farmers are experts in their own fields.

Depending on the type of crops, field school groups typically comprise 25 farmers. The groups come together during the growing season for 4–5 h on a weekly basis for rice and vegetables, or monthly in the case of bananas, for example. Field school groups establish a study field and carry out experiments and field trials to enhance understanding of the agroecosystem and compare 'regular' farmer practices with IPM practices. Groups are supported by a facilitator, usually an IPM-trained extensionist or a farmer-faciliator. IPM programs aim to reduce pesticide use to a minimum, to lessen the negative impact of agrochemicals on environment and health, and to decrease production costs. Through improved crop and pest management, IPM farmers may achieve substantial savings on pesticides while increasing or maintaining yields.

Field schools primarily aim to strengthen farmers' technical knowledge. But they also aim to enhance organizational, management and communication skills. Being able to access and process field and economic data independently, field school farmers learn how to communicate better and present findings to others. By documenting their field observations in writing and through drawings, farmers become aware of the knowledge they possess on (local) production constraints, including diseases and pests, water and soil problems and social impediments to intensification of production (labor shortages at peak periods, etc.). Groups critically review field findings by individual members of the group and expose farmers to knowledge and experiences of other farmers in a structured way. This, combined with their enhanced

[1] Examples include pesticide policy studies carried out by the University of Hanover/GTZ Pesticide Policy Project on Zimbabwe: GTZ Pesticide Policy Project (1999), Côte D'Ivoire: GTZ Pesticide Policy Project (1998), and a recently completed study on pesticide use in Mali: INSAH (2001).

[2] See among others documentation on FAO programs in Indonesia: FAO Technical Assistance Indonesia National IPM Program (1998).

[3] In this chapter the abbreviation IPM is used to mean 'Integrated Production and Pest Management'.

analytical capabilities, increases the self-confidence of farmers. It also makes them more outward looking and more critical towards externally imposed solutions for their problems.

IPM Development from Asia to Africa and Latin America

In 1980 a first FAO-supported IPM program on rice was set up in Asia. It was initiated in response to challenges created by highly pesticide and fertilizer intensive rice programs promoted as of the 1950s as a central element of the Green Revolution. Intensive fertilizer and synthetic pesticide use was promoted throughout Asia to maximize returns of high yielding, short duration rice varieties, allowing farmers to produce two crops a year. Though rice productivity greatly increased, negative effects such as the disruption of previously well-balanced rice ecosystems became increasingly present. The regular use of broad-spectrum insecticides reduced natural enemy populations favoring the resurgence of the brown planthopper, hitherto a minor pest in tropical rice.

This outbreak of brown planthopper affected numerous Asian countries and led to serious production losses. Besides economic implications, the brown planthopper outbreaks had a political impact, threatening for example Indonesia's self-sufficiency in rice. Research by IRRI revealed that the disruptive effect of insecticides on the rice ecosystem was the key factor in the vast and unprecedented brown planthopper outbreaks in Asia. Slowly, consensus emerged among the research and science community that chemical pest control should not be the sole form of pest management. It should instead be based on a thorough understanding of the agroecosystem and rely as much as possible on non-chemical control measures. This acknowledgment led to a more prominent place of IPM on the agenda of decision makers, community leaders and agricultural research. In 1986, the Government of Indonesia banned 57 insecticides for use on rice when the country was faced with a full-blown brown planthopper outbreak. The ban was based on the recognition of the potentially disrupting effects of insecticides on rice ecosystems. In addition, subsidies on pesticides were reduced and a major IPM farmer training program was started. In support of the decision by the Indonesian and other Governments to promote IPM, FAO's IPM inter-country program on rice was started in South and Southeast Asia. In 1989 the program, a collaborative effort among IPM practitioners, ecologists and sociologists, initiated its first field schools in Indonesia.

The inter-country program gradually included more Asian countries and was expanded to other crops, particularly cotton, cabbage, French beans and soybeans. From 1997 onward, support has been made available to countries in other continents[4].

Upon request of some West African countries, the FAO-based Global IPM Facility implemented rice IPM pilot projects in Ghana, Burkina Faso, Côte d'Ivoire and Mali. FAO's Technical Cooperation Unit financially supported these projects which were executed between 1995 and 1998. FFS in Ghana, Côte d'Ivoire and Burkina Faso showed savings for rice farmers of over US$ 90/ha, with yields maintained or increased. Profits increased by over 25%[5].

In 2000, in its third year of existence, the Facility had helped initiate IPM programs in over 12 African and several Latin American countries. Demand for technical assistance to develop IPM is enormous and growing. Evidence is now abundant that participatory

[4] With the establishment of the Global IPM Facility in 1997, FAO, in collaboration with UNDP, UNEP and the World Bank and financially supported by the Governments of The Netherlands, Switzerland and Norway, started supporting IPM initiatives in Africa and Latin America as well. In 1999, as a follow-up to several FAO-funded IPM pilot initiatives in Burkina Faso, Ghana, Côte d'Ivoire and Mali, the Facility helped initiate a pilot IPM project on rice in Mali's Office du Niger.

[5] Global IPM Facility Field Reports, unpublished.

IPM with and by farmers works, also in the African context. In most cases, national programs expanded IPM to incorporate production and soil issues, thus giving rise to the term 'integrated production and pest management (IPPM)'. Originally used by the Zimbabwe IPPM Program, other African and Latin American countries have adopted this term in recognition of the links between growing a healthy crop and pest management.

Global IPM Facility-assisted IPM Initiatives in West Africa

Ghana

Ghana was the first West African country which started an IPM program based on farmer participatory training through field schools. In 1995, three Asian IPM experts facilitated a so-called Training of Trainers (TOT) for 28 facilitators including 24 Ghanaians, three Ivorians and one Burkinabe. The training aimed at getting trainees and farmers acquainted with all stages of crop development. Ghana's first TOT was organized in the Dawhenya irrigation project, in the Greater Accra region combined with three field schools comprising about 75 farmers. Since this pilot project, the Ghanaian government has been able to secure funding for a national IPM program through the United Nations Development Program and GTZ, the German development agency. The program has expanded from rice to vegetables and plantain and has so far trained several thousands of farmers.

Côte d'Ivoire

From March to July 1996, two of the three Ivorians and the Burkinabe trained in Ghana facilitated a season-long training for facilitators and three field schools in the Sakassou irrigation project Center in Côte d'Ivoire. In total 17 Ivorians, four Burkinabe and four Malian extensionists were trained. The Ivorians trained in Sakassou subsequently facilitated field schools for 125 farmers in five different irrigated rice projects. In 1997, a national workshop was organized in Yamoussokro to review the main project results. With assistance of the Global IPM Facility and FAO's Investment Center, a national IPM proposal for rice and peri-urban vegetables was drafted. In August 1998, however, Agence Nationale d'Appui au Développement Rural with world bank encouragement, signed an agreement with Rhone Poulenc to establish pilot operations to demonstrate new agrochemicals on rice, cocoa and vegetables. This demonstration program effectively replaced the proposed IPM program.

Burkina Faso

FAO's Technical Cooperation Program (TCP) funding supported a pilot project in Burkina Faso from 1996 to 1997. It was hosted by the national Plant Protection Service of the Ministry of Agriculture. In 1996, a 4-month TOT and three field schools were organized in the Vallee du Kou irrigated rice scheme. The training included 20 participants of whom 19 were Burkinabe from across the country and one Malian. The following year, an inter-country field school program was run in seven irrigated rice schemes throughout the country: Bagre, Banzon, Dakiri, Karfiguela, Niassan, Tamasgo and Vallee du Kou. Each rice scheme accommodated one FFS. A total number of 213 farmers were trained during the Burkina Faso pilot project. In 1997, at the end of the project, a national workshop evaluated the activities and made several recommendations to further strengthen the achievements.

Mali

In 1996, FAO provided technical support to the Plant Protection Service to implement an IPM pilot project. In May 1996 a national workshop was organized to assess the status of IPM in the country and set up priorities for follow-up activities. Ten facilitators and 88 farmers were trained on rice IPM in Sélingue and Baguineda. In 1998, the Global IPM Facility helped to prepare a proposal for a national IPM program. This proposal was subsequently included in the World Bank program to support the rural development sector in Mali, the Programme d'Appui aux Services Agricoles et aux Organisations de Producteurs (PASAOP).

In anticipation of the forthcoming PASAOP, an IPM project on rice was implemented in Mali's Office du Niger, from June to November 1999[6]. Funding was provided by the Royal Netherlands' Embassy in Mali, the Office du Niger and the Global IPM Facility. With presently about 70,000 ha of irrigated land, the Office du Niger is West Africa's largest rice irrigation system. Though pesticide use on rice is typically limited to herbicides, farmers are increasing the use of agrochemicals, including insecticides due to the continuing intensification of irrigated rice farming. The IPM pilot project was set up to help farmers enhance their production and pest management skills, making alternative, non-chemical pest management strategies available to them. It trained 15 extension workers and about 575 farmers from 23 villages throughout Niono, Molodo and N'Debougou, three central areas of the Office du Niger. In 2000 and 2001, the Office du Niger continued and expanded IPM activities to all five administrative zones. From 2002 onwards, a 4-year consolidation and expansion project for IPM on rice and vegetables will provide continued support in the form of IPM training and awareness activities. IPM rice trials implemented in the Office du Niger have shown good to dramatic improvements in production, in many cases simultaneously reducing costs[7].

In June 2000, FAO's SPFS started six rice IPM field schools in the Mopti region with about 150 participating farmers. Training continued in 2001 and may be expanded to Kita and Kangaba, other SPFS intervention areas.

Follow-up to TCP Pilot Projects through the Sub-regional IPM Program for Burkina Faso, Mali and Senegal

One of the main objectives of the pilot IPM projects described above was to show national governments, parastatals such as the Office du Niger in Mali, farmers and farmers' associations, NGOs, research and donors intervening in West Africa, that by enhancing farmers' crop and production management skills pesticide use can be greatly reduced or avoided altogether. Institutionally, one of the objectives of the pilot initiatives was to build linkages with a wide range of stakeholders to ensure strong anchorage for follow-up IPM training programs.

To summarize, the main outputs of the pilot projects carried out in Burkina Faso, Côte d'Ivoire, Ghana and Mali were:

- Raised awareness on IPM and IPM-relevant issues (e.g. pesticide use and pesticide subsidies) through the organization of national workshops aimed at sensitizing the general public, political and technical decision-makers and donors.
- Strengthened capacities of the national agriculture extension system by training extension workers.
- Reduced use of chemical pesticides on IPM training plots. The experimentation showed a reduction of up to 20–30% of agrochemical products, in many cases with concomitant decrease in production costs.
- Overall improvement of the revenues of small-scale farmers through better management of the agroecosystem.
- More independent decision making by farmers resulting in changed behavior of farmers towards external actors or partners: farmers are known to become more critical when dealing with officials, extension educators, traders and chemical companies representatives

[6] Nacro (1999).
[7] Global IPM Facility, January 2000.

due to their enhanced technical competence, knowledge and analytical skills.
- Enhanced social cohesion of some farmer organizations.

In order to further strengthen national IPM training initiatives and regional expert networks, the Global IPM Facility assisted the Governments of Mali, Burkina Faso and Senegal in developing a sub-regional IPM program. Burkina Faso, Mali and Senegal decided to opt for a regional approach because the countries have many similarities. They share the same geographical region, the Sahel, characterized by cyclic droughts and have a common pest complex with chemical control being the first control tactic. In addition, agriculture is a major sector of their economies. The large majority of the actors in this sector are small-scale farmers who practice subsistence agriculture. Although some progress has been achieved in food production in the three countries in recent years, food security is still fragile. The degradation of natural resources is inexorably in progress not only because of the adverse climatic conditions that have weakened the natural ecosystems, but also because of man's action: extensive clearing, wild fires, abusive cutting of wood for cooking, irrational use of agropesticides and inappropriate production systems.

In anticipation of the forthcoming approval of the sub-regional program, the Royal Netherlands' Embassy in Dakar and the Global IPM Facility funded a regional vegetable IPM training. The training was organized by the Center for Ecotoxicological Research in the Sahel/Locustox Foundation and the Global IPM Facility in greater Dakar, from November 2000 to March 2001. Technical inputs and trainers were provided by the national IPM programs of Vietnam and Ghana. A total of 31 extension/training specialists from Senegal, Mali and Burkina Faso participated in the training and facilitated field schools for about 350 farmers, including both women and men from the Niaye zone. The training marked a diversification of the West African IPM programs from rice to vegetables. Technical results were encouraging with no chemical pesticides used on the IPM produce and IPM yields being equal to yields in regular farmer-practice plots.

In July 2001, the Royal Netherlands' Embassy in Dakar confirmed their contribution to the sub-regional IPM Program for Burkina Faso, Mali and Senegal. Each participating government will contribute by paying the salaries of facilitators/extension workers and make study plots and training facilities available to the program which has a duration of 3 to 5 years. The Global IPM Facility will provide overall technical and operational support.

The objectives of the sub-regional IPM program

The overall objective of the sub-regional program is to: 'Reinforce national extension and agricultural research systems through improved technical support to small farmers, particularly women, to allow them to enhance agricultural production in a sustainable manner to meet the objective of food security and increase revenues.'

Specific objectives of the program are to:

- Develop sub-regional IPM capacities by taking advantage of comparative advantages of each of the three countries involved in the program (rice in Burkina Faso, cotton in Mali and vegetables in Senegal).
- Create awareness among the general public, decision makers and development partners in all three countries concerned, through the implementation of activities such as national and regional workshops which reveal the impact of IPM and policy and regulatory constraints to the promotion of IPM (e.g. pesticide subsidies), food security, environment, health of producers and consumers and the export of agricultural produce.
- Promote the exchange of experiences among IPM experts of the three countries through the organization of study tours and sub-regional workshops.

The expected outputs of the program are:

- Approximately 25,000 farmers and 358 extension workers trained on rice, vegetables and cotton IPM.
- Various technical reports and policy documents on IPM and pesticide policy issues prepared.
- Awareness of the general public and policy makers on IPM and pesticide issues raised.
- Networks established among governmental and non-governmental organizations, farmers' associations, technical and research institutes and donor agencies.

Strategy of the sub-regional program

The sub-regional IPM program will use the comparative advantage of each of the three countries to organize training courses for each of the three crops targeted by the program: rice (Burkina Faso), cotton (Mali) and vegetables (Senegal). In each of the countries the program will build on national expertise in research, extension, plant protection, universities, non-governmental organizations, etc. Links will be established between countries to develop regional networks aimed at sharing information and experiences. Extension workers will facilitate the farmers' training, but the program will also provide additional technical training to some selected farmers to become farmer trainers. This facilitation of farmer-to-farmer training will be one of the gauges of the durability of the program and enable a smooth transfer of competencies and responsibilities to its main beneficiaries.

Whenever chemical control is requested, each country will only use pesticides authorized by the Sahelian Pesticides Committee. In this regard, all country members will take advantage of the expertise of the Regional Project of the Application of the International Code of Conduct and Use of Pesticides. This project, funded by the Netherlands government is hosted by the Sahel Institute, a specialized agency of the ISDMS. This strategy is in accordance with the regional approach to crop protection developed by the ISDMS which promotes changes in crop protection and the adoption of IPM. The methodology of this project is patterned after the specific objectives No. 2 (training of national IPM specialists) and No. 4 (development of participatory training and transfer of knowledge) of the national action plan prepared thanks to the ISDMS.

The sub-regional program is founded on the political will of the three countries to protect the environment through, among others, the suppression of pesticide subsidies. It offers alternatives to the classical production systems and pays special attention to women's participation. Women are a key interest group in the agricultural sector in these three countries. The program will organize season-long IPM training activities through TOT and farmer field schools, policy/awareness activities on IPM and pesticide use through a series of pesticide policy studies, policy seminars and related activities. It aims to increase agricultural production in the three countries and improve productivity of small-scale farmers. In other words, the program, while contributing to the increase of small-scale farmers' revenues, will also catalyze a change in farmers' behavior vis-à-vis the management of natural resources and the use of agricultural inputs, particularly pesticides.

Conclusion

The initial experiences on rice and vegetables IPM in West Africa are encouraging. Farmers have expressed strong interest in field schools as a participatory training methodology, and have taken initial responsibility for further dissemination of information on IPM and expansion of IPM training programs. The sub-regional IPM program aims at putting the responsibility for the implementation of IPM programs progressively into the hands of farmers, farmers' associations and local communities. The role of women, particularly in vegetables, is acknowledged by the IPM programs in place and is given high priority in new initiatives on, for example, cotton IPM.

The pilot projects carried out in West Africa have revealed that by reducing pesticide use, production costs can be significantly cut and yields maintained or even improved. With a range of local, regional and international partners, the Global IPM Facility has started looking at policy and institutional issues related to IPM and pesticide use. In Mali, a comprehensive analysis was made of socio-economic factors favoring pesticide use. The study revealed some policy distortions which are likely to be found in other countries in the sub-region as well. Thus, besides the need to further enhance efforts to expand field programs, targeted support is needed to address adverse policies (e.g. pesticide subsidies) and promote an enabling policy environment conducive to IPM.

References

FAO Technical Assistance Indonesia National IPM Programme (1998) Community IPM: Six Cases from Indonesia.

INSAH (2001) Etude Socio-économique de l'utilisation des pesticides au Mali. *Series Les Monographies Saheliennes*, No. 12.

Nacro, S. (1999) *Rapport Technique sur le Projet Intérimaire de Formation Participative en Gestion Integrée des Deprédateurs du Riz à Travers des Champs Ecoles de Producers en Zone Office du Niger.* Global IPM Facility Report, Rome.

ter Weel, P. and van der Wulp, H. (1999) *Participatory Pest Management.* DGIS Policy and Best Practice Document No. 3, The Hague.

University of Hanover/GTZ Pesticide Policy Project (1998) *Analyse Socio-économique de la Filière des Pesticides en Côte D'Ivoire.* Série de Publication No. 06/F.

University of Hanover/GTZ Pesticide Policy Project (1999) *Pesticide Policies in Zimbabwe.* Publication Series Special Issue No. 1.

Appendix 30.1.

Past IPM training activities and available IPM trainers and farmers in francophone West African countries (status as of July 2001).

Country	Pilot activity	Year	Crop	IPM resources
Burkina Faso	TCP	1995–1998	Rice	20 IPM trainers 213 IPM farmers
Burkina Faso	Participation in Regional Vegetable IPM training project Senegal	2000–2001	Vegetables	3 IPM trainers trained in Senegal
Mali	TCP and Interim IPM project Office du Niger	1997–2000	Rice	19 IPM trainers 889 IPM farmers 42 IPM farmer-trainers
Mali	Participation in Regional Vegetable IPM training project Senegal	2000–2001	Vegetables	3 IPM trainers trained in Senegal
Senegal	Pilot phase Vegetable IPM Senegal	2000–2001	Vegetables	26 IPM trainers 375 IPM farmers
Côte D'Ivoire	TCP	1995–1996	Rice	17 National IPM Trainers 200 IPM farmers

TCP, Technical Cooperation Program.

Appendix 30.2. IPM Web References

Global IPM Facility
http://www.fao.org/globalipmfacility/

Provides essential information on IPM programs assisted by the Global IPM Facility, with many useful IPM contacts worldwide.

Community IPM in ASIA
http://www.communityipm.org

This site is a source of information on Community IPM in Asia, providing details on IPM training activities in 12 countries with contact details of key IPM experts. Contains case studies on FFS processes and a virtual library of training materials, scientific papers and case studies related to IPM.

PAN
http://www.pan-international.org

A network of over 600 participating non-governmental organizations, institutions and individuals in over 60 countries working to replace hazardous pesticides with ecologically sound alternatives. Its projects and campaigns are coordinated by five autonomous Regional Centers: Africa, Asia/Pacific, Europe, Latin America and North America.

- **PAN Asia and Pacific:** www.poptel.org.uk/panap
- **PAN North America:** www.panna.org/panna/
- **PAN UK:** www.pan-uk.org/

CAB International
http://www.cabi.org/

An organization specialized in sustainable solutions for agricultural and environmental problems.

IPM in schools, EPA
http://www.epa.gov/pesticides/ipm/

To protect children's health from unnecessary exposure to pesticides.

OECD Pesticide Program
http://www1.oecd.org/ehs/pest_rr.htm

Focuses on chemical pesticides and biological pesticides (e.g. bacteria, pheromones, insects, plant extracts) which are used in agriculture, including horticulture and forestry and other settings (e.g. products used in houses, in swimming pools, on pests).

Chapter 31

Integrated Pest Management Collaborative Research Support Program (USAID – IPM CRSP): Highlights of its Global Experience

Brhane Gebrekidan

IPM CRSP, Virginia Tech, Blacksburg, Virginia, USA

Introduction

This chapter presents an overview of the Global Experience of the Integrated Pest Management Collaborative Research Support Program (IPM CRSP). The chapter covers the following aspects of the IPM CRSP: background, organizational and operational structure of the IPM CRSP, mode of collaboration, the Participatory Appraisal (PA) process, Annual Work Plan development and its implementation, developing IPM packages, examples of successful IPM packages in selected regions, technology transfer, Technical Assistance (TA), gender-related issues that affect IPM adoption, training and national capacity building, regionalization and globalization of IPM, information exchange, and finally mutuality of benefits to the USA and the host countries.

Background

The IPM CRSP was initiated in 1993 with the financial support of the USAID. The main mission of the CRSP remains to be to foster IPM through collaborative research between USA and developing host country institutions for their mutual benefit by improving their abilities to develop and implement economically and environmentally sound crop protection methods. The IPM CRSP, which is one of nine CRSPs supported and managed by the Global Bureau of USAID, has successfully completed its first 5-year phase and is in the middle of its second 5-year phase.

The purpose of the IPM CRSP is to develop and implement appropriate IPM techniques and strategies that will help reduce: (i) agricultural losses due to pests; (ii) damage to national ecosystems; and (iii) pollution and contamination of food and water supplies. The long-term goals of the CRSP are to develop improved IPM technologies and institutional changes that will reduce crop losses, increase farmer income, reduce pesticide use and pesticide residues on crop products, improve IPM research and education program capabilities, and increase the participation of women in IPM decision making and program design.

Working towards this goal the IPM CRSP follows the following specific objectives:

- Identify and describe the technical factors affecting pest management.

 Integrated Pest Management in the Global Arena (eds K.M. Maredia, D. Dakouo and D. Mota-Sanchez)

- Identify and describe the social, economic, political, and institutional factors affecting pest management.
- Work with participating groups to design, test, and evaluate appropriate participatory IPM strategies.
- Work with participating groups to promote training and information exchange on participatory IPM.
- Work with participating groups to foster policy and institutional changes.

The research activities of the IPM CRSP are based on close collaborations between scientists of the participating host countries and US institutions. The participating prime host country sites of this CRSP currently include Albania, Bangladesh, Ecuador, Guatemala, Jamaica, Mali, The Philippines, and Uganda. Among the active partner USA institutions are: University of Georgia, Lincoln University, Montana State University, Ohio State University, Penn State University, Purdue University, University of California – Davis and Riverside, University of Maryland – Eastern Shore, North Carolina A&T University, Florida A&M University, Fort Valley State University, USDA, and Virginia Tech (VT), with VT as the lead and the Management Entity (ME) institution.

Organizational and Operational Structure of the IPM CRSP

With some modifications, the IPM CRSP organizational and operational structure follows the general CRSP Guidelines provided by USAID to all CRSPs. Accordingly, the main entities in implementing the IPM CRSP are the ME, the Board of Directors, the Technical Committee (TC), the Site Committees (SC), the External Evaluation Panel (EEP) and the USAID Project Manager.

Virginia Polytechnic Institute and State University (Virginia Tech) is the ME for the IPM CRSP and is the primary grantee of USAID. The ME is accountable to USAID for all IPM CRSP programmatic and fiscal issues although certain site specific responsibilities are delegated by the ME to the participating USA and host country institutions. Collaborative research arrangements between participating USA and host country institutions are governed by a Memorandum of Understanding (MOU) between the lead host country institution and the IPM CRSP ME. The MOU creates the official environment in which participating US scientists and their host country partners can initiate and carry out collaborative research in the host country or region.

The Board of Directors deals with policy issues and advises the ME on these and other related matters. The TC reviews the research and training plans of the CRSP, participates in the development of the annual work plan and budget, and recommends them to the ME for implementation. The SC has the primary responsibility of developing and implementing collaborative IPM activities related to research, training and networking for its specific host country or region.

The EEP is charged by the USAID Global Bureau with the overall evaluation of the IPM CRSP, which includes program direction and research collaboration with the host countries. The USAID Project Manager of IPM CRSP and other appropriate members in the USAID Global Bureau advise and guide the ME, the Board, and other entities of the CRSP in areas of policy, technical and program management, collaborating host country coordination, budget management, and review.

Mode of Collaboration

The IPM CRSP operates in eight prime sites in five major regions, Africa, Asia, Latin America, the Caribbean, and Eastern Europe. The African programs focus on irrigated peri-urban horticulture as well as rain-fed cereals and legumes, both the Latin American and the Caribbean programs emphasize non-traditional agricultural exports (NTAEs), the Asian programs concentrate on vegetables grown in rice/vegetable cropping systems, while the Eastern European program deals with IPM of a single crop, olive. At each site, USA and host country scientists collaboratively and jointly plan, implement, and report research activities. A Site Chair from a US institution and a host country Site Coordinator from the lead collaborating institution in the host country take joint leadership in planning and implementing the IPM CRSP activities in the country. The current site chair and host country coordination leadership distribution by institutions is as given below:

- African Site in Mali: Virginia Tech (Site Chair), Institut d'Economie Rurale (IER) (Site Coordinator);
- African Site in Uganda: Ohio State University (Site Chair), Makerere University (Site Coordinator);
- Latin American Site in Ecuador: Virginia Tech (Site Chair), Instituto Nacional de Investigaciones Agropecuarias (INIAP) (Site Coordinator);
- Latin American Site in Guatemala: Purdue University (Site Chair), Universidad de Valle de Guatemala (Site Coordinator);
- Caribbean Site in Jamaica: Virginia Tech (Site Chair), Caribbean Agricultural Research and Development Institute (CARDI) (Site Coordinator);
- Asian Site in the Philippines: Ohio State University (Site Chair), PhilRice (Site Coordinator);
- Asian Site in Bangladesh: Virginia Tech (Site Chair), IRRI Dhaka/ Bangladesh Agricultural Research Institute (BARI) (Site Coordinator);
- Eastern Europe Site in Albania: Virginia Tech (Site Chair), Plant Protection Institute, Durres (Site Coordinator).

The Participatory Appraisal Process

The foundation of the IPM CRSP approach is the use of the PA in the determination of high priority crops, pests, and processes to follow in program implementation. Central to this approach is the involvement of the appropriate stakeholders such as scientists, extension personnel, farmers, policy makers, government officials, input suppliers, and NGOs in identifying the high priority pest problems at the site and the approaches to be used in solving these problems. Before initiating a program in a site, the IPM CRSP typically conducts PAs focusing on the identification of agroecosystems, baseline surveys, existing pest management practices, high priority crops or cropping systems and their key pests, and other related topics. The results of the PA are jointly analyzed and written as a reference document by USA and their host country partners and are used in defining the research, training and information exchange agenda of the IPM CRSP at the site. The priority crops and pests the IPM CRSP is working on at each of its sites have been determined through the PA process. The PA results from most of our sites have been published as IPM CRSP Working Papers and are available at the IPM CRSP Management Entity office.

Annual Work Plan Development and Its Implementation

Initial annual work plans and budgets under the IPM CRSP are prepared by a team of US and host country co-principal investigators and submitted to the host country site committee through the Site Coordinator. Under the leadership of the Site Chair, the site committee discusses each proposal and budget and recommends appropriate modifications to each team for revising the proposal. Eventually all proposals from a site are forwarded to the Site Chair who assembles the proposals received, prepares a site work plan and budget, and distributes the same to the site committee members for discussion and comments before submission to the IPM CRSP TC. The draft work plans and budgets from all sites are submitted to the TC for discussion and recommendations during its annual meeting. The final annual work plan is reviewed by the ME for consistency and uniformity across sites and submitted to USAID for its approval.

Developing IPM Packages

The approved annual work plan as described above is implemented at each host country site. The vast majority of the planned experiments are carried out on-farm with the direct participation of small-scale farmers where they contribute both land and labor. Through this scheme, farmers have the opportunity to be active partners in the implementation of the experiments, while simultaneously observing the results for themselves. The active participation of farmers in this manner facilitates the direct transfer of the experimental results to the participating farmers and their neighbors. As a complement to the on-farm trials, a lesser number of on-station, greenhouse, and laboratory experiments are also conducted. Based on experimental results, replicated over locations and seasons, suitable IPM packages are determined and tried out on farmers' fields, and eventually extended to a wider range of farmers in the host country and in the region.

Examples of Successful IPM Packages in Selected Regions

Over the last 7 years the IPM CRSP has developed successful IPM packages applicable to the various cropping systems where it is operating. These packages have been disseminated or are being disseminated to producers in the host countries. Selected examples from the four well-established regions are given below.

Asia – rice–vegetable cropping systems (the Philippines)

IPM CRSP has been working on vegetable IPM in the rice–vegetable cropping systems since 1994, initially in the Philippines and more recently in Bangladesh. The IPM approaches used have dealt with weeds, diseases, and insects. In the case of the rice–onion cropping systems in the Philippines, the most serious weed problem in onion production is the nut sedge (*Cyperus rotundus*). The IPM CRSP has developed an economical weed management system that is suitable for onion producers in the Philippines where the results of the IPM CRSP show that the cost of farmers' weed control practices can be reduced by 50% from one herbicide application followed by three hand weedings to one herbicide application and one hand weeding without reducing weed control efficacy and onion yields.

In addition, IPM CRSP research has shown that rice hull burning, which is practiced commonly by onion farmers in the Nueva Ecija region of the Philippines, could significantly reduce the soil population of pathogenic fungi where the onion pink root disease is common. Over several seasons, the incidence and severity of pink root infection in onion was lower and onion yields were higher in plots in which rice hulls were burned, compared with unburned plots. This practice is being recommended to a wider range of farmers who have access to economical rice hull.

Further, IPM CRSP research has shown that the use of (NPV) and Bt in onion production can be a viable alternative to chemical insecticides to control the larvae of the key onion insect, onion cutworm (*Spodoptera litura*). Thus, the use of NPV and Bt would greatly benefit onion farmers who are dependent on chemical insecticides for control of onion cutworms. Direct effects of this new technology are reduced pesticide use, better health of farmers and their families, and sustainable *Spodoptera* management. Farmers can, in fact, mass produce NPV themselves and use the technology on their fields, which will cut down on cost of crop protection by onion growers. As a result, the market quality of farmers' onion produce will be greatly enhanced by the low insecticide residue levels, thereby meeting the export requirements of foreign markets. The application of these technologies in an IPM package can greatly benefit onion producers in the Philippines.

Africa – maize–bean cropping systems (Uganda)

In one of its African sites, in Uganda, the IPM CRSP has been involved in developing IPM strategies for insect and disease control in the maize–bean cropping systems of eastern Uganda where *Chilo partellus*, the low altitude stem borer, is the predominant cereal pest. The results of the IPM CRSP on-farm trials for 3 years have confirmed that the introduced wasp parasitoid, *Cotesia flavipes*, a potential biological control, is effective in reducing significantly the stem borer damage on maize. The parasitoid which has now been established both in eastern and northern Uganda causes parasitism on the *Chilo* stem borer of up to 23%. This biological control agent was multiplied and released in a collaborative activity involving a graduate student, the IPM CRSP, and ICIPE.

Further, in the same region, IPM CRSP on-farm trials on the common bean *Phaseolus vulgaris* have confirmed that bean grain yields can be increased by as much as 150% with endosulfan seed dressing to control the bean fly (*Ophiomyia* sp.) and root rots (*Fusarium solani* and *Fusarium phaseoli*). Additionally, earthing-up or ridging at first weeding reduced bean fly damage and increased grain yield by about 35%. The combined use of the wasp parasitoid for maize stem borer control and seed dressing and earthing-up or ridging for bean fly control are being introduced to Ugandan farmers engaged in the maize–bean cropping system. To ensure that the technology is disseminated on a large scale, this technology has been passed on the USAID funded IDEA Project operating in Uganda. IDEA is now conducting nationwide demonstration trials for bean and maize growers. IPM CRSP and IDEA have also produced fact sheets and posters for use by extension workers. It is anticipated that adoption of this technology will boost bean and maize production in the region, leading to reduction in malnutrition and poverty.

Latin America – horticultural export crops (Guatemala)

In Guatemala, since 1994 the IPM CRSP has been working on developing IPM technologies for non-traditional agricultural export (NTAE) crops of which snow pea is the leading commodity. The key snow pea pest in Guatemala is the leaf miner, *Liriomyza huidobrensis*. A wide range of IPM component technologies for snow peas have been developed by the IPM CRSP and introduced to small-scale farmers. Snow pea producers participating in the IPM CRSP developed integrated pest management/integrated crop management programs composed of the use of certified seed, adequate fertilizer application, using wheat straw mulch, weekly scouting of pest levels, threshold based spraying of chemicals, and the use of the mobile yellow sticky traps, which reduced pesticide applications for the typical Guatemalan snow pea farmer from 13 to four in each cropping cycle.

The establishment of an effective quality-control program for snow peas that will guarantee: (i) the quality of the product to the final consumer; and (ii) the sustainability of the snow pea industry in Guatemala are main impacts of the IPM CRSP research. It is expected that in the next few years the majority of Guatemalan snow pea exporters will implement the IPM CRSP-generated ICM production programs. It is worth noting that most of the snow pea produced by small farmers in Guatemala is exported to the US market.

Another IPM CRSP tested crop management strategy effective for minimizing pest damage is strip cropping, which should become an attractive option for farmers in the highlands of Guatemala. The diversification of crops will favor the long-term sustainability of export crops and locally marketed vegetables as well. In addition, the higher diversity will promote the build-up of natural pest enemies and help maintain pests at manageable levels. The existence of a healthy agroecosystem will also prevent the emergence of new primary pests, and better natural control of existing pests. The ICM advocated by the IPM CRSP in Guatemala is applicable to multi-crop systems as well. The ICM strategy increases the economic benefits to the farmer, as profit margins are increased due to reduced usage of chemical pesticides.

Caribbean – sweet potato (Jamaica)

In the Caribbean program of the IPM CRSP, one of the high priority activities dealt with developing IPM strategies for sweet potato (*Ipomoea batatas*) production under the conditions of small-scale farmers. As a result of these activities, in replicated on-farm trials, several USDA-developed sweet potato clones and Caribbean varieties demonstrated good resistance to the sweet potato weevil *Cylus formicarius* which is an important insect constraint in sweet potato production in Jamaica. Other components of a sweet potato IPM strategy developed by the IPM CRSP in Jamaica included the use of sex pheromone baited traps, application of good cultural practices (field sanitation, removal of old sweet potato vines, optimum irrigation, timely harvesting, and crop rotation), and chemical spraying based on insect number scouting and predetermined threshold levels. The sweet potato IPM developed in Jamaica is being disseminated to farmers and is being regionalized to other Caribbean islands such as St Kitts and St Vincent.

Technology Transfer

The IPM CRSP works with national technology transfer agencies, cooperatives, NGOs, and other appropriate bodies to extend to producers the IPM technologies it has developed in a given site. Very often farmers are involved in on-farm testing of IPM technologies and demonstrations giving them the opportunity to observe and adopt results directly.

At the African site in Uganda, both the national extension system as well as the IDEA project are active in disseminating IPM CRSP results to farmers. At the Latin American site in Guatemala, pest management technology and information developed by IPM CRSP are transferred through grower workshops, technician seminars, and field demonstrations. It is estimated that in a typical year approximately 45% of Guatemala's NTAE producers were engaged in the use of pest management practices recommended by IPM CRSP. In Jamaica, technology transfer training sessions for sweet potato and hot pepper farmers were conducted in different communities. Topics covered in these sessions included pest identification, fertility management, and principles of IPM. Technology transfer sessions were not only geared towards production technology but also demonstrated the benefits of developing a sound marketing strategy, i.e. an integrated approach to hot pepper and sweet potato production and marketing. Demonstration plots were used in the technology-transfer process. Field days were often held in selected communities of sweet potato and hot pepper growers to demonstrate IPM systems to key farmers and extension officers. In the Philippines, technology-transfer activities were carried out in cooperation with the Training Division of PhilRice, in both organization of meetings and preparation of training materials for season-long training programs for extension officers and farmer leaders. IPM CRSP scientists from the host country were active participants in these training programs in technology transfer. Training manuals, fact sheets, brochures, single page flyers, flip charts and book marks on a number of diseases and pests of onion and aubergines were prepared, evaluated and disseminated to the participants.

Technical Assistance

In addition to its core activities in its prime sites, the IPM CRSP has set aside TA funds to respond to emergency pest situations arising in developing countries. Such funds are usually accessed through requests from national programs and the USAID Mission in the country. If the USAID Mission supports the request it must be prepared to contribute 50% of the funds to respond to the emergency situation.

In Mali, through a joint funding from the IMP CRSP TA funds and the USAID Mission in Mali, the IPM CRSP and Malian scientists initiated a collaborative work to strengthen the Environmental Quality Laboratory (EQL) of Mali. The TA program provided essential equipment and trained key Malian scientists at Virginia Tech in the use of special equipment and the application of good laboratory practices. Proper analysis of vegetables for pesticide residues was the main part of the training. Overall the Mali site of the IPM CRSP began collaboration with the EQL of Mali, to address needs in pesticide residue analysis, environmental monitoring, and development of a quality assurance program for agricultural products in Mali. The EQL, a part of the Central Veterinary Laboratory, has a comprehensive mandate that reflects the needs of Mali and the West African region for the analysis of pesticide residues on crops, food, and medicinal products, and in water, soil, and sediments.

At the beginning of 2000 the Ugandan NARO and the USAID Mission in Uganda requested the IPM CRSP to assist them in properly managing the emerging coffee wilt disease epidemic in the country. The CRSP responded positively with its TA funds along with 50% contribution from the USAID Uganda Mission. This TA is now underway and is expected to develop an IPM strategy to contain this serious disease of the leading export commodity of Uganda. This IPM CRSP TA project will focus on etiology, pathogenesis and epidemiology of coffee wilt disease (*Fusarium xylarioides* Steyaert). The main goal of the project is to control the epidemic of coffee wilt and restore coffee production in Uganda.

At the Caribbean site in Jamaica, the USAID Mission in Jamaica and the IPM CRSP contributed funds and technical assistance to strengthen the 'National Strategic Plan To Combat the Gall Midge Complex Affecting Hot Pepper.' The main goal of this activity has been to address the fundamental issues surrounding the emergence of new pests, the gall midges *Contarina lycopersci* and *Prodiplosis longifilia*, and their impact on the hot pepper export market. Among the expected outputs of this project are: (i) improvements in the quality and quantity of exportable and local hot peppers; (ii) development of IPM options for managing gall midge populations; (iii) improvements in the knowledgebase of farmers in pest management; (iv) increased number of farmers and extension agents trained in IPM; (v) knowledge of the relation of the gall midge with respect to phenology of the crop and agroecology.

In Guatemala, in 1996, a crisis emerged when Guatemalan snow peas were detained at US ports of entry because of infestation by an unknown leafminer species. The IPM CRSP responded as a TA by completing a taxonomic survey of the snow pea agromyzid leafminer species in the Guatemalan highlands. In the survey the CRSP found that *Liriomyza huidobrensis* is the major leafminer species found in snow peas and other export crops in the central highlands. This species also occurs in the USA. Results of this research were accepted by APHIS as a basis for removing the requirement for detention of Guatemalan snow peas found with minor leafminer infestations. The CRSP has subsequently introduced to small farmers a holistic crop management practice of which an effective IPM is an important component. The IPM CRSP has also been instrumental in getting growers and exporters in Guatemala to institute an effective pre-inspection procedure to ensure that snow peas exported to the USA are free from pests and pesticide residues.

Gender-related Issues that Affect IPM Adoption

The IPM CRSP was designed, and has been implemented, with a strong gender-equity component. The CRSP is committed to the equitable involvement of women as both program scientists and beneficiaries. As a research program, the focus of gender-equity programming has been on research into the likely outcomes of IPM research activities for women and men, and on involving women farmers in these activities, in order to ensure that women's livelihood strategies receive equal attention with those of men. With a view to sectoral growth as well as to equity issues, programming is designed to ensure that women who produce targeted crops are included in research and dissemination activities.

Among the gender equity issues addressed is whether the adoption of IPM is likely to alter the gendered division of labor and resources within households in ways that would disadvantage women. Findings indicate that IPM adoption would not disadvantage women. However, the potential benefits of IPM adoption may not be as available to women as to men, as women are less likely to receive relevant technical assistance or to be involved in technology development. Across the IPM CRSP sites, women appear to have less access than men to IPM-related extension. Recommendations for the inclusion of women farmers in technology development, as well as in related extension efforts, have been made for several sites, including those in Uganda, Mali, the Philippines, Guatemala, and Jamaica. For example, in Mali and Uganda, these recommendations have resulted in attention to women's crops and production constraints, and to technologies that improve women's food processing enterprises. Half of producer groups involved in IPM research in Uganda are female-oriented.

In the Philippines, household surveys and ethnographic research have demonstrated that even women who do not work in the fields often control funds used to purchase IPM technologies, and thus should be included in all information campaigns. Philippine women are somewhat more likely than their husbands to prioritize spending for pesticides, as they are less inclined to risk crop loss, but are also more likely to be receptive to cost-effective IPM practices.

The IPM CRSP has worked to ensure that both US and host-country women scientists are involved as investigators and women students are given opportunity for training. The IPM CRSP features an approach that is genuinely committed to gender equity issues. The commitment of the IPM CRSP to gender equity issues can be seen most clearly in the record of graduate training, host-country scientist participation, global gender-focused research, and increasing incorporation of women farmers in collaborative research and extension efforts.

Training and National Capacity Building

The IPM CRSP has given high priority to training of both host country and US students in the various disciplines contributing to IPM research and implementation. As of the end of the year 2000 there were a total of 58 trainees receiving partial or complete IPM CRSP financial support. These trainees have either completed or are undergoing their MS or PhD degree or short-term training objectives. Looking at the gender distribution, 42% of the trainees were females and the vast majority (72%) of the trainees were from the IPM CRSP collaborating host countries. The disciplines in which the trainees are involved cover the whole range of IPM-related topics including entomology, plant pathology, weed science, nematology, horticulture, agronomy, plant breeding, ecology, agricultural economics, statistics, geography, and sociology.

The human resource development strategy planned for all the IPM CRSP sites is long term in perspective, assuring a breadth of skills and capacities available for IPM research and implementation with emphasis on graduate student training at selected national universities such as the University of Mali, Makerere University in Uganda, and the University of the Philippines at Los Baños. Degree training of host-country students has also been done in US universities. Co-Principal Investigators (Co-PIs) from the USA often make several trips to the host countries each year to participate in both training and research. The IPM CRSP's emphasis on process, including research planning and farmer participation, and the locally recognized need to advance multi-institutional and disciplinary research have been recognized by host-country partners as a key contribution of the IPM CRSP to IPM-related research in each host country.

Graduate students sponsored and supervised by the IPM CRSP have made substantial contributions to on-farm data collection and to IPM CRSP activities in the host countries as a whole. Short-term training for national scientists on various aspects of IPM has been conducted both in the USA and the host countries. Several trips are made each year to the host country sites by USA based Co-PIs to participate in annual work plan development and to pursue collaborative research activities. Such visits and institutional relationships continue to be important in strengthening US and host country linkages and moving the IPM CRSP research, technology transfer, and information exchange efforts forward in each of our regions.

One of the objectives of the IPM CRSP training and human capacity building is the institutionalization of IPM into the national system and crop protection policy. There are positive indications to show that IPM is being institutionalized at the national level. Some examples of the institutionalization of IPM in the host countries are: (i) the creation of a new department of Crop Protection at Makerere University partially stimulated by the IPM CRSP; (ii) the establishment of the IPM network for the Caribbean through the participation of the IPM CRSP; (iii) the incorporation of IPM as an integral part of the snow pea production and export system of Guatemala; (iv) the adoption of IPM as part of the national policy in both the Philippines and Bangladesh.

Regionalization and Globalization of IPM

Although the IPM CRSP operates in strategically selected prime sites, regionalization and globalization of IPM technologies and information is a primary goal of the IPM CRSP. Each of our prime site programs embodies a regionalization effort to disseminate IPM technologies and information generated in a given prime site. In the Caribbean region for example, the IPM technologies developed for sweet potato pest management in Jamaica have been extended to other Caribbean islands such as St Kitts/Nevis and St Vincent. In Uganda, the utilization of the wasp parasitoid for stem borer biological control is being used in Kenya as well through the efforts of ICIPE. In both Mali and Uganda, an integrated *Striga* management strategy for cereals, comprising resistant varieties, modest nitrogen application, improved cultural practices, and crop rotation are being promoted. In the Philippines and Bangladesh the promotion of improved aubergine varieties along with grafting technology for bacterial wilt control are being introduced to farmers. The NTAE pre-inspection protocol which has served the Guatemalan horticultural exports effectively is being introduced in the neighboring Honduras. Regional workshops and IPM CRSP global symposia held at regular intervals contribute significantly to the regionalization and eventually the globalization efforts of this CRSP.

Information Exchange

IPM information exchange and dissemination nationally, regionally, and eventually globally is central to the mission of the IPM CRSP. To this end, the CRSP produces a wide range of publications such as refereed journal articles, books and book chapters, theses and dissertations, workshop proceedings, annual reports, working papers, websites, extension publications, fact sheets, newsletters, magazine and newspaper articles, trip reports, and bibliographic databases and distributes them widely. Most of these publications are available as hard copies at the ME office of the CRSP and are distributed to selected recipients globally. The IPM CRSP website http://www.ag.vt.edu/ipmcrsp/ contains full contents or lead information on most of these products.

One of our sites, the African site, maintains an active electronic communication network, IPM Electronic Communication in Africa: Africa IPM Link. The main goal of AIL is the promotion and use of ICTs in Africa for IPM and related topics. The AIL activities were partially supported by funds originating from USAID African Sustainable Development office but received through the Global Bureau of USAID. Hence the activities have been implemented as an integral part of the IPM CRSP. The principal objectives of AIL have been to provide electronic networking opportunities for African IPM practitioners, to initiate and promote electronic discussions among professionals with an interest in IPM related issues in and on Africa, to update content and improve the IPM CRSP and Africa IPM Link websites, and to provide training opportunities for African IPM practitioners in the effective use of e-mail and the Internet for IPM information sharing.

One of the main activities of AIL is running and maintaining an active listserv (Afrik-IPM) and e-mail discussion group on IPM-related issues on sub-Saharan Africa. The main purpose of the Afrik-IPM discussion list is to provide IPM practitioners in sub-Saharan Africa with a networking tool for quick, inexpensive, and effective IPM information exchange. The discussion list has been well established and is operating actively and effectively. List 'membership' is currently well over 100, representing a wide range of organizations and individuals from 29 countries, including 18 African ones.

Another main activity of AIL has been designing and maintaining a French and English bilingual Africa IPM Link website http://www.ag.vt.edu/ail/ The site currently contains links to over 200 English IPM sites, 100 French IPM sites, and some useful ICT sites. Information from the site is routinely accessed from several African countries by IPM practitioners.

As mentioned earlier the IPM CRSP maintains a main website http://www.ag.vt.edu/ipmcrsp which contains comprehensive information on the CRSP such as highlights of the IPM CRSP global activities, list of collaborating institutions, bibliographic services, documents, and relevant links. Among the documents available at the website are annual work plans, annual reports, trip reports, IPM CRSP Update Newsletters, working papers, IPM CRSP policy and operating guidelines.

Mutuality of Benefits to the USA and the Host Countries

One of the main aims of the CRSP programs in general is to show mutuality of benefits both to the USA and the partner host countries. A good number of the IPM CRSP activities contribute to this overall aim. For example, most of the pest problems addressed in our host country site activities are widespread throughout the various regions and also occur in other parts of the world. Strategies developed to manage these pests economically and in a sustainable manner can be applied to countries where the IPM CRSP is not present. IPM methods have been developed for managing pests of onion and aubergine in Asia, NTAE crops in Central America, potatoes in Ecuador and Uganda, sweet potato and hot pepper in the Caribbean. Working with such pests gives US scientists global experience on the status and management of these pests. Advance information on emerging pest situations which may threaten US agriculture could be obtained before the pests enter the USA. Availability of safe and pesticide-free imported foods, particularly from Latin America and the Caribbean, to the US consumer is also a major benefit to the USA. A broader concurrent benefit and feedback to the United States will be through: (i) the effects of economic growth in the IPM CRSP regions on trade and demand for US products in international markets; and (ii) improved relations with the various countries in politically sensitive areas of the world.

Lessons Learned

A wide range of lessons have been learnt by all the participating partners during the implementation of this global IPM program. The salient features of the lessons learned are highlighted below.

- Genuine collaboration and equal partnership between USA and host country scientists has laid the foundation for continued global IPM promotion, a core global IPM community has been established.
- PA has been the cornerstone in the identification of the high priority pests and crops on which the IPM CRSP is working, the results of the PA have guided the activities of the IPM CRSP.
- The participation of farmers as active partners in conducting on-farm IPM experiments has been essential for focusing on the actual problems of farmers who have been active research partners and lead adopters of the results.
- It has been necessary to develop or adopt specific component technologies at each site before formulating a comprehensive IPM package to disseminate to farmers, sometimes this has involved the use of laboratory and greenhouse experiments.
- Integration of IPM technologies and testing them at the farm and the community levels is essential before wide-scale dissemination of IPM packages, farmers are introduced to such packages in their own or their neighbors fields.
- The IPM CRSP has developed and disseminated a wide range of successful IPM packages appropriate for the various regions in which it has been working including Asia, Africa, Latin America, and the Caribbean.
- The availability and/or development of relevant IPM technology for each pest/cropping system situation is a prerequisite if a technology transfer activity is going to be successful with farmers, approaching and working with farmers without the necessary technologies does not cultivate their confidence and respect.
- Attention to gender-related issues in IPM technology development and extension activities is essential to address the special cases of women farmers and family members; women play crucially important roles in the production, marketing, and resource allocations of selected crops.
- Training of host country nationals and human capacity-building are the foundations for sustained growth and maintenance of a national, regional, and indeed global IPM programs, the IPM CRSP has given the highest priority to this effort.
- Institutionalization of IPM and accepting it as a national policy can ensure the long-term viability of IPM in a given country, some of our host countries have taken significant steps in this direction.
- Continuing regionalization and globalization efforts of IPM are essential to bring more of the global crop production under IPM, some IPM packages generated by IPM CRSP have been disseminated regionally.
- It is becoming increasingly clear that IPM is an information-intensive approach and multiple tactics have to be employed to disseminate IPM globally, the CRSP has disseminated such information through the print media and electronic media as well as workshops and field days.
- The IPM CRSP has demonstrated that there are mutual benefits both to the USA and the participating developing host countries in implementing a collaborative program such as this one, USA and host country scientists and students, host country farmers and US consumers are among the beneficiaries of this collaborative program.

Acknowledgments

The IPM CRSP is funded by the USAID under Grant No. LAG-G-00-93-00053-00 and the participating institutions. The contributions of the IPM CRSP investigators and the Management Entity are gratefully acknowledged.

Bibliography

Erbaugh, J.M., Kyamanywa, S., Epieru, G. and Mwanje, E. (1999) Farmer knowledge of pest management in Eastern Uganda: a baseline assessment. In: *African Crop Science Conference Proceedings* 4, 1–10.

IPM CRSP Management Entity, OIRD, Virginia Tech. (1998a) *IPM CRSP Fourth Annual Report, 1996–1997.* OIRD, Virginia Tech.

IPM CRSP Management Entity, OIRD, Virginia Tech. (1999a) *IPM CRSP Fifth Annual Report, 1997–1998.* OIRD, Virginia Tech.

IPM CRSP Management Entity, OIRD, Virginia Tech. (2000a) *IPM CRSP Sixth Annual Report, 1998–1999.* OIRD, Virginia Tech.

IPM CRSP Management Entity, OIRD, Virginia Tech. (1998b) *IPM CRSP Highlights, 1993–1998.* OIRD, Virginia Tech.

IPM CRSP Management Entity, OIRD, Virginia Tech. (1999b) *IPM CRSP Annual Highlights for Year 6 (1998–1999).* OIRD, Virginia Tech.

IPM CRSP Management Entity, OIRD, Virginia Tech. (2000b) *IPM CRSP Annual Highlights for Year 7 (1999–2000).* OIRD, Virginia Tech.

IPM CRSP Management Entity, OIRD, Virginia Tech. (1995–2000) *IPM CRSP Update Newsletter.* Published Quarterly.

IPM CRSP Management Entity, OIRD, Virginia Tech. (1998c) *Progress in IPM CRSP Research: Proceedings of the Third IPM CRSP Symposium. 15–18 May 1998,* Blacksburg, Virginia.

Lawrence, J.L., Edwards, C.A., Schroeder, M., Martin, R.D., McDonald, F.D. and Gold-Smith, J. (2000) An integrated approach for managing hot pepper pests in the Caribbean. *BCPC Conference – Pests & Diseases,* pp. 239–244.

Lefter Daku, Norton, G.W., Pfeiffer, D.G., Luther, G.C., Pitts, C.W., Taylor, D.B., Tedeschini, J. and Uka, R. (2000) Farmers' knowledge and attitudes towards pesticide use and olive pest management practices in Vlora, Albania: a baseline survey. *IPM CRSP Working Paper 00-1.* January 2000. IPM CRSP, Office of International Research and Development, Virginia Tech., Blacksburg, Virginia, 173 pp.

Norton, G.W., Rajotte, E.G. and Gapud, V. (1999) Participatory research in integrated pest management: Lessons from the IPM CRSP. *Agriculture and Human Values* 16, 431–439.

Sullivan, G.H., Sánchez, G.E., Weller, S.C. and Edwards, C.R. (1999) Sustainable development in Central America's non-traditional export crops sector through adoption of integrated pest management practices: Guatemalan case study. *Sustainable Development International* Launch edition, pp. 123–126, ICG Publishing, London.

Chapter 32

Bridging the Gap with the CGIAR Systemwide Program on Integrated Pest Management

B. James[1], P. Neuenschwander[1], R. Markham[2], P. Anderson[3], A. Braun[3], W.A. Overholt[4], Z.R. Khan[4], K. Makkouk[5] and A. Emechebe[6]

[1]Biological Control Center for Africa, IITA-Benin; [2]Formerly, The SP-IPM Secretariat, IITA, Ibadan, Nigeria; [3]CIAT, Cali, Colombia; [4]ICIPE, Nairobi, Kenya; [5]ICARDA, Aleppo, Syria; [6]IITA, Kano, Nigeria

Introduction

Demographic pressures and higher food demands cause changes in land use and agricultural production patterns. In resource-limited agriculture, the consequences frequently include decreasing fallow periods, declining soil fertility, reduction in amount of arable land per farmer, environmental degradation, growing pest[1] problems, falling food productivity, and increasing poverty. Africa, for example, accounts for 19% of the world's poor, and this figure is estimated to rise to 28% by the year 2005 (IITA, 1997; IITA, 2001). In the search for sustainable options to increase food security, IPM plays a key role, having the potential to increase the productivity of agricultural systems while minimizing threats to human health and the environment. IPM has evolved from pesticide abatement strategies (e.g. economic thresholds, scouting or forecasting systems to control application calendars) aimed at avoiding the 'pesticide treadmill' into analytical approaches to understand pest status within production ecologies (e.g. investigate, understand and utilize biodiversity, unravel trophic relationships and causes of population changes, determine effect of abiotic factors on crop/livestock and pest growth) in order to make informed decisions on appropriate options. In this regard, the CGIAR centers have pioneered outstanding contributions in IPM, notably in the areas of genetic resistance against pests of mandate crops (e.g. Osada and Fucikovsky, 1986; Chavez *et al.*, 1988; Hibino *et al.*, 1988; Efron *et al.*, 1989; Dahal *et al.*, 1990; Dixon *et al.*, 1992; Vuylsteke *et al.*, 1993; Mahungu *et al.*, 1994; Bandyopadhyay *et al.*, 1998; Huet *et al.*, 1999; Legg and Thresh, 2000; Singh *et al.*, 2000; Bellotti and Arias, 2001; Morales, 2001), biological control of alien invasive species (e.g. Herren and Neuenschwander, 1991; Neuenschwander *et al.*, 1994; van Thielen *et al.*, 1994; Yaninek *et al.*, 1998; Neuenschwander, 2001), substitution of inorganic

[1] Pests = arthropods, vertebrates, pathogens and weeds and other organisms detrimental to agricultural production.

 Integrated Pest Management in the Global Arena (eds K.M. Maredia, D. Dakouo and D. Mota-Sanchez)

pesticides with biopesticides (Lomer *et al.*, 2001), and development of holistic approaches (e.g. Markham *et al.*, 1994; Yaninek *et al.*, 1994).

The full potential of IPM to reduce shortfalls in food production is, however, poorly realized in developing countries. Many research initiatives focus largely on developing component technologies with minimal understanding of client-oriented approaches in the innovation process, and neglect the key role of the policy environment in IPM promotion. Research centers had also tended to concentrate efforts on genetic resistance, which invariably acquiesced dependence on agricultural inputs including pesticides. CGIAR center interactions with farmer support groups, such as NGOs, have indicated missed opportunities for joint activities to develop sustainable strategies to promote wider implementation of IPM (Kanoute *et al.*, 1999). In recognition of such obstacles, the CGIAR established in 1995 the SP-IPM to coordinate inter-center IPM partnerships and to ensure that IPM contributes decisively to food security and poverty reduction demands.

The Systemwide Program on IPM

The SP-IPM was launched in 1995 in recognition of the need for a radically different approach if IPM research is to be more responsive to sustainable agricultural development. The SP-IPM is also part of the CGIAR response to Agenda 21 – the action plan developed by the United Nations Conference on Environment and Development, convened in Rio de Janeiro in 1992, which identifies IPM as a key element of sustainable agricultural development. The SP-IPM is a global network (Table 32.1) of CG and other International Agricultural Research Centers in partnership with the Global IPM Facility, IPM Forum (housed by CABI Bioscience), Pesticide Action Network representing the CGIAR NGO Committee, CropLife International representing private crop protection industry, the International Association for the Plant Protection Sciences (IAPPS), the global Taxonomic network BioNET International and national agricultural research systems including extension services and NGOs in targeted countries in Africa, Asia and Latin America and the Caribbean. Beneficiaries of the program are farmers, national and international agricultural research organizations, extension programs and NGOs who benefit from exchange of expertise, information as well as genetic and other resources to increase their capacity to manage pest problems, stabilize productivity and income, and foster a pesticide-safe environment.

Table 32.1. Members of the Inter-institutional working group on IPM.

The SP-IPM partner organizations
Asian Vegetable Research and Development Center (AVRDC)
BioNET INTERNATIONAL
CABI Bioscience
Centro Internacional de Agricultura Tropical (CIAT)*
Centro Internacional de la Papa (CIP)*
Centro Internacional de Mejoramiento de Maiz y Trigo (CIMMYT)*
CropLife International
Global IPM Facility
International Association for the Plant Protection Sciences (IAPPS)
International Center for Agricultural Research in the Dry Areas (ICARDA)*
International Center for Research on Agroforestry (ICRAF)*
International Center of Insect Physiology and Ecology (ICIPE)
International Crops Research Institute for the Semi-Arid Tropics (ICRISAT)*
International Institute of Tropical Agriculture (IITA)*
International Rice Research Institute (IRRI)*
International Service for National Agricultural Research (ISNAR)*
IPM Forum
Pesticide Action Network (PAN) representing the CGIAR NGO Committee
West Africa Rice Development Association (WARDA)*

*Indicates CGIAR center.

Coordinating Mechanism

The SP-IPM target zones are in Africa, Asia and Latin America, but the program attracts technical inputs globally to:

- encourage better communication, and closer coordination among the CG/IARCs and their partners to ensure all IPM stakeholders develop a common perception and shared sense of purpose;
- strengthen multi-disciplinary and holistic research approaches to understand the production constraints, and the range of opportunities needed for tackling them;
- promote client-oriented partnerships to develop, test, adapt and disseminate novel IPM options, develop farmers' capacity to investigate production constraints more effectively, and to increase the impact of IPM;
- discourage harmful pesticide regimes in production systems of common concern;
- facilitate information exchange between stakeholder groups to increase awareness of the benefits of IPM, to create a favorable policy environment for widespread implementation.

These activities are planned, implemented and coordinated through four implementation structures.

Steering Committee (SC)

SC is the decision-making body and reports to partner stakeholder groups and the Interim Science Council of the CGIAR. The SC comprises institutionally nominated representatives of partner institutions (Table 32.1) who elect a Chair from any of the participating institutions. The SC ensures that proposed workplans and budgets guarantee inter-institutional collaboration, multidisciplinary approaches, use of appropriate scientific methodologies, and that the activities complement existing efforts and respond to clearly identified needs. The SC endorses amendments to the SP-IPM mission and policy statements and approves financial plans and statements, and nominates the program coordinator for appointment at the coordinating secretariat. The Chair serves for a renewable 3-year period and provides overall leadership, promotes collaborative linkages, advocacy/public relations, and fundraising.

Inter-Institutional Working Group (IIWG)

IIWG comprises institutionally designated representatives of partner institutions (Table 32.1) and coordinators of SP-IPM projects and thematic working groups. The IIWG meets annually for partners to discuss and agree on policy and vision (Box 32.1), identify problems for which an inter-institutional effort could make a difference, set priorities, agree on contractual obligations, program, process and budgetary issues to strengthen collaboration, promote networking, and review progress. Occasionally, special interest groups are invited to IIWG workshops where the meeting agenda require additional expert advice.

Box 32.1. CGIAR policy statement on IPM.

Definition

IPM is an approach to enhancing crop and livestock production, based on an understanding of ecological principles, that empowers farmers to promote the health of crops and animals within a well-balanced agroecosystem, making full use of available technologies, especially host resistance, biological control and cultural control methods. Chemical pesticides are used only when the above measures fail to keep pests below acceptable levels, and when assessment of associated risks and benefits (considering effects on human and environmental health, as well as profitability) indicates that the benefits of their use outweigh the costs. All interventions are need-based and are applied in ways that minimize undesirable side effects.

continued

Box 32.1. *Continued.*

Mission

The Mission of the CGIAR is, 'through research and related activities, . . . to contribute to sustainable improvements in the productivity of agriculture, forestry and fisheries in developing countries in ways that enhance nutrition and well-being, especially of low-income people.' In pursuance of this mission and in full accord with the articles of UNCED Agenda 21 and the Convention on Biological Diversity, the International Agricultural Research Centers (IARCs – which here include the CGIAR Centers plus AVRDC and ICIPE) recognize that IPM, as a system that contributes to productivity, prosperity and human well-being in an environmentally sound and equitable manner, has a key role to play in sustainable agricultural development. The IARCs therefore affirm that IPM is their preferred plant and animal health strategy and that, through their research and related activities, they will promote the adoption of IPM by farmers.

Guiding principles

- IPM research is interdisciplinary and pursues a holistic approach to management of agricultural and natural ecosystems.
- IPM maintains and utilizes biodiversity as the natural foundation for pest management in the context of sustainable agricultural development.
- IPM development is guided by farmer participation, from problem diagnosis, through component research, to on-farm validation.
- IPM adoption depends on the ability of farmers to make informed decisions, based on an understanding of ecological and economic principles. Farmer empowerment is achieved through participatory research and training methods that encourage the integration of traditional and 'science-based' knowledge.
- Success in implementing IPM is contingent on a favorable public policy environment.

Strategy

- The IARCs will further develop their existing comparative advantage in researching pest problems, developing IPM components, implementing pilot projects and assessing impact.
- The IARCs will maintain and expand their effective partnerships with NARS, NGOs, and other appropriate national, international and bilateral organizations to promote IPM research and implementation.
- In full-scale IPM implementation, the IARCs will play a supporting role to organizations such as national extension services, NGOs and IGOs.
- The IARCs will promote more effective communication between farmers, extensionists and researchers to ensure that research efforts are clearly focused on farmers' needs and provide direct support to implementation efforts.
- The IARCs will engage in direct dialogue with policy makers and provide information to the general public to raise awareness of the benefits of IPM and promote a policy environment more favorable to IPM implementation.
- The IARCs will collaborate with the private sector in developing biopesticides, semiochemicals, drugs and other products which can be used in an economically sound and environmentally responsible way within an IPM framework.
- The IARCs will explore the full potential of biotechnological tools (including tissue culture, marker-assisted selection, diagnostic tools and gene transfer) in developing IPM tactics. Genetically engineered products will be evaluated for their non-target effects before deployment within an IPM framework appropriate to the biophysical and socioeconomic environment.

Collaboration

The Systemwide Program on Integrated Pest Management (SP-IPM) was established in 1995 to realize the full potential of IPM within sustainable agricultural development. Guided by the principles set out above and through closer collaboration among the IARCs and their partners, the SP-IPM seeks to achieve synergies and greater impact in IPM research and implementation, and to ensure that these activities are fully responsive to the needs of IPM practitioners.

Thematic Working Groups (TWGs)

TWGs (previously task forces) have primary responsibility for scientific quality of SP-IPM projects and special initiatives. TWGs examine and analyze priority problems, develop coherent responses, and provide advisory services. Peer review by TWGs ensures that recommended responses complement existing efforts, promote synergy of efforts and respond to farmers' felt needs. Thematic groups submit proposals for approval by the SC. TWGs are multi-institutional in membership and include representatives of key stakeholders in national and international research and development organizations. There are currently eight TWGs (Table 32.2), which also assist to link up national institutions to crop germplasm, biological control agents, specialized information, expertise and other relevant resources available from other institutions.

Coordinating Secretariat

Coordinating Secretariat is managed by a Coordinator who serves as the global contact point to catalyze and facilitate approved activities, mobilize and disseminate resources, and facilitate communication between stakeholder groups and with donors. The coordinator serves as *ex officio* member of the SC, assists in fundraising, participates in planning and review meetings of thematic working groups, manages the program's budget as approved by the SC, prepares financial reports, and organizes external evaluation of the program in consultation with the SC.

Highlights

Box 32.1 presents the SP-IPM vision of IPM as defined by the IIWG. Current initiatives embodying this vision include TWGs and projects to develop biologically based IPM options, develop effective models for farmer participatory research and learning, establish pilot sites for IPM learning and adoption, advise on biopesticide regulation and registration, and provide relevant information and advocacy to the general public and donor communities.

Developing Biologically Based IPM Options

Certain pests cause severe damage and losses across cropping systems,

Table 32.2. The SP-IPM projects, thematic working groups and special initiatives.

Activity	Focal point: Institution	Focal point: Contact person
1. Projects		
Whiteflies and whitefly transmitted viruses	CIAT	F.Morales@cgiar.org
Farmer participatory research in IPM	CIAT	A.Braun@cgiar.org
2. Thematic working groups		
Alien invasive species	IITA	M.Tamo@cgiar.org
Crop loss and IPM impact assessment	CIP	O.Ortiz@cgiar.org
Farmer participatory research and participatory learning in IPM	CABI Bioscience	J.Vos@cabi.org
Functional agrobiodiversity	ICIPE	igordon@icipe.org
IPM policy research	FAO/GIF	Harry.vanderWulp@fao.org
Leaf miners	CIP	(to be named)
Soil biota (including termites)	CIAT	A.Bellotti@cgiar.org
Tropical whiteflies and viruses	CIAT	F.Morales@cgiar.org
3. Special initiatives		
Pilot sites initiatives for IPM learning and adoption	IITA	B.James@cgiar.org

agroecologies and geographic locations, and the problems they create simply refuse to go away. For example, whiteflies (principally *Trialeurodes* and *Bemisia* species) and *Bemisia*-transmitted viruses are major threats to the production of cassava in Africa (Dubern, 1994; Thresh *et al.*, 1994, 1998; Legg, 1999; Legg and Thresh, 2000; Neuenschwander *et al.*, 2001), common bean in Latin America (Morales, 2000), and vegetable crops throughout the tropics (Polston and Anderson, 1997; Morales and Anderson, 2001). Similarly, leaf miners (mainly *Liriomyza* species) cause havoc worldwide in a variety of cropping systems (e.g. Spencer, 1972; De Lima, 1979; Singh and Merrett, 1980; Bourdouxhe, 1982; Diekmann, 1982; Cardona *et al.*, 1985; ICARDA, 1987; Salas, 1992; Cisneros and Gregory, 1994; Barea *et al.*, 1995; Ujiye and Adachi, 1995; Uygen *et al.*, 1995; Kotze and Dennill, 1996; Singh and Weigand, 1996; Sharaf-El-Din *et al.*, 1997; Cardona *et al.*, 1998; de Souza *et al.*, 1998; Shepard *et al.*, 1998). Stemborers and parasitic weeds (e.g. *Striga* and *Orobanche*) are equally an area-wide problem in a variety of cereal–legume cropping systems in most of Africa (Sallé and Raynal-Roques, 1989; Sauerborn, 1991; Oikeh *et al.*, 1996). Weeds are generally of over-riding concern to production in many low-input farming systems, and soil biota (e.g. white grubs and nematodes) pose similar threats to food production. Furthermore, post-harvest IPM research is much neglected.

For many of these system-wide problems, farmers' coping strategies are ineffective, and would benefit greatly from properly coordinated researcher partnerships to develop sustainable strategies. Interconnectivity between research centers individually working on the problems would bring such benefits of scientific expertise to bear on food security and livelihoods threats posed by the pests. In this regard, the global project on whiteflies as pests and vectors of plant viruses in the tropics is the first SP-IPM initiative to tackle stubborn system-wide pest problems. In 1996, the SP-IPM whitefly Task Force proposed the global whitefly project (Table 32.3) that focused on *T. vaporariorum* as a direct feeding pest in the tropical highlands, *B. tabaci* as a vector of viruses in legume–vegetable mixed cropping systems of annuals in the tropical lowlands, and *B. tabaci* as a vector and pest in cassava. The primary project partners were five CG and other International Agricultural Research Centers, ten advanced research institutions in Australia, Denmark, Germany, New Zealand, the UK, and the USA working in partnership with national agricultural research systems in 30 countries of Latin America, the Caribbean, Africa and Asia.

Table 32.3. Sub-projects of the global project on whiteflies as pests and vectors of plant virus in the Tropics.

Sub-project	Focal point: Institution	Focal point: Contact person
Trialeurodes vaporariorum as a direct pest in the tropical highlands of Latin America	CIAT	C.Cardona@cgiar.org
Bemisia tabaci as a virus vector in mixed cropping systems of the Caribbean, Mexico and Central America	CIAT	F.Morales@cgiar.org
Bemisia tabaci as a virus vector in mixed cropping systems of eastern and southern Africa	ICIPE	Lisbeth@africaonline.co.ke
Bemisia tabaci as a virus vector in mixed cropping systems of SE Asia	AVRDC	p.hanson@cgnet.com
Bemisia tabaci as a virus vector in cassava and sweet potato in sub-Saharan Africa	IITA	Jlegg@infocom.co.ug
Whiteflies as direct pests on cassava in South America	CIAT	A.Bellotti@cgiar.org
Project coordination	CIAT	F.Morales@cgiar.org

In the first phase of activities which started in 1997, the project established an international network for researchers on whiteflies and whitefly-transmitted viruses in the tropics, and characterized agronomic, socioeconomic, and epidemiological features of whiteflies and whitefly transmitted viruses in cassava, legumes and sweet potato in targeted countries in Latin America, Africa, and the Caribbean and Mexico using standardized diagnostic survey methodologies. The results (Anderson and Markham, 2003) formed the basis of subsequent activities to improve understanding of whitefly and disease dynamics in critical target areas, define and develop whitefly IPM strategies, strengthen national research capacity, policy formation and IPM implementation, and to extend project activities to new geographical areas.

Developing Effective Models for Participatory Research

The success of IPM depends largely on how well farmers understand and combine knowledge of biological and ecological processes with their farming experience to develop/select options that reduce losses to pests, increase agricultural productivity, manage risk, and meet the demands of local and global markets. Globally, the IPM community is convinced that FPR ensures integration of scientific and indigenous knowledge to make research more understandable and useful. However, participation and FPR mean different things to different partners (Pretty, 1995; Lilja and Ashby, 1999; Ashby and Sperling, 2000; Braun and Hocdé, 2000). The label 'FPR' is applied to a diverse array of approaches involving different objectives and many types and levels of participation. These include facilitation of farmers' experiments (Haverkort *et al.*, 1991), farmer participation at different stages of formal plant breeding (Witcombe, 1996; Weltzien *et al.*, 2000), farmer testing of 'best bet' options generated by researchers (Snapp, 1999), and varied approaches involving interactive participation (Pretty, 1995; Braun and Hocdé, 2000), action-research and social learning (for examples see: Röling and Wagemakers, 1998; Ashby *et al.*, 2000; Defoer and Budelman, 2000). All aim at informed decision making by farmers to solve location-specific problems, respond to opportunities and cope with rapid change.

Likewise the definitions and objectives of approaches such as Participatory Learning, Participatory Extension, and FFS (van de Fliert, 1993; Hagmann 1999a,b; Connell, 2000; Ooi, 2000) also vary. One of these in particular, the FFS, was originally developed for the IPM context; nevertheless confusion about the similarities and differences among these, and about their relationship to participatory research approaches is commonplace (Braun *et al.*, 2000; van de Fliert *et al.*, 2000). In all these approaches, the impact in terms of reduced pesticide use, improved productivity, yields, or returns, less vulnerability to risk, better responsiveness to market forces, or sustainability of farming practices hinges largely on improving human capacities for analysis, decision making and facilitation (Groot and Maarleveld, 2000). A case study of FPR and participatory learning (PL) approaches across agroecologies and cultures can contribute to orient current and future projects and project managers towards available opportunities in participatory IPM. The potential of FPR and PL to increase IPM impact also needs to be assessed, with special focus on their complementarities.

In an initial response to challenges of this kind, the SP-IPM Task Force on FPR-IPM, in collaboration with the Systemwide Program on Participatory Research and Gender Analysis, FAO Global IPM Facility, CABI Biosciences, and other arms of the SP-IPM organized an international workshop in 1996 to clarify FPR concepts and develop a framework for action. The workshop output was translated into a project proposal mainly through discussions on the listserve FPR-IPM@cgiar.org, workshops and the SP-IPM annual meetings. The project proposed to undertake a series of mentored study-exchanges between contrasting pairs of successful projects on participatory research and learning to collate existing information, case studies on methodologies, practices, and processes for wider distribution; provide research planners and managers and policy makers with guidelines on key participatory principles and practices underpinning successful IPM projects, and encourage the incorporation of such approaches in modifications of existing IPM projects and in designing new ones. Table 32.4 lists the projects/programs that were invited to participate in the study tour and learning workshop because they have a track record either with an FPR or PL-based approach or have experience in combining and integrating the practice of the two approaches. To capitalize on FPR and PL experiences worldwide, the mentored exchange visits were followed by a learning workshop, at which project participants and representatives from other projects/programs conducted a cross-cutting analysis of the case studies generated by the study tour participants.

During the first phase of activities, 12 frontline project staff completed 7–10 days of mentored reciprocal exchange visits among the participating projects and programs. At the pioneer learning workshop (Anonymous, 2001a) 43 participants (IPM facilitators, researchers, extensionists, project managers) jointly analyzed study tour case studies and experiences with the view to develop an agreed-upon framework for an integrated participatory approach to research and dissemination/extension in IPM in the context of a broader innovation process. Elements of the joint analysis included identification of common elements and principles of FPR and PL processes, analysis of experimental processes, synthesis of success factors and lessons learnt and of the challenges faced by the projects/programs from the cross-country sharing and exposure visits. The workshop discussed differences in various approaches to advise on possible institutional arrangements for the development of strategies to influence policies and institutions for mainstreaming and operationalization of FPR/PL IPM policies and practices. Specifically, the participants developed a vision of FPR/PL in terms of what would need to be done differently at the level of farmers and community organizations, extensionists, national and CG/IARC researchers, and policy makers if FPR and PL were to be successful in IPM (Anonymous, 2001a).

The synthesis of success factors, challenges and lessons learned formed the basis of a conceptual framework for FPR and learning interventions. In this regard, the

Table 32.4. FPR-IPM project: study-tour projects/programs.

	Focal point	
Host project/program	Institution	Contact person
PROINPA: Foundation for the promotion and investigation of Andean products	PROINPA, Bolivia	egandari@proinpa.org
UPWARD: Users perspectives with agricultural research and development	UPWARD, Philippines	csb@laguna.net
CIP-ICM: Participatory development of potato and sweet potato ICM in Indonesia by CIP and its partners	CIP, Indonesia	E.vandeFliert@cgiar.org
CABI-IPPM: Sub-regional project on integrated production and pest management (IPPM)	CABI, Kenya	ffsproj@africaonline.co.ke
FAO-CIPM: FAO Community IPM program in Vietnam	FAO, Hanoi, Vietnam	matteson@fpt.vn
IPCA: Participatory research in Central America (Honduras)	Proyecto IPCA, Honduras	IPCA@laceiba.com
FPR-IPM project coordination	CIAT	A.Braun@cgiar.org

learning workshop identified cornerstones that need to be in place in a process of planning, executing and managing FPR/PL in IPM. The cornerstones were local organizational capacity; process facilitation capacity; a basket of technical options; quality participation; benefits for farmers; institutional capacity for support services; commitment to longer-term interventions; scaling-up strategies and approaches; research with and by farmers; farmer experimentation, learning and sharing; a vision beyond IPM; inclusion of marketing aspects; impact assessment and self-evaluation; supportive policies; interdisciplinary approach; institutional collaboration and networking; and funding and creative financing mechanisms. In follow-up activities, a sub-group of workshop participants will develop the fuller conceptual framework from these cornerstones. The consultative activities in this first phase of the project will lay a solid foundation for a longer-term process of training, advocacy, exposure and sharing of a variety of practices and practical methods, and of institutional change to promote more effective farmer participatory research and learning approaches among the partner organizations and beyond.

Pilot Sites for IPM Learning and Adoption

Although a number of promising IPM options are available, adoption of IPM at farm level is disappointingly slow in the developing countries. Poor communication between researchers and farmers is believed by many stakeholders in the agricultural development process to be a major constraint limiting IPM adoption. Effective partnerships to involve farmers in the research process to develop, test/adapt and disseminate IPM options are also uncommon. Additionally, organizational and policy obstacles and extension bottlenecks discourage the dissemination of proven options. In response to these constraints, SP-IPM introduced ‘pilot sites’ in 2000 as part of its implementation strategy to introduce ‘best-bet’ IPM options to farming communities, assist participating organizations to gain experience in developing effective farmer–scientist–extension (FSE) partnerships, and increase understanding, dissemination and adoption of IPM options by farmers.

Pilot sites (Table 32.5) were selected in major agroecologies where farmers had already identified pests as a principal concern. The selection criteria included prior characterization of other biophysical and socioeconomic features; availability of promising local IPM options; existence of research and development activities that provide a platform for pilot site development; opportunities for achieving new synergies by closer collaboration between IARCs and partners; the presence of strong partners to take primary responsibility for site activities; and identification of a wide extrapolation domain for the results of pilot activities. At each pilot site, FSE teams analyzed production problems, identified farmers’ coping strategies, and agreed on ‘best-bet’ options to evaluate.

In Kenya, pilot-site farmers have conducted field experiments to control

Table 32.5. The SP-IPM pilot sites initiative.

	Focal point	
Pilot site	Institution	Contact person
East Africa: mid-altitudes in Kenya	ICIPE	zkhan@MBITA.MIMCOM.NET
North Africa: irrigated ecologies in Egypt	ICARDA	A.Yahyaoui@cgiar.org
North Africa: rain-fed sub-humid ecologies in Morocco	ICARDA	A.Yahyaoui@cgiar.org
West Africa: Guinea savanna in Nigeria	IITA	A.Emechebe@cgiar.org
West Africa: Sahel in Mali and Burkina Faso	ICRISAT	O.Youm@cgiar.org
General coordination	IITA	B.James@cgiar.org

Striga hermonthica and stemborers by combining *Striga*-tolerant maize varieties with the fodder legume *Desmodium* in an intercropping and habitat management system that includes the use of Napier grass, a system developed by ICIPE (Khan *et al.*, 1997a,b, 2000). In this IPM strategy, *Desmodium* stimulates *Striga* germination, and suppresses haustorial growth (hence reduces the seed bank in the soil), repels stemborer moths and improves soil fertility, while Napier grass acts as a trap plant for the pest moths (whose offspring hardly survive in these grasses). More than 100 of the pilot-site farmers have experienced a 20% increase in maize yields, and a new market opportunity for sale of *Desmodium* and Napier grass for livestock. In Nigeria, 58 farmers, working with IITA and national program scientists, have initiated experimentation to manage *Striga* and improve soil fertility by integrating *Striga*-tolerant varieties in crop rotation systems with food legumes that stimulate suicidal germination of *S. hermonthica* (Parkinson *et al.*, 1987; Alibi *et al.*, 1994; Berner *et al.*, 1996a,b; Carsky *et al.*, 2000). In the field experiments, maize variety Acr.97TZL Comp.1-W suppressed *Striga* emergence by 63% compared with traditional varieties, and the suppression of *Striga* emergence was greater when the variety was rotated with soybean (variety TGX-1448-2E) than in the traditional practice of growing one maize crop after another. In the rotation schemes tested, the ability to suppress *Striga* emergence was greatest with soybean followed by cowpea (variety IT-93K-452-1) and groundnut (variety RMP 12) in that order.

In Egypt, the parasitic weed *Orobanche* and virus diseases seriously undermine faba bean production (Fam, 1983; El-Khouly *et al.*, 1997). The area under faba bean in Beni Suef Governorate, for example, had declined from 17,600 ha in 1991 to 800 in 2000, mostly because of epidemic spread of faba bean necrotic yellows virus that caused total crop failures in 1992 (Makkouk *et al.*, 1994). Through the pilot site initiative, 20 pilot-site farmers, working with ICARDA and national researchers increased their yields of faba bean by 60% compared with the yields of non-pilot-site farmers. The pilot-site farmers combined faba bean varieties tolerant to *Orobanche* with improved agronomic practices to minimize damage by the weed and reduce virus spread. In Morocco, the pilot site was in the Settat area in Central Morocco where rain-fed wheat and chickpea are the major crops grown in rotation. The major production problems are Hessian fly (*Mayetiola destructor*) in bread wheat, the fungal disease *Ascochyta* blight, and leaf miners (*Liriomyza cicerena*) in chickpea (Blaeser-Diekmann, 1982; Hamdaoui and Lahmar, 1996). In first season pilot site trials, the farmers, working with researchers from ICARDA and national researchers and extensionists, doubled wheat yield, largely through a combination of wheat variety resistant to Hessian fly, early planting and other improved agronomic practices. The farmers also doubled chickpea yield by integrating blight tolerant varieties with early planting and pre-emergence herbicide application, compared with traditional varieties and late planting.

Information Exchange and Advocacy

A key activity of the SP-IPM-underpinning its coordination role, is information dissemination to increase public and donor awareness of the benefits of IPM and to raise the profile of IPM within the communities. The program has produced and disseminated various information materials and activity results in print form (e.g. Anonymous, 1997), electronic form (e.g. Anonymous, 2001b), and through its website www.cgiar.org/spipm, special project websites (e.g. www.ciat.cgiar.org/fpr-ipm), and listserves fpr-ipm@cgiar.org and NGO-IPMnet@cgiar.org. The website www.cgiar.org/spipm features a database of IPM research and researchers at CG/IARCs. It consists of 'basic project data' and 'project planning information', which is the sort of data most likely of interest to research planners who start a new project or seek interaction with an existing one.

Within the framework of the 'IPM Information Partnership'[2], the SP-IPM has promoted IPM information exchange through the 'IPM Communications Workshop for Eastern & Southern Africa' at which diverse stakeholders in IPM explored how electronic mail and the Internet could enhance communication and knowledge transfer of IPM in sub-Saharan Africa (Shaefers and Krauss, 1998). Also, in collaboration with the CGIAR-NGO Committee, the SP-IPM co-organized an international workshop 'Towards More Effective Implementation of IPM in Africa' (Kanoute *et al.*, 1999) at which participants advocated concerted efforts by agricultural development agencies and research institutes to lobby governments to adopt IPM as a national policy. Also, the participants resolved to strengthen multidisciplinary research within the framework of FPR/PL approaches that allow research priorities to be determined with farmers, NGOs and other stakeholders. The immediate outcomes of the NGO-IPM workshop included an electronic network (NGO-IPMnet@cgiar.org), adaptation of scientific information into user-friendly format to facilitate farm-level understanding (James *et al.*, 2000a,b; Melifonwu *et al.*, 2000; Msikita *et al.*, 2000), and the promotion of aqueous leaf extracts as alternatives to inorganic pesticides in peri-urban agriculture (OBEPAB, 2000).

Conclusion

IPM researchers used to limit themselves to studying pests and developing more effective pest control technologies and packages. In recent years, the focus on IPM research is shifting to the better understanding of farmers' felt needs to intensify production for raising income and living standards. The SP-IPM partners are working with farmers to understand the biological, ecological and sociological processes that underpin agriculture and the problems that arise when these natural processes are disrupted. Among the future activities, pilot sites will be established in other key agroecologies around the tropical world to serve as focal points for developing and implementing new models of partnership, promote participatory approaches and introduce 'best-bet' options to farming communities. The major challenges at the sites include the need to develop non-chemical pest control and soil management options, sharpen farmer participatory approaches, implement cost-effective mechanisms to disseminate IPM options to a larger number of farmers, and assess the impact of pilot sites in IPM learning, promotion and adoption. Additionally, continuing problem analysis by TWGs explore opportunities to enhance global collaboration on production, quality control and regulation of biopesticides and the incorporation of biotechnology/transgenic crops into IPM. IPM options for problems associated with systemwide production constraints such as soil-biota, leaf miners, thrips, pod borers in chickpea, pigeon pea and cowpea, and post-harvest IPM are badly needed. Broader partnerships with the IPM-Forum will be explored to intensify information exchange and IPM advocacy at regional and global levels.

Acknowledgments

The core program of the SP-IPM was supported by the development agencies of Switzerland and Norway, and by The Word Bank/CGIAR. Donors to SP-IPM projects and activities include Australia, Denmark, New Zealand, Switzerland (complementary funding), the UK and the USA. The SP-IPM was established under the leadership of Dr Lukas Brader, Director-General, IITA.

---

[2] IPM Information Partnership was formed in 1996 between the Consortium for International Crop Protection (CICP), the IPM Forum, IPM Europe and the SP-IPM.

References

Alibi, M.O., Berner, D.K. and Olaniyan, G.O. (1994) Characterization of soybean cultivars for *Striga hermonthica* control. *Phytopathology* 84, 1151.

Anderson, P.K. and Markham, R. (eds) (2003) *Whiteflies and Geminiviruses in the Tropics: Building a Knowledge Base for Global Action.* International Center for Tropical Agriculture (CIAT), Cali, Colombia (in press).

Anonymous (1997) *Integrated Pest Management at the International Agricultural Research Centers.* The SP-IPM, IITA.

Anonymous (2001a) *Capitalizing on Experiences in Farmer Participatory Research and Learning for Integrated Pest Management.* Report on the FPR-IPM learning workshop, Chiang Mai, Thailand, 4–8 September, FPR-IPM project, CIAT.

Anonymous (2001b) *Breaking the Cycle.* The SP-IPM, IITA.

Ashby, J. and Sperling, L. (2000) What do we mean by participatory research in agriculture? In: Heinrich, G.M. (ed.) *Farmer Participatory Approaches.* Proceedings of a regional workshop on Farmer Participatory Approaches, 7–11 July 1997, Harare, Zimbabwe. ICRISAT, Bulawayo, Zimbabwe, pp. 11–22.

Ashby, J., Braun, A.R., Gracia, T., Guerrero, M.P., Hernandez, L.A., Quirós, C.A. and Roa, J.I. (2000) *Investing in Farmers as Researchers. Experience with Local Agricultural Research Committees in Latin America.* Centro Internacional de Agricultura Tropical, Cali, Colombia, 199 pp.

Bandyopadhyay, R., Frederickson, D.E., McLaren, N.W., Odvody, G.N. and Ryley, M.J. (1998) Ergot: a new disease threat to sorghum in the Americas and Australia. *Plant Disease* 82, 356–367.

Barea, O., Ramirez, O., Cubillo, D. and Hilje-L. (1995) Importancia economica de *Liriomyza huidobrensis* (Blanchard) en la papa en Cartago, Costa Rica. *Manejo Integrado de Plagas* 38, 1–7.

Bellotti, A.C. and Arias, B. (2001) Host plant resistance to whiteflies with emphasis on cassava as a case study. *Crop Protection* 20, 813–823.

Berner, D.K., Alabi, M.O., Di-Umba, U. and Ikie, F.O. (1996a) Proposed integrated control program for *Striga hermonthica* in Africa. In: Moreno, M.T., Cubero, J.I., Berner, D., Musselman, L.J. and Parker, C. (eds) *Advances in Parasitic Plant Research.* Proceedings of the 6th Parasitic Weeds Symposium, 16–18 April, 1996, Cordoba, Spain: Junta de Andalucia, Direction General de Investigation Agraria, pp. 817–825.

Berner, D.K., Carsky, R., Dashiell, K., Kling, J. and Manyong, V. (1996b) A land management based approach to integrated *Striga hermonthica* control in Sub-Saharan Africa. *Outlook on Agriculture* 25, 157–164.

Blaeser-Diekmann, M. (1982) Survey on pests and diseases of faba beans (*Vicia faba*) in Egypt, Morocco and Tunisia. *Fabis Newsletter* 4, 44–45.

Bourdouxhe, L. (1982) La mineuse nord-americaine des feuilles (*Liriomyza trifolii*, Diptera, Agromyzidae) sur cultures maraicheres au Senegal. *FAO Plant Protection Bulletin* 30(2), 81–82.

Braun, A.R. and Hocdé, H. (2000) Farmer participatory research in Latin America: four cases. In: Stur, W.W., Horne, P.M., Hacker, J.B. and Kerridge, P.C. (eds) *Working with Farmers: the Key to Adoption of Forage Technologies.* ACIAR Publication PR095, pp. 32–53. http://www.aciar.gov.au/publications/db/abstract.asp?pubsID=428

Braun, A.R., Thiele, G. and Fernández, M. (2000) Farmer field schools and local agricultural research committees: complementary platforms for integrated decision-making in sustainable agriculture. *AgREN Network Paper No. 105.* Agricultural Research and Extension Network, Overseas Development Institute, London, United Kingdom, 20 pp. http://www.odi.org.uk/agren/papers/agrenpaper_105.pdf

Cardona, C., Saxena, M.C. and Varma, S. (1985) Insect pests of faba beans, lentils, and chickpeas in North Africa and West Asia: a review of their economic importance. In: Saxena, M.C. and Varma, S. (eds) *Proceedings. Faba Beans, Kabuli Chickpeas, and Lentils in the 1980s.* An international workshop, 16–20 May 1983, pp. 159–167.

Cardona, C., Soto, O., Prada, P., de la Cruz, A.M. and Ramon, M.T. (1998) Influencia de la percepcion del agricultor en el manejo del minador del frijol, *Liriomyza huidobrensis* (Diptera: Agromyzidae), en la zona andina. *Revista Colombiana de Entomologia* 24, 1–2, 55–59.

Carsky, R.J., Berner, D.K., Oyewole, B.D., Dashiell, K. and Schulz, S. (2000) Reduction of *Striga hermonthica* parasitism on maize using soybean rotation. *International Journal of Pest Management* 46(2) 115–120.

Chavez, R., Schmieidieche, P.E., Jackson, M.T. and Raman, K.V. (1988) The breeding potential of wild potato species resistant to the potato tuber moth, *Phthorimaea operculella* (Zeller). *Euphytica* 39, 123–132.

Cisneros, F. and Gregory, P. (1994) Potato pest management. *Aspects of Applied Biology* 39, 113–124.

Connell, J.G. (2000) Scaling-up: the roles of participatory technology development and participatory extension approaches. In: Stur, W.W., Horne, P.M., Hacker, J.B. and Kerridge, P.C. (eds) *Working with Farmers: the Key to Adoption of Forage Technologies*. ACIAR Publication PR095, pp. 69–82 http://www.aciar.gov.au/publications/db/abstract.asp?pubsID=428

Dahal, G., Hibino, H., Cabunagan, R.C., Tiongco, E.R., Flores, Z.M. and Aguierro, V.M. (1990) Changes in cultivar reactions to tungro due to changes in 'virulence' of the leaf hopper vector. *Phytopathology* 80, 659–665.

De Lima, C.P.F. (1979) *Liriomyza trifolii* (Diptera: Agromyzidae), an important new leaf-miner pest in Kenya. *Kenya Entomologist's Newsletter* 10, 8.

de Souza, J.C., Salgado, L.O., de-o- Rigitano, R.L.O. and Reis, P.R. (1998) Danos causados pela mosca-minadora *Liriomyza huidobrensis* Blanchard, 1926 (Diptera: Agromyzidae) na cultura da batata (*Solanum tuberosum* L.), no plantio de inverno no sul de Minas Gerais, e eficiencia do aldicarb no seu controle. *Ciencia e Agrotecnologia* 22(1), 22–29.

Defoer, T. and Budelman, A. (eds) (2000) *Managing Soil Fertility in the Tropics: a Resource Guide for Participatory Learning and Action Research*. Royal Tropical Institute, Amsterdam, The Netherlands.

Diekmann, M. (1982) Survey on pests and diseases of faba bean (*Vicia faba*) in Egypt, Morocco and Tunisia. *FABIS Newsletter* 4, 44–46.

Dixon, A.G.O., Asiedu, R. and Hahn, S.K. (1992) Cassava germplasm enhancement at IITA. In: Akoroda, M.O. and Arene, O.B. (eds) *Promotion of root crop-based industries: an incentive for research and development*. Proceedings of the 4th Triennial Symposium of the International Society for Tropical Root Crops – Africa Branch (ISTRC-AB), Kinshasa, Zaire, 5–8 December 1989, pp. 83–87.

Dubern, J. (1994) Transmission of African cassava mosaic geminivirus by the whitefly (*Bemisia tabaci*). *Tropical Science* 34, 82–91.

Efron, Y., Kim, S.K., Fajemisin, J.M., Mareck, J.H., Tang, C.Y., Dabrowski, Z.T., Rossel, H.W. and Buddenhagen, I.W. (1989) Breeding for resistance to maize streak virus: a multidisciplinary team approach. *Plant Breeding* 103, 1–36.

El-Khouly, A.S., Metwally, M.M., Salem, R.M., Helal, H.A. and El-Mezaien, A.B. (1997) Field studies on broad bean leafminer, *Liriomyza congesta* (Beck.) A. Varietal resistance. B. Spatial distribution of larvae under conditions of Kafr El-Sheikh governorate, Egypt. *Egyptian Journal of Agricultural Research*, 75(4), 987–994.

Fam, E.Z. (1983) Reducing losses due to insects in faba beans in Egypt. *FABIS-Newsletter*, No. 7, 48–49.

Groot, A. and Maarleveld, M. (2000) *Demystifying Facilitation in Participatory Development*. International Institute for Environment and Development (IIED) Gatekeeper Series No. 89, 24 pp.

Hagmann, J. (1999a) *Learning Together for Change. Facilitating Innovation in Natural Resource Management through Learning Process Approaches in Rural Livelihoods in Zimbabwe*. Margraf Verlag, Wiekersheim, Germany, 310 pp.

Hagmann, J., Chuma, E., Murwira, K. and Connolly, M. (1999b) *Putting Process into Practice: Operationalising Participatory Extension*. In: ODI Agricultural Research and Extension (AGREN) Network Paper 94. http://www.odi.org.uk/agrenpaper_94.pde

Hamdaoui, F. and Lahmar, M. (1996) *Lutte contre la mineuse du pois chiche au Maroc*. Fiche technique, INRA, Maroc.

Haverkort, B., van der Kamp, J. and Waters-Bayer, A. (1991) *Joining Farmers' Experiments*. Intermediate Technology Publications, London, 269 pp.

Herren, H.R. and Neuenschwander, P. (1991) Biological control of cassava pests in Africa. *Annual Review of Entomology* 36, 257–283.

Hibino, H., Daquioag, R.D., Cabauatan, P.Q. and Dahal, G. (1988) Resistance to rice tungro virus in rice. *Plant Disease* 72, 843–847.

Huet, H., Mahendra, S., Wang, J., Sivamani, E., Ong, C.A., Chen, L., de Kochko, A., Beachy, R.N. and Fauquet, C. (1999) Near immunity to rice tungro spherical virus achieved in rice by a replicase-mediated resistance strategy. *Phytopathology* 89, 1022–1027.

ICARDA (1987) *Food Legume Improvement Program: Annual Report for 1986*. ICARDA, Aleppo, Syria.

IITA (1997) *Towards Sustainable Development in Africa*. IITA Medium-term Plan, 1998–2000, IITA, Ibadan, Nigeria.

IITA (2001) *IITA Strategic Plan 2001–2010, supporting document.* IITA, Ibadan, Nigeria, 41 pp.

James, B., Yaninek, J., Neuenschwander, P., Cudjoe, A., Modder, W., Echendu, N. and Toko, M. (2000a) *Lutte Contre les Ravaguers du Manioc.* Guide de la pratique de lutte intégrée a l'usage des vulgarisateurs. Modern Design Ltd, Lagos, Nigeria, 36 pp.

James, B., Yaninek, J., Tumanteh, A., Maroya, N., Dixon, A., Salawu, R. and Kwateng, J. (2000b) *Comment Démarrer un Champ de Manioc.* Guide de la pratique de lutte intégrée a l'usage des vulgarisateurs. Modern Design Ltd, Lagos, Nigeria, 24 pp.

Kanoute, A., James, B., Akatse, J., Kimani, M. and Markham, R. (1999) *Towards more effective implementation of IPM in Africa.* NGO-IPM workshop summary, Cotonou, Benin, 29 March to 3 April, 1999. IITA, Nigeria, 32 pp.

Khan, Z.R., Chiliswa, P., Ampong-Nyarko, K., Smart, L.E., Polaszek, A., Wandera, J. and Mulaa, M.A. (1997a) Utilisation of wild gramineous plants for the management of cereal stemborers in Africa. *Insect Science and its Application* 17, 143–150.

Khan, Z.R., Ampong-Nyarko, K., Chiliswa, P., Hassanali, A., Kimani, S., Lwande, W., Overholt, W.A., Pickett, J.P., Smart, L.E., Wadhams, L.J. and Woodcock, C.M. (1997b) Intercropping increases parasitism of pests. *Nature* 388, 631–632.

Khan, Z.R., Pickett, J.A., Van den Berg, J., Wadhams, L.J. and Woodcock, C.M. (2000) Exploiting chemical ecology and species diversity: stem borer and striga control for maize and sorghum in Africa. *Pest Management Science* 56, 957–962.

Kotze, D.J. and Dennill, G.B. (1996) The effect of *Liriomyza trifolii* (Burgess) (Dipt., Agromyzidae) on fruit production and growth of tomatoes, *Lycopersicon esculentum* (Mill) (Solanaceae). *Journal of Applied Entomology* 120(4), 231–235.

Legg, J.P. (1999) Emergence, spread and strategies for controlling the pandemic of cassava mosaic virus disease in east and central Africa. *Crop Protection* 18, 627–637.

Legg, J. and Thresh, J.M. (2000) Cassava mosaic virus disease in East Africa: a dynamic disease in a changing environment. *Virus Research* 71, 135–150.

Lilja, N. and Ashby, J.A. (1999) *Types of Participatory Research Based on Locus of Decision-making.* CGIAR Systemwide Program on Participatory Research and Gender Analysis Working Document No. 6, PRGA Program, Cali, Colombia, 10 pp.

Lomer, C.J., Bateman, R.P., Johnson, D.L., Langewald, J. and Thomas, M. (2001) Biological control of locusts and grasshoppers. *Annual Review of Entomology* 46, 667–702.

Makkouk, K.M., Rizkallah, L., Madkour, M., El-Sherbeiny, M., Kumari, S.G., Amriti A.W. and Solh, M.B. (1994) Survey of faba bean (*Vicia faba* L.) for viruses in Egypt. *Phytopathologia Mediterranea* 33, 207–211.

Mahungu, N.M., Dixon, A.G.O. and Mkumbira, J. (1994) Breeding cassava for multiple pest resistance in Africa. *African Crop Science Journal* 2, 539–552.

Markham, R.H., Bosque-Pérez, N.A., Borgemeister, C. and Meilke, W.G. (1994) Developing pest management strategies for the maize weevil, *Sitophilus zeamais* and the larger grain borer, *Prostephanus truncatus*, in the humid and sub-humid tropics. *FAO Plant Protection Bulletin* 42, 97–116.

Melifonwu, A., James, B., Aihou, K., Weise, S., Awah, E. and Gbaguidi, B. (2000) *Lutte Contre les Adventices dans les Champs de Manioc.* Guide de la pratique de lutte intégrée a l'usage des vulgarisateurs. Modern Design Ltd, Lagos, Nigeria, 28 pp.

Morales, F.J. (ed.) (2000) *Bean Golden Mosaic Virus and other Diseases of Common Bean caused by Whitefly-transmitted Geminiviruses in Latin America.* International Center for Tropical Agriculture (CIAT), Cali, Colombia, 169 pp.

Morales, F.J. (2001) Conventional breeding to *Bemisia tabaci*-transmitted geminiviruses. *Crop Protection* 20, 825–834.

Morales, F.J. and Anderson, P.K. (2001) The emergence and dissemination of whitefly-transmitted geminiviruses in Latin America. *Archives of Virology* 146(3), 415–441.

Msikita, W., James, B., Nnodu, E., Wydra, K. and Ogbe, F. (2000) *Lutte contre les maladies du manioc.* Guide de la pratique de lutte intégrée a l'usage des vulgarisateurs. Modern Design Ltd, Lagos, Nigeria, 26 pp.

Neuenschwander, P. (2001) Biological control of the cassava mealybug in Africa: a review. *Biological Control* 21, 214–229.

Neuenschwander, P., Boavida, C., Bokonon-Ganta, A., Gado, A.H. and Herren, H.R. (1994) Establishment and spread of *Gyranusoidea tebygi* Noyes and *Anagyrus mangicola* Noyes (Hymenoptera: Encyrtidae), two biological control agents released against the mango mealybug, *Rastrococcus invadens* Williams

(Homoptera: Pseudococcidae) in Africa. *Biocontrol Science and Technology* 4, 61–69.

Neuenschwander, P., Hughes, J.d'A., Ogbe, F., Ngatse, J.M. and Legg, J.P. (2001) The occurrence of the Uganda Variant of East African Cassava Mosaic Virus (EACMV-Ug) in western Democratic Republic of Congo and the Congo Republic defines the westernmost extent of the CMD pandemic in East/Central Africa. *Disease Reports* www.bspp.org.uk/ndr/reports.htm

OBEPAB (2000) *Formation de Maraîchers de Cotonou sur l'Utilisation des Feuilles de Neem et de Papayer pour la Protection des Cultures de Chou et de Concombre.* Rapport général. Organisation Béninoise pour la Promotion de l'Agriculture Biologique (OBEPAB), 15 pp.

Oikeh, S.O., Weber, G., Lagoke, S.T.O. and Awad, A.E. (1996) Assessment of yield losses from *Striga hermonthica* in farmers' fields in the northern Guinea savanna. *Nigerian Journal of Weed Science* 9, 1–6.

Ooi, P.A.C. (2000) From passive observer to pest management expert: science education and farmers. In: Guijt, I., Berdegue, J.A. and Loevinsohn, M. (eds) *Deepening the Basis of Rural Resource Management.* Proceedings 16–18 Feb 2000. ISNAR, The Hague, pp. 167–177.

Osada, K.S. and Fucikovsky, Z.L. (1986) Study on the germination of the ergot of maize and host–parasite interaction (*Zea mays* L. and *Claviceps gigantea* Fuentes *et al.*). *Agrociencia* 66, 57–70.

Parkinson, V., Kim, S.K., Efron, Y., Bello, L. and Dashiell, K. (1987) Trap crops as a cultural measure in *Striga* control in Africa. *FAO Plant Protection Bulletin* 35, 51–54.

Polston, J.E. and Anderson, P.K. (1997) The emergence of whitefly-transmitted geminiviruses in tomato in the Western Hemisphere. *Plant Disease* 81, 1358–1369.

Pretty, J.N. (1995) Participatory learning for sustainable agriculture. *World Development* 23(8), 1247–1263.

Röling, N. and Wagemakers, M.A.E. (eds) (1998) *Facilitating Sustainable Agriculture.* Cambridge University Press, 318 pp.

Salas, J. (1992) Integrated pest-insects management program for tomato crops (*Lycopersicon esculentum* Mill.) Lara State, Venezuela. *Acta Horticulturae* 301, 199–204.

Sallé, G. and Raynal-Roques, A. (1989) Le Striga. *La Recherche* 20, 44–52.

Sauerborn, J. (1991) The economic importance of the phytoparasites *Orobanche* and *Striga.* In: Ransom, J.K., Musselman, L.J., Worsham, A.D. and Parker, C. (eds) *Proceedings of the 5th International Symposium of Parasite Weeds, April, 1991.* CIMMYT, Nairobi, pp. 137–143.

Shaefers, G.A. and Krauss, R.E. (eds) (1998) *IPM Communications and Information Workshop for Eastern and Southern Africa (ICWESA).* Workshop proceedings. International Centre for Insect Physiology and Ecology (ICIPE), Nairobi, Kenya, 1–6 March 1998, 165 pp.

Sharaf-El-Din, N., Kannan, H., Babiker, A., Weigand, S. and Saxena, M.C. (1997) The leaf miner, *Liriomyza trifolii* Burgess (Diptera: Agromyzidae), a serious insect pest of faba bean (*Vicia faba* L.) in the Sudan. 1. Identification of the pest, natural enemies and abundance. *Arab Journal of Plant Protection* 15(1), 39–42.

Shepard, B.M. and Samsudin, Braun, A.R. (1998) Seasonal incidence of *Liriomyza huidobrensis* (Diptera: Agromyzidae) and its parasitoids on vegetables in Indonesia. *International Journal of Pest Management* 44(1), 43–47.

Singh, B.B. and Merrett, P.J. (1980) Leaf miner – a new pest of cowpeas. *Tropical Grain Legume Bulletin* 21, 15–17.

Singh, K.B. and Weigand, S. (1996) Registration of three leafminer-resistant chickpea germplasm lines: ILC 3800, ILC 5901, and ILC 7738. *Crop Science* 36(2), 472.

Singh, S.P., Morales, F.J., Milas, P.N. and Terán, H. (2000) Selection for bean golden resistance in intra- and interracial bean populations. *Crop Science* 40(6), 1565–1572.

Snapp, S. (1999) Mother and Baby Trials: a novel farmer participatory trial design and methodology being tried out in Malawi. *Target* 17, 8.

Spencer, K.A. (1972) Agromyzidae (Diptera) of economic importance. In: Linsley, E.G. and Chemsak, J.A. (eds) *Cerambycidae of North America. Part VI, no. 1. Taxonomy and classification of the subfamily Lepturinae. Series Entomologica* Vol. 9, 16 pp.

Thresh, J.M., Fishpool, L.D.C., Otim-Nape, G.W. and Fargette, D. (1994) African cassava mosaic virus disease: an under-estimated and unsolved problem. *Tropical Science* 34, 3–14.

Thresh, J.M., Otim-Nape, G.W., Thankappan, M. and Muniyappa, V. (1998) The mosaic disease of cassava in Africa and India caused by whitefly-borne geminiviruses. *Review of Plant Pathology* 77, 935–945.

Ujiye, T. and Adachi, I. (1995) Parasitoids of the citrus leafminer, *Phyllocnistis citrella*

Stainton (Lepidoptera: Phyllocnistidae) in Japan and Taiwan. *Bulletin of the Fruit Tree Research Station* 27, 79–102.

Uygen, N., Yumruktepe, R., Kersting, U., Karaca, I., Yigit, A., Tekeli, N.Z., Aytas, M., Ulusoy, M.R. and Canhilal, R. (1995) Turuncgillerde zararli Turuncgil yaprak galeriguvesi *Phyllocnistis citrella* Stainton (Lepidoptera, Gracillaridae). *Turkiye Entomoloji Dergisi* 19(4), 247–252.

van de Fliert, E. (1993) *Integrated Pest Management. Farmer Field Schools Generate Sustainable Practices: a Case Study in Central Java Evaluating IPM Training.* WU Papers 93-3. Wageningen Agricultural University, The Netherlands.

van de Fliert, E., Asmunati, R. and Tantowijoyo, W. (2000) Participatory approaches and scaling-up. In: Stur, W.W., Horne, P.M., Hacker, J.B. and Kerridge, P.C. (eds) *Working with Farmers: The Key to Adoption of Forage Technologies.* ACIAR Publication PR095, pp. 83–90. http://www.aciar.gov.au/publications/db/abstract.asp?pubsID=428

van Thielen, R., Ajuonu, O., Schade, V., Neuenschwander, P., Adité, A. and Lomer, C.J. (1994) Importation, release, and establishment of *Neochetina* spp. (Col.: Curculionidae) for biological control of water hyacinth, *Eichhornia crassipes* (Lil.: Pontederiaceae), in Benin, West Africa. *Entomophaga* 39, 179–188.

Vuylsteke, D.R., Swennen, R.L. and Rodomiro, O. (1993) Development and performance of black sigatoka-resistant tetraploid hybrids of plantain (*Musa* spp., AAB group). *Euphytica* 65, 33–42.

Weltzien, E., Smith, M.E., Meitzner, L.E. and Sperling, L. (2000) *Technical and Institutional Issues in Participatory Plant Breeding from the Perspective of Formal Plant Breeding: a Global Analysis of Issues, Results and Current Experience.* CGIAR Systemwide Program on Participatory Research and Gender Analysis Working Document No. 3, 106 pp. http://www.prgaprogram.org/

Witcombe, J.R. (1996) Decentralization versus farmer participation in plant breeding: some methodology issues. *New Frontiers in Participatory Research and Gender Analysis.* Proceedings of the International Seminar on Participatory Research and Gender Analysis for Technology Development 9–14 September. CIAT Publication No. 294, pp. 135–154.

Yaninek, J.S., James, B.D. and Bieler, P. (1994) Ecologically sustainable cassava plant protection (ESCaPP): a model for environmentally sound pest management in Africa. *African Crop Science* 2, 553–562.

Yaninek, J.S., Mégevand, B., Ojo, B., Cudjoe, A.R., Abole, E., Onzo, A. and Zannou, I. (1998) Establishment and spread of *Typhlodromalus manihoti* (Acari: Phytoseiidae), an introduced phytoseiid predator of *Mononychellus tanajoa* (Acari: Tetranychidae) in Africa. *Environmental Entomology* 27, 1496–1505.

Chapter 33
The World Bank and Pest Management[1]

T.W. Schillhorn van Veen
The World Bank, Washington, DC, USA

Introduction

The world's human population continues to grow. Feeding this increasing population has been a challenge for the world agricultural and livestock producers and has made considerable demand on the world's resources. A large part of the production increases will have to come from local or regional sources. This is especially a challenge for the farmers in Asia, Africa, and Latin America who will have to provide food for the still-increasing populations on their continents. This can be achieved through three major actions: (i) organization of food production and distribution; (ii) increasing production; and (iii) decreasing losses.

Dramatic increases in yields were obtained during the Green Revolution in the early 1970s when higher yielding cereal varieties were introduced in Asia thereby averting expected food shortages and famine. However, the newer varieties were also more demanding, and their use increased the need for inputs such as water, pesticides, fertilizers and extension. In some cases, these high input requirements lead to non-sustainable production systems and the 'health' of the land has become under threat of salinity, soil depletion and erosion. Moreover, the use of pesticides came with certain risks to farmers, farm workers as well as consumers. As such, the centralized approach (including state-managed procurement of inputs) was increasingly criticized, leading to a demand for a new paradigm of feeding the world through *sustainable* food production.

Development banks such as the World Bank, who supported the earlier systemic Green Revolution approach, have been learning lessons, and are increasingly exploring ways to invest in agricultural development with focused interventions that are economically and environmentally sustainable. They began to understand the risk and liabilities[2] of financing state-supported procurement of farmer inputs (Farah, 1994; Schillhorn van Veen *et al.*, 1997; Sheriff and Fleischer, 2003). These institutions are now trying to focus on policies and investments that will lead to more sustainable agriculture (including pest control), and let farmers, rather than the state, decide what farmers need.

---

[1] This note is prepared by T.W. Schillhorn van Veen of the World Bank (e-mail: tschillhornvanve@worldbank.org). The statements in this communication are those of the author and do not necessary represent the opinion of the World Bank and its affiliates.

[2] Prominent among these liabilities is the creation of obsolete pesticide stocks in State warehouses etc. and their costly removal and incineration process (see below).

 Integrated Pest Management in the Global Arena (eds K.M. Maredia, D. Dakouo and D. Mota-Sanchez)

The World Bank, therefore, is paying more attention to sustainable pest control practices and to promoting IPM. The Rural Development Department recently completed a study on IPM policies and implementation strategies of development agencies to be used as a background paper in its upcoming review of the Bank's Integrated Pest Management position paper (see Sorby *et al.*, 2003). In the Bank's view, IPM is not only about preventing environmental and health problems, but also about sustainable agriculture and improving farmer incomes through reduced production costs. The Bank is approaching this through two broad objectives: (i) the 'do-no-harm' principle; and (ii) the 'do-good' principle. The boundary between these two objectives is by no means firm, and often the Bank applies both approaches[3].

Do No Harm[4]

The *do no harm* principle is basically applied through internal (and sometimes external) peer review and through the enforcement of the Bank's policies and safeguards. The broad objective is to ensure that pest management activities in Bank projects are sustainable and that social, health and environmental risks are reviewed, minimized and are, or can, properly be managed by the user. Such activities include projects that directly or indirectly finance pesticides as well projects that do not finance pesticides but nevertheless indirectly increase pesticide use or affect pest management. For example, it is considered appropriate to set out clear targets for moving current practices towards IPM and to provide the necessary support for this process in projects where current practices in pest management are unsustainable, are not based on an IPM approach, and/or pose significant health and environmental risks. Immediate measures may be required to reduce risks associated with the handling and use of pesticides to a level that can safely be managed by the users.

Operational policy: pest management – some highlights of the text

General: In assisting borrowers to manage pests that affect either agriculture or public health, the Bank supports a strategy that promotes the use of biological or environmental control methods and reduces reliance on synthetic chemical pesticides. In Bank-financed projects, the borrower addresses pest management issues in the context of the project's environmental assessment.

In appraising a project that will involve pest management, the Bank assesses the capacity of the country's regulatory framework and institutions to promote and support safe, effective, and environmentally sound pest management. As necessary, the Bank and the borrower incorporate in the project components to strengthen such capacity.

Agricultural pest management: The Bank uses various means to assess pest management in the country and support integrated pest management (IPM) and the safe use of agricultural pesticides: economic and sector work, sectoral or project-specific environmental assessments, participatory IPM assessments, and adjustment or investment projects and components aimed specifically at supporting the adoption and use of IPM.

In Bank-financed agriculture operations, pest populations are normally controlled through IPM approaches, such as biological control, cultural practices, and the development and use of crop varieties that are resistant or tolerant to the pest. The Bank may finance the purchase of pesticides when their use is justified under an IPM approach.

Pest management in public health: In Bank-financed public health projects, the Bank supports controlling pests primarily through environmental methods. Where environmental methods alone are not effective, the Bank may finance the use of pesticides for control of disease vectors.

[3] Such investments, however, are not offsetting, i.e. 'do-good' investments do not diminish Bank staff and borrower vigilance to enforce the 'do-no-harm' objective.

[4] The do-no-harm and do-good concepts derive from the work of H.P. van der Wulp (Word Bank/FAO IPM Facility).

Safeguards

The Bank's Pest Management Policy was developed in the early 1980s. Since then, it has been changed and amended. However, its implementation/application in World Bank projects has been questioned (see Schillhorn van Veen *et al.*, 1997). Consequently, the language of the policy was strengthened in the mid-1990s with further focus on clearly enforceable and monitorable rules.

The strengthening of the Bank's environmental policies and their implementation more or less coincided with the recognition of the reputational risk to the institution in situations involving less than adequate adherence to the Bank's policies. The outcome of this recognition has resulted in stricter review and enforcement. The resulting '*safeguard*' policies mainly focus on environmental and social risks, covering eight specific fields[5], one of which is adherence to its pest management policy. The environmental and social units in the Bank's regional vice presidencies are responsible for the enforcement of, and oversight over, the implementation of the policies by the borrowers.

All Bank projects are now subject to review for possible environmental and social impacts, including the borrowers' regulatory framework, institutional capacity and enforcement. With respect to pest management, the safeguard is guided by the Bank's pest management policy (OP 4.09) that focuses on sustainable pest control methods with, where possible, a reduction on the reliance on chemical pesticides. In line with the objective of being 'implementable and enforceable', the policy is relatively simple (see Box). Its implementation is further worked out in Bank Procedures (BP 4.01 C) under the environmental policy[6]. The implementation, by regional operational staff, was originally supported and monitored by the agricultural department (Rural Development Department) which was seen by some as a conflict of interest. Since 2001, the quality management is located separately in the Quality Assurance and Compliance Unit that has the task to ensure environmental due diligence of Bank investments.

This environmental quality assurance has increased the burden in project preparation and review on borrowers as well as the Bank's project officers (project task leader) who have to verify adequate institutional oversight capability to ensure a sustainable (pest control) investment.

IPM Facility

The World Bank itself has limited relevant expertise to assist task leaders in encouraging inclusion of pest management interventions in Bank projects, or determining whether appropriate products and techniques used in Bank-financed projects are best practices (in IPM). The co-sponsorship of an IPM facility by the Bank with FAO, UNEP and UNDP has been a positive development by providing access to specialist expertise and to a network of relevant expert institutions including NGOs. The Facility, which was initially financed by the World Bank and FAO, is now largely supported by bilateral donors. It is located in Rome and aims to help countries in the development and implementation of pest management policies and IPM programs. It is also helping both developing countries and international funding agencies by initiating some pilot projects that serve as examples of sound investments in sustainable pest management, and can be mainstreamed through larger lending programs. It was envisaged that the Facility would lead to an increased number of improved quality lending operations in

[5] Others include social issues as well as technical environmental issues such as sustainable forestry, natural habitats and dam safety.

[6] The text of these policies is available on the Bank's external website: http://www.worldbank.org/whatwedo/policies.htm

IPM. It was seen as a possible cost-effective means to leverage a major impact in this sphere. The impact of the Facility has been mixed, as it concentrated on Asia and Africa, and on the application of specific extension techniques (i.e. the 'FFS') in implementing IPM. Its technical expertise, however, and the provision of IPM experts have been of value.

Do Good

The *do-good* principle calls for improving awareness, enhancing policy reform and strengthening the regulatory framework and institutional capacity for the implementation of IPM and the control of pesticide use and handling. The expected level of project involvement depends on the circumstances and the scope of the project. Relevant factors in this respect are the:

- status of pest management in a country (including parameters such as the amount of pesticides used on a per hectare or per capita basis, the health effects of farm workers, and the status and implementation capacity of pesticide laws);
- magnitude of the activity involving or affecting pest management;
- nature of the risks involved;
- size of the gap between actual practices and good practices;
- geographical scope of the project; and
- degree to which policy reform and capacity building fit in the project.

The tools vary from policy dialogue, advice and training to specific investments. The tools and policies do not only relate to agriculture but also to other sectors including health, energy and finance.

Overall, the Bank believes that sound macro- and micro-economic policies that encourage producers and consumers to make rational well-informed decisions will lead to efficient resource use and to sustainable production in agriculture. So far, environmental sustainability or sustainable pest management have not really been part of the World Bank's macro policy discussions or even of agricultural policy discussions or policy loans. This may change with the application of the safeguard policy, mentioned above.

Pest management continues to be part of project loans. Few loans are solely dealing with pest management, and the pest-management related activities are generally a component or subcomponent of the broader agricultural loan, with a mix of objectives and interventions that include the following.

1. **Governance/regulation.** Increasingly, Bank projects support the strengthening of monitoring capacity, by financing chemical analytical laboratories (residue testing, product quality) combined with improving regulations and enforcement capacity. Exporting countries, especially, are increasingly faced with international demands to certify the safety of agricultural products and adhere to the quality rules in their intended markets. This provision of laboratory equipment and expertise is often combined with policy dialogue. Recently, the Bank has also been requested to support the development of regulatory frameworks for organic agricultural production (in Romania).

2. **Introduction of IPM technologies.** Examples are: Bank support for the bio-control systems for cotton pests in Central Asia, for organic coffee growing in Mexico, for field-school based IPM training in Indonesia, for control of water hyacinth in East Africa, as well as the introduction of tsetse traps in the control of animal trypanosomiasis in central Africa. However, much more could be done.

Although many aspects of IPM consist of 'private goods', the mobility of pest populations often also requires a (national) collaborative effort[7], i.e. there are clear 'public goods' aspects. Moreover, in many countries

[7] Locust control is often used as an example. However, rather than only focusing on locust control interventions, other approaches, including insurance schemes, may have a higher benefit:cost ratio.

there is lack of awareness, and public sector support can be justified to correct this (again, because there may be considerable positive trade-offs). The introduction of IPM is complex, touching upon various technical and non-technical interventions, which complicates the calculation of the economic rate of return.

3. Research/extension. An example was the agricultural research support project in Turkey that specifically allocated some funds to research on IPM, resulting in greater attention of researchers for this subject matter and a number of applicable technologies. Extension services in many countries still involve themselves in providing goods rather than information and knowledge, and too often such goods include pesticides. This paradigm is changing only slowly. Bank projects generally are not to support extension systems that introduce pesticides (either as goods or as advice for 'calendar' spraying, etc.), without having exhausted the options of non-chemical control. The latter are numerous but there is a general lack of awareness among the agricultural staff in many countries[8].

4. Training. The best known example of training is the support for the farmer field schools in Indonesia, a concept earlier developed by UNDP and FAO and mainstreamed in the Bank-financed Indonesia IPM project. This (extension) technology has also been applied in other countries such as Vietnam and Ghana; so far mainly focusing on pest control in rice.

5. Financing pest control. The financing of pest control is slowly shifting away from pesticide procurements of the past. This is in part supported by the Bank's overall policy to stress the 'public good' and the 'private good' aspect of interventions, and limit investments to those with a clear public good benefit. Most on-farm pest control is a private good (albeit with some public good implications), which is not directly financed by the State[9] (and consequently by the Bank which lends only to governments). However, the Bank has been encouraging non-chemical pest control such as the cultivation, use and small-scale marketing of beneficial insects to control pests in, for example, cotton in Central Asia and coffee in Mexico.

6. Clean up. Although state-managed pest control and pesticide procurements can occasionally be useful in emergency situations, they tend to distort markets and the leftover ('obsolete') stocks are a persisting liability and expensive to remove. Most obsolete stocks[10], whether in the former Soviet Union or in developing countries, are results of previous pesticide procurements managed by the State. International donors and banks have been requested to participate in the financing of clean-up programs. However, as countries are reluctant to borrow from development banks for such clean-up, they prefer to appeal to donor organizations. For example, FAO managed such a clean-up program in some African countries financed by bilateral support from European countries. The World Bank is contemplating a follow-up that would finance the removal of all obsolete stocks in a proposed 'Africa Stockpiles Program.'

As previously mentioned, the Bank increasingly relies on others to develop and monitor the technical aspects of its lending program. This dependency increases the responsibility of local (technical) staff in the

[8] Too often farmers and agricultural staff still associate the use of pesticides with 'modern' agriculture, and are not aware of the advances made in non-chemical pest control in developing countries in recent years. Inexperienced extension staff still prefer providing tangible goods (fertilizer, pesticide) rather than sound advice.

[9] Indirectly the Bank could finance agricultural inputs through its on-lending (credit) projects. In this case, however, the Bank's safeguard policy requires review of the capacity in Government as well as in the participating banks to assure safe use.

[10] Donations or free supply of farm inputs such as agrochemicals, still a habit of a limited number of donors and governments, may impede the development of independent farmers (as government and donors decide, rather than farmers, what is needed), may consequently lead to poor choice of inputs and, with respect to pesticides, may lead to outdated or obsolete stocks.

planning and implementation of Bank loans, requiring training programs such as the IPM training program provided by Michigan State University (and many others).

Of course not everyone has been happy with the Bank's approach. Some in the pesticide industry see the pest management policy as restrictive and have made various efforts to neglect, change or circumvent the Bank's policy. The pressure on borrowers by such players in the industry has been increasing and was somewhat facilitated by the Bank's emphasis on private sector development and participation. There has been much debate on the biotechnical aspects of pest control. Currently, the Bank has no official position on genetically modified organisms as too little information is available to provide a well-informed opinion about these recent technological developments. After all, the World Bank, as a bank, is risk adverse.

On the other hand, many in the environmental community find the policy still weak and its implementation lacking. This community has a valuable role to play in pointing out to the Bank where its policies are not implemented. Indeed, a recent review[11] indicated that the application of the OP 4.09 was highly variable. This is in part related to the lack of awareness in borrowing countries, to the lack of knowledgeable staff (most agricultural specialists have left the Bank in the last decade), and to the persisting paradigm that associates 'modern' agriculture with high consumption of inputs, including pesticides. Bank-financed projects are, in principle, designed by the borrower[12] and, because of their lack of awareness about sustainable agriculture and pest management, excellent opportunities to introduce sustainable practices are too often missed.

Within the Bank there is debate about the economic returns of sustainable farming practices. The major issue is the type of tools required to measure the impact, and especially how to calculate and include the indirect spin-offs, i.e. effects such as human health (especially of farm workers) and well-being, environmental factors ranging from clean water to increasing fauna, and the loss or creation of local jobs[13].

All parties, including industry, international development organizations and the environmental community could play a constructive role and help to reduce the reliance on toxic pesticides by creating awareness and promoting a diversified approach to pest control in agriculture as well as in other sectors (for example, health) without unacceptable risks to the environment or human health and well-being.

References

Farah, J. (1994) Pesticide policies in developing countries: do they encourage excessive use. *Discussion Paper No. 238*, World Bank, Washington, DC.

Schillhorn van Veen, T.W., Forno, D.A. *et al.* (1997) Integrated Pest Management. Strategies and policies for effective implementation. *Environmental Sustainable Development Studies and Monographs No. 13*, World Bank, Washington, DC.

Sheriff, G. and Fleischer, G. (2003) Towards economically rational pesticide policies. *World Bank Research Observer*, World Bank, Washington, DC (in press).

Sorby, K., Fleischer, G. and Pehu, E. (2003) IPM in Development – a review of trends and implementation strategies. *Agriculture and Rural Development Working Paper No. 5*, World Bank, Washington.

Tozun, N. (2001) New policy, old patterns, a survey of IPM in World Bank projects. *Global Pesticide Campaigner* 11(1), Pesticide Network, San Francisco.

---

[11] Tozun (2001) and Sorby *et al.* (2003).

[12] Borrowers are often helped by, or leave the detailed design to, the project design unit in FAO financed under a cooperative agreement with the Bank, which also has few technical specialists, and often draws expertise from the private consulting community.

[13] Most pesticides are produced in developed countries. The use of chemical pesticides is often a labor-saving tool, the need of which can be questioned in developing countries where unemployment is high.

Chapter 34

Integrated Pest Management Case Studies from ICIPE

Z.R. Khan, W.A. Overholt and A. Ng'eny-Mengech
ICIPE, Nairobi, Kenya

Evolution of ICIPE

The ICIPE, with its global headquarters in Nairobi, Kenya, is an autonomous intergovernmental research institute devoted to alleviating poverty and ensuring food security and good health for the peoples of the tropics through the management of both harmful and useful insects and other arthropods (www.icipe.org). The Centre was founded in April 1970 by the Kenyan insect physiologist, Thomas R. Odhiambo (http://www.thp.org/prize/87/odhiambo.html), in direct response to the need for alternative pest and vector management strategies that are sustainable, selective, non-polluting, and affordable to resource-limited rural communities. The tropical African environment in which ICIPE is located contains a rich variety and abundance of arthropod life, which cause a number of serious developmental constraints on the one hand, but offer unique opportunities for research on the other, touching on all spheres of human, animal, plant and environmental life.

ICIPE's current Director General, Hans R. Herren, a Swiss entomologist, was appointed in 1994. Winner of the 1995 World Food prize (http://www.worldfoodprize.org/Laureates/Past/1995.html), Herren has introduced new programs that intrinsically link IPVM with the Centre's unique mandate to promote the use of insects as a valuable natural resource and raw material for enterprise development.

In its earlier years, ICIPE concentrated on basic research to acquire the critical information base needed for a holistic understanding of its target pests and vectors. Later, building on this base, ICIPE then began concentrating on application of this knowledge to the design of innovative methods of managing key pests. The third decade saw ICIPE's agenda evolving to address the immediate needs of its constituencies more directly, while at the same time being cognizant of regional and global concerns, by developing more products and services, as well as integrated packages for improved IPVM and commercial insect enterprise development. Now in its fourth decade, ICIPE is developing solutions based on its original mandate but very much taking into consideration the growing global concerns about such issues as biodiversity conservation and sustainable utilization, poverty reduction, and the role of women in development.

(eds K.M. Maredia, D. Dakouo and D. Mota-Sanchez)

Organizational Structure and Operation

ICIPE's organizational structure is based on a matrix. The lateral dimension is composed of four research divisions representing the '4-H' areas of operation, namely Plant **H**ealth, Human **H**ealth, Animal **H**ealth, and Environmental **H**ealth. The second, vertical dimension consists of two research departments in which ICIPE's core competency resides (the Population Ecology and Ecosystem Science Department and the Behavioral and Chemical Ecology Department) and the research and support units (Molecular Biology and Biotechnology, Biomathematics, Bioinformatics, Entomopathology, Arthropod Rearing and Quarantine, Biosystematics, Social Sciences, Information and Publications and a Technology Transfer unit). The third dimension is provided by Capacity Building, which permeates every program and activity, and serves to build the capabilities of developing-country individuals and institutions while forging a link to the future.

Research is conducted at the Centre's headquarters on the outskirts of Nairobi, as well as at the main field research and training facility, ICIPE-Mbita, on the shores of Lake Victoria in western Kenya. There are six other field sites in Kenya, altogether totaling 500 ha, representing all of the main agroecological zones found in the tropics. A field site on the Red Sea Coast of Sudan (for locust research) and a focus in Ethiopia are also sites of research and development activities. Through extensive collaborative agreements with national agricultural research and extension systems (NARES) and NGOs working at grassroots level, a much greater target area and population is reached. ICIPE currently collaborates with over 200 national, regional and international institutions and organizations in 64 countries around the world.

Impact of ICIPE's Achievements

In its 30-some years of operation, ICIPE has made some significant contributions to developing IPM components, particularly through work on pest-tolerant/resistant varieties, biological control, use of botanicals and behavior modification techniques. In the Centre's first 25 years, ICIPE agronomists concentrated on developing pest-tolerant 'IPM-fit' varieties of staple food crops such as maize, sorghum, cowpea and striga-resistant rice which are now being used by NARES in several African countries. A germplasm collection of over 175 banana cultivars was established by ICIPE researchers and also transferred to NARES in several African countries, and methods for multiplying clean planting materials devised. Cultural practices for reducing pest incidence in field crops and for improving the land equivalence ratio, such as by strip relay-intercropping, were developed. Horticultural crop pests research is a more recent addition to ICIPE's R&D agenda, but over the past several years, germplasm collections have been established and evaluated for pest resistance for several important export and local crops, including tomatoes, okra and aubergine. Cultural methods have also been devised for pest control in cabbage, French bean, tomato and indigenous vegetables. Prefacing the study of every new crop are biodiversity surveys of the pest complex and its natural enemies.

Botanical pest control agents, including neem (*Azadirachta indica*) have been successfully deployed by farmers against a variety of crop pests, including maize and sorghum stemborers, cowpea thrips, diamondback moth, aphids, termites, banana weevils, and banana and tobacco root-knot nematodes. The search for new botanicals with pest control properties continues, and of the several hundred indigenous plants screened, a score or more are proving promising in controlling not only crop pests, but also disease vectors such as mosquitoes and ticks. Several of these are being grown by local communities as income-generating activities, and have been registered and formulated on a small scale into commercial products via ICIPE's technology transfer sections.

ICIPE has been a leader in behavioral and chemical ecology research leading to pest control methods that rely on behavior modification. One of the best examples is the development of traps and odor baits to lure insects to their death. Thousands of the NGU tsetse traps made of cheap, locally available materials have been deployed by communities in several African countries (e.g. Kenya, Uganda, Rwanda, Ethiopia) and have been effective in ridding the areas of these vectors of animal and human sleeping sickness, all without the use of expensive and environmentally harmful chemical pesticides. Another trap design, the NZI trap, is proving popular for control of many different kinds of biting flies, including stable flies. A new bednet trap for mosquitoes is nearing completion, and fruit fly baiting stations have been successfully deployed on farms. Semiochemicals have been isolated and/or developed as insect attractants for use in traps (e.g. for fruit flies, stemborers, savanna and riverine tsetse), as repellents (e.g. for mosquitoes, tsetse) and as agents to disrupt locusts changing to the gregarious phase. ICIPE's Locusts and Migratory Pests research program in particular has made important progress in elucidating the chemical signals responsible for insect behavior, in this case the transformation of the desert locust (*Schistocerca gregaria*) from harmless solitary individuals into the damaging gregarious swarms.

In the area of biological control, several entomopathogens that have been developed against a wide range of arthropod pests are proving popular with project farmers: local Bt (*Bacillus thuringiensis*) strains effective against the diamondback moth (*Plutella xylostella*) in cabbage and kale and stemborers in maize; Bt strains active against mosquito larvae; fungal preparations such as strains of *Metarhizium anisopliae* against termites and thrips in onion, French bean and flowers; *Beauveria bassiana* and nematodes against ticks; and viruses against locusts and maize pests, among others.

This chapter describes one of ICIPE's main thrusts in classical biological control, against the exotic stemborer *Chilo partellus*, a major pest of gramineous crops. At the same time, the Centre is rapidly moving toward introduction of biological control agents for horticultural crop pests, with the first release of exotic *Diadegma* species in early 2002 to control the diamondback moth in Africa. Intensive searches for natural enemies of red spider mites (*Tetranychus* spp., a major pest of tomato) and the African fruitworm, *Helicoverpa armigera*, are also being conducted.

A relatively recent area in which ICIPE has made a substantial impact is in the development and introduction of small-scale enterprises based on apiculture and sericulture. The Commerical Insects program has developed full technology packages for beekeepers, from improved queen bees, to introduction of modern 10-frame Langstroth hives, and improved methods of harvesting and processing honey and other high-value hive products such as royal jelly, venom and propolis. Races of the domestic silkworm (*Bombyx mori*) have been developed for rearing under African conditions, and the full technology package from eggs to reeling, spinning and weaving of finished silk fabric introduced. Methods for rearing wild silkmoths, the source of tussar silk, and for conserving their forest habitats are also being established. These commercial insect technologies have already been successfully introduced to over 7000 farmers in East Africa, with requests to extend them to other parts of the continent and Latin America.

Closely related to efficient IPVM development is another of ICIPE's programs, that of Arthropod Biodiversity Conservation and Utilization, wherein the roles of insects in conservation biology (e.g. as pollinators) and in gene flow from genetically modified crops are studied. This program is also creating an African insect database (now about 60% complete) and biodiversity inventory, which will be useful for identification purposes and in developing decision-support tools.

As part of its vision and strategy for the next 10 years, ICIPE will continue to concentrate on developing pest and vector control technologies for use by the resource-poor smallholder farmers and urban communities of Africa, and other areas of the tropics with similar problems. Because ICIPE believes that true development can only take place in a healthy environment, special emphasis will be given to conserving and/or restoring environmental integrity. The basic approach will be one of more effective prevention – keeping an insect from ever becoming a pest in the first place – combined with 'smarter' cures. The latter will often be recommended as components of a more comprehensive development package, however, where both crop pests and disease vectors (e.g. mosquitoes, tsetse, ticks) will be tackled simultaneously, while introducing income-generating activities to help relieve poverty – the root cause of underdevelopment – rather than as an isolated solution for a single pest. This paradigm of 'total health' (i.e. the 4Hs) is being used, for example, in ICIPE's Kakamega Forest Conservation project, where active community participation at all levels is an intrinsic feature of the approach. The 'habitat management' project described later in this chapter is an example of such an integrated development approach, combining as it does several IPM components for pest control with environmental conservation and income generation. The 'push-pull' or 'stimulo-deterrent' concept used in this project is now being developed for other applications, such as in tsetse and tick control. ICIPE will also take advantage of the genomics and proteomics revolutions, for instance to identify active sites for altering pest or pesticide resistance, vectorial capacity and behavioral characteristics.

ICIPE's Role in Training and Capacity Building

As an intrinsic part of its mandate, ICIPE plays an important role in training and capacity building in insect science and its application. In 1983, in collaboration with several institutions of higher learning in Africa and with financial backing of several donors, ICIPE initiated a unique regional program: the ARPPIS (http://www.bu~wood.or~/agriforum/html/icipe.html). Today, with 27 participating universities in Africa and an equal number in another 15 countries, the ARPPIS network has become a model in cooperative high-level education for scientific leadership on the continent. The Programme has thus far a cumulative enrollment of more than 160 scholars, of whom 119 have completed their PhD degrees and are working in Africa. Students conduct their research at ICIPE under supervision of an ICIPE scientist and a staff from the degree-awarding university. Another 120 plus students have been trained at MSc level in three regional ARPPIS centers, at the University of Legon, Ghana; Addis Ababa University, Ethiopia; and the University of Zimbabwe; a fourth MSc center in francophone Africa is planned. A program that allows students from other continents to conduct research at ICIPE, called the Dissertation Research Internship Programme (DRIP), was established several years ago and has benefited 69 students, including 25 doctoral candidates. Professional development schemes for Postdoctoral Fellows, Visiting Scientists and Research Associates also exist. In addition to these higher-level training opportunities, ICIPE has served as a regional focus for training IPVM practitioners, through tailor-made short courses on specific themes; thus far over 9000 farmers have been trained as well as 1000-some extension agents and trainers. Future plans are to strengthen even further the Centre's capacity-building activities, for instance by inclusion of training in modern genomics and proteomics at postgraduate level and adoption of a FFS approach for training in IPVM at grassroots level. It is envisaged that ICIPE will be a focus of bioscience research and a facilitator of bioscience education in Africa.

ICIPE maintains close interactive contact with the African scientific community who are working to find solutions to the continent's pressing problems. This is another aspect of ICIPE's mandate: to serve as a regional focus for sharing, creating and documenting scientific information and promoting a science culture in Africa. The AAIS, founded by a group of eminent African scientists and nurtured over the years by ICIPE, is a vital link in this regard. The Association co-sponsors an international journal of tropical insect science, *Insect Science and its Application*, which is hosted by ICIPE (http://www.inasp.ora.uk/ajol/jou~als/isa/about.html). Now in its 23rd volume, the journal can be found on-line at http://www.bioline.org.br/ti. In an effort to keep in touch with alumni, the ARPPIS Scholars Association (ASA) was founded in 1998 to facilitate closer research networking across the African continent. ICIPE is also the focus of several other networks, including BiONET-Africa for the biosciences; the Africa Remote Sensing Databank (http://informatics.icipe.org) and the Africa IPM Forum (http://informatics.icipe.org/IPMAfrica).

Successful IPM Applications

Classical biological control of stemborers in eastern and southern Africa

Several lepidopteran stemborers (http://195.202.86.131/stemborers) attack maize, sorghum and millet in Africa, and often two or more species will attack a crop coincidentally in time and space. In eastern and southern Africa, two borers dominate and are responsible for most crop losses: *Busseola fusca* Fuller (Noctuidae) in higher elevation areas, and *Chilo partellus* (Swinhoe) (Crambidae, formerly Pyralidae) in lower- and mid-elevation areas. *Busseola fusca* is a native African borer, whereas *Ch. partellus* is an exotic species which was probably introduced into Africa from southern Asia in the early 1900s (Kfir *et al.*, 2002). Recent evidence suggests that *Ch. partellus* is displacing *B. fusca* and other native stemborers at some locations (Kfir, 1997; Ofomata, 1999).

In 1991, in collaboration with Wageningen Agricultural University in The Netherlands and the KARI, ICIPE initiated a classical biological control program aimed primarily against *Ch. partellus* (http://ipmworld.umn.edu/chapters/overholt.html). Pre-release field surveys conducted from 1991 to 1993 revealed that parasitism of *Ch. partellus* due to native natural enemies was typically very low (Overholt *et al.*, 1994a; Scovgard and Pats, 1996). Thus an exotic parasitoid from Pakistan, *Cotesia flavipes* Cameron (Hymenoptera: Braconidae), was introduced and released in southeastern Kenya where *Ch. partellus* regularly occurs in high densities. *Cotesia flavipes* is a gregarious koinobiont larval parasitoid which had previously been used to successfully control stemborers in sugarcane in several areas of the world (Polaszek and Walker, 1991). Releases were made in 1993 at three locations in the southern coastal area of Kenya (Overholt *et al.*, 1994b), and the parasitoid was recovered during the season of release from *Ch. partellus* and two native stemborers, *Ch. orichalcociliellus*, and *Sesamia calamistis* (Overholt *et al.*, 1997).

Cotesia flavipes was released at a fourth site in coastal Kenya during the non-cropping season (February) of 1994 in an area where the vegetation was dominated by a wild grass, *Sorghum arundinaceum* (Desv.) Stapf. Recoveries in both the wild habitat and in a nearby maize field during the following cropping season indicated that the parasitoid could sustain its population during the dry season in wild grasses, and then colonize maize fields during cropping seasons (Overholt *et al.*, 1997).

Other than recoveries at the wild sorghum site, only one stemborer parasitized by *Co. flavipes* was found in 1994, despite intensive sampling. In 1995 and 1996, a few recoveries were made, but parasitism was extremely low (Overholt *et al.*, 1997). In 1997, the number of recoveries increased dramatically and parasitism at 30 sites averaged about 6%. Parasitism continued to increase during the next 2 years with average parasitism of about 13% at 67 sites in 1999 (Zhou *et al.*, 2001).

Surveys in other maize growing areas of Kenya showed that *Co. flavipes* was present in the Eastern Province (Songa, 1999) and in the area bordering Lake Victoria in western Kenya (Omwega *et al.*, 1995). In the Eastern Province, which borders the Coast Province, *Co. flavipes* was found in low densities in 1996 and then released at three sites in 1997. Parasitism during the season following the releases was about 14% (Songa, 1999). As *Co. flavipes* was never released in western Kenya, Omwega *et al* (1995) speculated that the establishment was the result of insects which had escaped from a laboratory colony maintained at a research station in that region in 1992. However, parasitism in western Kenya has not increased to the levels observed in coastal Kenya nor in the Eastern Province (Ogedah, 1999). In western Kenya, four stemborers are common: *Ch. partellus*, *S. calamistis*, *B. fusca* and *Eldana saccharina* (Seshu Reddy, 1983). All of these are attractive and acceptable hosts for *Co. flavipes*, but two (*B. fusca* and *E. saccharina*) are not suitable for its development (Ngi-Song *et al.*, 1995; Overholt *et al.*, 1997). Overholt (1998) speculated that the presence of acceptable but unsuitable hosts in an area would depress population growth of *Co. flavipes*.

The impact of *Co. flavipes* on stemborer populations in coastal Kenya was recently investigated and showed a reduction in total stem borer density of about 26%. Reduction of *Ch. partellus* density was highest at approximately 50% (Fig. 34.1) (Zhou *et al.*, 2001). By coupling this information with stemborer yield loss estimates (Songa *et al.*, 2001), we believe that the establishment of *Co. flavipes* has resulted in an 8–10% increase in maize yields.

In addition to the work conducted in Kenya, a survey in 1995 in northern and central Tanzania recovered *Co. flavipes* at two locations near Lake Victoria in an area bordering southwestern Kenya (Omwega *et al.*, 1997). Based on surveys conducted before 1994, and on electrophoretic evidence, it was concluded that the most likely explanation was that *Co. flavipes* moved into Tanzania from Kenya (Omwega *et al.*, 1997). Zhou and Overholt (2001) have recently used modeling to produce an interpolated spatial distribution of *Co. flavipes* in Kenya.

Beginning in 1996, the program expanded to include several other countries. Releases of *Co. flavipes* were made in Mozambique in 1996 (Cugala *et al.*, 1999), and in Uganda and Somalia in 1997 (Overholt, 1998). Sampling in Mozambique in 1999 indicated that the parasitoid had established, but parasitism was low. In Uganda, Matama-Kauma (2000) reported that one year after its release, *Co. flavipes* had become the most common parasitoid attacking a complex of four stemborers, and

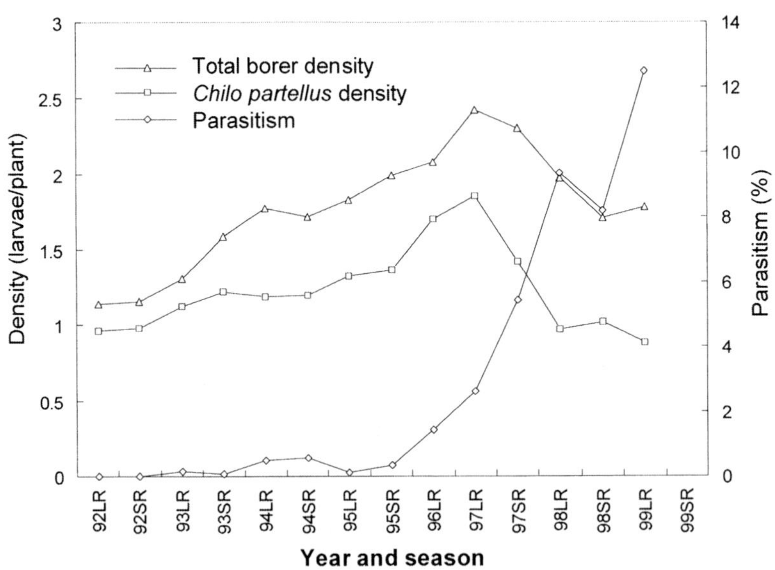

Fig. 34.1. Impact of *Cotesia flavipes* on total stemborer density and the density of the *Chilo partellus* in southeastern Kenya from the long rains (LR) of 1993 to the short rains (SR) of 1999.

that parasitism averaged about 20%. No post-release surveys have been conducted in Somalia, but recoveries in neighboring Ethiopia, where the parasitoid was never released (Degaga, 2002), strongly suggests that the parasitoid established in Somalia, and moved later into Ethiopia. Releases in Zimbabwe, Zambia, Zanzibar and Malawi were made in 1998/99, and so far, establishment has been confirmed in Malawi and Zanzibar (W.A. Overholt, unpublished information).

Future efforts on stemborer biological control will include the release of additional exotic natural enemies, such as the idiobiont pupal parasitoid, *Xanthopimpla stemmator* (Hymenoptera: Ichneumonidae), which is currently undergoing host range studies at ICIPE. We believe that *X. stemmator* will provide an additional level of mortality to stemborers, and not interfere with *Co. flavipes* nor native pupal parasitoids due to its unique attack strategy which involves drilling through plant stems with a stout ovipositor. No common African pupal parasitoids use this 'drill and sting' method. Additionally, greater emphasis will be placed on evaluating the impact of the classical biological control program on a regional basis, and finally, efforts will be made to enhance natural control through the integration of classical biological control and the habitat management approach (see below).

Habitat management for controlling stemborers and striga weed in maize-based farming systems

Stemborers and the parasitic striga weed are two major biotic constraints to increased cereal production in eastern and southern Africa. At least four species of stemborers (http://informatics.icipe.org/icwesa/proceedings/doc22.htm) infest maize in the region, causing reported yield losses of 20–40% of potential output, depending on the agroecological conditions, crop cultivar, agronomic practices and intensity of infestation (Ampofo, 1986; Seshu Reddy and Sum, 1992). Stemborers are difficult to control, largely because of the nocturnal habits of adult moths, and the cryptic feeding behavior by the larvae in plant stems. The main method of stemborer control that is recommended to farmers is chemical pesticides. However, chemical control of stemborers is uneconomical and impractical for many resource-poor, smallholder farmers.

Parasitic weeds (http://www.bio.vu.nl/vakgroepen/plantecologie/weeds/clmap.html) of the genus *Striga* threaten the lives of over 100 million people in Africa and infest 40% of arable land in the savanna region, causing an annual loss of US$7–13 billion (M'boob, 1989; Lagoke *et al.*, 1991; Musselman *et al.*, 1991). Striga infestation is associated with increased cropping intensity and declining soil fertility. Infestations by weeds of *Striga* spp. have resulted in the abandonment of much arable land by farmers in Africa. The problem is more widespread and serious in areas where both soil fertility and rainfall are low. Striga infestation continues to extend to new areas; another 40% of arable land may become infested in the next 10 years. Recommended control methods to reduce striga infestation include heavy applications of nitrogen fertilizer, crop rotation, use of trap crops and chemical stimulants to abort seed germination, hoeing and hand pulling, herbicide application and the use of resistant or tolerant crop varieties. All of these methods, including the most widely practised hoe weeding, have a serious limitation in the reluctance of farmers to accept them, for both biological and socioeconomic reasons (Lagoke *et al.*, 1991). Unfortunately, the burden of weeding striga, which is a time-consuming and labour-intensive activity, tends to fall on already-overworked African women. Reducing the losses caused by stemborers and striga through improved management strategies could significantly increase maize production, and result in better nutrition and increased purchasing power of many maize producers.

No single method of control has so far provided a solution to both the stemborer and striga problems (Berner *et al.*, 1995).

To put stemborer and striga control within the reach of African farmers, simple and relatively inexpensive measures that are tailored to the diversity of African farming systems need to be developed (Lagoke *et al.*, 1991). A sustainable solution would be an integrated approach that simultaneously addresses both of these major problems.

The 'push–pull' strategy

The 'push–pull', or 'stimulo-deterrent diversionary', strategy to control cereal stemborers was developed by ICIPE in 1997 together with several partner institutions, including KARI (http://www.kari.org), the Kenya Ministry of Agriculture, and the Institute of Arable Crops Research (IACR), Rothamsted, UK (http://www.iacr.bbsrc.ac.uk/iacr/tiacrhome.html). The strategy is based on a holistic approach to understanding and utilizing chemical ecology and agrobiodiversity for stemborer and striga management. The strategy involves the combined use of trap and repellent fodder plants of economic importance, whereby stemborers are repelled from the maize (main) crop and are simultaneously attracted to the trap crop. The fodder plants which are used in a push–pull strategy are Napier grass (*Pennisetum purpureum*) (http://www.blueplanetbiomes.org/elephant_grass.html); Sudan grass (*Sorghum vulgare sudanense*) (http://www.gov.on.ca/OMAFRA/english/crops/facts/98–043.html); molasses grass (*Melinis minutiflora*) (http://www.dpi.qld.gov.au/pastures/4557.html); and silverleaf desmodium (*Desmodium uncinatum*) (http://www.dpi.qld.gov.au/pastures/4493.html). Napier grass and Sudan grass have shown potential for use as trap plants, whereas molasses grass and silverleaf desmodium repel ovipositing stemborers. The trap plants used in the present push–pull strategy have the inherent property of not allowing development of the stemborers once they are trapped (Khan *et al.*, 2000).

The push–pull strategy also attempts to exploit the borers' natural enemies (Khan *et al.*, 1997a,b). Molasses grass, when intercropped with maize, not only reduces infestation of the maize by stemborers, but also increases stemborer parasitism by a natural enemy, *Cotesia sesamiae* (Khan *et al.*, 1997b). Desmodium, when intercropped with maize, not only repels stemborers but also inhibits striga (Khan *et al.*, 2000). Striga control appears to operate through a combination of mechanisms, including abortive germination of seeds that fail to develop and attach on the host.

Benefits of the push–pull strategy

The present push–pull strategy is quite unusual in the way that it has developed from research on the basic science to technology transfer, to farmer take-up and spontaneous technology transfer between farmers (http://plantprotection.org/news/NewsIIIOl.htm#1). By the end of 2001, more than 1000 farmers in six districts in Kenya reported that this approach is effective and results in significant reductions in stemborers and striga infestation and an increase in maize yields (Khan *et al.*, 2000, 2001). The following benefits of the strategy were recorded.

FOOD SECURITY The intercropping feature of the approach is amenable to the mixed farming conditions which are prevalent in eastern and southern Africa. Intercropping or mixed cropping of maize, grasses and fodder legumes has enabled farmers to increase crop yields, thus improving food security (Fig. 34.2). In recent years, this contribution to human nutrition has been particularly important in most of the farming areas around the shores of Lake Victoria. The water hyacinth menace has seriously and negatively affected the traditional fishing industry to the extent that families around the lake no longer have adequate protein in their diets nor a sound income base.

DAIRY AND LIVESTOCK PRODUCTION The push–pull strategy has contributed significantly to increased livestock production (for milk and meat) by providing more fodder and crop residues, especially on small farms where competition for land use is high. For example, in Suba District of Kenya, a

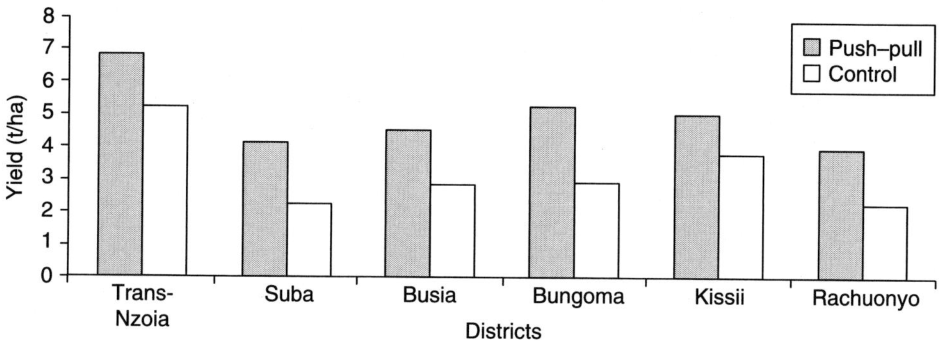

Fig. 34.2. Average increase in maize yields in 'push–pull' fields in six districts of Kenya in 2001. The control is maize monocrop.

milk-deficit region on the shores of Lake Victoria which produces only 6 million liters of milk against an estimated annual demand of 13 million l, the majority of cattle are of the indigenous Zebu breed. In this district, a major constraint to keeping improved dairy cattle for milk production is the inadequacy and seasonality of feed, which is often of low quality in any case. Adoption of the push–pull approach by 150 farmers in this district has resulted in an increase in livestock feed such that the number of dairy cattle being kept by farmers increased from four in 1997 to 370 in December 2001 with a concomitant increase in milk production of 1 million l annually.

SOIL CONSERVATION AND FERTILITY Soil erosion and low fertility are very common problems in eastern and southern Africa. The push–pull technology introduces practices for soil and water conservation that are already familiar to African farmers. For example, Napier grass is already being widely grown in eastern Africa as livestock fodder and for soil conservation. Similarly, the *Desmodium* spp. of nitrogen-fixing legumes have already been introduced into these regions as livestock fodder and for increasing soil fertility.

EXPLOITING BIODIVERSITY The push–pull approach embodies maintenance of species diversity by intercropping with different plants as a means of avoiding the pest problems inevitably encountered with a monoculture system. It is well established that wild host plants on uncultivated land adjacent to crop fields can provide an extremely important refuge for natural enemies, as well as sources of nectar, pollen, and host/alternate prey. The worsening of most pest problems is linked to the expansion of crop monoculture at the expense of natural vegetation. In ICIPE's push–pull program, it was observed that predators, mainly generalists, were significantly more abundant in push–pull fields than in maize monocrop fields. These predators included ants, spiders, earwigs and cockroaches. Other taxa were also recovered, although in relatively lower numbers, including coccinellids, staphylinids, reduviids, nabiids and gryliids. Stemborer populations were conspicuously lower in push–pull than in maize monocrop fields.

SUSTAINABILITY In the push–pull strategy, the full integration of several crop protection approaches such as the use of trap crops and increased parasitism of pests prevents the high selection pressure resulting from the use of a single approach. This works to create a sustainable system by obviating rapid development of resistance/adaptation by pests, a feature common to single-control measures such as the use of a pesticide or genetically based resistance.

PROTECTING FRAGILE ENVIRONMENTS The higher crop yields and improved livestock production resulting from habitat management strategies will support many rural households under existing socioeconomic and agroecological conditions. Thus, there will be less motivation for human migration to fragile environments in search of cultivable land.

INCOME GENERATION AND GENDER EMPOWERMENT Women's contribution to agricultural production in African countries is significant, often reaching 80% or higher. Despite variations across cultural and sociopolitical backgrounds, women make important contributions towards agricultural resource allocation decisions. Indeed, women engage in time-consuming and labor-intensive activities in many maize-based farming systems in Africa. The push–pull approach has contributed considerably towards improving farm incomes (Fig. 34.3) and gender empowerment through sale of farm grain surpluses, fodder and *Desmodium* seed, especially by women farmers and women's groups and by rural youth groups.

Transfer of push–pull technology

More than 100 farmers' groups and individual farmers in Kenya are multiplying forage crops for income generation and control of stemborers and striga weed. This number will increase to 250 in 2003. The Kagera Agricultural and Environmental Management Project (KAEMP) in collaboration with ICIPE has set up 50 push–pull plots in farmers' fields in Kagera, Tanzania. These plots will be increased to 100 in 2002. Parallel programs on development of push–pull strategies are being undertaken by the Agricultural Research Council of South Africa and the National Agricultural Research Organization of Uganda, both in collaboration with ICIPE. Additionally, funds are available to develop push–pull strategies in Ethiopia, Malawi and Mozambique in close collaboration with the national programs of South Africa and an NGO. Other complementary programs presently running are the CGIAR-funded 'Systemwide IPM Project' being conducted by ICIPE in Lambwe Valley (Kenya) on stemborer and striga management, and the BBSRC-funded project on chemical studies on striga suppression being conducted by IACR-Rothamsted. Another major ICIPE project, funded by the Global Environmental Facility of the World Bank to study grass–arthropod associations in Kenya, Ethiopia and Mali is looking into promoting practical applications of grasses and their associated arthropod life in sustainable agriculture (http//www.plantprotection.org/news/NewsJanuary02.html).

Although ICIPE's experience to date has been restricted to maize-based farming systems, the Centre believes that the general habitat management or push–pull approach is applicable to a much wider range of pest problems in a variety of crops (for instance in sorghum), and will be a model for other

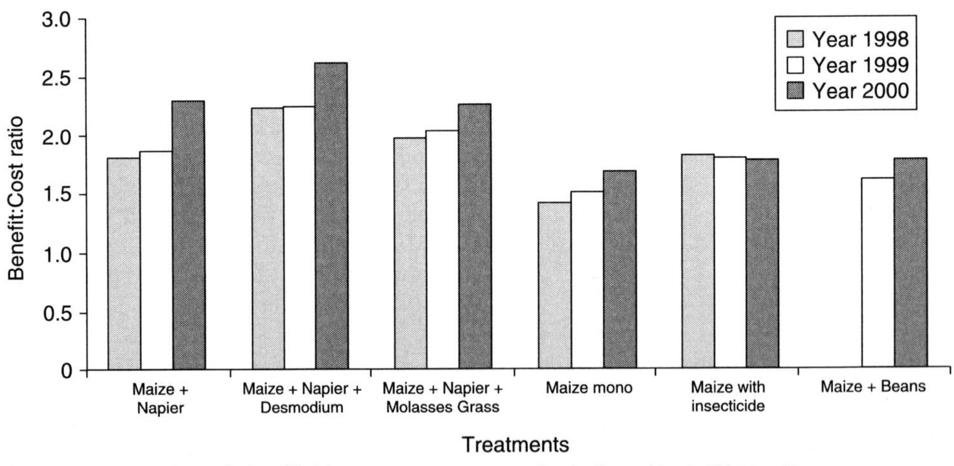

Fig. 34.3. Economic analysis of habitat management strategies in Trans Nzoia District, Kenya.

researchers in their efforts to minimize pest-induced yield losses in an economically and environmentally sustainable manner.

References

Ampofo, J.K.O. (1986) Maize stalk borer (Lepidoptera: Pyralidae) damage and plant resistance. *Environmental Entomology* 15, 1124–1129.

Berner, D.K., Kling, J.G. and Singh, B.B. (1995) *Striga* research and control: a perspective from Africa. *Plant Disease* 79, 652–660.

Cugala, D., Overholt, W.A., Giga, D. and Santos, L. (1999) Performance of *Cotesia sesamiae* and *Cotesia flavipes* (Hymenoptera: Braconidae) as biological control agents against cereal stemborers in Mozambique. *African Crop Science Journal* 7, 497–502.

Degaga, E.G. (2002) Ecological analysis of stem borers and their natural enemies under maize and sorghum based agroecosystems in Ethiopia. PhD thesis, Kenyatta University, Nairobi, Kenya.

Kfir, R. (1997) Competitive displacement of *Busseola fusca* (Lepidoptera: Noctuidae) by *Chilo partellus* (Lepidoptera: Pyralidae). *Annals of the Entomological Society of America* 90, 620–624.

Kfir, R., Overholt, W.A., Khan, Z.R. and Polaszek, A. (2002) Biology and management of economically important lepidopteran cereal stem borers in Africa. *Annual Review of Entomology* 47, 701–731.

Khan, Z.R., Chiliswa, P., Ampong-Nyarko, K., Smart, L.E., Polaszek, A., Wandera, J. and Mulaa, M.A. (1997a) Utilisation of wild gramineous plants for the management of cereal stemborers in Africa. *Insect Science and its Application* 17, 143–150.

Khan, Z.R., Ampong-Nyarko, K., Chiliswa, P., Hassanali, A., Kimani, S., Lwande, W., Overholt, W.A., Pickett, J.A., Smart, L.E., Wadhams, L.J. and Woodcock, C.M. (1997b) Intercropping increases parasitism of pests. *Nature* 388, 631–632.

Khan, Z.R., Pickett, J.A., Van den Berg, J., Wadhams, L.J. and Woodcock, C.M. (2000) Exploiting chemical ecology and species diversity: stem borer and striga control for maize and sorghum in Africa. *Pest Management Science* 56, 957–962.

Khan, Z.R., Pickett, J.A., Wadhams, L.J. and Muyekho, F. (2001) Habitat management strategies for control of cereal stemborers and striga weed in maize-based farming systems in Kenya. *Insect Science and its Application* 21, 375–380.

Lagoke, S.T.O., Parkinson, V. and Agunbiade, R.M. (1991) Parasitic weeds and control methods in Africa. In: Kim, S.K. (ed.) *Combating Striga in Africa.* Proceedings of an International Workshop organized by IITA, ICRISAT and IDRC, 22–24 August, 1988, IITA, Ibadan, Nigeria, pp. 3–14.

Matama-Kauma, T. (2000) Yield losses caused by stem borers on maize and establishment of *Cotesia flavipes* Cameron (Hymenoptera: Braconidae) in Eastern Uganda. MSc thesis, Makerere University, Kampala, Uganda.

M'Boob, S.S. (1989) A regional programme for West and Central Africa. In: Tobson, T.O. and Broad, H.R. (eds) *Striga Improved Management in Africa.* Proceedings of the FAO/OAU All African Government Consultation on Striga Control, 20–24 October, 1988, Maroua, Cameroon, pp. 190–194.

Musselman, L.J., Bhrathalakshmi, Safa, S.B., Knepper, D.A., Mohamed, K.I. and White, C.L. (1991) Recent research on biology of *Striga asiatica, S. gesnerioides* and *S. hermonthica.* In: Kim, S.K. (ed.) *Combating Striga in Africa.* Proceedings of an International Workshop organised by IITA, ICRISAT and IDRC, 22–24 August, 1988, IITA, Ibadan, Nigeria, pp. 31–41.

Ngi-Song, A.J., Overholt, W.A. and Ayertey, J.N. (1995) Suitability of African gramineous stemborers for the development of *Cotesia flavipes* and *Cotesia sesamiae* (Hymenoptera: Braconidae). *Environmental Entomology* 24, 978–984.

Ofomata, V.C., Overholt, W.A., Van Huis, A., Egwuatu, R.I. and Ngi-Song, A.J. (1999) Niche overlap and interspecific association between *Chilo partellus* and *Chilo orichalcociliellus* on the Kenya coast. *Entomologia Experimentalis et Applicata* 93, 141–148.

Ogedah, K.O. (1999) Evaluation of the impact of *Cotesia flavipes* Cameron (Hymenoptera: Braconidae) and indigenous parasitoids on stemborer populations in southwestern Kenya. MSc thesis, Kenyatta University, Nairobi, Kenya.

Omwega, C.O., Kimani, S.W., Overholt, W.A. and Ogol, C.K.P.O. (1995) Evidence of the establishment of *Cotesia flavipes* Cameron (Hymenoptera: Bracondiae) in continental Africa. *Bulletin of Entomological Research* 85, 525–530.

Omwega, C.O., Overholt, W.A., Mbapila, J.C. and Kimani-Njogu, S.W. (1997) Establishment and dispersal of *Cotesia flavipes* Cameron (Hyemenoptera: Braconidae), an exotic endoparasitoid of *Chilo partellus* Swinhoe (Lepidoptera: Pyralidae) in northern Tanzania. *African Entomology* 5, 71–75.

Overholt, W.A. (1998) Biological control. In: Polaszek, A. (ed.) *African Cereal Stem Borers: Economic Importance, Taxonomy, Natural Enemies and Control.* CAB International, Wallingford, UK, pp. 349–362.

Overholt, W.A., Ngi-Song, A.J., Kimani, S.W., Mbapila, J., Lammers, P. and Kioko, E. (1994a) Ecological considerations of the introduction of *Cotesia flavipes* Cameron (Hymenoptera: Braconidae) for biological control of *Chilo partellus* (Swinhoe) (Lepidoptera: Pyralidae) in Africa. *Biocontrol News and Information* 15, 19–24.

Overholt, W.A., Ochieng, J.O., Lammers, P.M. and Ogedah, K. (1994b) Rearing and field release methods for *Cotesia flavipes* Cameron (Hymenoptera: Braconidae), a parasitoid of tropical gramineous stemborers. *Insect Science and its Application* 15, 253–259.

Overholt, W.A., Ngi-Song, A.J., Omwega, C.O., Kimani-Njogu, S.W., Mbapila, J., Sallam, M.N. and Ofomata, V. (1997) A review of the introduction and establishment of *Cotesia flavipes* Cameron (Hymenoptera: Braconidae) in East Africa for biological control of cereal stemborers. *Insect Science and its Application* 17, 19–35.

Polaszek, A. and Walker, A.K. (1991) The *Cotesia flavipes* species complex: parasitoids of cereal stemborers in the tropics. *Redia* 74, 335–341.

Scovgard, H. and Pats, P. (1996) Effects of intercropping on maize stemborers and their natural enemies. *Bulletin of Entomological Research* 86, 599–607.

Seshu Reddy, K.V. (1983) Sorghum stem borers in eastern Africa. *Insect Science and its Application* 4, 3–10.

Seshu Reddy, K.V. and Sum, K.O.S. (1992) Yield-infestation relationship and determination of economic injury level of the stem borer, *Chilo partellus* (Swinhoe) in three varieties of maize, *Zea mays* L. *Maydica* 37, 371–376.

Songa, J.M. (1999) Distribution, importance and management of stemborers (Lepidoptera) in maize production systems of semi-arid eastern Kenya with emphasis on biological control. PhD thesis, Kenyatta University, Nairobi, Kenya.

Songa, J.M., Zhou, G. and Overholt, W.A. (2001) Relationships of stem borer damage and plant physical conditions to maize yield in a semi-arid zone of eastern Kenya. *Insect Science and its Application* 21, 243–249.

Zhou, G. and Overholt, W.A. (2001) Spatial–temporal population dynamics of *Cotesia flavipes* (Hymenoptera: Braconidae) in Kenya. *Environmental Entomology* 30, 869–876.

Zhou, G., Baumgartner, J. and Overholt, W.A. (2001) Impact of an exotic parasitoid on stemborer (Lepidoptera) population dynamics in Kenya. *Ecological Applications* 11, 1554–1562.

Chapter 35

Integrated Pest Management Experiences of CIRAD-France in Developing Countries

A. Ratnadass, X. Mourichon, M. Vaissayre, S. Quilici and J.-P. Deguine
Centre de Coopération Internationale en Recherche pour le Développement (CIRAD), Montpellier, France

Origins and Mission of CIRAD

CIRAD (Centre de Coopération Internationale en Recherche Agronomique pour le Développement or Center for International Cooperation in Agronomic Research for Development) is a French state-owned organization devoted to agricultural research with an emphasis on developing countries. It was formed in 1984 by the mergers of several existing agronomic research institutes. The mission of CIRAD is to aid economic development of countries in the tropics, subtropics and the Mediterranean through research, training and technology transfer. CIRAD is organized into seven departments with a staff of 1800 and an annual turnover of US$150,000,000. Field stations are distributed around the world, allowing CIRAD to address a broad range of agricultural systems and farmers' calendars. Some of the many crops studied under CIRAD include groundnuts, banana and plantain, cocoa, coffee, sugarcane, cotton, fruit, vegetables, horticultural crops, food crops (cereals, pulses, roots and tubers), rubber, oil palm, coconut and forest plantations. CIRAD's philosophy of crop protection is both flexible and realistic. Crop management programs are focused on achieving economically sustainable production for farmers, while minimizing the use of pesticides.

For much of its history, CIRAD has been primarily involved in treating existing pest problems, rather than prevention. More recently, the focus has shifted towards a preventive and ecologically sound approach to pest management. However, CIRAD acknowledges that some disturbance to the environment is a necessary part of agriculture. Many years of experience has shown that IPM alternatives are not always economically feasible in developing countries.

Communication between the field stations and CIRAD headquarters is critical to the development of integrated crop management systems. Field stations transfer management recommendations to growers, who in turn have an active role in selecting appropriate management strategies.

Scientific collaboration

The 100 scientists involved in Crop Protection at CIRAD are represented by 2PI (Scientific Delegation of CIRAD for Crop

(eds K.M. Maredia, D. Dakouo and D. Mota-Sanchez)

Protection), one of the seven trans-departmental and thematic delegations reporting to CIRAD's Scientific Management. Its mandate is scientific coordination and maintenance of high scientific standards which are coherent with the crop protection concept. These scientists, representing the academic fields of entomology/acridology, mycology, virology, bacteriology, weed science, nematology, molecular biology, phytopharmacy and socioeconomics, are posted either in mainland France, French overseas territories or in tropical countries abroad.

Areas of research include plant/pest interactions, population studies, taxonomy, quarantine, and integrated protection (including chemical control and pesticide resistance). The area of plant/pest interactions encompasses host plant resistance to pests, variability, pathogenicity, characterization and detection tools. Population studies include modeling, population dynamics, prediction of pest population changes etc. All specialists on pests, diseases and weeds work closely with scientists from related disciplines (particularly agronomy, sociology, plant breeding, technology and agricultural engineering).

Working partnerships

CIRAD has a history of active involvement in collaborative efforts to develop IPM approaches. CIRAD has developed partnerships with French and foreign universities, national, regional and international research institutes and networks including CILSS, INIBAP, IPM Facility, IPHYTROP[1], AFPP[2], IICA-PROMECAFE[3], international organizations, donors, the private sector and producers' organizations. CIRAD is also the French representative to IPM-Europe.

Case Studies of IPM Programs Developed and Implemented by CIRAD

Cotton IPM

In many tropical countries, cotton is grown by individual farmers on small plots of approximately 0.5–1 ha. Pest control under these conditions relies primarily on chemical pesticides, but CIRAD and NARS have been developing IPM strategies that can be successfully used on small farms. Several alternative control methods for cotton pests have been studied: cultural control, host plant resistance, natural enemies (predators and pathogens) and pheromones. Some of these approaches have proved efficient: early sowing, host plant hairiness, and pest detection with pheromone traps. However, biological control attempts using mass-releases of parasitoids or pathogens have not yet been successful.

CIRAD uses a three-step approach to planning IPM programs for small-scale cotton farmers: (i) prevention of pest damage; (ii) risk evaluation; and (iii) control measures. The prevention of insect damage requires a comprehensive assessment of pest and beneficial insects throughout the cotton cropping system, including identification, biology and population dynamics. Entomologists are involved at all stages of the pest-resistant cotton breeding programs (resistance screening, plant breeding and marker-assisted selection). Another major focus is preventing or delaying of pesticide resistance in the bollworm *Helicoverpa armigera*, the cotton aphid *Aphis gossypii* and the whitefly *Bemisia tabaci*.

Risk evaluation includes characterizing pest biotypes and damage, studying migration patterns and gene flow for key pests, modeling plant–insect relationships, quantifying damage at critical plant growth stages, and laying down economic thresholds.

[1] IPHYTROP is a phytopharmacy and IPM network based at the University of Gembloux, Belgium, http//www.fsagx.ac.be/ca/iphytrop.htm

[2] Association Française de Protection des Plantes.

[3] Programa Cooperativo Regional Para el Desarrollo. Technologico de la Caficultura en Centroamerica Republica Dominicana y Jamaica.

Sampling plans have been developed for key pests (bollworms and sap-sucking insects). Other important research areas include the dynamics of natural enemies and insect pathogens, the effects of pesticides on beneficial insects, and the epidemiology and impact of insect diseases.

Cultural control measures

Cultural practices are the first line of defense against pest damage. Cultural control strategies are geared to ensure that the cotton plant escapes the peak of pest pressure, and reduce opportunities for pests to find food or refuge. Cotton growth can be managed through the choice of sowing dates or practices such as higher plant densities. Intercropping and crop diversity favor the conservation of the natural enemy complex. Cultivation of alternate hosts for the main cotton pests is discouraged. Weed control reduces the number of hosts available as pest refuge. Destruction of crop residues at the end of the cropping season kills a large part of the wintering pest population. This can be achieved either mechanically, or, in small farming systems, by allowing cattle to feed on the green bolls remaining on the cotton plants, and burning the stems afterwards. Careful use of fertilizers is also required to control cotton growth: a late supply of nitrogen results in excessive foliage development, which favors the development of aphids and whiteflies at the end of the season and increases the risk of honeydew contamination.

Host plant resistance

Resistant host cultivars have been identified and are used throughout the cotton-producing countries in West Africa. Resistant traits include either morphological (hairiness, okra leaf, frego bracts, nectariless) or biochemical traits (gossypol and tannin contents). CIRAD recommends the use of 'hairy' cultivars in Africa and Southeast Asia, where leafhoppers are a major problem: cotton plants become less attractive to these insects, making early-season sprays unnecessary and allowing natural enemies to invade the crop. Other traits can be successfully introduced. Either they reduce insect attractancy (nectariless) or they disturb pest population dynamics (okra leaf). Biochemicals such as amino acids, sugars and gossypol affect the reproductive potential of pests.

Introduction of Bt toxins and protease inhibitors in cotton is an important step towards insect growth and development regulation without pesticides, but ecological effects of such an innovation must be carefully studied and managed.

Natural enemies

Inundative releases of natural enemies were first tested on cotton by CIRAD in Madagascar: *Trichogramma* wasps were used for the control of *Helicoverpa armigera* with the goal of delaying the first insecticide spray. An imported strain of *Trichogramma* was mass-produced locally on *Anagasta kuehniella* before release. Although results were disappointing in terms of reduction in pesticide use, similar experiments were conducted in Senegal (1979–1980), Togo and Cameroon (1982–1983) (Bournier, 1979; Couilloud, 1983, unpublished report). The conclusion was that results were not economically acceptable for use on a larger scale.

Insect pathogens

The importance of fungal diseases as methods of controlling insects has been reported in West Africa. After experiments using locally isolated viruses (Chad, Côte d'Ivoire), the possibility of using NPVs from other insect species (*Autographa californica* and *Mamestra brassicae*) has been investigated to reduce production costs. Insect viral diseases satisfactorily control cotton pests, provided adequate dosages (*c.* 10^{13} Pib/ha) are applied in a timely manner. However, field efficiency is negatively affected by factors such as UV and foliar exudates, ingestion-related activity, and a narrow spectrum of activity (species-specific). In several countries, synergism between NPV and a low dosage of pyrethroid insecticide has been demonstrated (Ferron *et al.*, 1983; Montaldo, 1991; Silvie *et al.*, 1993). However, mass production of viruses is difficult in tropical countries and current production costs are limiting to the wider application of microbial pest control.

Chemical attractants and mating disruption

CIRAD collaborated with the French Institut National de la Recherche Agronomique (INRA)'s Semio-Chemicals Laboratory on researching sexual attractants of *H. armigera*, *Cryptophlebia leucotreta*, *Earias* sp. and *Diparopsis watersi*.

Successful use of pheromone formulations for mating disruption in small-scale farming systems has only been observed for the pink bollworm *Pectinophora gossypiella* (Vaissayre, 1987).

Targeted staggered control

Development of sampling plans and action thresholds help farmers to shift from calendar-based spraying programs to more judicious use of pesticides. A new insecticide spraying program, known as 'Lutte Etagée Ciblée', or targeted staggered control (Fig. 35.1) has been developed to reduce production costs and harmful effects of pesticides. It involves calendar-based applications of reduced insecticide dosages along with periodic field scouting. This method was first tested in Cameroon (Deguine *et al.*, 1993). In 1993, 5 years after set-up, one third of the cotton-growing area was treated under this program (Deguine and Ekukole, 1994). In Mali pesticide use was reduced by 40–50% over conventional spraying programs. In 1998, 4 years after being set up in Mali, this new program covered more than 8000 ha (Michel *et al.*, 2000).

IPM of sorghum panicle pests in West Africa

Sorghum is the most important food crop in the savanna areas of West and Central Africa. It can be attacked by a complex of pests, particularly shoot-fly (*Atherigona soccata*), stemborers (*Busseola fusca*), sorghum midge (*Stenodiplosis sorghicola*), panicle-feeding bugs (*Eurystylus oldi*) and storage pests (*Rhyzopertha dominica*). Several preventive measures can be taken to reduce damage by these pests. Infestation by shoot-fly is negligible when sorghum planting is timely (early), synchronized, and dense. These practices also reduce infestation by the sorghum midge, particularly if pure seeds are used, contributing to synchronized flowering.

Using local cultivars

Using local Guinea sorghum cultivars reduces pest damage. High tillering ability,

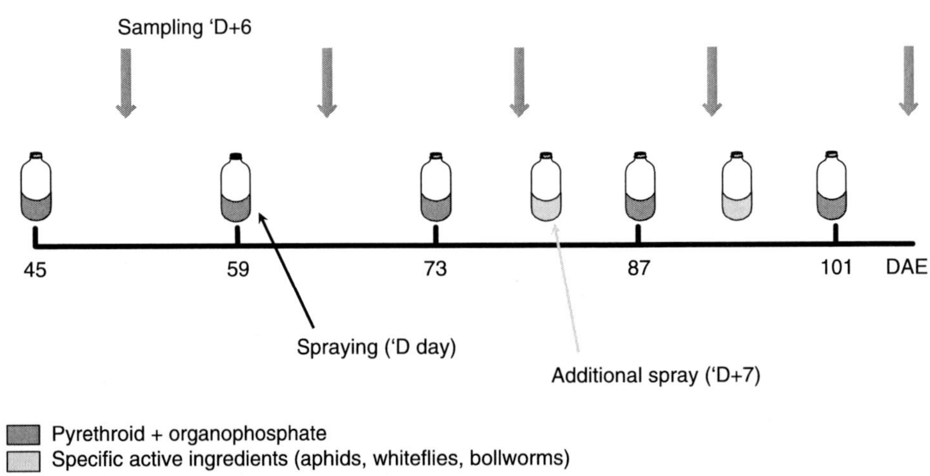

Fig. 35.1. IPM in cotton. Staggered and targeted control. DAE, days after emergence.

which characterizes local cultivars, particularly in response to dead-heart formation, helps the plant to recover from shoot-fly or stemborer attacks. The photoperiod sensitivity which is typical of Guinea cultivars results in uniform, synchronized flowering. In addition, Guineas are tolerant to head-bugs or grain molds in relation to their open panicles, level of grain coverage by glumes, and grain maturing under dry conditions.

Cultural control in field and storage

Management of crop residues and alternate host plants reduces pest populations and has been successfully used against *B. fusca* and *S. sorghicola* in the field. Several methods have been developed to reduce damage by storage pests for subsistence farming in West and Central Africa, particularly in Mali. Such methods include harvesting during the dry season, use of hard-grained cultivars, storage as unthreshed panicles, and using mud-brick or raised woven-grass granaries to prevent dampness.

New market pressures

However, the demand for improved high-yielding cultivars (particularly for short-cycled, non-photoperiod sensitive and compact-panicled caudatum cultivars) is likely to increase. Changes in the rainfall pattern are responsible in part. Market-driven changes in the end uses of sorghum have led to increased cultivation of varieties adapted to specific uses. This includes grain suitable for 'pre-cooked' food preparations for urban consumption, industrial use, brewing/malting, or dual-purpose sorghums with stems suitable for cattle feed. This change in preferred cultivars is likely to result in increased pest problems, in particular with head-bugs, grain molds, storage pests, and midges.

Need for IPM on caudatum cultivars

The increased adoption of caudatum culivars has been observed in parts of Africa, notably in Nigeria (particularly during the ban on imported cereals) and the Kolokani region north of Bamako, Mali. In this context, the mirid panicle-feeding bug *E. oldi*, grain molds, and sorghum midge have recently become major pests, respectively on short-cycled or early-planted cultivars, and on long-cycled or late-planted cultivars. Soft-grained caudatum cultivars are also more susceptible to storage insect pests. Although chemical control is effective, it is neither economical, nor safe for producers, consumers and the environment. CIRAD, in collaboration with ICRISAT and NARS participating in WCASRN, conducted studies from 1989 to 1999 on host plant resistance and other methods to increase sorghum production in a sustainable and environmentally friendly manner.

Host plant resistance

The status of the mirid panicle-feeding bug *E. oldi* as a major threat to the increase of sorghum production through the extension of high-yielding compact-headed varieties was confirmed in Mali. Recent work characterized its damages on various sorghum varieties and described its role as a factor increasing mold infection (Ratnadass *et al.*, 2001a). These studies resulted in the development of reliable screening techniques to identify sources of resistance to *E. oldi*. High, stable resistance in the compact-panicled sorghum cultivar Malisor 84-7 was confirmed (Fliedel *et al.*, 1996). Using pedigree selection, head-bug resistance was transferred from Malisor 84-7 to several advanced progenies such as 87W810, which combine reasonable head-bug tolerance with acceptable agronomic traits and have been field-tested in Mali for several years.

Genetic studies suggested an independent genetic system for head-bug resistance and midge resistance (Ratnadass *et al.*, 2002). Factors conferring resistance to sorghum midge or head-bugs can be brought together in lines that combine short glumes, rapid ovary development and quick endosperm hardening (Fliedel *et al.*, 1996). This was achieved by crossing Malisor 84-7 with ICSV 197. Progenies exhibiting multiple resistance were obtained, one of which is currently being field-tested in Burkina Faso (Ratnadass *et al.*, 2001b, and Table 35.1). A sorghum genetic map based on the cross between Malisor 84-7 and the head-bug susceptible cultivar S 34 was plotted (Deu *et al.*, 2001). Several significant QTLs were detected. This should accelerate the creation of head-bug resistant cultivars using marker-assisted selection.

Management of alternate hosts

Two-year field trials of evaluating the effect of castor bean management through sowing dates and sorghum genotypes on head-bug infestation and damages, demonstrated that castor bean was an alternate host and a significant source of sorghum infestation by head-bugs. This led to the prospect of reducing head-bug damage by management of castor beans (Ratnadass *et al.*, 2001c). IPM strategies based on castor management (by destruction of its spikes before sorghum flowering), using castor as a trap crop, and resistant cultivars are now being tested to determine their potential to perform well under farm conditions.

IPM of Sigatoka leaf spot diseases of banana

Sigatoka leaf spot diseases are the most devastating diseases in most banana producing areas of the world. Yield losses of up to 50% have been reported in cases of severe infection. Two related ascomycetous fungi are responsible for these diseases: *Mycosphaerella fijiensis*, causing black leaf streak disease (BLSD), and *M. musicola* causing Sigatoka disease. BLSD causes severe defoliation on a broad range of banana cultivars, affecting many cultivars that are resistant to Sigatoka disease, and the susceptible varieties' range is still expanding. Effective control can be achieved by fungicide application in commercial plantations. However, frequent sprays, as observed in major banana-producing countries, can result in problems such as resistance to some chemicals that has already been developed in both *M. fijiensis* and *M. musicola* (Romero and Sutton, 1997; Romero, 2000) and other undesirable effects of pesticide use.

CIRAD has focused its research on developing strategies to reduce the number of fungicide applications on commercial plantations; its goal is to develop suitable IPM methods for small farmers. Resultant control programs are based on a combination of cultural practices, host-plant resistance, and reliable forecasting systems for appropriately timing the use of pesticides.

Cultural practices

Cultural control measures reduce the inoculum level in the field. They are critical for limiting disease incidence on small farms and increasing the effectiveness of forecasting systems on commercial plantations. The critical point is to prevent production and/or dissemination of ascospores. Removing spotted leaves or leaf

Table 35.1. Agronomic performance of CIRAD 441 in seven on-farm tests conducted in the Eastern region of Burkina Faso in the 2000 rainy season (Ratnadass *et al.*, 2001b).

Cultivars	No. days to 50% flowering	Plant height (cm)	Grain yield (t/ha)	1000 grain mass (g)	Farmers' desirability score*
CIRAD 441	77 b†	186 c	1.3 a	15 b	1.9 a
BF 94-6/11-1K-1K	72 ab	130 a	0.8 b	15 b	2.4 b
CIRAD 440	68 a	159 b	1.0 ab	16 b	1.9 a
Farmers' local cultivars	86 c	330 d	0.6 b	20 a	2.3 b

*Mean score given by collaborating farmers based on 14 agronomic and grain and fodder quality parameters, on a 1–5 rating scale: 1 = excellent; 2 = good; 3 = acceptable; 4 = poor; 5 = bad.
†Means followed by the same letter in each column are not significantly different at $P = 0.05$ with the Newman–Keuls test.

areas with necrotic tissues reduces the inoculum pressure and interrupts the life cycle before the sexual stage (ascospores) occurs. Other practices such as the use of under-tree or drip irrigation (preferable to sprinkling systems) and plant densities which avoid overlapping of foliage also hinder pathogen development through the reduction of relative humidity inside the crop (Romero, 2000).

Using resistant cultivars

Breeding for resistance to BLSD and Sigatoka disease is relatively difficult. Considering the high genetic diversity within the two *Mycosphaerella* species, priority is given to the use of cultivars with partial resistance (polygenic-durable resistance). Resistance to BLSD is most important, especially for small farmers and for reducing the need for fungicide applications. Several resistant hybrids have been produced by CIRAD and are currently being evaluated under field conditions.

The forecasting system for Sigatoka disease

Forecasting involves the continuous analysis of two categories of data: (i) biological descriptors (field observations), of early symptoms of leaf infection; and (ii) climatic descriptors (evaporation and temperature), which affect the persistence of spray applications. Forecasting for banana leaf spot diseases has been successfully used to reduce the number of treatments (fungicide applications) and keep damage levels below economic threshold levels (Ganry and Laville, 1983; Bureau and Ganry, 1987; de Lapeyre *et al.*, 1997). Properly timed systemic fungicide sprays are both persistent and highly effective. However, fungicides with differing modes of action must be used to reduce the risk of fungicide resistance.

This forecasting system has provided efficient, sustainable control of Sigatoka disease for 25 years in the French West Indies (Guadeloupe and Martinique). The number of pesticide applications has been reduced to six per year since 1973, down from 35–40 applications/year in the 1950s. Ten to 20 treatments/year are applied in other countries (Ecuador, Suriname, Dominican Republic, Jamaica, Windward Islands). This reduction in the number of fungicide applications has effectively reduced control costs as well as adverse environmental effects and pesticide residues on exported fruits. Today, Sigatoka disease is satisfactorily controlled in these areas. The cost of disease control accounts for less than 3% of the total production cost.

Forecasting of BLSD

The same approach is used for the control of *Mycosphaerella fijiensis.* A biological forecasting system (combined with cultural practices) was successfully developed on plantain in Central America (Costa Rica (Lescot *et al.*, 1998), and Panama (Bureau, 1990)) and on commercial banana plantations in West Africa (Côte d'Ivoire) and Central Africa (Cameroon) (Fouré, 1988). This forecasting system makes it possible to control BLSD in commercial plantations with only *c.* 15–20 fungicide applications/year. In comparison, 30–40 treatments/year are applied in most Central American banana-producing countries.

IPM of fruit flies

Fruit flies (Diptera: Tephritidae) are mobile insects with high reproductive capacity. They can cause substantial economic losses, frequently preventing the sale of contaminated fruit from developing countries. The Fruit and Horticultural Crops Department of CIRAD develops IPM strategies to control these insects on the French island of Réunion (Quilici, 1994), in French Guiana (Cayol, 1999), and in New Caledonia.

On Réunion, seven fruit fly species cause considerable damage to fruit and vegetable crops. The Natal fruit fly, *Ceratitis rosa* and the Mediterranean fruit fly, *C. capitata*, are the major pests, with *C. rosa* being particularly harmful because of its widespread distribution on the island (from sea level to 1500 m elevation) and its polyphagy.

Cultivated *Solanaceae* are attacked by the tomato fly *Neoceratitis* (=*Trirhithromyia*) *cyanescens*, and *Cucurbitaceae* by *Bactrocera cucurbitae*, *Dacus ciliatus* and *Dacus demmerezi*. In French Guiana, fruit crops are attacked by *Anastrepha* spp. and by an exotic Asian species, the carambola fly *Bactrocera carambolae*, a rapidly spreading invasive species already found in several South American countries. In New Caledonia, fruits are attacked by several species of the *Bactrocera* genus.

Preventive chemical control has long been used against these flies. On Réunion, conventional chemical control consists of weekly preventive spraying with organophosphates, generally starting 3–7 weeks before the harvest. In addition, a synthetic pyrethroid is sprayed every 7–10 days, 1 week before the start of harvest and throughout the harvest period. Pyrethroids are highly effective, but must be used only during the harvest period because of high toxicity. In recent years, negative impacts on non-target species and high costs (of both pesticide and labor) have led researchers and producers to develop an alternative approach based on action thresholds.

Using action thresholds for Ceratitis *spp. in Réunion*

A chemical control method based on action thresholds has been proposed to control flies in citrus and mango production areas. Dry traps (Nadel type) combining trimedlure, a synthetic substance that selectively attracts *C. rosa* and *C. capitata* males, are used to monitor fly population and to detect when to start spraying (Fig. 35.2). Fly catches are recorded twice a week. Flies attracted into the trap are killed by a strip of insecticide, which remains effective for about 1 month. Traps are small plastic containers with yellow lids and four slot openings. A small dispenser ('Magnet', Agrisense, UK) containing the sexual attractant is placed under the lid of the trap. The attractant is released evenly for about 2 months. Four traps per hectare are installed in orchards by hanging in trees at about head height. Action thresholds vary according to the fruit species and varieties, climatic conditions and the type of trap used. For example, the threshold level is 20 fruit flies/trap/week for citrus and mango orchards in the lowland areas of Réunion (Quilici, 1989).

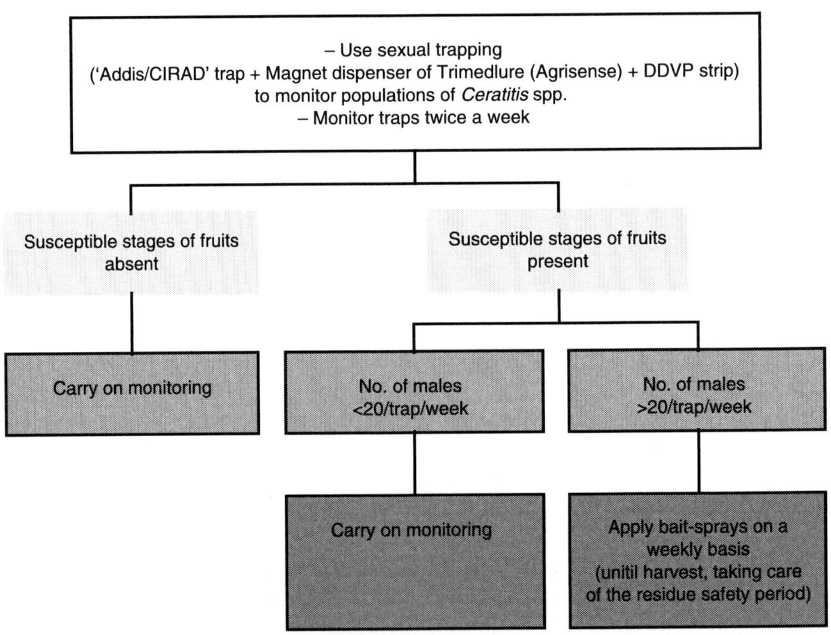

Fig. 35.2. Steps for monitoring of *Ceratitis* spp. populations and bait-spray application.

Bait spraying

On Réunion in 1991, farmers proposed and adopted a 'bait spraying' method of fruit fly control. A mixture of insecticide and food attractant (protein hydrolysate) is applied at the rate of 200 ml/tree to half of the trees/orchard. This strategy is reliable, economical and friendly to beneficial insects and the environment (Table 35.2). It has resulted in a sharp reduction in the amount of insecticide used for fruit fly control. Conventional mechanical and cultural control methods (destruction of fallen fruits and wild hosts near the orchards) remain useful (Quilici, 1993).

Biological control in Réunion and French Guiana

Current research focuses on ways of further reducing insecticide use, through preventive biological control and other biotechnology methods. Preventive biological control can reduce fly populations developing in non-cultivated areas before they migrate to crop areas. As part of this program, the establishment of *Psyttalia fletcheri*, a parasitoid of the melon fly *B. cucurbitae* originating in Hawaii, was successfully achieved in 1997 in Réunion, following the release of 200,000 insects in 1995–1997. Trials aiming at establishing other parasitoids are also under way: *Diachasmimorpha tryoni*, a parasitoid of *C. capitata*, was released in Réunion, and *D. longicaudata*, a parasitoid of *Anastrepha* spp., was released in French Guiana.

Eradication of B. carambolae *in French Guiana*

CIRAD has been involved in the Carambola Fruit Fly Program in French Guiana since 1999. This is a regional program seeking to eradicate *B. carambolae* from Suriname, Guyana, French Guiana and Brazil. The eradication program is based on a combination of Male Annihilation Technique (MAT), bait spraying, and other methods.

Table 35.2. Compared costs of bait-sprays vs. cover-sprays for control of fruit flies on citrus in Réunion Island (2002).

	Cover sprays (tractor with atomizer)	Cover sprays (tractor with manual hose)	Bait-sprays (backpack sprayer)	Bait-sprays (tractor with manual hose)
No. of sprays	5	5	8	8
Spraying duration	1 h	8 h	2 h	0.5 h
Labour cost + mechanised interventions	33.35 + 228.65 = 262 Euros	266.80 + 1829.20 = 2096 Euros	106.72 Euros	26.68 + 182.92 = 209.60 Euros
Products cost	fenthion[a] 167.70 Euros	fenthion[a] 167.70 Euros	protein hydrolysate[b] + lambdacyhalothrin[c] 7.32 + 16.77 = 24.09 Euros	protein hydrolysate[b] + lambdacyhalothrin[c] 7.32 + 16.77 = 24.09 Euros
Traps cost	37.20 Euros	37.20 Euros	37.20 Euros	37.20 Euros
Total cost	466.90 Euros	2300.90 Euros	168.01 Euros	270.89 Euros
Cost per kg of fruits (for a mean yield of 15 t/ha)	0.03 Euros	0.15 Euros	0.01 Euros	0.02 Euros

Note: Costs are calculated for a density of 400 trees/ha. Labour cost is based on the SMIC (minimum salary) at 6.67 Euros/h (salaries exempted from social insurance contribution but including paid holidays). Cost of mechanized interventions is based on the cooperatives (SICA) rate: 45.73 Euros/h. Commercial products and dosages: [a]Lebaycid (Bayer SA) 1 l/ha; [b]Buminal (Bayer SA) 0.8 l/ha; [c]Karaté Vert (Zeneca Sopra) 0.16 l/ha. The Citrus variety considered is Tangor Ortanique. Cost of trapping is based on a density of 2 traps/ha.
Source: D. Vincenot (Chambre d'Agriculture de la Réunion), 2002.

The MAT kills males that are highly sensitive to sexual attractants (e.g. male *B. carambolae* to methyleugenol). It is used in conjunction with bait spraying against *B. carambolae* in French Guiana (Malavasi, 2000). During the program, methyleugenol was found to have no observable effect on non-target insects.

This program was first tested in the Saint-Georges area, near the Brazilian border. Initial trapping studies determined the abundance of *B. carambolae* populations. The eradication project began in February 1999. By December 1999, *B. carambolae* was found only in three specific areas at very low density. The program is gradually being expanded to other areas of French Guiana in three stages: first, surveying to determine the distribution of the fly, then applying control measures, followed by careful monitoring to verify the absence of the fly. The program is currently progressing towards the Cayenne area. The high success rate may eventually result in eradication of this species from the entire region. Along with the *B. carambolae* eradication program, data are gathered on the local fruit flies and their host plant range, population fluctuations and parasitoid complexes (Cayol, 1999).

Other control methods include optimizing the effectiveness of female trapping systems. A female mass-trapping method is being developed for *Ceratitis* spp. Better knowledge of the response of females to visual or olfactory stimuli should result in more species-specific trapping systems, especially for those species for which no attractant is presently known (e.g. tomato fly) (Brevault and Quilici, 1999).

New Caledonia

Most of CIRAD's fruit fly research in the Pacific has been conducted in New Caledonia. From 1994 to 1999, a trapping network with 41 sites was used to monitor the seasonal abundance of the major fruit flies in New Caledonia. Three polyphagous species are of economic and quarantine importance: *Bactrocera tryoni*, *B. psidii* and *B. curvipennis*. Fruit fly populations generally showed strong seasonal variations in most regions. The most important factors influencing seasonal abundance were temperature, rainfall and host fruit availability. For example, low temperatures during the cool season, from June to August, were detrimental to fruit fly reproduction, as was the hot and dry season. In natural or rural habitats, host fruit availability reached a maximum during summer, corresponding to peaks of fly abundance.

Postharvest treatments

Although eradication or control programs are important, treatment of the fruit after harvest is sometimes necessary to allow export and prevent economic loss. A project on postharvest control of fruit flies in New Caledonia was carried out in collaboration with HortResearch (New Zealand). Host ranges and effective postharvest treatments for the major fruit fly species were determined to allow the export of fruit from quarantined areas to New Zealand. For example, the time–mortality response under exposure to hot (44–48°C) water was determined for the egg and larval stages of *B. tryoni*, *B. psidii* and *B. curvipennis* (Sales *et al.*, 1998). These results helped develop two generic heat treatments recently agreed upon by the New Zealand Ministry of Agriculture and Food. Fresh fruits and vegetables that can tolerate those treatments are now exportable to New Zealand. Further research should be conducted on the possibility of using cold treatments. Current programs are now focused on the development of integrated control methods, using the bait-spray technique on various fruit crops.

The AFFI and other collaborative projects

CIRAD collaborates in the AFFI, a program coordinated by the ICIPE in Nairobi (Kenya) with the goal of improving knowledge and control methods of fruit flies of economic importance in Africa. As part of this program, the biology and behavior of the Natal fruit fly *C. rosa* is being studied on Réunion. The CIRAD team in Réunion is also involved in FAO/International Atomic

Energy Agency-coordinated research projects. From 1996 to 1999, the team participated in a program on sexual behavior of *C. capitata*; during this program, comprehensive field cage and video studies were conducted for several *Ceratitis* spp. The team has also collaborated in a study on quality of mass-reared fruit flies. Recently, a program on development of new attractants for fruit flies has begun. The 3-components lure (putrescine, trimethylamine, ammonium acetate) developed by USDA will be tested in various conditions on the local fruit fly species of Réunion.

Technology Transfer

CIRAD and its collaborators worldwide have worked on IPM in a wide range of systems, resulting in a uniquely rich diversity of practical experience. The dissemination of technical knowledge to other international and national organizations is vitally important to the mission of CIRAD, especially to countries far from CIRAD stations. The goal of technology transfer is to allow as many users as possible to benefit from innovations and improvements in IPM strategies.

Educational software

CIRAD has most recently worked with the Crop Protection Service of Réunion to produce AdvenRun – a reference manual and a CD-ROM – providing guidelines to reduce the impact of weeds, which are estimated to cause 25% of all crop losses and account for 30–50% of farmers' time. The package provides assistance with weed identification and suggests effective weed management strategies (Le Bourgeois *et al.*, 2000). AdvenRun is the latest of a series of CD-ROMs dealing with IPM and pest identification. Others include: Adventrop, for identifying crop weeds in the Sudano-Sahelian zone of Africa (including an information database on weeds (Grard *et al.*, 1996; Le Bourgeois *et al.*, 2000)); CotonDoc, multimedia software on cotton insect pests and diseases in Africa (Girardot, 1994); EntoDoc, a bilingual (French–English) encyclopedia on the insect pests of food crops and sugarcane in Africa and the Indian Ocean region (Girardot, 1997); and D-CAS, a guide to sugarcane diseases, including software to aid in diagnosis and treatment of sugarcane diseases (Rott *et al.*, 2000).

Printed materials

CIRAD has printed fact sheets on several topics such as IPM on perennial trees, IPM of locusts and grasshoppers, and tools to evaluate the impact of IPM projects. Fact sheets on all areas of expertise of CIRAD will soon be available on the Web at http://www.cirad.fr. Books are also available for general use. Mariau (1999, 2001) recently summarized the results of 30 years of CIRAD research on IPM of tropical tree crops. A handbook by Michel and Bournier (1997) lists the most common species of beneficial natural enemies of tropical agricultural pests. The practical guide by Vaissayre and Cauquil (2000) enables rapid identification of cotton pests, diseases, and beneficial natural enemies. All CIRAD publications can be seen on the Internet at http://www.cirad.fr/publications/publications.shtml

Conclusion

Knowledge gained by research and experience in IPM has often made it possible to answer specific pest management questions and provide feasible and effective solutions. IPM research is uniquely focused on both gathering knowledge and utilizing it to create effective, sustainable management programs for end users. Continuing improvements will require additional efforts such as:

- the integration of disciplinary knowledge (pests and plant–insect interactions);

- organizing a multidisciplinary approach and integrating knowledge or tools from other disciplines;
- publication of scientific achievements;
- translating these and making information available directly to users.

CIRAD and its partners are continuing to meet these challenges.

References

Bournier, J.P. (1979) Note sur une souche nouvelle de Trichogramme parasitant les oeufs de *Diparopsis watersi* Rothshild au Sénégal. *Congrès sur la Lutte Contre les Insectes en Milieu Tropical*, 13–16 March, 1979, Marseille, France.

Brevault, T. and Quilici, S. (1999) Factors affecting behavioural responses to visual stimuli in the tomato fruit fly, *Neoceratitis cyanescens*. *Physiological Entomology* 24, 333–338.

Bureau, E. (1990) Adaptation of a forecasting system to control black Sigatoka (*Mycosphaerella fijiensis*) in plantain plantations of Panama. *Fruits* 45, 329–338.

Bureau, E. and Ganry, J. (1987) A climatic forecasting system to control banana Sigatoka (*M. musicola*) using sterol-biosynthesis inhibiting fungicides. *Fruits* 42, 199–205.

Cayol, J.P. (1999) *Rapport d'Activités Lutte Contre les Mouches des Fruits*. CIRAD-FLHOR Report (limited distribution), Montpellier, France, 8 pp.

Deguine, J.P. and Ekukole, G. (1994) A new cotton crop protection programme in Cameroon. *Agriculture et Développement* Special Issue, pp. 41–45.

Deguine, J.P., Ekukole, G. and Amiot, E. (1993) Targeted staggered control: a new insecticide programme on cotton in Cameroon. *Coton et Fibres Tropicaux* 48, 112–119.

De Lapeyre de Bellaire, L., Mourichon, X. and Ganry, J. (1997) A forecasting system for regulated control of Sigatoka disease of bananas in Guadeloupe. In: *Crop Protection and Good Quality*. University of Kent, Canterbury, pp. 189–196.

Deu, M., Ratnadass, A., Diabaté, M., Hamada, M.A., Noyer, J.L., Togola-Fane, S. and Chantereau, J. (2001) Quantitative trait loci for head-bug resistance in sorghum. *International Sorghum and Millets Newsletter*.

Ferron, P., Biache, G. and Aspirot, J. (1983) Synergisme entre Baculovirus à polyèdres nucleaires de lépidoptères Noctuidae et doses réduites de pyréthrinoïdes photostables. *Compte Rendus Académie des Sciences, Paris* 296 (III), 511–514.

Fliedel G., Ratnadass, A. and Yajid, M. (1996) Study of some physicochemical characteristics of developing sorghum grains in relation with head bug resistance. In: Dendy, D.A.V. (ed.) *Sorghum and Millets: Proceedings of the Symposia 7 May 1993, Pretoria, South Africa, and 9 June 1996, Porto Carras, Greece*. International Association for Cereal Science and Technology, Vienna, Austria, pp. 46–63.

Fouré, E. (1988) Stratégies de lutte contre la cercosporiose noire des bananiers et des plantains provoquée par *M. fijiensis*. L'avertissement biologique au Cameroun. Evaluation des possibilités d'amélioration. *Fruits* 43, 269–273.

Ganry, J. and Laville, E. (1983) Les cercosporioses du bananier et leurs traitements. Evolution des méthodes de traitement. 1 – traitements fongicides. 2 – avertissement. *Fruits* 38, 3–20.

Girardot, B. (1994) *Coton Doc, système multimédia sur le cotonnier et ses ennemis en Afrique francophone au sud du Sahara*. Colloques Universités francophones – Nouveaux supports. Compact Disc (CD-ROM). CIRAD Aupelf-Uref.

Girardot, B. (1997) *Ento Doc*. (CD-ROM) CIRAD-CA, Montpellier, France.

Grard, P., Le Bourgeois, T. and Merlier, H. (1996) *Adventrop Doc: les Adventices d'Afrique Soudano-sahélienne*. (CD-ROM) CIRAD-CA, Montpellier, France.

Le Bourgeois, T., Jeuffrault, E., Grard, P. and Carrara, A. (2000) *AdvenRun V.1.0. Principales mauvaises herbes de la Réunion*. (CD-ROM) CIRAD/ DAF-SPV, Monpellier, France.

Lescot, T., Simonot, H., Fages, O. and Escalant, J.V. (1998) Developing a biometeorological forecasting system to control leaf spot disease of plantains in Costa Rica. *Fruits* 53, 3–16.

Malavasi, A. (2000) An eradication programme of the Carambola Fruit Fly, *Bactrocera carambolae*, in the north of South America. In: Price, N.S. and Seewooruthun, S.I. (eds) *Proceedings of the Indian Ocean Commission Regional Fruit Fly Symposium, Flic-en-Flac, Mauritius, 5–9th June 2000*. Indian Ocean Commission, pp. 131–134.

Mariau, D. (ed.) (1999) *Integrated Pest Management of Tropical Perennial Crops*. Collection Repères. Science Publishers Montpellier, CIRAD and New Hampshire, 167 pp.

Mariau, D. (ed.) (2001) *Diseases of Tropical Tree Crops*. Collection Repères. CIRAD, Science Publishers, Oxford & IBH, 270 pp.

Michel, B. and Bournier, J.P. (1997) *Beneficials in Tropical Crops*. CIRAD, Montpellier, France, 88 pp.

Michel, B., Togola, M., Tereta, I. and Traoré, N.N. (2000) La lutte contre les ravageurs du cotonnier au Mali: problématique et évolution récente. *Cahiers Agriculture* 9, 109–115.

Montaldo, T. (1991) Microbiological control in cotton crops in North Cameroon: overview of experiments conducted from 1979 to 1988. *Coton et Fibres Tropicales* 46(3), 217–241.

Quilici, S. (1989) Aménagement de la lutte chimique contre les mouches des fruits à la Réunion. Fruit Flies of Economic Importance 87. In: Cavalloro, R. (ed.) *Proceedings of the CEC/IOBC International Symposium 7–10 April 1987*. A.A. Balkema Publishers, Rome, Italy, pp. 515–524.

Quilici, S. (1993) Protection phytosanitaire des agrumes: les ravageurs. In: Grisoni, M. (ed.) *La Culture des Agrumes à la Réunion*. CIRAD/FLHOR, Montpellier, France, pp. 55–89.

Quilici, S. (1994) Research and control programmes on fruit flies in Réunion. *Fruits* 49(5–6), 417–420.

Ratnadass, A., Butler, D.R., Marley, P.S., Ajayi, O., Bandyopadhyay, R., Hamada, M.A., Hess, D.E., Assamoi, F., Atokple, I.D.K., Beyo, J., Cissé, O., Dakouo, D., Diakité, M., Dossou-Yovo, S., Le Diambo, B., Sissoko, I., Vopeyande, M.B. and Akintayo, I. (2001a) Interacting effect of head bugs, molds and climate on sorghum grains in West and Central Africa. In: Akintayo, I. and Sedgo, J. (eds) *Towards Sustainable Sorghum Production, Utilization, and Commercialization in West and Central Africa*. Proceedings of a Technical Workshop of the West and Central Africa Sorghum Research Network, 19–22 April 1999, Lomé, Togo. ICRISAT, WCASRN/ROCARS, Bamako, Mali and ICRISAT Patancheru, Andhra Pradesh, India, pp. 120–140.

Ratnadass, A., Trouche, G., Chantereau, J., Dakouo, D., Hamada, M.A., Fliedel, G. and Luce, C. (2001b) Selection of a sorghum line CIRAD 441 combining productivity and resistance to midge and head-bugs. *International Sorghum and Millets Newsletter*.

Ratnadass, A., Ag Hamada, M., Traoré, S., Cissé, S. and Sidibé, B. (2001c) On-farm development and testing of IPM packages for control of sorghum head-bugs in Mali. *Mededelingen Faculteit Landbouwkundige & Toegepaste Biologische Wetenschappen*, Universiteit Gent 66/2a, 315–324.

Ratnadass, A., Chantereau, J., Coulibaly, M.F. and Cilas, C. (2002) Inheritance of resistance to the panicle-feeding bug *Eurystylus oldi* and the sorghum midge *Stenodiplosis sorghicola* in sorghum. *Euphytica* 123, 131–138.

Romero, R.A. (2000) Control of black leaf streak. In: Jones, D.R. (ed.) *Diseases of Banana, Abaca and Ensete*. CAB International, Wallingford, UK, pp. 72–79.

Romero, R.A. and Sutton, T.B. (1997) Sensitivity of *Mycosphaerella fijiensis*, causal agent of black Sigatoka of banana, to propiconazole. *Phytopathology* 87, 96–100.

Rott, P., Bailey, R.A., Comstock, J.C., Croft, B.J., Saumtally, S. and Krit, H. (2000) *D-CAS 1.2: A Guide to Sugarcane Diseases*. (CD-ROM) CIRAD, Montpellier, France.

Sales, F., Paulaud, D. and Maindonald, J. (1998) Comparison of egg and larval stage mortality of three fruit fly species (Diptera: Tephritidae) after immersion in hot water. In: Allwood, A.J. and Drew, R.A.I. (eds) *Management of Fruit Flies in the Pacific: a Regional Symposium*, 28–31 Oct 1996, Nadi, Fiji. *ACIAR Proceedings* 76, 247–250.

Silvie, P., Le Gall, P. and Sognigbe, B. (1993) Evaluation of a virus-insecticide combination for cotton pest control in Togo. *Crop Protection* 12, 591–596.

Vaissayre, M. (1987) Tentative d'éradication du ver rose *Pectinophora gossypiella* Saunders par la méthode dite de confusion sur la Station de Bouaké, Cote d'Ivoire. *Coton et Fibres tropicales* 42(4), 267–271.

Vaissayre, M. and Cauquil, J. (2000) *Main Pests and Diseases of Cotton in sub-Saharan Africa*. 2nd revised edn, CIRAD/CTA, Montpellier, 64 pp.

Chapter 36
IPMEurope, the European Group for Integrated Pest Management in Development Cooperation: Adding Value to Research Effort

Malcolm Iles
IPMEurope, Natural Resources Institute, Chatham Maritime, Kent, UK

IPMEurope, the European Group for Integrated Pest Management in Development Cooperation, has operated from 1992 to date. The wider goal of IPMEurope is to promote the impact and uptake of European IPM research output to manage pests of field and stored crops and livestock and thus improve livelihoods of the poor in the developing world. IPM provides economically, environmentally and ecologically sound food security through contributing to sustainable agriculture (SA). Actions will capitalize on the comparative advantage of the European resource base, harmonize European Union (EU) development action, promote coherence, and maximize benefits of development cooperation.

Specific objectives of IPMEurope are to:

- strengthen further a European approach to IPM and heighten international recognition of European capacity;
- promote and facilitate appropriate technology and social practices research to support participatory and farmer-oriented components of SA;
- explore the research–development interface and promote feedback to ensure relevance and uptake;
- improve availability and access to IPM-related projects and programs, and thus enhance cooperation;
- build further on the growing collaboration and trust among EU scientists by facilitating debate on technical, social and policy issues and by providing a framework for information exchange and management.

Activities

IPM is an approach to reducing losses through the use of an optimum combination of pest control techniques, and enabling farmers to make management decisions in full awareness of factors operating in their agroecosystems. However, the work of the Group is equally concerned with the importance of the policy dimension and the institutional and capacity-building context within which research and development activities take place.

1. **Guidance** on IPM policy and implementation, including:

- IPM in development cooperation: the role of Europe;

 Integrated Pest Management in the Global Arena (eds K.M. Maredia, D. Dakouo and D. Mota-Sanchez)

- European Policy Framework for IPM in international cooperation; and
- donor guidelines for IPM planning.

2. Information

- IPMEurope Projects Database (IPD) for investment record, analysis and interactive use;
- website providing general information on Group activities, and access through European Initiative for Agricultural Research for Development (EIARD) InfoSys to European and other agriculture and IPM-related information;
- website address: http://www.nri.org. IPMEurope/homepage.htm
- other information dissemination: activity reports and keynote presentations;
- the IPM Information Partnership – formed between many of the principal IPM networks to improve access to IPM information.

3. Task forces (TFs). A facility for European institutions to form TFs around important issues in order to improve understanding or delivery of development. The number and duration of TFs is not fixed. A Task Force will normally be convened by an IPMEurope country member or be supported by IPM/SA interests in that country. Otherwise there are no restrictions on membership. The main selection criteria are:

- relating to a key issue;
- European comparative advantage;
- drawn from an appropriate range of potential stakeholders with ability to contribute; and
- level of commitment.

4. Plenary meetings and workshops, at which key issues are reviewed in consultation with the Group's development partners, and future plans developed.

5. Promoting IPM research and development with regional groups in developing countries.

6. Access to European expertise for teams or individual assignments on project cycle investment consultation.

7. Representation of European interests in IPM and international fora.

Stakeholders

Beneficiaries

The main beneficiaries of the outputs are European institutions and their development partners dealing with IPM and sustainable agriculture research. Benefit to policy makers, researchers, those involved in dissemination and farmers is increasing as this aspect of the Group's work receives further emphasis.

In order to raise and stabilize developing country agricultural productivity in 'traditional' systems and to reduce pesticide use in high external input farming systems, IPM has received growing attention over the past five decades. Until the past decade, progress with adoption at the farm level was slower than expected, while there was unprecedented growth in the use of agrochemicals. The key players involved in agriculture, from those that fund research to environmental groups and increasingly the agrochemical industry itself, are concerned about this situation and its non-sustainability. This concern, recognized in UNCED Agenda 21, spurred the creation of the global IPMForum in 1990 to promote IPM implementation through coordinated international action.

Within this context, and with much of the expertise relating to pest management in the developing world residing within Europe, there was a need for a concerted European effort. With this in view, IPMEurope was set up in 1992 to enable Europe to play a more proactive and influential role on key issues; strengthen the policy, research and development environment; and promote European impact within the international effort.

Research partners

IPMEurope is a network for coordinating European support to IPM in research and development and concerted European effort on IPM policy and implementation in development cooperation. It involves institutions of the European Commission (EC), EU member states, Norway and Switzerland (the associate states) with an interest in promoting integrated pest management in developing countries. The *raison d'être* of the Group is concerted European effort on IPM research, policy and implementation in development cooperation.

The Group consults widely in the course of developing activities with its key development partners in: developing countries, regional fora, civil society, the CGIAR System and other IPM networks. Although some consultation is informal, it also draws on European expertise for consultancies to prepare resource papers for formal workshops, at which these issues are reviewed and future plans developed.

Donors and budgets

The EU member states provide joint support for the participation of their national institutions through contributions in kind and direct support when hosting meetings. GTZ, Netherlands Development Aid and NRI have also provided support for specific activities. Core support has been provided by the European Commission's Directorate General (DG) Research, special program for scientific cooperation with developing countries. Details are given in Table 36.1.

Project Results and Impact

Main results

Pan-European collaborative mechanism developed a project database, partnerships with other key stakeholders, a series of strategic and policy workshops and other advisory services (technical workshops planned). The IPMEurope functions as a sectoral network of the EU and associated states concerned with the promotion of IPM as a means of meeting the policy objectives identified under Agenda 21 and the International Convention on Biodiversity. Both identify IPM as the preferred approach to crop protection. This will be achieved through continued consultation between EU scientists, research partners and donor agencies on policy and technical matters; increased research collaboration; strengthened web-based information flow; further development of EU guidelines and standards on IPM and providing project cycle and program support for sustainable pest management.

Table 36.1. Estimated IPMEurope expenditure, October 1996–October 2002 (core support, contributions in kind).

	Euros	
Details	EC support	Member state contributions
Activities		
Annual plenary meetings and workshops	170,000	500,000
Information (database, website)	130,000	130,000
Studies		
Policy study		
Guidelines study		110,000
Task forces	40,000	40,000
Organization		
Steering committee and planning	80,000	140,000
Secretariat	200,000	150,000
Operating costs	60,000	60,000
Totals	680,000	1,170,000
Total	1,850,000	

Dissemination of results

Results are disseminated as hard copy, through the Internet, as publications and newsletters via mail, poster displays at workshops, and as presentations to key stakeholders. Novel means of dissemination to improve farmers' access to information were sought through the IPM Information Partnership.

Impact of the project

IPM is an interdisciplinary approach to reducing losses through the use of an optimum combination of pest control techniques. It has arisen out of the need to avoid the problems of pest resistance build-up, secondary pest outbreaks, human health problems, the high cost of pesticide control and environmental degradation caused by excessive and inappropriate chemical pesticide use. The approach combines the aims of agricultural productivity, environmental sustainability and cost effectiveness, enabling farmers to make management decisions in full awareness of factors operating their agroecosystems. It is a knowledge-intensive approach.

A key dimension of the work of IPMEurope and other IPM networks has been to stress the important effect non-technical factors have on adoption. Policy, institutional (principally research/ extension/farmer linkages), recognized as equally significant. This offers parallels for other sectors. The IPM philosophy is equally applicable to the crop protection, post-harvest, livestock and forestry sectors. With emphasis on making the best use of local and human resources, IPM encourages, wherever appropriate, the use of natural control mechanisms (such as enhancing the role of pest predators and parasites) and 'traditional' pest management techniques known to farmers. However, the adoption of practical alternatives to chemical methods of control may be constrained by the absence of technical solutions, the lack of resources, or socioeconomic and other factors. Where such constraints are severe, optimal IPM control could include alternative non-chemical control techniques and chemical pesticides.

During 2002, the Group was evaluated by an external team. They concluded that IPMEurope has achieved outputs that are directed to its broad objectives: a working network of IPM specialists; an accessible IPM database; active topical meetings both in Europe and in Africa; policy papers on IPM planning and strategy. Areas suggested for improvement in the future were: increased advocacy in bringing about policy changes; further strengthening of support to developing-country partners; increased promotion of the Group's work; greater management transparency and inclusiveness, and more diversified funding.

The impacts therefore include: widespread increased awareness of the IPM concept; significant increased incorporation of the IPM approach in European projects; several Projects benefiting improved planning of European IPM investment; several initiatives started to improve flow of IPM information to implementers in Africa; significant steps towards policy-level collaboration; significant steps towards field-level partnerships between European institutions and development partners (CGIAR Centers, NGOs, NARS, private sector, universities), and increased use of European expertise.

Partnership

Respective roles of the different stakeholders and coordination mechanisms

Project design

Shaping of Group activities takes place through a regular process centered on its meetings at which progress is reviewed and future plans agreed. The key elements of the structure are the annual plenary meetings, when all available members attend and agree on the program for the coming year. The steering committee and Secretariat, which execute the agreed plans, monitor progress and report to the next annual plenary. The

Group's development partners will participate in these annual meetings. These meetings are associated wherever possible with policy-related and technical workshops in areas where EU scientists have a clear international comparative advantage.

Project implementation

The Group has an advisory position within Europe. Although this is an increasingly important function, it does not implement projects per se.

Project management

Representation on the Group has been determined on a national basis with members of the INCO-DEV (International Committee for Development of the European Commission) Committee nominating institutions as national nodes. Active members have been provided by Belgium, France, Germany, Spain, The Netherlands, Portugal, Greece, Italy, UK and more recently Denmark, Finland, Norway, Sweden, Austria and Switzerland. Other member states have been represented occasionally. National representatives will take responsibility for data collection from and information flow to relevant bodies and individuals within their country. DG Research will nominate an appropriate representative associated with the INCO-DEV program and members of other Directorates General and of the Service Commun Relex will be invited, as at present, to participate.

The activities of IPMEurope have been funded jointly by DG Research and the member states. The UK's NRI, which hosts the Secretariat, undertakes contract management on behalf of the Group, taking responsibility for disbursement of funds associated with agreed activities, maintaining audited accounts and providing annual technical and financial reports. Consistent with IPMEurope policy to initiate activities centrally and decentralizing them to other institutions after inception whenever possible, the IPD is subcontracted to the International Agricultural Centre (IAC) at Wageningen which is responsible for integrating the national datasets within the Web-based Information System for Agricultural Research for Development (WISARD), which allows interactive data management by the national nodes.

The Secretariat is managed by an Executive Secretary, supported by a part-time assistant. Plans are in hand to provide the Secretariat with a Technical Secretary, who will take responsibility for the routine administrative tasks required to ensure efficient and timely operation. The Secretariat is responsible for maintaining the flow of communication between members, largely through electronic means based on a dynamic, and increasingly interactive, website. NRI also provides IT support and inputs from ICT specialists in the design and maintenance of the website hosted on the NRI server, and professional support in accounting and financial matters.

The decision-making body of IPMEurope is the 2-day Annual Plenary Meeting (APM), which insofar as is practicable is held in different member states each year. The APM elects a chairperson and a five person Steering Committee. The Chair assumes overall responsibility for the conduct of the APM and for the operation of the Steering Committee. The Steering Committee (each member 10 days/annum) meet three times a year and is responsible for oversight of the agreed annual program of activities, providing support and advice to the Technical Secretary and other activity leaders. The program of activities is determined by consensus at the APM. Officers of the Group may stand for re-election at the end of each 1-year period of service. The Chair, Secretary or another member of the Steering Committee represents IPMEurope as appropriate in other IPM fora. Wherever possible the management of particular core activities will be devolved to member institutions to encourage ownership and a decentralized mode of operation such that the Secretariat role is primarily one of coordination, facilitation and support.

Result dissemination

The basic functions of the Group comprise the maintenance of the website as a dynamic tool for information exchange, the projects database, a register of European IPM expertise, and a response capacity to access advisory and support activities in the broad field of IPM.

Added value of the partnership

IPMEurope adds value in three key areas.

Tackling challenges faced by developing countries

Developing countries must intensify the production of food and fiber and safeguard health while giving due regard to conserving the ecological base on which sustainable livelihoods depend. Humankind must manage agroecosystems effectively and wisely to minimize losses in crops, livestock and stored products without undue and often damaging effects on functional biodiversity. The causal organisms of such losses and those that transmit human and animal diseases are commonly described as pests; this term is currently used to describe all organisms with adverse effects on human well-being, including insects, nematodes, fungal and viral pathogens, weeds, and vertebrates such as rats and birds.

Enhancing productivity through reducing losses and maintaining the health of the human population is particularly important given the rising pressure on productive land and the trend towards inappropriate use and degradation of marginal areas. The sustainable management of natural resources can only be achieved through an integrated approach to all aspects of the production cycle and the systems that support it. IPM provides a key element of such an approach.

In the past chemical pesticides have provided a relatively simple and undeniably powerful means of controlling pest organisms. However, experience has shown that secondary effects have worsened the impact of existing pests, induced pest status in previously benign organisms and led to unforeseen effects on the underlying ecology of production systems and environmental quality through hazards to human health and well-being. Pesticide-induced crisis is well documented in a range of developing country crops, including cotton, coffee, maize and rice. IPM arose from the need to escape from pesticide dependence and to seek more environmentally benign alternatives.

IPM provides a 'basket of technologies' from which the user can select technical interventions relevant to the specific pest problem to be solved. The technologies may include genetic variation (e.g. resistant plant varieties); exotic biological control agents; enhancing the impact of indigenous pathogens, parasites and predators; cultural control through habitat management; interference with pest behavior; or judicious use of chemical pesticides. IPM is knowledge based and requires that the user is empowered to analyze the constraints operating in his or her system and take appropriate decisions. Making available existing knowledge and developing the capacity of developing country individuals and institutions to adopt such an approach is a powerful contribution to promoting independence and internalizing the processes that lead to sustainability.

The adoption of IPM leads to a new and more relevant basis for the derivation of research priorities in sustainable plant and animal production and will have impact far beyond the issues of protection that it addresses directly. An approach based on rational analysis, knowledge and user empowerment is broadly applicable to many of the challenges faced in both North and South.

Mobilizing the European science and technology community with developing country partners

The Group has already demonstrated considerable achievements in mobilizing the European IPM community and forging new links between individuals and institutions. It has also involved developing-country nationals in its deliberations.

IPMEurope was established in 1992 to promote European pest management research in international fora dominated by multilaterals. Europe has a considerable skill base and direct involvement in IPM-related development activities through donor funding and in project implementation. Improving the coherence of this approach was seen as a way of promoting the European position in sustainable development.

The initial focus of IPMEurope was on research and creating a framework for improved European collaboration through information exchange. During the first phase from 1992 to 1996 an encouraging start was made by:

- intensifying European efforts to promote interest in the IPM concept within the international development community;
- sustaining the costs of participation from the member state's institutions to match the core grant from EC DGXII;
- through the creation of growing confidence in collaboration between the member states' institutions;
- supporting a number of secondments from member institutions to the Secretariat;
- helping to steer the DGVIII-funded study on pesticide use and IPM policy;
- establishing the European Projects Database using the CDS-ISIS software developed for SPAAR;
- Establishing a node of the IPMNet for the exchange of IPM information on the Internet, in collaboration with the USA-based Consortium for International Crop Protection.

During the period 1996–2002, these achievements were extended by:

- establishment of a website carrying information about IPM, IPMEurope and its activities and integrated with international efforts through membership of the IPM Information Partnership with colleagues from the CGIAR, USA and other institutions;
- improvement and expansion of the projects database and its transfer to an interactive web-based format as part of a pilot with InfoSys to explore methods for managing European scientific information across sectors and institutions and providing access to development partners;
- holding two policy workshops with other IPM stakeholders that gave rise to a European Strategy for IPM as a contribution to sustainable development which was later endorsed by the EIARD;
- expanding the remit of the Group from research to cover IPM implementation and initiating dialogue with other Directorates General (DGVIII, DGI, DGXIII) on the potential for IPMEurope to take a wider role in EC development programs;
- cooperating with the DGVIII study on the development of guidelines for pesticide management and IPM for its officers;
- establishing National IPM Fora around the IPMEurope National Nodes to broaden involvement and increase national legitimacy;
- providing European representation to the FAO/World Bank IPM Facility, the CGIAR Systemwide Initiative on IPM and regional IPM meetings in Africa, Asia, the Caribbean and Latin America;
- co-sponsoring a regional meeting in East Africa on IPM information needs;
- preparing the European IPM Guidelines based on the Strategy Paper and drawing on the existing guidelines prepared by individual member states;
- sponsoring a meeting with IPM researchers and implementers in Africa to identify regional needs and assess the relevance of the IPM Guidelines;
- co-sponsoring a regional meeting in Asia to strengthen NGO–GO collaboration on IPM;
- launching the Task Force initiative for Food Safety and Quality, Biopesticides, Advanced Biotechnology, and Farmer Innovation;
- sponsoring a TF workshop 'Sharing responsibility for food quality standards and its implications for small scale producers in developing countries.'

Using research and technological development (RTD) cooperation to support EC development cooperation policy

The range of actions described above has had both direct and indirect effects in support of EC development policy. Enhanced communication, collaboration and information exchange between EU scientists has led to greater common understanding and a more consistent appreciation of the technical nature of IPM and the contribution it can make to sustainable management of renewable natural resources. This strengthens the coherence of action of EU scientists operating under nationally funded development programs and promotes consistency in their dealings with development partners such that European approaches are more readily identifiable. The joint undertaking in preparing a European IPM Strategy has further strengthened this position and a common position has now been accepted at the technical level, which will influence the nature of project proposals submitted to programs such as the forthcoming Framework 6 and individual member states R&D initiatives.

Conclusion

Next steps

There is a demonstrated need for IPM-Europe to continue. It is in a unique position to encourage IPM as a tool for sustainable agricultural development, which should be used to enhance the output of European institutions.

The Task Force facility is a priority mechanism for delivery of IPMEurope outputs and will play an important role in the future of the Group. It is anticipated that the direction taken by TFs will define the nature of the Group.

There is an open invitation to European institutions to form TFs around important issues in which their institutions are involved in improvment of understanding or delivery of development. Members should inform institutions in their country of this facility. They can be of a policy, institutional, social, economic, technical or methodological nature, in which European institutions have a comparative advantage. It is anticipated that at least three will operate at any one time with a start-up rate of one or two per year.

Sustainable agriculture

IPM makes a dual contribution as an approach to sustainable agriculture (through beneficiary empowerment and agroecology management) and development of appropriate pest management technologies. Improvements made by the Group through IPM are sustainable.

Partnership continuation

All Group activities are supported by the member states and the European Commission. The work of the Group is aligned to the broader information (InfoSys) and policy (EIARD) European initiatives, in turn providing European collaboration with a functioning model in a key theme area. The partnerships sponsored by the Group are independent of its functioning. The importance of joint support from a wide range of key stakeholders shows a willingness to see the benefits of the common good beyond the immediate needs of the organization.

IPMEurope continues to redefine its role and activities in response to growing European collaboration and strong international interest in IPM. Harmonization of European policy and implementation effort remains the priority. Current activities will continue; through strengthening links with local, national and regional organizations in the South; capacity development by promoting North–South partnerships; and needs identification and prioritization by agencies implementing IPM in developing countries.

The current structure and approach are expected to continue; through inter-institutional collaboration on specific activities with the Secretariat augmented by secondments from European institutions, and furthering the involvement of European development institutions. The Group continues to serve as a key sectoral network in association with the EIARD. It will work closely with international IPM networks and initiatives that have emerged, which focus on different aspects relating to pest management.

In its longer-term research perspective, the Group's work will continue to emphasize: IPM as an influence pathway, policy research, system research, alternative tools research through participatory technology development, food crops for resource poor groups, and perennial crops.

Additional Resources on IPMEurope

1. IPM in Development Cooperation: The role of Europe. IPMEurope Consultation Workshop, Friedberg, Germany. *IPMEurope Proceedings No. 1.*16–18 June, 1997.

2. *Integrated Pest Management Communications and Information Workshop for Eastern and Southern Africa.* ICIPE, Nairobi, Kenya. 1–6 March 1998. (ICWESA Publication).

3. *Concerted European Policy on IPM in International Cooperation: towards a Policy Framework.* IPMEurope Consultation Workshop, Accademia dei Georgofili, Florence, Italy. 8–9 June, 1998, IPMEurope, Chatham Maritime, UK.

4. *Concerted European Policy on IPM in International Cooperation: A Framework Towards a Strategy.* December 1998, IPMEurope, Chatham Maritime, UK.

5. *IPMEurope IPM Guidelines Workshop.* Biri, Norway. 7–8 June, 1999, IPMEurope, Chatham Maritime, UK.

6. *IPM Planning Meeting for East Africa: Reviewing the IPM Situation in East Africa, and Reviewing and Re-drafting European Guidelines for the Design and Appraisal of IPM projects.* New Arusha Hotel, Arusha, Tanzania. 23–26 January, 2000, IPMEurope, Chatham Maritime, UK.

7. European Group for Integrated Pest Management in Developing Countries. *Phase II Final Report.* 1996–2000. April, 2000, IPMEurope, Chatham Maritime, UK.

8. *Guidelines for IPM Planning for Donors.* For the harmonization of European support to developing countries in the use of IPM to improve agricultural sustainability. December, 2000, IPMEurope, Chatham Maritime, UK.

9. *IPME Annual Report 2000–2001.* June, 2001, IPMEurope, Chatham Maritime, UK.

10. *A Future for IPMEurope, Final Evaluation Report.* April, 2002, IPMEurope, Chatham Maritime, UK.

11. *IPME Annual Report 2001–2002.* June 2002, IPMEurope, Chatham Maritime, UK.

12. *Sharing Responsibility for Food Quality Standards and its Implications for Small Scale Producers in Developing Countries.* Food Quality and Safety Task Force; Workshop Report, Wageningen, The Netherlands, October 2002.

Chapter 37

Building IPM Programs in Central America: Experiences of CATIE

Charles Staver and Falguni Guharay
Regional program CATIE/IPM-AF (NORAD), Managua, Nicaragua

Background for IPM in Central America

Guatemala, Honduras, El Salvador, Nicaragua, Belize, and Costa Rica are traditionally known for exporting coffee, bananas, and sugar. More recently, export agriculture has diversified into vegetables, ornamentals, and tropical fruits. Although the percentage of the population dependent on agriculture has declined from 65% in 1950 to 26% in 2000 (FAOSTAT, 2002), agriculture continues to be an important segment of the economy and food supply.

In Central America, pesticide use became increasingly common in the 1960s, especially in export crops and vegetables (Hilje *et al.*, 2003). Pesticide use escalated in crops such as cotton and rice through the 1970s and 1980s. Low-cost credit programs contributed to the expansion of pesticide application in food grains such as maize and beans, although the increase in pesticide use did not significantly affect pest damage. However, economic changes in the late 1980s increased interest rates and the price of pesticides. At the same time, commodity prices became more volatile. Extreme weather conditions due to El Niño affected farming communities that often faced drought and hurricanes in a single year. These challenges called for a multidimensional approach to improving national IPM programs at the farmer level.

CATIE – a regional center dedicated to agriculture and natural resources

CATIE is a non-profit international association, whose mission is to improve the wellbeing of humanity through the application of scientific research and higher education to development, conservation and sustainable use of natural resources in tropical America. Established in 1942, CATIE is governed by 12 member countries in Central and South America and the Caribbean. CATIE administers programs in research, graduate education and regional outreach focused on improving sustainable management of tropical ecosystems. CATIE is internationally recognized in areas such as plant protection, crop genetics, agroforestry, and natural forest management.

The graduate education program offers Master of Science in ecological agriculture, agroforestry systems, watershed management, tropical natural forest and biodiversity management, and environmental

(eds K.M. Maredia, D. Dakouo and D. Mota-Sanchez)

economics. The PhD program, created in 1996, is a cooperative program with partner universities abroad. Current partners include the University of Gottingen (Germany), Colorado State University (USA), the University of Idaho (USA) and the University of Wales (UK). CATIE works with member countries to conduct regional outreach and development of human resources through donor-funded projects. Use of interactive methods for planning, monitoring and analyzing local needs form the core of CATIE's outreach strategy. CATIE gives priority to small and medium households and recognizes that the pernicious interaction between environmental degradation and poverty can only be tackled through sustainable management practices, human resource and institution strengthening, and appropriate policies.

CATIE Programs

Programs in plant protection

CATIE began work in plant protection in the late 1970s with a farming systems project in El Salvador, Honduras, Nicaragua, and Guatemala. This integrative project brought together scientists from several disciplines to improve crop production. Researchers worked with farmers to develop research priorities and plan experiments. Many of these scientists continued to work at CATIE in other projects in the 1980s.

In the 1980s USAID provided financial support for CATIE to establish a Plant Protection Unit in Central America with expertise in several disciplines of pest management. A Central American IPM Network was funded until 1991. CATIE staff worked with country coordinators in Costa Rica, Honduras, Panama, El Salvador, and Guatemala. Country coordinators carried out field experiments and tested demonstration IPM plots in crops characterized by high pesticide use. The results of this work were published in four IPM guides in vegetables and food crops (CATIE, 1990a,b,c, 1993). Another essential part of the IPM network included education and scientific outreach. Over 100 students completed master's degree training, and thousands participated in short courses. CATIE established the *IPM Journal* in 1986, which has become a leading scientific publication in the area (www.catie.ac.cr/informacion/rmip).

In the past decade, the number of scientists in the plant protection unit has declined, but a core staff in entomology, plant pathology, and biological control has continued to work in cocoa, plantain, food grains, timber trees, vegetables, biopesticides, and management of whiteflies. A whitefly network, established in 1993, is coordinated by CATIE with rotating annual meetings (www.catie.ac.cr/informacion/rmip). This network, operating without ongoing project funding, serves to keep its participants abreast of new developments.

IPM program in Nicaragua

In 1989, two agencies from Norway and Sweden (NORAD, Norwegian Agency for Development Cooperation and SIDA, Swedish International Development Agency) provided financing for CATIE to place a team of IPM specialists in Nicaragua. This project was designed to integrate Nicaragua into the Central American Plant Protection network. The Nicaraguan project, financed for 5 years, proposed to strengthen national capacity in IPM by following a stepwise approach. Crop losses were assessed in year 1, research on IPM components in years 2–4, and transfer of IPM packages in year 5. Key crops included cotton, tomato, and coffee. However, while the project team was still engaged in diagnostic studies and the organization of short courses to strengthen basic research skills, a change in government brought a 2–3 year period of reorganization and attempts to privatize the agricultural research and extension service. During the same period, the Central American plant protection network lost financing.

By early 1992 the project team had concluded that a new approach to IPM development was needed to make IPM practical at the farmer level. Later, NORAD provided continued financing for the CATIE IPM project in Nicaragua. The approach was modified to be more flexible and site-specific. Training programs at each level were coordinated by multi-institutional networks. Early on, the project team experimented with actively involving farmers in IPM development, prioritized crops to use as model systems (Table 37.1), and promoted the formation of multi-institutional and multi-disciplinary working groups.

After the initial phase, interactive training programs geared toward understanding of pest cycles, natural pest control mechanisms, and scouting techniques were held for both farmers and extension workers. The training sessions progressed into developing pest management skills based on an ecological understanding of the crop. The program also began to focus on sustaining small family-run farming operations. The program formed a national advisory committee, with local IPM coordinators meeting at collaborating institutions.

In 1998, CATIE began to take a different approach to financing IPM activities, by focusing on small projects developing skills in planning, monitoring and evaluation with coordinating institutions. CATIE staff and teams of collaborators worked together on year-long projects focusing on multi-institutional coordination, training for extension workers and farm families, and research. New learning tools were continually emphasized, such as discovery exercises, experimentation and testing of alternative practices, and analysis of pest and crop variability based on farmer data. CATIE has financed over 1000 small IPM projects with counterparts in Nicaragua (Table 37.2). CATIE's annual budget for IPM in Nicaragua of US$400,000 in small projects and a similar amount in technical advising is matched by a counterpart contribution in time, transportation, and facilities of over US$1 million.

This program is being tested in three other Central American countries with CATIE staff and country offices. The interactive methods used by the program have been incorporated into a SIDA-financed watershed project, a tomato IPM project in Costa Rica financed by The Netherlands, and a degraded pastures program under design with NORAD. The Nicaraguan

Table 37.1. Model crops used by CATIE's Nicaraguan IPM program.

	Production problems	Pest problems	Pesticide use	Extension learning routine with farmer groups
Coffee	Unproductive plant architecture, shade management, plant nutrition	Berry borer, disease complex, weeds	*Minimal to low* 30–50% of growers make occasional use of pesticides	Yearly cycle with meeting every 2 months
Vegetables	Poor variety selection and seed quality, plant nutrition, water management	Whiteflies or virus, insect pests of fruits, soil diseases, leaf diseases	*High* frequent applications of insecticides and fungicides	Six meetings during 4–5 month crop cycle
Maize and beans	Poor seed quality, low plant population, low value crop	Beans – virus complex, leaf diseases, weeds; maize – insects, weeds, stalk and ear rot; soil insects	*Minimal to low* 50–60% of planted area receives occasional use of pesticides	Five meetings during 4–5 month crop cycle
Plantains/ cooking bananas	Poor seed material	Insect and disease pest problems accumulate and reduce useful life of field	*Minimal to low*	Two or three meetings during year using fields of different ages

program serves as an example of this type of collaborative approach.

The CATIE IPM Program in Nicaragua

The CATIE IPM program in Nicaragua is composed of four key elements:

1. Farmer group-learning based on observation and experimentation at each crop stage.
2. Parallel training in ecology and methods for extension workers.
3. Multi-institutional working groups with a research agenda linked to ecologically sound pest management.
4. Planning and monitoring of national infrastructure and capacity for IPM implementation.

Farmer group learning

Although farmers have an intimate knowledge of their crops and local conditions, they tend to have a weaker understanding of pest life cycles, trophic relationships, and the causes of disease, and as a result often apply ineffective pest management practices. To strengthen their ability for accurate field observations, ecological reasoning, and pest management decision-making, farmers meet at key points in the crop cycle to discuss their experiences (Fig. 37.1 for coffee, but also adaptable for annual crops).

The farmer group-learning program begins before crops are planted. A meeting is held in which farmers discuss their pest management concerns and draw up a plan with extension agents to experiment with IPM methods. Regular meetings are held at

Table 37.2. Participants in CATIE-financed and co-managed training projects in Nicaragua.

	1999	2000	2001
Farmer groups in training	196	393	420
(Participants – % women)	(3750–29%)	(7814–21%)	(8400–30%)
Training processes for extension agents	16	14	12
(Participants – % women)	(196–15%)	(306–17%)	(317–15%)
Trainers of extension agents – % women	66–39%	67–40%	69–40%
Local IPM coordinating groups	5	5	5
(institutions)	(74)	(78)	(75)
IPM coordinating groups by crop/theme	5	7	6
(institutions)	(36)	(36)	(8)
National IPM committee	1	1	1
(institutions)	(6)	(7)	(12)

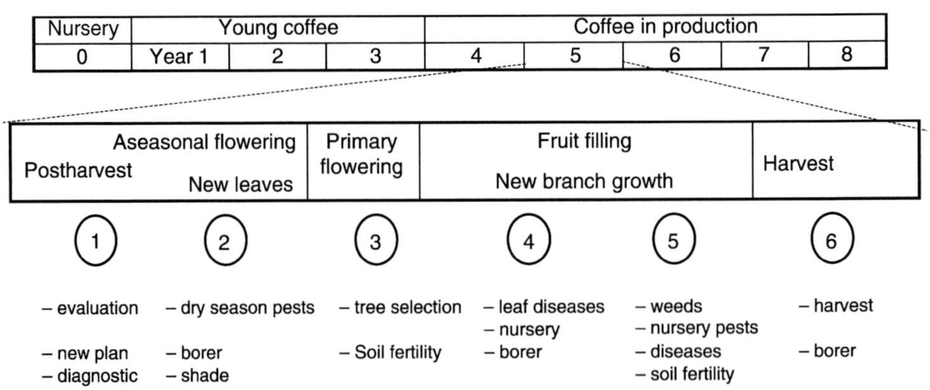

Fig. 37.1. Farmer group-learning and experimentation by crop stage centers on the pests, crop management, and decisions under conditions of weather, food web, and price uncertainty for successive moments in the crop cycle.

each successive stage of the crop where farmers discuss conditions in their fields and review pest management expenditures. They discuss alternatives for pest control measures, and how to render the crop system less favorable for pests and more suitable for natural enemies. Each meeting includes a field exercise to observe and quantify pest problems, crop vigor, and beneficial flora and fauna. After each meeting, farmers are encouraged to share their knowledge with their families and communities. Farmers scout their fields and report the results at the next meeting. They may also conduct simple learning exercises and experiments with alternative management practices in their own fields. At the end of the cycle farmers review crop vigor and pest problems during the crop cycle, analyze the effectiveness of their management decisions, and plan for the next crop cycle.

Extension agent training

Extension agents are typically familiar with a wide range of subjects. However, they may not have enough knowledge of pest biology to explain the basis for IPM strategies or assess specific problems in the field. They are accustomed to organizing short workshops for farmers, but often have little experience in planning a lengthy training process. To develop extension agents' ability to teach long-term management strategies to farmers, extension agents and farmers undergo a parallel training process (Fig. 37.2).

Before the crop is planted, a 2–3-day workshop for extension agents provides a technical and ecological overview of IPM for the crop, an introduction to interactive training methods, and a background in designing small IPM demonstration projects. Between the workshop and the first follow-up session, each extension agent holds a planning session with farmers and designs a small project. Extension agents meet to review their results and plan the next workshop with their farmer group. During each of two to four workshops, extension agents discuss the previous meeting, complete training exercises in the field, and plan their next session with farmers. At the last workshop, the extension workers analyze the overall results of the season's crop management, report the results from work with their farmer group, and develop a proposal for improved farmer training for the following cycle.

This parallel training process for farmer groups and extension agents also serves as a training format for the scientist instructors (Fig. 37.2), who are most accustomed to lecturing with material in an academic setting, rather than hands-on workshops. First, the group meets to build a curriculum based on the primary problems in local pest management in the given crop. Throughout the training process, the instructors strive to teach farmers ecological processes effectively through workshops and interactive exercises. At the close of the season, they report the impact of their training sessions and meet with other instructor groups to exchange results and upgrade the content and methods for the next cycle. In this way they build skills for a discovery-based, problem-solving approach to learning.

Multi-institutional working groups with a research agenda linked to ecologically sound pest management

For farmer and extension agent training to be effective, certain elements must be available to all instructors:

- an understanding of the variability in crop yields and food web dynamics;
- simple methods for scouting and decision making; and
- alternative management practices amenable to farmer resources.

Typically this information is incomplete and scattered. Collaboration between CATIE and other research and teaching institutions has shown that multi-institutional working groups can effectively assemble this information into an organized framework. Such groups meet regularly to

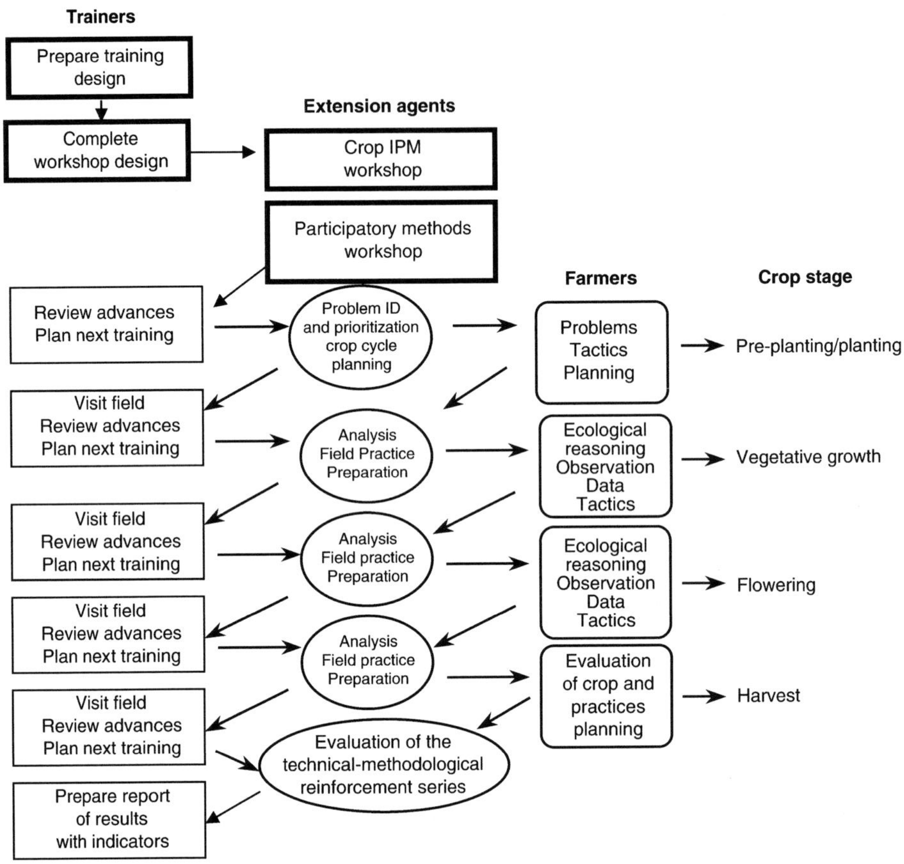

Fig. 37.2. Training by crop stage for farmers, extension agents and trainers ensures that training topics are relevant to field problems and that training participants practice what they learn. At each moment in training participants analyze what they have practiced, learn new elements and plan their next steps. The IPM/Agroforestry Program trains trainers and also collaborates with trainers in training extension agents.

develop a database summarizing the state of IPM knowledge among farmers and extension agents, a training curriculum, a research agenda, and links for scientific information exchange (Fig. 37.3). Each of these elements can be updated regularly with data on pest levels and crop yields reported by farmer groups, studies of training impacts, and research results.

Planning and monitoring of capacity for IPM implementation

Two factors are important for successful integration of IPM practices by farm families – high quality training programs that emphasize farmer learning, and a favorable policy environment (H. Waibel, Hannover, 2002, personal communication). This reflects the experience of the CATIE project in Nicaragua, although policy change has been slow and piecemeal. The focus of the Nicaraguan program is establishing a national infrastructure for IPM implementation. CATIE has worked with several collaborating institutions to develop a practical and effective IPM regime for farmers and disseminate the information throughout the country. This collaborative work has played a crucial role in ongoing improvement in training programs, by linking field-level trainers with institutional leaders (Fig. 37.4). Farmer and extension agent training are coordinated by regional IPM groups,

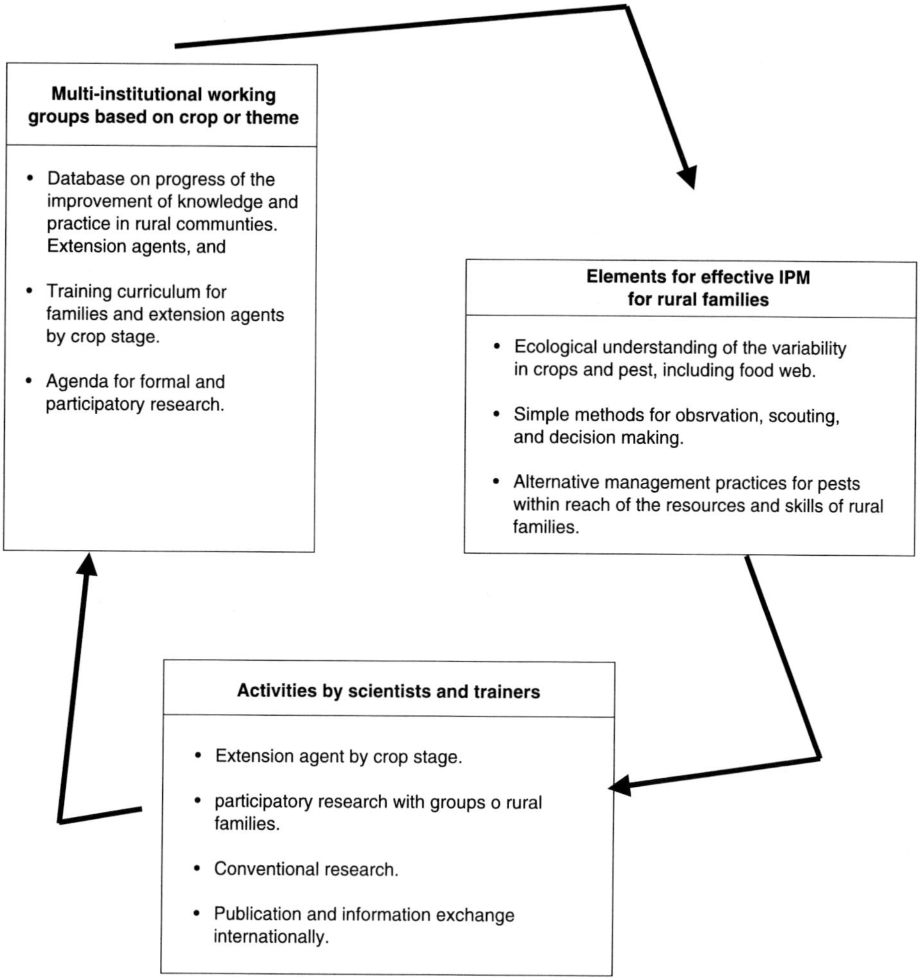

Fig. 37.3. The multi-institutional crop working groups achieve critical elements for effective use of IPM by farm families with group activities which strengthen and integrate individual and small group activities among scientists and trainers.

which are linked to multi-institutional crop working groups. A national IPM committee reports to each institution.

Impact of the CATIE IPM Program in Nicaragua

Measuring the impact of the CATIE IPM program has been the focus of several sustained and continually improving efforts. A DANIDA (Danish International Development Assistance)-financed project placed a high priority on improved institutional monitoring and evaluation. The annual small project routine provides a framework to measure increases in farmer knowledge and practices following training. Farmers complete a simple diagnostic workbook at the beginning and end of the training cycle to record activities carried out, IPM options tried, and results obtained. A visual prototype for testing pest identification and ecological reasoning has also been used with some groups (after Wiegel *et al.*, 2000). During farm visits, extension agents also document farmer activities. Similar

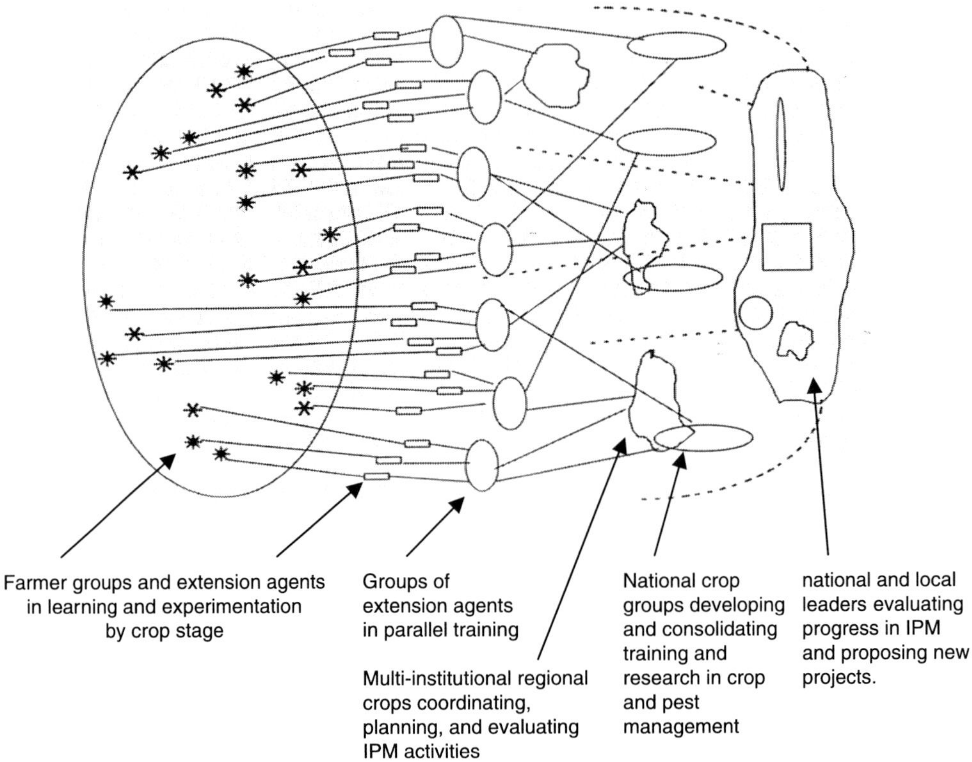

Fig. 37.4. Collaboration among national and local institutions and organizations at several levels is designed to strengthen national IPM capacity. Groups of farm families increasing their pest and crop management ability are the reference point for the system (represented as the * in the figure). The other levels in the system operate to make the work more effective with farm families. This system links decision makers through levels of specialists, trainers, and extension agents to put IPM in the hands of farm families.

procedures are used with extension agents and with scientist instructors. Each participant records their own activities; they are tested on their technical and ecological knowledge; and independent monitoring is used to verify actual practice.

For a deeper understanding of how and why certain training methods are most effective, thematic studies have been used in areas such as farmer perception of technologies and training approaches (Diestch and Kuan, 2002; Paredes, 2002); farmer practices vs. small project reports (van Aalsburg *et al.*, 2002); institutional perception of CATIE IPM working methods (Rodriguez and Meyrat, 2001; Paredes *et al.*, 2002); and the role of gender in training (Schibli, 2001). These primarily qualitative studies have involved participant observation and used a case study approach. A survey of over 1000 farm households in seven municipalities related progress in learning and practice among farm households in IPM training programs vs. households not in training (Dumazert, 2002).

The extent of the Nicaraguan IPM program

CATIE's collaboration with field-based institutions and organizations working in IPM is directed towards improving the quality of existing IPM programs rather than increasing the coverage too rapidly. The promotion of farmer-to-farmer communication contributes somewhat to increasing the reach of the IPM program. In a 2001 study, 22% of farm households reported currently receiving IPM training and technical assistance (Dumazert, 2002). A nationwide study reported 16% of households receiving IPM assistance (Programa Nacional de Transferencia de Tecnología Agropecuaria World Bank/IFAD/COSUDE (Swiss Agency for Development and Cooperation)/Government of Nicaragua). Approximately 21% of small-scale coffee growers, 35% of vegetable growers, and only 0.4% of maize and bean growers have participated in IPM training for more than two seasons. Farmer attendance at in-season training workshops is another issue. Although highly variable from one group to another, on average only about 50–70% of farmers participate in more than half of the farmer meetings during the crop cycle. Earlier studies have showed that farmers attending irregularly were much less likely to try alternative management practices (Wiegel *et al.*, 1997).

Improved farmer ability for pest identification and ecological reasoning

In two training cycles, farmers improved their recognition of most insect pests (Table 37.3). However, recognition levels were still low for less apparent problems such as diseases and soil imbalances. Over half of farmers correctly answered questions relating pest damage to a specific crop stage or to shade levels in the case of coffee pests (Table 37.4). Farmers with better skills tended to identify ecological understanding as important to decision making.

Increase in pest scouting

A study of 1000 farmers indicated that there was no difference in the use of pest scouting by farmers in training and those not in training with only 3.3% using scouting. However, in the small project reports, the percentage of farmers reporting use of pest scouting was much higher (Table 37.5). Farmers were motivated to try scouting during the training program, although they did not always continue afterwards. A follow-up visit to coffee farmers in Aranjuez and San Ramon found positive results:

> [Some of the coffee farmers] have continued to use some IPM practices, despite problems with declining coffee prices. They carry out routine observation of pest and disease levels, although in a much less rigorous way than during the training, and some substitute non-chemical methods for certain practices. The fact that farmers are still using the methods promoted by the Program despite the changes in the market is an important indicator of the effectiveness of the Program's work.
>
> (CABI Bioscience, 2001, p. 10)

Increase in non-chemical pest management practices

The study of 1000 farmers shows that IPM training contributes to increased use of IPM practices, soil conservation, agroforestry (where applicable), and crop diversification (Table 37.6). Specific practices such as

Table 37.3. Progress in the improvement of knowledge among men and women coffee farmers at the end of two cycles of group learning and experimentation 1999–2000 in Nicaragua. Values are per cent of farmers who could identify the pest correctly. (Source: small project final reports IPM/AF Program.)

	1999	2000				1999	2000		
Coffee pests	All farmers	Total *n* = 3665	Men *n* = 3020	Women *n* = 645	Vegetable pests	All farmers	Total *n* = 1682	Men *n* = 1196	Women *n* = 486
Berry borer	71	96	92	89	Whitefly	57	90	91	87
Rust	72	93	94	89	Pepper weevil	6	57	58	56
Brown leaf spot	55	77	79	69	Diamond-back moth	59	52	57	40
Leaf miner	43	54	55	48	Damping off	67	32	36	21
Nematodes	19	22	23	16	Tomato blight	21	62	67	49

Table 37.4. Farmer capacity for ecological reasoning measured by farmer understanding of pest relations to microhabitat conditions. (Source: small project final reports IPM/AF Program.)

	% farmers who identify shade effects on coffee pests 2000–2001		% farmers who can relate crop stages with pest damage 2000–2001	
	Men	Women	Men	Women
0 of 5 answers correct	1	2	4	6
1–2 of 5 answers correct	22	27	39	57
3–5 of 5 answers correct	77	71	56	38

Table 37.5. Farmer use of scouting in coffee, vegetables and food grains after two seasons of training. (Source: small project final reports IPM/AF Program.)

	% farm households carrying out pest scouting activities during 2000–2001		
Pest scouting activities	Coffee	Vegetables	Food-grains
Observe plots regularly	77	82	82
Do integrated pest counts	63	55	63
Do weed sampling	38	–	37
Note down data on a sheet	41	48	51
Teach others to do pest count	25	29	–

Table 37.6. Frequency of use of alternative IPM practices (IPM, soil conservation, crop diversification) among men and women farmers in group learning and experimentation ($n = 936$) and not in groups ($n = 360$) in three regions of Nicaragua (Dumazert, 2002).

Number of practices in use	Farmers in training groups (%)	Farmers not in groups (%)
0	9	32
1	9	18
2	13	13
3	23	18
4	24	13
5	14	5
6	7	1

botanical insecticides were used more frequently by farmers in training, although the increase was from 1% to only 7%. Farmers in training were three times more likely to use IPM practices as farmers not in training. Data from the small projects also indicated an increase in the use of specific IPM practices (Table 37.7). This indicates the impact of sustained participation in the process of learning and experimentation at each crop stage. Farmers may try an IPM method in a small area, on a larger scale test plot or they may apply it widely in their fields. The scale of practice has generally not been measured in any of the studies. Alternative pest management strategies include technical innovations such as covered nurseries, non-chemical soil disinfection, and botanical pesticides, and also manual techniques such as shade management, gleaning of infected coffee fruits, and the use of higher planting density.

A farmer perception study examined the benefits and motivation for IPM implementation. The information was generated through farm visits and interviews of 80 farmers selected by their institution as outstanding collaborators, frequently using from two to 16 different IPM practices. Most commonly used IPM strategies are home-made botanical pesticides, pest scouting, trap crops and organic fertilizer. Farmers were motivated to implement IPM to reduce cost, to discover new pest management methods, to protect their health and the environment, to have more effective pest suppression and to improve the produce quality. Higher yield was claimed by 82% of farmers and 96% reported improved product quality. A total of 73% have planted new crops in their farm over the last 2 years, 87%

had more trees in their farms and 93% consider that their farms now have better value. However, farmers stated that the availability of many IPM options is hindered by lack of financial resources and difficulties in obtaining the biological control agents. They report only about 50% of the farmers in their respective communities are employing some IPM practices.

Reduction in pesticide use

Overall, pesticide use declined after farmer training, depending on the crop and type of pesticide. The reduction in pesticide use was most dramatic in vegetables and food grains, with 20–30% of farmers reporting fewer pesticide applications. A 1997/98 study of five communities showed a sharp decline in current levels of pesticide use (Table 37.8). This was most apparent in food grains (71%), coffee (62%), and vegetables (37%). Of the 80 farmers in the perception study 78% claim to apply fewer synthetic pesticides, and some may have completely stopped pesticide applications. About 50% report less pest damage and 38% report observing more beneficial natural enemies.

Table 37.7. Implementation of non-chemical pest management options by farm households participating in program supported training. (Source: small project final reports IPM/AF Program.)

Crop	Alternative non-chemical pest management practices	% farm households implementing practices	
		1999/2000	2000/01
Coffee	Shade regulation	56	75
	Gleaning of fruits after harvest	55	67
	Removal of berry borer infested berries	61	74
	Sanitary pruning for anthracnose	42	47
	Selective weeding	39	62
Vegetables	Seed bed disinfection with lime	58	59
	Covered nurseries	10	35
	Barrier crops	10	22
	Use of yellow sticky traps	13	35
	Use of homemade botanical pesticides	25	33
	Intercropping	8	14
Food grains	Higher planting density	17	61
	Use of disease tolerant varieties	22	49
	Elimination of host plants of pests	57	61
	Use of homemade botanical pesticides	13	31
	Use of neem	8	22

Table 37.8. Reduction in synthetic pesticide use by farm households participating in program supported training.

	Coffee	Vegetables	Food grains
Pesticide use in 1997–98 (l/mz)[a]	4.5	7.5	2.8
Pesticide use in 2000–01 (l/mz)[b]	1.7	4.7	0.8
Number of participating farm households 2000–01	4,565	2,338	911
Average cropping area (mz)	2	1	2
Total saving/season (US$) 2000–01	383,000	98,000	35,000

[a]Source: study of 90 farm households in 5 communities.
[b]Source: small project final reports IPM/AF Program.
mz, manzana (equivalent to 7000 m^2 or 0.7 ha).

Improved teaching skills of extension agents

During 2000/01, 90% of extension workers used hands-on exercises as a method of learning. About 40–50% have begun to use small demonstration experiments, and 85% use interactive learning methods employing open questions, promoting exchange of experiences and stimulating discussion. About 75% of participating extension workers reported that they have increased their knowledge on the ecology of pests, crops and trees and have improved their capacity for planning, teaching and evaluating farmer group workshops.

Multi-institutional working groups

Over 50 professionals participate in collaborative working groups in coffee, vegetables, plantains and bananas. Most working groups meet regularly (Table 37.9). These professionals are also active in training extension agents. Self-evaluation revealed that while they had strengthened many areas, there are still many areas for improvement (Table 37.10). For example, the coffee program was strongest in ecology and management of coffee and its pests and weakest in addressing gender and family issues. Three measures are currently used to

Table 37.9. Work themes developed by multi-institutional crop or theme groups in Nicaragua. (Original data based on group reports 1999–2001.)

	Work themes established by groups directed towards capacity for field implementation of IPM					
Groups	Representative institutional membership	Regular work routine	Analysis of system capacity	Research agenda field-oriented	Training curriculum field-oriented	Regional and international links
Coffee	X	X	X	X	X	
Vegetable systems	X	X	X	X	X	X
Bananas/plantains	X	X	X	X	X	X
Food grains					X	
Gender and agriculture		X		X		

Table 37.10. Self-evaluation of improvement in knowledge and skills by trainers and scientists in coffee IPM and agroforestry in Nicaragua in collaboration with CATIE. (Original survey data 2001.)

	Coffee		Plantains/bananas	
	Before contact with CATIE IPM	Current state	Before contact with CATIE IPM	Current state
Themes	Rating on scale 1–10 (n = 14)		Rating on scale 1–10 (n = 11)	
Ecology and management of crops and pests	3	6.5	2.9	6.8
Participatory methods	2.8	6.6	3.7	7.5
Project formulation and evaluation	2.7	5.7	3.1	6.5
Gender and family	2.5	5.2	3.5	6.2
Writing of training materials for beginning readers	2.8	5.5	4.3	7.0
Multi-institutional coordination	2	5.7	2.8	6.0

evaluate progress towards greater national capacity for supporting effective IPM programs: (i) consolidation of working mechanisms of multi-institutional working groups; (ii) attitudes of institutional leaders; and (iii) continued development of new projects based on experience. CATIE is continuing to develop its evaluation criteria to reflect institutional and national capacity for IPM implementation (Paredes *et al.*, 2002).

The working groups (Table 37.9) have undertaken many different types of activities. However, these may not be related to their sustainability after the funded small projects come to an end. Since 1999 the five IPM coordinating groups in Nicaragua have coordinated their annual work plans through a joint planning process, organizing activities in work areas and projects. The level of execution of the work plans of different groups has varied. In 2000/01 it ranged from 43% to 97% with an average of 64%. Most of the activities coordinated by the groups have been financed through the CATIE program or Promipac, a COSUDE-financed IPM project. The national IPM committee was recently recognized as an official advisory body by the Ministry of Agriculture, which may create opportunities for an ongoing role in national IPM support.

Based on a survey conducted in 2000, the collaborating institutions have different concepts of IPM: use of biological control (28%); improving crop and pest management decisions (26%); using a combination of pest management practices (24%); and practicing organic agriculture (22%). Most institutions do see the season-long training process as important, rather than as isolated training events. The integration of gender issues is seen in different ways: more women participating in project activities (30%); more equity of access to information and income-sharing (28%); empowerment of women (20%); and planning and implementation of actions based on a better understanding of the different roles of family members (17%). More than 50% of the institutional decision makers express that IPM and agroforestry are important themes for their institutions because they contribute to a sustainable and organic agriculture.

Of the 117 member institutions participating in the five regional groups in Nicaragua, the majority incorporated IPM–agroforestry activities in their 2000/01 work plan, and 19 new projects were designed and financed by different sources for implementation of IPM with farm households (Table 37.11). A study of 23 institutions collaborating with CATIE in IPM in Nicaragua found that institutions take several years to develop their capacity to incorporate ecology, participation, gender–family focus and multi-institutional working procedures satisfactorily into their institutional routine. Longer-term partners formulate most of the new projects based on making IPM accessible for farmers.

Future of the CATIE IPM Program in Nicaragua

Expanded farmer education

A review of the IPM program found that the involvement and support of Nicaraguan

Table 37.11. Institutional uptake of IPM advances as shown in IPM planning framework and formulation of new projects. (Source: 2002 regional planning workshops.)

	Regions of Nicaragua				
	Nueva Segovias	Southern Pacific	Matagalpa Jinotega	Western Plains	South Central
Number of collaborating institutions	25	27	26	14	25
% of institutions with IPM–AF activities in their 2000–2001 work plan	84	60	100	66	31
Number of new IPM–AF projects approved during 2000–2001	2	7	4	5	4

organizations and donors was critical to realizing the potential of CATIE's work on development of agroecological management skills among farmers, extension agents, and scientist instructors. The review also recommended that the program be expanded to include organization, financial management, and marketing skills. The crisis created by low coffee prices had already generated pressure from farmer groups to incorporate these areas of knowledge in training programs. CATIE is working on programs for coffee quality and certification, small-scale capitalization, and financial management. These are in addition to continuing education on ecological processes and IPM strategies.

Improving outreach of the IPM program

Although the Nicaraguan IPM program reaches 22% of farm families in some areas, this percentage is often lower nationwide and in other Central American countries. Improving outreach to the farm sector is critical. CATIE, in its role as a regional organization in support of national and local institutions, prioritizes the development of programs that can be continued following CATIE involvement. The project team in Nicaragua is currently developing printed training materials to communicate training methods to new audiences. Materials on curriculum design, crop-specific guides for extension agents on the farmer group-learning approach, and farmer workbooks are currently underway. Summary guidebooks and videos are produced to communicate the importance of interactive season-long farmer training to institutional policy-makers and planners. The project team also continues to incorporate horizontal communication into the training model. Understanding how farmers, extension agents, and scientist instructors access information, build a personal network, and use their experiences for program improvement is essential to continued growth of the IPM program.

Incorporating agroecological concepts into university and technical school programs of studies

The CATIE IPM project in Nicaragua has invested approximately US$500 in training each extension agent, above and beyond logistical costs. Without major retooling of university and technical school programs throughout Central America, there will be an ongoing need for this expensive on-the-job training. CATIE and collaborators have started to explore approaches to incorporate agroecological concepts and learning approaches into university and high school curricula. Existing networks such as the Central American Education Network will help to promote these programs.

Rural community learning in an academic context

CATIE has a tradition of scientific research and academic training linked with national systems for agricultural research and extension. Also, CATIE has traditionally prioritized small farmers and natural resources. In recent decades, two contradictory forces have emerged in Central American agricultural development – an expansion of NGOs which promote participatory mechanisms, the use of local resources, and sustainable agriculture; and donor-financed private sector projects promoting genetically modified crops. The centralized government system for agricultural research and extension is still important, but is now supplemented by the efforts of other organizations.

The experience gained from field-level CATIE projects such as OLAFO (conservation for sustainable development in Central America project) in community management of natural resources (Ammour and Ramirez, 1999) and TRANSFORMA (technology transfer and formation of technical personnel for sustainable forest management project) in sustainable forest management (Galloway, 1997) have remained largely outside the graduate program teaching curriculum. CATIE academic programs face an important challenge in maintaining the scientific basis for the graduate programs, while incorporating interactive and farmer-driven approaches, cross-disciplinary analytical skills, and an understanding of the interface between social and ecological systems.

References

Ammour, T. and Ramirez, S. (1999) Desarrollo rural basado en el manejo de ecosistemas naturales: experiencias de investigación participativa en los proyectos OLAFO y Manglares. In: Prins, K., Galloway, G., Fasseart, C. and Nilsson, M. (eds) *II Taller de Investigación Participativa Buscando la Convergencia.* CATIE, Turrialba, Costa Rica. pp. 61–67.

CABI Bioscience (2001) *External Monitoring and Review of CATIE IPM/AF (NORAD) Program.* Final report (Contract CS/2000-231). CABI Bioscience, Egham, UK.

CATIE (1990a) Guia para el manejo integrado de plagas del cultivo de repollo. *Serie Técnica – Informe Técnico No. 150.* CATIE, Turrialba, Costa Rica.

CATIE (1990b) Guia para el manejo integrado de plagas del cultivo de tomate. *Serie Técnica – Informe Técnico No. 151.* CATIE, Turrialba, Costa Rica.

CATIE (1990c) Guia para el manejo integrado de plagas del cultivo de maíz. *Serie Técnica – Informe Técnico No. 152.* CATIE, Turrialba, Costa Rica.

CATIE (1993) Guia para el manejo integrado de plagas del cultivo de chile dulce. *Serie Técnica – Informe Técnico No. 201.* CATIE, Turrialba, Costa Rica.

Diestch, L. and Kuan, E. (2002) *Evaluación Campesina Sobre el Uso y la Efectividad de Prácticas MIP y Agrofoestería en Sistemas de Café, Hortalizas y Granos básicos en Nicaragua.* Universidad Centroamericana/CATIE/ PROMIPAC, Managua, Nicaragua.

Dumazert, P. (2002) *Evaluación Cuantitativa del Impacto de los Programas Participativos de MIP y Agroforestería en Café Implementados en Nicaragua por CATIE y Promipac.* CATIE/ MAGFOR/Promipac, Managua, Nicaragua.

FAOSTAT (2002) *FAOSTAT Statistics Database Agriculture.* Internet: http://apps.fao.org/ cgi-bin/nph-db

Galloway, G. (1997) Proyecto regional TRANSFORMA (CATIE/COSUDE): transferencia de tecnología y promoción de la formación profesional en manejo de bosques naturales. In: *Resúmenes de Ponencias, III Congreso Forestal Centroamericano.* San José, Costa Rica, pp. 188–190.

Hilje, L., Araya, C.M. and Valverde, B.E. (2003) Pest management in Mesoamerican agroecosystems. In: Vandermeer, J. (ed.) *Tropical Agroecosystems.* Advances in Agroecology Series. CRC Press, Boca Raton, Florida, pp. 59–93.

Paredes, M. (2002) Estudios de Aprendizaje Amplio – Definición y justificación de la selección de comunidades con niveles de impacto diferentes. *External Monitoring and Evaluation of CATIE IPM-AF (NORAD) Program Phase 2 – Output II.* CABI Bioscience, Egham, UK. (CABI Ref: XB 2980[a])

Paredes, M., Meir, C. and Stonehouse, J. (2002) Wider learning: CABI Consortium for the External Monitoring and Evaluation of the CATIE IPM/AF (NORAD) Program – conceptual Framework. *External Monitoring and Evaluation of CATIE IPM-AF (NORAD) Program Phase 1 – Output IV.* CABI Bioscience, Egham, UK.

Rodriguez, L. and Meyrat, M. (2001) *Incorporación de los Enfoques de Manejo Integrado de Plagas y Agroforestería en Café por Instituciones Colaboradores en Nicaragua.* Report presented to CATIE IPM Program, Managua, Nicaragua.

Schibli, C. (2001) Percepciones de familias productoras sobre el uso y manejo de sistemas agroforestales con café en el norte de Nicaragua. *Agroforestería de las Américas* 8(29), 8–14.

van Aalsburg, C., Kuan, E. and Guharay, F. (2002) Validación del sistema de evaluación participativa con familias productoras en proyectos de capacitación MIP en café en las zonas de Matagalpa y Nueva Segovia. *Sixth National Congress on IPM,* Managua, Nicaragua, 24–26 July, 2002.

Wiegel, J., Padilla, D., Gómez, D., Zeledón, R. and Peralta, A. (1997) Transferencia de ofertas tecnológicas MIP en tomate con metodología participativa en El Barro, Esquipulas. *V Congreso Nacional MIP,* León, Nicaragua, 26–28 November, 1997.

Wiegel, J., Staver, C., Mendez, E., Carcache, M. and Chaput, P. (2000) Una herramienta sencilla para evaluar conocimientos claves de extensionistas y productores/as para el manejo integrado de plagas. *Memoria VIII*

Congreso Latinoamericano y del Caribe de Manejo Integrado de Plagas, Panama, 22–24 November, 2000, p. 108.

Additional CATIE Publications

CATIE (2001) Manejo integrado de plagas en sistemas agroforestales con café en Nicaragua. *Agroforestería en las Americas* 8(29). www.catie.ac.cr/informacion/rafa/

Díaz, J., Guharay, F., Miranda, F., Molina, J., Zamora, M. and Zelendón, R. (1999) Manejo integrado de plagas en el cultivo de Repollo. *CATIE Serie Técnica – Manual Técnico No. 38.* Managua, Nicaragua.

Gómez, D., Prins, C. and Staver, C. (1999) Racionalidad en la toma de decisiones de MIP por pequeños y medianos caficultores de Nicaragua. *Manejo Integrado de Plagas (Costa Rica)* 53, 43–51.

Guharay, F., Monterrey, J., Monterroso, D. and Staver, C. (2000) Manejo integrado de plagas en café. *CATIE Serie Técnica – Manual Técnico No. 44.* Managua, Nicaragua.

Monterrey, J. and Guharay, F. (1997) Proceso investigación-transferencia participativa con comunidades de productores hortícolas: la experiencia del proyecto CATIE/INTA-MIP, Nicaragua. In: *Memoría del Taller Investigación, participación, generación e intercambio de conocimientos con familias campesinas.* CATIE, Turrialba, Costa Rica, pp. 58–65.

Nelson, K. (1996) Estudio comparativo de la generación de tecnología MIP en el cultivo de tomate, Nicaragua. *Manejo Integrado de Plagas (Costa Rica)* 41, 16–28.

Staver, C. (1998) Managing ground cover heterogeneity in perennial crops under trees: from replicated plots to farmer practice. In: Buck, L., Lassoie, J. and Fernandes, E. (eds) *Agroforestry in Sustainable Agricultural Systems.* CRC Press, Boca Raton, Florida, pp. 67–96.

Staver, C., Aguilar, A., Aguilar, V. and Somarriba, S. (1995) Selective weeding for soil conservation in coffee in Nicaragua. *ILEIA Newsletter* 11(3), 22–23.

Staver, C., Guharay, F., Monterroso, D. and Muschler, R. (2001) Designing pest-suppressive multi-strata perennial crop systems: shade-grown coffee in Central America as a case study. *Agroforestry Systems* 53, 151–170.

SIMAS/CATIE (1996) Nicaragua: Experiencias con el MIP en el Proyecto CATIE-INTA/MIP. In: Thrup, L. (ed.) *New Partnerships for Sustainable Agriculture.* World Resources Institute, Washington, DC.

Valenzuela, G. and Prins, K. (2002) Involucramiento de las mujeres en procesos participativos en manejo integrado de plagas en café en Nicaragua. *Manejo Integrado de Plagas (Costa Rica)* 63, 56–63.

Chapter 38

Integrated Pest Management at CAB International

D.R. Dent, M. Holderness and J.G.M. Vos
CABI Bioscience UK Centre, CAB International, Egham, Surrey, UK

Introduction

CABI, an intergovernmental not-for-profit enterprise, has been a leader in sustainable agriculture for 90 years. The issues that currently dominate the world agenda on agriculture are those concerned with: (i) the development and use of both new and established technologies (agbiotechnology, continuing problems with pesticide use); (ii) the need for sustainable production systems; (iii) farmer futures (setting the research agenda, access to markets, fair and global trade, knowledge generation and delivery); (iv) Small to Medium Enterprise development in the sector; (v) estate crops; and (vi) poverty alleviation among rural communities. Alongside all of this, pests still remain one of the major production constraints and have significant direct and indirect effects on produce quality and marketability. Aside from crop losses, the limits set by importing markets on pesticide residues and mycotoxin contamination create trade barriers to small producers, particularly those unable to invest in large-scale capital equipment to ensure quality. The agricultural sector and its many investors are looking for an approach to agriculture where the social, scientific, ethical and profit components can proactively and beneficially co-exist. CABI, through its science division CABI Bioscience, has such an approach in its delivery of IPM.

CABI Bioscience – a Global Reach with Local Impact

CABI includes both CABI Bioscience and CABI Publishing divisions. It has international centers in the UK, Pakistan, Kenya, Switzerland, Malaysia, and Trinidad and Tobago and country offices in Costa Rica, India, Vietnam and China. These centers provide us with a global reach in undertaking projects and initiatives in IPM. Over the last 5 years CABI has worked on IPM projects in Australia, Bangladesh, Bolivia, Canada, China, Costa Rica, Ethiopia, Ghana, India, Indonesia, Italy, Pakistan, the Philippines, Spain, South Africa, Trinidad, Uganda, UK, USA, Venezuela, and Vietnam to name but a few.

CABI's approach is based on the use of scientific knowledge by farmers to advance productivity, rather than production, an essential consideration in markets that are already prone to over-supply.

Knowledge empowers individuals, communities and nations to make effective and informed choices to sustain and develop

 Integrated Pest Management in the Global Arena (eds K.M. Maredia, D. Dakouo and D. Mota-Sanchez)

livelihoods. Knowledge is generated by CABI through research in areas relevant to IPM, while access to knowledge is ensured at a range of levels (from the community to the policy maker) through innovative methods of information compilation and delivery. CABI promotes knowledge use through farmer participatory training and research approaches, so that the research agenda is driven more by the actual demands of small producers and importantly directly engages small producers as key participants in development of appropriate and effective technologies.

Our focus in CABI is actively supporting the generation of biologically based technologies to overcome specific major pest constraints where pests are taken to mean insects, pathogens, weeds and nematodes. Many of the initiatives undertaken by CABI involve pest complexes emphasizing a crop-based approach that includes crops as diverse as coconut, oil palm, bananas, coffee, cocoa, fruit trees, cotton, vegetables (tomatoes, chilis), and brassicas (rape, cabbages, canola).

Global Plant Clinic: Diagnostic and Advisory Services

One of the key aspects of IPM is the ability to identify pest species and to exclude them wherever possible as a preventative measure. Exclusion of pests is one of the first lines of defense whether undertaken at the field level or at national boundary through quarantine sanitary and phytosanitary measures. With an increase in globalization and free trade the necessity for effective quarantine procedures are essential for countries to prevent invasion from imported alien species.

Imported species can become invasive, presenting a major threat to the sustainability of natural systems and agricultural productivity. Pests reduce yields and income through pest management costs. Existing plant health services are often outmoded and under-resourced to face the demands of free trade in a competitive and demanding new environment. National quarantine systems are sometimes poorly equipped to predict and manage the threat posed by alien invasive species. There is a need for sanitary and phytosanitary provisions in trade; a dire need for effective systems – reliable, balanced yet rapid – to support free trade and prevent the introduction of injurious pest species.

Global plant health depends on functioning quarantine systems and world class back-up diagnostic services. CABI pioneers a Global Plant Health Clinic that provides customers with direct diagnostic support for the identification of new problems, and advises on management and containment. The Clinic examines diseased specimens, potential pathogens and related microorganisms and analyzes the cause of ill-health in all crop plants and trees. For example, in a study of the epidemiology of *Phytophthora* diseases in Indonesia the breeding of coconuts for improved varieties represents a national priority. However, one of the improved hybrids PB121 was susceptible to a disease causing bud rot and premature nut fall and now rates as the most significant disease affecting coconut in the country. An understanding of the disease (its genetic structure, mating type, host specificity and distribution) was required to ensure appropriate management practices could be introduced. CABI identified the causal agent of the disease as *Phytophthora palmivora*. The isolates taken from the coconut were shown to be genetically distinct from those *P. palmivora* that affect cocoa. The information has been used to ensure that the next generation of coconut varieties planted in Indonesia are not susceptible to *P. palmivora*.

The Clinic forms the basis for much of CABI's efforts to prevent the transfer of crop diseases in trade and germplasm exchange, an area of increasing significance under the GATT/WTO Sanitary and Phytosanitary provisions for world trade. Correct diagnosis and characterization of the causal agents of crop diseases also provides an essential basis for subsequent research to develop disease management solutions. The establishment of farmer-based systems for pest monitoring and management is also a key focus, addressing both farmer awareness and assurance of Good Agricultural Practice.

Rational Pesticide Use

Conventional chemical pesticides have been used extensively to reduce crop losses to pests but they have posed threats to the environment and to human health. The WHO estimates that there are 25 million cases of acute chemical poisoning in developing countries each year. New chemicals with improved properties are becoming available but are often beyond the means of developing country farmers. Biopesticides are pesticides based on natural microbes such as fungi, bacteria, viruses and nematodes that attack pests and can be developed to control them. They can provide an environmentally friendly alternative to chemical pesticides but they face a number of constraints to their development, manufacture and use, one being a major development cost versus restricted market.

Despite the public concern about pesticide misuse, the total value of world sales has increased by 2.5 times in the last 20 years to US$30 billion (Bateman, 2003). Scientists, practitioners and policy makers involved in IPM have tended to view any activity associated with pesticides as belonging to the pesticide companies and often have been avoided. In turn the chemical companies, which often provide farmers with most of their information on products, are unlikely to develop or promote techniques that reduce pesticide use, although it is in their interests to promote practices that maintain the longer term viability of their products. Thus truly Rational Pesticide Use techniques have been ignored in the 'no-mans land' between the environmental and the agrochemical industry camps. CABI has sought to bridge the gap and provide farmers and practitioners with an alternative, pragmatic approach to pesticide use to:

- improve dose transfer of pesticides to the biological target (e.g. by precise spray application);
- enable better timing of application;
- develop and promote biologically specific and safe products.

Initiatives undertaken in this area include: improved use of chemical applications to fruit trees, tea, coffee, sugarcane and rice crops; preparation of various biopesticide formulations for commercial companies and donor-funded research projects; assessment of sprayer performance, especially droplet sizing; form commercial product development and registration.

Biological Control and Biopesticides

CABI has a long history of supporting and encouraging the use of biologically based agricultural technologies with many decades of experience in the development and use of biological control against pests and in understanding the complex interrelationships of microorganisms and plants in agriculture.

Invasive alien species pose a significant threat to human livelihoods and ecological systems, threatening economic productivity, ecological stability and biodiversity in agricultural systems. Introduced species cause unanticipated havoc and extensive costs. This problem is growing in severity and geographic extent as global trade and travel accelerate. Invasive weeds cause agricultural production losses and degrade water catchments, clog rivers and irrigation systems, while imported pests of livestock, crops and forests reduce yields drastically.

Moniliasis or frosty pod (*Moniliophthora roreri*) of cocoa causes losses of up to 100% in many Latin American countries (Evans *et al.*, 1998). Cocoa is typically produced in smallholdings, and many farmers have abandoned their cocoa because of the impact of this disease. In January 1997, CABI initiated a program in the Huallaga valley of Peru in which cultural disease control in cocoa was combined with biological control using local antagonists, which had been isolated from cocoa farms in the valley. The success of this program in Peru was marked – combining cultural and biological control, losses from moniliasis were reduced from

100% in abandoned plots and 78% in plots with cultural control alone to 36% in plots treated with biocontrol antagonists. The program has now been extended to Costa Rica and Panama.

Biopesticides provide environmentally friendly and safe alternatives to chemical pesticides for the control of insects, weeds, pathogens and nematodes. CABI has a multidisciplinary team who take biopesticides from concept through to commercialization and their use in developing countries. Biopesticides are being developed for control of cattle ticks, sheep scab, storage pests and white grubs. The most notable success has been with the development of the oil formulation mycoinsecticide called Green Muscle™, based on the naturally occurring fungus *Metarhizium anisopliae*. This product has been commercialized and is manufactured in South Africa, is purchased by international donor agencies and is now used against locusts and grasshoppers in Africa (Neethling and Dent, 1998).

The sugarcane froghopper *Aeneolamia saccharina* is a serious constraint to production in Trinidad and Tobago and can reduce sugarcane yields by as much as 30%. The estimated annual costs for froghopper control is approximately US$4.76 million.

An integrated approach utilizing the fungus *Metarhizium anisopliae* as a biopesticide has been successfully introduced by Caroni Ltd. CABI Bioscience staff have made recommendations for improving production and application methods for *M. anisopliae* utilizing new spore separation equipment developed by CABI.

Soil and Seed Health

Production and management of good seed is crucial to food security and agricultural sustainability. In developing countries 90–95% of staple crop seed is farm saved or farmer traded. Quality and health of informal sector seed are greatly neglected. Seed is one of the simplest and most effective means by which innovative crop production technologies can be made available to farmers. CABI, as part of a rice–wheat consortium in a project in Bangladesh examined the effects of conservation tillage practices on the microbial population of soils to determine the implications of changing agronomic practices on system sustainability. The results showed that the microbial diversity was not diminished by resource-conserving technologies and that disease regulating mechanisms and organic matter turnover were maintained in such systems. Application of these approaches has led to 8% mean increase in rice yield in Bangladesh and 13% increase in rice yield in Tanzania.

Potato is an increasingly important staple component of sub-Saharan African diets. Production is dominated by low input smallholder agriculture and rarely achieves attainable yields. Low yield has been attributed to near continuous potato production increasing the incidence of diseases. Capacity and or linkage constraints in certified seed production prevents smallholders planting good seed. In collaboration with KARI and CIP and the farming community in Njabini, Kenya a small-scale seed production system (SSPS) was established. After five seasons of trials, seed–tuber production per unit area of land has been shown to be some two to three times greater under the SSPS regime. In addition, the reduction in land needed for seed–tuber production freed land for production of other crops. The project has been extended to Uganda, South Africa and Bolivia.

Farmer Participatory Training and Research

Farmer participatory approaches are rapidly gaining acceptance as effective and sustainable methods towards developing more ecological crop and pest management strategies. Since the early 1990s, CABI Bioscience has been playing an important catalytic role in the development and support of such programs. The goal of the Farmer Participatory Training and Research initiative is to develop and strengthen farmer participatory approaches to training and research globally in order to increase the knowledge and decision-making skills of farmers.

Highlights of curriculum development efforts to date have been discovery learning manuals for vegetables in general and cabbage specifically, cotton, coffee and cocoa. Pilot IPM implementation projects have been implemented in collaboration with national stakeholders and others on cabbage in the highlands of Luzon in the Philippines, cotton in Asia, coffee in Kenya, rainfed rice in eastern India, and there are ongoing collaborative regional implementation IPM projects in eastern Africa and the Caribbean. FPR is becoming an increasingly important element in the CABI Bioscience portfolio with examples such as FPR on soil-borne vegetables in Vietnam, FPR on rice seed health in Bangladesh, and a case study on FPR in community forestry. Policy development and aid programs are influenced through project development and studies, such as the farmer decision-making study to inform the DFID crop protection program, a study on delivery of biological control products to farmers, and awareness raising through workshops and information products.

Technical Support Group to the Global IPM Facility

CABI Bioscience assisted FAO and others in the establishment of the Global IPM Facility, a multi-donor funded body, hosted by FAO, that was setup to capitalize on lessons learned in farmer participatory IPM in Southeast Asia and enhancing synergies with other actors on a global level. Since the establishment of the Global IPM Facility in 1997, high quality technical support is being provided by CABI Bioscience's Technical Support Group through a partnership program to support development and dissemination of farmer participatory IPM information products, develop and pilot new curricula and methodologies and improve the quality of IPM programs that are operated by government extension, NGOs, national and international research organizations, and food supply chains.

Under the Technical Support Group, over 60 IPM information products have been developed and disseminated to IPM practitioners, extension staff, scientists, donors, policymakers, etc. Support is given to the development of FPR in countries not introduced to the approach. An example of the support to the development of end-user linkages with delivery of biologically based products is the work to promote and utilize traditional knowledge on the use of local biodiversity in Vietnam, where the weaver ant is being used in traditional citrus orchards for pest management. Technical and methodological support in general is given through dissemination of information products and, if needed, consultancy visits to help in development of programs and activities.

Commodities

Smallholder producers face particular difficulties in producing and marketing commodity crops. Returns from globally traded commodities are subject to price fluctuations, quality controls and market forces beyond the control of small producers and in which they are considerably disadvantaged; global prices and local economic returns dictate the viability of production. CABI is applying its knowledge resources to specifically understand and support the needs of small producers in this competitive sphere and thus to support economic and social development in producer countries. CABI's core skills in knowledge generation, access and use are now engaged in the wider frame of support to grower and civil society organizations. This help will ensure smallholders are armed with the knowledge required to make informed and effective choices in production and trade. CABI has made significant inputs in this way to a number of commodity crops systems and communities worldwide, particularly coffee, cocoa, oil palm, bananas, cotton and sugarcane.

CABI has a major Coffee Commodities Program that works with the International

Coffee Organization, the International Fund for Commodities, fair trade organizations, national coffee research institutes and multinational coffee houses. Work carried out by CABI in Colombia for instance, involving extensive economic and anthropological studies of how farmers control the coffee berry borer, the costs and their attitudes led to the introduction of a new parasitoid *Phymastichus coffea* as a biological control agent. The parasitoid was successfully introduced and has become established.

Banana diseases including wilts, leaf spots and parasitic nematodes have been found to be major constraints to the production of both indigenous and exotic bananas in Uganda and a key contributor to the recent decline in production. CABI have assisted UNBRP with evaluation of new technologies to manage banana diseases including: selection and use of host plant resistance, improved use of organic fertilizers and related cultural treatments to improve plant vigor, use of clean planting material and use of break crops. A total of 128 farmers in 24 villages are participating in on-farm trials. Early impact of the trials is encouraging in terms of the excellent performance of the cultivars and the positive response of the farmers. This has been confirmed by considerable new demand for planting material of the hybrids tested and a shift in production to bananas back from annual crops.

Agbiotechnology

Agricultural biotechnology is relatively new and surrounded by confusion, misunderstanding and hyperbole. CABI Bioscience acknowledges these problems, the pitfalls and the gaps in knowledge. We acknowledge the sense of the unknown and the ethical issues, but we also acknowledge the potential of the technology.

Crop-related GM biotechnologies are just one of a set of options open to farmers to improve the sustainability of cropping systems. CABI tests agronomic merit on a case by case basis, looking at soil type, climate factors, yield, quality, farmer ergonomics, consumer preferences and fiscal margins. CABI evaluates disease or herbicide resistance and pesticide management regimes. Assessing environmental impact of transgenic crops is also of great importance. CABI programs conduct, advise and direct independent research on biodiversity impact of crop-related GM technologies, for example the long term effect on soil fertility, gene flow, pollen contamination, effects on non-target organisms and the development of pest and pathogen resistance.

CABI Bioscience in collaboration with partners, is assessing the suitability of cotton plants genetically modified to produce the insecticidal toxin Cry1Ac for control of fruit-feeding caterpillars in smallholder farming systems in China. CABI is involved in assessing the impact of the Bt cotton on non target organisms, working proactively on the likely evolution of resistance of Bt transgenic cotton to the key pest *Helicoverpa armigera* and providing all the relevant information necessary in recommending the deployment and management of transgenic cotton. CABI's intergovernmental status and its mandate from member governments allows for the provision of independent and impartial information relating to the issues of GM technologies as a service for decision makers within governments.

Information and Publications

Facilitating pest management information access and use in developing countries is being addressed through a range of global mechanisms. The Crop Protection Compendium developed by an international multi-stakeholder consortium of over 40 organizations led by CABI, has provided a powerful model of how information technologies and global scientific knowledge can be combined to create an encyclopedic resource providing rapid access to knowledge of a very wide range of pests (see www.cabicompendium.org/cpc). This core knowledgebase is being used to generate tools to assist decision making for policy makers, farmer advisors and those concerned with the spread of pests in trade. CABI Publishing has developed new Internet gateways to establish knowledgebases, for example for Integrated Crop Management (ICM Focus, see www.icmfocus.com). This gateway brings together a wide range of information through a common access point, allowing cross linkage between different types and sources of information. CABI produces an extensive range of publications relating to IPM including the CAB ABSTRACTS database (in printed abstracts journals, on-line and on CD), books, primary journals, other CD-ROMS, and informal publications, all aimed at addressing 'knowledge gaps' that constrain the uptake of beneficial approaches and technologies.

Conclusion

CABI Bioscience has a global base and has a knowledge for development agenda, which is strongly focused on field-based IPM and farmer participatory training and knowledge generation and transfer. As such, CABI will continue to play an innovative and significant role in IPM globally as a strategic ally of other global leaders, such as the Global IPM Facility and the CGIAR SP-IPM.

References

Bateman, R.P. (2003) Rational Pesticide Use: spatially and temporally targeted application of specific products. In: Wilson, M. (ed.) *Optimising Pesticide Use.* John Wiley & Sons, Chichester, UK.

Evans, H.C., Krauss, U., Rios, R.R., Zecevich, T.A. and Arevalo-Gardini, E. (1998) Cocoa in Peru. *Cocoa Grower's Bulletin* 51, 7–22.

Neethling, D.C. and Dent, D.R. (1998) *Metarhizium anisopliae*, isolate IMI 330189; a myco-insecticide for locust and grasshopper control. *BCPC Conference Proceedings* 16–19 November 1998, pp. 37–42.

Chapter 39

Making IPM Successful Globally: Research, Policy, Management and Networking Recommendations

Karim M. Maredia[1], Dona Dakouo[2], David Mota-Sanchez[3] and K.V. Raman[4]

[1]Institute of International Agriculture, Michigan State University, East Lansing, Michigan, USA; [2]INERA, Bobo-Dioulasso, Burkina Faso; [3]Center for Integrated Plant Systems, Michigan State University, East Lansing, Michigan, USA; [4]Department of Plant Breeding, Cornell University, Ithaca, New York, USA

IPM strategies have been successfully implemented in a few countries and regions of the world. In some cases this has helped in reducing the overuse and misuse of pesticides and in other cases it has promoted the use of biological control for sustainable pest management. Globalization of the agriculture and food system is increasingly demanding foods produced in a safe and environmentally friendly way. This means that food which is consumed or exported has tolerable pesticide residue limits as recommended by the CODEX commission of the FAO.

The experiences from Asia, Africa and Latin America indicate that specific IPM packages have been developed and successfully adopted for the management of a single pest in a specific crop. Developing and implementing IPM for multiple pest, disease and weed complexes affecting crop production in developing countries has been difficult and progress in this area has been limited. The general constraints to the development and implementation/adoption have been the following:

1. Lack of national IPM Policy. Many countries do not have a national IPM policy.
2. Lack of institutionalization of IPM to help develop and coordinate IPM programs.
3. Lack of a multi-disciplinary approach leading to inadequate problem identification/definition and poor project design for the development and implementation of IPM. Current IPM packages have been based on the management of a single pest, leading to the outbreak of the secondary pests.
4. Lack of appropriate research to develop technologies and integration of various tactics to be used in IPM programs. The majority of the IPM packages have focused on the use and integration of one or two tactics. This has exerted selection pressure on pests to develop resistance. There has not been a good integration of various pest management tactics.

5. Lack of well-trained human resources, research facilities, financial resources, and institutional linkages (lack of collaboration among various government departments and ministries). IPM is a multi-disciplinary approach. Plant protection specialists must collaborate and work hand-in-hand with breeders, agronomists, social scientists, and specialists from all other appropriate disciplines.

6. IPM is an information intensive strategy. Farmer participation in the design, development and implementation of IPM packages is critical. The research–extension–farmer linkages are critical for the information flow and successful implentation/adotion of IPM. Also, the linkage with the private sector is critical.

Based on our combined international experience and networking within the Global IPM arena, we are making the following general and specific suggestions/recommendations to various stakeholders involved in the design, development and implementation of IPM programs at national, regional and international level. These stakeholders include local and national governments, farmers/commodity organizations, academia, NGOs, private sector, international organizations/centers, and the donor community.

Education, Training and Capacity Building

Human resource development

Success in IPM relies on well-trained and skilled human resources. Education, training and development of appropriate human resources will allow the formation of multidisciplinary IPM teams for planning, designing, development and implementation of IPM packages for specific crops or ecosystems. A special emphasis should be given to train personnel in IPM project/program management, experiment station management, and IPM-related business development.

Farmer participation and empowerment in IPM

Scouting and monitoring of pests and beneficial organisms is the foundation of any IPM program. Government extension services or private agencies should conduct area-wide monitoring and provide the appropriate information on pest outbreaks to farmers on a regular basis. Training and education of farmers and extension workers in pest identification, monitoring and management approaches should be provided. Farmer participation and empowerment is critical for the adoption of IPM packages. The weather data should be utilized in the development of predictive models for forecasting the outbreaks of pests, especially migratory pests. New and emerging technologies such as mating disruption using sex pheromone technologies are increasingly utilized in commercial agriculture. However, the use of such technologies is complex and very information intensive. Therefore these technologies must accompany appropriate information on their use.

Botanicals and biological control enterprises

The local governments and rural development programs should encourage the development of cottage industries to mass produce beneficial organisms and botanical pesticides for use in IPM programs. This may include mass production and commercialization of entomopathogenic fungi and nematodes, parasitoids, predators, and botanical agents such as neem-based pesticides.

Role of international organizations/centers

The international organizations and programs have served as a very good platform for providing training and networking in IPM-related areas. The CGIAR international research centers have played a pivotal role in delivering improved germplasm to NARs. This improved germplasm has provided usable sources of resistance for developing new varieties and hybrids that are resistant to major pests and diseases of specific crops worldwide. These activities have helped tremendously in building the capacity of NARs worldwide. These efforts should be further supported and strengthened.

Environmental and pesticide use education

Pesticides have been and will remain an integral part of the IPM programs. However, pesticide-use education and proper monitoring of pests will reduce the misuse and overuse of pesticides. Farmers should be made aware of and educated on the negative effects/impacts of overuse and misuse of pesticides on the environment, human health and beneficial organisms such as natural enemies of pests, honey bees, etc. Also, environmental education using IPM as a vehicle should be promoted in primary, secondary and high schools.

Education on grades and standards

We live in a global marketplace. Education and understanding the requirements of the international markets (grades and standards; export regulations on pesticide residues and pest-free products) is becoming very important for food and fiber products grown for export markets.

Policy

IPM policy

National governments should encourage and support the development and implementation/enforcement of a national IPM policy. This policy may be a part of an alternative agriculture or sustainable agriculture policy and should encompass issues related to pesticide use, pesticide subsidies, environmental and human/animal health protection aspects. For example, the Government of Ghana has developed an IPM policy (see Chapter 11).

Intellectual property rights (IPR), biosafety and food safety

Many of the emerging biotechnologies are proprietary. Developing countries will have to develop or adjust their policies related to IPR, biosafety and food safety to access and commercialize these technologies. Public–private sector linkages will become critical to access and commercialize these technologies.

Institutionalization of IPM

In most countries, IPM programs function in isolation with very poor coordination among people working in different departments and ministries. IPM must be institutionalized to provide a better planning and coordination at both institutional and national level. For example, the Indian Government has established a National Center for IPM. Michigan State University has an IPM Program Office.

Financial support and technical assistance

Governments, national and international donors will have to make IPM a priority and provide financial resources for continual development and implementation of IPM programs. Coordination and education of donors that provide financial resources and technical assistance in IPM will be critical to utilize the limited financial resources efficiently.

Balance between Basic and Applied Research

Long-term ecological research

Prevention is better than cure. IPM research programs/projects should seek a balance between basic and applied research. Most IPM research programs are designed as short-term programs. Landscape-level long-term ecological research projects will give a better understanding of the biological and ecological interactions within the landscapes. A thorough understanding of biology and ecology of pests and natural enemies and their ecosystem may reveal totally new approaches for pest management specially for the preventive IPM programs. The new tools of geographic information systems and global positioning systems may assist in characterizing the temporal and spatial dynamics of landscapes.

Integration of IPM into ICM

IPM must be viewed as an integral part or component of ICM programs that promote sustainable agriculture. IPM packages should be developed and tested by the multidisciplinary teams including plant protection specialists, breeders, agronomists, social scientists, extension workers and farmers. The use of the farmer school approach in getting the farm community involved in pest and disease identification, monitoring and use of effective control components has been effective in Latin America and Asia in controlling potato and rice pests.

Postharvest pest management

Globally, food grains and food products are lost both in the field and in storage. The IPM packages must include postharvest pest management approaches. There is a need for basic and applied research programs related to the management of pests during the postharvest period. The use of appropriate grades and standards is a must for promoting international trade of food grains, vegetables and fruits.

Biotechnology and genetic engineering

Biotechnology will play an important role in future pest management programs worldwide. The major applications of biotechnology and genetic engineering thus far have been to develop and commercialize insect- and disease-resistant transgenic crops. Also, many of the emerging biotechnologies are developed by the private sector. The use of conventional biotechnology tools such as tissue culture, micropropagation and diagnostics needs to be further encouraged to make available pest- and disease-free planting materials to developing country farmers

Pest resistance management

Resistance to pesticides and other methods of pest control is a global problem. Education and easy to use methods and tools for the detection/diagnosis of resistance problems should be developed and utilized. Resistance management will also be critical for new and emerging biotechnologies.

Communication, Information, Linkages and Networking

Information

IPM is an information intensive strategy. A free flow (exchange) of information should be promoted at local, regional and international level. The recent advances in computer and satellite technologies now allow the storage and worldwide dissemination of large volumes of information through electronic media (websites, CDs).

Education and training programs combining both conventional methods and distance learning methods should be promoted.

A wealth of information and research data/results on IPM exist in different countries. Many of these research data and much information exists as raw data or as unpublished reports. Some of this information is published in country reports in local languages. The international community should encourage and provide support to bring out this wealth of knowledge and information in a format that can be used by IPM programs around the world.

Collaboration and networking

Regional and global cooperation and networking (linkages/partnerships) will become increasingly important to efficiently exchange and use the wealth of IPM information. Cooperation between the northern and southern hemispheres and within the southern hemisphere countries will be critical to maximize worldwide IPM implementation. (The experience of Farmers' Field Schools from Asia has been now tested successfully in West African countries.)

Communication with consumers

Consumers all over the world are increasingly demanding not only more food but better quality and safe food, water and environment. Communication with the consumers about the food produced through IPM practices will be important issues for the easy acceptance (e.g. eco-labeling).

Global Trends and Challenges

Trade and invasive pests

Globalization of food trade and increased human travel has accelerated the movement of species around the world. Invasive species pose a serious threat to agricultural productivity and human health. Capacity building in the development and implementation of quarantine regulations and policies to prevent the introduction of foreign pests is critical. This will require well-equipped quarantine facilities and trained personnel.

Pests and human–animal health interactions

Globalization with its increased spread of organisms and increases in emerging infectious diseases necessitates that the global community must be actively engaged at the interface of pest management and human medicine.

Conserving and using biodiversity

Less than 1% of all organisms on this earth are harmful pests to agriculture and human beings. The wealth of useful organisms and biodiversity should be preserved. Many cultures around the world use insects and other arthropods as food sources in their diets. Honeybees and silkworms provide food or fiber to mankind. Bees also serve as pollinators for many important crops.

Global climate change

There is strong evidence of global climate change affecting the geographic distribution and damage caused by pests. Insects and other organisms serve as an indicator of global climate change. There is a need to study the effect of global change on the pest populations as well as developing models to predict the impact of the global climate change.

Index

Actinomyces 36
Africa IPM Forum 6, 445
Africa IPM Link 416
African bollworm 133, 179, 443
see also Helicoverpa armigera
African rhinoceros beetle 150
see also Oryctes monoceros
Agenda 21 (UNCD) 3, 420
*Ag*NPV 291–293, 295
agricultural policy (South Africa) 34, 169–170
Agricultural Research Council (ARC–South Africa) 169
Agrobacterium tumefaciens 34
agrochemicals *see* chemical pesticides
Aleurocanthus woglumi 173, 273
amplified fragment length polymorphisms (AFLP) 36–37
Anaphes nitens 183
Anastrepha ludens 280
Anastrepha obliqua 280
Andean potato weevil 308
Anthonomus grandis 250, 314
Anthonomus vestitus 307
Anticarsia gemmatalis 288, 290, 291
Aonidiella aurantii 173, 174
Aphidius matricariae 363
aphids 334, 346, 363, 374, 441
Aphis gossypii 133, 329, 454
apiculture 442
Argentina 313–324
ARPPIS 443
Asian corn borer 200, 201, 203, 243
Aspongopus viduatus 138
Atherigona soccata 456
augmentation biological control 23–25
Aulacophora Africana 139
Australia 371–381
Azadiracta indica 114, 115
see also neem

Bacillus thuringiensis (Bt) 3, 16, 31–45, 67, 143, 184, 201, 220, 225, 244, 287, 306, 362, 366, 410, 442, 443, 455
Bactrocera carambolae 460–462
banana 441, 458, 498
Bangladesh 410, 415, 496
BBT 32
bean 158, 411
Beauveria bassiana 203, 304
beet armyworm 277
behavioral control 2
Bemisia argentifolii 69, 289, 306, 307
Bemisia tabaci 133, 295, 304, 307, 329, 334, 345, 346, 424, 454
benefit–cost analysis 87
biodiversity 449, 505
bioinformatics 38, 441
biological control 2, 19–28, 67–68, 138, 200, 202, 212, 213, 231, 276, 290, 293, 302, 324, 327, 330, 332, 334, 345, 357, 362–363, 371–374, 389, 445, 461, 495, 502
biological monitoring 79, 82
biological pest control in greenhouse 332
biological pest management 4
biomagnification 227
BioNET International 420
biopesticides 2, 3, 4, 36, 215, 324, 473, 495
biosafety 33, 38, 43, 45, 503
biotechnology 3, 31–45, 209, 441, 473, 498, 504
botanical extracts 134, 139, 442
see also botanical pesticides
botanical pesticides 2, 116, 139–140, 213, 225, 441, 486, 502

botanicals *see* botanical pesticides
Brazil 285–296
brown planthopper 22, 226–227, 229, 231, 232
Bt *see Bacillus thuringiensis*
Bt cotton 33, 67–68, 201, 202, 378–379
Bt crops 33, 41–42, 67–68, 379
Bt maize 33, 40, 177
Bt potato 67–68, 181
Bt transgenic plants 33–36, 40
Burkina Faso 109–117, 400–403
Bussecola fusca 115, 176, 177, 445, 446

cabbage 123, 212, 213, 233, 244
CABI 195, 493–499
CABI Bioscience 5, 420, 425, 485, 493–499
Calidea dregii 148, 149
capacity building 414, 417
  see also IPM training
cassava 25
Catalonia, Spain 341–354
CATIE 477–491
CD-ROM 10, 12, 463, 499
Central America 477–491
Ceratitis capitata 280, 304, 348, 349, 459–460
CGIAR 5, 349, 420, 470, 473, 502
CGIAR SP-IPM 5, 419–429, 499
chemical pesticides 2, 4,, 49, 50, 87, 88, 97, 119, 202, 209, 232, 362
chickpea 215, 428
Chilo partellus 176, 177, 411, 443, 445
Chilo sacchariphagus 177
Chilo zacconius 113
China 197–204
Chromaphis juglandicola 24
CICP 6
CILSS 112
Cinara cronartii 183
CIP 305, 496
CIRAD 453–464
CIRAD-France *see* CIRAD
citrus 301, 460, 461
  IPM 172–175
  thrips, *Scirtothrips aurantii* 173
Clavigralla tomentosicollis 114
climate change 505
Coconut IPM 150–151
CODEX Commission 501
codling moth 374
coffee 438, 485, 497
  coffee berry disease (CBD) 146
  coffee leaf rust (CLR) 146
Colorado potato beetle 35, 362
conservation biological control 21–23
constraints of IPM 415, 427, 496, 441–450
conventional agriculture *see* traditional agriculture
Costa Rica 479
cost–benefit ratio 117
Côte d'Ivoire 400
Cotesia flavipes 445
cotton 69, 71, 111, 131–137, 200, 201, 209, 214, 301, 304, 403, 438
cowpea 160, 428, 441
CPI/OUA 122
CPITT 6
CropLife International 70
Ctenopseustis obliquana 391
cultural control 2, 230, 455, 458
Cycloneda limbifera 363
Cydia pomonella 258, 322, 347, 348, 374
Cylus formicarius 412

DDT 135, 136, 145
Decision Support System 10–12
Desmodium uncinatum 448
Diadegma molliplum 180
diagnostic tools 3, 36, 45
diamondback moth 278–280, 302, 441, 442
Diatraea saccharalis 286, 302
DNA markers 37

eco-labeling 505
  see also labeling
ecological approach 227, 230
ecological world-view 73, 76, 84
economic threshold 92, 101, 133, 138, 176, 197, 419
Egypt 428
Eldana saccharina 115, 178, 179, 446
Empoasca 148
Empoasca lybica 135
Encarsia 363
Encarsia formosa 24, 335, 346
enzyme-linked immunosorbent assays (ELISA) 37
Ephestia kuehniella 25
Ervinia amylovora 363
Ethiopia 441, 443
eucalyptus tortoise beetle, *Trachymela tincticollis* 183
Europe 327–336
evaluation of IPM 254

faba bean 428
FAO 110, 120, 146, 195, 305, 397–406, 420, 437, 439, 473
farmer empowerment 5
farmer participation 502
farmer schools 4
  see also farmers' field schools

farmers' field schools (FFS) 21, 88–90, 94, 120–129, 134, 136, 234–236, 240, 242, 245, 398, 400, 403, 439, 443, 505
field water melon 138
Food Quality Protection Act (FQPA) 3, 256, 267
food safety 38, 43, 267, 473, 503
forestry IPM 182
French Guiana 461
fruit flies 280–282, 459
Fusarium 158, 162, 163, 329, 333, 344, 411
Fusarium oxysporum f.sp. *vasinfectum* 148

gender issues in FFS 128
genetic engineering 504
    see also biotechnology
genetically modified organisms (GMOs) 2, 36, 43, 70, 170
Geocoris 67
Geographic Information System (GIS) 4, 504
Ghana 119–129, 400, 443
Global IPM Facility 5, 397, 399, 402, 420, 425, 437, 497, 499
Global Plant Clinic 494
Global Plant Health Clinic *see* Global Plant Clinic
Gonipterus scutellatus 183
granulosis virus 374
Grapholita molesta 374
Green Revolution 31, 435
Greenhouse agriculture 344, 327–329
groundnut 131, 158, 428
Guatemala 411, 412, 414, 415

habitat management 20, 447
Helicotylenchus 115
Helicoverpa armigera 133, 148, 149, 179, 198, 209, 344, 345, 346, 378, 379, 443, 454, 455, 456, 498
Heliothis virescens 307
Heterodera 116
Hirschmanniella spinicaudata 116
Honduras 415
host plant resistance 2, 136, 212, 226, 231, 294, 455, 457
Hylastes angustatus 184

IARCs *see* CGIAR
ICARDA 428
Icerya purchasii 25, 173
ICIPE 6, 164, 257, 415, 420, 428, 441–450
ICRISAT 164, 457
IFP 388
IITA 114, 119, 126, 428
importation biological control 25–27
India 209–221
indigenous knowledge 159, 162
Indonesia 223–236, 439
insect pathogens 455
insect resistance 33
insect resistance management 69
Integrated Crop Management (ICM) 157, 499, 504
Integrated Pest and Vector Management (IPVM) 443
Integrated Plant Protection Center (IPPC) 11
Intellectual Property Rights (IPRs) 38–39, 43, 45, 503
International Organization of Biological Control 26
Internet 9–17
invasive pests 505
IPM adoption 97–105, 254, 412, 427
IPM CRSP 5, 407–418
IPM education 302, 343, 351, 489, 502–504
IPM Facility *see* Global IPM Facility
IPM farmers' field schools 120, 136
    see also farmers' field schools
IPM Forum 5
IPM Policy 372
    Australia 287
    Brazil 112
    Burkina Faso 421–422
    CGIAR 199
    China 125
    Ghana 211
    India 274–275
    Mexico 386
    New Zealand 302
    Peru 342–343
    Russia *see* IPPM policy, Russia
    Spain 132
    Sudan 436
    World Bank 501, 503
IPM program 320–324, 347–349, 366, 388
    apple 146–147
    coffee 147–150, 179–180, 305–307, 314–316, 378, 454
    cotton 114
    cowpea 175–176, 347–350, 366, 373–374
    fruit orchards 329–336
    greenhouse 386, 411
    horticultural crops 113
    kiwifruit 390
    millet 320–324, 349–350
    ornamentals 330
    pear 157, 159, 161–164
    pigeonpea 180–181, 306–309, 316–321
    potato 113
    rice 115, 176, 456
    sorghum 287–295
    soybean 177–179

IPM program *continued*
sugarcane 344–347, 367
tomato 330, 366
vegetable 181–182, 366
wheat 352, 438
IPM stakeholders 124
IPM training 352, 414, 417, 439, 443
IPMEurope 5, 467–475
IPPM 357–367
IPPM policy, Russia 362–363
ISAAA 71

Jacobiasca lybica 133
Jamaica 411, 412, 414

Kenya 415, 427, 496, 441–450
constraints of IPM 117, 151
KiwiGreen 390–393

labeling 41, 45
leaf miners 329
legumes 158, 165
Leptinotarsa decemlineata 35, 363, 360
licensing 41
Liriomyza huidobrensis 411, 413
Liriomyza trifolii 210
Locusta migratoria manilensis 197

Maecolaspis bridarolii 319, 320
maize 21, 34, 158, 160, 176, 200, 201, 202, 203, 212, 243–244, 411, 428, 441, 447
Malawi 157–165
Male Annihilation Technique (MAT) 461–462
Mali 400–403, 412, 414, 415, 457
Maliarpha separatella 113
mango 460
mating disruption 375, 456
Mayetiola destructor 428
mechanical control 2, 232
Mediterranean region 341–354
Meloidogyne 115, 318, 319
Meloidogyne incognita, M. hispanica 115
melon bug 138
Metarrhizium 164
Meteorus layphygmarum 138
Mexico 273–282, 438
Michigan State University 4, 262–267, 503
Microtermes thoracalis 135
Mononychellus tanajoa 25
Morocco 428
mungbean 230
Mycosphaerella fijiensis 458–459
Myzus persicae 68, 317, 329

National Coconut Development Program (NCDP-Tanzania) 150
National Committee of Pesticides (Burkina Faso) 112
natural enemies 330–332, 455, 487
neem 91, 139, 441
nematode management 318–319
nematodes 115–116, 123
see also nematode management
Nephotettix virescens 212, 226
New Caledonia 462
New Zealand 385–394
Nezara viridula 289
NGOs 112, 117, 125, 215, 305, 397, 401, 409, 420, 429, 437, 441, 450, 470, 473, 497
Nicaragua 478
Nigeria 428, 457
Nilaparvata lugens 22, 100, 198, 213, 226
Nomureae rileyi 288, 290
NSF 442

olives 302
Online resources 9–17
see also World Wide Web
Opuntia vulgaris 172
Orseolia oryzivora 113
Orthotomicus erosus 184
Oryctes monoceros see African rhinoceros beetle
Ostrinia furnacalis 200, 243
Ostrinia nubilalis 25, 35

Panonychus ulmi 389
parasitic nematodes of rice 116
Pauesia cinaravora 135, 183
Pectinophora gossypiella 135, 273, 306, 307, 456
PEDUNE 114
Perkinsiella saccharicida 177
Peru 301–310
pest resistance management 39, 504
Pesticide Policy, Indonesia 240–241
Pesticide Regulatory Committee (Ghana) 119
pesticide use
global 50, 495
India 210
Mali 456
New Zealand 389
Nicaragua 487
Russia 363–364
pesticides 97
health costs 57, 60

health hazards 226, 228
see also health risks
health risks 91–92
sales 53, 55–57
subsidies 49–62
Phenacoccus manihoti 105
Philippines 239–247, 308, 410, 414, 415
Phthorimaea operculella 35, 305
Physic nut, *Jatropha curcas* 115
Phytophthora palmivora 494
Phytoseiulus 180, 363
Plant Protection Research Institute (PPRI–South Africa) 170, 171
plantain weevils 123
Plutella xylostella 225, 244, 304, 443
Podagarica puncticollis 135
Podosphaera leucotricha 363
polymerase chain reaction (PCR) 36–37
postharvest pest management 504
potato tuber moth 35, 180
potatoes 34, 233, 302, 496
Pratylenchus 115
precision agriculture 4
predatory mites 374
Premnotrypes 305, 306
ARPPIS (ICIPE–Kenya) 444
Project GREEN 263
PRONAF 114
Puccinia striiformis 202
push–pull strategy 448–450
Pyricularia oryzae 113

Quadraspidiotus perniciosus 349

random amplified polymorphic DNA (RAPD) 36–37
rapeseed 214
resistance breeding 335
see also host plant resistance
resistance management strategy (RMS) 39–40, 42
restriction fragment length polymorphism (RFLP) 36
Réunion 461
Rhizoctonia solani 149
Rhopalosiphum maidis 133
rice 22, 34, 38, 71, 116, 131, 201, 202, 214, 216, 223, 224, 226, 229, 234, 239, 242, 245, 403, 410
rice IPM schools 401
Rodolia cardinalis 173, 257
Rural Womens' Schools (Sudan) 135
Russia 357–367
Russian wheat aphid (RWA) 180–182

Saissetia oleae 302
Schistocerca gregaria 443
Schizaphis graminum 133, 285
Scirpophaga excerptalia 213
Scirtothrips aurantii 173
Scutellonema 115
Senegal 401–403
sericulture 442
Sesamia calamistis 113
sex pheromone 115, 18
sigatoka leaf spot 458, 459
silk worm, *Bombyx mori* 443
Sirex noctilio 183
snow peas 412
social and economic considerations 49–62, 87–94
socioeconomic data 92–93
Sogatella furcifera 213
sorghum 131, 441
sorghum midge, *Stenodiplosis* (= *Contarinia*) *sorgicola* 115, 456–457
South Africa 169–186
soybean 34, 230, 233, 428
Spain 341–354
SP-IPM *see* CGIAR SP-IPM
Spodoptera frugiperda 5
Spodoptera littoralis 21
Spodoptera litura 329
staggered and targeted control 456
stem borers 176, 445, 447, 456
Sternechus subsignatus 294
stinkbugs 293
stored grain 286
Striga 126, 159–161, 424, 428, 447
Sudan 131–143, 441
sugarcane 213, 286, 302, 204, 496
sustainable agriculture 73, 77, 81, 473
sustainable development 73–84
sustainable production 435–436, 449
sweet potato 230, 411
Symmetrischema tangolias 308
systems approach 1

Tanzania 145–152
Tephrosia vogelii 160
termites 135
Tetranychus urticae 329, 330, 335, 374, 378
thrips 334, 441
tobacco 441
tomato fruitworm 276
tomato pinworm 276
tomatoes 212, 275–277, 286–287, 330, 344
Trachymela tincticollis 183
traditional agriculture 76, 77, 120–122, 126–128, 400, 403

training of trainers (TOT) 120
transgenic
 cotton 40
  see also Bt cotton
 crops 27, 363
  see also Bt crops
 maize 177
  see also Bt maize
 plants 2, 33
  see also transgenic crops; Bt transgenic plants
 potato 35, 181
  see also Bt potato
Trialeurodes vaporariorum 133
Trichoplusia ni 35, 138
Tricoderma 213, 333
Tricogramma spp. 24–25, 180, 201, 203, 211, 214, 219, 239, 243–244, 246, 274, 287, 306, 346, 363, 455
Trinidad and Tobago 496
tsetse traps 438, 442
Tuta absoluta 286
Typhlodromalus aripo 26
Typhlodromalus manihoti 26

Uganda 411, 412, 414, 415
ultra low volume (ULV) 363
UN–FAO Plant Protection Service 5
University of California 258–262
USA 249–268, 416
 Adkisson Project 252
 Haffaker Project 251–252
USAID 126, 478
USAID–IPM CRSP *see* IPM CRSP
USSR *see* Russia

vegetables 71, 134, 233, 403
velvetbean caterpillar 291
Venturia 363
Virginia Polytechnic Institute and State University 408
Virginia Tech *see* Virginia Polytechnic Institute and State University

watermelon 135, 138–140
websites *see* World Wide Web
weed management 319–320
West Africa 456
wheat 133, 201, 202, 285, 428
whiteflies 424
World Bank 435–440, 473, 485
World Food Prize 441
World Wide Web (WWW) 9–17, 20, 32, 118, 155, 185–186, 204, 246, 260, 266, 379–381, 388, 406, 415, 428, 441, 463, 498, 499

Xanthomonas campestris pv. *malvacearum* 148
Xanthomonas malvacearum 135

Zimbabwe 443